Lecture Notes in Computer Science

Founding Editors

Gerhard Goos
Juris Hartmanis

Editorial Board Members

Elisa Bertino, *Purdue University, West Lafayette, IN, USA*
Wen Gao, *Peking University, Beijing, China*
Bernhard Steffen, *TU Dortmund University, Dortmund, Germany*
Moti Yung, *Columbia University, New York, NY, USA*

The series Lecture Notes in Computer Science (LNCS), including its subseries Lecture Notes in Artificial Intelligence (LNAI) and Lecture Notes in Bioinformatics (LNBI), has established itself as a medium for the publication of new developments in computer science and information technology research, teaching, and education.

LNCS enjoys close cooperation with the computer science R & D community, the series counts many renowned academics among its volume editors and paper authors, and collaborates with prestigious societies. Its mission is to serve this international community by providing an invaluable service, mainly focused on the publication of conference and workshop proceedings and postproceedings. LNCS commenced publication in 1973.

Aleš Leonardis · Elisa Ricci · Stefan Roth ·
Olga Russakovsky · Torsten Sattler · Gül Varol
Editors

Computer Vision – ECCV 2024

18th European Conference
Milan, Italy, September 29–October 4, 2024
Proceedings, Part VI

Editors
Aleš Leonardis
University of Birmingham
Birmingham, UK

Stefan Roth
Technical University of Darmstadt
Darmstadt, Germany

Torsten Sattler
Czech Technical University in Prague
Prague, Czech Republic

Elisa Ricci
University of Trento
Trento, Italy

Olga Russakovsky
Princeton University
Princeton, NJ, USA

Gül Varol
École des Ponts ParisTech
Marne-la-Vallée, France

ISSN 0302-9743 ISSN 1611-3349 (electronic)
Lecture Notes in Computer Science
ISBN 978-3-031-72657-6 ISBN 978-3-031-72658-3 (eBook)
https://doi.org/10.1007/978-3-031-72658-3

© The Editor(s) (if applicable) and The Author(s), under exclusive license
to Springer Nature Switzerland AG 2025

This work is subject to copyright. All rights are solely and exclusively licensed by the Publisher, whether the whole or part of the material is concerned, specifically the rights of translation, reprinting, reuse of illustrations, recitation, broadcasting, reproduction on microfilms or in any other physical way, and transmission or information storage and retrieval, electronic adaptation, computer software, or by similar or dissimilar methodology now known or hereafter developed.
The use of general descriptive names, registered names, trademarks, service marks, etc. in this publication does not imply, even in the absence of a specific statement, that such names are exempt from the relevant protective laws and regulations and therefore free for general use.
The publisher, the authors and the editors are safe to assume that the advice and information in this book are believed to be true and accurate at the date of publication. Neither the publisher nor the authors or the editors give a warranty, expressed or implied, with respect to the material contained herein or for any errors or omissions that may have been made. The publisher remains neutral with regard to jurisdictional claims in published maps and institutional affiliations.

This Springer imprint is published by the registered company Springer Nature Switzerland AG
The registered company address is: Gewerbestrasse 11, 6330 Cham, Switzerland

If disposing of this product, please recycle the paper.

Foreword

Welcome to the proceedings of the European Conference on Computer Vision (ECCV) 2024, the 18th edition of the conference, which has been held biennially since its founding in France in 1990. We are holding the event in Italy for the third time, in a convention centre where we can accommodate nearly 7000 in-person attendees.

The journey to ECCV 2024 started at CVPR 2019, where the general chairs met at a sketchy bar in Long Beach to discuss ideas on how to organize an amazing future ECCV. After that, Covid came and changed conferences probably forever, but we maintained our excitement to organize an awesome in-person conference in 2024, particularly having seen the wonderful event that was ECCV 2022 in Tel Aviv. And here we are, 5 years later, delighted that we finally get to welcome all of you to Milan!

First and foremost, ECCV is about the exchange of scientific ideas, and hence the most important organizational task is the selection of the programme. To our Program Chairs, Aleš Leonardis, Elisa Ricci, Gül Varol, Olga Russakovsky, Stefan Roth, and Torsten Sattler, our entire community owe a huge debt of gratitude. As you will read in their Preface to the proceedings, they dealt with over 8,500 submissions, and their diligence, thoroughness, and sheer effort has been inspirational. From the early stages of designing the call for papers, through recruiting and selecting area chairs and reviewers, to ensuring every paper has a fair and thorough assessment, their professionalism, commitment, and attention to detail has been exemplary. From all the choices that we made to organize ECCV, choosing this amazing team of Program Chairs has been, without a doubt, our very best decision.

The Program Chairs were advised and supported by an Ethics Review Committee, Chloé Bakalar, Kate Saenko, Remi Denton, and Yisong Yue, who provided invaluable input on papers where reviewers or Area Chairs had raised ethical concerns.

The Publication Chairs, Jovita Lukasik, Michael Möller, François Brémond, and Mahmoud Ali, did great work in assembling the camera-ready papers into these Springer volumes, dealing with numerous issues that arose promptly and professionally.

With the ever-expanding importance of workshops, tutorials, and demos at our major conferences comes a corresponding increase in the challenge of organizing over 70 workshops and 9 tutorials attached to the main conference. The Workshop and Tutorial Chairs, Alessio Del Bue, Jordi Pont-Tuset, Cristian Canton, and Tatiana Tommasi, and Demo Chairs, Hyung Chung and Marco Cristani, worked tirelessly to coordinate the very varied demands and structures of these different events, yielding an extremely rich auxiliary program which will enhance the conference experience for everyone who attends.

The Industry Chairs, Shaogang Gong and Cees Snoek, working with Limor Urfaly and Lior Gelfand, managed to attract numerous companies and organizations from all around the world to the industrial Expo, making ECCV 2024 an international event that not only showcases excellent foundational research, but also presents how such research can be transformed in real products. It is notable that together with the big companies, several start-ups chose ECCV to present their business ideas and activities.

As usual in the recent main conference, we also organized Speed Mentoring and Doctoral Consortium events. While the latter is an institutional occasion allowing senior PhDs to show their work, the former is especially important to answer to the many doubts and uncertainties young scholars have while undertaking a career in Computer Vision and Artificial Intelligence. We warmly thank our Social Activities Chairs, Raffaella Lanzarotti, Simone Bianco and Giovanni Farinella, and the Doctoral Consortium Chairs, Cigdem Beyan and Or Litany, for their dedication to organize such events, so important for our young colleagues, as well as all the mentors who were available to share their experience.

As the conference becomes larger and larger, the overall budget increases to a level where uncertainties in estimates of attendance can result in significant losses, perhaps enough to significantly damage the prospects of running the conference in the future. This risk imposed the need to require authors to commit to full conference registrations, with the undesirable consequent possibility to cause hardship to authors, particularly student authors, from organizations where funding attendance is difficult. To mitigate this effect, we allocated travel grants to students who might otherwise have had difficulty in attending. Our Diversity Chairs, David Fouhey and Rita Cucchiara, put tremendous effort into a wonderfully careful and thoughtful programme, balancing diversity in all its forms with a changing budget landscape, allowing us to support the participation of many people who might not otherwise have been able to attend.

This year's conference also marks the move to a unified, multi-year conference website, based on that used for other leading computer vision and AI conferences. This was an effort that we decided to take on to make the ECCV website more familiar to the community and support the organization of the future ECCV editions. Our thanks go to Lee Campbell of Eventhosts who managed the entire design and development of the new website and provided responsive support and updates throughout the conference organization timeline.

Our Finance Chairs, Gérard Medioni and Nicole Finn, provided invaluable advice and assistance on all aspects of the financial planning of the conference, in conjunction with the professional conference organizers AIM Group, where Lavinia Ricci led a fantastic team that managed thousands of details from space planning to food to registrations. In particular, Elder Bromley and Mara Carletti were a tremendous support to the entire conference organization. The conference centre, Mico DMC, also provided an excellent service, showing considerable flexibility in adapting the venue and processes to the particular demands of an academic-focused conference.

Finally, we cannot forget to thank our Social Media Chair, Kosta Derpanis, supported by Abby Stylianou and Jia-Bin Huang, who, with their presence on social media, spread ECCV news and responded to small and big questions in a timely manner, helping the community to promptly receive information and fix their problems.

October 2024

Andrew Fitzgibbon
Laura Leal-Taixé
Vittorio Murino

Preface

ECCV 2024 saw a remarkable 48% increase in submissions, with 8585 valid papers submitted, compared to 5804 in ECCV 2022, 5150 in 2020, and 2439 in 2018. Of these, 2387 papers (27.9%) were accepted for publication, with 200 (2.3% overall) selected for oral presentations.

A total of 173 complete submissions were desk-rejected for various reasons. Common issues included revealing author identities in the paper or supplementary material (for example: by including the author names, referring to specific grants in the acknowledgments, including links to GitHub accounts with visible author information, etc.), exceeding the page limit or otherwise significantly altering the submission template, or posting an arXiv preprint explicitly mentioning the work as an ECCV 2024 submission. The desk-rejected submissions also include 14 papers identified as dual submissions to other concurrent conferences, such as ICML, NeurIPS, and MICCAI. Seven papers were rejected among the accepted papers due to plagiarism issues, after the Program Chairs conducted a thorough plagiarism check with the iThenticate software followed by manual inspection.

The double-blind review process was managed using the CMT system, ensuring anonymity between authors and Area Chairs/reviewers. Each paper received at least three reviews, amounting to over 26,000 reviews in total. To maintain the quality and the fairness of the reviews, we recruited 469 Area Chairs (ACs) and more than 8200 reviewers. Of the recruited reviewers, 7293 ultimately participated.

Area Chairs were chosen for their technical expertise and reputation; many of them had served in similar roles at other top conferences. We also recruited additional Area Chairs through an open call for self-nominations. Among the selected ACs, around 18% were women. Geographical distribution was taken into account during the selection process, with 31% of ACs working in Europe, 38% in the Americas, 27% in Asia, and 4% in the rest of the world. Despite our best efforts to consider diversity factors in the selection process, the gender diversity and geographic diversity of ACs is still below what we were hoping for.

To ensure a smooth decision-making process, we grouped the ACs into triplets. Upon accepting their role, ACs were requested to enter into the CMT system domain conflicts and subject areas, as well as information about their DBLP, TPMS (Toronto Paper Matching System), and OpenReview profiles. These were used to determine whether different subject areas were sufficiently covered, to detect conflicts, and to automatically form AC triplets. An AC triplet was a group of three ACs who worked together throughout the process, which helped to facilitate discussion, calibrate the decision-making process, and provide support for first-time ACs. AC triplets were manually refined by Program Chairs to ensure no triplet contained more than two major time zones to enable easier coordination for virtual AC meetings. Paper assignments were made such that no AC within the triplet was conflicted with any of the papers. Each AC was responsible for overseeing 18 papers on average.

In each triplet, we identified a Lead AC who was an experienced member of the computer vision community willing to take on additional administrative duties. The primary responsibilities of Lead ACs included coordinating the AC triplet to ensure deadlines were met and offering guidance to less experienced ACs. They also served as the main contact person with the Program Chairs and helped in handling papers if an original AC was unable to complete their tasks.

Reviewers were invited from reviewer pools of previous conferences and also selected from the pool of authors. We asked experienced ACs to recommend more potential reviewers. We further recruited reviewers through self-nominations, where researchers volunteered to review for ECCV 2024 via an open call. The self-nominations were then filtered, selecting those reviewers with at least two papers published in related conferences such as ECCV, ICCV, CVPR, ICLR, NeurIPS, etc.

Similarly to ACs, reviewers were also asked to update their CMT, Toronto Paper Matching System (TPMS), and OpenReview profiles. Reviewers had the option to classify themselves as either senior researchers or junior researchers and students, with the number of assigned papers varying accordingly (a quota of 6 for senior researchers with 3 or more times reviewing experience, and a quota of 4 otherwise). On average, each reviewer was assigned about 4 papers.

Conflicts of interest among authors, ACs, and reviewers were managed automatically via conflict domains and co-authorship information from DBLP. Paper assignments to ACs and reviewers were determined using an algorithm that combined subject-area affinity scores from CMT, TPMS, and OpenReview, along with additional criteria such as ensuring that every paper had at least one non-student reviewer. For AC assignments, we additionally incorporated a newly built affinity score based on embedding similarity with the papers on the Google Scholar profile of the ACs. Once the assignments were made, ACs and reviewers were asked to promptly report any undetected conflicts of interest or other concerns that would preclude them from handling the assigned papers to ensure immediate reassignment, if necessary.

Given the widespread use of Large Language Models (LLMs), ECCV 2024, like previous conferences, implemented a special policy regulating their use. Authors were permitted to use any tools, including LLMs, in preparing their papers, but they remained fully responsible for any misrepresentation, factual inaccuracies, or plagiarism. Reviewers were also allowed to use LLMs to refine the wording of their reviews; however, they were held accountable for the accuracy of their content and were strictly prohibited from inputting submissions into an LLM. One challenge we ran into was that tools that detect LLM-generated content were not able to distinguish between text polished vs. text written from scratch by an LLM. Thus, it became quite difficult to enforce the policy in practice.

Overall, the challenges encountered during the review process were consistent with those faced at previous computer vision conferences. For instance, a small number of reviewers ultimately did not submit their reviews or provided brief, uninformative reviews, necessitating last-minute reassignment to emergency reviewers. As the community expands and the number of submissions rises, securing enough qualified reviewers and performing a quality check of the received reviews becomes increasingly difficult.

The review process included a rebuttal phase where authors were given a week to address any concerns raised by the reviewers. Each response was limited to a single PDF page using a predefined template. Rebuttals with formatting or anonymity violations were removed by the Program Chairs. ACs then facilitated discussions with reviewers to evaluate the merits of each submission. While the goal was to reach a consensus, the final decision was left to the ACs in the triplet to ensure fairness. As is common, the ACs within each triplet were tasked with organizing meetings to discuss papers, especially borderline cases. The entire process was conducted online, with no in-person meetings. For each paper, a primary and a secondary AC were assigned from the AC triplet. The primary AC was responsible for entering the decisions and meta-reviews into the CMT system, which were then reviewed and approved by the secondary ACs. Besides making accept/reject decisions, the ACs were also required to provide recommendations regarding oral/poster presentations and award nominations.

Reviewers and ACs were asked to flag papers with potential ethical concerns. An ethics review committee, consisting of four leading researchers from both industry and academia, was appointed. The role of the committee was to evaluate papers escalated by the AC for further investigation. In total, 31 papers were flagged to the committee. The committee requested that the authors of 15 papers address specific concerns in their camera-ready submissions. The authors all complied with the requests.

Following the recent IEEE decision to ban the use of the photograph of Lena Forsén, we automatically identified all submissions that included this image. Although the ban did not officially apply to ECCV, we nevertheless shared the context with the authors of the affected papers and suggested that they remove the photograph in the camera-ready submission. All authors complied.

To help the ACs to track the overall progress of the reviewing and decision-making processes, we created AC triplet activity dashboards for each AC triplet (via Google sheets), which were updated on a regular basis. During the review process, the activity dashboard enabled ACs to track the progress of incoming reviews for each paper and promptly identify papers that necessitated attention, e.g., papers potentially requiring the invitation of emergency reviewers, or papers with suspiciously short reviews. During the decision-making process, the dashboard was also used by the Program Chairs to identify papers with incomplete review information, e.g., for which reviewers did not enter their final recommendation, the AC did not enter their final decision or meta-review, the secondary AC did not confirm the recommendation, etc.

At the end of the decision-making progress, the Program Chairs carefully examined the very small number of cases where ACs overruled a unanimous reviewer recommendation for acceptance or rejection. In such instances, it was crucial for the primary and secondary ACs to explain in detail the motivation behind the decision. To ensure a fair evaluation, an additional Area Chair, not originally part of the decision-making triplet, was appointed to provide an independent opinion and advise the Program Chairs in determining whether to uphold or reverse the overturn.

Inevitably, after decisions were released, some authors were dissatisfied. The authors were given the possibility to appeal the final decision in case of procedural issues encountered during the review process by filling out a form to directly contact the Program Chairs. Valid reasons for appealing a decision included policy errors (e.g., reviewers or ACs enforcing a non-existent policy), clerical errors (e.g., the meta-review clearly indicated an intention to accept a paper, but it was mistakenly rejected), and significant misunderstandings by the reviewers or ACs. In total, we received 59 appeals and upheld 6 of them. One was due to a clear policy error by an AC, and the paper decision changed from reject to accept following a successful appeal. The other 5 papers were already conditionally accepted – but the AC/reviewers wrongly flagged them as papers for which the associated dataset was required to be released prior to final acceptance; this condition was removed following the appeal.

After the camera-ready deadline, additional checks were conducted, particularly for papers contributing a new dataset and for those flagged for potential ethical concerns. Specifically, during the submission phase, authors were required to specify if the primary contribution of their papers was the release of a new dataset, and for these papers, a valid dataset link had to be provided by the camera-ready deadline. A total of 435 submissions included a dataset as part of their contribution, and all complied with the requirement to release the dataset by the camera-ready deadline.

We sincerely thank all the ACs and reviewers for their invaluable contributions to the review process. Their dedication and expertise were crucial in maintaining the high standards of the conference. We would also like to thank our Technical Program Chair, Sascha Hornauer, who did tremendous work behind the scenes, and to the entire CMT team for their prompt support with any technical issues.

October 2024

Aleš Leonardis
Elisa Ricci
Stefan Roth
Olga Russakovsky
Torsten Sattler
Gül Varol

Organization

General Chairs

Andrew Fitzgibbon   Graphcore, UK
Laura Leal-Taixé   NVIDIA, Italy
Vittorio Murino   University of Verona & University of Genoa, Italy

Program Chairs

Aleš Leonardis   University of Birmingham, UK
Elisa Ricci   University of Trento, Italy
Stefan Roth   Technical University of Darmstadt, Germany
Olga Russakovsky   Princeton University, USA
Torsten Sattler   Czech Technical University in Prague, Czech Republic
Gül Varol   Ecole des Ponts Paris Tech, France

Technical Program Chair

Sascha Hornauer   Mines Paris – PSL, France

Industrial Liaison Chairs

Cees Snoek   University of Amsterdam, Netherlands
Shaogang Gong   Queen Mary University of London, UK

Publication Chairs

Francois Bremond   Inria, France
Mahmoud Ali   Inria, France
Jovita Lukasik   University of Siegen, Germany
Michael Moeller   University of Siegen, Germany

Poster Chairs

Aljoša Ošep Carnegie Mellon University, USA
Zuzana Kukelova Czech Technical University in Prague, Czech Republic

Diversity Chairs

David Fouhey New York University, USA
Rita Cucchiara Università di Modena e Reggio Emilia, Italy

Local Chairs

Raffaella Lanzarotti Università degli Studi di Milano, Italy
Simone Bianco University of Milano-Bicocca, Italy

Conference Ombuds

Georgia Gkioxari California Institute of Technology, USA
Greg Mori Borealis AI & Simon Fraser University, Canada

Ethics Review Committee

Chloé Bakalar Meta, USA
Kate Saenko Boston University, USA
Remi Denton Google Research, USA
Yisong Yue Caltech, Asari AI & Latitude AI, USA

Workshop and Tutorial Chairs

Alessio Del Bue Istituto Italiano di Tecnologia, Italy
Cristian Canton Meta AI, USA
Jordi Pont-Tuset Google DeepMind, Switzerland
Tatiana Tommasi Politecnico di Torino, Italy

Finance Chairs

Gerard Medioni Amazon, USA
Nicole Finn c to c events, USA

Demo Chairs

Hyung Chang University of Birmingham, UK
Marco Cristani University of Verona, Italy

Publicity and Social Media Chairs

Konstantinos Derpanis York University & Samsung AI Centre Toronto, Canada
Jia-Bin Huang University of Maryland College Park, USA
Abby Stylianou Saint Louis University, USA

Social Activities Chairs

Giovanni Maria Farinella University of Catania, Italy
Raffaella Lanzarotti Università degli Studi di Milano, Italy
Simone Bianco University of Milano-Bicocca, Italy

Doctoral Consortium Chairs

Cigdem Beyan University of Verona, Italy
Or Litany NVIDIA & Technion, USA

Web Developer

Lee Campbell Eventhosts, USA

Area Chairs

Ehsan Adeli	Stanford University, USA
Aishwarya Agrawal	University of Montreal & Mila & DeepMind, Canada
Naveed Akhtar	University of Western Australia, Australia
Yagiz Aksoy	Simon Fraser University, Canada
Karteek Alahari	Inria, France
Jose Alvarez	NVIDIA, USA
Djamila Aouada	Interdisciplinary Centre for Security, Reliability and Trust, University of Luxembourg, Luxembourg
Andre Araujo	Google, Brazil
Pablo Arbelaez	Universidad de los Andes, Colombia
Iro Armeni	Stanford University, USA
Yuki Asano	University of Amsterdam, Netherlands
Karl Åström	Lund University, Sweden
Mathieu Aubry	Ecole des Ponts ParisTech, France
Shai Bagon	Weizmann Institute of Science, Israel
Song Bai	ByteDance, Singapore
Xiang Bai	Huazhong University of Science and Technology, China
Yang Bai	Tencent, China
Lamberto Ballan	University of Padova, Italy
Linchao Bao	University of Birmingham, UK
Ronen Basri	Weizmann Institute of Science, Israel
Sara Beery	Massachusetts Institute of Technology, USA
Vasileios Belagiannis	University of Erlangen–Nuremberg, Germany
Serge Belongie	University of Copenhagen, Denmark
Sagie Benaim	Hebrew University of Jerusalem, Israel
Rodrigo Benenson	Google, Switzerland
Cigdem Beyan	University of Bergamo, Italy
Bharat Bhatnagar	Meta Reality Labs Research, Switzerland
Tolga Birdal	Imperial College London, UK
Matthew Blaschko	KU Leuven, Belgium
Federica Bogo	Meta Reality Labs Research, Switzerland
Timo Bolkart	Google, Switzerland
Katherine Bouman	California Institute of Technology, USA
Lubomir Bourdev	Primepoint, USA
Edmond Boyer	Inria, France
Yuri Boykov	University of Waterloo, Canada
Eric Brachmann	Niantic, Germany

Wieland Brendel	University of Tübingen, Germany
Gabriel Brostow	University College London, UK
Michael Brown	York University, Canada
Andrés Bruhn	University of Stuttgart, Germany
Andrei Bursuc	valeo.ai, France
Benjamin Busam	Technical University of Munich, Germany
Holger Caesar	TU Delft, Netherlands
Jianfei Cai	Monash University, Australia
Simone Calderara	University of Modena and Reggio Emilia, Italy
Octavia Camps	Northeastern University, Boston, USA
Joao Carreira	DeepMind, UK
Silvia Cascianelli	University of Modena and Reggio Emilia, Italy
Ayan Chakrabarti	Google Research, USA
Tat-Jen Cham	Nanyang Technological University, Singapore
Antoni Chan	City University of Hong Kong, China
Manmohan Chandraker	University of California, San Diego, USA
Xiaojun Chang	University of Technology Sydney, Australia
Devendra Singh Chaplot	Mistral AI, USA
Liang-Chieh Chen	TikTok, USA
Long Chen	Hong Kong University of Science and Technology, China
Mei Chen	Microsoft, USA
Shizhe Chen	Inria, France
Shuo Chen	RIKEN, Japan
Xilin Chen	Institute of Computing Technology, Chinese Academy of Sciences, China
Anoop Cherian	Mitsubishi Electric Research Labs, USA
Tat-Jun Chin	University of Adelaide, Australia
Minsu Cho	Pohang University of Science and Technology, Korea
Hisham Cholakkal	Mohamed bin Zayed University of Artificial Intelligence, United Arab Emirates
Yung-Yu Chuang	National Taiwan University, Taiwan
Ondrej Chum	Czech Technical University in Prague, Czech Republic
Marcella Cornia	University of Modena and Reggio Emilia, Italy
Marco Cristani	University of Verona, Italy
Cristian Canton	Meta AI, USA
Zhaopeng Cui	Zhejiang University, China
Angela Dai	Technical University of Munich, Germany
Jifeng Dai	Tsinghua University, China
Yuchao Dai	Northwestern Polytechnical University, China

Dima Damen	University of Bristol, UK
Kostas Daniilidis	University of Pennsylvania, USA
Antitza Dantcheva	Inria, France
Raoul de Charette	Inria, France
Shalini De Mello	NVIDIA Research, USA
Tali Dekel	Weizmann Institute of Science, Israel
Ilke Demir	Intel Corporation, USA
Joachim Denzler	Friedrich Schiller University Jena, Germany
Konstantinos Derpanis	York University, Canada
Jose Dolz	École de technologie supérieure Montreal, Canada
Jiangxin Dong	Nanjing University of Science and Technology, China
Hazel Doughty	Leiden University, Netherlands
Iddo Drori	Boston University and Columbia University, USA
Enrique Dunn	Stevens Institute of Technology, USA
Thibaut Durand	Borealis AI, Canada
Sayna Ebrahimi	Google, USA
Mohamed Elhoseiny	King Abdullah University of Science and Technology, Saudi Arabia
Bin Fan	University of Science and Technology Beijing, China
Yi Fang	New York University, USA
Giovanni Maria Farinella	University of Catania, Italy
Ryan Farrell	Brigham Young University, USA
Alireza Fathi	Google, USA
Christoph Feichtenhofer	Meta & FAIR, USA
Rogerio Feris	MIT-IBM Watson AI Lab & IBM Research, USA
Basura Fernando	Agency for Science, Technology and Research (A*STAR), Singapore
David Fouhey	New York University, USA
Katerina Fragkiadaki	Carnegie Mellon University, USA
Friedrich Fraundorfer	Graz University of Technology, Austria
Oren Freifeld	Ben-Gurion University, Israel
Huan Fu	University of Sydney, China
Yasutaka Furukawa	Simon Fraser University, Canada
Andrea Fusiello	University of Udine, Italy
Fabio Galasso	Sapienza University, Italy
Jürgen Gall	University of Bonn, Germany
Orazio Gallo	NVIDIA Research, USA
Chuang Gan	MIT-IBM Watson AI Lab, USA
Lianli Gao	University of Electronic Science and Technology of China, China

Shenghua Gao	University of Hong Kong, China
Sourav Garg	University of Adelaide, Australia
Efstratios Gavves	University of Amsterdam, Netherlands
Peter Gehler	Zalando, Germany
Theo Gevers	University of Amsterdam, Netherlands
Golnaz Ghiasi	Google DeepMind, USA
Rohit Girdhar	Carnegie Mellon University, USA
Xavier Giro-i-Nieto	Universitat Politecnica de Catalunya, Spain
Ross Girshick	Allen Institute for Artificial Intelligence, USA
Zan Gojcic	NVIDIA, Switzerland
Vladislav Golyanik	Max Planck Institute for Informatics, Germany
Stephen Gould	Australian National University, Australia
Liangyan Gui	University of Illinois Urbana-Champaign, USA
Fatma Guney	Koc University, Turkey
Yulan Guo	Sun Yat-sen University, China
Yunhui Guo	University of Texas at Dallas, USA
Mohit Gupta	University of Wisconsin-Madison, USA
Minh Ha Quang	RIKEN Center for Advanced Intelligence Project, Japan
Bumsub Ham	Yonsei University, Korea
Bohyung Han	Seoul National University, Korea
Kai Han	University of Hong Kong, China
Xiaoguang Han	Shenzhen Research Institute of Big Data & Chinese University of Hong Kong (Shenzhen), China
Tatsuya Harada	University of Tokyo & RIKEN, Japan
Tal Hassner	Weir AI, USA
Xuming He	ShanghaiTech University, China
Felix Heide	Princeton & Algolux, USA
Anthony Hoogs	Kitware, USA
Timothy Hospedales	Edinburgh University, UK
Mahdi Hosseini	Concordia University, Canada
Peng Hu	College of Computer Science, Sichuan University, China
Gang Hua	Wormpex AI Research, USA
Jia-Bin Huang	University of Maryland College Park, USA
Qixing Huang	University of Texas at Austin, USA
Sharon Xiaolei Huang	Pennsylvania State University, USA
Junhwa Hur	Google, USA
Nazli Ikizler-Cinbis	Hacettepe University, Turkey
Eddy Ilg	Saarland University, Germany
Ahmet Iscen	Google, France

Nathan Jacobs	Washington University in St. Louis, USA
Varun Jampani	Stability AI, USA
C. V. Jawahar	International Institute of Information Technology - Hyderabad, India
Laszlo Jeni	Carnegie Mellon University, USA
Jiaya Jia	Chinese University of Hong Kong, China
Qin Jin	Renmin University of China, China
Jungseock Joo	NVIDIA & University of California, Los Angeles, USA
Fredrik Kahl	Chalmers, Sweden
Yannis Kalantidis	NAVER LABS Europe, France
Vicky Kalogeiton	Ecole Polytechnique, IP Paris, France
Evangelos Kalogerakis	University of Massachusetts Amherst, USA
Hiroshi Kawasaki	Kyushu University, Japan
Margret Keuper	University of Mannheim & Max Planck Institute for Informatics, Germany
Salman Khan	Mohamed bin Zayed University of Artificial Intelligence, United Arab Emirates
Anna Khoreva	Bosch Center for Artificial Intelligence, Germany
Gunhee Kim	Seoul National University, Korea
Junmo Kim	Korea Advanced Institute of Science and Technology, Korea
Min H. Kim	Korea Advanced Institute of Science and Technology, Korea
Seon Joo Kim	Yonsei University, Korea
Tae-Kyun Kim	Korea Advanced Institute of Science and Technology & Imperial College London, Korea
Benjamin Kimia	Brown University, USA
Hedvig Kjellström	KTH Royal Institute of Technology, Sweden
Laurent Kneip	ShanghaiTech University, China
A. Sophia Koepke	University of Tübingen, Germany
Iasonas Kokkinos	University College London, UK
Wai-Kin Adams Kong	Nanyang Technological University, Singapore
Piotr Koniusz	Data61/CSIRO & Australian National University, Australia
Jana Kosecka	George Mason University, USA
Adriana Kovashka	University of Pittsburgh, USA
Hilde Kuehne	University of Bonn, Germany
Zuzana Kukelova	Czech Technical University in Prague, Czech Republic
Ajay Kumar	Hong Kong Polytechnic University, China
Kiriakos Kutulakos	University of Toronto, Canada

Suha Kwak	Pohang University of Science and Technology, Korea
Zorah Laehner	University of Bonn, Germany
Shang-Hong Lai	National Tsing Hua University, Taiwan
Iro Laina	University of Oxford, UK
Jean-Francois Lalonde	Université Laval, Canada
Loic Landrieu	École des Ponts ParisTech, France
Diane Larlus	Naver Labs Europe, France
Viktor Larsson	Lund University, Sweden
Stéphane Lathuilière	Telecom-Paris, France
Gim Hee Lee	National University of Singapore, Singapore
Yong Jae Lee	University of Wisconsin-Madison, USA
Bastian Leibe	RWTH Aachen University, Germany
Ales Leonardis	University of Birmingham, UK
Vincent Lepetit	Ecole des Ponts ParisTech, France
Fuxin Li	Oregon State University, USA
Hongdong Li	Australian National University, Australia
Xi Li	Zhejiang University, China
Yin Li	University of Wisconsin-Madison, USA
Yu Li	International Digital Economy Academy, Singapore
Xiaodan Liang	Sun Yat-sen University, China
Shengcai Liao	Inception Institute of Artificial Intelligence, United Arab Emirates
Jongwoo Lim	Seoul National University, Korea
Ser-Nam Lim	University of Central Florida, USA
Stephen Lin	Microsoft Research, China
Tsung-Yi Lin	NVIDIA Research, USA
Yen-Yu Lin	National Yang Ming Chiao Tung University, Taiwan
Yutian Lin	Wuhan University, China
Zhe Lin	Adobe Research, USA
Haibin Ling	Stony Brook University, USA
Or Litany	NVIDIA, USA
Jiaying Liu	Peking University, China
Jun Liu	Lancaster University, UK
Miaomiao Liu	Australian National University, Australia
Si Liu	Beihang University, China
Sifei Liu	NVIDIA, USA
Wei Liu	Tencent, China
Yanxi Liu	Pennsylvania State University, USA
Yaoyao Liu	Johns Hopkins University, USA

Yebin Liu	Tsinghua University, China
Zicheng Liu	Microsoft, USA
Ziwei Liu	Nanyang Technological University, Singapore
Ismini Lourentzou	University of Illinois, Urbana-Champaign, USA
Chen Change Loy	Nanyang Technological University, Singapore
Feng Lu	Beihang University, China
Le Lu	Alibaba Group, USA
Oisin Mac Aodha	University of Edinburgh, UK
Luca Magri	Politecnico di Milano, Italy
Michael Maire	University of Chicago, USA
Subhransu Maji	University of Massachusetts, Amherst, USA
Atsuto Maki	KTH Royal Institute of Technology, Sweden
Yasushi Makihara	Osaka University, Japan
Massimiliano Mancini	University of Trento, Italy
Kevis-Kokitsi Maninis	Google Research, Switzerland
Kenneth Marino	Google DeepMind, USA
Renaud Marlet	Ecole des Ponts ParisTech & Valeo, France
Niki Martinel	University of Udine, Italy
Jiri Matas	Czech Technical University in Pragu, Czech Republic
Yasuyuki Matsushita	Osaka University, Japan
Simone Melzi	University of Milano-Bicocca, Italy
Thomas Mensink	Google Research, Netherlands
Michele Merler	IBM Research, USA
Pascal Mettes	University of Amsterdam, Netherlands
Ajmal Mian	University of Western Australia, Australia
Krystian Mikolajczyk	Imperial College London, UK
Ishan Misra	Facebook AI Research, USA
Niloy Mitra	University College London, UK
Anurag Mittal	Indian Institute of Technology Madras, India
Davide Modolo	Amazon, USA
Davide Moltisanti	University of Bath, UK
Philippos Mordohai	Stevens Institute of Technology, USA
Francesc Moreno	Institut de Robotica i Informatica Industrial, Spain
Pietro Morerio	Istituto Italiano di Tecnologia, Italy
Greg Mori	Simon Fraser University & Borealis AI, Canada
Roozbeh Mottaghi	FAIR @ Meta, USA
Yadong Mu	Peking University, China
Vineeth N. Balasubramanian	Indian Institute of Technology, India
Hajime Nagahara	Osaka University, Japan
Vinay Namboodiri	University of Bath, UK
Liangliang Nan	Delft University of Technology, Netherlands

Michele Nappi	University of Salerno, Italy
Srinivasa Narasimhan	Carnegie Mellon University, USA
P. J. Narayanan	International Institute of Information Technology, Hyderabad, India
Nassir Navab	Technical University of Munich, Germany
Ram Nevatia	University of Southern California, USA
Natalia Neverova	Facebook AI Research, France
Juan Carlos Niebles	Salesforce & Stanford University, USA
Michael Niemeyer	Google, Germany
Simon Niklaus	Adobe Research, USA
Ko Nishino	Kyoto University, Japan
Nicoletta Noceti	MaLGa-DIBRIS & Università degli Studi di Genova, Italy
Matthew O'Toole	Carnegie Mellon University, USA
Jean-Marc Odobez	Idiap Research Institute, École polytechnique fédérale de Lausanne, Switzerland
Francesca Odone	Università di Genova, Italy
Takayuki Okatani	Tohoku University & RIKEN Center for Advanced Intelligence Project, Japan
Carl Olsson	Lund University, Sweden
Vicente Ordonez	Rice University, USA
Aljosa Osep	Technical University of Munich, Germany
Martin R. Oswald	University of Amsterdam, Netherlands
Andrew Owens	University of Michigan, USA
Tomas Pajdla	Czech Technical University in Prague, Czechia
Manohar Paluri	Meta, USA
Jinshan Pan	Nanjing University of Science and Technology, China
In Kyu Park	Inha University, Korea
Jaesik Park	Seoul National University, Korea
Despoina Paschalidou	Stanford, USA
Georgios Pavlakos	University of Texas at Austin, USA
Vladimir Pavlovic	Rutgers University, USA
Marcello Pelillo	University of Venice, Italy
David Picard	École des Ponts ParisTech, France
Hamed Pirsiavash	University of California, Davis, USA
Bryan Plummer	Boston University, USA
Thomas Pock	Graz University of Technology, Austria
Fabio Poiesi	Fondazione Bruno Kessler, Italy
Marc Pollefeys	ETH Zurich & Microsoft, Switzerland
Jean Ponce	Ecole normale supérieure - PSL, France
Gerard Pons-Moll	University of Tübingen, Germany

Jordi Pont-Tuset	Google, Switzerland
Fatih Porikli	Qualcomm R&D, USA
Brian Price	Adobe, USA
Guo-Jun Qi	Westlake University, USA
Long Quan	Hong Kong University of Science and Technology, China
Venkatesh Babu Radhakrishnan	Indian Institute of Science, India
Hossein Rahmani	Lancaster University, UK
Deva Ramanan	Carnegie Mellon University, USA
Vignesh Ramanathan	Facebook, USA
Vikram V. Ramaswamy	Princeton University, USA
Nalini Ratha	University at Buffalo, State University of New York, USA
Yogesh Rawat	University of Central Florida, USA
Hamid Rezatofighi	Monash University, Australia
Helge Rhodin	University of British Columbia, Canada
Stephan Richter	Apple, Germany
Anna Rohrbach	Technical University of Darmstadt, Germany
Marcus Rohrbach	Technical University of Darmstadt, Germany
Gemma Roig	Goethe University Frankfurt, Germany
Negar Rostamzadeh	Google, Canada
Paolo Rota	University of Trento, Italy
Stefan Roth	Technical University of Darmstadt, Germany
Carsten Rother	University of Heidelberg, Germany
Subhankar Roy	University of Trento, Italy
Michael Rubinstein	Google, USA
Christian Rupprecht	University of Oxford, UK
Olga Russakovsky	Princeton University, USA
Bryan Russell	Adobe Research, USA
Michael Ryoo	Stony Brook University & Google, USA
Reza Sabzevari	Delft University of Technology, Netherlands
Mathieu Salzmann	École polytechnique fédérale de Lausanne, Switzerland
Dimitris Samaras	Stony Brook University, USA
Aswin Sankaranarayanan	Carnegie Mellon University, USA
Sudeep Sarkar	University of South Florida, USA
Yoichi Sato	University of Tokyo, Japan
Manolis Savva	Simon Fraser University, Canada
Simone Schaub-Meyer	Technical University of Darmstadt, Germany
Bernt Schiele	Max Planck Institute for Informatics, Germany
Cordelia Schmid	Inria & Google, France
Nicu Sebe	University of Trento, Italy

Laura Sevilla-Lara	University of Edinburgh, UK
Fahad Shahbaz Khan	Mohamed bin Zayed University of Artificial Intelligence, United Arab Emirates
Qi Shan	Apple, USA
Shiguang Shan	Institute of Computing Technology, Chinese Academy of Sciences, China
Viktoriia Sharmanska	University of Sussex & Imperial College London, UK
Eli Shechtman	Adobe Research, USA
Evan Shelhamer	DeepMind, UK
Lu Sheng	Beihang University, China
Boxin Shi	Peking University, China
Jianbo Shi	University of Pennsylvania, USA
Mike Zheng Shou	National University of Singapore, Singapore
Abhinav Shrivastava	University of Maryland, USA
Aliaksandr Siarohin	Snap, USA
Kaleem Siddiqi	McGill University, Canada
Leonid Sigal	University of British Columbia, Canada
Oriane Siméoni	valeo.ai, France
Krishna Kumar Singh	Adobe Research, USA
Richa Singh	Indian Institute of Technology Jodhpur, India
Yale Song	Facebook AI Research, USA
Concetto Spampinato	University of Catania, Italy
Srinath Sridhar	Brown University, USA
Bjorn Stenger	Rakuten, Japan
Abby Stylianou	Saint Louis University, USA
Akihiro Sugimoto	National Institute of Informatics, Japan
Chen Sun	Brown University, USA
Deqing Sun	Google, USA
Jian Sun	Xi'an Jiaotong University, China
Min Sun	National Tsing Hua University, Taiwan
Qianru Sun	Singapore Management University, Singapore
Yu-Wing Tai	Dartmouth College, China
Ayellet Tal	Technion, Israel
Ping Tan	Hong Kong University of Science and Technology, China
Robby Tan	National University of Singapore, Singapore
Chi-Keung Tang	Hong Kong University of Science and Technology, China
Makarand Tapaswi	International Institute of Information Technology, Hyderabad & Wadhwani AI, India
Justus Thies	Technical University of Darmstadt, Germany

Radu Timofte	University of Wurzburg & ETH Zurich, Germany
Giorgos Tolias	Czech Technical University in Prague, Czechia
Federico Tombari	Google & Technical University of Munich, Switzerland
Tatiana Tommasi	Politecnico di Torino, Italy
James Tompkin	Brown University, USA
Matthew Trager	AWS AI Labs, USA
Anh Tran	VinAI, Vietnam
Du Tran	Google, USA
Shubham Tulsiani	Carnegie Mellon University, USA
Sergey Tulyakov	Snap, USA
Dimitrios Tzionas	University of Amsterdam, Netherlands
Maria Vakalopoulou	CentraleSupelec, France
Joost van de Weijer	Computer Vision Center, Spain
Laurens van der Maaten	Meta, USA
Jan van Gemert	Delft University of Technology, Netherlands
Sebastiano Vascon	Ca' Foscari University of Venice & European Centre for Living Technology, Italy
Nuno Vasconcelos	University of California, San Diego, USA
Mayank Vatsa	Indian Institute of Technology Jodhpur, India
Javier Vazquez-Corral	Autonomous University of Barcelona, Spain
Andrea Vedaldi	Oxford University, UK
Olga Veksler	University of Waterloo, Canada
Luisa Verdoliva	University Federico II of Naples, Italy
Rene Vidal	University of Pennsylvania, USA
Bo Wang	CtrsVision, USA
Heng Wang	TikTok, USA
Jingdong Wang	Baidu, China
Jingya Wang	ShanghaiTech University, China
Ruiping Wang	Institute of Computing Technology, Chinese Academy of Sciences, China
Shenlong Wang	University of Illinois, Urbana-Champaign, USA
Ting-Chun Wang	NVIDIA, USA
Wenguan Wang	Zhejiang University, China
Xiaolong Wang	University of California, San Diego, USA
Xiaoqian Wang	Purdue University, USA
Xiaoyu Wang	Hong Kong University of Science and Technology, USA
Xin Wang	Microsoft Research, USA
Xinchao Wang	National University of Singapore, Singapore
Yang Wang	Concordia University, Canada
Yiming Wang	Fondazione Bruno Kessler, Italy

Yu-Chiang Frank Wang	National Taiwan University, Taiwan
Yu-Xiong Wang	University of Illinois, Urbana-Champaign, USA
Zhangyang Wang	University of Texas at Austin, USA
Olivia Wiles	DeepMind, UK
Christian Wolf	Naver Labs Europe, France
Michael Wray	University of Bristol, UK
Jiajun Wu	Stanford University, USA
Jianxin Wu	Nanjing University, China
Shangzhe Wu	Stanford University, USA
Ying Wu	Northwestern University, USA
Yu Wu	Wuhan University, China
Jianwen Xie	Baidu Research, USA
Saining Xie	New York University, USA
Weidi Xie	Shanghai Jiao Tong University, China
Dan Xu	Hong Kong University of Science and Technology, China
Huijuan Xu	Pennsylvania State University, USA
Xiangyu Xu	Xi'an Jiaotong University, China
Yan Yan	Illinois Institute of Technology, USA
Jiaolong Yang	Microsoft Research, China
Kaiyu Yang	California Institute of Technology, USA
Linjie Yang	ByteDance AI Lab, USA
Ming-Hsuan Yang	University of California, Merced, USA
Yanchao Yang	University of Hong Kong, China
Yi Yang	Zhejiang University, China
Angela Yao	National University of Singapore, Singapore
Mang Ye	Wuhan University, China
Kwang Moo Yi	University of British Columbia, Canada
Xi Yin	Facebook, USA
Chang D. Yoo	Korea Advanced Institute of Science and Technology, Korea
Shaodi You	University of Amsterdam, Netherlands
Jingyi Yu	Shanghai Tech University, China
Ning Yu	Netflix Eyeline Studios, USA
Qi Yu	Rochester Institute of Technology, USA
Stella Yu	University of Michigan, USA
Yizhou Yu	University of Hong Kong, China
Junsong Yuan	University at Buffalo, State University of New York, USA
Lu Yuan	Microsoft, USA
Sangdoo Yun	NAVER AI Lab, Korea
Hongbin Zha	Peking University, China

Jing Zhang	Australian National University, Australia
Lei Zhang	Hong Kong Polytechnic University, China
Ning Zhang	Facebook AI, USA
Richard Zhang	Adobe, USA
Tianzhu Zhang	University of Science and Technology of China, China
Hengshuang Zhao	University of Hong Kong, China
Qi Zhao	University of Minnesota, USA
Sicheng Zhao	Tsinghua University, China
Liang Zheng	Australian National University, Australia
Wei-Shi Zheng	Sun Yat-sen University, China
Zhun Zhong	University of Nottingham, UK
Bolei Zhou	University of California, Los Angeles, USA
Kaiyang Zhou	Hong Kong Baptist University, China
Luping Zhou	University of Sydney, Australia
S. Kevin Zhou	University of Science and Technology of China, China
Lei Zhu	Hong Kong University of Science and Technology (Guangzhou), China
Linchao Zhu	Zhejiang University, China
Andrew Zisserman	University of Oxford, UK
Daniel Zoran	DeepMind, UK
Wangmeng Zuo	Harbin Institute of Technology, China

Technical Program Committee

Sathyanarayanan N. Aakur
Andrea Abate
Wim Abbeloos
Amos L. Abbott
Rameen Abdal
Salma Abdel Magid
Rabab Abdelfattah
Ahmed Abdelkader
Eslam Mohamed Abdelrahman
Ahmed Abdelreheem
Akash Abdu Jyothi
Kfir Aberman
Shahira Abousamra
Yazan Abu Farha
Shady Abu-Hussein

Abulikemu Abuduweili
Hanno Ackermann
Chandranath Adak
Aikaterini Adam
Lukáš Adam
Kamil Adamczewski
Donald Adjeroh
Mohammed Adnan
Mahmoud Afifi
Triantafyllos Afouras
Akshay Agarwal
Madhav Agarwal
Nakul Agarwal
Sharat Agarwal
Shivang Agarwal
Vatsal Agarwal

Abhinav Agarwalla
Shivam Aggarwal
Sara Aghajanzadeh
Shashank Agnihotri
Harsh Agrawal
Antonio Agudo
Shai Aharon
Saba Ahmadi
Waqar Ahmed
Byeongjoo Ahn
Chanho Ahn
Jaewoo Ahn
Namhyuk Ahn
Seoyoung Ahn
Nilesh A. Ahuja
Yihao Ai

Abhishek Aich
Emanuele Aiello
Noam Aigerman
Kiyoharu Aizawa
Kenan Emir Ak
Reza Akbarian Bafghi
Sai Aparna Aketi
Derya Akkaynak
Ibrahim Alabdulmohsin
Yuval Alaluf
Md Ferdous Alam
Stephan Alaniz
Badour A. Sh AlBahar
Paul Albert
Isabela Albuquerque
Emanuel Aldea
Luca Alessandrini
Emma Alexander
Konstantinos P.
　Alexandridis
Stamatis Alexandropoulos
Saghir Alfasly
Mohsen Ali
Waqar Ali
Daniel Aliaga
Sadegh Aliakbarian
Garvita Allabadi
Thiemo Alldieck
Antonio Alliegro
Jurandy Almeida
Gal Alon
Hadi Alzayer
Ibtihel Amara
Giuseppe Amato
Sameer Ambekar
Niki Amini-Naieni
Shir Amir
Roberto Amoroso
Dongsheng An
Jie An
Junyi An
Liang An
Xiang An
Yuexuan An
Zhaochong An

Zhulin An
Saket Anand
Pavan Kumar Anasosalu
　Vasu
Codruta O. Ancuti
Cosmin Ancuti
Fernanda Andaló
Connor Anderson
Juan Andrade-Cetto
Bruno Andreis
Alexander Andreopoulos
Bjoern Andres
Shivangi Aneja
Dragomir Anguelov
Damith Kawshan
　Anhettigama
Aditya Annavajjala
Michel Antunes
Tejas Anvekar
Saeed Anwar
Evlampios Apostolidis
Srikar Appalaraju
Relja Arandjelović
Nikita Araslanov
Alexandre Araujo
Helder Araujo
Eric Arazo
Dawit Mureja Argaw
Anurag Arnab
Federica Arrigoni
Alexey Artemov
Aditya Arun
Nader Asadi
Kumar Ashutosh
Vishal Asnani
Mahmoud Assran
Amir Atapour-Abarghouei
Nikos Athanasiou
Ali Athar
ShahRukh Athar
Rowel O. Atienza
Benjamin Attal
Matan Atzmon
Nicolas Audebert
Romaric Audigier

Tristan
　Aumentado-Armstrong
Yannis Avrithis
Muhammad Awais
Md Rabiul Awal
Görkay Aydemir
Mehmet Aygün
Melika Ayoughi
Ali Ayub
Kumar Ayush
Mehdi Azabou
Basim Azam
Mohammad Farid
　Azampour
Hossein Azizpour
Vivek B. S.
Yunhao Ba
Mohammadreza Babaee
Deepika Bablani
Sudarshan Babu
Roman Bachmann
Inhwan Bae
Jeonghun Baek
Kyungjune Baek
Seungryul Baek
Piyush Nitin Bagad
Mohammadhossein Bahari
Gaetan Bahl
Sherwin Bahmani
Bing Bai
Haoyue Bai
Jiawang Bai
Jinbin Bai
Lei Bai
Qingyan Bai
Shutao Bai
Xuyang Bai
Yalong Bai
Yuanchao Bai
Yue Bai
Yunpeng Bai
Ziyi Bai
Daniele Baieri
Sungyong Baik
Ramon Baldrich

Coloma Ballester
Irene Ballester
Sonat Baltaci
Biplab Banerjee
Monami Banerjee
Sandipan Banerjee
Snehasis Banerjee
Soumya Banerjee
Sreya Banerjee
Jihwan Bang
Ankan Bansal
Nitin Bansal
Siddhant Bansal
Chong Bao
Hangbo Bao
Runxue Bao
Wentao Bao
Yiwei Bao
Zhipeng Bao
Zhongyun Bao
Amir Bar
Omer Bar Tal
Manel Baradad Jurjo
Lorenzo Baraldi
Lorenzo Baraldi
Tal Barami
Danny Barash
Daniel Barath
Adrian Barbu
Nick Barnes
Alina J. Barnett
German Barquero
Joao P. Barreto
Jonathan T. Barron
Rodrigo C. Barros
Luca Barsellotti
Myles Bartlett
Kristijan Bartol
Hritam Basak
Dina Bashkirova
Chaim Baskin
Lennart C. Bastian
Abhipsa Basu
Soumen Basu
Peyman Bateni

Anil Batra
David Bau
Zuria Bauer
Eduard Gabriel Bazavan
Federico Becattini
Nathan A. Beck
Jan Bednarik
Peter A. Beerel
Ardhendu Behera
Harkirat Singh Behl
Jens Behley
William Beksi
Soufiane Belharbi
Giovanni Bellitto
Ismail Ben Ayed
Abdessamad Ben Hamza
Yizhak Ben-Shabat
Guy Ben-Yosef
Robert Benavente
Assia Benbihi
Yasser Benigmim
Urs Bergmann
Amit H Bermano
Jesus Bermudez-Cameo
Florian Bernard
Stefano Berretti
Gabriele Berton
Petra Bevandić
Matthew Beveridge
Lalit Bhagat
Yash Bhalgat
Homanga Bharadhwaj
Skanda Bharadwaj
Ambika Saklani Bhardwaj
Goutam Bhat
Gaurav Bhatt
Neel P. Bhatt
Deblina Bhattacharjee
Anish Bhattacharya
Rajarshi Bhattacharya
Uttaran Bhattacharya
Apratim Bhattacharyya
Anand Bhattad
Binod Bhattarai
Snehal Bhayani

Brojeshwar Bhowmick
Aritra Bhowmik
Neelanjan Bhowmik
Amran Bhuiyan
Ankan Kumar Bhunia
Ayan Kumar Bhunia
Jing Bi
Qi Bi
Xiuli Bi
Michael Bi Mi
Hao Bian
Jia-Wang Bian
Shaojun Bian
Simone Bianco
Pia Bideau
Xiaoyu Bie
Roberto Bigazzi
Mario Bijelic
Bahri Batuhan Bilecen
Guillaume-Alexandre
 Bilodeau
Alexander Binder
Massimo Bini
Niccolò Bisagno
Carmen Bisogni
Anthony R. Bisulco
Sandika Biswas
Sanket Biswas
Ksenia Bittner
Kevin Blackburn-Matzen
Nathaniel Blanchard
Hunter Blanton
Emile Blettery
Hermann Blum
Deyu Bo
Janusz Bobulski
Marius Bock
Lea Bogensperger
Simion-Vlad Bogolin
Moritz Böhle
Piotr Bojanowski
Alexey Bokhovkin
Georg Bökman
Ladislau Boloni
Daniel Bolya

Lorenzo Bonicelli
Gianpaolo Bontempo
Vivek Boominathan
Adrian Bors
Damian Borth
Matteo Bortolon
Davide Boscaini
Laurie Nicholas Bose
Shirsha Bose
Mark Boss
Andrea Bottino
Larbi Boubchir
Alexandre Boulch
Terrance E. Boult
Amine Bourki
Walid Bousselham
Fadi Boutros
Nicolas C. Boutry
Thierry Bouwmans
Richard S. Bowen
Kevin W. Bowyer
Ivaylo Boyadzhiev
Aidan Boyd
Joseph Boyd
Aljaz Bozic
Behzad Bozorgtabar
Manuel Brack
Samarth Brahmbhatt
Nikolas Brasch
Cameron E. Braunstein
Toby P. Breckon
Mathieu Bredif
Romain Brégier
Carlo Bretti
Joel R. Brogan
Clifford Broni-Bediako
Andrew Brown
Ellis L. Brown
Thomas Brox
Qingwen Bu
Tong Bu
Shyamal Buch
Alvaro Budria
Ignas Budvytis
José M. Buenaposada

Kyle R. Buettner
Marcel C. Bühler
Nhat-Tan Bui
Adrian Bulat
Valay M. Bundele
Nathaniel J. Burgdorfer
Ryan D. Burgert
Evgeny Burnaev
Pietro Buzzega
Marco Buzzelli
Kwon Byung-Ki
Fabian Caba
Felipe Cadar
Martin Cadik
Bowen Cai
Changjiang Cai
Guanyu Cai
Haoming Cai
Hongrui Cai
Jiarui Cai
Jinyu Cai
Kaixin Cai
Lile Cai
Minjie Cai
Mu Cai
Pingping Cai
Rizhao Cai
Ruisi Cai
Ruojin Cai
Shen Cai
Shengqu Cai
Weidong Cai
Weiwei Cai
Wenjia Cai
Wenxiao Cai
Yancheng Cai
Yuanhao Cai
Yujun Cai
Yuxuan Cai
Zhaowei Cai
Zhipeng Cai
Zhixi Cai
Zhongang Cai
Zikui Cai
Salvatore Calcagno

Roberto Caldelli
Akin Caliskan
Lilian Calvet
Necati Cihan Camgoz
Tommaso Campari
Dylan Campbell
Guglielmo Camporese
Yigit Baran Can
Shaun Canavan
Gemma Canet Tarrés
Marco Cannici
Ang Cao
Anh-Quan Cao
Bing Cao
Bingyi Cao
Chenjie Cao
Dongliang Cao
Hu Cao
Jiahang Cao
Jiale Cao
Jie Cao
Jiezhang Cao
Jinkun Cao
Meng Cao
Mingdeng Cao
Peng Cao
Pu Cao
Qinglong Cao
Qingxing Cao
Shengcao Cao
Song Cao
Weiwei Cao
Xiangyong Cao
Xiao Cao
Yu Cao
Xu Cao
Yan-Pei Cao
Yang Cao
Yichao Cao
Ying Cao
Yuan Cao
Yuhang Cao
Yukang Cao
Yulong Cao
Yun-Hao Cao

Yunhao Cao
Yutong Cao
Zhangjie Cao
Zhiguo Cao
Zhiwen Cao
Ziang Cao
Giacomo Capitani
Luigi Capogrosso
Andrea Caraffa
Jaime S. Cardoso
Chris Careaga
Zachariah Carmichael
Giuseppe Cartella
Vincent Cartillier
Paola Cascante-Bonilla
Lucia Cascone
Pol Caselles Rico
Angela Castillo
Lluis Castrejon
Modesto
    Castrillón-Santana
Francisco M. Castro
Pedro Castro
Jacopo Cavazza
Oya Celiktutan
Luigi Celona
Jiazhong Cen
Fernando J. Cendra
Fabio Cermelli
Duygu Ceylan
Eunju Cha
Junbum Cha
Rohan Chabra
Rohan Chacko
Julia Chae
Nicolas Chahine
Junyi Chai
Lucy Chai
Tianrui Chai
Wenhao Chai
Zenghao Chai
Anish Chakrabarty
Goirik Chakrabarty
Anirban Chakraborty
Deep Chakraborty

Souradeep Chakraborty
Punarjay Chakravarty
Jacob Chalk
Loick Chambon
Chee Seng Chan
David Chan
Dorian Y. Chan
Kin Sun Chan
Matthew Chan
Adrien Chan-Hon-Tong
Sampath Chanda
Shivam Chandhok
Prashanth Chandran
Kenneth Chaney
Chirui Chang
Di Chang
Dongliang Chang
Hong Chang
Hyung Jin Chang
Ju Yong Chang
Matthew Chang
Ming-Ching Chang
MingFang Chang
Nadine Chang
Qi Chang
Seunggyu Chang
Wei-Yi Chang
Xiaojun Chang
Yakun Chang
Yi Chang
Zheng Chang
Sumohana S.
    Channappayya
Hanqing Chao
Pradyumna Chari
Korawat Charoenpitaks
Dibyadip Chatterjee
Moitreya Chatterjee
Pratik Chaudhari
Muawiz Chaudhary
Ritwick Chaudhry
Ruchika Chavhan
Sofian Chaybouti
Zhengping Che
Gullal Singh Cheema

Kunal Chelani
Rama Chellappa
Allison Y. Chen
Anpei Chen
Aozhu Chen
Bin Chen
Bin Chen
Bin Chen
Bohong Chen
Bor-Chun Chen
Bowei Chen
Brian Chen
Chang Wen Chen
Changan Chen
Changhao Chen
Changhuai Chen
Chaofan Chen
Chaofan Chen
Chaofeng Chen
Chen Chen
Cheng Chen
Chengkuan Chen
Chieh-Yun Chen
Chong Chen
Chun-Fu Richard Chen
Congliang Chen
Cuiqun Chen
Da Chen
Daoyuan Chen
Defang Chen
Dian Chen
Dian Chen
Ding-Jie Chen
Dingfan Chen
Dong Chen
Dongdong Chen
Fangyi Chen
Guang-Yong Chen
Guangyao Chen
Guangyi Chen
Guanying Chen
Guikun Chen
Guo Chen
Haibo Chen
Haiwei Chen

Hansheng Chen
Hanting Chen
Hanzhi Chen
Hao Chen
Haodong Chen
Haokun Chen
Haoran Chen
Haoxin Chen
Haoyu Chen
Haoyu Chen
Hong-You Chen
Honghua Chen
Honglin Chen
Hongming Chen
Huanran Chen
Hui Chen
Hwann-Tzong Chen
Jiacheng Chen
Jiajing Chen
Jianqiu Chen
Jiaqi Chen
Jiaxin Chen
Jiayi Chen
Jie Chen
Jie Chen
Jie Chen
Jieneng Chen
Jierun Chen
Jinghui Chen
Jingjing Chen
Jintai Chen
Jinyu Chen
Joya Chen
Jun Chen
Jun Chen
Jun Chen
Jun-Cheng Chen
Jun-Kun Chen
Junjie Chen
Kai Chen
Kai Chen
Kaifeng Chen
Kefan Chen
Kejiang Chen
Kuan-Wen Chen

Kunyao Chen
Li Chen
Liang Chen
Lianggangxu Chen
Liangyu Chen
Ling-Hao Chen
Linghao Chen
Liyan Chen
Min Chen
Min-Hung Chen
Minghao Chen
Minghui Chen
Nenglun Chen
Peihao Chen
Pengguang Chen
Ping Chen
Qi Chen
Qi Chen
Qi Chen
Qiang Chen
Qimin Chen
Qingcai Chen
Rongyu Chen
Rui Chen
Runjian Chen
Shang-Fu Chen
Shen Chen
Sherry X. Chen
Shi Chen
Shiming Chen
Shiyan Chen
Shizhe Chen
Shoufa Chen
Shu-Yu Chen
Shuai Chen
Si Chen
Siheng Chen
Simin Chen
Sixiang Chen
Sizhe Chen
Songcan Chen
Tao Chen
Tianshui Chen
Tianyi Chen
Tsai-Shien Chen

Wei Chen
Weidong Chen
Weihua Chen
Weijie Chen
Weikai Chen
Wuyang Chen
Xi Chen
Xi Chen
Xiang Chen
Xiangyu Chen
Xianyu Chen
Xiao Chen
Xiao Chen
Xiaohan Chen
Xiaokang Chen
Xiaotong Chen
Xin Chen
Xin Chen
Xin Chen
Xinghao Chen
Xingyu Chen
Xingyu Chen
Xinlei Chen
Xiuwei Chen
Xuanbai Chen
Xuanhong Chen
Xuesong Chen
Xuweiyi Chen
Xuxi Chen
Yanbei Chen
Yang Chen
Yaofo Chen
Yen-Chun Chen
Yi-Ting Chen
Yi-Wen Chen
Yilun Chen
Yinbo Chen
Yingcong Chen
Yingwen Chen
Yiting Chen
Yixin Chen
Yixin Chen
Yu Chen
Yuanhong Chen
Yuanqi Chen

Yuedong Chen
Yueting Chen
Yuhao Chen
Yuhao Chen
Yujin Chen
Yukang Chen
Yun-Chun Chen
Yunjin Chen
Yunlu Chen
Yuntao Chen
Yushuo Chen
Yuxiao Chen
Zehui Chen
Zerui Chen
Zewen Chen
Zeyuan Chen
Zeyuan Chen
Zhang Chen
Zhao-Min Chen
Zhaoliang Chen
Zhaoxi Chen
Zhaoyu Chen
Zhaozheng Chen
Zhaozheng Chen
Zhe Chen
Zhen Chen
Zhen Chen
Zheng Chen
Zhenghao Chen
Zhengyu Chen
Zhenyu Chen
Zhi Chen
Zhibo Chen
Zhimin Chen
Zhineng Chen
Zhiwei Chen
Zhixiang Chen
Zhiyang Chen
Zhiyuan Chen
Zhuangzhuang Chen
Zhuo Chen
Zhuoxiao Chen
Zihan Chen
Zijiao Chen
Ziwen Chen

Ziyang Chen
An-Chieh Cheng
Bowen Cheng
De Cheng
Erkang Cheng
Feng Cheng
Guangliang Cheng
Hao Cheng
Hao Cheng
Harry Cheng
Ho Kei Cheng
Jingchun Cheng
Jun Cheng
Ka Leong Cheng
Kelvin B. Cheng
Kevin Ho Man Cheng
Lechao Cheng
Shuo Cheng
Silin Cheng
Ta-Ying Cheng
Tianhang Cheng
Weihao Cheng
Wencan Cheng
Wensheng Cheng
Xi Cheng
Xize Cheng
Xuxin Cheng
Yen-Chi Cheng
Yi Cheng
Yihua Cheng
Yu Cheng
Zesen Cheng
Zezhou Cheng
Zhanzhan Cheng
Zhen Cheng
Zhi-Qi Cheng
Ziang Cheng
Ziheng Cheng
Soon Yau Cheong
Anoop Cherian
Daniil Cherniavskii
Ajad Chhatkuli
Cheng Chi
Haoang Chi
Hyung-gun Chi

Seunggeun Chi
Zhixiang Chi
Naoya Chiba
Julian Chibane
Aditya Chinchure
Eleni Chiou
Mia Chiquier
Chiranjeev Chiranjeev
Kashyap Chitta
Hsu-kuang Chiu
Wei-Chen Chiu
Chee Kheng Chng
Shin-Fang Chng
Donghyeon Cho
Gyusang Cho
Hanbyel Cho
Hoonhee Cho
Jae Won Cho
Jang Hyun Cho
Jungchan Cho
Junhyeong Cho
Nam Ik Cho
Seokju Cho
Seungju Cho
Suhwan Cho
Sunghyun Cho
Sungjun Cho
Sungmin Cho
Wonwoong Cho
Nathaniel E. Chodosh
Jaesung Choe
Junsuk Choe
Changwoon Choi
Chiho Choi
Ching Lam Choi
Dongyoung Choi
Dooseop Choi
Hongsuk Choi
Hyesong Choi
Jaewoong Choi
Janghoon Choi
Jin Young Choi
Jinwoo Choi
Jinyoung Choi
Jonghyun Choi

Jongwon Choi
Jooyoung Choi
Kanghyun Choi
Myungsub Choi
Seokeon Choi
Wonhyeok Choi
Bhavin Choksi
Jaegul Choo
Baptiste Chopin
Ayush Chopra
Gene Chou
Shih-Han Chou
Divya Choudhary
Siddharth Choudhary
Sunav Choudhary
Rohan Choudhury
Vasileios Choutas
Ka-Ho Chow
Pinaki Nath Chowdhury
Sanjoy Chowdhury
Sammy Christen
Fu-Jen Chu
Qi Chu
Ruihang Chu
Sanghyeok Chu
Wei-Ta Chu
Wen-Hsuan Chu
Wen-Hsuan Chu
Wei Qin Chuah
Andrei Chubarau
Ilya Chugunov
Sanghyuk Chun
Se Young Chun
Hyungjin Chung
Inseop Chung
Jaeyoung Chung
Jihoon Chung
Jiwan Chung
Joon Son Chung
Thomas A. Ciarfuglia
Umur A. Ciftci
Antonio E. Cinà
Ramazan Gokberk Cinbis
Anthony Cioppa
Javier Civera

Christopher A. Clark
James J. Clark
Ronald Clark
Brian S. Clipp
Mark Coates
Federico Cocchi
Felipe Codevilla
Cecília Coelho
Danny Cohen-Or
Forrester Cole
Julien Colin
John Collomosse
Marc Comino-Trinidad
Nicola Conci
Marcos V. Conde
Alexandru Paul
  Condurache
Runmin Cong
Tianshuo Cong
Wenyan Cong
Yuren Cong
Alessandro Conti
Andrea Conti
Enric Corona
John R. Corring
Jaime Corsetti
Jason J. Corso
Theo W. Costain
Guillaume Couairon
Robin Courant
Davide Cozzolino
Donato Crisostomi
Francesco Croce
Florinel Alin Croitoru
Ioana Croitoru
James L. Crowley
Steve Cruz
Gabriela Csurka
Alfredo Cuesta Infante
Aiyu Cui
Hainan Cui
Jiali Cui
Jiequan Cui
Justin Cui
Qiongjie Cui

Ruikai Cui
Yawen Cui
Ying Cui
Yuning Cui
Zhen Cui
Zhenyu Cui
Ziteng Cui
Benjamin J. Culpepper
Xiaodong Cun
Federico Cunico
Claudio Cusano
Sebastian Cygert
Guido Maria D'Amely di
  Melendugno
Moreno D'Incà
Feipeng Da
Mosam Dabhi
Rishabh Dabral
Konstantinos M. Dafnis
Matthew L. Daggitt
Anders B. Dahl
Bo Dai
Bo Dai
Haixing Dai
Mengyu Dai
Peng Dai
Peng Dai
Pingyang Dai
Qi Dai
Qiyu Dai
Wenrui Dai
Wenxun Dai
Zuozhuo Dai
Nicola Dall'Asen
Nneor Damci
Bo Dang
Meghal Dani
Duolikun Danier
Donald G. Dansereau
Quan Dao
Son Duy Dao
Trung Tuan Dao
Mohamed Daoudi
Ahmad Darkhalil
Rangel Daroya

Abhijit Das
Abir Das
Anurag Das
Ayan Das
Nilotpal Das
Partha Das
Soumi Das
Srijan Das
Sukhendu Das
Swagatam Das
Ananyananda Dasari
Avijit Dasgupta
Gourav Datta
Achal Dave
Ishan Rajendrakumar Dave
Andrew Davison
Aram Davtyan
Axel Davy
Youssef Dawoud
Laura Daza
Francesca De Benetti
Daan de Geus
Alfredo S. De Goyeneche
Riccardo de Lutio
Maria De Marsico
Christophe De
 Vleeschouwer
Tonmoay Deb
Sayan Deb Sarkar
Dale Decatur
Thanos Delatolas
Fabien Delattre
Mauricio Delbracio
Ginger D. Delmas
Andong Deng
Bailin Deng
Bin Deng
Boyang Deng
Chaorui Deng
Cheng Deng
Congyue Deng
Jiacheng Deng
Jiajun Deng
Jiankang Deng
Jingjing Deng
Jinhong Deng
Kangle Deng
Kenan Deng
Lei Deng
Liang-Jian Deng
Shijian Deng
Weijian Deng
Xiaoming Deng
Xueqing Deng
Ye Deng
Yongjian Deng
Youming Deng
Yu Deng
Zhijie Deng
Zhongying Deng
Matteo Denitto
Mohammad Mahdi
 Derakhshani
Alakh Desai
Kevin Desai
Nishq P. Desai
Sai Vikas Desai
Jean-Emmanuel Deschaud
Frédéric Devernay
Rahul Dey
Arturo Deza
Helisa Dhamo
Amaya Dharmasiri
Ankit Dhiman
Vikas Dhiman
Yan Di
Zonglin Di
Ousmane A. Dia
Renwei Dian
Haiwen Diao
Xiaolei Diao
Gonçalo José Dias Pais
Abdallah Dib
Juan Carlos Dibene
 Simental
Sebastian Dille
Christian Diller
Mariella Dimiccoli
Anastasios Dimou
Or Dinari
Caiwen Ding
Changxing Ding
Choubo Ding
Fangqiang Ding
Guiguang Ding
Guodong Ding
Henghui Ding
Hui Ding
Jian Ding
Jian-Jiun Ding
Jianhao Ding
Li Ding
Liang Ding
Lihe Ding
Ming Ding
Runyu Ding
Shouhong Ding
Shuangrui Ding
Shuxiao Ding
Tianjiao Ding
Tianyu Ding
Wenhao Ding
Xiaohan Ding
Xinpeng Ding
Yaqing Ding
Yong Ding
Yuhang Ding
Yuqi Ding
Zheng Ding
Zhengming Ding
Zhipeng Ding
Zhuangzhuang Ding
Zihan Ding
Zihan Ding
Ziluo Ding
Anh-Dung Dinh
Markos Diomataris
Christos Diou
Alish Dipani
Alara Dirik
Ajay Divakaran
Abdelaziz Djelouah
Nemanja Djuric
Thanh-Toan Do
Tien Do

Anh-Dzung Doan
Bao Gia Doan
Khoa D. Doan
Carl Doersch
Andreea Dogaru
Nehal Doiphode
Csaba Domokos
Bowen Dong
Haoye Dong
Jiahua Dong
Jianfeng Dong
Jinpeng Dong
Junhao Dong
Junting Dong
Li Dong
Minjing Dong
Nanqing Dong
Peijie Dong
Qiaole Dong
Qihua Dong
Qiulei Dong
Runpei Dong
Shichao Dong
Shuting Dong
Sixun Dong
Siyan Dong
Tian Dong
Wei Dong
Wei Dong
Weisheng Dong
Xiaoyi Dong
Xiaoyu Dong
Xingping Dong
Yinpeng Dong
Zhenxing Dong
Jonathan C. Donnelly
Naren Doraiswamy
Gianfranco Doretto
Michael Dorkenwald
Muskan Dosi
Huanzhang Dou
Shuguang Dou
Yiming Dou
Zhiyang Frank Dou
Zi-Yi Dou

Arthur Douillard
Michail C. Doukas
Matthijs Douze
Amil Dravid
Vincent Drouard
Dawei Du
Jia-Run Du
Jinhao Du
Ruoyi Du
Weiyu Du
Xiaodan Du
Ye Du
Yingjun Du
Yong Du
Yuanqi Du
Yuntao Du
Zexing Du
Zhuoran Du
Radhika Dua
Haodong Duan
Haoran Duan
Huiyu Duan
Jiafei Duan
Jiali Duan
Peiqi Duan
Ye Duan
Yuchen Duan
Yueqi Duan
Zhihao Duan
Amanda Cardoso Duarte
Shiv Ram Dubey
Anastasia Dubrovina
Daniel Duckworth
Timothy Duff
Nicolas Dufour
Rahul Duggal
Shivam Duggal
Bardienus P. Duisterhof
Felix Dülmer
Matteo Dunnhofer
Chi Nhan Duong
Elona Dupont
Nikita Durasov
Zoran Duric
Mihai Dusmanu

Titir Dutta
Ujjal Kr Dutta
Isht Dwivedi
Sai Kumar Dwivedi
Sebastian Dziadzio
Thomas Eboli
Ramin Ebrahim Nakhli
Benjamin Eckart
Takeharu Eda
Johan Edstedt
Alexei A. Efros
Nikos Efthymiadis
Bernhard Egger
Max Ehrlich
Iván Eichhardt
Farshad Einabadi
Martin Eisemann
Peter Eisert
Mohamed El Banani
Mostafa El-Khamy
Alaa El-Nouby
Randa I. M. Elanwar
James Elder
Abdelrahman Eldesokey
Ismail Elezi
Ehsan Elhamifar
Moshe Eliasof
Sara Elkerdawy
Qing En
Ian Endres
Andreas Engelhardt
Francis Engelmann
Martin Engilberge
Kenji Enomoto
Chanho Eom
Guy Erez
Linus Ericsson
Maria C. Escobar
Marcos Escudero-Viñolo
Sedigheh Eslami
Valter Estevam
Ali Etemad
Martin Nicolas Everaert
Ivan Evtimov
Ralph Ewerth

Fevziye Irem Eyiokur
Cristobal Eyzaguirre
Gabriele Facciolo
Mohammad Fahes
Masud ANI Fahim
Fabrizio Falchi
Alex Falcon
Aoxiang Fan
Baojie Fan
Bin Fan
Chao Fan
David Fan
Haoqiang Fan
Hehe Fan
Heng Fan
Hongyi Fan
Huijie Fan
Junsong Fan
Lei Fan
Lei Fan
Lifeng Fan
Lue Fan
Mingyuan Fan
Qi Fan
Rui R. Fan
Xiran Fan
Yifei Fan
Yue Fan
Zejia Fan
Zhaoxin Fan
Zhenfeng Fan
Zhentao Fan
Zhipeng Fan
Zhiwen Fan
Zicong Fan
Sean Fanello
Chaowei Fang
Chuan Fang
Gongfan Fang
Guian Fang
Han Fang
Jiansheng Fang
Jiemin Fang
Jin Fang
Jun Fang

Kun Fang
Pengfei Fang
Rui Fang
Sheng Fang
Xiang Fang
Xiaolin Fang
Xiuwen Fang
Zhaoyuan Fang
Zhiyuan Fang
Eros Fanì
Matteo Farina
Farzan Farnia
Aiman Farooq
Azade Farshad
Mohsen Fayyaz
Arash Fayyazi
Ben Fei
Jingjing Fei
Xiaohan Fei
Zhengcong Fei
Michael Felsberg
Berthy T. Feng
Bin Feng
Brandon Y. Feng
Chao Feng
Chen Feng
Chengjian Feng
Chun-Mei Feng
Fan Feng
Guorui Feng
Hao Feng
Jianjiang Feng
Lan Feng
Lei Feng
Litong Feng
Mingtao Feng
Qianyu Feng
Ruicheng Feng
Ruili Feng
Ruoyu Feng
Ryan T. Feng
Shiwei Feng
Songhe Feng
Tuo Feng
Wei Feng

Weitao Feng
Weixi Feng
Xuelu Feng
Yanglin Feng
Yao Feng
Yue Feng
Zhanxiang Feng
Zhenhua Feng
Zunlei Feng
Clara Fernandez
Victoria Fernandez
  Abrevaya
Miguel-Ángel
  Fernández-Torres
Sira Ferradans
Claudio Ferrari
Ethan Fetaya
Mustansar Fiaz
Guénolé Fiche
Panagiotis P. Filntisis
Marco Fiorucci
Kai Fischer
Tobias Fischer
Tom Fischer
Volker Fischer
Robert B. Fisher
Alessandro Flaborea
Corneliu O. Florea
Georgios Floros
Alejandro Fontan
Tomaso Fontanini
Lin Geng Foo
Wolfgang Förstner
Niki M. Foteinopoulou
Simone Foti
Simone Foti
Victor Fragoso
Gianni Franchi
Jean-Sebastien Franco
Emanuele Frascaroli
Moti Freiman
Rafail Fridman
Sara Fridovich-Keil
Felix Friedrich
Stanislav Frolov

Andre Fu
Bin Fu
Changcheng Fu
Changqing Fu
Jianlong Fu
Jie Fu
Jingjing Fu
Keren Fu
Lan Fu
Lele Fu
Minghao Fu
Qichen Fu
Rao Fu
Taimeng Fu
Tianwen Fu
Yanwei Fu
Ying Fu
Yonggan Fu
Yun Fu
Yuqian Fu
Zehua Fu
Zhenqi Fu
Zipeng Fu
Wolfgang Fuhl
Yasuhisa Fujii
Yuki Fujimura
Katsuki Fujisawa
Kent Fujiwara
Takuya Funatomi
Christopher Funk
Antonino Furnari
Ryo Furukawa
Ryosuke Furuta
David Futschik
Akshay Gadi Patil
Siddhartha Gairola
Rinon Gal
Adrian Galdran
Silvio Galesso
Meirav Galun
Rofida Gamal
Weijie Gan
Yiming Gan
Yulu Gan
Vineet Gandhi

Kanchana Vaishnavi Gandikota
Aditya Ganeshan
Aditya Ganeshan
Swetava Ganguli
Sujoy Ganguly
Hanan Gani
Alireza Ganjdanesh
Harald Ganster
Roy Ganz
Angela F. Gao
Bin-Bin Gao
Changxin Gao
Chen Gao
Chongyang Gao
Daiheng Gao
Daoyi Gao
Difei Gao
Hang Gao
Hongchang Gao
Huiyu Gao
Junyu Gao
Junyu Gao
Kaifeng Gao
Katelyn Gao
Kuofeng Gao
Lin Gao
Ling Gao
Maolin Gao
Quankai Gao
Ruopeng Gao
Shanghua Gao
Shangqi Gao
Shangqian Gao
Shaobing Gao
Shenyuan Gao
Xiang Gao
Xiangjun Gao
Yang Gao
Yipeng Gao
Yuan Gao
Yue Gao
Yunhe Gao
Zhanning Gao
Zhi Gao

Zhihan Gao
Zhitong Gao
Zhongpai Gao
Ziteng Gao
Nicola Garau
Elena Garces
Noa Garcia
Nuno Cruz Garcia
Ricardo Garcia Pinel
Guillermo Garcia-Hernando
Saurabh Garg
Mathieu Garon
Risheek Garrepalli
Pablo Garrido
Quentin Garrido
Stefano Gasperini
Vincent Gaudilliere
Chandan Gautam
Shivam Gautam
Paul Gavrikov
Jonathan Z. Gazak
Chongjian Ge
Liming Ge
Runzhou Ge
Shiming Ge
Songwei Ge
Weifeng Ge
Wenhang Ge
Yanhao Ge
Yixiao Ge
Yunhao Ge
Yuying Ge
Zhiqi Ge
Zongyuan Ge
James Gee
Chen Geng
Daniel Geng
Xue Geng
Zhengyang Geng
Zichen Geng
Kyle Genova
Georgios Georgakis
Mariana-Iuliana Georgescu

Yiangos Georgiou
Markos Georgopoulos
Stamatios Georgoulis
Michal Geyer
Mina Ghadimi Atigh
Abhijay Ghildyal
Reza Ghoddoosian
Mohsen Gholami
Peyman Gholami
Enjie Ghorbel
Soumya Suvra Ghosal
Anindita Ghosh
Anurag Ghosh
Arnab Ghosh
Arthita Ghosh
Aurobrata Ghosh
Shreya Ghosh
Soham Ghosh
Sreyan Ghosh
Andrea Giachetti
Paris Giampouras
Alexander Gielisse
Andrew Gilbert
Guy Gilboa
Nate Gillman
Jhony H. Giraldo
Roger Girgis
Sharath Girish
Mario Valerio Giuffrida
Francesco Giuliari
Nikolaos Gkanatsios
Ioannis Gkioulekas
Alexi Gladstone
Derek Gloudemans
Aurele T. Gnanha
Hyojun Go
Akos Godo
Danilo D. Goede
Arushi Goel
Kratarth Goel
Rahul Goel
Shubham Goel
Vidit Goel
Tejas Gokhale
Vignesh Gokul

Aditya Golatkar
Micah Goldblum
S. Alireza Golestaneh
Gabriele Goletto
Lluis Gomez
Guillermo Gomez-Trenado
Alex Gomez-Villa
Nuno Gonçalves
Biao Gong
Boqing Gong
Chen Gong
Chengyue Gong
Dayoung Gong
Dong Gong
Jingyu Gong
Kaixiong Gong
Kehong Gong
Liyu Gong
Mingming Gong
Ran Gong
Rui Gong
Tao Gong
Xiaojin Gong
Xuan Gong
Xun Gong
Yifan Gong
Yu Gong
Yuanhao Gong
Yuning Gong
Yunye Gong
Cristina I. González
Adam Goodge
Nithin Gopalakrishnan Nair
Cameron Gordon
Dipam Goswami
Gaurav Goswami
Hanno Gottschalk
Chenhui Gou
Jianping Gou
Shreyank N. Gowda
Mohit Goyal
Julia Grabinski
Helmut Grabner
Patrick L. Grady

Alexandros Graikos
Eric Granger
Douglas R. Gray
Michael Alan Greenspan
Connor Greenwell
David Griffiths
Artur Grigorev
Alexey A. Gritsenko
Ivan Grubišić
Geonmo Gu
Jianyang Gu
Jiaqi Gu
Jiayuan Gu
Jinwei Gu
Peiyan Gu
Qiao Gu
Tianpei Gu
Xianfan Gu
Xiang Gu
Xiangming Gu
Xiao Gu
Xiuye Gu
Yanan Gu
Yiwen Gu
Yuchao Gu
Yuming Gu
Yun Gu
Zeqi Gu
Zhangxuan Gu
Banglei Guan
Junfeng Guan
Qingji Guan
Shanyan Guan
Tianrui Guan
Tongkun Guan
Ziqiao Guan
Ziyu Guan
Denis A. Gudovskiy
Antoine Guédon
Paul Guerrero
Ricardo Guerrero
Liangke Gui
Sadaf Gulshad
Manuel Günther
Chen Guo

Chenqi Guo
Chuan Guo
Guodong Guo
Haiyun Guo
Heng Guo
Hengtao Guo
Hongji Guo
Jia Guo
Jianwei Guo
Jianyuan Guo
Jiayi Guo
Jie Guo
Jie Guo
Jingcai Guo
Jingyuan Guo
Jinyang Guo
Lanqing Guo
Meng-Hao Guo
Mengxi China Guo
Minghao Guo
Pengfei Guo
Pengsheng Guo
Pinxue Guo
Qi Guo
Qin Guo
Qing Guo
Qingpei Guo
Saidi Guo
Shi Guo
Shuxuan Guo
Song Guo
Taian Guo
Tiantong Guo
Tianyu Guo
Wen Guo
Wenzhong Guo
Xianda Guo
Xiaobao Guo
Xiefan Guo
Yangyang Guo
Yanhui Guo
Yao Guo
Yichen Guo
Yong Guo
Yuan-Chen Guo

Yuliang Guo
Yuxiang Guo
Yuyu Guo
Zixian Guo
Ziyu Guo
Zujin Guo
Aarush Gupta
Akash Gupta
Akshita Gupta
Ankush Gupta
Anshul Gupta
Anubhav Gupta
Anup Kumar Gupta
Honey Gupta
Kunal Gupta
pravir singh gupta
Rohit Gupta
Saumya Gupta
Corina Gurau
Yeti Z. Gurbuz
Saket Gurukar
Siddharth Gururani
Vladimir Guzov
Matthew A. Gwilliam
Hyunho Ha
Ryo Hachiuma
Isma Hadji
Armin Hadzic
Daniel Haehn
Nicolai Haeni
Ronny Haensch
Meera Hahn
Oliver Hahn
Nguyen Hai
Yuval Haitman
Levente Hajder
Moayed Haji-Ali
Sina Hajimiri
Alexandros Haliassos
Peter M. Hall
Bumsub Ham
Ryuhei Hamaguchi
Abdullah J. Hamdi
Max Hamilton
Lars Hammarstrand

Hasan Abed Al Kader Hammoud
Beining Han
Boran Han
Dongyoon Han
Feng Han
Guangxing Han
Guangzeng Han
Hu Han
Jiaming Han
Jin Han
Jungong Han
Junlin Han
Kai Han
Kang Han
Kun Han
Ligong Han
Longfei Han
Mei Han
Mingfei Han
Muzhi Han
Sungwon Han
Tengda Han
Tengda Han
Wencheng Han
Xinran Han
Xintong Han
Yizeng Han
Yufei Han
Zhenjun Han
Zhi Han
Zhongyi Han
Zongbo Han
Zongyan Han
Ankur Handa
Tiankai Hang
Yucheng Hang
Asif Hanif
Param Hanji
Joëlle Hanna
Niklas Hanselmann
Nicklas A. Hansen
Fusheng Hao
Shaozhe Hao
Shengyu Hao

Shijie Hao
Tianxiang Hao
Yanbin Hao
Yu Hao
Zekun Hao
Takayuki Hara
Mehrtash Harandi
Sanjay Haresh
Haripriya Harikumar
Adam Harley
David M. Hart
Md Yousuf Harun
Md Rakibul Hasan
Connor Hashemi
Atsushi Hashimoto
Khurram Azeem Hashmi
Nozomi Hata
Ali Hatamizadeh
Ryuichiro Hataya
Timm Haucke
Joakim Bruslund Haurum
Stephen Hausler
Mohammad Havaei
Hideaki Hayashi
Zeeshan Hayder
Chengan He
Conghui He
Dailan He
Fan He
Fazhi He
Gaoqi He
Haoyu He
Hongliang He
Jiangpeng He
Jianzhong He
Jiawei He
Ju He
Jun-Yan He
Junfeng He
Lihuo He
Liqiang He
Nanjun He
Ruian He
Runze He
Ruozhen He
Shengfeng He
Shuting He
Songtao He
Tao He
Tianyu He
Tong He
Wei He
Xiang He
Xiangteng He
Xiangyu He
Xiaoxiao He
Xin He
Xingyi He
Xingzhe He
Xinlei He
Xinwei He
Yang He
Yang He
Yannan He
Yeting He
Yinan He
Ying He
Yisheng He
Yuhang He
Yuhang He
Zecheng He
Zewei He
Zexin He
Zhenyu He
Ziwen He
Ramya S. Hebbalaguppe
Peter Hedman
Deepti B. Hegde
Sindhu B. Hegde
Matthias Hein
Mattias Paul Heinrich
Aral Hekimoglu
Philipp Henzler
Byeongho Heo
Jae-Pil Heo
Miran Heo
Stephane Herbin
Alexander Hermans
Pedro Hermosilla Casajus
Jefferson E. Hernandez
Monica Hernandez
Charles Herrmann
Roei Herzig
Fabian Herzog
Georg Hess
Mauricio Hess-Flores
Robin Hesse
Chamin P. Hewa
  Koneputugodage
Masatoshi Hidaka
Richard E. L. Higgins
Mitchell K. Hill
Carlos Hinojosa
Tobias Hinz
Yusuke Hirota
Elad Hirsch
Shinsaku Hiura
Chih-Hui Ho
Man M. Ho
Tuan N. A. Hoang
Jennifer Hobbs
Jiun Tian Hoe
David T. Hoffmann
Derek Hoiem
Yannick Hold-Geoffroy
Lukas Höllein
Gregory I. Holste
Hiroto Honda
Cheeun Hong
Cheng-Yao Hong
Danfeng Hong
Deokki Hong
Fa-Ting Hong
Fangzhou Hong
Feng Hong
Guan Zhe Hong
Je Hyeong Hong
Ji Woo Hong
Jie Hong
Joanna Hong
Lanqing Hong
Lingyi Hong
Sangwoo Hong
Sungeun Hong
Sunghwan Hong

Susung Hong
Weixiang Hong
Xiaopeng Hong
Yan Hong
Yi Hong
Yuchen Hong
Julia Hornauer
Maxwell C. Horton
Eliahu Horwitz
Mir Rayat Imtiaz Hossain
Tonmoy Hossain
Mehdi Hosseinzadeh
Kazuhiro Hotta
Andrew Z. Hou
Fei Hou
Hao Hou
Ji Hou
Jian Hou
Jinghua Hou
Qibin Hou
Qiqi Hou
Ruibing Hou
Saihui Hou
Tingbo Hou
Xinhai Hou
Yuenan Hou
Yunzhong Hou
Zhi Hou
Lukas Hoyer
Petr Hruby
Jun-Wei Hsieh
Chih-Fan Hsu
Gee-Sern Hsu
Hung-Min Hsu
Yen-Chi Hsu
Anthony Hu
Benran Hu
Bo Hu
Dapeng Hu
Dongting Hu
Guosheng Hu
Haigen Hu
Hanjiang Hu
Hanzhe Hu
Hengtong Hu
Hezhen Hu
Jian Hu
Jian-Fang Hu
Jie Hu
Juan Hu
Junjie Hu
Kai Hu
Lianyu Hu
Lily Hu
MengShun Hu
Minghui Hu
Mu Hu
Panwen Hu
Panwen Hu
Ruizhen Hu
Shengshan Hu
Shiyu Hu
Shizhe Hu
Shu Hu
Tao Hu
Tao Hu
Vincent Tao Hu
Wenbo Hu
Xiaodan Hu
Xiaolin Hu
Xiaoling Hu
Xiaotao Hu
Xiaowei Hu
Xinting Hu
Xinting Hu
Xixi Hu
Xuefeng Hu
Yang Hu
Yaosi Hu
Yinlin Hu
Yueyu Hu
Zeyu Hu
Zhanhao Hu
zhengyu hu
Zhenzhen Hu
Zhiming Hu
Zhiming Hu
Zhongyun Hu
Zijian Hu
Andong Hua
Hang Hua
Miao Hua
Wei Hua
Mengdi Huai
Binbin Huang
Bo Huang
Buzhen Huang
Chao Huang
Chao-Tsung Huang
Chaoqin Huang
Ching-Chun Huang
Cong Huang
Danqing Huang
Di Huang
Feihu Huang
Haibin Huang
Haiyang Huang
Hao Huang
Hsin-Ping Huang
Huaibo Huang
Ian Y. Huang
Jiabo Huang
Jiahui Huang
Jiancheng Huang
Jiangyong Huang
Jiaxing Huang
Jin Huang
Jinfa Huang
Jing Huang
Jonathan Huang
Junkai Huang
Junwen Huang
Kejie Huang
Kuan-Chih Huang
Lei Huang
Libo Huang
Lin Huang
Linjiang Huang
Linyan Huang
linzhi Huang
Lun Huang
Luojie Huang
Mingzhen Huang
Qidong Huang
Qiusheng Huang

Runhui Huang
Ruqi Huang
Shaofei Huang
Sheng Huang
Sheng-Yu Huang
Shiyuan Huang
Shuaiyi Huang
Shuangping Huang
Siteng Huang
Siyu Huang
Siyuan Huang
Tao Huang
Thomas E. Huang
Tianyu Huang
Wenjian Huang
Wenke Huang
Xiaohu Huang
Xiaohua Huang
Xiaoke Huang
Xiaoshui Huang
Xiaoyang Huang
Xijie Huang
Xinyu Huang
Xuhua Huang
Yan Huang
Yaping Huang
Ye Huang
Yi Huang
Yi Huang
Yi Huang
Yi-Hua Huang
Yifei Huang
Yihao Huang
Ying Huang
Yizhan Huang
You Huang
Yue Huang
Yufei Huang
Yuge Huang
Yukun Huang
Yuyao Huang
Zaiyu Huang
Zehao Huang
Zeyi Huang
Zhe Huang

Zhewei Huang
Zhida Huang
Zhiqi Huang
Zhiyu Huang
Zhongzhan Huang
Ziling Huang
Zilong Huang
Ziqi Huang
Zixuan Huang
Ziyuan Huang
Jaesung Huh
Ka-Hei Hui
Le Hui
Tianrui Hui
Xiaofei Hui
Ahmed Imtiaz Humayun
Thomas Hummel
Jing Huo
Shuwei Huo
Yuankai Huo
Julio Hurtado
Rukhshanda Hussain
Mohamed Hussein
Noureldien Hussein
Chuong Minh Huynh
Tran Ngoc Huynh
Muhammad Huzaifa
Inwoo Hwang
Jaehui Hwang
Jenq-Neng Hwang
Sehyun Hwang
Seunghyun Hwang
Sukjun Hwang
Sunhee Hwang
Wonjun Hwang
Jeongseok Hyun
Sangeek Hyun
Ekaterina Iakovleva
Tomoki Ichikawa
A. S. M. Iftekhar
Masaaki Iiyama
Satoshi Ikehata
Sunghoon Im
Woobin Im
Nevrez Imamoglu

Abdullah Al Zubaer Imran
Tooba Imtiaz
Nakamasa Inoue
Naoto Inoue
Eldar Insafutdinov
Catalin Ionescu
Radu Tudor Ionescu
Nasar Iqbal
Umar Iqbal
Go Irie
Muhammad Zubair Irshad
Francesco Isgrò
Yasunori Ishii
Berivan Isik
Syed M. S. Islam
Mariko Isogawa
Vamsi Krishna K. Ithapu
Koichi Ito
Leonardo Iurada
Shun Iwase
Sergio Izquierdo
Sarah Jabbour
David Jacobs
David E. Jacobs
Yasamin Jafarian
Azin Jahedi
Achin Jain
Ayush Jain
Jitesh Jain
Kanishk Jain
Yash Jain
Nikita Jaipuria
Tomas Jakab
Muhammad Abdullah
    Jamal
Hadi Jamali-Rad
Stuart James
Nataraj Jammalamadaka
Donggon Jang
Sujin Jang
Young Kyun Jang
Youngjoon Jang
Steeven Janny
Paul Janson
Maximilian Jaritz

Ronnachai Jaroensri
Guillaume Jaume
Sajid Javed
Saqib Javed
Bhavin Jawade
Hirunima Jayasekara
Senthilnath Jayavelu
Guillaume Jeanneret
Pranav Jeevan
Rohit Jena
Tomas Jenicek
Simon Jenni
Hae-Gon Jeon
Jeimin Jeon
Seogkyu Jeon
Boseung Jeong
Jongheon Jeong
Yonghyun Jeong
Yoonwoo Jeong
Koteswar Rao Jerripothula
Artur Jesslen
Nikolay Jetchev
Ankit Jha
Sumit K. Jha
I-Hong Jhuo
Deyi Ji
Ge-Peng Ji
Hui Ji
Jianmin Ji
Jiayi Ji
Jingwei Ji
Naye Ji
Pengliang Ji
Shengpeng Ji
Wei Ji
Xiang Ji
Xiaopeng Ji
Xiaozhong Ji
Xinya Ji
Yu Ji
Yuanfeng Ji
Zhanghexuan Ji
Baoxiong Jia
Ding Jia
Jinrang Jia
Menglin Jia
Shan Jia
Shuai Jia
Tong Jia
Wenqi Jia
Xiaojun Jia
Xiaosong Jia
Xu Jia
Yiren Jian
Bo Jiang
Borui Jiang
Chen Jiang
Guangyuan Jiang
Hai Jiang
Haiyang Jiang
Haiyong Jiang
Hanwen Jiang
Hanxiao Jiang
Hao Jiang
Haobo Jiang
Huaizu Jiang
Huajie Jiang
Jianmin Jiang
Jianwen Jiang
Jiaxi Jiang
Jin Jiang
Jing Jiang
Kaixun Jiang
Kui Jiang
Li Jiang
Liming Jiang
Meirui Jiang
Ming Jiang
Ming Jiang
Peng Jiang
Peng-Tao Jiang
Runqing Jiang
Siyang Jiang
Tianjian Jiang
Tingting Jiang
Weisen Jiang
Wen Jiang
Xingyu Jiang
Xueying Jiang
Xuhao Jiang
Yanru Jiang
Yi Jiang
Yifan Jiang
Yingying Jiang
Yue Jiang
Yuming Jiang
Zeren Jiang
Zheheng Jiang
Zhenyu Jiang
Zhongyu Jiang
Zi-Hang Jiang
Zixuan Jiang
Ziyu Jiang
Zutao Jiang
Zutao Jiang
Jianbin Jiao
Jianbo Jiao
Licheng Jiao
Ruochen Jiao
Shuming Jiao
Wenpei Jiao
Zequn Jie
Dongkwon Jin
Gaojie Jin
Haian Jin
Haibo Jin
Kyong Hwan Jin
Lei Jin
Lianwen Jin
Linyi Jin
Peng Jin
Peng Jin
Sheng Jin
SouYoung Jin
Weiyang Jin
Xiao Jin
Xiaojie Jin
Xin Jin
Xin Jin
Yeying Jin
Yi Jin
Ying Jin
Zhenchao Jin
Chenchen Jing
Junpeng Jing

Mengmeng Jing
Taotao Jing
Jeonghee Jo
Yeonsik Jo
Younghyun Jo
Cameron A. Johnson
Faith M. Johnson
Maxwell Jones
Michael J. Jones
R. Kenny Jones
Ameya Joshi
Shantanu H. Joshi
Brendan Jou
Chen Ju
Wei Ju
Xuan Ju
Yan Ju
Felix Juefei-Xu
Florian Jug
Yohan Jun
Kim Jun-Seong
Masum Shah Junayed
Chanyong Jung
Claudio R. Jung
Hoin Jung
Hyunyoung Jung
Sangwon Jung
Steffen Jung
Sacha Jungerman
Hari Chandana K.
Prajwal K. R.
Berna Kabadayi
Anis Kacem
Anil Kag
Kumara Kahatapitiya
Bernhard Kainz
Ivana Kajic
Ioannis Kakogeorgiou
Niveditha Kalavakonda
Mahdi M. Kalayeh
Ajinkya Kale
Anmol Kalia
Sinan Kalkan
Jayateja Kalla
Tarun Kalluri

Uday Kamal
Sandesh Kamath
Chandra Kambhamettu
Meina Kan
Sai Srinivas Kancheti
Takuhiro Kaneko
Cuicui Kang
Dahyun Kang
Guoliang Kang
Hyolim Kang
Jaeyeon Kang
Juwon Kang
Li-Wei Kang
Mingon Kang
MinGuk Kang
Minjun Kang
Weitai Kang
Zhao Kang
Yash Mukund Kant
Yueying Kao
Saarthak Kapse
Aupendu Kar
Oğuzhan Fatih Kar
Ozgur Kara
Tamás Karácsony
Srikrishna Karanam
Neerav Karani
Mert Asim Karaoglu
Laurynas Karazija
Navid Kardan
Amirhossein Kardoost
Nour Karessli
Michelle Karg
Mohammad Reza Karimi
  Dastjerdi
Animesh Karnewar
Arjun M. Karpur
Shyamgopal Karthik
Korrawe Karunratanakul
Tejaswi Kasarla
Satyananda Kashyap
Yoni Kasten
Marc A. Kastner
Hirokatsu Kataoka
Isinsu Katircioglu

Kai Katsumata
Ilya Kaufman
Manuel Kaufmann
Chaitanya Kaul
Prannay Kaul
Prakhar Kaushik
Isaak Kavasidis
Ryo Kawahara
Yuki Kawana
Justin Kay
Evangelos Kazakos
Jingcheng Ke
Lei Ke
Tsung-Wei Ke
Wei Ke
Zhanghan Ke
Nikhil V. Keetha
Thomas Kehrenberg
Marilyn Keller
Rohit Keshari
Janis Keuper
Daniel Keysers
Hrant Khachatrian
Seyran Khademi
Wesley A. Khademi
Taras Khakhulin
Samir Khaki
Hasam Khalid
Umar Khalid
Amr Khalifa
Asif Hussain Khan
Faizan Farooq Khan
Muhammad Haris Khan
Naimul Khan
Qadeer Khan
Zeeshan Khan
Bishesh Khanal
Pulkit Khandelwal
Ishan Khatri
Muhammad Uzair Khattak
Vahid Reza Khazaie
Vaishnavi M. Khindkar
Rawal Khirodkar
Pirazh Khorramshahi
Sahil S. Khose

Ronnachai Jaroensri
Guillaume Jaume
Sajid Javed
Saqib Javed
Bhavin Jawade
Hirunima Jayasekara
Senthilnath Jayavelu
Guillaume Jeanneret
Pranav Jeevan
Rohit Jena
Tomas Jenicek
Simon Jenni
Hae-Gon Jeon
Jeimin Jeon
Seogkyu Jeon
Boseung Jeong
Jongheon Jeong
Yonghyun Jeong
Yoonwoo Jeong
Koteswar Rao Jerripothula
Artur Jesslen
Nikolay Jetchev
Ankit Jha
Sumit K. Jha
I-Hong Jhuo
Deyi Ji
Ge-Peng Ji
Hui Ji
Jianmin Ji
Jiayi Ji
Jingwei Ji
Naye Ji
Pengliang Ji
Shengpeng Ji
Wei Ji
Xiang Ji
Xiaopeng Ji
Xiaozhong Ji
Xinya Ji
Yu Ji
Yuanfeng Ji
Zhanghexuan Ji
Baoxiong Jia
Ding Jia
Jinrang Jia

Menglin Jia
Shan Jia
Shuai Jia
Tong Jia
Wenqi Jia
Xiaojun Jia
Xiaosong Jia
Xu Jia
Yiren Jian
Bo Jiang
Borui Jiang
Chen Jiang
Guangyuan Jiang
Hai Jiang
Haiyang Jiang
Haiyong Jiang
Hanwen Jiang
Hanxiao Jiang
Hao Jiang
Haobo Jiang
Huaizu Jiang
Huajie Jiang
Jianmin Jiang
Jianwen Jiang
Jiaxi Jiang
Jin Jiang
Jing Jiang
Kaixun Jiang
Kui Jiang
Li Jiang
Liming Jiang
Meirui Jiang
Ming Jiang
Ming Jiang
Peng Jiang
Peng-Tao Jiang
Runqing Jiang
Siyang Jiang
Tianjian Jiang
Tingting Jiang
Weisen Jiang
Wen Jiang
Xingyu Jiang
Xueying Jiang
Xuhao Jiang

Yanru Jiang
Yi Jiang
Yifan Jiang
Yingying Jiang
Yue Jiang
Yuming Jiang
Zeren Jiang
Zheheng Jiang
Zhenyu Jiang
Zhongyu Jiang
Zi-Hang Jiang
Zixuan Jiang
Ziyu Jiang
Zutao Jiang
Zutao Jiang
Jianbin Jiao
Jianbo Jiao
Licheng Jiao
Ruochen Jiao
Shuming Jiao
Wenpei Jiao
Zequn Jie
Dongkwon Jin
Gaojie Jin
Haian Jin
Haibo Jin
Kyong Hwan Jin
Lei Jin
Lianwen Jin
Linyi Jin
Peng Jin
Peng Jin
Sheng Jin
SouYoung Jin
Weiyang Jin
Xiao Jin
Xiaojie Jin
Xin Jin
Xin Jin
Yeying Jin
Yi Jin
Ying Jin
Zhenchao Jin
Chenchen Jing
Junpeng Jing

Mengmeng Jing
Taotao Jing
Jeonghee Jo
Yeonsik Jo
Younghyun Jo
Cameron A. Johnson
Faith M. Johnson
Maxwell Jones
Michael J. Jones
R. Kenny Jones
Ameya Joshi
Shantanu H. Joshi
Brendan Jou
Chen Ju
Wei Ju
Xuan Ju
Yan Ju
Felix Juefei-Xu
Florian Jug
Yohan Jun
Kim Jun-Seong
Masum Shah Junayed
Chanyong Jung
Claudio R. Jung
Hoin Jung
Hyunyoung Jung
Sangwon Jung
Steffen Jung
Sacha Jungerman
Hari Chandana K.
Prajwal K. R.
Berna Kabadayi
Anis Kacem
Anil Kag
Kumara Kahatapitiya
Bernhard Kainz
Ivana Kajic
Ioannis Kakogeorgiou
Niveditha Kalavakonda
Mahdi M. Kalayeh
Ajinkya Kale
Anmol Kalia
Sinan Kalkan
Jayateja Kalla
Tarun Kalluri

Uday Kamal
Sandesh Kamath
Chandra Kambhamettu
Meina Kan
Sai Srinivas Kancheti
Takuhiro Kaneko
Cuicui Kang
Dahyun Kang
Guoliang Kang
Hyolim Kang
Jaeyeon Kang
Juwon Kang
Li-Wei Kang
Mingon Kang
MinGuk Kang
Minjun Kang
Weitai Kang
Zhao Kang
Yash Mukund Kant
Yueying Kao
Saarthak Kapse
Aupendu Kar
Oğuzhan Fatih Kar
Ozgur Kara
Tamás Karácsony
Srikrishna Karanam
Neerav Karani
Mert Asim Karaoglu
Laurynas Karazija
Navid Kardan
Amirhossein Kardoost
Nour Karessli
Michelle Karg
Mohammad Reza Karimi
  Dastjerdi
Animesh Karnewar
Arjun M. Karpur
Shyamgopal Karthik
Korrawe Karunratanakul
Tejaswi Kasarla
Satyananda Kashyap
Yoni Kasten
Marc A. Kastner
Hirokatsu Kataoka
Isinsu Katircioglu

Kai Katsumata
Ilya Kaufman
Manuel Kaufmann
Chaitanya Kaul
Prannay Kaul
Prakhar Kaushik
Isaak Kavasidis
Ryo Kawahara
Yuki Kawana
Justin Kay
Evangelos Kazakos
Jingcheng Ke
Lei Ke
Tsung-Wei Ke
Wei Ke
Zhanghan Ke
Nikhil V. Keetha
Thomas Kehrenberg
Marilyn Keller
Rohit Keshari
Janis Keuper
Daniel Keysers
Hrant Khachatrian
Seyran Khademi
Wesley A. Khademi
Taras Khakhulin
Samir Khaki
Hasam Khalid
Umar Khalid
Amr Khalifa
Asif Hussain Khan
Faizan Farooq Khan
Muhammad Haris Khan
Naimul Khan
Qadeer Khan
Zeeshan Khan
Bishesh Khanal
Pulkit Khandelwal
Ishan Khatri
Muhammad Uzair Khattak
Vahid Reza Khazaie
Vaishnavi M. Khindkar
Rawal Khirodkar
Pirazh Khorramshahi
Sahil S. Khose

Kourosh Khoshelham
Sena Kiciroglu
Benjamin Kiefer
Kotaro Kikuchi
Mert Kilickaya
Benjamin Killeen
Beomyoung Kim
Boah Kim
Bumsoo Kim
Byeonghwi Kim
Byoungjip Kim
Changhoon Kim
Changick Kim
Chanho Kim
Chanyoung Kim
Dahun Kim
Dahyun Kim
Diana S. Kim
Dong-Jin Kim
Donggun Kim
Donghyun Kim
Dongkeun Kim
Dongwan Kim
Dongyoung Kim
Eun-Sol Kim
Eunji Kim
Euyoung Kim
Geeho Kim
Giseop Kim
Guisik Kim
Gwanghyun Kim
Hakyeong Kim
Hanjae Kim
Hanjung Kim
Heewon Kim
Howon Kım
Hyeongwoo Kim
Hyeongwoo Kim
Hyo Jin Kim
Hyung-Il Kim
Hyunwoo J. Kim
Insoo Kim
Jae Myung Kim
Jeongsol Kim
Jihwan Kim

Jinkyu Kim
Jinwoo Kim
Jongyoo Kim
Joonsoo Kim
Jung Uk Kim
Junho Kim
Junho Kim
Junsik Kim
Kangyeol Kim
Kunhee Kim
Kwang In Kim
Manjin Kim
Minchul Kim
Minji Kim
Minjung Kim
Namil Kim
Namyup Kim
Nayeong Kim
Sanghyun Kim
Seong Tae Kim
Seungbae Kim
Seungryong Kim
Seungwook Kim
Soo Ye Kim
Soohwan Kim
Sungnyun Kim
Sungyeon Kim
Sunnie S. Y. Kim
Tae Hyun Kim
Tae Hyung Kim
Taehoon Kim
Taehun Kim
Taehwan Kim
Taekyung Kim
Taeoh Kim
Taewoo Kim
Won Hwa Kim
Wonjae Kim
Woo Jae Kim
Young Min Kim
YoungBin Kim
Youngeun Kim
Youngseok Kim
Youngwook Kim
Akisato Kimura

Andreas Kirsch
Nikita Kister
Furkan Osman Kınlı
Marcus Klasson
Florian Kleber
Tzofi M. Klinghoffer
Jan P. Klopp
Florian Kluger
Hannah Kniesel
David M. Knigge
Byungsoo Ko
Dohwan Ko
Jongwoo Ko
Takumi Kobayashi
Muhammed Kocabas
Yeong Jun Koh
Kathlén Kohn
Subhadeep Koley
Nick Kolkin
Soheil Kolouri
Jacek Komorowski
Deying Kong
Fanjie Kong
Hanyang Kong
Hui Kong
Kyeongbo Kong
Lecheng Kong
Lingdong Kong
Linghe Kong
Lingshun Kong
Naejin Kong
Quan Kong
Shu Kong
Xianghao Kong
Xiangtao Kong
Xiangwei Kong
Xiaoyu Kong
Xin Kong
Youyong Kong
Yu Kong
Zhenglun Kong
Aishik Konwer
Nicholas C. Konz
Gwanhyeong Koo
Juil Koo

Julian F. P. Kooij
George Kopanas
Sanjeev J. Koppal
Bruno Korbar
Giorgos Kordopatis-Zilos
Dimitri Korsch
Adam Kortylewski
Divya Kothandaraman
Suraj Kothawade
Iuliia Kotseruba
Sasikanth Kotti
Alankar Kotwal
Shashank Kotyan
Alexandros Kouris
Petros Koutras
Rama Kovvuri
Dilip Krishnan
Praveen Krishnan
Ranganath Krishnan
Rohan M. Krishnan
Georg Krispel
Alexander Krull
Tianshu Kuai
Haowei Kuang
Zhengfei Kuang
Andrey Kuehlkamp
David Kügler
Arjan Kuijper
Anna Kukleva
Jonas Kulhanek
Peter Kulits
Akshay R. Kulkarni
Ashutosh C. Kulkarni
Kuldeep Kulkarni
Nilesh Kulkarni
Abhinav Kumar
Abhishek Kumar
Akash Kumar
Avinash Kumar
B. V. K. Vijaya Kumar
Chandan Kumar
Pulkit Kumar
Ratnesh Kumar
Sateesh Kumar
Satish Kumar

Suryansh Kumar
Yogesh Kumar
Nupur Kumari
Sudhakar Kumawat
Nilakshan Kunananthaseelan
Rohit Kundu
Souvik Kundu
Meng-Yu Jennifer Kuo
Weicheng Kuo
Shuhei Kurita
Yusuke Kurose
Takahiro Kushida
Uday Kusupati
Alina Kuznetsova
Jobin K. V.
Henry Kvinge
Ho Man Kwan
Hyeokjun Kweon
Donghyeon Kwon
Gihyun Kwon
Heeseung Kwon
Hyoukjun Kwon
Myung-Joon Kwon
Taein Kwon
YoungJoong Kwon
Cameron Kyle-Davidson
Christos Kyrkou
Jorma Laaksonen
Patrick Labatut
Yann Labbé
Manuel Ladron de Guevara
Florent Lafarge
Jean Lahoud
Bolin Lai
Farley Lai
Jian-Huang Lai
Shenqi Lai
Xin Lai
Yu-Kun Lai
Yung-Hsuan Lai
Zeqiang Lai
Zhengfeng Lai
Barath Lakshmanan
Rohit Lal

Rodney LaLonde
Hala Lamdouar
Meng Lan
Yushi Lan
Federico Landi
George V. Landon
Chunbo Lang
Jochen Lang
Nico Lang
Georg Langs
Raffaella Lanzarotti
Dong Lao
Yixing Lao
Yizhen Lao
Zakaria Laskar
Alexandros Lattas
Chun Pong Lau
Shlomi Laufer
Justin Lazarow
Svetlana Lazebnik
Duy Tho Le
Hieu Le
Hoang Le
Hoang Le
Thi-Thu-Huong Le
Trung Le
Trung-Nghia Le
Tung Thanh Le
Hoàng-Ân Lê
Herve Le Borgne
Guillaume Le Moing
Erik Learned-Miller
Tim Lebailly
Byeong-Uk Lee
Byung-Kwan Lee
Cheng-Han Lee
Chul Lee
Daeun Lee
Dogyoon Lee
Dong Hoon Lee
Eugene Eu Tzuan Lee
Eung-Joo Lee
Gyuseong Lee
Hsin-Ying Lee
Hwee Kuan Lee

Hyeongmin Lee
Hyungtae Lee
Jae Yong Lee
Jaeho Lee
Jaeseong Lee
Jaewon Lee
Jangho Lee
Jangwon Lee
Jihyun Lee
Jiyoung Lee
Jong-Seok Lee
Jongho Lee
Jongmin Lee
Joo-Ho Lee
Joon-Young Lee
Joonseok Lee
Jungbeom Lee
Jungho Lee
Jungwoo Lee
Junha Lee
Junhyun Lee
Junyong Lee
Kibok Lee
Kuan-Ying Lee
Kwonjoon Lee
Kwot Sin Lee
Kyungmin Lee
Kyungmoon Lee
Minhyeok Lee
Minsik Lee
Pilhyeon Lee
Saehyung Lee
Sangho Lee
Sanghyeok Lee
Sangmin Lee
Sehun Lee
Sehyung Lee
Seon-Ho Lee
Seong Hun Lee
Seongwon Lee
Seung Hyun Lee
Seung-Ik Lee
Seungho Lee
Seunghun Lee
Seungmin Lee
Seungyong Lee
Sohyun Lee
Suhyeon Lee
Sungho Lee
Sungmin Lee
Suyoung Lee
Taehyun Lee
Wooseok Lee
Yao-Chih Lee
Yi-Lun Lee
Yonghyeon Lee
Youngwan Lee
Leonidas Lefakis
Bowen Lei
Chenyang Lei
Chenyi Lei
Jiahui Lei
Na Lei
Qinqian Lei
Yinjie Lei
Thomas Leimkuehler
Abe Leite
Abdelhak Lemkhenter
Jiaxu Leng
Luziwei Leng
Zhiying Leng
Hendrik P. A. Lensch
Jan E. Lenssen
Ted Lentsch
Simon Lepage
Stefan Leutenegger
Filippo Leveni
Axel Levy
Ailin Li
Aixuan Li
Baiang Li
Baoxin Li
Bin Li
Bing Li
Bing Li
Bo Li
Bowen Li
Boying Li
Changlin Li
Changlin Li
Chao Li
Chenghong Li
Chenglin Li
Chenglong Li
Chengze Li
Chun-Guang Li
Daiqing Li
Dasong Li
Dian Li
Dong Li
Fangda Li
Feiran Li
Fenghai Li
Gen Li
Guanbin Li
Guangrui Li
Guihong Li
Guorong Li
Haifeng Li
Han Li
Hang Li
Hangyu Li
Hanhui Li
Hao Li
Hao Li
Haoang Li
Haoran Li
Haoxiang Li
Haoxin Li
He Li
Heng Li
Hengduo Li
Hongshan Li
Hongwei Bran Li
Hongxiang Li
Hongyang Li
Hongyu Li
Huafeng Li
Huan Li
Hui Li
Jiacheng Li
Jiahao Li
Jialu Li
Jiaman Li
Jiangmeng Li

Jiangtong Li
Jiangyuan Li
Jianing Li
Jianwei Li
Jianwu Li
Jiaqi Li
Jiaqi Li
Jiatong Li
Jiaxuan Li
Jiazhi Li
Jichang Li
Jie Li
Jin Li
Jinglun Li
Jingzhi Li
Jingzong Li
Jinlong Li
Jinlong Li
Jinpeng Li
Jinxing Li
Jun Li
Jun Li
Junbo Li
Juncheng Li
Junxuan Li
Junyi Li
Kai Li
Kaican Li
Kailin Li
Ke Li
Kehan Li
Keyu Li
Kun Li
Kunchang Li
Kunpeng Li
Lei Li
Lei Li
Li Li
Li Erran Li
Liang Li
Lin Li
Lincheng Li
Liulei Li
Liunian Harold Li
Lujun Li

Manyi Li
Maomao Li
Meng Li
Mengke Li
Mengtian Li
Mengtian Li
Ming Li
Ming Li
Minghan Li
Mingjie Li
Nannan Li
Nianyi Li
Peike Li
Peizhao Li
Peng Li
Pengpeng Li
Pengyu Li
Ping Li
Puhao Li
Qiang Li
Qing Li
Qingyong Li
Qiufu Li
Qizhang Li
Ren Li
Rong Li
Rongjie Li
Ru Li
Rui Li
Ruibo Li
Ruihui Li
Ruilong Li
Ruining Li
Ruixuan Li
Runze Li
Ruoteng Li
Shaohua Li
Shasha Li
Shigang Li
Shijie Li
Shikun Li
Shile Li
Shuai Li
Shuai Li
Shuang Li

Shuwei Li
Si Li
Siyao Li
Siyuan Li
Siyuan Li
Taihui Li
Tianye Li
Wanhua Li
Wanqing Li
Wei Li
Wei Li
Wei Li
Wei-Hong Li
Weihao Li
Weijia Li
Weiming Li
Wenbin Li
Wenbo Li
Wenhao Li
Wenjie Li
Wenshuo Li
Wentong Li
Wenxi Li
Xiang Li
Xiang Li
Xiang Li
Xiang Li
Xiang Li
Xiangtai Li
Xiangyang Li
Xianzhi Li
Xiao Li
Xiao Li
Xiaoguang Li
Xiaomeng Li
Xiaoming Li
Xiaoqi Li
Xiaoqiang Li
Xiaotian Li
Xiaoyu Li
Xin Li
Xin Li
Xin Li
Xinghui Li
Xingyi Li

Xingyu Li
Xinjie Li
Xinyu Li
Xiu Li
Xiujun Li
Xuan Li
Xuanlin Li
Xuelong Li
Xuelu Li
Xueqian Li
Ya-Li Li
Yanan Li
Yang Li
Yang Li
Yangyan Li
Yanjing Li
Yansheng Li
Yanwei Li
Yanyu Li
Yaohui Li
Yaowei Li
Yawei Li
Yi Li
Yi Li
Yicong Li
Yicong Li
Yifei Li
Yijin Li
Yijun Li
Yijun Li
Yikang Li
Yimeng Li
Yiming Li
Yiming Li
Yingwei Li
Yiting Li
Yixuan Li
Yize Li
Yizhuo Li
Yong Li
Yong-Lu Li
Yongjie Li
Yuanman Li
Yuanming Li
Yuelong Li

Yuexiang Li
Yuezun Li
Yuhang Li
Yuheng Li
Yulin Li
Yumeng Li
Yunfan Li
Yunheng Li
Yunqiang Li
Yunsheng Li
Yuyan Li
Yuyang Li
Zejian Li
Zekun Li
Zekun Li
Zhangheng Li
Zhangzikang Li
Zhaoshuo Li
Zhaowen Li
Zhe Li
Zhe Li
Zhen Li
Zhen Li
Zhen Li
Zheng Li
Zhengqin Li
Zhengyuan Li
Zhenyu Li
Zhichao Li
Zhihao Li
Zhihao Li
Zhiheng Li
Zhiqi Li
Zhixuan Li
Zhong Li
Zhuoling Li
Zhuowan Li
Zhuowei Li
Zhuoxiao Li
Zihan Li
Ziqiang Li
Wen Qiao Li
Dongze Lian
Long Lian
Qing Lian

Ruyi Lian
Zhouhui Lian
Jin Lianbao
Chao Liang
Chia-Kai Liang
Dingkang Liang
Feng Liang
Gongbo Liang
Hanxue Liang
Hao Liang
Hui Liang
Jiadong Liang
Jiajun Liang
Jian Liang
Jingyun Liang
Jinxiu S. Liang
Junwei Liang
Kaiqu Liang
Ke Liang
Kevin J. Liang
Luming Liang
Mingfu Liang
Pengpeng Liang
Siyuan Liang
Xiaoxiao Liang
Xinran Liang
Xiwen Liang
Yang Liang
Yixun Liang
Yongqing Liang
Youwei Liang
Yuanzhi Liang
Zhexin Liang
Zhihao Liang
Zhixuan Liang
Kang Liao
Liang Liao
Minghui Liao
Ting-Hsuan Liao
Wei Liao
Xin Liao
Yinghong Liao
Yue Liao
Zhibin Liao
Ziwei Liao

Benedetta Liberatori
Daniel J. Lichy
Maiko Lie
Qin Likun
Isaak Lim
Teck Yian Lim
Bannapol Limanond
Baijiong Lin
Beibei Lin
Cheng Lin
Chenhao Lin
Chia-Wen Lin
Chieh Hubert Lin
Chuang Lin
Chung-Ching Lin
Chunyu Lin
Ci-Siang Lin
Di Lin
Fanqing Lin
Feng Lin
Fudong Lin
Guangfeng Lin
Haotong Lin
Haozhe Lin
Hubert Lin
Hui Lin
Jason Lin
Jianxin Lin
Jiaqi Lin
Jiaying Lin
Jiehong Lin
Jierui Lin
Jintao Lin
Kai-En Lin
Ke Lin
Kevin Lin
Kevin Qinghong Lin
Kuan Heng Lin
Kun-Yu Lin
Kunyang Lin
Kwan-Yee Lin
Lijian Lin
Liqiang Lin
Liting Lin
Luojun Lin

Mingyuan Lin
Qiuxia Lin
Shaohui Lin
Shih-Yao Lin
Sihao Lin
Siyou Lin
Tiancheng Lin
Tsung-Yu Lin
Wanyu Lin
Wei Lin
Wei Lin
Wen-Yan Lin
Wenbin Lin
Xiangbo Lin
Xianhui Lin
Xiaofan Lin
Xiaofeng Lin
Xin Lin
Xudong Lin
Xue Lin
Xuxin Lin
Ya-Wei Eileen Lin
Yan-Bo Lin
Yancong Lin
Yi Lin
Yijie Lin
Yiming Lin
Yiqi Lin
Yiqun Lin
Yongliang Lin
Yu Lin
Yuanze Lin
Yuewei Lin
Zhi-Hao Lin
Zhiqiu Lin
ZhiWei Lin
Zinan Lin
Ziyi Lin
David B. Lindell
Philipp Lindenberger
Jingwang Ling
Jun Ling
Yongguo Ling
Zhan Ling
Alexander Liniger

Stefan P. Lionar
Phillip Lippe
Lahav O. Lipson
Joey Litalien
Ron Litman
Mattia Litrico
Dor Litvak
Aishan Liu
Ajian Liu
Akide L. Y. Liu
Andrew Liu
Ao Liu
Bang Liu
Benlin Liu
Bin Liu
Bin Liu
Bing Liu
Binghao Liu
Bingyuan Liu
Bo Liu
Bo Liu
Bo Liu
Boning Liu
Bowen Liu
Boxiao Liu
Chang Liu
Chang Liu
Chang Liu
Chang Liu
Chao Liu
Chengxin Liu
Chengxu Liu
Chih-Ting Liu
Chuanjian Liu
Chun-Hao Liu
Daizong Liu
Decheng Liu
Di Liu
Difan Liu
Dong Liu
Dongnan Liu
Fang Liu
Fang Liu
Fangyi Liu
Feng Liu

Fengbei Liu
Fenglin Liu
Fengqi Liu
Furui Liu
Fuxiao Liu
Haisong Liu
Han Liu
Hanwen Liu
Hanyuan Liu
Hao Liu
Haolin Liu
Haotian Liu
Haozhe Liu
Heshan Liu
Hong Liu
Hongbin Liu
Hongfu Liu
Hongyu Liu
Hsueh-Ti Derek Liu
Huidong Liu
Isabella Liu
Ji Liu
Jia-Wei Liu
Jiachen Liu
Jiaheng Liu
Jiahui Liu
Jiaming Liu
Jiancheng Liu
Jiang Liu
Jianmeng Liu
Jiashuo Liu
Jiawei Liu
Jiawei Liu
Jiayang Liu
Jiayi Liu
Jie Liu
Jie Liu
Jie Liu
Jihao Liu
Jing Liu
Jing Liu
Jing Liu
Jingyuan Liu
Jingyuan Liu
Jiuming Liu
Jiyuan Liu
Jun Liu
Kang-Jun Liu
Kangning Liu
Kenkun Liu
Kunhao Liu
Li Liu
Lijuan Liu
Lingbo Liu
Lingqiao Liu
Liu Liu
Liyang Liu
Meng Liu
Mengchen Liu
Miao Liu
Ming Liu
Minghao Liu
Minghua Liu
Mingxuan Liu
Mingyuan Liu
Nan Liu
Nian Liu
Ning Liu
Peidong Liu
Peirong Liu
Peiye Liu
Pengju Liu
Ping Liu
Qi Liu
Qiankun Liu
Qihao Liu
Qing Liu
Qingjie Liu
Richard Liu
Risheng Liu
Rui Liu
Ruicong Liu
Ruoshi Liu
Ruyu Liu
Shaohui Liu
Shaoteng Liu
Shaowei Liu
Sheng Liu
Shenglan Liu
Shikun Liu
Shilong Liu
Shuaicheng Liu
Shuaicheng Liu
Shuming Liu
Songhua Liu
Tao Liu
Tian Yu Liu
Tianci Liu
Tianshan Liu
Tongliang Liu
Tyng-Luh Liu
Wei Liu
Weifeng Liu
Weixiao Liu
Weiyu Liu
Wen Liu
Wenxi Liu
Wenyu Liu
Wenyu Liu
Wu Liu
Xian Liu
Xianglong Liu
Xianpeng Liu
Xiao Liu
Xiaohong Liu
Xiaoyu Liu
Xiaoyu Liu
Xihui Liu
Xin Liu
Xin Liu
Xinchen Liu
Xingtong Liu
Xingyu Liu
Xinhang Liu
Xinhui Liu
Xinpeng Liu
Xinwei Liu
Xinyu Liu
Xiulong Liu
Xiyao Liu
Xu Liu
Xubo Liu
Xudong Liu
Xueting Liu
Xueyi Liu

Yan Liu
Yanbin Liu
Yang Liu
Yang Liu
Yang Liu
Yang Liu
Yang Liu
Yanwei Liu
Yaojie Liu
Ye Liu
Yi Liu
Yihao Liu
Yingcheng Liu
Yingfei Liu
Yipeng Liu
Yipeng Liu
Yixin Liu
Yizhang Liu
Yong Liu
Yong Liu
Yonghuai Liu
Yongtuo Liu
Yu Liu
Yu-Lun Liu
Yu-Shen Liu
Yuan Liu
Yuang Liu
Yuanpei Liu
Yuanpeng Liu
Yuanwei Liu
Yuchen Liu
Yuchen Liu
Yuchi Liu
Yueh-Cheng Liu
Yufan Liu
Yuhao Liu
Yuliang Liu
Yun Liu
Yun Liu
Yun Liu
Yunfan Liu
Yunfei Liu
Yunze Liu
Yupei Liu
Yuqi Liu

Yuyang Liu
Yuyuan Liu
Zhaoqiang Liu
Zhe Liu
Zhe Liu
Zhen Liu
Zheng Liu
Zhenguang Liu
Zhi Liu
Zhihua Liu
Zhijian Liu
Zhili Liu
Zhuoran Liu
Ziquan Liu
Ziyi Liu
Zuxin Liu
Zuyan Liu
Josep Llados
Ling Lo
Shao-Yuan Lo
Liliana Lo Presti
Sylvain Lobry
Yaroslava Lochman
Fotios Logothetis
Suhas Lohit
Marios Loizou
Vishnu Suresh Lokhande
Cheng Long
Chengjiang Long
Fuchen Long
Guodong Long
Rujiao Long
Shangbang Long
Teng Long
Xiaoxiao Long
Zijun Long
Ivan Lopes
Vasco Lopes
Adrian Lopez-Rodriguez
Javier Lorenzo-Navarro
Yujing Lou
Brian C. Lovell
Weng Fei Low
Changsheng Lu
Chun-Shien Lu

Daohan Lu
Dongming Lu
Erika Lu
Fan Lu
Guangming Lu
Guo Lu
Hao Lu
Hao Lu
Hongtao Lu
Jiachen Lu
Jiaxin Lu
Jiwen Lu
Lewei Lu
Liying Lu
Quanfeng Lu
Shenyu Lu
Shun Lu
Tao Lu
Xiangyong Lu
Xiankai Lu
Xin Lu
Xuanchen Lu
Xuequan Lu
Yan Lu
Yang Lu
Yanye Lu
Yawen Lu
Yifan Lu
Yongchun Lu
Yongxi Lu
Yu Lu
Yu Lu
Yuzhe Lu
Zhichao Lu
Zhihe Lu
Zijia Lu
Tianyu Luan
Pauline Luc
Simon Lucey
Timo Lüddecke
Jonathon Luiten
Jovita Lukasik
Ao Luo
Cheng Luo
Chuanchen Luo

Donghao Luo
Fangzhou Luo
Gen Luo
Gongning Luo
Hongchen Luo
Jiahao Luo
Jiebo Luo
Jinqi Luo
Jinqi Luo
Jun Luo
Katie Z. Luo
Kunming Luo
Lei Luo
Mandi Luo
Mi Luo
Ruotian Luo
Sihui Luo
Tiange Luo
Wenhan Luo
Xiao Luo
Xiaotong Luo
Xiongbiao Luo
Xu Luo
Yadan Luo
Yawei Luo
Ye Luo
Yisi Luo
Yong Luo
You-Wei Luo
Yuanjing Luo
Zelun Luo
Zhengxiong Luo
Zhengyi Luo
Zhiming Luo
Zhipeng Luo
Zhongjin Luo
Zilin Luo
Ziyang Luo
Tung M. Luu
Diogo C. Luvizon
Jun Lv
Pei Lv
Yunqiu Lv
Zhaoyang Lv
Gengyu Lyu

Jiancheng Lyu
Jipeng Lyu
Junfeng Lyu
Mengyao Lyu
Mingzhi Lyu
Weimin Lyu
Xiaoyang Lyu
Xinyu Lyu
Yiwei Lyu
Youwei Lyu
Ailong Ma
Andy J. Ma
Benteng Ma
Bingpeng Ma
Chao Ma
Chuofan Ma
Cong Ma
Cuixia Ma
Fan Ma
Fangchang Ma
Fei Ma
Guozheng Ma
Haoyu Ma
Hengbo Ma
Huimin Ma
Jiahao Ma
Jianqi Ma
Jiawei Ma
Jiayi Ma
Kai Ma
Kede Ma
Lei Ma
Li Ma
Lin Ma
Liqian Ma
Lizhuang Ma
Mengmeng Ma
Ning Ma
Qianli Ma
Rui Ma
Shijie Ma
Shiqiang Ma
Shiqing Ma
Shuailei Ma
Sizhuo Ma

Tao Ma
Teli Ma
Wenxuan Ma
Wufei Ma
Xianzheng Ma
Xiaoxuan Ma
Xinyin Ma
Xinzhu Ma
Xu Ma
Yeyao Ma
Yifeng Ma
Yuexiao Ma
Yuexin Ma
Yunsheng Ma
Zhan Ma
Zhanyu Ma
Ziping Ma
Ziqiao Ma
Muhammad Maaz
Anish Madan
Neelu Madan
Spandan Madan
Sai Advaith Maddipatla
Rishi Madhok
Filippo Maggioli
Simone Magistri
Marcus Magnor
Sabarinath Mahadevan
Shweta Mahajan
Aniruddha Mahapatra
Sarthak Kumar Maharana
Behrooz Mahasseni
Upal Mahbub
Arif Mahmood
Kaleel Mahmood
Mohammed Mahmoud
Tanvir Mahmud
Jinjie Mai
Helena de Almeida Maia
Josef Maier
Shishira R. Maiya
Snehashis Majhi
Orchid Majumder
Sagnik Majumder
Ilya Makarov

Sina Malakouti
Hashmat Shadab Malik
Mateusz Malinowski
Utkarsh Mall
Srikanth Malla
Clement Mallet
Dimitrios Mallis
Abed Malti
Yunze Man
Oscar Mañas
Karttikeya Mangalam
Fabian Manhardt
Ioannis Maniadis Metaxas
Fahim Mannan
Rafal Mantiuk
Dongxing Mao
Jiageng Mao
Wei Mao
Weian Mao
Weixin Mao
Ye Mao
Yongsen Mao
Yunyao Mao
Yuxin Mao
Zhiyuan Mao
Emanuela Marasco
Matthew Marchellus
Alberto Marchisio
Diego Marcos
Alina E. Marcu
Riccardo Marin
Manuel J. Marín-Jiménez
Octave Mariotti
Dejan Markovic
Imad Eddine Marouf
Valerio Marsocci
Diego Martin Arroyo
Ricardo Martin-Brualla
Brais Martinez
Renato Martins
Damien Martins Gomes
Tetiana Martyniuk
Pierre Marza
David Masip
Carlo Masone

Timothée Masquelier
André G. Mateus
Minesh Mathew
Yusuke Matsui
Bruce A. Maxwell
Christoph Mayer
Prasanna Mayilvahanan
Amir Mazaheri
Amrita Mazumdar
Pratik Mazumder
Alessio Mazzucchelli
Amarachi B. Mbakwe
Scott McCloskey
Naga Venkata Kartheek Medathati
Henry Medeiros
Guofeng Mei
Haiyang Mei
Jie Mei
Jieru Mei
Kangfu Mei
Lingjie Mei
Xiaoguang Mei
Dennis Melamed
Luke Melas-Kyriazi
Iaroslav Melekhov
Yifang Men
Ricardo A. Mendoza-León
Depu Meng
Fanqing Meng
Jingke Meng
Lingchen Meng
Qier Meng
Qingjie Meng
Quan Meng
Yanda Meng
Zibo Meng
Otniel-Bogdan Mercea
Pablo Mesejo
Safa Messaoud
Nico Messikommer
Nando Metzger
Christopher Metzler
Vasileios Mezaris
Liang Mi

Zhenxing Mi
S. Mahdi H. Miangoleh
Bo Miao
Changtao Miao
Jiaxu Miao
Zichen Miao
Bjoern Michele
Christian Micheloni
Marko Mihajlovic
Zoltán Á. Milacski
Simone Milani
Leo Milecki
Roy Miles
Christen Millerdurai
Monica Millunzi
Chaerin Min
Cheol-Hui Min
Dongbo Min
Hyun-Seok Min
Jie Min
Juhong Min
Kyle Min
Yifei Min
Yuecong Min
Zhixiang Min
Matthias Minderer
Di Ming
Qi Ming
Xiang Ming
Riccardo Miotto
Aymen Mir
Pedro Miraldo
Parsa Mirdehghan
Seyed Ehsan Mirsadeghi
Muhammad Jehanzeb Mirza
Ashkan Mirzaei
Dmytro Mishkin
Anand Mishra
Ashish Mishra
Samarth Mishra
Shlok K. Mishra
Diganta Misra
Abhay Mittal
Gaurav Mittal

Surbhi Mittal
Trisha Mittal
Taiki Miyanishi
Daisuke Miyazaki
Hong Mo
Kaichun Mo
Sangwoo Mo
Sicheng Mo
Sicheng Mo
Zhipeng Mo
Michael Moeller
Peyman Moghadam
Hadi Mohaghegh Dolatabadi
Salman Mohamadi
Mirgahney H. Mohamed
Deen Dayal Mohan
Fnu Mohbat
Satyam Mohla
Tony C. W. Mok
Liliane Momeni
Pascal Monasse
Ajoy Mondal
Anindya Mondal
Mathew Monfort
Tom Monnier
Yusuke Monno
Eduardo F. Montesuma
Gyeongsik Moon
Taesup Moon
WonJun Moon
Dror Moran
Julie R. C. Mordacq
Deeptej S. More
Arthur Moreau
Davide Morelli
Luca Morelli
Pedro Morgado
Alexandre Morgand
Henrique Morimitsu
Matteo Moro
Lia Morra
Matteo Mosconi
Ali Mosleh
Sayed Mohammad Mostafavi Isfahani
Saman Motamed
Chong Mou
Dana Moukheiber
Pierre Moulon
Ramy A. Mounir
Théo Moutakanni
Fangzhou Mu
Jiteng Mu
Yao Mark Mu
Manasi Muglikar
Yasuhiro Mukaigawa
Amitangshu Mukherjee
Avideep Mukherjee
Prerana Mukherjee
Tanmoy Mukherjee
Anirban Mukhopadhyay
Soumik Mukhopadhyay
Yusuke Mukuta
Ravi Teja Mullapudi
Lea Müller
Norman Müller
Chaithanya Kumar Mummadi
Muhammad Akhtar Munir
Subrahmanyam Murala
Sanjeev Muralikrishnan
Ana C. Murillo
Nils Murrugarra-Llerena
Mohamed Adel Musallam
Damien Muselet
Josh David Myers-Dean
Byeonghu Na
Taeyoung Na
Muhammad Ferjad Naeem
Sauradip Nag
Pravin Nagar
Rajendra Nagar
Varun Nagaraja
Tushar Nagarajan
Seungjun Nah
Shu Nakamura
Gaku Nakano
Yuta Nakashima
Kiyohiro Nakayama
Mitsuru Nakazawa
Krishna Kanth Nakka
Yuesong Nan
Karthik Nandakumar
Paolo Napoletano
Syed S. Naqvi
Dinesh Reddy Narapureddy
Supreeth Narasimhaswamy
Kartik Narayan
Sriram Narayanan
Fabio Narducci
Erickson R. Nascimento
Muzammal Naseer
Kamal Nasrollahi
Lakshmanan Nataraj
Vishwesh Nath
Avisek Naug
Alexander Naumann
K. L. Navaneet
Pablo Navarrete Michelini
Shah Nawaz
Nazir Nayal
Niv Nayman
Amin Nejatbakhsh
Negar Nejatishahidin
Reyhaneh Neshatavar
Pedro C. Neto
Lukáš Neumann
Richard Newcombe
Alejandro Newell
Evonne Ng
Kam Woh Ng
Trung T. Ngo
Tuan Duc Ngo
Anh Nguyen
Anh Duy Nguyen
Cuong Cao Nguyen
Duc Anh Nguyen
Hoang Chuong Nguyen
Huy Hong Nguyen
Khai Nguyen
Khanh-Binh Nguyen

Khanh-Duy Nguyen
Khoi Nguyen
Khoi D. Nguyen
Kiet A. Nguyen
Ngoc Cuong Nguyen
Pha Nguyen
Phi Le Nguyen
Phong Ha Nguyen
Rang Nguyen
Tam V. Nguyen
Thao Nguyen
Thuan Hoang Nguyen
Toan Tien Nguyen
Trong-Tung Nguyen
Van Nguyen Nguyen
Van-Quang Nguyen
Thuong Nguyen Canh
Thien Trang Nguyen Vu
Haomiao Ni
Jiangqun Ni
Minheng Ni
Yao Ni
Zhen-Liang Ni
Zixuan Ni
Dong Nie
Hui Nie
Jiahao Nie
Lang Nie
Liqiang Nie
Qiang Nie
Ying Nie
Yinyu Nie
Yongwei Nie
Aditya Nigam
Kshitij N. Nikhal
Nick Nikzad
Jifeng Ning
Rui Ning
Xuefei Ning
Li Niu
Muyao Niu
Shuaicheng Niu
Wei Niu
Xuesong Niu
Yi Niu
Yulei Niu
Zhenxing Niu
Zhong-Han Niu
Shohei Nobuhara
Jongyoun Noh
Junhyug Noh
Nadhira Noor
Parsa Nooralinejad
Sotiris Nousias
Tiago Novello
Gal Novich
David Novotny
Slawomir Nowaczyk
Ewa M. Nowara
Evangelos Ntavelis
Valsamis Ntouskos
Leonardo Nunes
Oren Nuriel
Zhakshylyk Nurlanov
Simbarashe Nyatsanga
Lawrence O'Gorman
Anton Obukhov
Michael Oechsle
Ferda Ofli
Changjae Oh
Dongkeun Oh
Junghun Oh
Seoung Wug Oh
Youngtaek Oh
Hiroki Ohashi
Takehiko Ohkawa
Takeshi Oishi
Takahiro Okabe
Fumio Okura
Daniel Olmeda Reino
Suguru Onda
Trevine S. J. Oorloff
Michael Opitz
Roy Or-El
Jose Oramas
Jordi Orbay
Tribhuvanesh Orekondy
Evin Pınar Örnek
Alessandro Ortis
Magnus Oskarsson
Julian Ost
Daniil Ostashev
Mayu Otani
Naima Otberdout
Hatef Otroshi Shahreza
Yassine Ouali
Amine Ouasfi
Cheng Ouyang
Wanli Ouyang
Wenqi Ouyang
Xu Ouyang
Poojan B. Oza
Milind G. Padalkar
Johannes C. Paetzold
Gautam Pai
Anwesan Pal
Simone Palazzo
Avinash Paliwal
Cristina Palmero
Chengwei Pan
Fei Pan
Hao Pan
Jianhong Pan
Junting Pan
Liang Pan
Lili Pan
Linfei Pan
Liyuan Pan
Tai-Yu Pan
Xichen Pan
Xingjia Pan
Xinyu Pan
Yingwei Pan
Zhaoying Pan
Zhihong Pan
Zixuan Pan
Zizheng Pan
Rohit Pandey
Saurabh Pandey
Bo Pang
Guansong Pang
Lu Pang
Meng Pang
Tianyu Pang
Youwei Pang

Ziqi Pang
Omiros Pantazis
Juan J. Pantrigo
Hsing-Kuo Kenneth Pao
Marina Paolanti
Joao P. Papa
Samuele S. Papa
Dim P. Papadopoulos
Symeon Papadopoulos
George Papandreou
Toufiq Parag
Chethan Parameshwara
Foivos Paraperas
  Papantoniou
Shaifali Parashar
Alejandro Pardo
Jason R. Parham
Kranti K. Parida
Rishubh Parihar
Chunghyun Park
Daehee Park
Dongmin Park
Dongwon Park
Eunbyung Park
Eunhyeok Park
Eunil Park
Geon Yeong Park
Gyeong-Moon Park
Hyoungseob Park
Jae Sung Park
JaeYoo Park
Jin-Hwi Park
Jinhyung Park
Jinyoung Park
Jongwoo Park
JoonKyu Park
JungIn Park
Junheum Park
Kiru Park
Kwanyong Park
Seongsik Park
Seulki Park
Song Park
Sungho Park
Sungjune Park

Taesung Park
Yeachan Park
Gaurav Parmar
Paritosh Parmar
Maurizio Parton
Magdalini Paschali
Vito Paolo Pastore
Or Patashnik
Gaurav Patel
Maitreya Patel
Diego Patino
Suvam Patra
Viorica Patraucean
Badri Narayana Patro
Danda Pani Paudel
Angshuman Paul
Sneha Paul
Soumava Paul
Sudipta Paul
Sujoy Paul
Rémi Pautrat
Ioannis Pavlidis
Svetlana Pavlitska
Raju Pavuluri
Kim Steenstrup Pedersen
Marco Pedersoli
Adithya Pediredla
Pieter Peers
Jiju Peethambaran
Sen Pei
Wenjie Pei
Yuru Pei
Simone Alberto Peirone
Chantal Pellegrini
Latha Pemula
Abhirama Subramanyam
  V. B. Penamakuri
Adrian Penate-Sanchez
Baoyun Peng
Bo Peng
Can Peng
Cheng Peng
Chi-Han Peng
Chunlei Peng
Jie Peng

Jingliang Peng
Kebin Peng
Kunyu Peng
Liang Peng
Liangzu Peng
Pai Peng
Peixi Peng
Sida Peng
Songyou Peng
Wei Peng
Wen-Hsiao Peng
Xi Peng
Xiaojiang Peng
Yi-Xing Peng
Yuxin Peng
Zhiliang Peng
Ziqiao Peng
Matteo Pennisi
Or Perel
Gabriel Perez
Gustavo Perez
Juan C. Perez
Andres Felipe Perez
  Murcia
Eduardo Pérez-Pellitero
Neehar Peri
Skand Peri
Gabriel J. Perin
Federico Pernici
Chiara Pero
Elia Peruzzo
Marco Pesavento
Dmitry M. Petrov
Ilya A. Petrov
Mathis Petrovich
Vitali Petsiuk
Tomas Pevny
Shubham Milind Phal
Chau Pham
Hai X. Pham
Khoi Pham
Long Hoang Pham
Trong Thang Pham
Trung X. Pham
Tung Pham

Hoang Phan
Huy Phan
Minh Hieu Phan
Julien Philip
Stephen Phillips
Cheng Perng Phoo
Hao Phung
Shruti S. Phutke
Weiguo Pian
Yongri Piao
Luigi Piccinelli
A. J. Piergiovanni
Sara Pieri
Vipin Pillai
Wu Pingyu
Silvia L. Pintea
Francesco Pinto
Maura Pintor
Giovanni Pintore
Vittorio Pippi
Robinson Piramuthu
Fiora Pirri
Leonid Pishchulin
Francesca Pistilli
Francesco Pittaluga
Fabio Pizzati
Edward Pizzi
Benjamin Planche
Iuliia Pliushch
Chiara Plizzari
Ryan Po
GIovanni Poggi
Matteo Poggi
Kilian Pohl
Chandradeep Pokhariya
Ashwini Pokle
Matteo Polsinelli
Adrian Popescu
Teodora Popordanoska
Nikola Popović
Ronald Poppe
Samuele Poppi
Andrea Porfiri Dal Cin
Angelo Porrello
Pedro Porto Buarque de Gusmão
Rudra P. K. Poudel
Kossar Pourahmadi Meibodi
Hadi Pouransari
Ali Pourramezan Fard
Omid Poursaeed
Anish J. Prabhu
Mihir Prabhudesai
Aayush Prakash
Aditya Prakash
Shraman Pramanick
Mantini Pranav
B. H. Pawan Prasad
Meghshyam Prasad
Prateek Prasanna
Ekta Prashnani
Bardh Prenkaj
Derek S. Prijatelj
Véronique Prinet
Malte Prinzler
Victor Adrian Prisacariu
Federica Proietto Salanitri
Sergey Prokudin
Bill Psomas
Dongqi Pu
Mengyang Pu
Nan Pu
Shi Pu
Rita Pucci
Kuldeep Purohit
Senthil Purushwalkam
Waqas A. Qazi
Charles R. Qi
Chenyang Qi
Haozhi Qi
Jiaxin Qi
Lei Qi
Mengshi Qi
Peng Qi
Xianbiao Qi
Xiangyu Qi
Yuankai Qi
Zhangyang Qi
Guocheng Qian
Hangwei Qian
Jianing Qian
Qi Qian
Rui Qian
Shengsheng Qian
Shengyi Qian
Shenhan Qian
Wen Qian
Xuelin Qian
Yaguan Qian
Yijun Qian
Yiming Qian
Zhenxing Qian
Wenwen Qiang
Feng Qiao
Fengchun Qiao
Xiaotian Qiao
Yanyuan Qiao
Yi-Ling Qiao
Yu Qiao
Hangyu Qin
Haotong Qin
Jie Qin
Peiwu Qin
Siyang Qin
Wenda Qin
Xuebin Qin
Xugong Qin
Yang Qin
Yipeng Qin
Yongqiang Qin
Yuzhe Qin
Zequn Qin
Zeyu Qin
Zheng Qin
Zhenyue Qin
Ziheng Qin
Jiaxin Qing
Congpei Qiu
Haibo Qiu
Hang Qiu
Heqian Qiu
Jiayan Qiu
Jielin Qiu

Longtian Qiu
Mufan Qiu
Ri-Zhao Qiu
Weichao Qiu
Xuchong Qiu
Xuerui Qiu
Yuda Qiu
Yuheng Qiu
Zhongxi Qiu
Maan Qraitem
Chao Qu
Linhao Qu
Yanyun Qu
Kha Gia Quach
Ruijie Quan
Fabio Quattrini
Yvain Queau
Faisal Z. Qureshi
Rizwan Qureshi
Hamid R. Rabiee
Paolo Rabino
Ryan L. Rabinowitz
Petia Radeva
Bhaktipriya Radharapu
Krystian Radlak
Bodgan Raducanu
M. Usman Rafique
Francesco Ragusa
Sahar Rahimi Malakshan
Tanzila Rahman
Aashish Rai
Arushi Rai
Shyam Nandan Rai
Zobeir Raisi
Amit Raj
Kiran Raja
Sachin Raja
Deepu Rajan
Jathushan Rajasegaran
Gnana Praveen Rajasekhar
Ramanathan Rajendiran
Marie-Julie Rakotosaona
Gorthi Rama Krishna Sai
    Subrahmanyam
Sai Niranjan
    Ramachandran
Santhosh Kumar
    Ramakrishnan
Srikumar Ramalingam
Michaël Ramamonjisoa
Ravi Ramamoorthi
Shanmuganathan Raman
Mani Ramanagopal
Ashish Ramayee Asokan
Andrea Ramazzina
Jason Rambach
Sai Saketh Rambhatla
Sai Saketh Rambhatla
Clément Rambour
Francois Bernard Julien
    Rameau
Visvanathan Ramesh
Adín Ramírez Rivera
Haoxi Ran
Xuming Ran
Aakanksha Rana
Srinivas Rana
Kanchana N. Ranasinghe
Poorva G. Rane
Aneesh Rangnekar
Harsh Rangwani
Viresh Ranjan
Anyi Rao
Sukrut Rao
Yongming Rao
ZhiBo Rao
Carolina Raposo
Hanoona Abdul Rasheed
Amir Rasouli
Deevashwer Rathee
Christian Rathgeb
Avinash Ravichandran
Bharadwaj Ravichandran
Arijit Ray
Dripta S. Raychaudhuri
Sonia Raychaudhuri
Haziq Razali
Daniel Rebain
William T. Redman
Albert W. Reed
Aniket Rege
Christoph Reich
Christian Reimers
Simon Reiß
Konstantinos Rematas
Tal Remez
Davis Rempe
Bin Ren
Chao Ren
Chuan-Xian Ren
Dayong Ren
Dongwei Ren
Jiawei Ren
Jiaxiang Ren
Jing Ren
Mengwei Ren
Pengfei Ren
Pengzhen Ren
Qibing Ren
Shuhuai Ren
Sucheng Ren
Tianhe Ren
Weihong Ren
Wenqi Ren
Xuanchi Ren
Yanli Ren
Yihui Ren
Yixuan Ren
Yufan Ren
Zhenwen Ren
Zhihang Ren
Zhiyuan Ren
Zhongzheng Ren
Jose Restom
George Retsinas
Ambareesh Revanur
Ferdinand Rewicki
Manuel Rey Area
Md Alimoor Reza
Farnoush Rezaei Jafari
Hamed Rezazadegan
    Tavakoli
Rafael S. Rezende
Wonjong Rhee

Anthony D. Rhodes
Daniel Riccio
Alexander Richard
Christian Richardt
Luca Rigazio
Benjamin Risse
Dominik Rivoir
Luigi Riz
Mamshad Nayeem Rizve
Antonino M. Rizzo
Wes J. Robbins
Damien Robert
Jonathan Roberts
Joseph Robinson
Antonio Robles-Kelly
Mrigank Rochan
Chris Rockwell
Chris Rockwell
Ivan Rodin
Erik Rodner
Ranga Rodrigo
Andres C. Rodriguez
Cristian Rodriguez
Carlos Rodriguez-Pardo
Antonio J.
  Rodriguez-Sanchez
Barbara Roessle
Paul Roetzer
Alina Roitberg
Javier Romero
Meitar Ronen
Keran Rong
Xuejian Rong
Yu Rong
Marco Rosano
Bodo Rosenhahn
Gabriele Rosi
Candace Ross
Andreas Rössler
Giulio Rossolini
Mohammad Rostami
Edward Rosten
Daniel Roth
Karsten Roth
Mark S. Rothermel

Matthias Rottmann
Anastasios Roussos
Aniket Roy
Anirban Roy
Debaditya Roy
Shuvendu Roy
Sudipta Roy
Ahana Roy Choudhury
Amit Roy-Chowdhury
Aruni RoyChowdhury
Dávid Rozenberszki
Denys Rozumnyi
Lixiang Ru
Lingyan Ruan
Shulan Ruan
Viktor Rudnev
Daniel Rueckert
Nataniel Ruiz
Ewelina Rupnik
Evgenia Rusak
Chris Russell
Marc Rußwurm
Fiona Ryan
Dawid Damian Rymarczyk
DongHun Ryu
Sari Saba-Sadiya
Robert Sablatnig
Mohammad Sabokrou
Ragav Sachdeva
Ali Sadeghian
Arka Sadhu
Sadra Safadoust
Bardia Safaei
Ryusuke Sagawa
Avishkar Saha
Gobinda Saha
Oindrila Saha
Aditya Sahdev
Lakshmi Babu Saheer
Aadarsh Sahoo
Pritish Sahu
Aneeshan Sain
Nirat Saini
Saurabh Saini
Kuniaki Saito

Shunsuke Saito
Rahul Sajnani
Fumihiko Sakaue
Parikshit V. Sakurikar
Riccardo Salami
Soorena Salari
Mohammadreza Salehi
Leonard Salewski
Driton Salihu
Benjamin Salmon
Cristiano Saltori
Joel Saltz
Tim Salzmann
Sina Samangooei
Babak Samari
Nermin Samet
Fawaz Sammani
Leo Sampaio Ferraz
  Ribeiro
Shailaja Keyur Sampat
Alessio Sampieri
Jorge Sanchez
Pedro Sandoval-Segura
Nong Sang
Shengtian Sang
Patsorn Sangkloy
Depanshu Sani
Juan C. Sanmiguel
Hiroaki Santo
Joshua Santoso
Bikash Santra
Soubhik Sanyal
Hitesh Sapkota
Ayush Saraf
Nikolaos Sarafianos
István Sárándi
Kyle Sargent
Andranik Sargsyan
Josip Šarić
Mert Bulent Sariyildiz
Abhijit Sarkar
Anirban Sarkar
Chayan Sarkar
Michel Sarkis
Paul-Edouard Sarlin

Sara Sarto
Josua Sassen
Srikumar Sastry
Imari Sato
Takami Sato
Shin'ichi Satoh
Ravi Kumar Satzoda
Jack Saunders
Corentin Sautier
Mattia Savardi
Bogdan Savchynskyy
Mohamed Sayed
Marin Scalbert
Gianluca Scarpellini
Gerald Schaefer
Guilherme G. Schardong
David Schinagl
Phillip Schniter
Patrick Schramowski
Matthias Schubert
Peter Schüffler
Samuel Schulter
René Schuster
Klamer Schutte
Luca Scofano
Jesse Scott
Marcel Seelbach Benkner
Karthik Seemakurthy
Mattia Segù
Santi Seguí
Sinisa Segvic
Constantin Marc Seibold
Roman Seidel
Lorenzo Seidenari
Taiki Sekii
Yusuke Sekikawa
Matan Sela
Pratheba Selvaraju
Agniva Sengupta
Ahyun Seo
Jinhwan Seo
Junyoung Seo
Kwanggyoon Seo
Seonguk Seo
Seunghyeon Seo

Jinseok Seol
Hongje Seong
Ana F. Sequeira
Dario Serez
Dario Serez
David Serrano-Lozano
Pratinav Seth
Francesco Setti
Giorgos Sfikas
Mohammad Amin Shabani
Faisal Shafait
Anshul Shah
Chintan Shah
Jay Shah
Ketul Shah
Mubarak Shah
Viraj Shah
Mohamad Shahbazi
Muhammad Bilal B. Shaikh
Abdelrahman M. Shaker
Greg Shakhnarovich
Md Salman Shamil
Fahad Shamshad
Caifeng Shan
Dandan Shan
Hongming Shan
Xiaojun Shan
Chong Shang
Fanhua Shang
Jinghuan Shang
Lei Shang
Sifeng Shang
Wei Shang
Yuzhang Shang
Yuzhang Shang
Sukrit Shankar
Dian Shao
Mingwen Shao
Rui Shao
Ruizhi Shao
Shuai Shao
Shuwei Shao
Ron A. Shapira Weber
S. M. A. Sharif

Aashish Sharma
Avinash Sharma
Charu Sharma
Prafull Sharma
Prasen Kumar Sharma
Allam Shehata
Mark Sheinin
Sumit Shekhar
Oleksandr Shekhovtsov
Chuanfu Shen
Fei Shen
Fengyi Shen
Furao Shen
Hui-liang Shen
Jiajun Shen
Jianghao Shen
Jiangrong Shen
Jiayi Shen
Li Shen
Li-Yong Shen
Linlin Shen
Maying Shen
Qiu Shen
Qiuhong Shen
Shuai Shen
Shuhan Shen
Siqi Shen
Tianwei Shen
Tong Shen
Xiaolong Shen
Xiaoqian Shen
Yan Shen
Yanqing Shen
Yilin Shen
Ying Shen
Yiqing Shen
Yuan Shen
Yucong Shen
Yuhan Shen
Yunhang Shen
Zehong Shen
Zengming Shen
Zhijie Shen
Zhiqiang Shen
Hualian Sheng

Tao Sheng
Yichen Sheng
Zehua Sheng
Shivanand Venkanna
  Sheshappanavar
Ivaxi Sheth
Baoguang Shi
Botian Shi
Dachuan Shi
Daqian Shi
Haizhou Shi
Hengcan Shi
Jia Shi
Jing Shi
Jingang Shi
QingHongYa Shi
Ruoxi Shi
Tianyang Shi
Weishi Shi
Wu Shi
Wuxuan Shi
Xiaodan Shi
Xiaoshuang Shi
Xiaoyu Shi
Xingjian Shi
Xinyu Shi
Xuepeng Shi
Yichun Shi
Yujiao Shi
Zhenbo Shi
Zheng Shi
Zhensheng Shi
Zhenwei Shi
Zhihao Shi
Zifan Shi
Takashi Shibata
Meng-Li Shih
Yichang Shih
Dongseok Shim
Wataru Shimoda
Ilan Shimshoni
Changha Shin
Gyungin Shin
Hyungseob Shin
Inkyu Shin

Seungjoo Shin
Ukcheol Shin
Yooju Shin
Young Min Shin
Koichi Shinoda
Kaede Shiohara
Suprosanna Shit
Palaiahnakote
  Shivakumara
Sindi Shkodrani
Michal
  Shlapentokh-Rothman
Debaditya Shome
Hyounguk Shon
Sulabh Shrestha
Aman Shrivastava
Ayush Shrivastava
Gaurav Shrivastava
Aleksandar Shtedritski
Dong Wook Shu
Han Shu
Jun Shu
Xiangbo Shu
Xiujun Shu
Yang Shu
Bing Shuai
Hong-Han Shuai
Qing Shuai
Changjian Shui
Pushkar Shukla
Mustafa Shukor
Hubert P. H. Shum
Nina Shvetsova
Chenyang Si
Jianlou Si
Zilin Si
Mennatullah Siam
Sven Sickert
Désiré Sidibé
Ioannis Siglidis
Alberto Signoroni
Karan Sikka
Pedro Silva
Julio Silva-Rodríguez
Hyeonjun Sim

Jae-Young Sim
Chonghao Sima
Christian Simon
Martin Simon
Alessandro Simoni
Enis Simsar
Abhishek Singh
Apoorv Singh
Ashish Singh
Bharat Singh
Jasdeep Singh
Jaskirat Singh
Krishnakant Singh
Manish Kumar Singh
Mannat Singh
Nikhil Singh
Pravendra Singh
Rajat Vikram Singh
Simranjit Singh
Darshan Singh S.
Utkarsh Singhal
Dipika Singhania
Vasu Singla
Abhishek Kumar Sinha
Animesh Sinha
Sanjana Sinha
Saptarshi Sinha
Sudipta Sinha
Sophia A.
  Sirko-Galouchenko
Josef Sivic
Elena Sizikova
Geri Skenderi
Gregory Slabaugh
Habib Slim
Dmitriy Smirnov
James S. Smith
William Smith
Noah Snavely
Kihyuk Sohn
Bolivar E. Solarte
Mattia Soldan
Sobhan Soleymani
Samik Some
Nagabhushan Somraj

Jeany Son
Seung Woo Son
Byung Cheol Song
Chen Song
Guanglu Song
Jie Song
Jifei Song
Li Song
Liangchen Song
Lin Song
Luchuan Song
Mingli Song
Ran Song
Sibo Song
Sifan Song
Siyang Song
Weilian Song
Weinan Song
Wenfeng Song
Xiangchen Song
Xibin Song
Xinhang Song
Yafei Song
Yang Song
Yi-Yang Song
Yizhi Song
Yue Song
Zeen Song
Zhenbo Song
Zikai Song
Ekta Sood
Tomáš Souček
Rajiv Soundararajan
Albin Soutif-Cormerais
Jeremy Speth
Indro Spinelli
Jon Sporring
Manogna Sreenivas
Arvind Krishna Sridhar
Deepak Sridhar
Balaji Vasan Srinivasan
Pratul Srinivasan
Anuj Srivastava
Astitva Srivastava
Dhruv Srivastava

Koushik Srivatsan
Pierre-Luc St-Charles
Ioannis Stamos
Anastasis Stathopoulos
Colton Stearns
Jan Steinbrener
Jan-Martin O. Steitz
Sinisa Stekovic
Federico Stella
Michael Stengel
Alexandros Stergiou
Gleb Sterkin
Rainer Stiefelhagen
Noah Stier
Timo N. Stoffregen
Vladan Stojnić
Nick O. Stracke
Ombretta Strafforello
Julian Straub
Nicola Strisciuglio
Vitomir Struc
Yannick Strümpler
Joerg Stueckler
Chi Su
Hang Su
Hang Su
Kun Su
Rui Su
Shaolin Su
Sitong Su
Xingzhe Su
Xiu Su
Yao Su
Yiyang Su
Yongyi Su
Zhaoqi Su
Zhixun Su
Zhuo Su
Zhuo Su
Iago Suárez
Arulkumar Subramaniam
Sanjay Subramanian
A. Subramanyam
Swathikiran Sudhakaran
Yusuke Sugano

Masanori Suganuma
Yumin Suh
Mohammed Suhail
Xiuchao Sui
Yang Sui
Yao Sui
Heung-Il Suk
Pavel Suma
Baigui Sun
Baochen Sun
Bin Sun
Bo Sun
Changchang Sun
Che Sun
Cheng Sun
Chong Sun
Chunyi Sun
Gan Sun
Guofei Sun
Guoxing Sun
Haifeng Sun
Hanqing Sun
Haoliang Sun
He Sun
Heming Sun
Hongbin Sun
Huiming Sun
Jennifer J. Sun
Jian Sun
Jiande Sun
Jianhua Sun
Jiankai Sun
Jipeng Sun
Keqiang Sun
Lei Sun
Lichao Sun
Long Sun
Mingjie Sun
Peize Sun
Pengzhan Sun
Qiyue Sun
Shangquan Sun
Shanlin Sun
Shuyang Sun
Tao Sun

Tiancheng Sun
Wei Sun
Weiwei Sun
Weixuan Sun
Xianfang Sun
Xiaohang Sun
Xiaoshuai Sun
Xiaoxiao Sun
Ximeng Sun
Xuxiang Sun
Yanan Sun
Yasheng Sun
Yihong Sun
Ying Sun
Yixuan Sun
Yu Sun
Yuan Sun
Yuchong Sun
Zeren Sun
Zhanghao Sun
Zhaodong Sun
Zhaohui H. Sun
Zhicheng Sun
Zhicheng Sun
Haomiao Sun
Varun Sundar
Shobhita Sundaram
Minhyuk Sung
Kalyan Sunkavalli
Yucheng Suo
Indranil Sur
Saksham Suri
Naufal Suryanto
Vadim Sushko
David Suter
Roman Suvorov
Fnu Suya
Teppei Suzuki
Kunal Swami
Archana Swaminathan
Gurumurthy Swaminathan
Robin Swanson
Eran Swears
Alexander Swerdlow
Sirnam Swetha

Tabish A. Syed
Tanveer Syeda-Mahmood
Stanislaw K. Szymanowicz
Sethuraman T. V.
Calvin-Khang T. Ta
The-Anh Ta
Babak Taati
Samy Tafasca
Andrea Tagliasacchi
Haowei Tai
Yuan Tai
Francesco Taioli
Peng Taiying
Keita Takahashi
Naoya Takahashi
Jun Takamatsu
Nicolas Talabot
Hugues G. Talbot
Hossein Talebi
Davide Talon
Gary Tam
Toru Tamaki
Dipesh Tamboli
Andong Tan
Bin Tan
Cheng Tan
David Joseph New Tan
Fuwen Tan
Guang Tan
Jianchao Tan
Jing Tan
Jingru Tan
Lei Tan
Mingkui Tan
Mingxing Tan
Shuhan Tan
Shunquan Tan
Weimin Tan
Xin Tan
Zhentao Tan
Zhentao Tan
Masayuki Tanaka
Chen Tang
Chengzhou Tang
Chenwei Tang

Fan Tang
Feng Tang
Hao Tang
Haoran Tang
Jiajun Tang
Jiapeng Tang
Jiaxiang Tang
Jie Tang
Junshu Tang
Keke Tang
Luming Tang
Luyao Tang
Lv Tang
Ming Tang
Quan Tang
Shengji Tang
Sheyang Tang
Shitao Tang
Shixiang Tang
Tao Tang
Weixuan Tang
Xu Tang
Yang Tang
Yansong Tang
Yehui Tang
Yu-Ming Tang
Zheng Tang
Zhipeng Tang
Zitian Tang
Md Mehrab Tanjim
Julian Tanke
An Tao
Chaofan Tao
Chenxin Tao
Jiale Tao
Junli Tao
Keda Tao
Ming Tao
Ran Tao
Wenbing Tao
Xinhao Tao
Jean-Philippe G. Tarel
Laia Tarres
Laia Tarrés
Enzo Tartaglione

Keisuke Tateno
SaiKiran K. Tedla
Antonio Tejero-de-Pablos
Bugra Tekin
Purva Tendulkar
Minggui Teng
Ruwan Tennakoon
Andrew Beng Jin Teoh
Konstantinos Tertikas
Piotr Teterwak
Piotr Teterwak
Anh Thai
Kartik Thakral
Nupur Thakur
Sadbhawna Thakur
Balamurugan Thambiraja
Vikas Thamizharasan
Kevin Thandiackal
Sushil Thapa
Daksh Thapar
Jonas Theiner
Christian Theobalt
Spyridon Thermos
Fida Mohammad Thoker
Christopher L. Thomas
Diego Thomas
William Thong
Mamatha Thota
Mukund Varma
  Thottankara
Changyao Tian
Chunwei Tian
Jinyu Tian
Kai Tian
Lin Tian
Tai-Peng Tian
Xin Tian
Xinyu Tian
Yapeng Tian
Yu Tian
Yuan Tian
Yuesong Tian
Yunjie Tian
Yuxin Tian
Zhuotao Tian
Mert Tiftikci
Javier Tirado-Garín
Garvita Tiwari
Lokender Tiwari
Anastasia Tkach
Andrea Toaiari
Sinisa Todorovic
Pavel Tokmakov
Tri Ton
Adam Tonderski
Jinguang Tong
Peter Tong
Xin Tong
Zhan Tong
Francesco Tonini
Alessio Tonioni
Alessandro Torcinovich
Marwan Torki
Lorenzo Torresani
Fabio Tosi
Matteo Toso
Anh T. Tran
Hung Tran
Linh-Tam Tran
Minh-Triet Tran
Ngoc-Trung Tran
Phong Tran
Jonathan Tremblay
Alex Trevithick
Aditay Tripathi
Subarna Tripathi
Felix Tristram
Gabriele Trivigno
Emanuele Trucco
Prune Truong
Thanh-Dat Truong
Tomasz Trzcinski
Fu-Jen Tsai
Yu-Ju Tsai
Michael Tschannen
Tze Ho Elden Tse
Ethan Tseng
Yu-Chee Tseng
Shahar Tsiper
Hanzhang Tu
Rong-Cheng Tu
Yuanpeng Tu
Zhengzhong Tu
Zhigang Tu
Narek Tumanyan
Anil Osman Tur
Haithem Turki
Mehmet Ozgur Turkoglu
Daniyar Turmukhambetov
Victor G. Turrisi da Costa
Tinne Tuytelaars
Bartlomiej Twardowski
Radim Tylecek
Christos Tzelepis
Seiichi Uchida
Hideaki Uchiyama
Vishaal Udandarao
Mostofa Rafid Uddin
Kohei Uehara
Tatsumi Uezato
Nicolas Ugrinovic
Youngjung Uh
Norimichi Ukita
Amin Ullah
Markus Ulrich
Ardian Umam
Mesut Erhan Unal
Mathias Unberath
Devesh Upadhyay
Paul Upchurch
Shagun Uppal
Yoshitaka Ushiku
Anil Usumezbas
Yuzuko Utsumi
Roy Uziel
Anil Vadathya
Sharvaree Vadgama
Pratik Vaishnavi
Gregory Vaksman
Matias A. Valdenegro Toro
Lucas Valença
Eduardo Valle
Ernest Valveny
Laurens van der Maaten
Wouter Van Gansbeke

Nanne van Noord
Max W. F. van Spengler
Lorenzo Vaquero
Farshid Varno
Cristina Vasconcelos
Francisco Vasconcelos
Igor Vasiljevic
Florin-Alexandru Vasluianu
Subeesh Vasu
Arun Balajee Vasudevan
Vaibhav S. Vavilala
Kyle Vedder
Vijay Veerabadran
Ronny Xavier Velastegui Sandoval
Senem Velipasalar
Andreas Velten
Raviteja Vemulapalli
Deepika Vemuri
Edward Vendrow
Jonathan Ventura
Lucas Ventura
Jakob Verbeek
Dor Verbin
Eshan Verma
Manisha Verma
Monu Verma
Sahil Verma
Constantin Vertan
Eli Verwimp
Noranart Vesdapunt
Jordan J. Vice
Sara Vicente
Kavisha Vidanapathirana
Dat Viet Thanh Nguyen
Sudheendra Vijayanarasimhan
Sujal T. Vijayaraghavan
Deepak Vijaykeerthy
Elliot Vincent
Yael Vinker
Duc Minh Vo
Huy V. Vo
Khoa H. V. Vo

Romain Vo
Antonin Vobecky
Michele Volpi
Riccardo Volpi
Igor Vozniak
Nicholas Vretos
Vibashan V. S.
Ngoc-Son Vu
Tuan-Anh Vu
Khiem Vuong
Mårten Wadenbäck
Neal Wadhwa
Sophia J. Wagner
Muntasir Wahed
Nobuhiko Wakai
Devesh Walawalkar
Jacob Walker
Matthew Walmer
Matthew R. Walter
Bo Wan
Guancheng Wan
Jia Wan
Jin Wan
Jun Wan
Qiyang Wan
Renjie Wan
Wei Wan
Xingchen Wan
Yecong Wan
Zhexiong Wan
Ziyu Wan
Karan Wanchoo
Alex Jinpeng Wang
Angtian Wang
Baoyuan Wang
Benyou Wang
Biao Wang
Bin Wang
Bing Wang
Binghui Wang
Binglu Wang
Can Wang
Ce Wang
Changwei Wang
Chao Wang

Chaoyang Wang
Chen Wang
Chen Wang
Chen Wang
Chengrui Wang
Chien-Yao Wang
Chu Wang
Chuan Wang
Congli Wang
Dadong Wang
Di Wang
Dong Wang
Dong Wang
Dongdong Wang
Dongkai Wang
Dongqing Wang
Dongsheng Wang
X. Wang
Fan Wang
Fangfang Wang
Fangjinhua Wang
Fei Wang
Feng Wang
Feng Wang
Fu-Yun Wang
Gaoang Wang
Guangcong Wang
Guangming Wang
Guangrun Wang
Guangzhi Wang
Guanshuo Wang
Guo-Hua Wang
Guoqing Wang
Guoqing Wang
Haixin Wang
Haiyan Wang
Han Wang
Hanjing Wang
Hanyu Wang
Hao Wang
Hao Wang
Haobo Wang
Haochen Wang
Haochen Wang
Haohan Wang

Haoqi Wang
Haoran Wang
Haotao Wang
Haoxuan Wang
HaoYu Wang
Hengkang Wang
Hengli Wang
Hengyi Wang
Hesheng Wang
Hong Wang
Hongjun Wang
Hongxiao Wang
Hongyu Wang
Hongzhi Wang
Hua Wang
Huafeng Wang
Huan Wang
Huijie Wang
Huiyu Wang
Jiadong Wang
Jiahao Wang
Jiahao Wang
Jiahao Wang
Jiakai Wang
Jialiang Wang
Jiamian Wang
Jian Wang
Jiang Wang
Jiangliu Wang
Jianjia Wang
Jianyi Wang
Jianyuan Wang
Jiaqi Wang
Jiashun Wang
Jiayi Wang
Jiaze Wang
Jin Wang
Jinfeng Wang
Jingbo Wang
Jinghua Wang
Jingkang Wang
Jinglong Wang
Jinglu Wang
Jinpeng Wang
Jinqiao Wang

Jue Wang
Jun Wang
Jun Wang
Junjue Wang
Junke Wang
Junxiao Wang
Kai Wang
Kai Wang
Kai Wang
Kai Wang
Kaihong Wang
Kewei Wang
Keyan Wang
Kun Wang
Kup Wang
Lan Wang
Lanjun Wang
Le Wang
Lei Wang
Lei Wang
Lezi Wang
Liansheng Wang
Liao Wang
Lijuan Wang
Lijun Wang
Limin Wang
Lin Wang
Linwei Wang
Lishun Wang
Lixu Wang
Liyuan Wang
Lizhen Wang
Lizhi Wang
Longguang Wang
Luozhou Wang
Luting Wang
Mang Wang
Manning Wang
Mei Wang
Mengjiao Wang
Mengmeng Wang
Miaohui Wang
Min Wang
Naiyan Wang
Nannan Wang

Ning-Hsu Wang
Pei Wang
Peihao Wang
Peiqi Wang
Peng Wang
Pengfei Wang
Pengkun Wang
Pichao Wang
Pu Wang
Qi Wang
Qian Wang
Qiang Wang
Qiang Wang
Qiangchang Wang
Qianqian Wang
Qifei Wang
Qilong Wang
Qin Wang
Qing Wang
Qingzhong Wang
Qitong Wang
Qiufeng Wang
Ronggang Wang
Rui Wang
Rui Wang
Rui Wang
Ruibin Wang
Ruisheng Wang
Ruoyu Wang
Sai Wang
Sen Wang
Sen Wang
Shan Wang
Shaoru Wang
Sheng Wang
Sheng-Yu Wang
Shengze Wang
Shida Wang
Shijie Wang
Shipeng Wang
Shiping Wang
Shiyu Wang
Shizun Wang
Shuhui Wang
Shujun Wang

Shunli Wang
Shunxin Wang
Shuo Wang
Shuo Wang
Shuo Wang
Shuzhe Wang
Siqi Wang
Siwei Wang
Song Wang
Song Wang
Su Wang
Tan Wang
Tao Wang
Taoyue Wang
Teng Wang
Tengfei Wang
Tiancai Wang
Tianqi Wang
Tianyang Wang
Tianyu Wang
Tong Wang
Tsun-Hsuan Wang
Tuanfeng Wang
Tuanfeng Y. Wang
Wei Wang
Weihan Wang
Weikang Wang
Weimin Wang
Weiqiang Wang
Weixi Wang
Weiyao Wang
Weiyun Wang
Wen Wang
Wenbin Wang
Wenhao Wang
Wenjing Wang
Wenqian Wang
Wentao Wang
Wenxiao Wang
Wenxuan Wang
Wenzhe Wang
Xi Wang
Xi Wang
Xiang Wang
Xiao Wang
Xiao Wang
Xiao Wang
Xiaobing Wang
Xiaofeng Wang
Xiaohan Wang
Xiaosen Wang
Xiaosong Wang
Xiaoxing Wang
Xiaoyang Wang
Xijun Wang
Xijun Wang
Xinggang Wang
Xinghan Wang
Xinjiang Wang
Xinshao Wang
Xintong Wang
Xizi Wang
Xu Wang
Xuan Wang
Xuanhan Wang
Xue Wang
Xueping Wang
Xuyang Wang
Yali Wang
Yan Wang
Yan Wang
Yang Wang
Yangang Wang
Yangtao Wang
Yaohui Wang
Yaoming Wang
Yaxing Wang
Yaxiong Wang
Yi Wang
Yi Ru Wang
Yidong Wang
Yifan Wang
Yifeng Wang
Yifu Wang
Yikai Wang
Yilin Wang
Yilun Wang
Yin Wang
Yinggui Wang
Yingheng Wang
Yingqian Wang
Yipei Wang
Yiqun Wang
Yiran Wang
Yiwei Wang
Yixu Wang
Yizhi Wang
Yizhou Wang
Yizhou Wang
Yong Wang
Yu Wang
Yu-Shuen Wang
Yuan-Gen Wang
Yuchen Wang
Yude Wang
Yue Wang
Yuesong Wang
Yufei Wang
Yufu Wang
Yuguang Wang
Yuhan Wang
Yujia Wang
Yulin Wang
Yunke Wang
Yuting Wang
Yuxi Wang
YuXin Wang
Yuzheng Wang
Ze Wang
Zedong Wang
Zehan Wang
Zengmao Wang
Zeyu Wang
Zeyu Wang
Zhao Wang
Zhaokai Wang
Zhaowen Wang
Zhe Wang
Zhen Wang
Zhen Wang
Zhendong Wang
Zheng Wang
Zheng Wang
Zheng Wang
Zhengyi Wang

Zhennan Wang
Zhenting Wang
Zhenyi Wang
Zhenyu Wang
Zhenzhi Wang
Zhepeng Wang
Zhi Wang
Zhibo Wang
Zhihao Wang
Zhihui Wang
Zhijie Wang
Zhikang Wang
Zhixiang Wang
Zhiyong Wang
Zhongdao Wang
Zhonghao Wang
Zhouxia Wang
Zhu Wang
Zian Wang
Zifu Wang
Zihao Wang
Zijian Wang
Ziqiang Wang
Ziqin Wang
Ziqing Wang
Zirui Wang
Zirui Wang
Ziwei Wang
Ziyan Wang
Ziyang Wang
Ziyi Wang
Ziyun Wang
Frederik Warburg
Syed Talal Wasim
Daniel Watson
Jamie Watson
Ethan Weber
Silvan Weder
Jan Dirk Wegner
Chen Wei
Donglai Wei
Fangyin Wei
Fangyun Wei
Guoqiang Wei
Jia Wei
Jiacheng Wei
Kaixuan Wei
Kun Wei
Longhui Wei
Megan Wei
Mian Wei
Mingqiang Wei
Pengxu Wei
Ping Wei
Qiuhong Anna Wei
Shikui Wei
Tianyi Wei
Wei Wei
Wenqi Wei
Xian Wei
Xin Wei
Xing Wei
Xinyue Wei
Xiu-Shen Wei
Yi Wei
Yixuan Wei
Yunchao Wei
Yuxiang Wei
Yuxiang Wei
Zeming Wei
Zhipeng Wei
Zihao Wei
Zimian Wei
Jean-Baptiste Weibel
Luca Weihs
Martin Weinmann
Michael Weinmann
Bihan Wen
Bowen Wen
Chao Wen
Chenglu Wen
Chuan Wen
Congcong Wen
Jie Wen
Jing Wen
Qiang Wen
Rui Wen
Sijia Wen
Song Wen
Xiang Wen
Xin Wen
Yilin Wen
Youpeng Wen
Yuanbo Wen
Yuxin Wen
Chung-Yi Weng
Junwu Weng
Shuchen Weng
Wenming Weng
Yijia Weng
Zhenzhen Weng
Thomas Westfechtel
Christopher Johannes
 Wewer
Spencer Whitehead
Tobias Jan Wieczorek
Thaddäus Wiedemer
Julian Wiederer
Kevin Tirta Wijaya
Asiri Wijesinghe
Kimberly Wilber
Jeffrey R. Willette
Bryan M. Williams
Williem Williem
Christian Wilms
Benjamin Wilson
Richard Wilson
Felix Wimbauer
Vanessa Wirth
Scott Wisdom
Calden Wloka
Alex Wong
Chau-Wai Wong
Chi-Chong Wong
Ka Wai Wong
Kelvin Wong
Kok-Seng Wong
Kwan-Yee K. Wong
Yongkang Wong
Sangmin Woo
Simon S. Woo
Markus Worchel
Scott Workman
Marcel Worring
Safwan Wshah

Aming Wu
Bo Wu
Bojian Wu
Boxi Wu
Changguang Wu
Chaoyi Wu
Chen Henry Wu
Cheng-En Wu
Chenming Wu
Chenyan Wu
Chenyun Wu
Cho-Ying Wu
Chongruo Wu
Cong Wu
Dayan Wu
Di Wu
Dongming Wu
Fangzhao Wu
Fuxiang Wu
Gaojie Wu
Guanyao Wu
Guile Wu
Haiping Wu
Haiwei Wu
Haiyu Wu
Han Wu
Haoning Wu
Haoning Wu
Haotian Wu
Hefeng Wu
Huisi Wu
Jane Wu
Jay Zhangjie Wu
Jhih-Ciang Wu
Ji-Jia Wu
Jialian Wu
Jiaye Wu
Jimmy Wu
Jing Wu
Jing Wu
Jinjian Wu
Jiqing Wu
Jun Wu
Junfeng Wu
Junlin Wu

Junru Wu
Junyang Wu
Junyi Wu
Letian Wu
Lifang Wu
Lin Yuanbo Wu
Liwen Wu
Min Wu
Minye wu
Peng Wu
Penghao Wu
Qian Wu
Qiangqiang Wu
Qianyi Wu
Qingbo Wu
Rongliang Wu
Rui Wu
Rundi Wu
Shuang Wu
Shuzhe Wu
Tao Wu
Tao Wu
Tao Wu
Te-Lin Wu
Tianfu Wu
Tianhao Wu
Tianhao Wu
Ting-Wei Wu
Tong Wu
Tong Wu
Tsung-Han Wu
Tz-Ying Wu
Weibin Wu
Weijia Wu
Xian Wu
Xiao Wu
Xiaodong Wu
Xiaohe Wu
Xiaoqian Wu
Xiaoyang Wu
Xindi Wu
Xingjiao Wu
Xinxiao Wu
Xiuzhe Wu
Yang Wu

Yangzheng Wu
Yanze Wu
Yanzhao Wu
Yawen Wu
Yicheng Wu
Ying Nian Wu
Yingwen Wu
Yong Wu
Yuanwei Wu
Yue Wu
Yue Wu
Yuqun Wu
Yushu Wu
Yushuang Wu
Zhe Wu
Zheng Wu
Zhi-Fan Wu
Zhihao Wu
Zhijie Wu
Zhiliang Wu
Zhonghua Wu
Zijie Wu
Ziyi Wu
Zizhao Wu
Zongwei Wu
Zongyu Wu
Zongze Wu
Stefanie Wuhrer
Jamie M. Wynn
Monika Wysoczańska
Jianing Xi
Teng Xi
Bin Xia
Changqun Xia
Haifeng Xia
Jiaer Xia
Jiahao Xia
Kun Xia
Mingxuan Xia
Shihong Xia
Weihao Xia
Xiaobo Xia
Yan Xia
Ye Xia
Yifei Xia

Zhaoyang Xia
Zhihao Xia
Zhihua Xia
Zhuofan Xia
Zimin Xia
Chuhua Xian
Wenqi Xian
Yongqin Xian
Donglai Xiang
Jinhai Xiang
Liuyu Xiang
Tian-Zhu Xiang
Tiange Xiang
Wangmeng Xiang
Xiaoyu Xiang
Yuanbo Xiangli
Anqi Xiao
Aoran Xiao
Bei Xiao
Chunxia Xiao
Fanyi Xiao
Han Xiao
Jiancong Xiao
Jimin Xiao
Jing Xiao
Jing Xiao
Jun Xiao
Junbin Xiao
Junfei Xiao
Mingqing Xiao
Qingyang Xiao
Ruixuan Xiao
Taihong Xiao
Yang Xiao
Yanru Xiao
Yao Xiao
Yijun Xiao
Yuting Xiao
Zehao Xiao
Zeyu Xiao
Zihao Xiao
Binhui Xie
Chaohao Xie
Chi Xie
Christopher Xie
Chuanlong Xie
Fei Xie
Guo-Sen Xie
Haozhe Xie
Hongtao Xie
Jiahao Xie
Jiaxin Xie
Jin Xie
Jinheng Xie
Jiu-Cheng Xie
Jiyang Xie
Junyu Xie
Liuyue Xie
Ming-Kun Xie
Mingyang Xie
Qian Xie
Tingting Xie
Weicheng Xie
Xianghui Xie
Xiaohua Xie
Xudong Xie
Yichen Xie
Yiming Xie
You Xie
Yuan Xie
Yusheng Xie
Yutong Xie
Zeke Xie
Zhenda Xie
Zhenyu Xie
ZiYang Xie
Chaoyue Xing
Fuyong Xing
Jinbo Xing
Xiaoyan Xing
Xiaoying Xing
XiMing Xing
Xin Xing
Yazhou Xing
Yifan Xing
Yun Xing
Zhen Xing
Hongkai Xiong
Jingjing Xiong
Jinhui Xiong
Junwen Xiong
Peixi Xiong
Wei Xiong
Weihua Xiong
Yu Xiong
Yuanhao Xiong
Yuanjun Xiong
Yuwen Xiong
Zhexiao Xiong
Zhiwei Xiong
Yuliang Xiu
Alessio Xompero
An Xu
Angchi Xu
Baixin Xu
Bicheng Xu
Bo Xu
Chao Xu
Chenfeng Xu
Chenshu Xu
Chenxin Xu
Chi Xu
Dejia Xu
Dongli Xu
Feng Xu
Gangwei Xu
Haiming Xu
Haiyang Xu
Han Xu
Haofei Xu
Haohang Xu
Haoran Xu
Hongbin Xu
Hongmin Xu
Jiale Xu
Jianjin Xu
Jiaqi Xu
Jie Xu
Jilan Xu
Jinglin Xu
Jingyi Xu
Jun Xu
Kai Xu
Katherine Xu
Ke Xu

Kele Xu
Lan Xu
Lian Xu
Liang Xu
Linning Xu
Lumin Xu
Manjie Xu
Mengde Xu
Mengdi Xu
Mengmeng Frost Xu
Min Xu
Ming Xu
Mutian Xu
Peiran Xu
Peng Xu
Qi Xu
Qiang Xu
Qiangeng Xu
Qingshan Xu
Qingyang Xu
Qiuling Xu
Ran Xu
Renzhe Xu
Ruikang Xu
Runsen Xu
Runsheng Xu
Shichao Xu
Sirui Xu
Tongda Xu
Wanting Xu
Wei Xu
Weiwei Xu
Wenjia Xu
Wenju Xu
Wenqiang Xu
Xiang Xu
Xianghao Xu
Xiangyu Xu
Xiangyu Xu
Xiaogang Xu
Xiaohao Xu
Xin Xu
Xin Xu
Xin-Shun Xu
Xing Xu

Xinli Xu
Xinyu Xu
Xiuwei Xu
Xiyan Xu
Xudong Xu
Xuemiao Xu
Xun Xu
Yan Xu
Yan Xu
Yan Xu
Yangyang Xu
Yanwu Xu
Yating Xu
Yi Xu
Yi Xu
Yi Xu
Yihong Xu
YiKun Xu
Yinghao Xu
Yingyan Xu
Yinshuang Xu
Yiran Xu
Yixing Xu
Yongchao Xu
Yue Xu
Yufei Xu
Yunqiu Xu
Zexiang Xu
Zhan Xu
Zhe Xu
Zhengqin Xu
Zhenlin Xu
Zhiqiu Xu
Zhiyuan Xu
Zhongcong Xu
Zhuoer Xu
Zipeng Xu
Ziyue Xu
Zongyi Xu
Ziwei Xuan
Danna Xue
Fanglei Xue
Fei Xue
Feng Xue
Han Xue

Jianru Xue
Le Xue
Lixin Xue
Mingfu Xue
Nan Xue
Qinghan Xue
Shangjie Xue
Xiangyang Xue
Zihui Xue
Abhay Yadav
Amit Kumar Singh Yadav
Takuma Yagi
Tomas F Yago Vicente
I. Zeki Yalniz
Kota Yamaguchi
Shin'ya Yamaguchi
Burhaneddin Yaman
Toshihiko Yamasaki
Kohei Yamashita
Lee Juliette Yamin
Chaochao Yan
Hongyu Yan
Jiexi Yan
Kai Yan
Pei Yan
Qingan Yan
Qingsen Yan
Qingsong Yan
Rui Yan
Shaoqi Yan
Shi Yan
Siming Yan
Siming Yan
Siyuan Yan
Weilong Yan
Wending Yan
Xiangyi Yan
Xinchen Yan
Xingguang Yan
Xueting Yan
Yan Yan
Yichao Yan
Zhaoyi Yan
Zhiqiang Yan
Zhiyuan Yan

Zike Yan
Zizheng Yan
Keiji Yanai
Pinar Yanardag
Anqi Yang
Anqi Joyce Yang
Bangbang Yang
Baoyao Yang
Bin Yang
Binwei Yang
Bo Yang
Bo Yang
Boyu Yang
Changdi Yang
Chao Yang
Charig Yang
Cheng-Fu Yang
Cheng-Yen Yang
Chenhongyi Yang
Chuanguang Yang
De-Nian Yang
Dingcheng Yang
Dingkang Yang
Dong Yang
Erkun Yang
Fan Yang
Fan Yang
Fan Yang
Fan Yang
Fan Yang
Feng Yang
Fengting Yang
Fengxiang Yang
Fengyuan Yang
Fu-En Yang
Gang Yang
Gengshan Yang
Guandao Yang
Guanglei Yang
Haitao Yang
Hanqing Yang
Heran Yang
Honghui Yang
Huanrui Yang
Huiyuan Yang

Huizong Yang
Hunmin Yang
Jiange Yang
Jiaqi Yang
Jiawei Yang
Jiayu Yang
Jiazhi Yang
Jie Yang
Jie Yang
Jiewen Yang
Jihan Yang
Jing Yang
Jingkang Yang
Jinhui Yang
Jinlong Yang
Jinrong Yang
Jinyu Yang
Kaicheng Yang
Kailun Yang
Lan Yang
Le Yang
Lehan Yang
Lei Yang
Lei Yang
Lei Yang
Li Yang
Lihe Yang
Ling Yang
Lingxiao Yang
Linlin Yang
Lixin Yang
Longrong Yang
Lu Yang
Luwei Yang
Michael Ying Yang
Min Yang
Ming Yang
MingKun Yang
Mouxing Yang
Muli Yang
Peiyu Yang
Qi Yang
Qian Yang
Qiushi Yang
Ren Yang

Rui Yang
Ruihan Yang
Sejong Yang
Shan Yang
Shangrong Yang
Shiqi Yang
Shuai Yang
Shuai Yang
Shuang Yang
Shuo Yang
Shusheng Yang
Sibei Yang
Siwei Yang
Siyuan Yang
Siyuan Yang
Song Yang
Songlin Yang
Tianyu Yang
Tong Yang
Wankou Yang
Wenhan Yang
Wenhan Yang
Wenjie Yang
Wenqi Yang
William Yang
Xi Yang
Xi Yang
Xiangpeng Yang
Xiao Yang
Xiaofeng Yang
Xiaoshan Yang
Xin Jeremy Yang
Xingyi Yang
Xinlong Yang
Xitong Yang
Xiulong Yang
Xu Yang
Xuan Yang
Xue Yang
Xuelin Yang
Xun Yang
Yan Yang
Yan Yang
Yang Yang
Yaokun Yang

Yezhou Yang
Yiding Yang
Yijun Yang
Yijun Yang
Yin Yang
Yinfei Yang
Yixin Yang
Yongqi Yang
Yongqi Yang
Yue Yang
Yuewei Yang
Yuezhi Yang
Yujiu Yang
Yung-Hsu Yang
Yuwei Yang
Ze Yang
Ze Yang
Zetong Yang
Zhangsihao Yang
Zhaoyuan Yang
Zhen Yang
Zhenpei Yang
Zhibo Yang
Zhiwei Yang
Zhiwen Yang
Zhiyuan Yang
Zhuoqian Yang
Ziyan Yang
Ziyun Yang
Zongxin Yang
Zuhao Yang
Chengtang Yao
Cong Yao
Hantao Yao
Jiawen Yao
Lina Yao
Mingde Yao
Mingshuai Yao
Qingsong Yao
Shunyu Yao
Taiping Yao
Ting Yao
Xincheng Yao
Xinwei Yao
Xu Yao
Xufeng Yao
Yao Yao
Yazhou Yao
Yue Yao
Ziwei Yao
Sudhir Yarram
Rajeev Yasarla
Mohsen Yavartanoo
Botao Ye
Dengpan Ye
Fei Ye
Hanrong Ye
Jianglong Ye
Jiarong Ye
Jin Ye
Jingwen Ye
Jinwei Ye
Junjie Ye
Keren Ye
Maosheng Ye
Meng Ye
Meng Ye
Muchao Ye
Nanyang Ye
Peng Ye
Qi Ye
Qian Ye
Qinghao Ye
Qixiang Ye
Ruolin Ye
Shuquan Ye
Tian Ye
Vickie Ye
Wenqian Ye
Xinchen Ye
Yufei Ye
Moon Ye-Bin
Yousef Yeganeh
Chun-Hsiao Yeh
Raymond Yeh
Yu-Ying Yeh
Florence Yellin
Sriram Yenamandra
Tarun Yenamandra
Promod Yenigalla
Chandan Yeshwanth
Dong Yi
Hongwei Yi
Kai Yi
Ran Yi
Renjiao Yi
Xinyu Yi
Alper Yilmaz
Jonghwa Yim
Aoxiong Yin
Fei Yin
Fukun Yin
Jia-Li Yin
Ming Yin
Nan Yin
Ruihong Yin
Tianwei Yin
Wenzhe Yin
Xiaoqi Yin
Yingda Yin
Yu Yin
Yufeng Yin
Zhenfei Yin
Xianghua Ying
Xiaowen Ying
Naoto Yokoya
Chen YongCan
ByungIn Yoo
Innfarn Yoo
Jinsu Yoo
Sungjoo Yoo
Hee Suk Yoon
Jae Shin Yoon
Jihun Yoon
Sangwoong Yoon
Sejong Yoon
Sung Whan Yoon
Sung-Hoon Yoon
Sunjae Yoon
Youngho Yoon
Youngseok Yoon
Youngseok Yoon
Yuichi Yoshida
Ryota Yoshihashi
Yusuke Yoshiyasu

Chenyu You
Haoran You
Haoxuan You
Shan You
Yang You
Yingxuan You
Yurong You
Chan-Hyun Youn
Kim Youwang
Nikolaos-Antonios Ypsilantis
Baosheng Yu
Bei Yu
Bruce X. B. Yu
Chaohui Yu
Chunlin Yu
Cunjun Yu
Dahai Yu
En Yu
En Yu
Fenggen Yu
Gang Yu
Haibao Yu
Hanchao Yu
Hang Yu
Hao Yu
Hao Yu
Haojun Yu
Heng Yu
Hong-Xing Yu
Houjian Yu
Jianhui Yu
Jiashuo Yu
Jing Yu
Jiwen Yu
Jiyang Yu
Kaicheng Yu
Lei Yu
Lidong Yu
Lijun Yu
Mulin Yu
Peilin Yu
Qian Yu
Qihang Yu
Qing Yu
Rui Yu
Ruixuan Yu
Runpeng Yu
Shaozuo Yu
Shuzhi Yu
Sihyun Yu
Tan Yu
Tao Yu
Tianjiao Yu
Wei Yu
Weihao Yu
Wenwen Yu
Xi Yu
Xiaohan Yu
Xin Yu
Xin Yu
Xuehui Yu
Yingchen Yu
Yongsheng Yu
Yunlong Yu
Zehao Yu
Zhaofei Yu
Zhengdi Yu
Zhengdi Yu
Zhixuan Yu
Zhongzhi Yu
Zhuoran Yu
Zitong Yu
Chun Yuan
Chunfeng Yuan
Hangjie Yuan
Haobo Yuan
Jiakang Yuan
Jiangbo Yuan
Liangzhe Yuan
Maoxun Yuan
Shanxin Yuan
Shengming Yuan
Shuai Yuan
Shuaihang Yuan
Wentao Yuan
Xiaoding Yuan
Xiaoyun Yuan
Xin Yuan
Xin Yuan
Yixuan Yuan
Yu-Jie Yuan
Yuan Yuan
Yuan Yuan
Yuhui Yuan
Zheng Yuan
Zhuoning Yuan
Mehmet Kerim Yücel
Dongxu Yue
Haixiao Yue
Kaiyu Yue
Tao Yue
Xiangyu Yue
Zihao Yue
Zongsheng Yue
Heeseung Yun
Jooyeol Yun
Juseung Yun
Kimin Yun
Se-Young Yun
Sukwon Yun
Tian Yun
Raza Yunus
Ekim Yurtsever
Eloi Zablocki
Riccardo Zaccone
Martin Zach
Muhammad Zaigham Zaheer
Ilya Zakharkin
Egor Zakharov
Abhaysinh S. Zala
Pierluigi Zama Ramirez
Eduard Sebastian Zamfir
Amir Zamir
Luca Zancato
Yuan Zang
Yuhang Zang
Zelin Zang
Pietro Zanuttigh
Giacomo Zara
Samira Zare
Olga Zatsarynna
Denis Zavadski
Vitjan Zavrtanik

Jan Zdenek
Yanjie Ze
Bernhard Zeisl
John Zelek
Oliver Zendel
Ailing Zeng
Chong Zeng
Dan Zeng
Fangao Zeng
Haijin Zeng
Huimin Zeng
Jia Zeng
Jiabei Zeng
Kuo-Hao Zeng
Libing Zeng
Ling-An Zeng
Ming Zeng
Pengpeng Zeng
Runhao Zeng
Tieyong Zeng
Wei Zeng
Yan Zeng
Yanhong Zeng
Yawen Zeng
Yuyuan Zeng
Zilai Zeng
Ziyao Zeng
Kaiwen Zha
Ruyi Zha
Yaohua Zha
Bohan Zhai
Qiang Zhai
Runtian Zhai
Wei Zhai
YiKui Zhai
Yuanhao Zhai
Yunpeng Zhai
De-Chuan Zhan
Fangneng Zhan
Guanqi Zhan
Huangying Zhan
Huijing Zhan
Kun Zhan
Xueying Zhan
Aidong Zhang

Baochang Zhang
Baoheng Zhang
Baoming Zhang
Biao Zhang
Bingfeng Zhang
Binjie Zhang
Bo Zhang
Bo Zhang
Borui Zhang
Bowen Zhang
Can Zhang
Ce Zhang
Chang-Bin Zhang
Chao Zhang
Chao Zhang
Chen-Lin Zhang
Cheng Zhang
Cheng Zhang
Chenghao Zhang
Chenyangguang Zhang
Chi Zhang
Chongyang Zhang
Chris Zhang
Chuhan Zhang
Chunhui Zhang
Chuyu Zhang
Congyi Zhang
Daichi Zhang
Dan Zhang
Daoan Zhang
Daoqiang Zhang
David Junhao Zhang
Dexuan Zhang
Dingwen Zhang
Dingyuan Zhang
Dongsu Zhang
Fan Zhang
Fan Zhang
Fang-Lue Zhang
Feilong Zhang
Frederic Z. Zhang
Fuyang Zhang
Gang Zhang
Gengwei Zhang
Gengyu Zhang

Gengyuan Zhang
Gongjie Zhang
GuiXuan Zhang
Guofeng Zhang
Guozhen Zhang
Hang Zhang
Hang Zhang
Hanwang Zhang
Hao Zhang
Hao Zhang
Hao Zhang
Haokui Zhang
Haonan Zhang
Haotian Zhang
Hengrui Zhang
Hongguang Zhang
Hongrun Zhang
Hongyuan Zhang
Howard Zhang
Huaidong Zhang
Huaiwen Zhang
Hui Zhang
Hui Zhang
Jason Y. Zhang
Ji Zhang
Jiahui Zhang
Jiakai Zhang
Jiaming Zhang
Jian Zhang
Jianfu Zhang
Jiangning Zhang
Jianhua Zhang
Jianming Zhang
Jianpeng Zhang
Jianping Zhang
Jianrong Zhang
Jichao Zhang
Jie Zhang
Jie Zhang
Jie Zhang
Jimuyang Zhang
Jing Zhang
Jing Zhang
Jinghao Zhang
Jingyi Zhang

Jinlu Zhang
Jiqing Zhang
Jiyuan Zhang
Junbo Zhang
Junge Zhang
Junyi Zhang
Juyong Zhang
Kai Zhang
Kai Zhang
Kaidong Zhang
Kaihao Zhang
Kaipeng Zhang
Kaiyi Zhang
Ke Zhang
Ke Zhang
Kui Zhang
Le Zhang
Le Zhang
Lefei Zhang
Lei Zhang
Leo Yu Zhang
Li Zhang
Lianbo Zhang
Liang Zhang
Liangpei Zhang
Lin Zhang
Linfeng Zhang
Liqing Zhang
Lu Zhang
Malu Zhang
Manyuan Zhang
Mengmi Zhang
Mengqi Zhang
Mi Zhang
Min Zhang
Min-Ling Zhang
Mingda Zhang
Mingfang Zhang
Minghui Zhang
Mingjin Zhang
Mingyuan Zhang
Minjia Zhang
Ni Zhang
Pan Zhang
Peiyan Zhang
Pengze Zhang
Pingping Zhang
Qi Zhang
Qi Zhang
Qian Zhang
Qiang Zhang
Qijian Zhang
Qiming Zhang
Qing Zhang
Qing Zhang
Renrui Zhang
Rongyu Zhang
Ruida Zhang
Ruimao Zhang
Ruixin Zhang
Runze Zhang
Sanyi Zhang
Shan Zhang
Shanghang Zhang
Shaofeng Zhang
Sheng Zhang
Shengping Zhang
Shengyu Zhang
Shimian Zhang
Shiwei Zhang
Shizhou Zhang
Shu Zhang
Shuo Zhang
Siwei Zhang
Song-Hai Zhang
Tao Zhang
Tianyun Zhang
Ting Zhang
Tong Zhang
Weixia Zhang
Wendong Zhang
Wenlong Zhang
Wenqiang Zhang
Wentao Zhang
Wentian Zhang
Wenxiao Zhang
Wenxuan Zhang
Xi Zhang
Xiang Zhang
Xiang Zhang
Xianling Zhang
Xiao Zhang
Xiaohan Zhang
Xiaoming Zhang
Xiaoran Zhang
Xiaowei Zhang
Xiaoyun Zhang
Xikun Zhang
Xin Zhang
Xinfeng Zhang
Xingchen Zhang
Xingguang Zhang
Xingxuan Zhang
Xiong Zhang
Xiuming Zhang
Xu Zhang
Xuanyang Zhang
Xucong Zhang
Xuying Zhang
Yabin Zhang
Yabo Zhang
Yachao Zhang
Yahui Zhang
Yan Zhang
Yan Zhang
Yanan Zhang
Yang Zhang
Yanghao Zhang
Yawen Zhang
Yechao Zhang
Yi Zhang
Yi Zhang
Yi Zhang
Yi-Fan Zhang
Yifan Zhang
Yifei Zhang
Yifeng Zhang
Yihao Zhang
Yihua Zhang
Yimeng Zhang
Yiming Zhang
Yin Zhang
Yinan Zhang
Yinda Zhang
Ying Zhang

Yingliang Zhang
Yitian Zhang
Yixin Zhang
Yiyuan Zhang
Yongfei Zhang
Yonggang Zhang
Yonghua Zhang
Youjian Zhang
Youmin Zhang
Youshan Zhang
Yu Zhang
Yu Zhang
Yuan Zhang
Yuechen Zhang
Yuexi Zhang
Yufei Zhang
Yuhan Zhang
Yuhang Zhang
Yunchao Zhang
Yunhe Zhang
Yunhua Zhang
Yunpeng Zhang
Yunzhi Zhang
Yuxin Zhang
Yuyao Zhang
Zaixi Zhang
Zeliang Zhang
Zewei Zhang
Zeyu Zhang
Zhang Zhang
Zhao Zhang
Zhaoxiang Zhang
Zhen Zhang
Zheng Zhang
Zheng Zhang
Zhenyu Zhang
Zhenyu Zhang
Zheyuan Zhang
Zhicheng Zhang
Zhilu Zhang
Zhishuai Zhang
Zhitian Zhang
Zhiwei Zhang
Zhixing Zhang
Zhiyuan Zhang

Zhong Zhang
Zhongping Zhang
Zhongqun Zhang
Zicheng Zhang
Zicheng Zhang
Zihao Zhang
Ziming Zhang
Ziqi Zhang
Qilong Zhangli
Bin Zhao
Bingchen Zhao
Bingyin Zhao
Cairong Zhao
Can Zhao
Chen Zhao
Dong Zhao
Dongxu Zhao
Fang Zhao
Feng Zhao
Fuqiang Zhao
Gangming Zhao
Ganlong Zhao
Guiyu Zhao
Haimei Zhao
Hanbin Zhao
Handong Zhao
Jian Zhao
Jiaqi Zhao
Jie Zhao
Kai Zhao
Kaifeng Zhao
Lei Zhao
Liang Zhao
Lirui Zhao
Long Zhao
Luxi Zhao
Mingyang Zhao
Minyi Zhao
Na Zhao
Nanxuan Zhao
Pu Zhao
Qi Zhao
Qian Zhao
Qibin Zhao
Qingsong Zhao

Qingyu Zhao
Qinyu Zhao
Rongchang Zhao
Rui Zhao
Rui Zhao
Rui Zhao
Ruiqi Zhao
Shanshan Zhao
Shihao Zhao
Shiyu Zhao
Shizhen Zhao
Shuai Zhao
Siheng Zhao
Tianchen Zhao
TianHao Zhao
Tiesong Zhao
Wang Zhao
Wangbo Zhao
Weichao Zhao
Weiyue Zhao
Wenda Zhao
Wenliang Zhao
Xiangyun Zhao
Xiaoming Zhao
Xiaonan Zhao
Xiaoqi Zhao
Xin Zhao
Xin Zhao
Xingyu Zhao
Xu Zhao
Yajie Zhao
Yang Zhao
Yifan Zhao
Ying Zhao
Yiqun Zhao
Yiqun Zhao
Yizhou Zhao
Yizhou Zhao
Yucheng Zhao
Yue Zhao
Yunhan Zhao
Yuyang Zhao
Yuzhi Zhao
Zelin Zhao
Zengqun Zhao

Zhen Zhao
Zhenghao Zhao
Zhengyu Zhao
Zhou Zhao
Zibo Zhao
Zimeng Zhao
Ziwei Zhao
Ziwei Zhao
Zixiang Zhao
Jin Zhe
Anling Zheng
Ce Zheng
Chaoda Zheng
Chengwei Zheng
Chuanxia Zheng
Duo Zheng
Ervine Zheng
Guangcong Zheng
Haitian Zheng
Haiyong Zheng
Hao Zheng
Haotian Zheng
Haoxin Zheng
Huan Zheng
Huan Zheng
Jia Zheng
Jian Zheng
Jian-Qing Zheng
Jianqiao Zheng
Jianwei Zheng
Jin Zheng
Jingxiao Zheng
Kaiwen Zheng
Kecheng Zheng
Léon Zheng
Meng Zheng
Naishan Zheng
Peng Zheng
Qi Zheng
Qian Zheng
Rongkun Zheng
Shen Zheng
Shuai Zheng
Shuhong Zheng
Shunyuan Zheng

Siming Zheng
Tianhang Zheng
Wenting Zheng
Wenzhao Zheng
Xiaozheng Zheng
Xiawu Zheng
Xu Zheng
Yajing Zheng
Yalin Zheng
Yang Zheng
Ye Zheng
Yinglin Zheng
Yinqiang Zheng
Yu Zheng
Zangwei Zheng
Zehan Zheng
Zerong Zheng
Zhaoheng Zheng
Zhedong Zheng
Zhuo Zheng
Zilong Zheng
Shuaifeng Zhi
Tiancheng Zhi
Bineng Zhong
Fangcheng Zhong
Fangwei Zhong
Guoqiang Zhong
Nan Zhong
Yaoyao Zhong
Yijie Zhong
Yiqi Zhong
Yiran Zhong
Yiwu Zhong
Yunshan Zhong
Zichun Zhong
Ziming Zhong
Brady Zhou
Chong Zhou
Chu Zhou
Chunluan Zhou
Da-Wei Zhou
Dawei Zhou
Dewei Zhou
Dingfu Zhou
Donghao Zhou

Dongzhan Zhou
Fan Zhou
Hang Zhou
Hang Zhou
Hao Zhou
Hao Zhou
Haoyi Zhou
Hong-Yu Zhou
Honglu Zhou
Huayi Zhou
Jiahuan Zhou
Jiaming Zhou
Jian Zhou
Jianan Zhou
Jiantao Zhou
Jianxiong Zhou
JIngkai Zhou
Junsheng Zhou
Kailai Zhou
Kaiyang Zhou
Keyang Zhou
Kun Zhou
Lei Zhou
Mingyang Zhou
Mingyi Zhou
Mingyuan Zhou
Mo Zhou
Pan Zhou
Peng Zhou
Peng Zhou
Qianyi Zhou
Qianyu Zhou
Qihua Zhou
Qin Zhou
Qinqin Zhou
Qunjie Zhou
Sheng Zhou
Shenglong Zhou
Shijie Zhou
Shuchang Zhou
Sihang Zhou
Tao Zhou
Tianfei Zhou
Xiangdong Zhou
Xiaoqiang Zhou

Xin Zhou
Xingyi Zhou
Yan-Jie Zhou
Yang Zhou
Yang Zhou
Yanqi Zhou
Yi Zhou
Yi Zhou
Yichao Zhou
Yichen Zhou
Yin Zhou
Yiyi Zhou
Yu Zhou
Yucheng Zhou
Yufan Zhou
Yunsong Zhou
Yuqian Zhou
Yuxiao Zhou
Yuxuan Zhou
Zhenyu Zhou
Zijian Zhou
Zikun Zhou
Ziqi Zhou
Zixiang Zhou
Zongwei Zhou
Alex Z. Zhu
Benjin Zhu
Bin Zhu
Bin Zhu
Chenming Zhu
Chenyang Zhu
Deyao Zhu
Dongxiao Zhu
Fangrui Zhu
Fei Zhu
Feida Zhu
Fengqing Maggie Zhu
Guibo Zhu
Haidong Zhu
Hanwei Zhu
Hao Zhu
Hao Zhu
Heming Zhu
Jiachen Zhu
Jianke Zhu

Jiawen Zhu
Jiayin Zhu
Jinjing Zhu
Junyi Zhu
Kai Zhu
Ke Zhu
Lanyun Zhu
Lin Zhu
Linchao Zhu
Liyuan Zhu
Meilu Zhu
Muzhi Zhu
Qingtian Zhu
Ronghang Zhu
Rui Zhu
Rui Zhu
Rui-Jie Zhu
Ruizhao Zhu
Shengjie Zhu
Sijie Zhu
Siyu Zhu
Tyler Zhu
Wang Zhu
Weicheng Zhu
Wenwu Zhu
Xiangyu Zhu
Xiaofeng Zhu
Xiaoguang Zhu
Xiaosu Zhu
Xiaoyu Zhu
Xingkui Zhu
Xinxin Zhu
Xiyue Zhu
Yangguang Zhu
Yanjun Zhu
Yao Zhu
Ye Zhu
Feida Zhu
Fengqing Maggie Zhu
Guibo Zhu
Haidong Zhu
Hanwei Zhu
Hao Zhu
Hao Zhu
Heming Zhu

Jiachen Zhu
Jianke Zhu
Jiawen Zhu
Jiayin Zhu
Jinjing Zhu
Junyi Zhu
Kai Zhu
Ke Zhu
Lanyun Zhu
Lin Zhu
Linchao Zhu
Liyuan Zhu
Meilu Zhu
Muzhi Zhu
Qingtian Zhu
Ronghang Zhu
Rui Zhu
Rui Zhu
Rui-Jie Zhu
Ruizhao Zhu
Shengjie Zhu
Sijie Zhu
Siyu Zhu
Tyler Zhu
Wang Zhu
Weicheng Zhu
Wenwu Zhu
Xiangyu Zhu
Xiaofeng Zhu
Xiaoguang Zhu
Xiaosu Zhu
Xiaoyu Zhu
Xingkui zhu
Xinxin Zhu
Xiyue Zhu
Yangguang Zhu
Yanjun Zhu
Yao Zhu
Ye Zhu
Ye Zhu
Yichen Zhu
Yingying Zhu
Yousong Zhu
Yuansheng Zhu
Yurui Zhu

Zhen Zhu
Zhenwei Zhu
Zhenyao Zhu
Zhifan Zhu
Zhigang Zhu
Zhihong Zhu
Zihan Zhu
Zixin Zhu
Zunjie Zhu
Bingbing Zhuang
Jia-Xin Zhuang
Jiafan Zhuang
Peiye Zhuang
Wanyi Zhuang
Weiming Zhuang
Yihong Zhuang
Yixin Zhuang
Mingchen Zhuge
Tao Zhuo
Wei Zhuo

Yaoxin Zhuo
Bartosz Zieliński
Wojciech Zielonka
Filippo Ziliotto
Karel Zimmermann
Primo Zingaretti
Nikolaos Zioulis
Liu Ziyin
Mohammad Zohaib
Yongshuo Zong
Zhuofan Zong
Maria Zontak
Gaspard Zoss
Changqing Zou
Chuhang Zou
Danping Zou
Dongqing Zou
Haoming Zou
Longkun Zou
Shihao Zou

Xingxing Zou
Xueyan Zou
Yang Zou
Yuexian Zou
Yuli Zou
Yuliang Zou
Yunhao Zou
Zhiming Zou
Zihang Zou
Silvia Zuffi
Idil Esen Zulfikar
Maria A. Zuluaga
Ronglai Zuo
Xingxing Zuo
Xinxin Zuo
Yifan Zuo
Yiming Zuo
Reyer Zwiggelaar
Vlas Zyrianov

Contents – Part VI

Raindrop Clarity: A Dual-Focused Dataset for Day and Night Raindrop Removal 1
    Yeying Jin, Xin Li, Jiadong Wang, Yan Zhang, and Malu Zhang

Unsupervised Moving Object Segmentation with Atmospheric Turbulence 18
    Dehao Qin, Ripon Kumar Saha, Woojeh Chung, Suren Jayasuriya, Jinwei Ye, and Nianyi Li

AccDiffusion: An Accurate Method for Higher-Resolution Image Generation 38
    Zhihang Lin, Mingbao Lin, Meng Zhao, and Rongrong Ji

Uncertainty-Driven Spectral Compressive Imaging with Spatial-Frequency Transformer 54
    Lintao Peng, Siyu Xie, and Liheng Bian

CaesarNeRF: Calibrated Semantic Representation for Few-Shot Generalizable Neural Rendering 71
    Haidong Zhu, Tianyu Ding, Tianyi Chen, Ilya Zharkov, Ram Nevatia, and Luming Liang

MapTracker: Tracking with Strided Memory Fusion for Consistent Vector HD Mapping 90
    Jiacheng Chen, Yuefan Wu, Jiaqi Tan, Hang Ma, and Yasutaka Furukawa

Image Demoiréing in RAW and sRGB Domains 108
    Shuning Xu, Binbin Song, Xiangyu Chen, Xina Liu, and Jiantao Zhou

LiDAR-Event Stereo Fusion with Hallucinations 125
    Luca Bartolomei, Matteo Poggi, Andrea Conti, and Stefano Mattoccia

X-Former: Unifying Contrastive and Reconstruction Learning for MLLMs 146
    Swetha Sirnam, Jinyu Yang, Tal Neiman, Mamshad Nayeem Rizve, Son Tran, Benjamin Yao, Trishul Chilimbi, and Mubarak Shah

Learning Anomalies with Normality Prior for Unsupervised Video Anomaly Detection 163
    Haoyue Shi, Le Wang, Sanping Zhou, Gang Hua, and Wei Tang

Revisiting Supervision for Continual Representation Learning 181
  Daniel Marczak, Sebastian Cygert, Tomasz Trzciński,
  and Bartłomiej Twardowski

FLAT: Flux-Aware Imperceptible Adversarial Attacks on 3D Point Clouds 198
  Keke Tang, Lujie Huang, Weilong Peng, Daizong Liu, Xiaofei Wang,
  Yang Ma, Ligang Liu, and Zhihong Tian

MMBench: Is Your Multi-modal Model an All-Around Player? 216
  Yuan Liu, Haodong Duan, Yuanhan Zhang, Bo Li, Songyang Zhang,
  Wangbo Zhao, Yike Yuan, Jiaqi Wang, Conghui He, Ziwei Liu,
  Kai Chen, and Dahua Lin

Implicit Filtering for Learning Neural Signed Distance Functions from 3D
Point Clouds 234
  Shengtao Li, Ge Gao, Yudong Liu, Ming Gu, and Yu-Shen Liu

Unsupervised Exposure Correction 252
  Ruodai Cui, Li Niu, and Guosheng Hu

Anytime Continual Learning for Open Vocabulary Classification 269
  Zhen Zhu, Yiming Gong, and Derek Hoiem

External Knowledge Enhanced 3D Scene Generation from Sketch 286
  Zijie Wu, Mingtao Feng, Yaonan Wang, He Xie, Weisheng Dong,
  Bo Miao, and Ajmal Mian

G3R: Gradient Guided Generalizable Reconstruction 305
  Yun Chen, Jingkang Wang, Ze Yang, Sivabalan Manivasagam,
  and Raquel Urtasun

DreamScene360: Unconstrained Text-to-3D Scene Generation
with Panoramic Gaussian Splatting 324
  Shijie Zhou, Zhiwen Fan, Dejia Xu, Haoran Chang, Pradyumna Chari,
  Tejas Bharadwaj, Suya You, Zhangyang Wang, and Achuta Kadambi

Frequency-Spatial Entanglement Learning for Camouflaged Object
Detection 343
  Yanguang Sun, Chunyan Xu, Jian Yang, Hanyu Xuan, and Lei Luo

VisionTrap: Vision-Augmented Trajectory Prediction Guided by Textual
Descriptions 361
  Seokha Moon, Hyun Woo, Hongbeen Park, Haeji Jung,
  Reza Mahjourian, Hyung-gun Chi, Hyerin Lim, Sangpil Kim,
  and Jinkyu Kim

Occluded Gait Recognition with Mixture of Experts: An Action Detection
Perspective 380
*Panjian Huang, Yunjie Peng, Saihui Hou, Chunshui Cao, Xu Liu,
Zhiqiang He, and Yongzhen Huang*

EDTalk: Efficient Disentanglement for Emotional Talking Head Synthesis 398
Shuai Tan, Bin Ji, Mengxiao Bi, and Ye Pan

Groma: Localized Visual Tokenization for Grounding Multimodal Large
Language Models 417
Chuofan Ma, Yi Jiang, Jiannan Wu, Zehuan Yuan, and Xiaojuan Qi

On the Utility of 3D Hand Poses for Action Recognition 436
*Md Salman Shamil, Dibyadip Chatterjee, Fadime Sener, Shugao Ma,
and Angela Yao*

DG-PIC: Domain Generalized Point-In-Context Learning for Point Cloud
Understanding 455
*Jincen Jiang, Qianyu Zhou, Yuhang Li, Xuequan Lu, Meili Wang,
Lizhuang Ma, Jian Chang, and Jian Jun Zhang*

Operational Open-Set Recognition and PostMax Refinement 475
Steve Cruz, Ryan Rabinowitz, Manuel Günther, and Terrance E. Boult

Author Index 493

Raindrop Clarity: A Dual-Focused Dataset for Day and Night Raindrop Removal

Yeying Jin[1], Xin Li[2], Jiadong Wang[1], Yan Zhang[1], and Malu Zhang[3] (✉)

[1] National University of Singapore, Singapore, Singapore
jinyeying@u.nus.edu
[2] University of Science and Technology of China, Hefei, China
xin.li@ustc.edu.cn
[3] University of Electronic Science and Technology of China, Chengdu, China
maluzhang@uestc.edu.cn

Abstract. Existing raindrop removal datasets have two shortcomings. First, they consist of images captured by cameras with a focus on the background, leading to the presence of blurry raindrops. To our knowledge, none of these datasets include images where the focus is specifically on raindrops, which results in a blurry background. Second, these datasets predominantly consist of daytime images, thereby lacking nighttime raindrop scenarios. Consequently, algorithms trained on these datasets may struggle to perform effectively in raindrop-focused or nighttime scenarios. The absence of datasets specifically designed for raindrop-focused and nighttime raindrops constrains research in this area. In this paper, we introduce a large-scale, real-world raindrop removal dataset called Raindrop Clarity. Raindrop Clarity comprises 15,186 high-quality pairs/triplets (raindrops, blur, and background) of images with raindrops and the corresponding clear background images. There are 5,442 daytime raindrop images and 9,744 nighttime raindrop images. Specifically, the 5,442 daytime images include 3,606 raindrop- and 1,836 background-focused images. While the 9,744 nighttime images contain 4,834 raindrop- and 4,906 background-focused images. Our dataset will enable the community to explore background-focused and raindrop-focused images, including challenges unique to daytime and nighttime conditions (Our data and code are available at. https://github.com/jinyeying/RaindropClarity).

Keywords: Raindrop-focused · Background-focused · Nighttime

1 Introduction

Adherent raindrops on lenses or windscreens can significantly reduce visibility. Raindrop removal is crucial for surveillance, self-driving cars, object detection,

(a) Qian (b) RainDS Quan (c) RobotCar Porav (d) Windshield

Fig. 1. Existing raindrop datasets (e.g., [24–26,30]) exhibit two limitations: first, they do not include raindrop-focused images (Fig. 2 right); and second, they lack night raindrops (Fig. 2 bottom).

Fig. 2. Motivation for Raindrop Clarity: real images sourced from the Internet feature scenarios overlooked by existing datasets, including raindrop-focused (right) and nighttime raindrops (bottom). In this figure, Rd = Raindrop, Bg = Background.

outdoor photography, filmmaking, augmented reality, *etc.*. Recently, a few raindrop datasets (e.g., [24–26,30]) have been proposed. In Fig. 1, these datasets primarily contain daytime background scenes with blurry raindrops, as the images are taken with the camera's focus on the background. They have impacts in the raindrop removal area, inspiring several successful methods [24,25,27].

While these datasets are valuable, they have two limitations. First, a significant issue arises when the camera is in automatic focus mode. In Fig. 2 (right), the camera may unintentionally focus on raindrops that are attached to the lens, windshield, or glass, resulting in images with sharp raindrops and a blurry background. Consequently, methods designed for images with blurry raindrops and a sharp background may not perform optimally. Second, existing datasets consist only of daytime images. Therefore, existing methods, especially those reliant on training data, may struggle to effectively handle nighttime raindrop images because of the inherent domain gap between day and night conditions.

In Fig. 2, we observe that besides daytime images with blurry raindrops and a sharp background, it is common to come across images with sharp raindrops and a blurry background (right), as well as nighttime raindrops (bottom). Thus, we collected varied raindrop images as shown in Fig. 3: the top-left with daytime blurry raindrops against a sharp background, the top-right with daytime sharp raindrops and a blurry background, the bottom-left with nighttime blurry

Fig. 3. Raindrop Clarity: Examples of our pairs $(\tilde{x}, b_0)$ and triplets $(\tilde{x}, x_0, b_0)$ of a raindrop image $\tilde{x}$, the corresponding blurry background image x_0 and the corresponding clear background image b_0, in day and night. For the pairs $(\tilde{x}, b_0)$, there is no x_0, since in this case $x_0 = b_0$. The reason we have the blurry background image x_0 in triplets is to have the raindrop difference maps $\tilde{m}$ (see Fig. 8).

raindrops and a clear background, and the bottom-right with nighttime sharp raindrops with a blurry background.

Physical solutions like rain covers or wipers can reduce the presence of raindrops, but their effectiveness is limited. In certain situations where wipers cannot be used, or wind blows rain onto the lens, these physical measures become impractical. Physics-based methods [1,8,23,44,45] remove raindrops by leveraging the physics of raindrops or utilizing generated synthetic data. However, the effectiveness of methods trained on synthetic data relies on the quality of raindrop rendering models. Currently, raindrop rendering models commonly employ surface modeling and ray tracing [8,28,29]. Unfortunately, these methods may not fully consider factors such as scene depth, lighting conditions, and the 3D shapes of raindrops [23,28], which are crucial for realistic raindrop rendering. Additionally, raindrops can introduce refraction effects [7,44], distorting the shapes and positions of objects, which further complicates accurate raindrop rendering. To the best of our knowledge, developing a rendering approach that can produce physically accurate raindrops adaptable to various scenes is challenging.

Motivated by the aforementioned challenges and the observation that existing raindrop datasets predominantly consist of background-focused images while neglecting raindrop-focused images and nighttime raindrops, we collect a novel dataset, called "Raindrop Clarity". Our dataset comprises 6,742 (1,836 daytime, 4,906 nighttime) pairs and 8,444 (3,606 daytime, 4,838 nighttime) triplets of raindrop-focused images and background-focused images. All of them have the corresponding clear background images. The triplets include one more raindrop-free blurry background images. Figure 3 shows samples and explanation on the meaning of our pairs and triplets. From the 15,186 raindrops pairs/triplet, 5,442 are captured during daytime and the 9,744 at nighttime. Raindrops Clarity exhibits diverse sizes and shapes, and raindrop appearance varies based on the scene and lighting conditions. To our knowledge, Raindrop Clarity is the largest real-world raindrop dataset that contains both the pair and triplets, day and night. Our main contributions are summarized as follows:

- We present the Raindrop Clarity dataset, a large-scale real-world dataset that encompasses both raindrop-focused triplets and background-focused pairs. For the pairs, we provide images with blurry raindrops and their corresponding clear backgrounds. For the triplets, we additionally provide the raindrop-free blurry background.
- Raindrop Clarity includes images collected under both daytime and nighttime. It aims to remove raindrops and recover clear backgrounds regardless of day or night, whether the raindrops or background are blurry.
- Our experimental results indicate that even with the application of state-of-the-art raindrop removal methods, there are instances where these methods fail. This suggests the existence of unresolved challenges in the field of raindrop removal, which await further attention and investigation.

2 Related Work

Raindrop Dataset. Qian *et al.* [25] introduce a raindrop dataset consisting of raindrop and corresponding ground truths. Porav *et al.* [24] propose a stereo RobotCar dataset, using a rain-making device that causes one lens to be affected by raindrops, while the other lens remains clear. Hao *et al.* [8] create a synthetic dataset of adherent raindrops with pixel-level masks for training deraining algorithms. Quan *et al.* [26] introduce RainDS, a dataset that includes raindrops, rain steaks, and their corresponding ground truths. Soboleva and Shipitko [30] introduce Windshields, a dataset containing raindrop images on both camera lenses and windshields, annotated with binary masks representing the raindrop areas. However, all of these datasets primarily consist of daytime images taken by cameras focusing on the background scenes and thus have blurry raindrops.

Single Image Raindrop Removal. Most image deraining methods have primarily focused on removing rain streaks rather than raindrops. However, recent learning-based methods have been proposed specifically for single-image raindrop removal. Eigen *et al.* [6] introduce a supervised CNN-based method to remove raindrops, using a shallow CNN. Qian *et al.* [25] propose an AttentiveGAN for raindrop removal, that relies on attention maps but has a limited focus on local spatial information. Quan *et al.* [27] propose RaindropAttention, which employs an attention mechanism to generate attention maps based on mathematical raindrop shapes, yet it lacks exploration of long-range information. Liu *et al.* [18] demonstrate a dual residual network, by leveraging the complementary nature of paired residual blocks. Hao *et al.* [8] propose a raindrop synthesis algorithm, used their synthetic dataset to train a multi-task network for raindrop detection and removal. Porav *et al.* [24] propose a Pix2PixHD-based deraining network architecture. Subsequent work by Quan *et al.* [26] propose a cascaded network architecture for simultaneous removal of rain streaks and raindrops, achieved through a neural architecture search. Yasarla and Patel [41] propose the uncertainty guided multi-scale residual learning network to effectively remove rain from images. Zhang *et al.* [47] propose the Dual Attention-in-Attention Model to remove rain streaks and raindrops.

Fig. 4. Pairs $(\tilde{\mathbf{x}}, \boldsymbol{b}_0)$ show various sizes, shapes, and densities.

Fig. 5. The triplets $(\tilde{\mathbf{x}}, \boldsymbol{x}_0, \boldsymbol{b}_0)$ cover diverse scenes and shapes.

Restoration Backbone. There are image restoration and enhancement tasks that utilize transformer-based models, taking advantage of the long-range dependencies, including IPT [2] (vanilla transformer), Uformer [36], SwinIR [17] and Restormer [46]. Xiao et al. [40] propose an image deraining transformer (IDT) for rain streak and raindrop removal with window-based and spatial-based dual Transformer. RaindropDiffusion [21] develops a UNet-based diffusion framework, which demonstrates robust generation capabilities for raindrops, rain streaks, and snow. RainDiffusion [38], an unsupervised diffusion model, utilizes unpaired real-world data instead of weakly adversarial training. DiT [22] is a transformer-based diffusion model.

3 Raindrop Clarity Dataset

3.1 Data Collection

To create Raindrop Clarity, we employed a sphere pan-tilt platform to keep our camera stationary. The images were captured using Sony FDR-AX33 4K Ultra HD Handycam Camcorder, Sony alpha 7R III digital camera, the back camera of an iPhone 14 Pro and iPhone 15 Pro Max, ensuring the highest quality possible. The 4K Camcorder captured RGB images at a resolution 3840×2164.

Using the optical refraction model (raindrops appear upside-down), we projected raindrops by spraying water on glass and adjusting the distance, angle, speed, and patterns of the nozzle/wipers. We also captured in rain conditions.

Our collection procedure was designed to ensure diversity in the raindrop (such as focuses, shapes and sizes), background scenes and lighting conditions. First, we fixed the camera position on a tripod and installed a glass plate. The distance between the camera and the glass plate ranged from 5 cm to 25 cm. Second, we sprayed waterdrops/raindrops onto the glass plate and focused the camera on the raindrops. Third, we sprayed the waterdrops/raindrops to capture images of them under different shapes and sizes, ranging from sparse to dense. The captured images include elliptical raindrops and raindrop flow traces ($\tilde{\mathbf{x}}$). Fourth, we removed the glass plate, allowing the camera to focus on the very close plane. The duration is kept very short, within milliseconds, to minimize lighting changes. In this way, we obtained raindrop-free blurry background images ($\boldsymbol{x}_0$). Finally, we adjusted the camera focus to the background, directing the camera to focus on the distant background ($\boldsymbol{b}_0$).

3.2 Characteristics of Raindrop Clarity

Raindrop Clarity is a new dataset consisting of around 15,186 high-quality image raindrop pairs and triplets. In Fig. 3, our dataset introduces a unique element of triplet data, which includes raindrop-focused images, corresponding blurry raindrop-free background images, and their clear background equivalents, distinguishing it from existing datasets. The second feature that distinguishes the Raindrop Clarity from the existing ones is the collection of night raindrops.

Objectives. Our dataset is designed to support research in the elimination of raindrops and the restoration of a clear background, irrespective of whether the camera's focus is on the raindrops (leading to a blurry background) or on the background (resulting in blurry raindrops). This holds true regardless of whether it is daytime or nighttime.

Raindrops and Diversity. Our Raindrop Clarity dataset offers an extensive coverage of diverse raindrop sizes, shapes, and occlusion types in various scenes and lighting conditions, making it an exceptionally comprehensive resource for advancing raindrop removal research. In Fig. 4 and Fig. 5, the dataset captures raindrops and rain flows of different sizes and shapes, including heavy raindrops that result in unpredictable water flows and blurry backgrounds when they hit

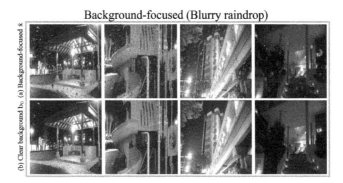

Fig. 6. The pairs $(\tilde{\mathbf{x}}, b_0)$ under different lighting conditions.

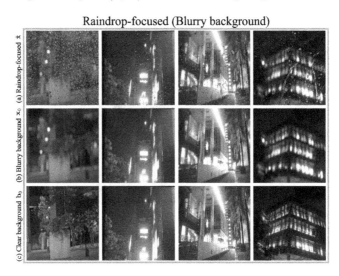

Fig. 7. Triplets $(\tilde{\mathbf{x}}, x_0, b_0)$ under different lighting conditions.

window panes or car windshields. Two primary types of occlusions caused by raindrops, namely partial and complete occlusions, are included in our dataset. Both types present unique challenges for raindrop removal algorithms

Scenes and Large-Scale. The dataset encompasses a wide range of background scenes, such as urban, countryside, campus, wilderness, road, and aerial views, covering both day and night. This variation allows for the exploration of different lighting and weather conditions in different scenes. With approximately 38,816 images in total, including 15,186 high-quality pairs/triplets of images, the Raindrop Clarity dataset is one of the largest datasets of its kind.

Lighting Conditions. The appearance of raindrops highly depends on the surrounding environment. A distinctive feature of our dataset is its comprehensive coverage of various environmental conditions, setting it apart from other

Table 1. Summary of comparisons between our Raindrop Clarity and existing datasets. Bg-focus = Background-focused, Rd-focus = Raindrop-focused, RD = Raindrop. Raindrop Clarity uniquely features raindrop-focused triplets and the nighttime raindrops.

Datasets	Components	N. Images	Syn/Real	Bg-focus Day	Bg-focus Night	Rd-focus Day	Rd-focus Night
Our RD Clarity (Day)	3,606 Triplets+1,836 Pairs	14,490	Real	✓	×	✓	×
Our RD Clarity (Night)	4,838 Triplets+4,906 Pairs	24,326	Real	×	✓	×	✓
Raindrop Qian [25]	919 Pairs	1,838	Real	✓	×	×	×
RainDS-Real [26]	248 Quadruplets	992	Real	✓	×	×	×
RainDS-Syn [26]	1,200 Quadruplets	4,800	Syn	✓	×	×	×
RobotCar Porav [24]	4,818 Pairs	9,636	Real	✓	×	×	×
Windshield [30]	3,390	3,390	Real	✓	×	×	×

Fig. 8. We show $(\tilde{x}, x_0, \tilde{m}, b_0)$, difference map $\tilde{m}$ is derived by distinguishing raindrop-focused $\tilde{x}$ from blurry background x_0.

datasets. The dataset's variation in day and night conditions makes it valuable for developing raindrop removal algorithms.

During the daytime, raindrop images exhibit a diverse array of appearances influenced by multiple factors such as sunlight direction, intensity, and shadowing. In Figs. 4 and 5, our dataset captures this variety, with images taken at different times of the day and under varying weather conditions (such as light rain, heavy rain, and rain combined with wind), providing rich information for algorithms to learn.

Unlike daytime, nighttime raindrops present unique challenges due to artificial lighting conditions and low light. The appearance of raindrops can drastically change under street lights, headlights, or neon lights. In Fig. 6 and Fig. 7, our dataset includes night images with varied artificial lights, contributing to the complexity and comprehensiveness of the dataset. These night images are challenging due to the higher contrast between raindrops and the background, and potential overexposure or underexposure issues.

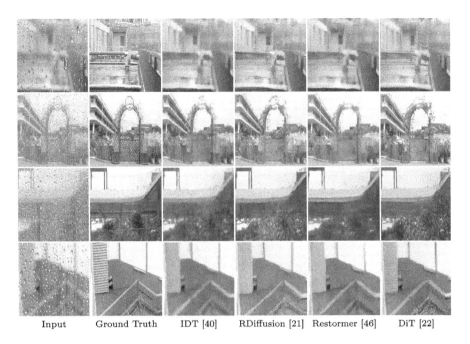

Fig. 9. Comparison of existing methods on daytime raindrop-focused images, covering both raindrop algorithms and restoration backbones.

Background- and Raindrop-focused. When the raindrop image $\tilde{\mathbf{x}}$ is background-focused, we decompose a raindrop degraded image $\tilde{\mathbf{x}}$ into a clear background image b_0 by following this model [25]: $\tilde{\mathbf{x}} = (1 - \mathbf{M}) \odot x_0 + \mathbf{D}$, where $\mathbf{D}$ is the effect of the blurry raindrops, referring to the localized blurry imagery formed by the light reflected from the surroundings. The operator $\odot$ denotes element-wise multiplication. When binary mask $\mathbf{M}(\mathbf{x})$ equals 1, it signifies that the pixel x belongs to the area affected by raindrops. Conversely, if $\mathbf{M}(\mathbf{x})$ does not equal 1, it indicates that the pixel is within the background areas. In the case where $\tilde{\mathbf{x}}$ is background-focused raindrop images, x_0 is identical to the clear, raindrop-free image b_0, namely $x_0 = b_0$. Therefore, we use $(\tilde{\mathbf{x}}, b_0)$ to stand for background-focused (resulting in blurry raindrops) pairs, shown in the left part of Fig. 3.

In the right part of Fig. 3 and Fig. 8, when $\tilde{\mathbf{x}}$ is the raindrop-focused image, x_0 is the corresponding raindrop-free blurry background image, while b_0 is the clear background image. In this case, $x_0 \neq b_0$. We can calculate the pixel-level difference map $\tilde{\mathbf{m}}$ between the raindrop-free blurry background x_0 and the raindrop-focused $\tilde{\mathbf{x}}$ as: $\tilde{\mathbf{m}} = \tilde{\mathbf{x}} - x_0$. Therefore, we use $(\tilde{\mathbf{x}}, x_0, b_0)$ to stand for raindrop-focused (resulting in blurry background) triplets.

Since the camera is stationary, neither the pairs $(\tilde{\mathbf{x}}, b_0)$ nor the triplets $(\tilde{\mathbf{x}}, x_0, b_0)$ have alignment problems. This is also evident in the raindrop difference maps $\tilde{\mathbf{m}}$, which accurately assign values to raindrop regions across multiple

Table 2. Quantitative evaluation of daytime and nighttime raindrop data with a benchmark for raindrop algorithms and restoration backbones.

Data\Method		Input	Raindrop Algorithms					Restoration Backbones		
			AtGAN [25]	Robot [24]	RainDS [26]	IDT [40]	RDdiff [21]	Restor. [46]	Uform. [36]	DiT [22]
Day	PSNR↑	21.81	23.00	23.01	24.31	24.82	24.47	24.63	24.52	**25.32**
	SSIM↑	0.573	0.654	0.658	0.706	0.716	0.712	0.717	0.704	**0.753**
	LPIPS↓	0.241	0.202	0.196	0.154	0.150	0.120	0.144	0.159	**0.105**
Night	PSNR↑	24.78	24.08	25.17	25.41	**26.81**	26.48	26.08	25.26	26.23
	SSIM↑	0.726	0.762	0.796	0.820	**0.851**	0.831	0.827	0.816	0.826
	LPIPS↓	0.209	0.190	0.174	0.139	0.125	0.112	0.135	0.142	**0.111**

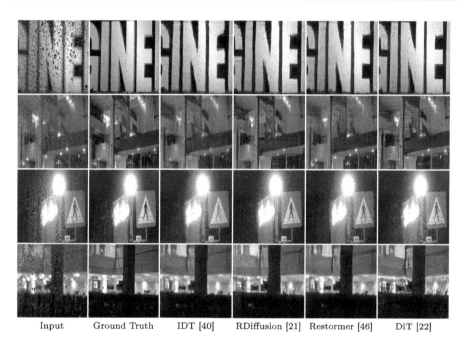

    Input   Ground Truth   IDT [40]   RDiffusion [21]   Restormer [46]   DiT [22]

Fig. 10. Comparison of methods on nighttime raindrop-focused images, covering both raindrop algorithms and restoration backbones.

images. Specifically, for each pixel **x** in the input image, any discrepancies in the map's assigned values for the static background would indicate alignment issues; however, these are not observed, as $\tilde{\mathbf{m}}$ indicates only the raindrop regions. Since the pairs/triplets of images are captured in quick succession, there are no problems with changes in illumination.

The captured images do not contain any personal information such as phone numbers, or human faces. We will release the dataset publicly with car license plates obscured, as we believe that our Raindrop Clarity dataset will benefit the progress of the raindrop removal field.

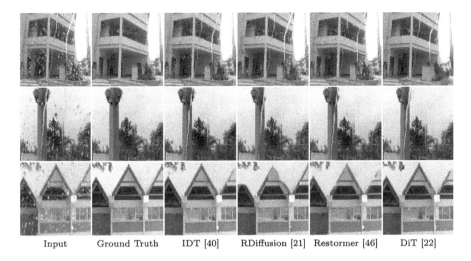

Fig. 11. Comparison of methods on daytime background-focused images, covering both raindrop algorithms and restoration backbones.

4 Experiments

Experimental Setup. Our benchmark includes two parts: Daytime and Nighttime Raindrop Removal. As shown in Table 1, our daytime raindrop dataset contains a total of 5,442 paired/triplets images, where 4,713 paired images are used for the training set, and the remaining 729 paired images are regarded as the test set. Our nighttime raindrop dataset consists of 9,744 paired/triplets images, where 8,655 paired images are selected as the training set and the rest 1,089 paired images are used for evaluation.

Baselines. To create the benchmark, we retrain existing state-of-the-art methods on our daytime and nighttime raindrop datasets. Our benchmark consists of five raindrop algorithms [21, 24–26, 40] and three restoration methods [22, 36, 46]. Specifically, IDT [40], RDiffusion [21], RainDS [26], Robot Car [24] and AtGAN [25] are raindrop algorithms, while Restormer [46], Uformer [36] and DiT [22] are restoration backbones. For a fair comparison, we conduct experiments using the original implementation of these methods. All experiments are implemented in PyTorch and are trained on four NVIDIA RTX A5000 GPUs.

Comparison with Existing Datasets. The difference between our Raindrop Clarity dataset and the existing datasets is shown in Table 1 and Fig. 2.

The stereo RobotCar dataset [24] includes 4,818 image pairs, randomly selected for training, validation, and testing. However, the raindrops in this dataset are disproportionately large and lack realism.

The raindrop dataset proposed by Qian *et al.* [25] consists of 861 training pairs and 58 testing pairs. While it features common raindrop shapes, it lacks the diversity of real-world scenes [44]. Also, the backgrounds between the raindrop images and the ground truth are not perfectly aligned [39].

Input   Ground Truth   IDT [40]   RDiffusion [21]   Restormer [46]   DiT [22]

Fig. 12. Comparison of methods on nighttime background-focused images, covering both raindrop algorithms and restoration backbones.

The RainDS-Real dataset, proposed by Quan *et al.* [26], comprises 248 quadruplets, including raindrops, rain streaks, and their corresponding ground truths. The RainDS-Synthetic dataset has a domain gap with the real raindrop images. Moreover, the real-world raindrops are consistently small, obscuring only a fraction of the image details, making recovery less challenging.

The Windshield dataset, proposed by Soboleva and Shipitko [30], was captured by a camera attached to the vehicle during its movement in urban areas and highway environments. The Windshield does not provide the corresponding clean pairs, making this dataset not available for robust raindrop removal method testing. Additionally, frequent moving objects further complicate the dataset.

Existing raindrop datasets, though valuable, have certain limitations. They do not include raindrop-focused and nighttime raindrops. In addition, we have noticed reflection errors in the datasets from Qian [25] and RainDS-Real [26].

Comparison with Synthetic Raindrops. In Fig. 13, we show the existing state-of-the-art synthetic physical model [23], although it can generate raindrops on blurry and clear backgrounds (Fig. 13 d generated on b, Fig. 13 e generated on c). It still has a gap with real-world collected raindrops. Real-world collected raindrops display diverse appearances, including various shapes such as round, elliptical, and long strips. Also, the transparency and glare effects of raindrops collected in the real-world varies.

Quantitative Comparison. To effectively evaluate the results on our proposed datasets, we use three popular metrics PSNR, SSIM [37], and LPIPS [48] in generation tasks [3–5, 9–16, 19, 20, 31–35, 42, 43].

The qualitative results are shown in Table 2. For the daytime dataset, the PSNR and SSIM of the original input images are 21.81 and 0.573. Raindrop algorithm IDT [40] achieves the best performance (24.82 PSNR of and 0.716 of SSIM) compared with other raindrop algorithms. For the restoration backbones, DiT [22] achieves a PSNR of 25.32 and a SSIM of 0.753, outperforming IDT by 0.5 PSNR and 0.037 SSIM.

It can be observed that, for the nighttime dataset, PSNR and SSIM of the input images are 24.78 and 0.726. For the raindrop algorithms, IDT [40] achieves better performance than other methods. The PSNR and SSIM of IDT are 26.81 and 0.851, respectively. For the restoration backbones, DiT [22] achieves a PSNR of 26.23 and a SSIM of 0.826, which outperforms other backbones.

Qualitative Comparison. Figures 9, 10, 11 and 12 show the qualitative results of state-of-the-art methods. To be specific, Fig. 9 and Fig. 11 show the results of daytime raindrop removal. The input images in Fig. 9 and Fig. 11 are daytime raindrop-focused and background-focused, respectively. Figure 10 and Fig. 12 show the results of nighttime raindrop-focused and background-focused removal, respectively.

It can be found that, for daytime and nighttime raindrop-focused removal in Fig. 9 and Fig. 10, existing state-of-the-art methods fail to remove raindrops and recover the background. This is because most of them are designed for background-focused scenes and thus neglect the blurry problem. To our knowledge, Restormer [46] and Uformer [36] are specifically designed to address deblurring problem. However, they still struggle with recovering details and textures. As a result, these methods are struggling to simultaneously remove raindrops and recover the background.

In Fig. 11, for daytime background-focused inputs, we can observe that existing state-of-the-art methods can handle most raindrops. However, they are struggling to address long strips of raindrops. This is because their methods may not detect long strips of raindrops and thus fail to remove them. Furthermore, in Fig. 12, we find that existing state-of-the-art methods are struggling to handle nighttime background-focused samples. To be specific, raindrops in these samples cannot be totally removed and details of the restored images are not clear. This is because nighttime raindrops usually reflect artificial light and thus are more complex than daytime raindrops. Meanwhile, existing state-of-the-art methods are designed for daytime raindrop removal and neglect the effect of complex nighttime conditions. Thus, existing state-of-the-art methods cannot achieve promising results.

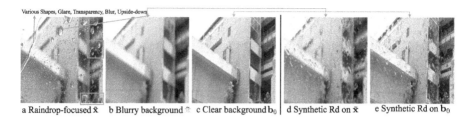

Fig. 13. We show the differences between the synthetic raindrops [23] and real-world raindrops from Raindrop Clarity.

5 Conclusion

In this paper, we propose a new raindrop dataset, named Raindrop Clarity. It includes a total of 15,186 high-quality pairs/triplets of images with raindrops and the corresponding clear background images. Moreover, our dataset is collected under various conditions, including different scenes, times (daytime and nighttime), and focuses (raindrop- and background-focused). To be specific, 3,606 daytime raindrop-focused, 1,836 daytime background-focused, 4,838 nighttime raindrop-focused and 4,906 nighttime background-focused images are collected. With the proposed dataset, we re-analyze existing state-of-the-art methods, including raindrop algorithms and restoration backbones, and point out several open problems: raindrop-focus raindrop removal and nighttime raindrop removal. We wish to draw the attention of the community to the open problems.

6 Social Impacts

Existing raindrop datasets focus on daytime background-focused raindrop removal. However, in the real world, raindrop images may focus on raindrops, resulting in a blurry background. Also, raindrops may occur in nighttime images. To draw the attention of the community to the open problems, we propose a new raindrop dataset, named Raindrop Clarity. It mainly collected from four conditions: daytime raindrop-focused, daytime background-focused, nighttime raindrop-focused and nighttime background-focused. Experimental results on our proposed dataset show that existing state-of-the-art methods cannot effectively remove raindrops under complex conditions. As a result, we believe that further exploration of our Raindrop Clarity dataset can help the community to address more complex raindrops.

Acknowledgments. This research is supported by the National Research Foundation, Singapore, under its AI Singapore Programme (AISG Award No: AISG2-PhD/2022–01-037[T]). Xin Li's research is supported by NSFC under Grant 623B2098. Malu Zhang's work is supported by the National Science Foundation of China under Grant 62106038, and in part by the Sichuan Science and Technology Program under Grant 2023YFG0259. We thank Robby T. Tan and Beibei Lin for their valuable discussions, revisions, and feedback.

References

1. [EB/OL]. http://liuyunfei.net/Projs/PBRR/, title = PBRR: Physically Based Raindrop Rendering, author = Yunfei Liu, Zhixiang Hao, Shadi You, Yu Li, and Feng Lu
2. Chen, H., et al.: Pre-trained image processing transformer. In: Proceedings of the IEEE/CVF Conference on Computer Vision and Pattern Recognition, pp. 12299–12310 (2021)
3. Chen, S., Ye, T., Bai, J., Chen, E., Shi, J., Zhu, L.: Sparse sampling transformer with uncertainty-driven ranking for unified removal of raindrops and rain streaks. In: Proceedings of the IEEE/CVF International Conference on Computer Vision, pp. 13106–13117 (2023)
4. Chen, S., Ye, T., Liu, Y., Chen, E.: Snowformer: Context interaction transformer with scale-awareness for single image desnowing. arXiv preprint arXiv:2208.09703 (2022)
5. Chen, S., et al.: MSP-former: Multi-scale projection transformer for single image desnowing. In: ICASSP 2023-2023 IEEE International Conference on Acoustics, Speech and Signal Processing (ICASSP), pp. 1–5. IEEE (2023)
6. Eigen, D., Krishnan, D., Fergus, R.: Restoring an image taken through a window covered with dirt or rain. In: Proceedings of the IEEE International Conference on Computer Vision, pp. 633–640 (2013)
7. Halimeh, J.C., Roser, M.: Raindrop detection on car windshields using geometric-photometric environment construction and intensity-based correlation. In: 2009 IEEE Intelligent Vehicles Symposium, pp. 610–615. IEEE (2009)
8. Hao, Z., You, S., Li, Y., Li, K., Lu, F.: Learning from synthetic photorealistic raindrop for single image raindrop removal. In: Proceedings of the IEEE/CVF International Conference on Computer Vision Workshops (2019)
9. Jin, Y., Lin, B., Yan, W., Yuan, Y., Ye, W., Tan, R.T.: Enhancing visibility in nighttime haze images using guided APSF and gradient adaptive convolution. In: Proceedings of the 31st ACM International Conference on Multimedia, pp. 2446–2457 (2023)
10. Jin, Y., Sharma, A., Tan, R.T.: Dc-shadownet: single-image hard and soft shadow removal using unsupervised domain-classifier guided network. In: Proceedings of the IEEE/CVF International Conference on Computer Vision, pp. 5027–5036 (2021)
11. Jin, Y., Yang, W., Tan, R.T.: Unsupervised night image enhancement: when layer decomposition meets light-effects suppression. In: European Conference on Computer Vision, pp. 404–421. Springer (2022). https://doi.org/10.1007/978-3-031-19836-6_23
12. Jin, Y., Ye, W., Yang, W., Yuan, Y., Tan, R.T.: Des3: adaptive attention-driven self and soft shadow removal using ViT similarity. In: Proceedings of the AAAI Conference on Artificial Intelligence. vol. 38, pp. 2634–2642 (2024)
13. Li, B., et al.: SED: semantic-aware discriminator for image super-resolution. In: Proceedings of the IEEE/CVF Conference on Computer Vision and Pattern Recognition (2024)
14. Li, X., et al.: Learning disentangled feature representation for hybrid-distorted image restoration. In: ECCV, pp. 313–329. Springer (2020). https://doi.org/10.1007/978-3-030-58526-6_19
15. Li, X., Li, B., Jin, X., Lan, C., Chen, Z.: Learning distortion invariant representation for image restoration from a causality perspective. In: Proceedings of the

IEEE/CVF Conference on Computer Vision and Pattern Recognition, pp. 1714–1724 (2023)
16. Li, X., et al.: Diffusion models for image restoration and enhancement–a comprehensive survey. arXiv preprint arXiv:2308.09388 (2023)
17. Liang, J., Cao, J., Sun, G., Zhang, K., Van Gool, L., Timofte, R.: Swinir: Image restoration using swin transformer. arXiv preprint arXiv:2108.10257 (2021)
18. Liu, X., Suganuma, M., Sun, Z., Okatani, T.: Dual residual networks leveraging the potential of paired operations for image restoration. In: Proceedings of the IEEE/CVF Conference on Computer Vision and Pattern Recognition, pp. 7007–7016 (2019)
19. Luo, J., et al.: Intrinsicdiffusion: joint intrinsic layers from latent diffusion models. In: ACM SIGGRAPH 2024 Conference Papers. SIGGRAPH '24, Association for Computing Machinery, New York, NY, USA (2024). https://doi.org/10.1145/3641519.3657472, https://doi.org/10.1145/3641519.3657472
20. Luo, J., Zhao, N., Li, W., Richardt, C.: Crefnet: learning consistent reflectance estimation with a decoder-sharing transformer. IEEE Trans. Visual. Comput. Graph. (2023)
21. Özdenizci, O., Legenstein, R.: Restoring vision in adverse weather conditions with patch-based denoising diffusion models. IEEE Trans. Pattern Anal. Mach. Intell. (2023)
22. Peebles, W., Xie, S.: Scalable diffusion models with transformers. arXiv preprint arXiv:2212.09748 (2022)
23. Pizzati, F., Cerri, P., de Charette, R.: Physics-informed guided disentanglement in generative networks. IEEE Trans. Pattern Anal. Mach. Intell. (2023)
24. Porav, H., Bruls, T., Newman, P.: I can see clearly now: Image restoration via deraining. In: 2019 International Conference on Robotics and Automation (ICRA), pp. 7087–7093. IEEE (2019)
25. Qian, R., Tan, R.T., Yang, W., Su, J., Liu, J.: Attentive generative adversarial network for raindrop removal from a single image. In: Proceedings of the IEEE Conference on Computer Vision and Pattern Recognition, pp. 2482–2491 (2018)
26. Quan, R., Yu, X., Liang, Y., Yang, Y.: Removing raindrops and rain streaks in one go. In: Proceedings of the IEEE/CVF Conference on Computer Vision and Pattern Recognition, pp. 9147–9156 (2021)
27. Quan, Y., Deng, S., Chen, Y., Ji, H.: Deep learning for seeing through window with raindrops. In: Proceedings of the IEEE/CVF International Conference on Computer Vision, pp. 2463–2471 (2019)
28. Roser, M., Geiger, A.: Video-based raindrop detection for improved image registration. In: 2009 IEEE 12th International Conference on Computer Vision Workshops, ICCV Workshops, pp. 570–577. IEEE (2009)
29. Roser, M., Kurz, J., Geiger, A.: Realistic modeling of water droplets for monocular adherent raindrop recognition using Bezier curves. In: Asian conference on computer vision, pp. 235–244. Springer (2010). https://doi.org/10.1007/978-3-642-22819-3_24
30. Soboleva, V., Shipitko, O.: Raindrops on windshield: Dataset and lightweight gradient-based detection algorithm. In: 2021 IEEE Symposium Series on Computational Intelligence (SSCI), pp. 1–7. IEEE (2021)
31. Wang, C., Pan, J., Lin, W., Dong, J., Wang, W., Wu, X.M.: Selfpromer: self-prompt dehazing transformers with depth-consistency. In: Proceedings of the AAAI Conference on Artificial Intelligence. vol. 38, pp. 5327–5335 (2024)
32. Wang, C., et al.: Promptrestorer: a prompting image restoration method with degradation perception. Adv. Neural. Inf. Process. Syst. **36**, 8898–8912 (2023)

33. Wang, C., et al.: Correlation matching transformation transformers for UHD image restoration. In: Proceedings of the AAAI Conference on Artificial Intelligence. vol. 38, pp. 5336–5344 (2024)
34. Wang, J., Pan, Z., Zhang, M., Tan, R.T., Li, H.: Restoring speaking lips from occlusion for audio-visual speech recognition. In: Proceedings of the AAAI Conference on Artificial Intelligence. vol. 38, pp. 19144–19152 (2024)
35. Wang, J., Qian, X., Zhang, M., Tan, R.T., Li, H.: Seeing what you said: talking face generation guided by a lip reading expert. In: Proceedings of the IEEE/CVF Conference on Computer Vision and Pattern Recognition, pp. 14653–14662 (2023)
36. Wang, Z., Cun, X., Bao, J., Zhou, W., Liu, J., Li, H.: Uformer: a general u-shaped transformer for image restoration. In: Proceedings of the IEEE/CVF Conference on Computer Vision and Pattern Recognition, pp. 17683–17693 (2022)
37. Wang, Z., Bovik, A.C., Sheikh, H.R., Simoncelli, E.P.: Image quality assessment: from error visibility to structural similarity. IEEE Trans. Image Process. **13**(4), 600–612 (2004)
38. Wei, M., Shen, Y., Wang, Y., Xie, H., Wang, F.L.: Raindiffusion: When unsupervised learning meets diffusion models for real-world image deraining. arXiv preprint arXiv:2301.09430 (2023)
39. Wen, Q., Wu, Y., Chen, Q.: Video waterdrop removal via spatio-temporal fusion in driving scenes. arXiv preprint arXiv:2302.05916 (2023)
40. Xiao, J., Fu, X., Liu, A., Wu, F., Zha, Z.J.: Image de-raining transformer. IEEE Trans. Pattern Anal. Mach. Intell. 1–18 (2022)
41. Yasarla, R., Patel, V.M.: Uncertainty guided multi-scale residual learning-using a cycle spinning CNN for single image de-raining. In: Proceedings of the IEEE/CVF Conference on Computer Vision and Pattern Recognition, pp. 8405–8414 (2019)
42. Ye, T., et al.: Adverse weather removal with codebook priors. In: Proceedings of the IEEE/CVF International Conference on Computer Vision, pp. 12653–12664 (2023)
43. Ye, T., et al.: Perceiving and modeling density for image dehazing. In: European conference on computer vision, pp. 130–145. Springer (2022). https://doi.org/10.1007/978-3-031-19800-7_8
44. You, S., Tan, R.T., Kawakami, R., Ikeuchi, K.: Adherent raindrop detection and removal in video. In: Proceedings of the IEEE Conference on Computer Vision and Pattern Recognition, pp. 1035–1042 (2013)
45. You, S., Tan, R.T., Kawakami, R., Mukaigawa, Y., Ikeuchi, K.: Adherent raindrop modeling, detection and removal in video. IEEE Trans. Pattern Anal. Mach. Intell. **38**(9), 1721–1733 (2015)
46. Zamir, S.W., Arora, A., Khan, S., Hayat, M., Khan, F.S., Yang, M.H.: Restormer: efficient transformer for high-resolution image restoration. In: CVPR (2022)
47. Zhang, K., Li, D., Luo, W., Ren, W.: Dual attention-in-attention model for joint rain streak and raindrop removal. IEEE Trans. Image Process. **30**, 7608–7619 (2021)
48. Zhang, R., Isola, P., Efros, A.A., Shechtman, E., Wang, O.: The unreasonable effectiveness of deep features as a perceptual metric. In: CVPR (2018)

Unsupervised Moving Object Segmentation with Atmospheric Turbulence

Dehao Qin[1](✉), Ripon Kumar Saha[2], Woojeh Chung[2], Suren Jayasuriya[2], Jinwei Ye[3], and Nianyi Li[1]

[1] Clemson University, Clemson, USA
dehaoq@g.clemson.edu
[2] Arizona State University, Tempe, USA
[3] George Mason University, Virginia, USA
https://turb-research.github.io/segment_with_turb

Abstract. Moving object segmentation in the presence of atmospheric turbulence is highly challenging due to turbulence-induced irregular and time-varying distortions. In this paper, we present an unsupervised approach for segmenting moving objects in videos downgraded by atmospheric turbulence. Our key approach is a detect-then-grow scheme: we first identify a small set of moving object pixels with high confidence, then gradually grow a foreground mask from those seeds to segment all moving objects. This method leverages rigid geometric consistency among video frames to disentangle different types of motions, and then uses the Sampson distance to initialize the seedling pixels. After growing per-frame foreground masks, we use spatial grouping loss and temporal consistency loss to further refine the masks in order to ensure their spatio-temporal consistency. Our method is unsupervised and does not require training on labeled data. For validation, we collect and release the first real-captured long-range turbulent video dataset with ground truth masks for moving objects. Results show that our method achieves good accuracy in segmenting moving objects and is robust for long-range videos with various turbulence strengths.

Keywords: Unsupervised learning · Object segmentation · Turbulence

1 Introduction

Moving object segmentation is critical for motion understanding with an important role in numerous vision applications such as security surveillance [24,35], remote sensing [39,66], and environmental monitoring [5,18]. Although tremendous success has been achieved in motion segmentation and analysis [6,26], the

Fig. 1. Our method robustly segments moving objects under various turbulence strengths, while state-of-the-art methods may fail under strong turbulence (2nd video).

problem becomes highly challenging for long-range videos captured with ultra-telephoto lenses (e.g., focal length > 800 mm). These videos often suffer from perturbations caused by atmospheric turbulence which can geometrically shift or warp pixels in images. When mixed with rigid motions in a dynamic scene, they break down the underlying assumption of most motion analysis algorithms: the intensity structures of local regions are constant under motion. Further, when averaged over time, the turbulent perturbation also yields blurriness in images, which blurs out moving object edges and makes it challenging to maintain the spatio-temporal consistency of segmentation masks.

Most motion segmentation algorithms solely consider static backgrounds and assume rigid body movement. Under these premises, learning-based approaches [6,26,53], either supervised or unsupervised, have achieved remarkable success in motion segmentation. However, these algorithms' performance significantly downgrades when applied to videos with turbulence effects (see failure examples in Fig. 6). Supervised methods [19,28,53], even trained on extensive labeled data, cannot generalize well on turbulent videos. Unsupervised methods [6,26], on the other hand, typically rely on optical flow, which becomes inaccurate when rigid motion is perturbed by turbulence. Furthermore, imaging at long distance makes videos highly susceptible to camera shake and motion due to the limited field of view when zooming, further complicating the segmentation task.

In this paper, we present an unsupervised approach for segmenting moving objects in long-range videos affected by air turbulence. The unsupervised nature of our approach is highly desirable, since real-captured turbulent video datasets are scarce and difficult to acquire. Our method directly takes in a turbulent video and outputs per-frame masks that segment all moving objects without the need for data supervision. The overall pipeline of our approach is illustrated in Fig. 2.

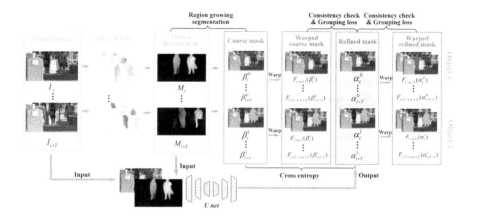

Fig. 2. Overall pipeline of our unsupervised motion segmentation method. We first generate motion feature maps by applying a geometry-based consistency check on optical flows. We then adopt a region-growing scheme to generate coarse segmentation masks. Finally, we refine the masks using cross entropy-based consistency losses to enforce their spatio-temporal consistency.

Our method starts with calculating bidirectional optical flow. To disentangle actual object motion from turbulent motion, we use a novel epipolar geometry-based consistency check to generate motion feature maps that only preserve object motions. We then adopt a region-growing scheme that generates per-object motion segmentation masks from a small set of seed pixels. Finally, we develop a U-Net [42] trained by our proposed bidirectional consistency losses and a pixel grouping function to improve the spatio-temporal consistency of estimated motion segmentation masks.

For evaluation, we collect and release a turbulent video dataset captured with an ultra-telephoto lens called Dynamic Object Segmentation in Turbulence (DOST), and manually annotate ground-truth masks for moving objects, which, to our knowledge, is the *first* dataset in this application domain. We benchmark our approach, as well as other state-of-the-art segmentation algorithms, on this real dataset. Our method is able to handle multiple objects in a dynamic scene and is robust to videos captured with various turbulence strengths (see Fig. 1). More details about DOST datasest can be found in https://turb-research.github.io/DOST/. Our key contributions include:

- A rigid geometry-based and consistency enhanced framework for motion disentanglement in long-range videos.
- Region-growing scheme for generating robust spatio-temporal consistent masks with tight object boundaries.
- A refinement pipeline with novel training losses, which improve the spatio-temporal consistency for segmenting dynamic objects in videos.
- The *first* real-captured long-range turbulent video dataset with ground-truth motion segmentation masks.

2 Related Work

Unsupervised Motion Segmentation. Early approaches tackle this problem using handcrafted features such as objectness [63], saliency [52], and motion trajectory [34]. Recent learning-based unsupervised methods largely rely on either optical flow or feature alignment for locating moving objects. Common optical flow models are RAFT [49], PWC-Net [48], and FLow-Net [8,14]. MP-Net [50] uses optical flow as the only cue for motion segmentation. Many other methods combine optical flow with other information, such as appearance features [21,36,57], long-term temporal information [41,64], and boundary similarity [25,68], to guide motion segmentation. Our method also uses optical flow but segregates different types of motion using a geometry-based consistency check to overcome turbulence-induced errors.

Another class of methods [10,27,58] rely more on appearance features and reduce their dependency on optical flow, so to achieve robust performance regardless of motion. The recent TMO [6] achieves state-of-the-art performance on object segmentation by prioritizing feature alignment and fusion. However, all these methods face challenges when the input video is recorded under air turbulence. Further, the literature on supervised motion segmentation (e.g., [22,29,61]) is not effective due to the lack of large, labelled turbulent video datasets.

Geometric Constraints for Motion Analysis. Geometric constraints are extensively studied for understanding spatial information, such as depth estimation, pose estimation, camera calibration [40,62,70], and 3D reconstruction [47,69]. They can also be useful in understanding motion. Valgaerts et al. [51] propose a variational model to estimate the optical flow, along with the fundamental matrix. Wedel et al. [54] use the fundamental matrix as a weak constraint to guide optical flow estimation. Yamaguchi et al. [56] convert the problem of optical flow estimation into a 1D search problem by using precomputed fundamental matrices with small motion assumptions. Wulff et al. [55] use semantic information to separate dynamic objects from static backgrounds and apply strong geometric constraints to the static backgrounds. Zhong et al. [67] integrate global geometric constraints into network learning for unsupervised optical flow estimation. More recently, Ye et al. [59] use the Sampson error, which measures the consistency of epipolar geometry, to model a loss function for layer decomposition in videos.

Inspired by these works, our method adopts a geometry-based consistency check to separate rigid motion from other types of motion (i.e., turbulent motion and camera motion). Similar to [59], we use the Sampson distance to measure the geometric consistency between neighboring video frames.

Turbulent Image/Video Restoration and Segmentation. Analyzing images or videos affected by air turbulence has been a challenging problem in computer vision due to distortions and blur caused by turbulence. Most techniques have focused on turbulent image and video restoration. Early physics-based approaches [12,20,38] investigate the physical modeling of turbulence (e.g., the Kolmogorov model [20]), and then invert the model to restore clear images.

A popular class of methods [3,9] uses "lucky patches" to reduce turbulent artifacts, although they typically assume static scenes. Motion cues, such as optical flow, are also used for turbulent image restoration [33,46,71]. Notably, Mao et al. [31] adopt an optical-flow guided lucky patch technique to restore images of dynamic scenes. More recently, neural networks have been used to restore turbulent images or videos. Mao et al. [32] introduce a physics-inspired transformer model for restoration, and Zhang et al. [65] improve this thread of work to demonstrate state-of-the video restoration. Li et al. [23] propose an unsupervised network to mitigate the turbulence effect using deformable grids. Jiang et al. [16] extend this deformable grid model to handle more realistic turbulence effects.

In contrast, object segmentation with atmospheric turbulence is relatively understudied. Cui and Zhang [7] propose a supervised network for semantic segmentation with turbulence. They generate a physically-realistic simulated training dataset, but the method cannot handle scene motion and suffers from real-world domain generalization. Saha et al. [45] use simple optical flow-based segmentation in their turbulent video restoration pipeline. Unlike these existing methods, our approach aims to segment moving objects without restoring or enhancing the turbulent video. In this way, our method is able to generate segmentation masks that are consistent with the actual turbulent video.

3 Unsupervised Moving Object Segmentation

Algorithm Overview. Given an input turbulent video: $\{I_t | t = 1, 2, \ldots, T\}$ (where T is the total number of frames, and I_t represents a frame in the video), we first calculate its bidirectional optical flow: $\mathcal{O}_t = \{F_{t \to t \pm i} | i = 1, \ldots, B\}$ (where B is the maximum number of frames used for calculation; $F_{t \to t+i}$ is the forward flow, and $F_{t \to t-i}$ is the backward flow). We then perform an epipolar geometry-based consistency check to disentangle rigid object motion from turbulence-induced motions and camera motions (Sect. 3.1). We output per-frame motion feature maps: $\{M_t | t = 1, 2, \ldots, T\}$ to characterize candidate motion regions. Next, we leverage a detect-then-grow strategy, named "region growing", to generate motion segmentation masks: $\{\beta_t^m | t = 1, 2, \ldots, T\}$ for every moving object m, from a small set of seedling pixels selected from $\{M_t\}_{t=1}^{T}$ (Sect. 3.2). Finally, we further refine the masks by using a U-Net trained with our proposed bidirectional spatial-temporal consistency losses and pixel grouping loss (Sect. 3.3). The final output is a set of per-frame, per-object binary masks: $\{\alpha_t^m | t = 1, 2, \ldots, T\}$ segmenting each moving object in a dynamic scene.

3.1 Epipolar Geometry-based Motion Disentanglement

We first tackle the problem of motion disentanglement, which is a major challenge posed by turbulence perturbation in rigid motion analysis. Our key idea is to check on the rigid geometric consistency among video frames: pixels on moving objects do not obey the geometric consistency constraint posed by the image formation model. Specifcially, we use the Sampson distance, which measures geometric consistency with a given epipolar geometry, to improve the spatial-temporal consistency among video frames. We first average the optical flow between adjacent frames to stabilize the direct estimations $\{\mathcal{O}_t\}$, since they are susceptible to turbulence perturbation. We then calculate the Sampson distance using fundamental matrices estimated from the averaged optical flow. Next, we merge the Sampson distance maps as the motion feature maps $\{M_t | t = 1, 2, \ldots, T\}$. We use the motion feature map values as indicators of how likely a pixel has rigid motion (the higher the value, the higher the likelihood). Our epipolar geometry-based motion disentanglement pipeline is shown in Fig. 3.

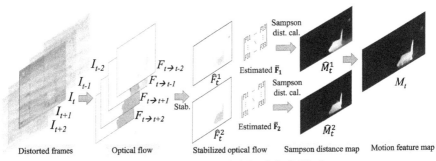

Fig. 3. Pipeline of epipolar geometry-based motion disentanglement. Since the raw optical flows are downgraded by turbulence, we apply a geometry-based consistency check to generate motion feature maps that only preserve object motion.

Optical Flow Stabilization. Since atmospheric turbulence causes erratic pixel shifts in video frames [31], estimating optical flow is prone to error. We address this by first stabilizing the optical flow estimations: we assume consistent object motion during a short period of time, and then average the optical flow within a small time step to reduce the error caused by turbulence perturbation without losing features of the actual rigid motion. Specifically, given a sequence of bi-directional optical flow $\mathcal{O}_t = \{F_{t \to t \pm i}\}_{i=0}^{B}$, we calculate a sequence of per-frame stabilized flow $\hat{\mathcal{O}}_t = \{\hat{F}_t^j | j = 1, 2, ..., A\}$ (where A is the total number of stabilized flows for each frame) by averaging the original sequence within a short interval:

$$\hat{F}_t^j = \frac{1}{|\mathcal{K}^j|} \sum_{i \in \mathcal{K}^j} \frac{F_{t \to t+i}}{i}, \qquad (1)$$

where $\mathcal{K}^j$ is the temporal interval used for calculating $\hat{F}_t^j$, namely the subset of $\{x \mid x \in \mathbb{Z}, -B \leq x \leq B\}$

Geometric Consistency Check. Our fundamental assumption is that pixels on moving objects have larger geometric consistency errors compared with static background, when mapping a frame to its neighboring time frame using foundamental matrix. The Sampson distance [13] measures rigid geometric consistency by calculating the distance between a frame pair in video, constrained by epipolar geometry. Since moving objects do not obey the epipolar geometry that assumes a static scene, their correspondences will have a large Sampson distance. Although the turbulent perturbation also breaks down the epipolar geometry, the errors that they introduce are much smaller and more random, and thus easily eliminated through averaging.

Given a stabilized optical flow $\hat{F}_t^j$, we calculate its Sampson distance map $\hat{M}_t^j$ as:

$$\hat{M}_t^j(\mathbf{p}_1, \mathbf{p}_2) = \frac{(\mathbf{p}_2^\top \bar{\mathbf{F}} \mathbf{p}_1)^2}{(\bar{\mathbf{F}} \mathbf{p}_1)_1^2 + (\bar{\mathbf{F}} \mathbf{p}_1)_2^2 + (\bar{\mathbf{F}}^\top \mathbf{p}_2)_1^2 + (\bar{\mathbf{F}}^\top \mathbf{p}_2)_2^2}, \qquad (2)$$

where $\mathbf{p}_1$ and $\mathbf{p}_2$ are the homogeneous coordinates of a pair of corresponding points in two neighboring frames. We determine the correspondence using the stabilized optical flow: $\mathbf{p}_2 = \mathbf{p}_1 + \hat{F}_t^j(\mathbf{p}_1)$. $\bar{\mathbf{F}}$ is the fundamental matrix between the two frames estimated by Least Median of Squares (LMedS) regression [44].

Distorted frame I_t Original optical flow F_t Stabilized optical flow $\hat{F}_t^j$ Sampson distance map $\hat{M}_t$ Motion feature map M_t

Fig. 4. Step-by-step intermediate results for motion feature map estimation.

We further average all available Sampson distance maps $\{\hat{M}_t^j | j = 1, 2, ..., A\}$ for a frame I_t to obtain the per-frame motion feature map: $M_t = \frac{1}{A} \sum_{j=1}^{A} \hat{M}_t^j$. This map indicates how likely a pixel is to belong to a moving object. Figure 4 compares the intermediate results when generating the motion feature map. Note that the original optical flow map is corrupted by turbulence and cannot resolve the airplane. After our stabilization and consistency check, the final motion feature map preserves the airplane's shape including the highlighted wheels.

3.2 Region Growing-Based Segmentation

Next, based on the motion feature maps, we adopt a "region growing" scheme to generate segmentation masks for moving objects (see Fig. 5). Note that while motion feature maps effectively characterize object motions, they tend to be non-binary and exhibit fuzziness at the object boundaries.

Initial Seed Selection. We first select a small set of seedling pixels that have high confidence of being on a moving object. This selection is based on the motion feature map which encodes how likely a pixel is to be in motion. Note that since we apply the sliding window on the motion map M_t, the size and appearance of the object would not affect the seed selection. Specifically, our sliding windows

$\{W_k\}_{k=1}^{K}$ are of size $D \times D$, where D is adaptively chosen based on the input resolution, to scan through M_t to determine the initial seed of each moving object k. A seed is detected when pixels within the search window are of similar large values, indicating large Sampson distances in this area. Specifically, for a search window W_k, we consider it has found a seed when it satisfies two criteria: (1) its average value $\bar{M}_t(W_k)$ is greater than a threshold δ_1; and (2) its variance $\sigma^2(M_t(W_k))$ is less than a small threshold δ_2. Each selected seedling region is assigned an integer mask ID to uniquely identify multiple moving objects.

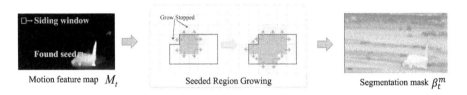

Fig. 5. Pipeline of region growing-based segmentation. We select seeds on the motion feature map using a sliding window. We then grow the seeds to full segmentation masks.

Seeded Region Growing. We then grow full segmentation masks from the initialized seedling pixels. We gradually expand each seeded region outwards to the nearest neighbors of boundary pixels using this criteria for pixel inclusion:

$$|M_t(\mathbf{p}_{new}) - M_t(\mathbf{p}_{seed})| < \delta_{seed}, \qquad (3)$$

where M_t is the motion feature map; $\mathbf{p}_{new}$ is the pixel under consideration; $\mathbf{p}_{seed}$ is the seed pixel that we grow from; and $\delta_{seed} = 0.2 \times M_t(\mathbf{p}_{seed})$ is the threshold for stopping the growth. Note that this threshold value depends on turbulence strength and needs to be adjusted for extreme cases. For stronger turbulence, we prefer larger δ_{seed} and increase the multiplier from 0.2 to 0.3. For weak turbulence, we decrease the multiplier to 0.1. The object's border tends to get blurred for videos with severe turbulence. Meanwhile, a larger threshold makes the region grow harder, resulting in more reliable results.

When growing from multiple seeds, we skip the pixels already examined so that different object masks are non-overlapping. We thus obtain a set of per-frame segmentation masks $\{\beta_t^m\}_{t=1}^{T}$.

Mask ID Unification. When a scene has multiple moving objects (say K), our region-growing algorithm will generate K segmentation masks with IDs ranging from 1 to K for each frame. Since the region-growing module is applied to each frame independently, the mask ID among different frames may be inconsistent with respect to objects. There is a need to unify the mask IDs across frames, so that the same mask ID always maps to the same object. We propose a K-means-based filtering technique to do so.

We first represent each mask region by its centroid: $\mathbf{c}_t^m = \text{mean}(\mathbf{p}_t^m)$ (where $\mathbf{p}_t^m$ is the coordinates of all foreground pixels in the mask). We take the mask centroids for all frames and all objects, and optimize K K-means cluster centroids μ^m:

$$\operatorname*{argmin}_{\mu^m} \sum_{m=1}^{K} \sum_{t=1}^{T} ||\mathbf{c}_t^m - \mu^m||^2, \qquad (4)$$

where K is the total number of objects, and T is the total number of frames.

We then re-ID the masks for each frame by comparing their centroids to the K-means cluster centroids. The K-means cluster IDs act as a global reference for all frames. Each mask is assigned the ID m of a K-means cluster that it is closest to: $m^* = \operatorname*{argmin}_{m} ||\mathbf{c}_t^m - \mu^m||$. After this re-assignment, the mask IDs for all frames are consistent with each ID uniquely identifying a moving object in the scene. This allows our method to handle multiple moving objects in a scene.

3.3 Spatio-temporal Refinement

Finally, we further refine the masks to improve their spatio-temporal consistency. We develop a Refine-Net Φ_θ with a U-Net [43] backbone to refine the masks generated by region growing.

Parameter Initialization. We concatenate the video frame I_t with its motion feature map M_t. The concatenated tensor is fed as input to the Refine-Net Φ_θ: $\alpha_t^m = \Phi_\theta(I_t, M_t)$ (here α_t^m is the output of Φ_θ, which is a refined mask). We use the following loss function for initializing the parameters of Φ_θ:

$$\mathcal{L}_{ini} = \gamma_1 \mathcal{L}_1 + \gamma_2 \sum_g \mathcal{L}_2^g + \gamma_3 \sum_g \mathcal{L}_3^g, \qquad (5)$$

where γ_1, γ_2, and γ_3 are balancing weights for each loss term. We run 20–30 epochs for initialization. Below, we describe our loss terms in detail.

$\mathcal{L}_1$ is a pixel-wise cross-entropy loss that enforces consistency between the refined output mask α_t^m and coarse input mask β_t^m. It is calculated as:

$$\mathcal{L}_1 = \frac{1}{\Omega} \sum_{\mathbf{p} \in \Omega} (-\alpha_t^m(\mathbf{p}) \log \alpha_t^m(\mathbf{p}) + \beta_t^m(\mathbf{p}) \log \beta_t^m(\mathbf{p})), \qquad (6)$$

where $\mathbf{p}$ is the pixel coordinates, and Ω represents the spatial domain.

$\mathcal{L}_2^g$ is a bidirectional consistency loss that enforces flow consistency between α_{t+g}^m and the optical flow-warped input mask: $\hat{\beta}_t^m = F_{t \to t+g}(\beta_t^m)$. We also use the cross-entropy for comparison, and $\mathcal{L}_2^g$ is written as:

$$\mathcal{L}_2^g = \frac{1}{\Omega} \sum_{\mathbf{p} \in \Omega} (-\alpha_{t+g}^m(\mathbf{p}) \log \alpha_{t+g}^m(\mathbf{p}) + \hat{\beta}_t^m(\mathbf{p}) \log \hat{\beta}_t^m(\mathbf{p})). \qquad (7)$$

$\mathcal{L}_3^g$ is another bidirectional consistency loss that enforces flow consistency between α_{t+g}^m and the optical flow-warped version of itself: $\hat{\alpha}_t^m = F_{t \to t+g}(\alpha_t^m)$. $\mathcal{L}_3^g$ is written as:

$$\mathcal{L}_3^g = \frac{1}{\Omega} \sum_{\mathbf{p} \in \Omega} (-\alpha_{t+g}^m(\mathbf{p}) \log \alpha_{t+g}^m(\mathbf{p}) + \hat{\alpha}_t^m(\mathbf{p}) \log \hat{\alpha}_t^m(\mathbf{p})). \qquad (8)$$

After initialization, the output mask α_t^m is aligned with the input mask β_t^m and has improved temporal consistency with the two bidirectional losses.

Iterative Refinement Using Grouping Function. To further improve the mask quality and consistency, we adopt an iterative refinement constrained by a grouping function. The refinement runs for 10 epochs. We update the input reference mask β_t^m using a K-means-based grouping function for every 3 epochs. Specifically, we concatenate each pixel's mask values $\{\beta_t^m(\mathbf{p})\}_{t=1}^T$, with its coordinates $\mathbf{p} = (x, y)$, to form a new tensor $\{T_t^m(\mathbf{p})\}_{t=1}^T$ that combines both the motion and spatial information. The combined tensor $T_t^m(\mathbf{p})$ of all pixels is used to optimize two K-means cluster centroids θ^1, θ^2: one for foreground cluster (θ^1) and the other for background (θ^2). The centroids are optimized using the following function:

$$\mathop{\mathrm{argmin}}_{\theta^1, \theta^2} \sum_{i=1}^{2} \sum_{\mathbf{p} \in \Omega} \|T_t^i - \theta_t^i\|^2, \tag{9}$$

where Ω is the domain of all pixels. We then re-assign the values of β_t^m: if $T_t^m(\mathbf{p})$ is closer to θ^1, we assign the mask value as 1; otherwise, we consider the pixel as background and assign the mask value as 0.

The loss function we use for network optimization is the same as the initialization step, but in this refinement step, β_t^m is updated using the K-means-based grouping every 3 epochs. Without this grouping-based refinement, the output mask tends to have gaps or other spatial inconsistent artifacts.

Table 1. Comparison of existing datasets on turbulent images or videos.

Dataset	Source	Format	Purpose	Num. of video	Availability
Turb Pascal VOC [7]	Synthetic	Image	Segmentation	✗	✗
Turb ADE20K [7]	Synthetic	Image	Segmentation	✗	✗
TSRW-GAN [17]	Real	Video	Restoration	27*	✓
OTIS [11]	Real	Video	Restoration	5	✓
BVI-CLEAR [2]	Real	Video	Restoration	3	✓
DOST (ours)	Real	Video	Segmentation	38	✓

* The actual number of videos in TSRW-GAN is greater than 27, but many of the videos have large overlaps. The number of unique (or non-overlapped) videos is 27.

4 Experiments

4.1 DOST Dataset

We capture a long-range turbulent video dataset, which we call *Dynamic Object Segmentation in Turbulence (DOST)*, to evaluate our method. DOST consists of 38 videos, all collected outdoors in hot weather using long focal length settings. All videos contain instances of moving objects, such as vehicles, aircraft, and pedestrians. We manually annotate the video to provide per-frame ground truth masks for segmenting moving objects. Specifically, we use a Nikon Coolpix P1000 to capture the videos. The camera has an adjustable focal length of up to 539mm (125× optical zoom), which is equivalent to 3000mm focal length in 35mm sensor format. We record videos with a resolution of 1920 × 1080. In total, our dataset has 38 videos with 1719 frames. We annotate moving objects in each video frame using the Computer Vision Annotation Tool (CVAT) [1]. Our dataset is the first of its kind to provide a ground truth moving object segmentation mask in the context of long-range turbulent video. DOST is designed for motion segmentation, but can be used for other tasks (e.g., turbulent video restoration).

Table 1 compares DOST with existing datasets designed for turbulent image or video processing. Since real images/videos with air turbulence are very difficult to acquire, real datasets are scarce, and existing ones are usually small in size. Cui and Zhang [7] synthesize large turbulent image datasets using standard image datasets (Pascal VOC 2012 [7] and ADE20K [7]) and a turbulence simulator. However, there is a domain gap between simulated data and real data. Further, their datasets only contain single images and cannot be used for studying motion. Other real turbulent video datasets [2,11,17] are all designed for the restoration task. Although some [11,17] have bounding box annotations, none provide object-tight segmentation masks.

Table 2. Quantitative comparisons with state-of-the-art unsupervised methods on DOST w.r.t. various turbulence strengths.

Model	$\mathcal{J}$			$\mathcal{F}$			$\mathcal{G}$
	Normal turb.	Severe turb.	Overall	Normal turb.	Severe turb.	Overall	Overall
TMO [6]	0.643	0.235	0.439	0.757	0.315	0.536	0.487
DSprites [60]	0.427	0.101	0.264	0.772	0.203	0.374	0.319
DS-net [26]	0.361	0.191	0.276	0.422	0.232	0.327	0.302
Ours	**0.851** ↑	**0.557** ↑	**0.703** ↑	**0.812** ↑	**0.634** ↑	**0.723** ↑	**0.713** ↑

4.2 Implementation Details

We implemented our network using PyTorch on a supercomputing node equipped with an NVIDIA GTX A100 GPU. The input frames are resized to a lower resolution of 240×432 for faster optical flow calculation and network training. We use RAFT [49] for optical flow estimation with a maximum frame interval of 4. The stabilized optical flow and Sampson distance maps are subsequently calculated based on the RAFT output. The region-growing algorithm's stopping threshold δ_{seed} (see Eq. 3) is dependent on the turbulence strength, and we need to adjust this parameter for varied strength turbulent videos (see more details in the supplementary material). For the bidirectional consistency losses, we compare with four neighboring time frames, $g \in \{-2, -1, 1, 2\}$ in Eqs. 7–8.

Evaluation Metrics. Given a ground-truth mask, we evaluate the accuracy of the estimated segmentation mask using two standard metrics [4]: (1) Jaccard's

Index ($\mathcal{J}$) that calculates the intersection over the union of two sets (also known as intersection-over-union or IoU measure); and (2) F1-Score ($\mathcal{F}$) that calculates the harmonic mean of precision and recall (also known as dice coefficient). We also calculate the average of $\mathcal{J}$ and $\mathcal{F}$, and denote this overall metric as $\mathcal{G}$.

4.3 Comparison with State-of-the-Art Methods

We compare our method against recent state-of-the-art *unsupervised* methods for motion segmentation, including TMO [6], Deformable Sprites [60], and DS-Net [26]. TMO [6] achieves high accuracy in object segmentation regardless of motion. It therefore has certain advantages in handling turbulent videos, although it is not specifically designed for this purpose. Deformable Sprites [60] integrates appearance features with optical flow, and further enforces consistency with optical flow-guided grouping loss and warping loss. DS-Net [26] uses multi-scale spatial and temporal features for segmentation and is able to achieve good performance when the input is noisy. For all three methods, we use the code implementations provided by authors and train their networks using settings described in papers. For methods that need optical flow, we use RAFT to estimate optical flow between consecutive frames.

Quantitative Comparisons. We show quantitative comparison results in Table 2. All methods are evaluated on our DOST dataset. We organize videos into two sets, "normal turb." and "severe turb.", according to their exhibited turbulence strength. Our method significantly outperforms these state-of-the-art on motion segmentation accuracy under various turbulence strengths. In normal cases, some can still achieve decent performance, whereas our method scores much higher in all metrics. Compared to TMO, whose overall score is the highest among the three state-of-the-art, our accuracy is increasing by 60.1% in $\mathcal{J}$ and 34.9% in $\mathcal{F}$. In severe cases, the performance of all state-of-the-art significantly downgrades, with all $\mathcal{J}$ values lower than 0.25 and $\mathcal{F}$ lower than 0.35. In contrast, our method is relatively robust to strong turbulence.

We also experiment on a larger synthetic dataset to evaluate our robustness with respect to turbulence strength. We use a physics-based turbulence simulator [30] and the DAVIS 2016 dataset [37] to synthesize videos with various strengths of turbulence. We test the three state-of-the-art methods on the synthetic data as well. $\mathcal{J}$ score (or IoU) plots with respect to turbulence strength are shown in Fig. 7b. The results demonstrate that our method is robust to various turbulence strengths. Note that when there is no turbulence in the scene (strength = 0), TMO has a slightly higher $\mathcal{J}$ score, as it incorporates more visual cues for segmentation, whereas our method focuses on turbulence artifacts.

Qualitative Comparisons. We show visual comparisons with state-of-the-art in Fig. 6. We can see that the optical estimations are largely affected by turbulence, especially in severe cases. Our method is able to generate segmentation masks that are tight to the object and works well for multiple moving objects. State-of-the-art methods face challenges when the turbulence strength is strong, or the moving object is too small. Their segmentation masks are incomplete in many cases. In the airplane scenes, DS-Net and DSprites fail to detect moving

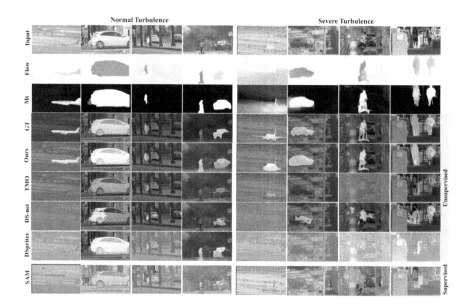

Fig. 6. Qualitative comparisons with state-of-the-art methods on DOST w.r.t. various turbulence strengths. Here we also show the raw optical flows ("Flow"), our motion feature maps ("M_t"), and our ground truth masks ("GT").

objects. Notably, our method achieves the highest robustness to turbulent distortions, and camera shakes, as shown in Fig. 7. We analyzed the effects of various camera motions, including complex multi-directional shake, on segmentation results in video sequences, as illustrated in Fig. 7a. Quantitatively, our method achieved an average IoU score of 0.712 on videos featured by camera shaking

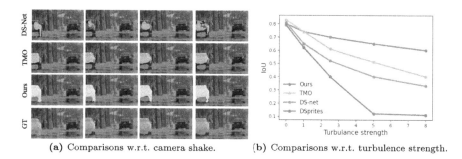

(a) Comparisons w.r.t. camera shake.   (b) Comparisons w.r.t. turbulence strength.

Fig. 7. Example results of handling videos (a) with camera motion and (b) impacted by different turbulence strength. (a) Our results achieves robust performance on different time frames (2nd to 4th columns) in videos suffering from significant camera shake. (b) We can also tell that our method achieves the best robustness across different turbulence strengths, even when it is very strong.

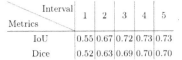

Variants	$\mathcal{L}_2^g + \mathcal{L}_3^g$	$\mathcal{L}_{GR}$	$\mathcal{J}$
A	✗	✗	0.663
B	✓	✗	0.690
C	✓	✓	**0.703**

(a) Ablation studies on our key components.

Interval Metrics	1	2	3	4	5
IoU	0.55	0.67	0.72	0.73	0.73
Dice	0.52	0.63	0.69	0.70	0.70

(b) Ablation study on optical flow interval.

Fig. 8. Ablation studies. (a) Variants of our network. A: Simply using region-growing; B: Full pipeline without grouping loss; C: Our full pipeline. (b) The interval $i = 1...5$ is the maximum temporal gap we used to stablize optical flow in Eq. 1.

in our dataset, compared with TMO/0.305, DS-Net/0.267, and DSprites/0.235, respectively.

We notice that supervised segmentation methods can hardly generalize to DOST, since these methods were trained solely using turbulence-free data. For instance, the foundation model for segmentation, i.e., SAM [19], which has been trained on 11 million images, and over 1B masks, still fails at segmenting whole objects under strong turbulence, as shown in Fig. 6 (last row). Additionally, SAM does not interpret motion, and SAM requires a user to click or prompt the algorithm, whereas we only segment moving objects without any need for user input. More comparison results can be found in supplemental material.

4.4 Ablation Studies

We perform ablation studies to evaluate individual components of our method. All experiments are performed on DOST. We test on three variants of our method: A only employs the region-growing algorithm (with Refine-Net excluded); B uses both region-growing and Refine-Net, but excludes the grouping loss for refinement; and C is implemented as our full approach. Their $\mathcal{J}$ score comparison results are shown in Fig. 8a. Each component is clearly effective, and our full model achieves the best performance. We also evaluate the influence of the optical flow interval (i.e., the number of temporal frames used for optical flow calculation). Accuracy scores with respect to the interval length are shown in Fig. 8b. We can see that the score achieves the plateau when the interval is greater than 4. Therefore, we set the interval to 4.

We also evaluated the effectiveness of our optical flow stabilization and geometric consistency check. Without the optical flow stabilization step, the IoU immediately drops to 0.354; without the geometric consistency check step, the IoU is 0.685, compared with IoU of 0.703 in our full pipeline.

5 Conclusions

In summary, we present an unsupervised approach for segmenting moving objects in videos affected by air turbulence. Our method uses a geometry-based consistency check to disambiguate motions and a region-growing scheme to generate tight segmentation masks. The masks are further refined with spatio-temporal consistency losses. Our method significantly outperforms the existing state-of-the-art in terms of accuracy and robustness when handling turbulent videos. We also contribute the first long-range turbulent video dataset designed for motion segmentation. Nevertheless, due to the unsupervised nature of our method, the current version can only achieve a latency of 0.95 FPS. Our future work will focus on optimizing the approach to reduce this latency by embedding results from foundations models such as SAM. Our method also has limited performance in separating overlapping moving objects in the videos. To address this, we plan to integrate additional visual cues, such as appearance and saliency.

Acknowledgements. This research is based upon work supported in part by the Office of the Director of National Intelligence (ODNI), Intelligence Advanced Research Projects Activity (IARPA), via 2022–21102100003 and NSF IIS-2232298/2232299/2232300. The views and conclusions contained herein are those of the authors and should not be interpreted as necessarily representing the official policies, either expressed or implied, of ODNI, IARPA, NSF or the U.S. Government. The U.S. Government is authorized to reproduce and distribute reprints for governmental purposes, notwithstanding any copyright annotation therein. We also wish to thank ASU Research Computing for providing GPU resources [15] to support this research.

References

1. Computer vision annotation tool (2024). https://www.cvat.ai/
2. Anantrasirichai, N., Achim, A., Kingsbury, N.G., Bull, D.R.: Atmospheric turbulence mitigation using complex wavelet-based fusion. IEEE Trans. Image Process. **22**(6), 2398–2408 (2013)
3. Aubailly, M., Vorontsov, M.A., Carhart, G.W., Valley, M.T.: Video enhancement through automated lucky region fusion from a stream of atmospherically distorted images. In: Frontiers in Optics 2009/Laser Science XXV/Fall 2009 OSA Optics & Photonics Technical Digest. p. CThC3. Optica Publishing Group (2009). https://doi.org/10.1364/COSI.2009.CThC3
4. Chen, L., Wu, Y., Stegmaier, J., Merhof, D.: Sortedap: rethinking evaluation metrics for instance segmentation. In: 2023 IEEE/CVF International Conference on Computer Vision Workshops (ICCVW), pp. 3925–3931. IEEE Computer Society, Los Alamitos, CA, USA (2023). https://doi.org/10.1109/ICCVW60793.2023.00424
5. Chen, X., et al.: Moving object segmentation in 3D lidar data: a learning-based approach exploiting sequential data. IEEE Rob. Autom. Lett. **6**, 6529–6536 (2021). https://doi.org/10.1109/LRA.2021.3093567
6. Cho, S., Lee, M., Lee, S., Park, C., Kim, D., Lee, S.: Treating motion as option to reduce motion dependency in unsupervised video object segmentation. In: Proceedings of the IEEE/CVF Winter Conference on Applications of Computer Vision, pp. 5140–5149 (2023)
7. Cui, L., Zhang, Y.: Accurate semantic segmentation in turbulence media. IEEE Access **7**, 166749–166761 (2019)
8. Dosovitskiy, A., et al.: Flownet: learning optical flow with convolutional networks. In: Proceedings of the IEEE International Conference on Computer Vision, pp. 2758–2766 (2015)
9. Fried, D.L.: Probability of getting a lucky short-exposure image through turbulence. JOSA **68**(12), 1651–1658 (1978)

10. Garg, S., Goel, V.: Mask selection and propagation for unsupervised video object segmentation. In: Proceedings of the IEEE/CVF Winter Conference on Applications of Computer Vision, pp. 1680–1690 (2021)
11. Gilles, J., Ferrante, N.B.: Open turbulent image set (OTIS). Pattern Recogn. Lett. **86**, 38–41 (2017)
12. Gutierrez, D., Seron, F.J., Munoz, A., Anson, O.: Simulation of atmospheric phenomena. Comput. Graph. **30**(6), 994–1010 (2006)
13. Hartley, R., Zisserman, A.: Multiple View Geometry in Computer Vision, 2nd edn. Cambridge University Press, USA (2003)
14. Ilg, E., Mayer, N., Saikia, T., Keuper, M., Dosovitskiy, A., Brox, T.: Flownet 2.0: evolution of optical flow estimation with deep networks. In: Proceedings of the IEEE Conference on Computer Vision and Pattern Recognition, pp. 2462–2470 (2017)
15. Jennewein, D.M., et al.: The sol supercomputer at Arizona state university. In: Practice and Experience in Advanced Research Computing. pp. 296–301. PEARC '23, Association for Computing Machinery, New York, NY, USA (2023). https://doi.org/10.1145/3569951.3597573
16. Jiang, W., Boominathan, V., Veeraraghavan, A.: Nert: implicit neural representations for unsupervised atmospheric turbulence mitigation. In: Proceedings of the IEEE/CVF Conference on Computer Vision and Pattern Recognition, pp. 4235–4242 (2023)
17. Jin, D., et al.: Neutralizing the impact of atmospheric turbulence on complex scene imaging via deep learning. Nat. Mach. Intell. **3**(10), 876–884 (2021)
18. Johnson, B.A., Ma, L.: Image segmentation and object-based image analysis for environmental monitoring: recent areas of interest, researchers' views on the future priorities. Remote Sens. **12**(11) (2020). https://doi.org/10.3390/rs12111772, https://www.mdpi.com/2072-4292/12/11/1772
19. Kirillov, A., et al.: Segment anything. arXiv **2304**, 02643 (2023)
20. Kolmogorov, A.N.: Dissipation of energy in locally isotropic turbulence. Akademiia Nauk SSSR Doklady **32**, 16 (1941)
21. Lee, M., Cho, S., Lee, S., Park, C., Lee, S.: Unsupervised video object segmentation via prototype memory network. In: Proceedings of the IEEE/CVF Winter Conference on Applications of Computer Vision, pp. 5924–5934 (2023)
22. Li, H., Chen, G., Li, G., Yu, Y.: Motion guided attention for video salient object detection. In: Proceedings of the IEEE/CVF international conference on computer vision, pp. 7274–7283 (2019)
23. Li, N., Thapa, S., Whyte, C., Reed, A., Jayasuriya, S., Ye, J.: Unsupervised non-rigid image distortion removal via grid deformation. In: 2021 IEEE/CVF International Conference on Computer Vision (ICCV), pp. 2502–2512 (2021). https://doi.org/10.1109/ICCV48922.2021.00252
24. Ling, Q., Yan, J., Li, F., Zhang, Y.: A background modeling and foreground segmentation approach based on the feedback of moving objects in traffic surveillance systems. Neurocomputing **133**, 32–45 (2014). https://doi.org/10.1016/j.neucom.2013.11.034, https://www.sciencedirect.com/science/article/pii/S0925231214000654
25. Liu, D., Yu, D., Wang, C., Zhou, P.: F2net: learning to focus on the foreground for unsupervised video object segmentation. In: Proceedings of the AAAI Conference on Artificial Intelligence. vol. 35, pp. 2109–2117 (2021)
26. Liu, J., Wang, J., Wang, W., Su, Y.: Ds-net: Dynamic spatiotemporal network for video salient object detection. Digit. Signal Process.

27. Lu, X., Wang, W., Danelljan, M., Zhou, T., Shen, J., Van Gool, L.: Video object segmentation with episodic graph memory networks. In: Computer Vision–ECCV 2020: 16th European Conference, Glasgow, UK, August 23–28, 2020, Proceedings, Part III 16, pp. 661–679. Springer (2020). https://doi.org/10.1007/978-3-030-58580-8_39
28. Lu, X., Wang, W., Shen, J., Crandall, D.J., Van Gool, L.: Segmenting objects from relational visual data. IEEE Trans. Pattern Anal. Mach. Intell. **44**(11), 7885–7897 (2022). https://doi.org/10.1109/TPAMI.2021.3115815
29. Mahadevan, S., Athar, A., Ošep, A., Hennen, S., Leal-Taixé, L., Leibe, B.: Making a case for 3D convolutions for object segmentation in videos. arXiv preprint arXiv:2008.11516 (2020)
30. Mao, Z., Chimitt, N., Chan, S.H.: Accelerating atmospheric turbulence simulation via learned phase-to-space transform. In: 2021 IEEE/CVF International Conference on Computer Vision (ICCV), pp. 14739–14748. IEEE Computer Society, Los Alamitos, CA, USA (2021). https://doi.org/10.1109/ICCV48922.2021.01449, https://doi.ieeecomputersociety.org/10.1109/ICCV48922.2021.01449
31. Mao, Z., Chimitt, N., Chan, S.H.: Image reconstruction of static and dynamic scenes through anisoplanatic turbulence. IEEE Trans. Comput. Imaging **6**, 1415–1428 (2020). https://doi.org/10.1109/TCI.2020.3029401
32. Mao, Z., Jaiswal, A., Wang, Z., Chan, S.H.: Single frame atmospheric turbulence mitigation: a benchmark study and a new physics-inspired transformer model. In: European Conference on Computer Vision, pp. 430–446. Springer (2022). https://doi.org/10.1007/978-3-031-19800-7_25
33. Nieuwenhuizen, R., Dijk, J.S.K.D.T.M.F.L.R.I.I.T.P.O.L.M.O.: Dynamic turbulence mitigation for long-range imaging in the presence of large moving objects. In: EURASIP J Image Video Process, pp. 1–8 (2019). https://doi.org/10.1186/s13640-018-0380-9
34. Ochs, P., Brox, T.: Object segmentation in video: a hierarchical variational approach for turning point trajectories into dense regions. In: 2011 International Conference on Computer Vision, pp. 1583–1590 (2011). https://doi.org/10.1109/ICCV.2011.6126418
35. Osorio, R., López, I., Peña, M., Lomas, V., Lefranc, G., Savage, J.: Surveillance system mobile object using segmentation algorithms. IEEE Lat. Am. Trans. **13**, 2441–2446 (2015). https://doi.org/10.1109/TLA.2015.7273810
36. Pei, G., Shen, F., Yao, Y., Xie, G.S., Tang, Z., Tang, J.: Hierarchical feature alignment network for unsupervised video object segmentation. In: Computer Vision–ECCV 2022: 17th European Conference, Tel Aviv, Israel, October 23–27, 2022, Proceedings, Part XXXIV, pp. 596–613. Springer (2022). https://doi.org/10.1007/978-3-031-19830-4_34
37. Perazzi, F., Pont-Tuset, J., McWilliams, B., Van Gool, L., Gross, M., Sorkine-Hornung, A.: A benchmark dataset and evaluation methodology for video object segmentation. In: Computer Vision and Pattern Recognition (2016)
38. Potvin, G., Forand, J., Dion, D.: A parametric model for simulating turbulence effects on imaging systems. DRDC Valcartier TR **2006**, 787 (2007)
39. Rai, M., Al-Saad, M., Darweesh, M., Al-Mansoori, S., Al-Ahmad, H., Mansoor, W.: Moving objects segmentation in infrared scene videos. In: 2021 4th International Conference on Signal Processing and Information Security (ICSPIS), pp. 17–20 (2021). https://doi.org/10.1109/icspis53734.2021.9652436

40. Ranjan, A., et al.: Competitive collaboration: Joint unsupervised learning of depth, camera motion, optical flow and motion segmentation. In: 2019 IEEE/CVF Conference on Computer Vision and Pattern Recognition (CVPR), pp. 12232–12241. IEEE Computer Society, Los Alamitos, CA, USA (2019). https://doi.org/10.1109/CVPR.2019.01252, https://doi.ieeecomputersociety.org/10.1109/CVPR.2019.01252
41. Ren, S., Liu, W., Liu, Y., Chen, H., Han, G., He, S.: Reciprocal transformations for unsupervised video object segmentation. In: Proceedings of the IEEE/CVF Conference on Computer Vision and Pattern Recognition, pp. 15455–15464 (2021)
42. Ronneberger, O., Fischer, P., Brox, T.: U-net: Convolutional networks for biomedical image segmentation (2015)
43. Ronneberger, O., Fischer, P., Brox, T.: U-net: convolutional networks for biomedical image segmentation. In: Navab, N., Hornegger, J., Wells, W.M., Frangi, A.F. (eds.) Medical Image Computing and Computer-Assisted Intervention - MICCAI 2015, pp. 234–241. Springer International Publishing, Cham (2015)
44. Rousseeuw, P.J.: Least median of squares regression. J. Am. Stat. Assoc. **79**(388), 871–880 (1984). http://www.jstor.org/stable/2288718
45. Saha, R.K., Qin, D., Li, N., Ye, J., Jayasuriya, S.: Turb-seg-res: a segment-then-restore pipeline for dynamic videos with atmospheric turbulence. In: Proceedings of the IEEE/CVF Conference on Computer Vision and Pattern Recognition (2024)
46. Shimizu, M., Yoshimura, S., Tanaka, M., Okutomi, M.: Super-resolution from image sequence under influence of hot-air optical turbulence. In: 2008 IEEE Conference on Computer Vision and Pattern Recognition, pp. 1–8 (2008). https://doi.org/10.1109/CVPR.2008.4587525
47. Shin, D., Ren, Z., Sudderth, E., Fowlkes, C.: 3D scene reconstruction with multi-layer depth and epipolar transformers. In: 2019 IEEE/CVF International Conference on Computer Vision (ICCV), pp. 2172–2182. IEEE Computer Society, Los Alamitos, CA, USA (2019)
48. Sun, D., Yang, X., Liu, M.Y., Kautz, J.: PWC-net: CNNS for optical flow using pyramid, warping, and cost volume. In: Proceedings of the IEEE Conference on Computer Vision and Pattern Recognition, pp. 8934–8943 (2018)
49. Teed, Z., Deng, J.: Raft: Recurrent all-pairs field transforms for optical flow. In: European Conference on Computer Vision, pp. 402–419. Springer (2020). https://doi.org/10.1007/978-3-030-58536-5_24
50. Tokmakov, P., Alahari, K., Schmid, C.: Learning motion patterns in videos. In: 2017 IEEE Conference on Computer Vision and Pattern Recognition (CVPR), pp. 531–539. IEEE Computer Society, Los Alamitos, CA, USA (2017). https://doi.org/10.1109/CVPR.2017.64, https://doi.ieeecomputersociety.org/10.1109/CVPR.2017.04
51. Valgaerts, L., Bruhn, A., Weickert, J.: A variational model for the joint recovery of the fundamental matrix and the optical flow. In: Rigoll, G. (ed.) Pattern Recogn., pp. 314–324. Springer, Berlin Heidelberg, Berlin, Heidelberg (2008)
52. Wang, W., Shen, J., Porikli, F.: Saliency-aware geodesic video object segmentation. In: 2015 IEEE Conference on Computer Vision and Pattern Recognition (CVPR), pp. 3395–3402 (2015). https://doi.org/10.1109/CVPR.2015.7298961
53. Wang, W., Zhou, T., Porikli, F., Crandall, D.J., Gool, L.: A survey on deep learning technique for video segmentation. IEEE Trans. Pattern Anal. Mach. Intell. **45**, 7099–7122 (2021). https://doi.org/10.1109/TPAMI.2022.3225573
54. Wedel, A., Cremers, D., Pock, T., Bischof, H.: Structure- and motion-adaptive regularization for high accuracy optic flow. In: 2009 IEEE 12th International Con-

ference on Computer Vision, pp. 1663–1668 (2009). https://doi.org/10.1109/ICCV.2009.5459375
55. Wulff, J., Sevilla-Lara, L., Black, M.J.: Optical flow in mostly rigid scenes. In: 2017 IEEE Conference on Computer Vision and Pattern Recognition (CVPR), pp. 6911–6920 (2017). https://doi.org/10.1109/CVPR.2017.731
56. Yamaguchi, K., McAllester, D., Urtasun, R.: Robust monocular epipolar flow estimation. In: 2013 IEEE Conference on Computer Vision and Pattern Recognition, pp. 1862–1869 (2013). https://doi.org/10.1109/CVPR.2013.243
57. Yang, S., Zhang, L., Qi, J., Lu, H., Wang, S., Zhang, X.: Learning motion-appearance co-attention for zero-shot video object segmentation. In: Proceedings of the IEEE/CVF International Conference on Computer Vision, pp. 1564–1573 (2021)
58. Yang, Z., Wang, Q., Bertinetto, L., Hu, W., Bai, S., Torr, P.H.: Anchor diffusion for unsupervised video object segmentation. In: Proceedings of the IEEE/CVF International Conference On Computer Vision, pp. 931–940 (2019)
59. Ye, V., Li, Z., Tucker, R., Kanazawa, A., Snavely, N.: Deformable sprites for unsupervised video decomposition. In: 2022 IEEE/CVF Conference on Computer Vision and Pattern Recognition (CVPR), pp. 2647–2656 (2022). https://doi.org/10.1109/CVPR52688.2022.00268
60. Ye, V., Li, Z., Tucker, R., Kanazawa, A., Snavely, N.: Deformable sprites for unsupervised video decomposition. In: Proceedings of the IEEE/CVF Conference on Computer Vision and Pattern Recognition, pp. 2657–2666 (2022)
61. Yin, Y., Xu, D., Wang, X., Zhang, L.: Agunet: annotation-guided u-net for fast one-shot video object segmentation. Pattern Recogn. **110**, 107580 (2021)
62. Yin, Z., Shi, J.: Geonet: Unsupervised learning of dense depth, optical flow and camera pose. In: 2018 IEEE/CVF Conference on Computer Vision and Pattern Recognition (CVPR), pp. 1983–1992. IEEE Computer Society, Los Alamitos, CA, USA (2018). https://doi.org/10.1109/CVPR.2018.00212, https://doi.ieeecomputersociety.org/10.1109/CVPR.2018.00212
63. Zhang, D., Javed, O., Shah, M.: Video object segmentation through spatially accurate and temporally dense extraction of primary object regions. In: 2013 IEEE Conference on Computer Vision and Pattern Recognition, pp. 628–635 (2013). https://doi.org/10.1109/CVPR.2013.87
64. Zhang, K., Zhao, Z., Liu, D., Liu, Q., Liu, B.: Deep transport network for unsupervised video object segmentation. In: Proceedings of the IEEE/CVF International Conference on Computer Vision, pp. 8781–8790 (2021)
65. Zhang, X., Mao, Z., Chimitt, N., Chan, S.H.: Imaging through the atmosphere using turbulence mitigation transformer. IEEE Trans. Comput. Imaging **10**, 115–128 (2024). https://doi.org/10.1109/tci.2024.3354421, http://dx.doi.org/10.1109/TCI.2024.3354421
66. Zheng, Z., Zhong, Y., Wang, J., Ma, A.: Foreground-aware relation network for geospatial object segmentation in high spatial resolution remote sensing imagery. In: 2020 IEEE/CVF Conference on Computer Vision and Pattern Recognition (CVPR), pp. 4095–4104 (2020). https://doi.org/10.1109/CVPR42600.2020.00415
67. Zhong, Y., Ji, P., Wang, J., Dai, Y., Li, H.: Unsupervised deep epipolar flow for stationary or dynamic scenes. In: Proceedings of the IEEE/CVF Conference on Computer Vision and Pattern Recognition (2019)
68. Zhou, Y., Xu, X., Shen, F., Zhu, X., Shen, H.T.: Flow-edge guided unsupervised video object segmentation. IEEE Trans. Circuits Syst. Video Technol. **32**(12), 8116–8127 (2021)

69. Zhou, Z., Tulsiani, S.: Sparsefusion: Distilling view-conditioned diffusion for 3D reconstruction. In: CVPR (2023)
70. Zou, Y., Luo, Z., Huang, J.B.: DF-Net: unsupervised Joint Learning Of Depth And Flow Using Cross-task Consistency. In: Proceedings of the European Conference on Computer Vision (ECCV), pp. 38–55. Springer International Publishing (2018). https://doi.org/10.1007/978-3-030-01228-1_3, http://dx.doi.org/10.1007/978-3-030-01228-1_3
71. Çaliskan, T., Arica, N.: Atmospheric turbulence mitigation using optical flow. In: 2014 22nd International Conference on Pattern Recognition, pp. 883–888 (2014). https://doi.org/10.1109/ICPR.2014.162

AccDiffusion: An Accurate Method for Higher-Resolution Image Generation

Zhihang Lin[1], Mingbao Lin[2], Meng Zhao[3], and Rongrong Ji[1(✉)]

[1] Key Laboratory of Multimedia Trusted Perception and Efficient Computing, Ministry of Education of China, Xiamen University, Xiamen, China
zhihanglin@stu.xmu.edu.cn, rrji@xmu.edu.cn
[2] Skywork AI, Singapore, Singapore
[3] Tencent Youtu Lab., Shanghai, China

Abstract. This paper attempts to address the object repetition issue in patch-wise higher-resolution image generation. We propose AccDiffusion, an accurate method for patch-wise higher-resolution image generation without training. An in-depth analysis in this paper reveals an identical text prompt for different patches causes repeated object generation, while no prompt compromises the image details. Therefore, our AccDiffusion, for the first time, proposes to decouple the vanilla image-content-aware prompt into a set of patch-content-aware prompts, each of which serves as a more precise description of an image patch. Besides, AccDiffusion also introduces dilated sampling with window interaction for better global consistency in higher-resolution image generation. Experimental comparison with existing methods demonstrates that our AccDiffusion effectively addresses the issue of repeated object generation and leads to better performance in higher-resolution image generation. Our code is released at https://github.com/lzhxmu/AccDiffusion.

Keywords: Image Generation · High Resolution · Diffusion Model

1 Introduction

Diffusion models have garnered significant attention and made notable advancements with the emergence of works such as DDPM [10], DDIM [28], ADM [3], and LDMs [21], owing to their outstanding generative ability and wide range of applications. However, stable diffusion models entail tremendous training costs primarily due to the large number of timestamps required and the quadratic relationship between computing costs and resolution. Consequently, it is common to limit the resolution to a relatively low level, such as 512^2 for SD 1.5 [20] and 1024^2 for SDXL [17], during training. Even at such low resolution, stable diffusion

Supplementary Information The online version contains supplementary material available at https://doi.org/10.1007/978-3-031-72658-3_3.

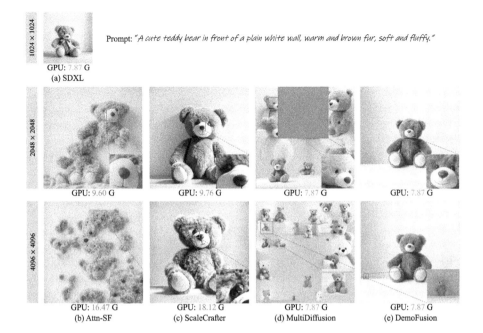

Fig. 1. Comparison of image quality and GPU overhead for existing higher-resolution generation methods. The GPU memory of Attn-SF [12] and ScaleCrafter [6] significantly increases with resolution, while patch-wise denoising methods, *e.g.*, MultiDiffusion [1] and DemoFusion [4] suffer object repetition issue. Best viewed zoomed in.

1.5 still entails over 20 d of training on 256 A100 GPUs [20]. Nonetheless, high-resolution generation finds widespread application in real-life scenarios, such as advertisements. The demand for generating high-resolution images clashes with the expensive training costs involved.

Therefore, researchers have shifted their focus to training stable diffusion models with low resolution and subsequently applying fine-tuning [30,33] or training-free [1,4,6,14] methods to achieve image generation extrapolation. A naive approach is to directly use pre-trained stable diffusion models to generate higher-resolution images. However, the resulting images from this approach are proved to suffer from issues such as object repetition and inconsistent object structures [4,12]. Previous methods attempted to achieve image generation extrapolation from the perspectives of attention entropy [12] or the receptive field of stable diffusion model [6]. However, these methods have been proven to be less practical in two folds, as shown in Fig. 1(b, c): (1) a substantial increase in GPU memory consumption [33] as the resolution rises and (2) poor quality of the generated images [4]. Thanks to stable diffusion's outstanding local detail generation ability, recent works [1,4,14] have started conducting higher-resolution image generation in a patch-wise fashion for the sake of less GPU memory consumption. Previous works MultiDiffusion [1] and SyncDiffusion [14] fuse multiple

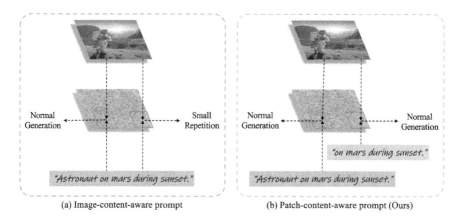

Fig. 2. Image-content-aware prompt *v.s.* Patch-content-aware prompt.

overlapped patch-wise denoising results to generate higher-resolution panoramic images without a seam. However, the direct application of these approaches to generate higher-resolution object-centric images leads to repeated and distorted results lacking global semantic coherence, as shown in Fig. 1(d). Recently, DemoFusion [4] has introduced global semantic information into the patch-wise higher-resolution image generation through residual connection and dilated sampling. It only partially solves the problem of repeated object generation and still exhibits small object repetition in ultra-high image generation as depicted in Fig. 1(e). How to resolve the issue of repeated object generation completely in patch-wise higher-resolution image generation remains an unresolved problem.

In this paper, our in-depth analysis of DemoFusion [4] indicates, as illustrated in Fig. 2(a), small object repetition generation is the adversarial outcome of an identical text prompt on all patches, encouraging to generate repeated objects, and global semantic information from residual connection and dilated sampling, suppressing the generation of repeated objects. To address the above issues, we propose AccDiffusion, an accurate method for higher-resolution image generation, with its major novelty in two folds:

(1) To completely solve small object repetition, as illustrated in Fig. 2(b), we propose to decouple the vanilla image-content-aware prompt into a set of patch-content-aware substrings, each of which serves as a more precise prompt to describe the patch contents. Specifically, we utilize the cross-attention map from the low-resolution generation process to determine whether a word token should serve as the prompt for a patch. If a word token has a high response in the cross-attention map region corresponding to the patch, it should be included in the prompt, and vice versa.

(2) Through visualization, we observe that the dilated sampling operation in DemoFusion generates globally inconsistent and noisy information, disrupting the generation of higher-resolution images. Such inconsistency stems from the independent denoising of dilation samples without interaction. To

address this, we employ a position-wise bijection function to enable interaction between the noise from different dilation samples. Experimental results show that our dilated sampling with interaction leads to the generation of smoother global semantic information (see Fig. 3(c, d)).

We have conducted extensive experiments to verify the effectiveness of AccDiffusion. The qualitative results demonstrate that AccDiffusion effectively addresses the issue of repeated object generation in higher-resolution image generation. And the quantitative results show that AccDiffusion achieves state-of-the-art performance in training-free image generation extrapolation.

2 Related Work

2.1 Diffusion Models

Diffusion models [3,10,21,28] are generative probabilistic models that transform Gaussian noise into samples through gradual denoising steps. DDPM [10] is a pioneering model that demonstrates impressive image generation capabilities using Markovian forward and reverse processes. Based on DDPM, DDIM [28] utilizes non-Markovian reverse processes to decrease sampling time effectively. Furthermore, LDMs [21] incorporate the diffusion process into the latent space, resulting in efficient training and inference. Subsequently, a series of LDMs-based stable diffusion models are open-sourced and achieve state-of-the-art image synthesis capability. This has led to widespread applications in various downstream generative tasks, including images [3, 10, 16, 22, 28], audio [5, 11], video [9, 26] and 3D objects [15, 18, 31], *etc.*

2.2 Training-Free Higher-Resolution Image Generation

Although stable diffusion demonstrates impressive results, its training cost limits low-resolution training and thus generates low-fidelity images when the inference resolution differs from the training resolution [4, 6, 12]. Recent works [1, 4, 6, 12] have attempted to utilize pre-trained diffusion models for generating higher-resolution images. These works [1, 4, 6, 12] can be broadly categorized into two categories: direct generation [6, 12] and indirect generation [1, 4]. Direct generation methods scale the input of the diffusion models to the target resolution and then perform forward and reverse processes directly on the target resolution. These kinds of methods require modifications to the fundamental architecture, such as adjusting the attention scale factor [12] and the receptive field of convolutional kernels [6], to prevent repetition generation. However, the generated images fail to yield the higher-resolution detail desired. Additionally, direct generation methods encounter out-of-memory errors when generating ultra-high resolution images (*e.g.* 8K) on consumer-grade GPUs, due to the quadratic increase in memory overhead as the latent space size grows. Indirect generation methods generate higher-resolution images through multiple overlapped denoising paths of LDMs and are capable of generating images of any resolution on consumer-grade GPUs. However, these mothods [1,14] suffer from local repetition and structural distortion. Du et al. [4] tried to address repeated generation by introducing global structural information from lower-resolution image.

3 Method

3.1 Backgrounds

Latent Diffusion Models (LDMs). LDMs [3] apply an autoencoder $\mathcal{E}$ to encode an image $\mathbf{x}_0 \in \mathbb{R}^{H \times W \times 3}$ into a latent representation $\mathbf{z}_0 = \mathcal{E}(\mathbf{x}_0) \in \mathbb{R}^{h \times w \times c}$, where the regular diffusion process is constructed as:

$$\mathbf{z}_t = \sqrt{\bar{\alpha}_t}\mathbf{z}_0 + \sqrt{1-\bar{\alpha}_t}\varepsilon, \quad \varepsilon \sim \mathcal{N}(0, \mathbf{I}), \tag{1}$$

where $\{\alpha_t\}_{t=1}^{T}$ is a set of prescribed variance schedules and $\bar{\alpha}_t = \Pi_{i=1}^{t}\alpha_i$. To perform conditional sequential denoising, a network ε_θ is trained to predict added noise, constrained by the following training objective:

$$\min_{\theta} \mathbb{E}_{\mathcal{E}(x_0),\varepsilon \sim \mathcal{N}(0,1), t \sim \text{Uniform}(1,T)} \left[\|\varepsilon - \varepsilon_\theta(\mathbf{z}_t, t, \tau_\theta(y))\|_2^2 \right], \tag{2}$$

in which $\tau_\theta(y) \in \mathbb{R}^{M \times d_\tau}$ is an intermediate representation of condition y and M is the number of word tokens in the prompt y. The $\tau_\theta(y)$ is then mapped to keys and values in cross-attention of U-Net ε_θ:

$$\begin{aligned} Q = W_Q \cdot \varphi(z_t), \quad K &= W_K \cdot \tau_\theta(y), \quad V = W_V \cdot \tau_\theta(y), \\ \mathcal{M} = \text{Softmax}(\frac{QK^T}{\sqrt{d}}&), \quad \text{Attention}(Q, K, V) = \mathcal{M} \cdot V. \end{aligned} \tag{3}$$

Here, for simplicity, we omit the expression of multi-head cross-attention and $\varphi(z_t) \in \mathbb{R}^{N \times d_\epsilon}$ denotes an intermediate representation of noise in the U-Net. Here $N = h \times w$ represents the pixel number of the latent noise z_t. $W_Q \in \mathbb{R}^{d \times d_\epsilon}, W_K \in \mathbb{R}^{d \times d_\tau}$, and $W_V \in \mathbb{R}^{d \times d_\tau}$ are learnable projection matrices. $\mathcal{M} \in \mathbb{R}^{N \times M}$ is the cross-attention maps.

In contrast, the denoising process aims to recover the cleaner version $\mathbf{z}_{t-1}$ from $\mathbf{z}_t$ by estimating the noise, which can be expressed as:

$$\mathbf{z}_{t-1} = \sqrt{\frac{\alpha_{t-1}}{\alpha_t}}\mathbf{z}_t + \left(\sqrt{\frac{1}{\alpha_{t-1}} - 1} - \sqrt{\frac{1}{\alpha_t} - 1} \right) \cdot \varepsilon_\theta(\mathbf{z}_t, t, \tau_\theta(y)). \tag{4}$$

During inference, a decoder $\mathcal{D}$ is employed at the end of the denoising process to reconstruct the image from the latent representation $\mathbf{x}_0 = \mathcal{D}(\mathbf{z}_0)$.

Patch-wise Denoising. MultiDiffusion [1] achieve higher-resolution image generation by fusing multiple overlapped denoising patches. In simple terms, given a latent representation $\mathcal{Z}_t \in \mathbb{R}^{h' \times w' \times c}$ of higher-resolution image with $h' > h$ and $w' > h$, MultiDiffusion utilizes a shifted window to sample patches from

Fig. 3. Results of higher-resolution image generation. (a) The result of DemoFusion without text prompt. (b) The result of DemoFusion without residual connection and dilated sampling. (c) The result of dilated sampling without window interaction. (d) The result of our dilated sampling with window interaction.

$\mathcal{Z}_t$ and results in a series of patch noise $\{\mathbf{z}_t^i\}_{i=1}^{P_1}$, where $\mathbf{z}_t^i \in \mathbb{R}^{h \times w \times c}$ and $P_1 = (\frac{h'-h}{d_h} + 1) \times (\frac{w'-w}{d_w} + 1)$ is the total number of patches, d_h and d_w is the vertical and horizontal stride, respectively. Then, MultiDiffusion performs patch-wise denoising via Eq. (4) and obtains $\{\mathbf{z}_{t-1}^i\}_{i=1}^{P_1}$. Then $\{\mathbf{z}_{t-1}^i\}_{i=1}^{P_1}$ is reconstructed to get $\mathcal{Z}_{t-1}$, where the overlapped parts take the average. Eventually, a higher-resolution image can be obtained by directly decoding $\mathcal{Z}_0$ into image $\mathbf{X}_0$. Based on MultiDiffusion, DemoFusion [4] additionally introduces: 1) progressive upscaling to gradually generate higher-resolution images; 2) residual connection to maintain global consistency with the lower-resolution image by injecting the intermediate noise-inversed representation. 3) dilated sampling to enhance global semantic information of higher-resolution images.

3.2 In-Depth Analysis of Small Object Repetition

DemoFusion demonstrates the possibility of using pre-trained LDMs to generate higher-resolution images. However, as shown in Fig. 1(e), small object repetition continues to challenge the performance of DemoFusion.

Delving into an in-depth analysis, we respectively: 1) remove the text prompt during higher-resolution generation of DemoFusion and the resulting Fig. 3(a) indicates the disappearance of repeated objects but more degradation in details. 2) remove the operations of residual connection & dilated sampling in DemoFusion and the resulting Fig. 3(b) denotes severe large object repetition. Therefore, we can make a safe conclusion that small object repetition is the adversarial outcome of an identical text prompt on all patches and operations of residual connection & dilated sampling. The former encourages to generate repeated objects while the latter suppresses the generation of repeated objects. Consequently, DemoFusion tends to generate small repeated objects.

Overall, text prompts play a significant role in image generation. It is not a viable solution to address small object repetition by removing text prompts during the higher-resolution generation, as it would lead to a decline in image quality. Instead, we require more accurate prompts specifically tailored for each

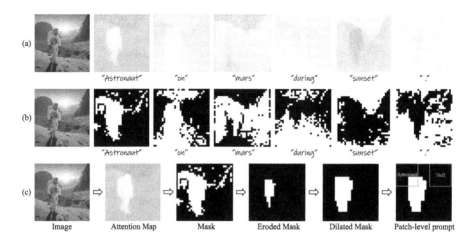

Fig. 4. Visualization of averaged attention map from the up blocks and down blocks in U-Net. We reshape the attention map into a 2D shape before visualization. (a) Cross-attention map visualization using open source code [7]. (b) Highly responsive regions of each word. (c) The illustration of the patch-level prompt generation process, including morphological operations to eliminate small connected areas. Here we use the word "Astronaut" as an example. All words in the prompt will go through the above process.

patch. That is, if an object is not present in a patch, the corresponding word in the text prompts should not serve as a prompt for that patch.

To this end, in Sect. 3.3, we eliminate the restriction of having an identical text prompt for all patches in previous patch-wise generation approaches. Instead, we generate more precise patch-content-aware prompts that adapt to the content of different patches. In Sect. 3.4, we introduce how to enhance the global structure information to generate higher-resolution images without repetition.

3.3 Patch-Content-Aware Prompts

Considering the significance of text prompt in higher-resolution generation, we explore patch-content-aware substring set $\{\gamma^i\}_{i=1}^{P_1}$ of the entire text prompt, each of which is responsible for injecting a condition to the corresponding patch. In general, it is challenging to know in advance what content a patch generates, but in DemoFusion [4], the global information from low-resolution image is injected into the high-resolution image generation through residual connections. Therefore, the structure of the generated higher-resolution image is similar to that of the low-resolution image. This inspires us to decide patch contents from the low-resolution image. A direct but cumbersome approach is to manually observe the patch content of low-resolution image and then set the prompt for each patch, which undermines the usability of stable diffusion. Another approach is to use SAM [13] to segment the upscaled low-resolution image and determine whether each object appears in the patch, introducing huge storage and computational costs of the segmentation model. How to automatically generate patch-content-aware prompts without external models is the key to success.

Inspired by image editing [7], instead we consider the cross-attention maps in low-resolution generation $\mathcal{M} \in \mathbb{R}^{N \times M}$, to determine patch-content-aware prompts. Recall N represents the pixel number of the latent noise z_t and M denotes the number of word tokens in the prompt y. Thus, the column $\mathcal{M}_{:,j}$ represents the attentiveness of latent noise to the j-th word token. The basic principle lies in that the attentiveness ($\mathcal{M}_{i,j}$) of image regions is mostly higher than others if it is attended by the j-th word token, as shown in Fig. 4(a). To find the highly relevant region of each word token, we convert the attention map $\mathcal{M}$ into a binary mask $\mathcal{B} \in \mathbb{R}^{N \times M}$ as:

$$\mathcal{B}_{i,j} = \begin{cases} 1, & \text{if } \mathcal{M}_{i,j} > \overline{\mathcal{M}_{:,j}}, \\ 0, & \text{otherwise,} \end{cases} \quad (5)$$

where i and j enumerate N and M, respectively. The threshold $\overline{\mathcal{M}_{:,j}}$ is the mean of $\mathcal{M}_{:,j}$, which design is elaborated in Sect. 4.4. Regions with values above the threshold are considered highly responsive, while regions with values below the threshold are considered less responsive.

Next, we obtain word-level masks $\{\mathcal{B}_j\}_{j=1}^{M}$ using the following equation:

$$\hat{\mathcal{B}}_j = \text{Reshape}(\mathcal{B}_{:,j}, (h_a, w_a)), \quad (6)$$

where $h_a = \frac{h}{s}$ and $w_a = \frac{w}{s}$ represent the height and width of the attention map, respectively. Recall h and w represent the height and width of the noise, respectively. The "s" corresponds to the down-sampling scale in the corresponding block of the U-Net model. The mask $\mathcal{B}_j$ oriented for the j-th word token is reshaped into a 2d shape for further processing.

After obtaining the highly responsive regions for each word, we observe that they contain many small connected areas, as shown in Fig. 4(b). To alleviate the influence of these small connected areas, we apply the opening operation $\mathcal{O}(\cdot)$ from mathematical morphology [27], resulting in the final mask for each word, as shown in Fig. 4(c). The processed mask $\{\tilde{\mathcal{B}}_j\}_{j=1}^{M}$ can be formulated as:

$$\tilde{\mathcal{B}}_j = \mathcal{O}(\hat{\mathcal{B}}_j) = \omega(\delta(\hat{\mathcal{B}}_j)), \quad (7)$$

where $\delta(\cdot)$ and $\omega(\cdot)$ is erosion operation and dilation operation, respectively. Next, we interpolate $\tilde{\mathcal{B}}_j \in \mathbb{R}^{h_a \times w_a}$ to $\tilde{\mathcal{B}}'_j \in \mathbb{R}^{h'_a \times w'_a}$, where $h'_a = \frac{h'}{s}$ and $w'_a = \frac{w'}{s}$. Recall h' and w' are the size of higher-resolution latent representation as defined in Sect. 3.1. Inspired by MultiDiffusion [1], we use a shifted window to sample patches from $\tilde{\mathcal{B}}'_j$, resulting in a series of patch masks $\{\{\mathbf{m}_j^i\}_{i=1}^{P_1}\}_{j=1}^{M}$, where $\mathbf{m}_j^i \in \mathbb{R}^{h_a \times w_a}$ and P_1 is the total number of patches. It is important to note that each $\mathbf{m}_i^j$ corresponds to a specific patch noise $\mathbf{z}_t^i$.

Recall if an object is not present in a patch, the corresponding word token in the text prompts should not serve as a prompt for that patch. So, we can

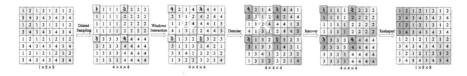

Fig. 5. Illustration of dilated sampling with window interaction: 8×8 higher-resolution and 4×4 low-resolution. The number $\{1, 2, 3, 4\}$ represent the different positions within the same window (same color). The interaction operation is conducted in the window.

determine the patch-content-aware prompt γ^i, a sub-sequence of prompt y, for each patch z_t^i using the following formulation:

$$\begin{cases} y_j \in \gamma^i, & \text{if } \frac{\sum (\mathbf{m}_j^i)_{:,:}}{h_a \times w_a} > c, \\ y_j \notin \gamma^i, & \text{otherwise,} \end{cases} \quad (8)$$

where j and i enumerates M and P_1, respectively. The pre-given hyper-parameter $c \in (0, 1)$ determines whether a highly responsive region's proportion of a word y_j exceeds the threshold for inclusion in the prompts of patch z_t^i. We then concatenate all words that should appear in a patch together, resulting in patch-content-aware prompts $\{\gamma^i\}_{i=1}^{P_1}$ for noise patches $\{z_t^i\}_{i=1}^{P_1}$ during patch-wise denoising.

3.4 Dilated Sampling with Window Interaction

Recall $\mathcal{Z}_t \in \mathbb{R}^{h' \times w' \times c}$ stands for the latent representation of a higher-resolution image in Sect. 3.1. In this section, we continue proposing dilated sampling with window interaction, for a set of patch samples $\{\mathcal{D}_t^k\}_{k=1}^{P_2}$, to improve the global semantic information in the latent representation $\mathcal{Z}_t$. In DemoFusion [4], each sample $\mathcal{D}_t^k$ is a subset of the latent representation $\mathcal{Z}_t$, formulated as:

$$\mathcal{D}_t^k = (\mathcal{Z}_t)_{i::h_s, j::w_s, :}, \quad (9)$$

where $k = i \times w_s + j + 1$, and k ranges from 1 to P_2. The variables i and j range from 0 to $h_s - 1$ and $w_s - 1$, respectively. The sampling stride is determined by $h_s = \frac{h'}{h}$ and $w_s = \frac{w'}{w}$. Recall $\{h', w'\}$ and $\{h, w\}$ are the height and width of higher and low resolution latent representation. DemoFusion independently performs denoising on $\mathcal{D}_t$ via Eq. (4) and obtains $\mathcal{D}_{t-1} \in \mathbb{R}^{P_2 \times h \times w \times c}$, where $P_2 = h_s \times w_s$. Then $\{\mathcal{D}_{t-1}^k\}_{k=1}^{P_2}$ is reconstructed as $G_{t-1} \in \mathbb{R}^{h' \times w' \times c}$ and added to patch-wise denoised latent representation $\mathcal{Z}_{t-1}$ using:

$$\hat{\mathcal{Z}}_{t-1} = (1 - \eta) \cdot \mathcal{Z}_{t-1} + \eta \cdot G_{t-1}, \quad (10)$$

where $(G_{t-1})_{i::h_s, j::w_s, :} = \mathcal{D}_{t-1}^k$ and η decreases from 1 to 0 using a cosine schedule. Due to the lack of interaction between different samples during the denoising process, the global semantic information is non-smooth, as depicted in Fig. 3(c). The sharp global semantic information disturbs the higher-resolution generation.

To solve above issue, as illustrated in Fig. 5, we enable window interaction between different samples before each denoising process through bijective function:

$$\mathcal{D}_t^{k,h,w} = \mathcal{D}_t^{f_t^{h,w}(k),h,w}, \quad f_t^{h,w} : \{1,2,\cdots,P_2\} \Rightarrow \{1,2,\cdots,P_2\}, \quad (11)$$

where $f_t^{h,w}$ is bijective function, and the mapping varies based on the position or time step. We then perform normal denoising progress on $\{\mathcal{D}_t^k\}_{k=1}^{P_2}$ to obtain $\{\mathcal{D}_{t-1}^k\}_{k=1}^{P_2}$. Before applying Eq. (10) to $\{\mathcal{D}_{t-1}^k\}_{k=1}^{P_2}$, we use the inverse mapping $(f_t^{h,w})^{-1}$ of $f_t^{h,w}$ to recover the position as:

$$\mathcal{D}_{t-1}^{k,h,w} = \mathcal{D}_{t-1}^{(f_t^{h,w})^{-1}(k),h,w}, \quad (f_t^{h,w})^{-1} : \{1,2,\cdots,P_2\} \Rightarrow \{1,2,\cdots,P_2\}. \quad (12)$$

4 Experimentation

4.1 Experimental Setup

AccDiffusion is a plug-and-play extension to stable diffusion without additional training costs. We mainly validate the feasibility of AccDiffusion using the pre-trained SDXL [17]. More results for other stable diffusion variants are in the supplementary material. AccDiffusion follows the pipeline of DemoFusion [4] (SOTA higher-resolution generation) and uses the patch-content-aware prompts during the progress of higher-resolution image generation. Additionally, AccDiffusion enhances dilated sampling with window interaction. For fairness, we adhere to the default setting of DemoFusion, as described in the supplementary material. In Sect. 4.2, the hyper-parameter c in Eq. (8) is set to 0.3. Considering the training-free nature of AccDiffusion, the methods we compare include: **SDXL-DI** [17], **Attn-SF** [12], **ScaleCrafter** [6], **MultiDiffusion** [1], and **DemoFusion** [4]. We do not compare with image super-resolution methods [23,29,32] which take images as input and have been proven to lack texture details [4,6].

4.2 Quantitative Comparison

For quantitative comparison, we use three widely-recognized metrics: FID (Frechet Inception Distance) [8], IS (Inception Score) [24], and CLIP Score [19]. Specifically, FID_r measures the Frechet Inception Distance between generated high-resolution images and real images. IS_r represents the Inception Score of generated high-resolution images. Given that FID_r and IS_r necessitate resizing images to 299^2, which may not be ideal for assessing high-resolution images. Motivated by [2,4], we crop 10 local patches at 1× resolution from each generated high-resolution image and subsequently resize them to calculate FID_c and IS_c. The CLIP score measures the cosine similarity between image embedding and text prompts. We randomly selected 10,000 images from the Laion-5B [25] dataset as our real images set and randomly chose 1,000 text prompts from Laion-5B as inputs for AccDiffusion to generate a set of high-resolution images.

Table 1. Comparison of quantitative metrics between different training-free image generation extrapolation methods. We use **bold** to emphasize the best result and underline to emphasize the second best result.

Resolusion	Method	FID$_r$ ↓	IS$_r$ ↑	FID$_c$ ↓	IS$_c$ ↑	CLIP ↑	Time
1024 × 1024 (1×)	SDXL-DI	58.49	17.39	58.08	25.38	33.07	<1 min
2048 × 2048 (4×)	SDXL-DI	124.40	11.05	88.33	14.64	28.11	1 min
	Attn-SF	124.15	11.15	88.59	14.81	28.12	1 min
	MultiDiffusion	81.46	12.43	44.80	20.99	31.82	2 min
	ScaleCrafter	99.47	12.52	74.64	15.42	28.82	1 min
	DemoFusion	<u>60.46</u>	<u>16.45</u>	<u>38.55</u>	<u>24.17</u>	<u>32.21</u>	3 min
	AccDiffusion	**59.63**	**16.48**	**38.36**	**24.62**	**32.79**	3 min
3072 × 3072 (9×)	SDXL-DI	170.61	7.83	112.51	12.59	24.53	3 min
	Attn-SF	170.62	7.93	112.46	12.52	24.56	3 min
	MultiDiffusion	101.11	8.83	51.95	17.74	29.49	6 min
	ScaleCrafter	131.42	9.62	105.79	11.91	27.22	7 min
	DemoFusion	<u>62.43</u>	<u>16.41</u>	<u>47.45</u>	<u>20.42</u>	<u>32.25</u>	11 min
	AccDiffusion	**61.40**	**17.02**	**46.46**	**20.77**	**32.82**	11 min
4096 × 4096 (16×)	SDXL-DI	202.93	6.13	119.54	11.32	23.06	9 min
	Attn-SF	203.08	6.26	119.68	11.66	23.10	9 min
	MultiDiffusion	131.39	6.56	61.45	13.75	26.97	10 min
	ScaleCrafter	139.18	9.35	116.90	9.85	26.50	20 min
	DemoFusion	<u>65.97</u>	<u>15.67</u>	<u>59.94</u>	<u>16.60</u>	<u>33.21</u>	25 min
	AccDiffusion	**63.89**	**16.05**	**58.51**	**16.72**	**33.79**	26 min

As shown in Table 1, AccDiffusion achieves the best results and obtains state-of-the-art performance. Since the implementation of AccDiffusion is based on DemoFusion [4], it exhibits similar quantitative results and inference times with DemoFusion. However, AccDiffusion outperforms DemoFusion due to its more precise patch-content-aware prompt and more accurate global information introduced by dilated sampling with interaction, especially in high-resolution generation scenarios. Compared to other training-free image generation extrapolation methods, the quantitative results of AccDiffusion are closer to quantitative results calculated at pre-trained resolutions (1024 × 1024), demonstrating the excellent image generation extrapolation capabilities of AccDiffusion. Note that FID, IS, and CLIP-Score do not intuitively reflect the degree of repeated generation in the resulting images, so we conduct a qualitative comparison to validate the effectiveness of AccDiffusion in eliminating repeated generation.

4.3 Qualitative Comparison

In Fig. 6, AccDiffusion is compared with other training-free text-to-image generation extrapolation methods, such as MultiDiffusion [1], ScaleCrafter [6], and

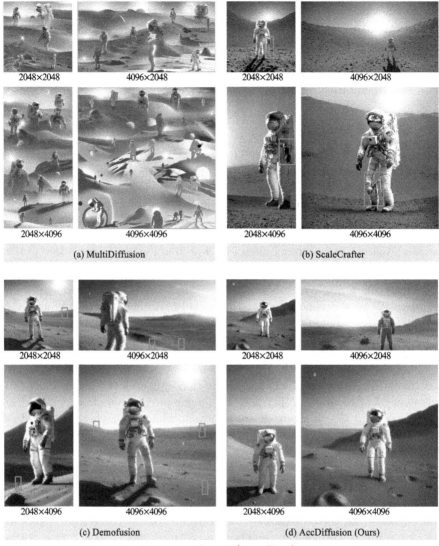

Fig. 6. Qualitative comparison of our AccDiffusion with existing training-free image generation extrapolation methods [1,4,6]. We draw a red box upon the generated images to highlight the repeated objects. Best viewed zoomed in. (Color figure online)

DemoFusion [4]. We provide more results in supplementary material. MultiDiffusion can generate seamless images but suffers severe repeated and distorted generation. ScaleCrafter, while avoiding the repetition of astronauts, suffers from structural distortions as highlighted in the red box, resulting in local repetition and a lack of coherence. DemoFusion tends to generate small, repeated astronauts, with the frequency of repetition escalating with image resolution, thereby significantly degrading image quality. Conversely, AccDiffusion demonstrates superior performance in generating high-resolution images without such repetitions. As Attn-SF [12] and SDXL-DI [17] cannot alleviate the repetition issue, their qualitative results are not compared here.

4.4 Ablation Study

In this section, we first perform ablation studies on the two core modules proposed in this paper, and then discuss the settings of the threshold for the binary mask in Eq. (5) and the threshold c for deciding patch-content-aware prompt in Eq. (8). All experiments are carried out at a resolution of 4096^2 (16×). Considering the fact that existing quantitative metrics are unable to accurately reflect the extent of object repetition, we choose to provide visualizations to demonstrate the effectiveness of our core modules in preventing repeated generation.

Ablations on Core Modules. As illustrated in Fig. 7, the absence of any module leads to a decline in generation quality. Without patch-content-aware prompts, the resulting image contains numerous repeated small objects, emphasizing the importance of patch-content-aware prompts in preventing the generation of repetitive elements. Conversely, without our window interaction in dilated sampling, the generated small object becomes unrelated to the image, indicating that dilated sampling with window interaction enhances the image's semantic consistency and suppresses repetition. The maximum number of repeated objects is produced when both modules are removed, while employing both modules simultaneously generates an image free of repetitions. This implies that the two modules work together to effectively alleviate repetitive objects.

Ablations on Hyper-Parameters. As depicted in Table 2, there is a significant variation in the range of different cross-attention maps $\mathcal{M}_j$. When using a fixed threshold for these maps, two scenarios may occur. If the threshold is too high, some words will not have highly responsive regions in their corresponding attention maps, resulting in their absence from the patch-content-aware prompt. Conversely, if the threshold is too low, the entire attention map consists of highly responsive regions, causing those words to consistently appear in the patch-content-aware prompt. By considering the average $\overline{\mathcal{M}}_{:,j}$, we can ensure that each word has suitable highly responsive regions, as demonstrated in Fig. 4(b).

Recall in Eq. (8), the c determines whether the proportion of a highly responsive region for a word y_j surpasses the threshold required for inclusion in the prompts of patch z_t^i. A very small value of c leads to more words being included in the patch prompt, potentially causing object repetition. Conversely, a very large value of c simplifies the patch prompt, which may lead to degradation of details. Our analysis is demonstrated in Fig. 8. It should be noted that this is a user-specific hyper-parameter, adjustable to suit different application scenarios.

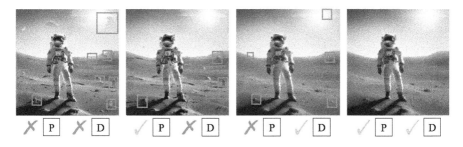

Fig. 7. Ablations of Patch-content-aware prompts (P) and Dilated sampling with window interaction (D). The "✗"/"✓" denotes removing/preserving the component. The repeated objects are highlighted by a red box. Best viewed zoomed in.

Table 2. Statistics of cross-attention maps $\mathcal{M}$ using prompt $y =$ "Astronaut on mars during sunset." as an example. Each word $\{y_j\}_{j=1}^{6}$ has a cross-attention map $\{\mathcal{M}_{:,j}\}_{j=1}^{6}$.

Statistics	"Astronaut" ($j=1$)	"on" ($j=2$)	"mars" ($j=3$)	"during" ($j=4$)	"sunset" ($j=5$)	"." ($j=6$)
Min($\mathcal{M}_{:,j}$)	0.1274	0.0597	0.2039	0.0457	0.0921	0.0335
Mean($\mathcal{M}_{:,j}$)	0.1499	0.0676	0.2533	0.0521	0.1189	0.0386
Max($\mathcal{M}_{:,j}$)	0.2096	0.0779	0.2979	0.0585	0.1499	0.0419

Fig. 8. Visual results of different threshold c, prompted by "A cute corgi on the lawn." The repeated objects are highlighted with a red box and the detail degradation is stressed with a blue box. Best viewed zoomed in. (Color figure online)

5 Limitations and Future Work

AccDiffusion is limited in: (1) As it follows the DemoFusion pipeline, similar drawbacks arise such as inference latency from progressive upscaling and overlapped patch-wise denoising. (2) AccDiffusion focuses on image generation extrapolation, meaning the fidelity of high-resolution images depends on the pre-trained diffusion model. (3) Relying on LDMs' prior knowledge of cropped images, it may produce local irrational content in sharp close-up image generation.

Future studies could explore the possibility of developing non-overlapped patch-wise denoising techniques for efficiently generating high-resolution images.

6 Conclusion

In this paper, we propose AccDiffusion to address the object-repeated generation issue in higher-resolution image generation without training. AccDiffusion first introduces patch-content-aware prompts, which makes the patch-wise denoising more accurate and can avoid repeated generation from the root. And then we further propose dilated sampling with window interaction to enhance the global consistency during higher-resolution image generation. Extensive experiments, including qualitative and quantitative results, show that AccDiffusion can successfully conduct higher-resolution image generation without object repetition.

Acknowledgements. This work was supported by National Science and Technology Major Project (No. 2022ZD0118202), the National Science Fund for Distinguished Young Scholars (No. 62025603), the National Natural Science Foundation of China (No. U21B2037, No. U22B2051, No. U23A20383, No. 62176222, No. 62176223, No. 62176226, No. 62072386, No. 62072387, No. 62072389, No. 62002305 and No. 62272401), and the Natural Science Foundation of Fujian Province of China (No. 2022J06001).

References

1. Bar-Tal, O., Yariv, L., Lipman, Y., Dekel, T.: Multidiffusion: fusing diffusion paths for controlled image generation. In: ICML (2023)
2. Chai, L., Gharbi, M., Shechtman, E., Isola, P., Zhang, R.: Any-resolution training for high-resolution image synthesis. In: ECCV, pp. 170–188 (2022)
3. Dhariwal, P., Nichol, A.: Diffusion models beat gans on image synthesis. In: NeurIPS, pp. 8780–8794 (2021)
4. Du, R., Chang, D., Hospedales, T., Song, Y.Z., Ma, Z.: Demofusion: democratising high-resolution image generation with no $$$. In: CVPR, pp. 6158–6168 (2024)
5. Ghosal, D., Majumder, N., Mehrish, A., Poria, S.: Text-to-audio generation using instruction-tuned llm and latent diffusion model. In: ACM MM (2023)
6. He, Y., et al.: Scalecrafter: tuning-free higher-resolution visual generation with diffusion models. In: ICLR (2024)
7. Hertz, A., Mokady, R., Tenenbaum, J., Aberman, K., Pritch, Y., Cohen-Or, D.: Prompt-to-prompt image editing with cross attention control. In: ICLR (2023)
8. Heusel, M., Ramsauer, H., Unterthiner, T., Nessler, B., Hochreiter, S.: Gans trained by a two time-scale update rule converge to a local nash equilibrium. In: NeurIPS (2017)
9. Ho, J., et al.: Imagen video: High definition video generation with diffusion models (2022). arXiv preprint arXiv:2210.02303
10. Ho, J., Jain, A., Abbeel, P.: Denoising diffusion probabilistic models. In: NeurIPS (2020)

11. Huang, R., et al.: Make-an-audio: text-to-audio generation with prompt-enhanced diffusion models. In: ICML (2023)
12. Jin, Z., Shen, X., Li, B., Xue, X.: Training-free diffusion model adaptation for variable-sized text-to-image synthesis. In: NeurIPS (2023)
13. Kirillov, A., et al.: Segment anything. In: ICCV (2023)
14. Lee, Y., Kim, K., Kim, H., Sung, M.: Syncdiffusion: coherent montage via synchronized joint diffusions. In: NeurIPS (2023)
15. Lin, C.H., et al.: Magic3d: high-resolution text-to-3d content creation. In: CVPR (2023)
16. Nichol, A.Q., Dhariwal, P.: Improved denoising diffusion probabilistic models. In: ICML (2021)
17. Podell, D., et al.: Sdxl: Improving latent diffusion models for high-resolution image synthesis. In: ICLR (2024)
18. Poole, B., Jain, A., Barron, J.T., Mildenhall, B.: Dreamfusion: text-to-3d using 2d diffusion. In: ICLR (2023)
19. Radford, A., et al.: Learning transferable visual models from natural language supervision. In: ICML (2021)
20. Robin Rombach, P.E.: Stable diffusion v1-5 model card. https://huggingface.co/runwayml/stable-diffusion-v1-5
21. Rombach, R., Blattmann, A., Lorenz, D., Esser, P., Ommer, B.: High-resolution image synthesis with latent diffusion models. In: CVPR (2022)
22. Saharia, C., et al.: Photorealistic text-to-image diffusion models with deep language understanding. In: NeurIPS (2022)
23. Saharia, C., Ho, J., Chan, W., Salimans, T., Fleet, D.J., Norouzi, M.: Image super-resolution via iterative refinement. TPAMI (2022)
24. Salimans, T., Goodfellow, I., Zaremba, W., Cheung, V., Radford, A., Chen, X.: Improved techniques for training gans. In: NeurIPS (2016)
25. Schuhmann, C., et al.: Laion-5b: an open large-scale dataset for training next generation image-text models. In: NeurIPS (2022)
26. Singer, U., et al.: Make-a-video: text-to-video generation without text-video data. In: ICLR (2023)
27. Soille, P., et al.: Morphological image analysis: principles and applications, vol. 2. Springer (1999)
28. Song, J., Meng, C., Ermon, S.: Denoising diffusion implicit models. In: ICLR (2021)
29. Wang, J., Yue, Z., Zhou, S., Chan, K.C., Loy, C.C.: Exploiting diffusion prior for real-world image super-resolution (2023). arXiv preprint arXiv:2305.07015
30. Xie, E., et al.: Difffit: unlocking transferability of large diffusion models via simple parameter-efficient fine-tuning. In: ICCV (2022)
31. Xu, J., et al.: Dream3d: zero-shot text-to-3d synthesis using 3d shape prior and text-to-image diffusion models. In: CVPR (2023)
32. Zhang, K., Liang, J., Van Gool, L., Timofte, R.: Designing a practical degradation model for deep blind image super-resolution. In: CVPR (2021)
33. Zheng, Q., et al.: Any-size-diffusion: toward efficient text-driven synthesis for any-size hd images. In: AAAI (2024)

Uncertainty-Driven Spectral Compressive Imaging with Spatial-Frequency Transformer

Lintao Peng, Siyu Xie, and Liheng Bian(✉)

Beijing Institute of Technology, Beijing 100081, China
bian@bit.edu.cn

Abstract. Recently, learning-based Hyperspectral image (HSI) reconstruction methods have demonstrated promising performance. However, existing learning-based methods still face two issues. 1) They rarely consider both the spatial sparsity and inter-spectral similarity priors of HSI. 2) They treat all image regions equally, ignoring that texture-rich and edge regions are more difficult to reconstruct than smooth regions. To address these issues, we propose an uncertainty-driven HSI reconstruction method termed Specformer. Specifically, we first introduce a frequency-wise self-attention (FWSA) module, and combine it with a spatial-wise local-window self-attention (LWSA) module in parallel to form a Spatial-Frequency (SF) block. LWSA can guide the network to focus on the regions with dense spectral information, and FWSA can capture the inter-spectral similarity. Parallel design helps the network to model cross-window connections, and expand its receptive fields while maintaining linear complexity. We use SF-block as the main building block in a multi-scale U-shape network to form our Specformer. In addition, we introduce an uncertainty-driven loss function, which can reinforce the network's attention to the challenging regions with rich textures and edges. Experiments on simulated and real HSI datasets show that our Specformer outperforms state-of-the-art methods with lower computational and memory costs. The code is available at https://github.com/bianlab/Specformer.

Keywords: Hyperspectral Imaging · Spatial-Frequency Transformer · Uncertainty-Driven Learning

1 Introduction

Compared with RGB images, hyperspectral images (HSI) have more spectral bands, which makes them able to store richer spectral information and delineate more detailed characteristics of the target scene. Benefiting from this property,

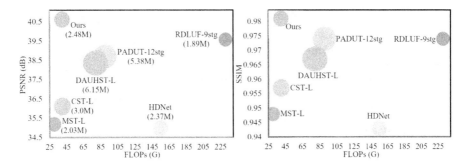

Fig. 1. PSNR-Params-FLOPs and SSIM-Params-FLOPs comparisons with exsiting HSI reconstruction methods. The circle radius represents the parameter numbers of the model. The reported Specformer technique outperforms state-of-the-art methods while requiring fewer FLOPs and Params.

HSIs have been widely used in multiple computer vision tasks, such as object detection [22,40], remote sensing [1,33] and medical image processing [31,36]. Conventional hyperspectral imaging systems scan the scene along the spatial or spectral dimension to obtain the HSI cubes. This scanning strategy is time-consuming, making it unsuitable for capturing and measuring dynamic scenes. Compared with traditional push-broom scanning spectral imaging techniques, snapshot compressive imaging (SCI) technology only requires one time of exposure to obtain a complete HSI cube. These SCI systems compress both spatial and spectral information into a single 2D measurement [52], and then use algorithms to reconstruct an HSI data cube [6,13,30,47,48]. The Coded Aperture Snapshot Spectral Imaging (CASSI) [47] technique stands out as a representative SCI technique.

In recent years, researchers have proposed multiple reconstruction algorithms to reconstruct 3D HSI cubes from 2D measurements. Traditional model-based methods use handcrafted priors such as sparsity [23,27,47], total variability [49,51], and non-local similarity [28,50,53] to guide the reconstruction process. However, these methods rely on manually tuned parameters and often require different parameters for different scenes, which leads to poor generality and slow reconstruction speed. Compared to model-based methods, CNN-based HSI reconstruction methods [20,34,35,37,38] do not require manual parameter tuning, and produce improvements on generalization performance and reconstruction speed. However, CNN-based methods show limitations in capturing non-local self-similarity and long-range dependencies of HSI, which results in unsatisfactory HSI reconstruction quality.

Recently, Transformer [46] has shown promising performance in image processing. The self-attention mechanism in Transformer can model long-range dependencies and non-local similarities. These advantages offer the possibility to address the shortcomings of CNN-based methods. Based on the Transformer technique, researchers have reported performance-leading HSI reconstruction methods such as MST [3], CST [2], DAUHST [4], and PADUT [25]. However, the existing Transformer-based methods still face the following issues. **First**, in the global attention Transformer, the computational complexity is quadratic to the spatial size. This burden is non-trivial and sometimes unaffordable. The local-window self-attention (LWSA) [29] module can effectively reduce computational complexity, but the receptive field of LWSA module is quite limited. **Second**, Transformer-based methods often ignore the spatial sparsity of HSI. These methods intensively collect all tokens and calculate global self-attention, causing a lot of computing resources wasting in areas with sparse spectral information. **Third**, the original Transformer method [12] learns to capture the long-range dependencies spatially, but the representations of HSIs are spectrally highly self-similar. In this case, the inter-spectral similarities are not well modeled.

To address these issues, we propose an uncertainty-driven HSI reconstruction method termed Specformer. Specifically, inspired by the spatial sparsity and inter-spectral similarity nature of HSIs, we first introduce frequency-wise self-attention (FWSA), a conceptually simple but computationally efficient architecture. It consists of the fast Fourier transform (FFT) [39], the learnable global filter, and the inverse fast Fourier transform (IFFT) [45]. FWSA module can calculate self-attention along the spectral dimension with linear complexity, modeling the inter-spectral long-distance dependencies and capturing the inter-spectral similarity of HSI. Next, we use a parallel design to combine it with the spatial-wise LWSA module to form a basic block termed spatial-frequency (SF) block. LWSA module can model the spatial sparsity and guide the network to focus on the spatial regions with dense spectral information. Parallel design helps the SF-block to model cross-window connections, and expand its receptive field while maintaining linear complexity. We insert the SF-block as the main building block in a U-shape architecture [44] to form our Specformer. In addition, considering that texture-rich and edge regions in HSIs are more difficult to reconstruct, we introduce an uncertainty-driven self-adaptive loss function to reinforce the network's attention on those regions, thus improving the HSI reconstruction quality. The contribution of this paper is summarised as follows,

- We propose a novel Specformer technique for HSI reconstruction. To the best of our knowledge, it is the first attempt to embed both the spatial sparsity and inter-spectral similarity of HSIs into learning-based reconstruction.
- We introduce a novel self-attention module termed FWSA, and combine it with the LWSA module in a parallel design to form an SF-block. It can model the spatial sparsity and inter-spectral similarity of HSIs.
- We introduce an uncertainty-driven self-adaptive loss to enhance the HSI reconstruction quality in texture-rich and edge regions.
- Our Specformer outperforms state-of-the-art methods in both quantitative evaluation and visual comparison, with lower computational and memory costs.

2 Related Work

2.1 HSI Reconstruction

Traditional model-based methods [23,27,47,49,51] recover 3D HSI cubes from 2D measurements based on hand-crafted priors. For example, Ref. [15] solved the sparse HSI reconstruction problem using the gradient projection algorithm based on the spatial sparse prior of HSI. In Ref. [51], the nonlocal self-similarity and low-rank properties of HSIs have been exploited to solve HSI reconstruction problems. These model-based methods rely on hand-crafted parameters, and are difficult to adapt to different scenes, leading to poor generalization ability and slow recovery speed.

Compared with model-based methods, CNN-based methods show improvements in generalization performance and reconstruction speed. The CNN-based techniques can be categorized into end-to-end (E2E) methods, deep unfolding methods, and plag-and-paly (PnP) methods. The E2E methods [16,35,38] aim to learn a mapping function from 2D measurements to 3D HSI cubes. The deep unfolding methods [11,20,32,34] employ multi-stage CNNs trained to map the measurements into the desired signal. Each stage contains two parts, i,e , linear projection and passing the signal through a CNN functioning as a denoiser. The PnP methods [7,43] insert the pre-trained CNN denoiser into a model-based optimization framework for HSI reconstruction. Despite of the developments, CNN-based methods still have limitations in capturing long-distance dependencies and modeling non-local similarities.

Recently, the Global Vision Transformer has achieved great success in image classification [12]. However, for the dense image processing tasks such as HSI reconstruction, the computational complexity of the Transformer technique is quadratic with the image size, making it unable to be directly applied for HSI reconstruction. Moreover, the existing Transformer-based methods [5,19] generally ignore the spatial sparsity of HSI, and intensively collect all tokens and calculate global self-attention, causing a lot of computing resources waste in the areas with sparse spectral information. The MST [3] method circumvents the computational complexity and spatial sparsity problems by calculating self-attention in the spectral domain, but this makes it unable to effectively model spatial-wise local and non-local features. The CST [2] technique embeds the HSI spatial sparsity into the learning process through a coarse-to-fine learning scheme, which effectively reduces the computational complexity of the Transformer, but ignores the inter-spectral similarity.

2.2 Uncertainty-Driven Loss

For the HSI reconstruction task, texture-rich and edge regions are more difficult to reconstruct compared to smooth regions, and the reconstruction quality of these regions is decisive for the final reconstruction quality. In previous HSI reconstruction methods [2,3,5], researchers tend to improve the reconstruction quality by designing deeper, larger, and more complex networks, where all pixels are still treated equally. However, treating each pixel equally during the training process is not the optimal choice for the HSI reconstruction task. Intuitively, we need a spatially self-adaptive loss function.

Recently, the uncertainty loss function [8,17,21] has attracted certain attention. The uncertainty in deep learning can be roughly divided into two categories [10]. Epistemic/model uncertainty describes how much the model is uncertain about its predictions. Another type is aleatoric/data uncertainty which refers to noise inherent in observation data. The GRAM [24] technique analyses the effect of aleatoric/data uncertainty on image reconstruction. By decreasing the loss attenuation of large variance pixels, GRAM achieves better results than directly applying the above uncertainty loss to image enhancement. In these tasks, pixels with a high degree of uncertainty are considered unreliable pixels that will suffer loss attenuation. However, this contradicts the intuition that the regions with rich textures and edges should be given priority in the HSI reconstruction task. In this regard, different from the above methods, we propose a novel uncertainty-driven spatially self-adaptive loss function, which can assign larger training weights to the texture-rich and edge regions of HSI, thus improving the HSI reconstruction performance.

2.3 CASSI Model

CASSI [47] is a mature and widely used spectral imaging technology. All experiments in this paper are based on CASSI. Figure 2(c) shows the principle of a single-dispenser CASSI. We denote the 3D HSI cube as $\mathbf{F} \in \mathbb{R}^{H \times W \times N_\lambda}$, where H, W, and N_λ refer to the HSI's height, width, and number of wavelengths, respectively. $\mathbf{F}$ is first collected by the objective lens and spatially encoded along the channel dimension by a coded aperture $\mathbf{M}^* \in \mathbb{R}^{H \times W}$, which is denoted as

$$\mathbf{F}'(:,:,n_\lambda) = \mathbf{F}(:,:,n_\lambda) \odot \mathbf{M}^*. \tag{1}$$

Among them, $\mathbf{F}'$ represents the signal modulated by the coded aperture, $n_\lambda \in [1,\ldots,N_\lambda]$ represents different spectral wavelengths, and $\odot$ represents element-wise multiplication. The $\mathbf{F}'$ passes through the disperser and becomes tilted, which can be considered as sheared along the y-axis. Assuming λ_c is the reference wavelength, then the dispersion can be formulated as

$$\mathbf{F}''(u,v,n_\lambda) = \mathbf{F}'(x, y + d(\lambda_n - \lambda_c), n_\lambda), \tag{2}$$

where $\mathbf{F}'' \in R^{H \times (W+d(N_\lambda-1)) \times N_\lambda}$ is the signal after dispersion, d refers to the step of spatial shifting, (u,v) locates the coordinate on the sensing detector, λ_n represents the wavelength of the n_λ-th channel, and $d(\lambda_n - \lambda_c)$ refers to the spatial shifting offset of the n_λ-th channel on $\mathbf{F}''$. Eventually, the data cube is compressed into a 2D measurement $\mathbf{Y} \in \mathbb{R}^{H \times (W+d(N_\lambda-1))}$ by integrating all the channels as

$$\mathbf{Y} = \sum_{n_\lambda=1}^{N_\lambda} \mathbf{F}''(:,:,n_\lambda) + \mathbf{G}, \tag{3}$$

where $\mathbf{G} \in \mathbb{R}^{H \times (W + d(N_\lambda - 1))}$ represents the random noise during the imaging process. The core task of HSI reconstruction is to recover the 3D HSI cube $\mathbf{F}$ from the 2D measurement $\mathbf{Y}$.

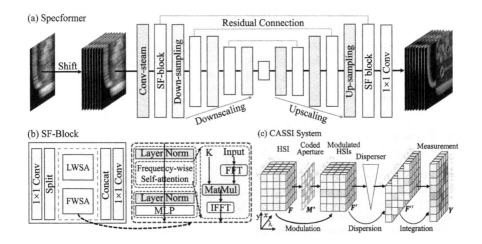

Fig. 2. (a) The overall structure of Specformer. (b) Detailed structure of the Spatial-Frequency (SF) block. We use a parallel design to combine the local-window self-attention (LWSA) module with the frequency-wise self-attention (FWSA) module to form the SF-block. LWSA can guide the network to focus on the regions with dense spectral information. FWSA can capture the inter-spectral similarity. It consists of the fast Fourier transform (FFT) [39], the learnable global filter $\mathbf{K}$ and the inverse fast Fourier transform (IFFT) [45]. Parallel design helps the specformer to model cross-window connections, and enlarge the receptive fields while maintaining linear complexity. (c) A schematic diagram of the CASSI model.

3 Method

3.1 Overall Reconstruction Architecture

The overall architecture of the reported Specformer technique is shown in Fig. 2(a), which consists of an encoder and a decoder built based on the SF-block. First, we reverse the dispersion process and shift back the measurement to obtain the initialized signal $\mathbf{H} \in \mathbb{R}^{H \times W \times N_\lambda}$ as

$$\mathbf{H}(\mathrm{x}, \mathrm{y}, n_\lambda) = \mathbf{Y}(\mathrm{x}, \mathrm{y} - \mathrm{d}(\lambda_n - \lambda_c)). \tag{4}$$

Next, we send $\mathbf{H}$ into Specformer. In Specformer, the input feature $\mathbf{H}$ is first processed by the conv-steam layer to convert the number of channels into C (we set C to be 32 in this work), and then obtain the preprocessing feature $\mathbf{X_0} \in \mathbb{R}^{H \times W \times C}$. Then, $\mathbf{X_0}$ is fed into the Specformer's encoder. There are four scales of encoding modules in the encoder, each of which contains several SF-blocks and a downsampling layer. Consequently, the output feature of the i-th stage of the encoder is denoted as $\mathbf{X}_i^e \in R^{\frac{H}{2^i} * \frac{W}{2^i} * 2^i C}$. Except for being input to the next encoding module, the output of the i-th encoding module is directly input to the decoding module with the same scale through residual connections. The decoder also contains four scales of decoding modules, each of which contains several SF-blocks and an upsampling layer. Similarly, the output of the i-th stage of the decoder is denoted as $\mathbf{X}_i^d \in R^{\frac{H}{2^i} * \frac{W}{2^i} * 2^i C}$. The final output features of the decoder are fed into a 1×1 convolutional layer to convert the number of channels into N_λ, and the final reconstruction result is $\mathbf{H}' \in R^{H * W * N_\lambda}$.

3.2 Spatial-Frequency (SF) Block

The detailed SF-block structure is shown in Fig. 2(b). Assuming that the inputs of SF-block are feature maps $\mathbf{X_{in}} \in R^{\frac{H}{2^i} * \frac{W}{2^i} * 2^i C}$ at different scales. To be specific, for an input feature map $\mathbf{X_{in}}$, it is first passed through a 1×1 convolution and split evenly into two feature maps $\mathbf{X_1}$ and $\mathbf{X_2}$ as

$$\mathbf{X_1}, \mathbf{X_2} = \text{Split}(\text{Conv1} \times 1(\mathbf{X_{in}})). \tag{5}$$

Next, $\mathbf{X_1}$ is fed into the LWSA module for further processing. In the LWSA module, the feature map $\mathbf{X_1}$ is linearly mapped to generate a one-dimensional feature sequence $\mathbf{S_1} \in R^{2^i C * d_i}(d = \frac{HW}{2^{2i}})$, which is then multiplied by the learnable weight matrices $W_Q \in R^{2^i C * d_i}$, $W_K \in R^{2^i C * d_i}$, and $W_V \in R^{2^i C * d_i}(d = \frac{HW}{2^{2i}})$ to generate $\mathbf{Q}$, $\mathbf{K}$ and $\mathbf{V}$, respectively, after layer normalization. The above calculation process is denoted as

$$\mathbf{Q} = \mathbf{S_1} W_Q; \mathbf{K} = \mathbf{S_1} W_K; \mathbf{V} = \mathbf{S_1} W_V. \tag{6}$$

After obtaining $\mathbf{Q}$, $\mathbf{K}$ and $\mathbf{V}$, we implement the following calculation based on the self-attentive mechanism as

$$\mathbf{Y}'_1 = \text{SoftMax}(\text{IN}(\frac{\mathbf{Q}^\mathbf{T}\mathbf{K}}{\sqrt[2]{2^i C}}))\mathbf{V}, \tag{7}$$

where IN represents the instance normalization. The 1D feature sequence $\mathbf{Y}'_1$ is then resized into 2D features $\mathbf{Y_1} \in R^{\frac{H}{2^i} * \frac{W}{2^i} * 2^i C}$ using feature remapping.

Similarly, $\mathbf{X_2}$ is sent to the FWSA module. As shown in Fig. 2(b), in the FWSA module, we propose to use a global learnable filter $\mathbf{K}$ as an alternative to the self-attention mechanism to interchange information globally among the Fourier domain tokens. For the input feature $\mathbf{X_2}$, we first perform layer normalization (LN), and then use 2D FFT [26] to convert $\mathbf{X_2}$ to the Fourier domain as

$$\mathbf{X_F} = \mathcal{F}(\text{LN}(\mathbf{X_2})), \tag{8}$$

where $\mathcal{F}(\cdot)$ denotes the 2D FFT operation. The output feature $\mathbf{X_F}$ represents the Fourier spectrum of $\mathbf{X_2}$. We can then modulate the spectrum by multiplying a filter $\mathbf{K} \in \mathbb{R}^{\frac{H}{2^i}*\frac{W}{2^i}*2^iC}$ to $\mathbf{X_F}$ as

$$\mathbf{Y_F} = \mathbf{K} \odot \mathbf{X_F}, \tag{9}$$

where $\odot$ is element-wise multiplication. The filter $\mathbf{K}$ is called the global filter since it has the same dimension as $\mathbf{X_F}$, which represents a learnable frequency filter for different hidden dimensions [42]. Finally, we adopt the inverse FFT (IFFT) operation [45] to transform the modulated spectrum $\mathbf{Y_F}$ back to the spatial domain and update the tokens as

$$\mathbf{Y_2} = \mathcal{F}^{-1}(\mathbf{Y_F}), \tag{10}$$

where $\mathcal{F}^{-1}(\cdot)$ denotes the 2D IFFT.

Before concatenating, $\mathbf{Y_1}$ and $\mathbf{Y_2}$ need to be processed by a feed-forward layer respectively, which consists of an LN layer and an MLP layer [46]. Finally, $\mathbf{Y_1}$ and $\mathbf{Y_2}$ are concatenated as the input of a 1×1 convolution which has a residual connection with the input $\mathbf{X_{in}}$. As such, the final output of SF-block is given by

$$\mathbf{X_{out}} = \mathrm{Conv1 \times 1}(\mathrm{Concat}(\mathbf{Y_1}, \mathbf{Y_2})) + \mathbf{X_{in}}. \tag{11}$$

3.3 Uncertainty-Driven Loss

To reinforce the network's attention on the texture-rich and edge regions, as shown in Fig. 3, we divide the training of the network into two stages. In the first stage, the network estimates both the HSI cube and the uncertainty map. In the second stage, the uncertainty values are used to generate a spatially adaptive loss to guide the network to prioritize the pixels in the regions with rich textures and edges. To better quantify the arbitrary uncertainty in HSI reconstruction, as shown in Fig. 3, we use x_i, y_i to denote the measurement and the corresponding ground truth, respectively. Let $f(.)$ denote an arbitrary HSI reconstruction network, and the aleatoric uncertainty is denoted by an additive term $\boldsymbol{\theta}_i$. The overall HSI reconstruction model can be formulated as

$$y_i = f(x_i) + \varepsilon \boldsymbol{\theta}_i, \tag{12}$$

where ε represents the Laplace distribution with zero-mean and unit-variance.

For a given input measurement x_i and corresponding HSI y_i, a Laplace distribution is assumed for characterizing the likelihood function as

$$\mathrm{p}(y_i, \boldsymbol{\theta}_i \mid x_i) = \frac{1}{2\boldsymbol{\theta}_i} \exp\left(-\frac{\|y_i - f(x_i)\|_1}{\boldsymbol{\theta}_i}\right), \tag{13}$$

where $f(x_i)$ denotes the reconstruction results, and $\boldsymbol{\theta}_i$ denotes the uncertainty (variance) which are learned by the network. Then, the log-likelihood can be formulated as

$$\ln \mathrm{p}(y_i, \boldsymbol{\theta}_i \mid x_i) = -\frac{\|y_i - f(x_i)\|_1}{\boldsymbol{\theta}_i} - \ln \boldsymbol{\theta}_i - \ln 2. \tag{14}$$

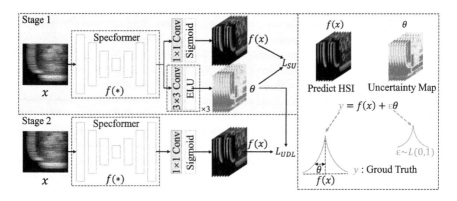

Fig. 3. The overview of the two-stage training strategy. The uncertainty estimation θ serves as the bridge connecting two steps, i.e., it is the output of the first stage, and is passed to the second stage as the guidance required for calculating Loss_{UDL}.

For numerical stability, we train the Specformer network to estimate the uncertainty variance $s_i = \ln(\theta_i)$ as shown in Fig. 3. Finally, the maximum likelihood estimate of Eq. 14 can be reformulated as minimizing the following loss function to estimate the uncertainty in the reconstruction process

$$\text{Loss}_U = \frac{1}{N}\sum_{i=1}^{N}\exp(-s_i)\|y_i - f(x_i)\|_1 + s_i. \tag{15}$$

The loss function Loss_U includes two terms. The first term is associated with reconstruction fidelity, and the second one prevents the network from predicting infinite uncertainty for all pixels. Based on the spatial sparsity nature of HSI and its uncertainty map, we propose to impose Jeffrey's prior[14] $p(w) \propto \frac{1}{w}$ on uncertainty θ_i as

$$p(y_i, \theta_i | x_i) = p(y_i | x_i, \theta_i)p(\theta_i) \propto \frac{1}{2\theta_i^2}\exp(-\frac{\|y_i - f(x_i)\|_1}{\theta_i}). \tag{16}$$

Then the log likelihood and loss function Loss_{SU} (SU represents sparse uncertainty) can be separately formulated as

$$\ln p(y_i | x_i) = -\frac{\|y_i - f(x_i)\|_1}{\theta_i} - 2\ln\theta_i - \ln 2, \tag{17}$$

$$\text{Loss}_{SU} = \frac{1}{N}\sum_{i=1}^{N}\exp(-s_i)\|y_i - f(x_i)\|_1 + 2s_i. \tag{18}$$

The above shows the loss function of the first training stage. When the Loss_{SU} converges, the trained Specformer network can be used to estimate the reconstruction uncertainty. Then, we construct a monotonically increasing function $\hat{s}_i = \ln(1 + e^{s_i})$ to prioritize the uncertainty values, and then use the ranked uncertainty as spatially adaptive weight to multiply with the HSI reconstruction loss function in the second stage as

$$\text{Loss}_{UDL} = \frac{1}{N} \sum_{i=1}^{N} \hat{s}_l \left(\text{Loss}_{L_1} + \text{Loss}_{SSIM} \right), \tag{19}$$

where Loss_{L_1} and Loss_{SSIM} denote the Mean Absolute Error (MAE) loss and Structure Similarity Index Measure (SSIM) loss [18], respectively. In Loss_{UDL}, the texture and edge pixels with higher uncertainty tend to have greater weights than smooth regions.

4 Experiments

4.1 Experiment Setup

Datasets. For simulation comparison, we employed the CAVE [41] and the KAIST [9] dataset. The CAVE dataset consists of 32 HSIs with a spatial size of 512×512 pixels. The KAIST dataset contains 30 HSIs of spatial size 2704×3376. Following the settings of DGSMP [20], 28 wavelengths from 450nm to 650nm were derived by spectral interpolation. The CAVE dataset was adopted as the training set, while 10 scenes from the KAIST dataset were selected for testing. For the real data experiment, five real HSIs collected in TSA-Net [35] were used for evaluation. Each testing sample has 28 channels, with a spatial size of 660×660 pixels.

Implementation Details. We implemented Specformer on ubuntu20 using the PyTorch framework, and trained it using the Adam optimization algorithm on NVIDIA RTX3090. During training, the batchsize was set to 4, and the Specformer was trained for a total of 600 epochs with a learning rate of 0.0001 (400 epochs for state 1, 200 epochs for stage 2). When conducting simulation comparison, patches at a spatial size of 256×256 cropped from the 3D cubes were fed into the networks. As for real HSI reconstruction, the patch size was set to 660×660 to match the real-world measurements. The shifting step d in dispersion is set to 2 pixels. Consequently, the measurement size was 256×310 and 660×714 for simulation and real data experiment, respectively.

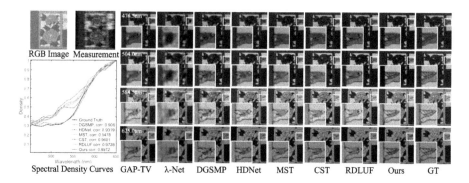

Fig. 4. Simulation reconstruction comparison of an exemplar Scene 7 with 4 out of 28 spectral channels. The results of 7 SOTA algorithms and our Specformer are presented. The spectral curves (bottom-left) correspond to the marked regions of the green box on the RGB image. (Color figure online)

Table 1. Quantitative reconstruction comparison on 10 scenes. We adopt PSNR and SSIM [18] as the metrics to evaluate the HSI reconstruction performance. The best results were marked in bold.

Methods	S1	S2	S3	S4	S5	S6	S7	S8	S9	S10	Avg
GAP-TV [34]	26.82	22.89	26.31	30.65	23.64	21.85	23.76	21.98	22.63	23.10	24.36
	0.754	0.610	0.802	0.852	0.703	0.663	0.688	0.655	0.682	0.584	0.669
DeSCI [28]	27.13	23.04	26.62	34.96	23.94	22.38	24.45	22.03	24.56	23.59	25.27
	0.748	0.620	0.818	0.897	0.706	0.683	0.743	0.673	0.732	0.587	0.721
λ-net [38]	30.10	28.49	27.73	37.01	26.19	28.64	26.47	26.09	27.50	27.13	28.53
	0.849	0.805	0.870	0.934	0.817	0.853	0.806	0.831	0.826	0.816	0.841
TSA-Net [35]	32.03	31.00	32.25	39.19	29.39	31.44	30.32	29.35	30.01	29.59	31.46
	0.892	0.858	0.915	0.953	0.884	0.908	0.878	0.888	0.890	0.874	0.894
DGSMP [20]	33.26	32.09	33.06	40.54	28.86	33.08	30.74	31.55	31.66	31.44	32.63
	0.915	0.898	0.925	0.964	0.882	0.937	0.886	0.923	0.911	0.925	0.917
HDNet [19]	35.14	35.67	36.03	42.30	32.69	34.46	33.67	32.48	34.89	32.38	34.97
	0.935	0.940	0.943	0.969	0.946	0.952	0.926	0.941	0.942	0.937	0.943
MST [3]	35.40	35.87	36.51	42.27	32.77	34.80	33.66	32.67	35.39	32.50	35.18
	0.941	0.944	0.953	0.973	0.947	0.955	0.925	0.948	0.949	0.941	0.948
CST [2]	35.96	36.85	38.16	42.44	33.25	35.72	34.86	34.34	36.51	33.09	36.12
	0.949	0.955	0.962	0.975	0.955	0.963	0.944	0.961	0.957	0.945	0.957
DAUHST [5]	37.25	39.02	41.05	46.15	35.80	37.08	37.57	35.10	40.02	34.59	38.36
	0.958	0.967	0.971	0.983	0.969	0.970	0.963	0.966	0.970	0.956	0.967
PADUT [25]	37.36	40.43	42.38	46.62	36.26	37.27	37.83	35.33	40.86	34.55	38.89
	0.962	0.978	0.979	0.990	0.974	0.974	0.966	0.974	0.978	0.963	0.974
RDLUF[11]	37.94	40.95	43.25	**47.83**	37.11	37.47	38.58	35.50	41.83	35.23	39.57
	0.966	0.977	0.979	**0.990**	0.976	0.975	0.969	0.970	0.978	0.962	0.974
Specformer	**38.82**	**41.93**	**43.98**	47.77	**38.78**	**38.61**	**39.91**	**36.72**	**42.82**	**36.73**	**40.61**
	0.973	**0.982**	**0.983**	0.989	**0.983**	**0.982**	**0.977**	**0.982**	**0.985**	**0.969**	**0.981**

4.2 Quantitative Results

We compared the HSI reconstruction performance of our Specformer with other 11 SOTA methods, including 2 model-based methods (GAP-TV [34], and DeSCI [28]), 3 CNN-based methods (λ-net [38], TSA-Net [35], and DGSMP [20]), and 6 recent Transformer-based methods (HDNet [19], MST [3], CST [2], DAUHST [5], PADUT [25] and RDLUF[11]). All the techniques were trained using the same datasets and evaluated under the same settings as DGSMP [20]. The quantitative reconstruction results on the 10 scenes of the KAIST dataset are presented in Table 1. Compared to DGSMP [20], MST [3], CST [2], DAUHST [5], and RDLUF[11], the reported Specformer method achieved PSNR improvements of 7.98 dB, 5.43dB, 4.49dB, 2.25dB and 1.04dB on average, respectively. Additionally, the reported method requires lower memory and computational costs as shown in Fig. 1. This demonstrates the effectiveness of simultaneously embedding the spatial sparsity and inter-spectral similarity of HSI into the learning process. It also validates that the parallel design can help the SF-block to model cross-window connections, and enlarge the receptive fields while maintaining linear complexity.

4.3 Qualitative Results

Simulation HSI Reconstruction. Fig. 4 shows the visualized HSI reconstruction results of 7 SOTA methods and our Specformer technique. From the reconstructed HSIs and the zoom-in patches of the selected yellow boxes, we can see that the reconstruction performance of previous methods in texture-rich regions and edge regions is unsatisfactory. They either produce overly smooth results, sacrificing fine-grained structural content and textural detail, or introduce undesirable color artifacts and speckled textures. In contrast, our Specformer can accurately reconstruct the details of texture-rich regions and edge regions, as well as preserve the spatial smoothness of the homogeneous regions. This is because the uncertainty-driven self-adoptive loss can reinforce the network's attention on the regions with rich textures and edges, thus improving the HSI reconstruction quality of these regions. In addition, we plot the spectral density curves (bottom-left) corresponding to the picked region of the green box in the RGB image (top-left). The highest correlation and coincidence between our curve and the ground truth demonstrate the spectral-wise consistency restoration effectiveness of our Specformer. This is because FWSA can accurately model the spectral-wise long-distance dependencies and capture the inter-spectral similarities.

Real HSI Reconstruction. We further applied the approaches to real HSI reconstruction. Similar to DGSMP[20], we retrained all the networks on all scenes of CAVE [41] and KAIST [9] datasets. The reconstruction results on real measurements are presented in Fig. 5, from which we can see that compared with existing methods, our Specformer technique produced higher reconstruction quality in texture-rich and edge regions. Meanwhile, Specformer also produced better performance in suppressing noise.

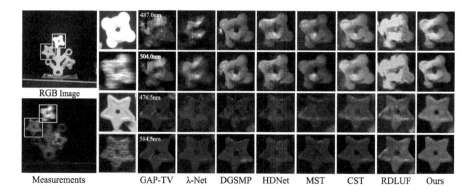

Fig. 5. Real reconstruction comparison of an exemplar Scene 1 with 4 out of 28 spectral channels. The results of 7 SOTA algorithms and our Specformer are presented. Please zoom in for a better view.

5 Ablation Study

To validate the effectiveness of each component in the Specformer network, we conducted a series of ablation studies on the CAVE [41] and KAIST [9] datasets. We consider several factors including the FWSA and LWSA modules in the SF-block, and the uncertainty-driven loss (UDL) function. The comparison results are presented in Table 2. The baseline (BL) model is derived by removing SF-block and UDL from Specformer. The "Serial" and "Parallel" represent combining the LWSA and FWSA modules using serial and parallel design, respectively.

Effectiveness of SF-Blcok. From Table 2, we can see that the absence of any component in the SF-block will result in performance degradation, which demonstrates the effectiveness of each component in the SF-block and the effectiveness of their combination. Moreover, the reconstruction performance of "BL+FWSA" is better than that of "BL+LWSA". This is because spectral representations are spatially sparse and spectrally highly self-similar. Hence, capturing spatial interactions may be less effective than modeling inter-spectra dependencies. However, only using FWSA cannot model the spatial sparsity, which is why the reconstruction performance of "BL+FWSA" is not as good as "BL+LWSA+FWSA". Furthermore, we can see that using a parallel design to combine FWSA and LWSA modules results in better reconstruction performance than using a serial design. Moreover, the computational and memory costs of parallel design (2.48M, 39.85G) are also lower than that of serial design (2.82M, 46.73G). This is because parallel design helps the SF-block to model cross-window connections, enlarges the receptive fields while maintaining linear complexity, and enables the complementary fusion of frequency-wise and spatial-wise features.

Effectiveness of UDL. We further investigated the contribution of the UDL function. The visualization results in Fig. 6 demonstrate the UDL's ability to enhance the reconstruction quality in texture-rich and edge regions. From

Table 2. Break-down ablation study. The models of different combinations were trained on the CAVE dataset and tested on the KAIST dataset. The first row is the baseline model.

BL	LWSA	FWSA	Serial	Parallel	UDL	PSNR	SSIM	Params	GFLOPs
✓						31.08	0.874	1.42M	14.87
✓	✓					33.34	0.927	1.89M	23.14
✓		✓				35.97	0.958	1.67M	19.79
✓	✓	✓	✓			37.58	0.966	2.82M	46.73
✓	✓	✓		✓		39.82	0.978	2.48M	39.85
✓	✓	✓		✓	✓	**40.61**	**0.981**	2.48M	39.85

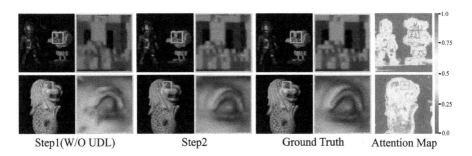

Step1(W/O UDL)　　　Step2　　　Ground Truth　　　Attention Map

Fig. 6. Ablation study of the UDL. Step1 represents the HSI reconstructed by the Specformer that has not been trained by $Loss_{UDL}$, and step2 represents the HSI reconstructed by the specformer that has been trained by $Loss_{UDL}$.

Table 2, we can also see that by enhancing the HSI reconstruction quality for the regions with rich textures and edges, the PSNR and SSIM of Specformer's reconstruction results are increased. Moreover, we can see from Table 2 that the improvements achieved by UDL do not bring any additional memory (Params) and computational (GFLops) costs during testing. This is because the two-stage training strategy of UDL does not change the structure of the final model, but only increases the training time by about 30%. However, owing to the low memory and computational costs of Specformer, we can still complete training within 6 h on a single RTX 3090 GPU.

6 Conclusion

In this work, we explored how to simultaneously embed the spatial sparsity and inter-spectral similarity nature of HSI into the learning-based reconstruction process. To this end, we proposed a novel uncertainty-driven method, termed Specformer, for HSI reconstruction. Specifically, we first introduced FWSA module, and used a parallel design to combine it with an LWSA module to form an SF-block. LWSA can guide the network to focus on the image regions with dense spectral information. FWSA can model the inter-spectral similarity. Parallel design helps the SF-block to model cross-window connections, and enlarge its receptive fields while maintaining linear complexity. We inserted the SF-block as the main building block in a U-shape encoder-decoder architecture to form Specformer. In addition, considering that texture-rich and edge regions in HSI are more difficult to reconstruct than smooth regions, we designed an uncertainty-driven self-adaptive loss function, which assigns greater priority to the regions with rich textures and edges during training, thus improving the reconstruction quality of these regions. Extensive quantitative and qualitative experiments demonstrate that the reported Specformer technique outperforms other SOTA methods with lower computational and memory costs.

Acknowledgements. This work was supported by the National Natural Science Foundation of China (Nos. 62322502, 61827901, 62131003), the Guangdong Province Key Laboratory of Intelligent Detection in Complex Environment of Aerospace, Land and Sea (2022KSYS016), and the Guangdong Cross domain Intelligent Detection and Information Processing Innovation Team (2023KCXT044).

References

1. Borengasser, M., Hungate, W.S., Watkins, R.: Hyperspectral remote sensing: principles and applications. CRC press (2007)
2. Cai, Y., et al.: Coarse-to-fine sparse transformer for hyperspectral image reconstruction. In: ECCV, pp. 686–704 (2022)
3. Cai, Y., et al.: Mask-guided spectral-wise transformer for efficient hyperspectral image reconstruction. In: CVPR, pp. 502–511 (2022)
4. Cai, Y., et al.: Degradation-aware unfolding half-shuffle transformer for spectral compressive imaging. NIPS **35**, 37749–37761 (2022)
5. Cai, Y., et al: Degradation-aware unfolding half-shuffle transformer for spectral compressive imaging. In: NIPS, vol. 35, pp. 749–761 (2022)
6. Cao, X., et al.: Computational snapshot multispectral cameras: toward dynamic capture of the spectral world. IEEE Signal Process. Mag. **33**(5), 95–108 (2016)
7. Chan, S.H., Wang, X., Elgendy, O.A.: Plug-and-play ADMM for image restoration: fixed-point convergence and applications. IEEE Trans. Comput. Imag. **3**(1), 84–98 (2016)
8. Chang, J., Lan, Z., Cheng, C., Wei, Y.: Data uncertainty learning in face recognition. In: CVPR. pp. 5710–5719 (2020)
9. Choi, I., Kim, M., Gutierrez, D., Jeon, D., Nam, G.: High-quality hyperspectral reconstruction using a spectral prior. Tech. rep. (2017)
10. Der Kiureghian, A., Ditlevsen, O.: Aleatory or epistemic? Does it matter? Struct. Saf. **31**(2), 105–112 (2009)
11. Dong, Y., Gao, D., Qiu, T., Li, Y., Yang, M., Shi, G.: Residual degradation learning unfolding framework with mixing priors across spectral and spatial for compressive spectral imaging. In: CVPR, pp. 22262–22271 (2023)
12. Dosovitskiy, A., et al.: An image is worth 16 × 16 words: Transformers for image recognition at scale (2020). arXiv preprint arXiv:2010.11929
13. Du, H., Tong, X., Cao, X., Lin, S.: A prism-based system for multispectral video acquisition. In: ICCV, pp. 175–182. IEEE (2009)
14. Figueiredo, M.: Adaptive sparseness using Jeffreys prior. Adv. Neural Inf. Process. Syst. **14** (2001)

15. Figueiredo, M.A., Nowak, R.D., Wright, S.J.: Gradient projection for sparse reconstruction: application to compressed sensing and other inverse problems. IEEE J-STSP **1**(4), 586–597 (2007)
16. Fu, Y., Zhang, T., Wang, L., Huang, H.: Coded hyperspectral image reconstruction using deep external and internal learning. IEEE TPAMI **44**(7), 3404–3420 (2021)
17. Gu, Y., Jin, Z., Chiu, S.C.: Active learning combining uncertainty and diversity for multi-class image classification. IET Comput. Vision **9**(3), 400–407 (2015)
18. Horé, A., Ziou, D.: Image quality metrics: PSNR vs. SSIM. In: ICPR, pp. 2366–2369 (2010). https://doi.org/10.1109/ICPR.2010.579
19. Hu, X., et al.: Hdnet: High-resolution dual-domain learning for spectral compressive imaging. In: CVPR, pp.17542–17551 (2022)
20. Huang, T., Dong, W., Yuan, X., Wu, J., Shi, G.: Deep gaussian scale mixture prior for spectral compressive imaging. In: CVPR, pp. 16216–16225 (2021)
21. Kendall, A., Gal, Y.: What uncertainties do we need in Bayesian deep learning for computer vision? In: NIPS **30** (2017)
22. Kim, M.H., et al.: 3D imaging spectroscopy for measuring hyperspectral patterns on solid objects. TOG **31**(4), 1–11 (2012)
23. Kittle, D., Choi, K., Wagadarikar, A., Brady, D.J.: Multiframe image estimation for coded aperture snapshot spectral imagers. Appl. Opt. **49**(36), 6824–6833 (2010)
24. Lee, C., Chung, K.S.: Gram: Gradient rescaling attention model for data uncertainty estimation in single image super resolution. In: ICMLA, pp. 8–13. IEEE (2019)
25. Li, M., fu, Y., Liu, J., Zhang, Y.: Pixel adaptive deep unfolding transformer for hyperspectral image reconstruction. In: ICCV, pp. 12959–12968 (2023)
26. Li, S., et al.: Falcon: a fourier transform based approach for fast and secure convolutional neural network predictions. In: CVPR, pp. 8705–8714 (2020)
27. Lin, X., Liu, Y., Wu, J., Dai, Q.: Spatial-spectral encoded compressive hyperspectral imaging. ACM Trans. Graph. (TOG) **33**(6), 1–11 (2014)
28. Liu, Y., Yuan, X., Suo, J., Brady, D.J., Dai, Q.: Rank minimization for snapshot compressive imaging. IEEE TPAMI **41**(12), 2990–3006 (2018)
29. Liu, Z., et al.: Swin transformer: hierarchical vision transformer using shifted windows. In: ICCV, pp. 10012–10022 (2021)
30. Llull, P., et al.: Coded aperture compressive temporal imaging. Opt. Express **21**(9), 10526–10545 (2013)
31. Lu, G., Fei, B.: Medical hyperspectral imaging: a review. J. Biomed. Opt. **19**(1), 010901–010901 (2014)
32. Ma, J., Liu, X.Y., Shou, Z., Yuan, X.: Deep tensor admm-net for snapshot compressive imaging. In: ICCV, pp. 10223–10232 (2019)
33. Melgani, F., Bruzzone, L.: Classification of hyperspectral remote sensing images with support vector machines. IEEE Trans. Geosci. Remote Sens. **42**(8), 1778–1790 (2004)
34. Meng, Z., Jalali, S., Yuan, X.: Gap-net for snapshot compressive imaging (2020). arXiv preprint arXiv:2012.08364
35. Meng, Z., Ma, J., Yuan, X.: End-to-end low cost compressive spectral imaging with spatial-spectral self-attention. In: ECCV, pp. 187–204. Springer (2020)
36. Meng, Z., Qiao, M., Ma, J., Yu, Z., Xu, K., Yuan, X.: Snapshot multispectral endomicroscopy. Opt. Lett. **45**(14), 3897–3900 (2020)
37. Meng, Z., Yu, Z., Xu, K., Yuan, X.: Self-supervised neural networks for spectral snapshot compressive imaging. In: ICCV, pp. 2622–2631 (2021)
38. Miao, X., Yuan, X., Pu, Y., Athitsos, V.: l-net: reconstruct hyperspectral images from a snapshot measurement. In: ICCV, pp. 4059–4069 (2019)

39. Nussbaumer, H.J., Nussbaumer, H.J.: The fast Fourier transform. Springer (1982)
40. Pan, Z., Healey, G., Prasad, M., Tromberg, B.: Face recognition in hyperspectral images. IEEE TPAMI **25**(12), 1552–1560 (2003)
41. Park, J.I., Lee, M.H., Grossberg, M.D., Nayar, S.K.: Multispectral imaging using multiplexed illumination. In: ICCV, pp. 1–8. IEEE (2007)
42. Pitas, I.: Digital Image Processing Algorithms and Applications. Wiley (2000)
43. Qiao, M., Liu, X., Yuan, X.: Snapshot spatial-temporal compressive imaging. Opt. Lett. **45**(7), 1659–1662 (2020)
44. Ronneberger, O., Fischer, P., Brox, T.: U-net: convolutional networks for biomedical image segmentation. In: MICCAI, pp. 234–241. Springer (2015)
45. Vaibhav, V.: Fast inverse nonlinear fourier transform. Phys. Rev. E **98**(1), 013304 (2018)
46. Vaswani, A., et al.: Attention is all you need. In: NIPS **30** (2017)
47. Wagadarikar, A., John, R., Willett, R., Brady, D.: Single disperser design for coded aperture snapshot spectral imaging. Appl. Opt. **47**(10), B44–B51 (2008)
48. Wagadarikar, A.A., Pitsianis, N.P., Sun, X., Brady, D.J.: Video rate spectral imaging using a coded aperture snapshot spectral imager. Opt. Express **17**(8), 6368–6388 (2009)
49. Wang, L., Xiong, Z., Gao, D., Shi, G., Wu, F.: Dual-camera design for coded aperture snapshot spectral imaging. Appl. Opt. **54**(4), 848–858 (2015)
50. Wang, L., Xiong, Z., Shi, G., Wu, F., Zeng, W.: Adaptive nonlocal sparse representation for dual-camera compressive hyperspectral imaging. IEEE TPAMI **39**(10), 2104–2111 (2016)
51. Yuan, X.: Generalized alternating projection based total variation minimization for compressive sensing. In: ICIP, pp. 2539–2543. IEEE (2016)
52. Yuan, X., Brady, D.J., Katsaggelos, A.K.: Snapshot compressive imaging: theory, algorithms, and applications. IEEE Signal Process. Mag. **38**(2), 65–88 (2021)
53. Zhang, S., Wang, L., Fu, Y., Zhong, X., Huang, H.: Computational hyperspectral imaging based on dimension-discriminative low-rank tensor recovery. In: ICCV, pp. 10183–10192 (2019)

CaesarNeRF: Calibrated Semantic Representation for Few-Shot Generalizable Neural Rendering

Haidong Zhu[1], Tianyu Ding[2(✉)], Tianyi Chen[2], Ilya Zharkov[2], Ram Nevatia[1], and Luming Liang[2]

[1] University of Southern California, Los Angeles, USA
{haidongz,nevatia}@usc.edu
[2] Microsoft, Washington, USA
{tianyuding,tiachen,zharkov,lulian}@microsoft.com

Abstract. Generalizability and few-shot learning are key challenges in Neural Radiance Fields (NeRF), often due to the lack of a holistic understanding in pixel-level rendering. We introduce CaesarNeRF, an end-to-end approach that leverages scene-level **CA**librat**E**d **S**em**A**ntic **R**epresentation along with pixel-level representations to advance few-shot, generalizable neural rendering, facilitating a holistic understanding without compromising high-quality details. CaesarNeRF explicitly models pose differences of reference views to combine scene-level semantic representations, providing a calibrated holistic understanding. This calibration process aligns various viewpoints with precise location and is further enhanced by sequential refinement to capture varying details. Extensive experiments on public datasets, including LLFF, Shiny, mip-NeRF 360, and MVImgNet, show that CaesarNeRF delivers state-of-the-art performance across varying numbers of reference views, proving effective even with a single reference image.

Keywords: Few-view rendering · Generalizable NeRF · Neural rendering

1 Introduction

Rendering a scene from a novel camera position is essential in view synthesis [5,11,62]. The recent advancement of Neural Radiance Field (NeRF) [44] has shown impressive results in creating photo-realistic images from novel viewpoints. However, conventional NeRF methods are either typically scene-specific, necessitating retraining for novel scenes [15,16,44,67,74], or require a large number of reference views as input for

---

H. Zhu and T. Ding—Equal contribution.
This work was done when Haidong Zhu was an intern at Microsoft.

Supplementary Information The online version contains supplementary material available at https://doi.org/10.1007/978-3-031-72658-3_5.

Fig. 1. Novel view synthesis for novel scenes using ONE reference view on Shiny [67], LLFF [42], and MVImgNet [77] (top to bottom). Each triplet of images corresponds to the results from GNT [60] (left), CaesarNeRF (middle) and groundtruth (right).

generalizing to novel scenarios [6,58,60,76]. These constraints highlight the complexity of the few-shot generalizable neural rendering, which aims to render unseen scenes from novel viewpoints with a limited number of reference images.

Generalizing NeRF to novel scenes often involves using pixel-level feature embeddings encoded from input images, as seen in existing methods [59,76]. These methods adapt NeRF to novel scenes by separating the scene representation from the model through an image encoder. However, relying solely on pixel-level features has its drawbacks: it requires highly precise epipolar geometry and often overlooks occlusion in complex scenes. Moreover, employing pixel-level features ignores the inherent interconnections within objects in the scene, treating the prediction of each pixel independently. Prior attempts to utilize scene-level representation either suffers from style mismatches [35] during scene-wide rendering or have been limited to specific object categories [23,39,69]. With few input reference under generalizable settings, these limitations become more pronounced, exacerbating the ambiguity in predictions due to biases from camera viewpoints. Although recent diffusion-based models [18,56] attempt to address these issues through generative approaches, they struggle to effectively use the input views as contextual references for specific scenes.

We present CaesarNeRF, a method that advances the generalizability of NeRF by incorporating calibrated semantic representation. This enables rendering from novel viewpoints using as few as one input reference view, as depicted in Fig. 1. Our approach combines semantic scene-level representation with per-pixel features, enhancing consistency across different views of the same scene. The encoder-generated scene-level representations capture both semantic features and biases linked to specific camera poses. When reference views are limited, these biases can introduce uncertainty in the rendered images. To counter this, CaesarNeRF integrates camera pose transformations into the semantic representation, hence the term *calibrated*. By isolating pose-specific information from the scene-level representation, our model harmonizes features across input views, mitigating view-specific biases and, in turn, reducing ambiguity. In addition, CaesarNeRF introduces a sequential refinement process, which equips the model with varying levels of detail needed to enhance the semantic features. Extensive experiments on datasets such as LLFF [42], Shiny [67], mip-NeRF 360 [4], and the newly released MVImgNet [77] demonstrate that CaesarNeRF outperforms current state-of-the-art methods with limited reference views available, proving effective in generalizable settings with as few as one reference view. The project can be found here.

In summary, our contributions are as follows:

- We introduce CaesarNeRF, which utilizes scene-level calibrated semantic representation to achieve few-shot, generalizable neural rendering. This innovation leads to coherent and high-quality renderings.
- We integrate semantic scene context with pixel-level details, in contrast to existing methods that rely solely on pixel-level features. We also address view-specific biases by modeling camera pose transformations and enhance the scene understanding through the sequential refinement of semantic features.
- We demonstrate through extensive experiments that CaesarNeRF consistently outperforms state-of-the-art generalizable NeRF methods across a variety of datasets. Furthermore, integrating the Caesar pipeline into other baseline methods leads to consistent performance gains, highlighting its effectiveness and adaptability.

2 Related Work

Neural Radiance Field (NeRF) implicitly captures the density and appearance of points within a scene or object [43,44] and enables rendering from novel camera positions. In recent years, NeRF has witnessed improvements in a wide range of applications, such as photo-realistic novel view synthesis for large-scale scenes [40,68,80], dynamic scene decomposition and deformation [24,30,34,46–48,50,84,85], occupancy or depth estimation [66,71,83], scene generation and editing [1,21,31,33,41,49, 72,73,79], and so on. Despite these advances, most methods still rely on the original NeRF and require retraining or fine-tuning for novel scenes not covered in the training data.

Generalizable NeRF aims to adapt a single NeRF model to multiple scenes by separating the scene representation from the model. This field has seen notable advancements, with efforts focused on avoiding the need for retraining [8,20,38,64,65,75,76]. PixelNeRF [76] and GRF [59] pioneered the application of an image encoder to transform images into per-pixel features, with NeRF functioning as a decoder for predicting density and color from these features. MVSNeRF [6] introduces the use of a cost volume from MVSNet [75] to encode 3-D features from multiple views. Recognizing the intrinsic connection between points along a ray, IBRNet [64] employs self-attention to enhance point density predictions. Transformer-based [61] networks like GNT [10,60], GeoNeRF [25], and GPNR [58] are explored as alternatives to volume rendering, concentrating on pixel and patch-level representations. Additionally, InsertNeRF [2] utilizes hypernet modules to adapt parameters for novel scenes efficiently.

These methods primarily depend on image encoders to extract pixel-aligned features from reference views. As a result, many of them lack a comprehensive understanding of the entire scene. Furthermore, with few reference views, the features become intertwined with view-specific details, compromising the quality of the rendering results.

Few-Shot Neural Radiance Field aims to render novel views using a limited number of reference images. To this end, various methods have been developed, incorporating additional information such as normalization-flow [78], semantic constraints [17,22], depth cues [13,52], geometry consistency [3,29,45,63,70], and frequency content [74]. Others [6,9] emphasize pretraining on large-scale datasets.

While these methods offer reasonable reconstructions with few inputs, they typically still require training or fine-tuning for specific scenes. Moreover, these methods usually require at least three reference images. With fewer than three, view-specific biases lead to ambiguity, complicating the rendering. Diffusion-based [12,35,57,82] and other generative methods [27,86] have been explored for single-view synthesis or generative rendering, yet they are mostly limited to single-object rendering and generally fall short for complex scenes, which often result in a style change.

CaesarNeRF confronts the above challenges by leveraging calibrated semantic representations that exploit scene geometry and variations in camera viewpoints. As a result, CaesarNeRF overcomes the limitations of pixel-level features and reduces dependency on external data or extensive pretraining, delivering high-quality renderings in few-shot and generalizable settings.

3 The Proposed Method

We first outline the general framework of existing generalizable NeRF in Sect. 3.1. Then, we present our proposed CaesarNeRF, as illustrated in Fig. 2. This model integrates elements of semantic representation, calibration, and sequential refinement, detailed in Sect. 3.2, 3.3, and 3.4, respectively. The training objective is given in Sect. 3.5.

3.1 NeRF and Generalizable NeRF

Neural Radiance Field (NeRF) [43,44] aims to render 3D scenes by predicting both the density and RGB values at points where light rays intersect the radiance field. For a query point $x \in \mathbb{R}^3$ and a viewing direction d on the unit sphere $\mathbb{S}^2$ in 3D space, the NeRF model $\mathcal{F}$ is defined as:

$$\sigma, c = \mathcal{F}(x, d). \tag{1}$$

Here, $\sigma \in \mathbb{R}$ and $c \in \mathbb{R}^3$ denote the density and the RGB values, respectively. After computing these values for a collection of discretized points along each ray, volume rendering techniques are employed to calculate the final RGB values for each pixel, thus reconstructing the image.

However, traditional NeRF models $\mathcal{F}$ are limited by their requirement for scene-specific training, making it unsuitable for generalizing to novel scenes. To overcome this, generalizable NeRF models, denoted by $\mathcal{F}_G$, are designed to render images of novel scenes without per-scene training. Given N reference images $\{I_n\}_{n=1}^N$, an encoder-based generalizable NeRF model $\mathcal{F}_G$ decouples the object representation from the original NeRF by using an encoder to extract per-pixel feature maps $\{F_n\}_{n=1}^N$ from the input images. To synthesize a pixel associated with a point x along a ray in direction

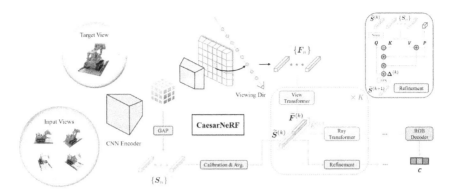

Fig. 2. Overview of CaesarNeRF. CaesarNeRF employs a shared encoder to capture two types of features from input views, including scene-level semantic representation $\{S_n\}$ and pixel-level feature representation $\{F_n\}$. We use the same encoder for both the scene-level semantic representation and the pixel-level embeddings. Following calibration and aggregation of $\{S_n\}$ from various views, we concatenate it with the pixel-level fused feature, processed by the view transformer. Subsequent use of the ray-transformer, coupled with sequential refinement, enables us to render the final RGB values for each pixel in the target view. The output features serve as the input for the next stage, indicated by matching line colors.

d, it projects $\{F_n\}_{n=1}^{N}$ from nearby views and aggregates this multi-view pixel-level information using techniques such as average pooling [76] or cost volumes [6]. This results in a fused feature embedding $\widetilde{F}$, allowing $\mathcal{F}_G$ to predict density σ and RGB values c for each point along the ray, as expressed by:

$$\sigma, c = \mathcal{F}_G(x, d, \widetilde{F}). \tag{2}$$

In our method, we adopt the recently introduced fully attention-based generalizable NeRF method, GNT [60], as both the backbone and the baseline. GNT shares a similar paradigm with (2) but employs transformers [61] to aggregate pixel-level features into $\widetilde{F}$. It uses a *view transformer* to fuse projected pixel-level features from reference views, and a *ray transformer* to combine features from different points along a ray, eliminating the need for volume rendering. Further details about GNT can be found in [60]. We also demonstrate that our approach can be extended to other generalizable NeRF models, as discusses in Sect. 4.2.

3.2 Scene-Level Semantic Representation

Both encoder-based generalizable NeRF models [6,59,76] and their attention-based counterparts [58,60] mainly rely on pixel-level feature representations. While effective, this approach restricts their capability for a holistic scene understanding, especially when reference views are scarce. This limitation also exacerbates challenges in resolving depth ambiguities between points along the rays, a problem that becomes more pronounced with fewer reference views.

To address these challenges, we introduce semantic representations aimed at enriching the scene-level understanding. We utilize a shared CNN encoder and apply a Global

(a) "L: green, R: red" (b) "L: red, R: green"

Fig. 3. An illustration of conflicting semantic meanings from multiple viewpoints of the same object. When observing the cup from distinct angles, the features extracted after pooling retain spatial information but are inconsistent in the scene-level semantic understanding, leading to conflicts across various reference images after aggregation.

Average Pooling (GAP) to its C-dimensional output feature map, generating N global feature vectors $\{S_n\}_{n=1}^N$ corresponding to each input view. These feature vectors are then averaged to form a unified scene-level representation S, i.e.,

$$S = \frac{1}{N}\sum_{n=1}^{N} S_n \in \mathbb{R}^C. \quad (3)$$

In GNT [60], which uses a view transformer to aggregate pixel-level features into an L-dimensional vector $\widetilde{F}$, we extend this by concatenating $\widetilde{F}$ with S to construct a *global-local* embedding E, as formulated by:

$$E = \text{Concat}(\widetilde{F}, S) \in \mathbb{R}^{L+C}. \quad (4)$$

This combined embedding E is then subjected to the standard self-attention mechanism used in GNT [60]. This approach enables the scene-level semantic representation (S) to integrate with per-point features ($\widetilde{F}$), offering a more nuanced understanding at both levels. It also allows each point to selectively draw from the scene-level information. To maintain dimensional consistency across the input and output layers of multiple transformer modules, we employ a two-layer MLP to project the enhanced features back to the original dimension L of the per-point embedding $\widetilde{F}$.

3.3 Calibration of Semantic Representation

The integration of the scene-level semantic representation S, generated through simple averaging of global feature vectors as in (3), improves rendering quality. However, this approach has limitations when dealing with multiple views. As illustrated in Fig. 3, viewing the same object from distinct angles may retain spatial attributes but can lead to conflicting semantic meanings. Merely averaging these global feature vectors without accounting for camera positions can result in a distorted scene-level understanding.

To mitigate this inconsistency, we propose a semantic calibration technique using feature rotation. This adjustment aligns the semantic representation across different

Fig. 4. Visualization of decoded feature maps for "orchid" in LLFF dataset, produced by *ray transformers* [60] at different stages. From left to right, the transformer stages increase in depth.

camera poses. Our inspiration comes from the use of camera pose projection in computing the fused pixel-level feature $\widetilde{F}$ and is further motivated by [51], which demonstrates that explicit rotation operations in feature spaces are feasible. Unlike point clouds in [51] that inherently lack a defined canonical orientation, NeRF explicitly encodes differences between camera viewpoints, thereby enabling precise calibration between the reference and target images.

Building on this observation, we calculate calibrated semantic representations $\{\widetilde{S}_n\}_{n=1}^N$ from the N original semantic representations $\{S_n\}_{n=1}^N$ derived from the reference views. We accomplish this by leveraging their respective rotation matrices $\{T_n\}_{n=1}^N$ to model the rotational variations between each input view and the target view. The alignment of the original semantic features is performed as follows:

$$\widetilde{S}_n = \mathcal{P}(T_n \cdot \mathcal{P}^{-1}(S_n)), \quad \text{where } T_n = T_{\text{out}}^{\text{w2c}} \cdot T_n^{\text{c2w}}. \tag{5}$$

Here, T_n^{c2w} is the inverse of the extrinsic matrix used for I_n, and $T_{\text{out}}^{\text{w2c}}$ is the extrinsic matrix for the target view. $\mathcal{P}(\cdot)$ and $\mathcal{P}^{-1}(\cdot)$ are the flattening and inverse flattening operations, which reshape the feature to a 1D vector of shape 1-by-C and a 2D matrix of shape 3-by-$\frac{C}{3}$, respectively.

Note that for the extrinsic matrix, we consider only the top-left 3×3 submatrix that accounts for rotation. Using GAP to condense feature maps of various sizes into a 1-by-C feature vector eliminates the need for scaling parameters in the semantic representation. As a result, modeling the intrinsic matrix is unnecessary, assuming no skewing, making our approach adaptable to different camera configurations.

With the calibrated semantic features $\{\widetilde{S}_n\}_{n=1}^N$ for each reference view, we average these, similar to (3), to obtain the calibrated scene-level semantic representation $\widetilde{S}$, i.e.,

$$\widetilde{S} = \frac{1}{N} \sum_{n=1}^N \widetilde{S}_n \in \mathbb{R}^C. \tag{6}$$

Finally, akin to (4), we concatenate the pixel-level fused feature $\widetilde{F}$ with the calibrated scene-level semantic representation $\widetilde{S}$ to form the final global-local embedding $\widetilde{E}$:

$$\widetilde{E} = \text{Concat}(\widetilde{F}, \widetilde{S}) \in \mathbb{R}^{L+C}. \tag{7}$$

This unified embedding then feeds into ray transformers, passing through standard self-attention mechanisms. In the original GNT [60], multiple view transformers and ray transformers are stacked alternately for sequential feature processing. The last ray transformer integrates features from multiple points along a ray to yield the final RGB value. We denote the corresponding feature representations at stage k as $\widehat{F}^{(k)}$ and $\widetilde{E}^{(k)}$. Notably, the calibrated semantic representation $\widetilde{S}$ remains constant across these stages.

3.4 Sequential Refinement

While leveraging $\widetilde{S}$ improves consistency, a single, uniform $\widetilde{S}$ may not be adequate for deeper layers that demand more nuanced details. In fact, we find that deeper transformers capture finer details compared to shallower ones, as shown in Fig. 4. To address this limitation, we introduce a sequential semantic feature refinement module that progressively enriches features at each stage. Specifically, we learn the residual $\Delta^{(k)}$ to update $\widetilde{S}$ at each stage k as follows:

$$\widetilde{S}^{(k+1)} \leftarrow \widetilde{S}^{(k)} + \Delta^{(k)}. \tag{8}$$

Here, $\Delta^{(k)}$ is calculated by first performing specialized cross-attentions between $\widetilde{S}^{(k)}$ and the original, uncalibrated per-frame semantic features $\{S_n\}_{n=1}^N$ (see Fig. 2), followed by their summation. Our goal is to fuse information from different source views to enrich the scene-level semantic representation with features from each reference frame. With this sequential refinement, we combine $\widetilde{S}^{(k)}$ with $\widehat{F}^{(k)}$ at each stage, yielding a stage-specific global-local embedding $\widetilde{E}^{(k)}$, which completes our approach.

Discussion. In scenarios with few reference views, especially when limited to just one, the primary issue is inaccurate depth estimation, resulting in depth ambiguity [12]. This compromises the quality of images when rendered from novel viewpoints. Despite this, essential visual information generally remains accurate across different camera poses. Incorporating our proposed scene-level representation improves the understanding of the overall scene layout [7], distinguishing our approach from existing generalizable NeRF models that predict pixels individually. The advantage of our approach is its holistic view; the semantic representation enriches per-pixel predictions by providing broader context. This semantic constraint ensures that fewer abrupt changes between adjacent points. Consequently, it leads to more reliable depth estimations, making the images rendered from limited reference views more plausible.

3.5 Training Objectives

During training, we employ three different loss functions:

MSE Loss. The Mean Square Error (MSE) loss is the standard photometric loss used in NeRF [43]. It computes the MSE between the actual and predicted pixel values.

Central Loss. Since we project the view-level features from different input camera poses to the shared target view, we introduce a central loss to ensure frame-wise calibrated semantic features $\{\widetilde{S}_n\}_{n=1}^N$ are consistent when projected onto the same target view, which is defined as:

$$\mathcal{L}_{\text{central}} = \frac{1}{N} \sum_{n=1}^{N} \left\| \widetilde{S}_n - \widetilde{S} \right\|_1. \tag{9}$$

Table 1. Results for generalizable scene rendering on LLFF with few reference views. GeoNeRF, MatchNeRF, and MVSNeRF necessitate variance as input, defaulting to 0 for single-image cases, hence their results are not included for 1-view scenarios.

Method	1 reference view			2 reference views			3 reference views		
	PSNR (↑)	LPIPS (↓)	SSIM (↑)	PSNR (↑)	LPIPS (↓)	SSIM (↑)	PSNR (↑)	LPIPS (↓)	SSIM (↑)
PixelNeRF [76]	9.32	0.898	0.264	11.23	0.766	0.282	11.24	0.671	0.486
GPNR [58]	15.91	0.527	0.400	18.79	0.380	0.575	21.57	0.288	0.695
NeuRay [38]	16.18	0.584	0.393	17.71	0.336	0.646	18.26	0.310	0.672
GeoNeRF [25]	-	-	-	18.76	0.473	0.500	23.40	0.246	0.766
MatchNeRF [8]	-	-	-	21.08	0.272	0.689	22.30	0.234	0.731
MVSNeRF [6]	-	-	-	19.15	0.336	0.704	19.84	0.314	0.729
IBRNet [64]	16.85	0.542	0.507	21.25	0.333	0.685	23.00	0.262	0.752
GNT [60]	16.57	0.500	0.424	20.88	0.251	0.691	23.21	0.178	0.782
Ours	18.31	0.435	0.521	21.94	0.224	0.736	23.45	0.176	0.794

Point-Wise Perceptual Loss. During the rendering of a batch of pixels in a target view, we inpaint the ground-truth image by replacing the corresponding pixels with the predicted ones. Then, a perceptual loss [26] is computed between the inpainted image and the target image to guide the training process at the whole-image level.

The final loss function is formulated as follows:

$$\mathcal{L} = \mathcal{L}_{\text{MSE}} + \lambda_1 \mathcal{L}_{\text{central}} + \lambda_2 \mathcal{L}_{\text{perc}}. \tag{10}$$

Empirically, we set $\lambda_1 = 1$ and $\lambda_2 = 0.001$, following [32].

4 Experiments

In this section, we first discuss our experimental setups and other details required for our experiments in Sect. 4.1, followed by our results for different experimental settings, along with ablation studies and analysis in Sect. 4.2.

4.1 Experimental Setups

Datasets. Firstly, following [60], we construct our training data from both synthetic and real data. This collection includes scanned models from Google Scanned Objects [14], RealEstate10K [81], and handheld phone captures [64]. For evaluation, we utilize real data encompassing complex scenes from sources such as LLFF [42], Shiny [67], and mip-NeRF 360 [4]. Additionally, we train and test our model using the recently released MVImgNet dataset [77]. We adhere to the official split, focusing on examples from the *containers* category, and select 2,500 scenes for training. During inference, we choose 100 scenes, using their first images as target views and the spatially nearest images as references. Since MVImgNet does not provide camera poses, we utilize COLMAP [54, 55] to deduce the camera positions within these scenes.

Implementation Details. CaesarNeRF is built upon GNT [60], for which we maintain the same configuration, setting the *ray* and *view* transformers stack number (K) to 8

Table 2. Results for generalizable scene rendering on Shiny with few reference views.

Method	1 reference view			2 reference views			3 reference views		
	PSNR (↑)	LPIPS (↓)	SSIM (↑)	PSNR (↑)	LPIPS (↓)	SSIM (↑)	PSNR (↑)	LPIPS (↓)	SSIM (↑)
MatchNeRF [8]	-	-	-	20.28	0.278	0.636	20.77	0.249	0.672
MVSNeRF [6]	-	-	-	17.25	0.416	0.577	18.55	0.343	0.645
IBRNet [64]	14.93	0.625	0.401	18.40	0.400	0.595	21.96	0.281	0.710
GNT [60]	15.99	0.548	0.400	20.42	0.327	0.617	22.47	0.247	0.720
Ours	17.57	0.467	0.472	21.47	0.293	0.652	22.74	0.241	0.723

Table 3. Results for generalizable scene rendering on mip-NeRF 360 with few reference views.

Method	1 reference view			2 reference views			3 reference views		
	PSNR (↑)	LPIPS (↓)	SSIM (↑)	PSNR (↑)	LPIPS (↓)	SSIM (↑)	PSNR (↑)	LPIPS (↓)	SSIM (↑)
MatchNeRF [8]	-	-	-	17.00	0.566	0.392	17.26	0.551	0.407
MVSNeRF [6]	-	-	-	14.23	0.681	0.366	14.29	0.674	0.406
IBRNet [64]	14.12	0.682	0.283	16.24	0.618	0.360	17.70	0.555	0.420
GNT [60]	13.48	0.630	0.314	15.21	0.559	0.370	15.59	0.538	0.395
Ours	15.20	0.592	0.350	17.05	0.538	0.403	17.55	0.512	0.430

for generalizable setting and 4 for single-scene setting. The feature encoder extracts bottleneck features, applies GAP, and then uses a fully connected (FC) layer to reduce the input dimension C to 96. Training involves 500,000 iterations using the Adam optimizer [28], with learning rate set at 0.001 for the feature encoder and 0.0005 for CaesarNeRF, halving them every 100,000 iterations. Each iteration samples 4,096 rays from a single scene. In line with [60], we randomly choose between 8 to 10 reference views for training, and 3 to 7 views when using the MVImgNet [77].

Baseline Methods. We compare CaesarNeRF with several state-of-the-art methods suited for generalizable NeRF applications, including earlier works such as MVSNeRF [6], PixelNeRF [76], and IBRNet [64], alongside more recent ones, including GPNR [58], NeuRay [38], GNT [60], GeoNeRF [25] and MatchNeRF [8].

4.2 Results and Analysis

We compare results in two settings: a generalizable setting, where the model is trained on multiple scenes without fine-tuning during inference for both few and all reference view cases, and a single-scene setting where the model is trained and evaluated on just one scene. Following these comparisons, we conduct ablation studies and test the generalizability of our method with other state-of-the-art approaches.

Generalizable Rendering. In the generalizable setting, we adopt two training strategies. First, we train the model on multiple datasets as described in Sect. 4.1 and evaluate on LLFF [42], Shiny [67] and mip-NeRF 360 [4] datasets. In addition, the model is trained and tested on the MVImgNet [77] for object-centric generalizability.

(a) LLFF, Shiny, and mip-NeRF 360. The results for few-reference view scenarios on these datasets are shown in Table 1, 2 and 3, respectively. Methods like MatchN-

Table 4. Results on MVImgNet across varying numbers of reference views. 'C.' represents the use of calibration before averaging.

Method	1 reference view			2 reference views			3 reference views			4 reference views			5 reference views		
	PSNR (↑)	LPIPS (↓)	SSIM (↑)	PSNR (↑)	LPIPS (↓)	SSIM (↑)	PSNR (↑)	LPIPS (↓)	SSIM (↑)	PSNR (↑)	LPIPS (↓)	SSIM (↑)	PSNR (↑)	LPIPS (↓)	SSIM (↑)
IBRNet	19.14	0.458	0.595	24.38	0.266	0.818	25.53	0.203	0.858	25.99	0.190	0.867	26.12	0.188	0.867
GNT	22.22	0.433	0.678	26.94	0.236	0.850	27.41	0.206	0.870	27.51	0.197	0.875	27.51	0.194	0.876
Ours w/o C.	23.61	0.371	0.718	26.34	0.274	0.817	27.10	0.228	0.850	27.30	0.210	0.862	27.34	0.203	0.865
Ours	24.28	0.334	0.747	27.34	0.215	0.856	27.82	0.190	0.875	27.92	0.181	0.881	27.92	0.179	0.882

eRF [8], MVSNeRF [75], and GeoNeRF [25] require at least two reference views. On the LLFF dataset, all methods experience a performance decline as the number of views decreases. CaesarNeRF, however, consistently outperforms others across varying reference view numbers, with the performance gap becoming more significant with fewer views. For example, with 3 views, while IBRNet [64] and GNT [60] have comparable PSNRs, CaesarNeRF demonstrates a more substantial lead in LPIPS and SSIM metrics.

Similar patterns are observed on the Shiny [67] and mip-NeRF 360 [4] datasets. We apply the highest-performing methods from the LLFF evaluations and report the results for those that produce satisfactory outcomes with few reference views. CaesarNeRF maintains superior performance throughout. Notably, for complex datasets like mip-NeRF 360 [4], which have sparse camera inputs, the quality of rendered images generally decreases with fewer available reference views. Nonetheless, CaesarNeRF shows the most robust performance compared to the other methods.

Table 5. Results of per-scene optimization on LLFF, in comparison with state-of-the-art methods.

Method	LLFF [42]	NeRF [44]	NeX [67]	GNT [60]	Ours
PSNR (↑)	23.27	26.50	27.26	27.24	27.64
LPIPS (↓)	0.212	0.250	0.179	0.087	0.081
SSIM (↑)	0.798	0.811	0.904	0.889	0.904

Table 6. Results on LLFF for few-shot generalization after adapting Caesar to other baselines.

Method	1 reference view			2 reference views			3 reference views		
	PSNR (↑)	SSIM (↑)	LPIPS (↓)	PSNR (↑)	SSIM (↑)	LPIPS (↓)	PSNR (↑)	SSIM (↑)	LPIPS (↓)
MatchNeRF [8]	–	–	–	20.59	0.775	0.276	22.43	0.805	0.244
+ Caesar	–	–	–	21.55	0.782	0.268	22.98	0.824	0.242
IBRNet [64]	16.85	0.507	0.542	21.25	0.685	0.333	23.00	0.752	0.262
+ Caesar	17.76	0.543	0.500	22.39	0.740	0.275	23.67	0.772	0.242

(b) MVImgNet. We extend our comparison of CaesarNeRF with GNT [60] and IBRNet [64] on the MVImgNet dataset, focusing on object-centric scenes, as shown

Table 7. Ablations on the semantic representation length R, sequential refinement (Seq.) and calibration (Cali.). 'Ext.' denotes the extension of per-pixel feature to a length of 64 in GNT.

Model Variations			PSNR (↑)	LPIPS (↓)	SSIM (↑)
R len.	Seq.	Cali.			
(Baseline GNT)			20.93	0.185	0.731
Ext.			20.85	0.173	0.735
+32			21.43	0.152	0.763
+64			21.49	0.149	0.766
+96			21.46	0.150	0.766
+128			21.49	0.147	0.763
+96	✓		21.53	0.146	0.770
+96		✓	21.51	0.147	0.769
+96	✓	✓	21.67	0.139	0.781

in Table 4. We examine a variant of CaesarNeRF where semantic calibration is substituted with simple feature averaging from multiple frames. While the performance of all methods improves with more views, CaesarNeRF consistently outperforms GNT and IBRNet. Notably, CaesarNeRF with feature averaging surpasses GNT in 1-view case but lags with additional views, implying that the absence of calibration lead to ambiguities when rendering from multiple views.

Per-Scene Optimization. Beyond the multi-scene generalizable setting, we demonstrate per-scene optimization results in Table 5. We calculate the average performance over 8 categories from the LLFF dataset [42]. CaesarNeRF consistently outperforms nearly all state-of-the-art methods in the comparison, across all three metrics, showing a significant improvement over our baseline method, GNT [60].

Adaptability. To test the adaptability of our Caesar pipeline, we apply it to two other state-of-the-art methods that use *view transformers*, namely MatchNeRF [8] and IBRNet [64]. We demonstrate in Table 6 that our enhancements in scene-level semantic understanding significantly boost the performance of these methods across all metrics. This indicates that the Caesar framework is not only beneficial in our CaesarNeRF, which is based on GNT [60], but can also be a versatile addition to other NeRF pipelines with view transformers to aggregate different input views.

Ablation Analysis. We conduct ablation studies on the "orchid" scene from the LLFF dataset, with findings detailed in Table 7. Testing variations in representation and the impact of the sequential refinement and calibration modules, we find that increasing the latent size in GNT yields marginal benefits. However, incorporating even a modest semantic representation size distinctly improves results. The length of the semantic representation has a minimal impact on quality. Our ablation studies indicate that while sequential refinement and calibration each offer slight performance gains, their

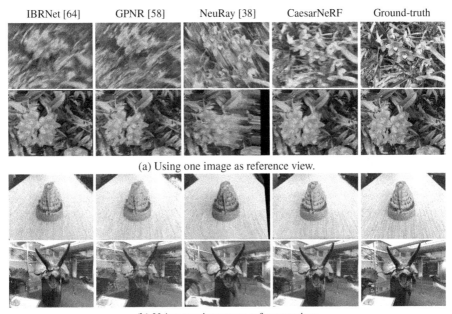

Fig. 5. Comparative visualization of our proposed method against other state-of-the-art methods.

Fig. 6. Depth estimation prediction using one reference view (first row) and two reference views (second row) as input from LLFF comparing CaesarNeRF with GNT.

combined effect is most significant. In a single-scene context, semantic information is effectively embedded within the representation, making the benefits of individual modules subtler. Together, however, they provide a framework where sequential refinement can leverage calibrated features for deeper insights.

Visualizations. We present our visualization results in Fig. 5, where we compare our method with others using one or two views from the LLFF dataset. Additional visual comparisons are provided in the supplementary materials. These visualizations highlight that in scenarios with few views, our method significantly surpasses other generalizable NeRF models, particularly excelling when only a single view is available. CaesarNeRF demonstrates rendering with sharper boundaries and more distinct objects.

Depth Estimation. We extend our evaluation to depth prediction within the LLFF [42] dataset, focusing on challenges presented by few reference views, such as scenarios with just one or two images. In a comparison between CaesarNeRF and GNT [60], we observe in Fig. 6 that GNT struggles to accurately capture the relative positions of objects when reference images are sparse. For instance, with only a single view of a flower, CaesarNeRF precisely indicates the flower's proximity to the camera compared to the leaves in the background, a distinction that GNT fails to make. Furthermore, the depth estimations provided by CaesarNeRF are consistently more reliable. In the horn example involving two views, CaesarNeRF offers better boundary delineation, showing particular strength in handling reflective surfaces such as grass in the background.

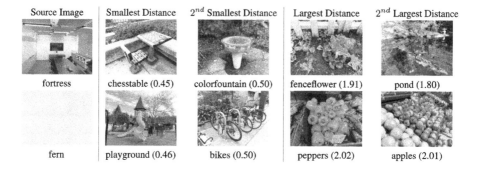

Fig. 7. Largest and smallest distances for two examples from LLFF test split when matching with training scenes. Numbers ($\times 10^{-2}$) denote the L-2 distance to the source image within the semantic feature space.

Semantic Analysis for $\tilde{S}$. To assess whether the calibrated semantic representation $\tilde{S}$ truly captures semantic details of the scene, we analyze the highest and lowest response based on the L-2 distance between features from two different scenes, interpreting smaller distances as greater similarity. We consider the first image from each category in the LLFF dataset as a reference and extract the scene-level representation for the ten closest views to this reference image using CaesarNeRF.

In Fig. 7, we present two examples, "room" and "fortress", from the LLFF dataset, representing the first image of these categories alongside images corresponding to the top-2 highest and lowest responses. This analysis reveals that the scene-level representation predominantly emphasizes structural information and objects of similar categories. For instance, with the source image "room", the highest responses correlate with images featuring table-like structures, indicating an object-centric focus. In contrast, the lowest responses include images of flowers or ponds found in open areas, which diverge significantly from the structural and object content in the source images.

Comparison with Generative Methods. Single-view scenarios are often addressed by generative methods [35–37] that employ diffusion models [19,53]. While these models can produce reasonable results for object-centric renderings, they struggle with scene-level renderings from novel viewpoints. We show two LLFF examples in Fig. 8 using zero 1-to-3 [35], where the left image is the input, and the right one is the output. Images rendered with zero 1-to-3 [35] suffer from style difference. Unlike NeRF-based approaches that reconstruct images from observed pixels, generative models synthesize an entire image from its semantic representation but may not maintain the style.

(a) leaves　　　　　　　　　　　　(b) orchid

Fig. 8. Synthetic results for two examples from LLFF [42], "leaves" and "orchid", using zero 1-to-3 [35] with one reference image as input and 2 °C of vertical shift. The left image of each pair is the input, and the right one is the output of zero 1-to-3 [35].

5 Conclusion and Limitation

In this paper, we introduce CaesarNeRF, a few-shot and generalizable NeRF pipeline that combines scene-level semantic with per-pixel feature representations, aiding in rendering from novel camera positions with limited reference views. We calibrate the semantic representations across different input views and employ a sequential refinement network to offer distinct semantic representations at various levels. Our method has been extensively evaluated on a broad range of datasets, exhibiting state-of-the-art performance in both generalizable and single-scene settings.

Limitations and Potential Negative Impact. Instead of a generative approach, CaesarNeRF relies on NeRF's scheme, using input images to render the target view. This approach restricts its ability to render parts of the scene that are not present in the reference images. CaesarNeRF may create fake images for authentication.

References

1. Bao, C., et al.: Sine: Semantic-driven image-based nerf editing with prior-guided editing field. In: CVPR, pp. 20919–20929 (2023)
2. Bao, Y., Ding, T., Huo, J., Li, W., Li, Y., Gao, Y.: Insertnerf: instilling generalizability into nerf with hypernet modules. In: ICLR (2024)
3. Bao, Y., Li, Y., Huo, J., Ding, T., Liang, X., Li, W., Gao, Y.: Where and how: Mitigating confusion in neural radiance fields from sparse inputs. arXiv preprint arXiv:2308.02908 (2023)
4. Barron, J.T., Mildenhall, B., Verbin, D., Srinivasan, P.P., Hedman, P.: Mip-nerf 360: unbounded anti-aliased neural radiance fields. In: CVPR, pp. 5470–5479 (2022)
5. Buehler, C., Bosse, M., McMillan, L., Gortler, S., Cohen, M.: Unstructured lumigraph rendering. In: SIGGRAPH, pp. 425–432 (2001)

6. Chen, A., et al.: Mvsnerf: fast generalizable radiance field reconstruction from multi-view stereo. In: ICCV, pp. 14124–14133 (2021)
7. Chen, W., et al.: Beyond appearance: a semantic controllable self-supervised learning framework for human-centric visual tasks. In: CVPR (2023)
8. Chen, Y., Xu, H., Wu, Q., Zheng, C., Cham, T.J., Cai, J.: Explicit correspondence matching for generalizable neural radiance fields. arXiv preprint arXiv:2304.12294 (2023)
9. Chibane, J., Bansal, A., Lazova, V., Pons-Moll, G.: Stereo radiance fields (srf): learning view synthesis for sparse views of novel scenes. In: CVPR, pp. 7911–7920 (2021)
10. Cong, W., et al.: Enhancing nerf akin to enhancing llms: generalizable nerf transformer with mixture-of-view-experts. In: ICCV (2023)
11. Debevec, P.E., Taylor, C.J., Malik, J.: Modeling and rendering architecture from photographs: a hybrid geometry-and image-based approach. In: SIGGRAPH, pp. 11–20 (1996)
12. Deng, C., et al.: Nerdi: single-view nerf synthesis with language-guided diffusion as general image priors. In: CVPR, pp. 20637–20647 (2023)
13. Deng, K., Liu, A., Zhu, J.Y., Ramanan, D.: Depth-supervised nerf: fewer views and faster training for free. In: CVPR, pp. 12882–12891 (2022)
14. Downs, L., et al.: Google scanned objects: a high-quality dataset of 3d scanned household items. In: ICRA, pp. 2553–2560 (2022)
15. Fridovich-Keil, S., Meanti, G., Warburg, F.R., Recht, B., Kanazawa, A.: K-planes: explicit radiance fields in space, time, and appearance. In: CVPR, pp. 12479–12488 (2023)
16. Fu, Y., Misra, I., Wang, X.: Multiplane nerf-supervised disentanglement of depth and camera pose from videos. In: ICML (2022)
17. Gao, Y., Cao, Y.P., Shan, Y.: Surfelnerf: neural surfel radiance fields for online photorealistic reconstruction of indoor scenes. In: CVPR, pp. 108–118 (2023)
18. Gu, J., Liu, L., Wang, P., Theobalt, C.: Stylenerf: A style-based 3d-aware generator for high-resolution image synthesis. arXiv preprint arXiv:2110.08985 (2021)
19. Ho, J., Jain, A., Abbeel, P.: Denoising diffusion probabilistic models. In: NeurIPS, pp. 6840–6851 (2020)
20. Irshad, M.Z., et al.: Neo 360: neural fields for sparse view synthesis of outdoor scenes. In: ICCV, pp. 9187–9198 (2023)
21. Jain, A., Mildenhall, B., Barron, J.T., Abbeel, P., Poole, B.: Zero-shot text-guided object generation with dream fields. In: CVPR, pp. 867–876 (2022)
22. Jain, A., Tancik, M., Abbeel, P.: Putting nerf on a diet: semantically consistent few-shot view synthesis. In: ICCV, pp. 5885–5894 (2021)
23. Jang, W., Agapito, L.: Codenerf: disentangled neural radiance fields for object categories. In: ICCV, pp. 12949–12958 (2021)
24. Jiang, Y., et al.: Alignerf: high-fidelity neural radiance fields via alignment-aware training. In: CVPR, pp. 46–55 (2023)
25. Johari, M.M., Lepoittevin, Y., Fleuret, F.: Geonerf: generalizing nerf with geometry priors. In: CVPR, pp. 18365–18375 (2022)
26. Johnson, J., Alahi, A., Fei-Fei, L.: Perceptual losses for real-time style transfer and super-resolution. In: Leibe, B., Matas, J., Sebe, N., Welling, M. (eds.) ECCV 2016. LNCS, vol. 9906, pp. 694–711. Springer, Cham (2016). https://doi.org/10.1007/978-3-319-46475-6_43
27. Kania, A., Kasymov, A., Zięba, M., Spurek, P.: Hypernerfgan: Hypernetwork approach to 3d nerf gan. arXiv preprint arXiv:2301.11631 (2023)
28. Kingma, D.P., Ba, J.: Adam: A method for stochastic optimization. arXiv preprint arXiv:1412.6980 (2014)
29. Kwak, M., Song, J., Kim, S.: Geconerf: Few-shot neural radiance fields via geometric consistency. arXiv preprint arXiv:2301.10941 (2023)
30. Li, Z., Wang, Q., Cole, F., Tucker, R., Snavely, N.: Dynibar: neural dynamic image-based rendering. In: CVPR, pp. 4273–4284 (2023)

31. Lin, C.H., et al.: Magic3d: high-resolution text-to-3d content creation. In: CVPR, pp. 300–309 (2023)
32. Lin, H., Peng, S., Xu, Z., Yan, Y., Shuai, Q., Bao, H., Zhou, X.: Efficient neural radiance fields for interactive free-viewpoint video. In: SIGGRAPH Asia 2022 Conference Papers pp. 1–9 (2022)
33. Lin, Y., et al.: Componerf: Text-guided multi-object compositional nerf with editable 3d scene layout. arXiv preprint arXiv:2303.13843 (2023)
34. Liu, L., Habermann, M., Rudnev, V., Sarkar, K., Gu, J., Theobalt, C.: Neural actor: Neural free-view synthesis of human actors with pose control. TOC **40**(6), 1–16 (2021)
35. Liu, R., Wu, R., Van Hoorick, B., Tokmakov, P., Zakharov, S., Vondrick, C.: Zero-1-to-3: Zero-shot one image to 3d object. In: ICCV (2023)
36. Liu, X., Kao, S.h., Chen, J., Tai, Y.W., Tang, C.K.: Deceptive-nerf: Enhancing nerf reconstruction using pseudo-observations from diffusion models. arXiv preprint arXiv:2305.15171 (2023)
37. Liu, Y., et al.: Syncdreamer: Generating multiview-consistent images from a single-view image. arXiv preprint arXiv:2309.03453 (2023)
38. l; Liu, Y., et al.: Neural rays for occlusion-aware image-based rendering. In: CVPR, pp. 7824–7833 (2022)
39. Mariotti, O., Mac Aodha, O., Bilen, H.: Viewnerf: Unsupervised viewpoint estimation using category-level neural radiance fields. arXiv preprint arXiv:2212.00436 (2022)
40. Martin-Brualla, R., et al.: Nerf in the wild: Neural radiance fields for unconstrained photo collections. In: CVPR, pp. 7210–7219 (2021)
41. Metzer, G., Richardson, E., Patashnik, O., Giryes, R., Cohen-Or, D.: Latent-nerf for shape-guided generation of 3d shapes and textures. In: CVPR, pp. 12663–12673 (2023)
42. Mildenhall, B., et al.: Local light field fusion: Practical view synthesis with prescriptive sampling guidelines. TOG **38**(4), 1–14 (2019)
43. Mildenhall, B., Srinivasan, P.P., Tancik, M., Barron, J.T., Ramamoorthi, R., Ng, R.: NeRF: representing scenes as neural radiance fields for view synthesis. In: Vedaldi, A., Bischof, H., Brox, T., Frahm, J.-M. (eds.) ECCV 2020. LNCS, vol. 12346, pp. 405–421. Springer, Cham (2020). https://doi.org/10.1007/978-3-030-58452-8_24
44. Mildenhall, B., et al.: Nerf: Representing scenes as neural radiance fields for view synthesis. Commun. ACM **65**(1), 99–106 (2021)
45. Niemeyer, M., Barron, J.T., Mildenhall, B., Sajjadi, M.S., Geiger, A., Radwan, N.: Regnerf: regularizing neural radiance fields for view synthesis from sparse inputs. In: CVPR, pp. 5480–5490 (2022)
46. Noguchi, A., Sun, X., Lin, S., Harada, T.: Neural articulated radiance field. In: ICCV, pp. 5762–5772 (2021)
47. Park, K., et al.: Nerfies: deformable neural radiance fields. In: ICCV, pp. 5865–5874 (2021)
48. Peng, S., et al.: Animatable neural radiance fields for modeling dynamic human bodies. In: ICCV, pp. 14314–14323 (2021)
49. Poole, B., Jain, A., Barron, J.T., Mildenhall, B.: Dreamfusion: Text-to-3d using 2d diffusion. arXiv preprint arXiv:2209.14988 (2022)
50. Pumarola, A., Corona, E., Pons-Moll, G., Moreno-Noguer, F.: D-nerf: neural radiance fields for dynamic scenes. In: CVPR, pp. 10318–10327 (2021)
51. Qi, C.R., Su, H., Mo, K., Guibas, L.J.: Pointnet: deep learning on point sets for 3d classification and segmentation. In: CVPR, pp. 652–660 (2017)
52. Roessle, B., Barron, J.T., Mildenhall, B., Srinivasan, P.P., Nießner, M.: Dense depth priors for neural radiance fields from sparse input views. In: CVPR, pp. 12892–12901 (2022)
53. Rombach, R., Blattmann, A., Lorenz, D., Esser, P., Ommer, B.: High-resolution image synthesis with latent diffusion models. In: CVPR, pp. 10684–10695 (2022)

54. Schönberger, J.L., Frahm, J.M.: Structure-from-motion revisited. In: CVPR (2016)
55. Schönberger, J.L., Zheng, E., Frahm, J.-M., Pollefeys, M.: Pixelwise view selection for unstructured multi-view stereo. In: Leibe, B., Matas, J., Sebe, N., Welling, M. (eds.) ECCV 2016. LNCS, vol. 9907, pp. 501–518. Springer, Cham (2016). https://doi.org/10.1007/978-3-319-46487-9_31
56. Schwarz, K., Liao, Y., Niemeyer, M., Geiger, A.: Graf: generative radiance fields for 3d-aware image synthesis. NeurIPS **33**, 20154–20166 (2020)
57. Shue, J.R., Chan, E.R., Po, R., Ankner, Z., Wu, J., Wetzstein, G.: 3d neural field generation using triplane diffusion. In: CVPR, pp. 20875–20886 (2023)
58. Suhail, M., Esteves, C., Sigal, L., Makadia, A.: Generalizable patch-based neural rendering. In: ECCV, pp. 156–174 (2022). https://doi.org/10.1007/978-3-031-19824-3_10
59. Trevithick, A., Yang, B.: Grf: learning a general radiance field for 3d representation and rendering. In: ICCV, pp. 15182–15192 (2021)
60. Varma, M., Wang, P., Chen, X., Chen, T., Venugopalan, S., Wang, Z.: Is attention all that nerf needs? In: ICLR (2023)
61. Vaswani, A., et al.: Attention is all you need. NeurIPS **30** (2017)
62. Waechter, M., Moehrle, N., Goesele, M.: Let there be color! large-scale texturing of 3D reconstructions. In: Fleet, D., Pajdla, T., Schiele, B., Tuytelaars, T. (eds.) ECCV 2014. LNCS, vol. 8693, pp. 836–850. Springer, Cham (2014). https://doi.org/10.1007/978-3-319-10602-1_54
63. Wang, G., Chen, Z., Loy, C.C., Liu, Z.: Sparsenerf: distilling depth ranking for few-shot novel view synthesis. In: ICCV (2023)
64. Wang, Q., et al.: Ibrnet: learning multi-view image-based rendering. In: CVPR, pp. 4690–4699 (2021)
65. Wang, T., et al.: Rodin: a generative model for sculpting 3d digital avatars using diffusion. In: CVPR, pp. 4563–4573 (2023)
66. Wei, Y., Liu, S., Rao, Y., Zhao, W., Lu, J., Zhou, J.: Nerfingmvs: guided optimization of neural radiance fields for indoor multi-view stereo. In: CVPR, pp. 5610–5619 (2021)
67. Wizadwongsa, S., Phongthawee, P., Yenphraphai, J., Suwajanakorn, S.: Nex: real-time view synthesis with neural basis expansion. In: CVPR, pp. 8534–8543 (2021)
68. Xiangli, Y., et al.: Bungeenerf: progressive neural radiance field for extreme multi-scale scene rendering. In: ECCV (2022). https://doi.org/10.1007/978-3-031-19824-3_7
69. Xie, C., Park, K., Martin-Brualla, R., Brown, M.: Fig-nerf: figure-ground neural radiance fields for 3d object category modelling. In: 3DV, pp. 962–971 (2021)
70. Xu, D., Jiang, Y., Wang, P., Fan, Z., Shi, H., Wang, Z.: Sinnerf: training neural radiance fields on complex scenes from a single image. In: ECCV, pp. 736–753 (2022). https://doi.org/10.1007/978-3-031-20047-2_42
71. Xu, Q., Xu, Z., Philip, J., Bi, S., Shu, Z., Sunkavalli, K., Neumann, U.: Point-nerf: point-based neural radiance fields. In: CVPR, pp. 5438–5448 (2022)
72. Yang, B., et al.: Learning object-compositional neural radiance field for editable scene rendering. In: ICCV, pp. 13779–13788 (2021)
73. Yang, H., Hong, L., Li, A., Hu, T., Li, Z., Lee, G.H., Wang, L.: Contranerf: Generalizable neural radiance fields for synthetic-to-real novel view synthesis via contrastive learning. In: CVPR, pp. 16508–16517 (2023)
74. Yang, J., Pavone, M., Wang, Y.: Freenerf: Improving few-shot neural rendering with free frequency regularization. In: CVPR, pp. 8254–8263 (2023)
75. Yao, Y., Luo, Z., Li, S., Fang, T., Quan, L.: MVSNet: depth inference for unstructured multi-view stereo. In: Ferrari, V., Hebert, M., Sminchisescu, C., Weiss, Y. (eds.) ECCV 2018. LNCS, vol. 11212, pp. 785–801. Springer, Cham (2018). https://doi.org/10.1007/978-3-030-01237-3_47

76. Yu, A., Ye, V., Tancik, M., Kanazawa, A.: pixelnerf: neural radiance fields from one or few images. In: CVPR, pp. 4578–4587 (2021)
77. Yu, X., et al.: Mvimgnet: a large-scale dataset of multi-view images. In: CVPR, pp. 9150–9161 (2023)
78. Zhang, J., Yang, G., Tulsiani, S., Ramanan, D.: Ners: neural reflectance surfaces for sparse-view 3d reconstruction in the wild. NeurIPS **34**, 29835–29847 (2021)
79. Zhang, J., Li, X., Wan, Z., Wang, C., Liao, J.: Text2nerf: Text-driven 3d scene generation with neural radiance fields. arXiv preprint arXiv:2305.11588 (2023)
80. Zhenxing, M., Xu, D.: Switch-nerf: learning scene decomposition with mixture of experts for large-scale neural radiance fields. In: ICLR (2023)
81. Zhou, T., Tucker, R., Flynn, J., Fyffe, G., Snavely, N.: Stereo magnification: Learning view synthesis using multiplane images. arXiv preprint arXiv:1805.09817 (2018)
82. Zhou, Z., Tulsiani, S.: Sparsefusion: distilling view-conditioned diffusion for 3d reconstruction. In: CVPR, pp. 12588–12597 (2023)
83. Zhu, H., et al.: Multimodal neural radiance field. In: ICRA, pp. 9393–9399 (2023)
84. Zhu, H., Zheng, Z., Zheng, W., Nevatia, R.: Cat-nerf: constancy-aware tx2former for dynamic body modeling. In: CVPRW, pp. 6618–6627 (2023)
85. Zhuang, Y., Zhu, H., Sun, X., Cao, X.: Mofanerf: morphable facial neural radiance field. In: ECCV, pp. 268–285 (2022)
86. Zimny, D., Trzciński, T., Spurek, P.: Points2nerf: Generating neural radiance fields from 3d point cloud. arXiv preprint arXiv:2206.01290 (2022)

MapTracker: Tracking with Strided Memory Fusion for Consistent Vector HD Mapping

Jiacheng Chen[1]([✉]), Yuefan Wu[1], Jiaqi Tan[1], Hang Ma[1], and Yasutaka Furukawa[1,2]

[1] Simon Fraser University, Burnaby, Canada
cjc0722hz@gmail.com
[2] Wayve, London, UK

Abstract. This paper presents a vector HD-mapping algorithm that formulates the mapping as a tracking task and uses a history of memory latents to ensure consistent reconstructions over time. Our method, *MapTracker*, accumulates a sensor stream into memory buffers of two latent representations: 1) Raster latents in the bird's-eye-view (BEV) space and 2) Vector latents over the road elements (i.e., pedestrian-crossings, lane-dividers, and road-boundaries). The approach borrows the query propagation paradigm from the tracking literature that explicitly associates tracked road elements from the previous frame to the current, while fusing a subset of memory latents selected with distance strides to further enhance temporal consistency. A vector latent is decoded to reconstruct the geometry of a road element. The paper further makes benchmark contributions by 1) Improving processing code for existing datasets to produce consistent ground truth with temporal alignments and 2) Augmenting existing mAP metrics with consistency checks. MapTracker significantly outperforms existing methods on both nuScenes and Agroverse2 datasets by over 8% and 19% on the conventional and the new consistency-aware metrics, respectively. The code and models are available on our project page: https://map-tracker.github.io.

1 Introduction

Humans forget, so do neural networks. A robust memory is crucial for online systems to produce consistent outputs. Vector HD mapping, a task of reconstructing vector road geometries from vehicle sensor data, has made dramatic progress. A consistent vector HD mapping system, capable of reconstructing a

---

J. Chen, Y. Wu and J. Tan—Equal contribution.

Supplementary Information The online version contains supplementary material available at https://doi.org/10.1007/978-3-031-72658-3_6.

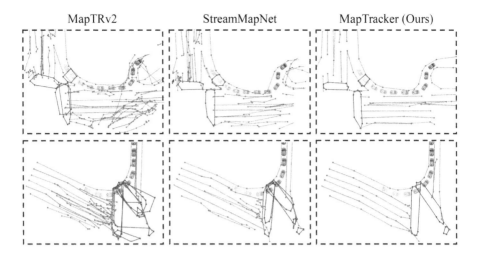

Fig. 1. MapTracker produces high-quality and temporally consistent vector HD maps, which are progressively merged into a global vector HD map by a simple online algorithm. The current state-of-the-art methods, MapTRv2 [19] and StreamMapNet [41], fail to produce consistent reconstructions, leading to very noisy global maps. The figure shows two challenging scenarios (cars are turning) from the nuScenes [2] dataset.

consistent HD map of a city from a single drive-through (See Fig. 1 for examples), would have a tremendous impact on our society, reducing the cost of HD map creation for tens of thousands of cities in the world and enhancing the safety and stability of self-driving cars.

Existing vector HD mapping methods [16,18,19,24,41] focus on per-frame reconstruction via detection-style transformer networks [4]. They detect road elements anew in every frame without consistency enforcement, potentially guided by reconstructions from the previous frame. Furthermore, a standard recurrent latent embedding is often the choice for memory mechanism [41], where accumulating the entire history in a single latent memory proves challenging, especially for cluttered environments with numerous vehicles obscuring road structures.

Towards ultimate temporal consistency, this paper presents MapTracker with two key design elements. First, tracking instead of detection becomes the formulation, specifically borrowing the query propagation paradigm from the tracking literature that explicitly associates tracked road elements across frames. Second, a sequence of memory latents from past frames serves as the memory mechanism. Concretely, we retain memory buffers for two latent representations from the past frames: 1) Raster latents in the bird's-eye-view (BEV) space and 2) Vector latents over the tracked road elements, while using a subset of memory latents based on distance strides for effective information fusion. A vector latent reconstructs a road element geometry.

To prepare the tracking labels and measure the consistency of HD map reconstructions, this paper introduces a new benchmark based on nuScenes [2] and Agroverse2 [37] datasets. Specifically, we improve the processing code of the two datasets to produce consistent ground truth data with temporal alignments, then propose a consistency-aware mean average precision (mAP) metric.

We have made extensive comparative evaluations based on the traditional and the new mAP metrics. MapTracker significantly outperforms the competing methods by over 8% on the conventional distance-based mAP, reaching 76.1 mAP on nuScenes and 76.9 mAP on Argoverse2. With the new consistency-aware metrics, MapTracker demonstrates superior temporal consistency and improves the StreamMapNet baseline by over 19%.

To summarize, this paper makes three contributions: 1) A novel vector HD mapping algorithm that formulates HD mapping as tracking and leverages the history of memory latents in two representations to achieve temporal consistency; 2) An improved vector HD mapping benchmark with temporally consistent ground truth and a consistency-aware mAP metric; and 3) SOTA performance with significant improvements over the current best methods on traditional and new metrics. The code and the new benchmark data will be available.

2 Related Work

This paper tackles consistent vector HD mapping by 1) borrowing an idea from the visual object tracking literature and 2) devising a new memory mechanism. The section first reviews recent trends in visual object tracking with transformers and memory designs in vision-based autonomous driving. Lastly, we discuss competing vector HD mapping methods.

Visual Object Tracking with Transformers. Visual object tracking [40] has a long history, where end-to-end transformer [35] methods become a recent trend due to the simplicity. TrackFormer [27], TransTrack [34], and MOTR [42,45] leverage the attention mechanism with *track queries* to explicitly associate instances across frames. MeMOT [3] and MeMOTR [8] further extend the tracking transformers with memory mechanisms for better long-term consistency. This paper formulates vector HD mapping as a tracking task by incorporating track queries with a more robust memory mechanism.

Memory Designs in Autonomous Driving. Single-frame self-driving systems have difficulty in handling occlusion, sensor failure, or complex environments. Temporal modeling with memories offers promising complements [9,10,12,13,22,23,36,39,41]. Many memory designs exist for the raster BEV features [17,29], which form the foundation of most autonomous-driving tasks [15, 26]. BEVDet4D [12] and BEVFormerv2 [39] stack features of multiple past frames as a memory, but the computation scales linearly with history length, struggling to capture long-term information. VideoBEV [10] propagates BEV raster queries across frames to accumulate information recurrently. In the vector domain, Sparse4Dv2 [22] employs a similar RNN-style memory for object

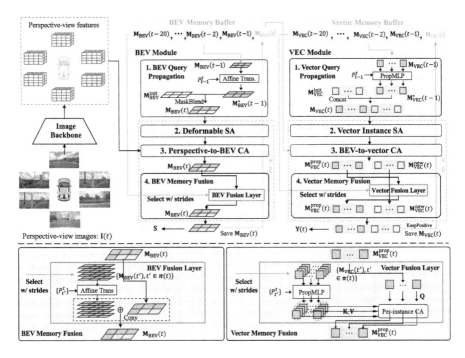

Fig. 2. (**Top**) The overall architecture of MapTracker. (**Bottom**) The close-up views of the BEV and the Vector fusion layers.

queries, while Sparse4Dv3 [23] further uses temporal denoising for robust temporal learning. These ideas have been partially incorporated by vector HD mapping approaches [36,41]. This paper proposes a new memory design for both the raster BEV latents and the vector latents of road elements.

Vector HD Mapping. Traditionally HD maps are reconstructed offline with SLAM-based methods [32,33,44], followed by human curation, requiring high maintenance costs. Online vector HD mapping algorithms are gaining more interest over their offline counterparts as their accuracy and efficiency improve, which would simplify the production pipeline and handle map changes. HDMapNet [16] turns raster map segmentation into vector map instances via post-processing and has established the first Vector HD mapping benchmark. VectorMapNet [24] and MapTR [18] both leverage DETR-based [4] transformers for end-to-end prediction. The former predicts the vertices of each detected curve autoregressively, while the latter uses hierarchical queries and matching loss to predict all the vertices simultaneously. MapTRv2 [19] further complements MapTR with auxiliary tasks and network modifications. Curve representation [7,30,46], network design [38], and training paradigm [5,43] are the focus of other works. StreamMapNet [41] steps towards consistent mapping by borrowing the streaming idea from BEV perception. The idea accumulates the past information into memory latents and passes as a condition (i.e., a conditional detection framework). SQD-MapNet [36] proposes temporal curve denoising to facilitate temporal learning, mimicking DN-DETR [14].

3 MapTracker

A robust memory mechanism is the core of MapTracker, accumulating a sensor stream into latent memories of two representations: 1) Bird's-eye-view (BEV) memory of a region around a vehicle in the top-down BEV coordinate frame as a latent image; and 2) Vector (VEC) memory of road elements (i.e., pedestrian-crossings, lane-dividers, and road-boundaries) as a set of latent vectors.

Two simple ideas with the memory mechanism achieve consistent mapping. The first idea is to use a buffer of memories from the past instead of a single memory at the current frame [10,17,41]. A single memory should hold the entire past information but is susceptible to memory loss, especially in cluttered environments with numerous vehicles obscuring road structures. Concretely, we select a subset of the past latent memories for fusion at each frame based on the vehicle motions for efficiency and coverage. The second idea is to formulate the online HD mapping as a tracking task. The VEC memory mechanism maintains a sequence of memory latents with each road element and makes this formulation straightforward by borrowing a query propagation paradigm from the tracking literature. The rest of the section explains our neural architectures (See Fig. 2 and Fig. 3), consisting of the BEV and VEC memory buffers and their corresponding network modules, and then presents the training details.

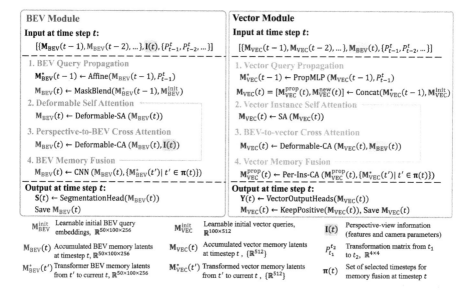

Fig. 3. The architecture details of the BEV and the Vector modules. The BEV-related representations are in green, while the vector-related representations are in cyan. Details of the attention layers are described in Sect. 3.

3.1 Memory Buffers

A BEV memory, $\mathbf{M}_{\text{BEV}}(t) \in \mathbb{R}^{50 \times 100 \times 256}$, is a 2D latent image in the BEV coordinate frame centered and oriented with the vehicle at frame t. The spatial dimension (i.e., 50×100) covers a rectangular area, 15 m left/right and 30 m front/back. Each memory latent accumulates the entire past information, while the buffer holds such memory latents for the last 20 frames, making the memory mechamism redundant but robust.

A VEC memory, $\mathbf{M}_{\text{VEC}}(t) \in \{\mathbb{R}^{512}\}$, is a set of vector latents, each of which accumulates information of an active road element up to frame t. The number of active elements varies per frame. The buffer holds the latent vectors of the past 20 frames and their correspondences across frames (i.e., a sequence of vector latents corresponding to the same road element).

3.2 BEV Module

Inputs are 1) CNN features of the onboard perspective images processed by the image backbone (the official ResNet50 model [11] pretrained on ImageNet [6]) and their camera parameters $\mathbf{I}(t)$; 2) the BEV memory buffer $\{\mathbf{M}_{\text{BEV}}(t-1), \mathbf{M}_{\text{BEV}}(t-2), ...\}$; and 3) the vehicle motions $\{P_{t-1}^{t}, P_{t-2}^{t}, ...\}$, where $P_{t_1}^{t_2} \in \mathbb{R}^{4 \times 4}$ is the affine transformation of the vehicle coordinate frame from frame t_1 to t_2. The following explains the four components of the BEV module architecture and its outputs.

[1. BEV Query Propagation]. A BEV memory is a 2D latent image in a vehicle coordinate frame. An affine transformation P_{t-1}^{t} and a bilinear interpolation initialize the current BEV memory $\mathbf{M}_{\text{BEV}}(t)$ with the previous one $\mathbf{M}_{\text{BEV}}(t-1)$. For pixels that fall outside the latent image after the transformation, per-pixel learnable embedding vectors $\mathbf{M}_{\text{BEV}}^{\text{init}} \in \mathbb{R}^{50 \times 100 \times 256}$ are the initialization instead, whose operation is denoted as "MaskBlend" in Fig. 3.

[2. Deformable Self-attention]. A deformable self-attention layer [47] enriches the BEV memory $\mathbf{M}_{\text{BEV}}(t)$.

[3. Perspective-to-BEV Cross-Attention]. Similar to StreamMapNet [41], a spatial deformable cross-attention layer from BEVFormer [17] injects the perspective-view information $\mathbf{I}(t)$ into $\mathbf{M}_{\text{BEV}}(t)$, followed by a standard feed-forward network (FFN) layer [35].

[4. BEV Memory Fusion]. The memory latents in the buffer are fused to enrich $\mathbf{M}_{\text{BEV}}(t)$. Using all the memories is computationally expensive and redundant. We use a strided selection of four memories without repetition, whose vehicle positions are the closest to (1 m/5 m/10 m/15 m) from the current position. An affine transformation and a bilinear interpolation align the coordinate frames of the selected memories to the current: $\{\mathbf{M}_{\text{BEV}}^{*}(t'), t' \in \boldsymbol{\pi}(t)\}$, where $\boldsymbol{\pi}(t)$ denotes the selected times. We concatenate $\mathbf{M}_{\text{BEV}}(t)$ with the aligned memories and use a lightweight residual block with two convolution layers to update

$\mathbf{M}_{\text{BEV}}(t)$. The last three components of the BEV module repeat twice without weight sharing.

Outputs are 1) the final memory $\mathbf{M}_{\text{BEV}}(t)$ saved to the buffer and passed to the VEC module; and 2) the rasterized road element geometries $\mathbf{S}(t)$ which is inferred by a segmentation head and used for a loss calculation (See Sect. 3.4). The segmentation head is a linear projection module that projects each pixel in the memory latent to a 2×2 segmentation mask, thus producing a 100×200 mask.

3.3 VEC Module

Inputs are 1) the BEV memory $\mathbf{M}_{\text{BEV}}(t)$; 2) the vector memory buffer $\{\mathbf{M}_{\text{VEC}}(t-1), \mathbf{M}_{\text{VEC}}(t-2), ...\}$; and 3) the vehicle motions $\{P^t_{t-1}, P^t_{t-2}, ...\}$.

[1. Vector Query Propagation]. A vector memory is a set of latent vectors of the active road elements. Borrowing the query propagation paradigm from transformer-based tracking approaches [27,34,42], we initialize the vector memory as $\mathbf{M}_{\text{VEC}}(t) = [\mathbf{M}^{\text{prop}}_{\text{VEC}}(t), \mathbf{M}^{\text{new}}_{\text{VEC}}(t)]$. $\mathbf{M}^{\text{new}}_{\text{VEC}}(t)$ denotes 100 latent vectors for 100 new road element candidates, which are initialized with 100 learnable embeddings $\mathbf{M}^{\text{init}}_{\text{VEC}}$. $\mathbf{M}^{\text{prop}}_{\text{VEC}}(t)$ denotes latent vectors for the currently tracked road elements, which are initialized with the corresponding latent vectors in the previous memory $\mathbf{M}_{\text{VEC}}(t-1)$ after using a two-layer MLP to align the coordinate frame. Concretely, we turn P^t_{t-1} into a 4D vector of rotation quaternion and 3D vector of translation parameters, represent with their positional encodings [35], concatenate them with each vector latent in $\mathbf{M}_{\text{VEC}}(t-1)$, and apply an MLP. We call it PropMLP, which handles the temporal propagation.

[2. Vector Instance Self Attention]. Similar to StreamMapNet, a standard self-attention layer enriches the vector latents in the memory $\mathbf{M}_{\text{VEC}}(t)$.

[3. BEV-to-Vector Cross Attention]. The Multi-Point Attention from StreamMapNet, which is an extension of the vanilla deformable cross-attention [47], injects the BEV information from $\mathbf{M}_{\text{BEV}}(t)$ into $\mathbf{M}_{\text{VEC}}(t)$.

[4. Vector Memory Fusion]. For each latent vector in the current memory $\mathbf{M}_{\text{VEC}}(t)$, latent vectors in the buffer associated with the same road element are fused to enrich its representation. The same strided frame-selection chooses four latent vectors, where the selected frames $\pi(t)$ would be different and fewer for some road elements with a short tracking history. For example, an element that has been tracked for two frames has only two latents in the buffer. A standard cross-attention followed by an FFN layer injects the selected latents after aligning their coordinate frames by the same PropMLP module $\{\mathbf{M}^*_{\text{VEC}}(t'), t' \in \pi(t)\}$. To be precise, a query is a latent in $\mathbf{M}^{\text{prop}}_{\text{VEC}}(t)$, where a key/value is a latent in $\{\mathbf{M}^*_{\text{VEC}}(t'), t' \in \pi(t)\}$. In Fig. 3, we omit the element index for the "Per-Ins-CA" operation, which acts on each element independently. The last three components of the VEC module repeat six times without weight sharing.

Outputs are 1) the final memory $\mathbf{M}_{\text{VEC}}(t)$ for "positive" road elements that pass the classification test by a single fully connected layer from $\mathbf{M}_{\text{VEC}}(t)$; and 2) vector road geometries of the positive road elements, regressed by the 3-layer MLP from $\mathbf{M}_{\text{VEC}}(t)$. The threshold of the classification test is 0.4 for the first frame, and 0.5/0.6 for the propagated/new road elements for subsequent frames. $\mathbf{Y}(t) = \{(V_i, p_i)\}$ denotes the outputs. Following the prior convention [18], each element geometry $V_i = [(x_1, y_1), \ldots, (x_{20}, y_{20})]$ is a polygonal curve with 20 points in the BEV coordinate frame. p_i is the class probability score.

3.4 Training

The ground-truth road element geometries are denoted as $\hat{\mathbf{Y}}(t) = \{\hat{Y}_i\}$. $\hat{Y}_i = (\hat{V}_i, \hat{c}_i)$, where V_i has 20 points interpolated from the raw ground-truth vector. $\hat{c}_i$ is the class label. Standard OpenCV and PIL libraries rasterize $\hat{\mathbf{Y}}(t)$ on an empty BEV canvas to obtain the ground-truth segmentation image $\hat{\mathbf{S}}(t)$

BEV loss. We employ per-pixel Focal loss [21] and per-class Dice loss [28] on the BEV outputs $\mathbf{S}(t)$, which are common auxiliary losses in vector HD mapping approaches [19,31]. The loss is defined by

$$\mathcal{L}_{\text{BEV}} = \lambda_1 \mathcal{L}_{\text{focal}}(\mathbf{S}(t), \hat{\mathbf{S}}(t)) + \lambda_2 \mathcal{L}_{\text{dice}}(\mathbf{S}(t), \hat{\mathbf{S}}(t)) \quad (1)$$

VEC Loss. Inspired by MOTR [42], an end-to-end transformer for multi-object tracking, we extend the matching-based loss [18,41] to explicitly consider ground-truth tracks (See Sect. 4 for ground-truth processing). For each frame t, $\hat{\mathbf{Y}}(t)$ consists of two disjoint subsets: new elements $\hat{\mathbf{Y}}_{\text{new}}(t)$ and tracked elements $\hat{\mathbf{Y}}_{\text{track}}(t)$. For the vector outputs, we denote the results from the propagated latents $\mathbf{M}_{\text{VEC}}^{\text{prop}}(t)$ as $\mathbf{Y}_{\text{track}}(t)$, and results from the new latents $\mathbf{M}_{\text{VEC}}^{\text{new}}(t)$ as $\mathbf{Y}_{\text{new}}(t)$. Note that to make the VEC module robust to potential errors in pose estimation, we randomly perturb the transformation matrix P_{t-1}^t by adding a Gaussian noise during training. We train the module with a tracking loss that explicitly considers the temporal alignments. Section 4 explains the ground-truth preparation. The optimal instance-level label assignment for new elements is defined as:

$$\boldsymbol{\omega}_{\text{new}}(t) = \arg\min_{\boldsymbol{\omega}_{\text{new}}(t) \in \Omega(t)} \mathcal{L}_{\text{match}}(\hat{\mathbf{Y}}_{\text{new}}(t)|_{\boldsymbol{\omega}_{\text{new}}(t)}, \mathbf{Y}_{\text{new}}(t)). \quad (2)$$

$\Omega(t)$ is the space of all bipartite matches. $\mathcal{L}_{\text{match}}$ is the hierarchical matching cost similar to the one proposed in MapTR [18], consisting of a focal loss $\mathcal{L}_{\text{focal}}(\{\hat{c}_i\}|_{\boldsymbol{\omega}_{\text{new}}(t)}, \{p_i\})$ and a permutation-invariant line coordinate loss $\mathcal{L}_{\text{line}}(\{\hat{V}_i\}|_{\boldsymbol{\omega}_{\text{new}}(t)}, \{V_i\})$. The label assignments $\boldsymbol{\omega}(t)$ between all outputs and ground truth is then defined inductively:

$$\boldsymbol{\omega}(t) = \boldsymbol{\omega}_{\text{track}}(t) \cup \boldsymbol{\omega}_{\text{new}}(t); \quad \boldsymbol{\omega}_{\text{track}}(t) = \begin{cases} \varnothing, & \text{if } t = 0 \\ \boldsymbol{\omega}(t-1), & \text{if } t > 0 \end{cases}. \quad (3)$$

$\omega_{\text{track}}(t)$ is the label assignments between $\mathbf{Y}_{\text{track}}(t)$ and $\hat{\mathbf{Y}}_{\text{track}}(t)$. The tracking-style loss for the vector outputs is:

$$\mathcal{L}_{\text{track}} = \lambda_3 \mathcal{L}_{\text{focal}}(\{\hat{c}_i\}|_{\omega(t)}, \{p_i\}) + \lambda_4 \mathcal{L}_{\text{line}}(\{\hat{V}_i\}|_{\omega(t)}, \{V_i\}). \quad (4)$$

Transformation Loss. We borrow the transformation loss L_{trans} from StreamMapNet [41] to train the PropMLP, which enforces that the query transformation in the latent space maintains the vector geometry and class type. Full details are provided in the Appendix. The final training loss is

$$\mathcal{L} = \mathcal{L}_{\text{BEV}} + \mathcal{L}_{\text{track}} + \lambda_5 \mathcal{L}_{\text{trans}}. \quad (5)$$

Training Details. For each training sample, we randomly choose 4 out of the previous 10 frames to compose a training clip with a length of 5. We freeze the image backbone for the first four training frames to reduce the memory cost for the clip-based training. The training of the system has three stages: 1) Pre-train the image backbone and BEV encoder with only $\mathcal{L}_{\text{BEV}}$; 2) Warm up the vector decoder while freezing all other parameters with $\mathcal{L}$, where the vector memory is turned on after 500 warmup iterations; 3) Jointly train all parameters with $\mathcal{L}$. The second stage warms up the vector module with a large batch size to facilitate initial convergence, as we cannot afford in the joint training. The loss weights are $\lambda_1 = 10.0, \lambda_2 = 1.0, \lambda_3 = 5.0, \lambda_4 = 50.0, \lambda_5 = 0.1$. We use an AdamW [25] optimizer with an initial learning rate 5e-4 and the weight decay is set to 0.01. A cosine learning rate scheduler is used with a final learning rate of 1.5e–6.

4 Consistent Vector HD Mapping Benchmarks

The section makes existing HD mapping benchmarks consistency-aware by 1) Improving pre-processing to generate temporally consistent ground truth with "track" labels (Sect. 4.1); and 2) Augmenting the standard mAP metric with consistency checks (Sect. 4.2).

4.1 Consistent Ground Truth

MapTR [18,19] created vector HD mapping benchmark from nuScenes and Agroverse2 datasets, adopted by many follow-ups [5,19,30,43,46]. However, pedestrian crossings are merged naively and inconsistent across frames. Divider lines are also inconsistent (for Argoverse2) with the failures of its graph tracing process.

StreamMapNet [41] inherited code from VectorMapNet [24] and created a benchmark with better ground truth, which has been used in the workshop challenge [1]. However, there are still issues. For Argoverse2, divider lines are sometimes split into shorter segments. For nuScenes, large pedestrian crossings sometimes split out small loops, whose inconsistencies arise randomly per frame, leading to temporarily inconsistent representations. We provide visualizations for the issues of existing benchmarks in the Appendix.

We improve processing code from existing benchmarks to (1) enhance per-frame ground-truth geometries, then (2) compute their correspondences across frames, forming ground-truth "tracks".

(1) Enhancing Per-Frame Geometries. We inherit and improve the MapTR codebase, which has been popular in the community, while making two changes: Replace the pedestrian-zone processing with the one in StreamMapNet and further improve the quality by more geometric constraints; and Enforce temporal consistency in the divider processing by augmenting the graph tracing algorithm to handle noises of raw annotations (only for Argoverse2).

(2) Forming Tracks. Given per-frame road element geometries, we solve an optimal bipartite matching problem between every pair of adjacent frames to establish correspondences of road elements. Pairwise correspondences are chained to form tracks of road elements. The matching score between a pair of road elements is defined as follows. A road-element geometry is either a polygonal curve or a loop. We transform an element geometry in an older frame to the newer one based on the vehicle motion, then rasterize both curves/loops with a certain thickness into instance masks. Their intersection over union is the matching score. Please refer to Appendix for the full algorithmic details.

4.2 Consistency-Aware mAP Metric

The standard mean average precision (mAP) metric does not penalize temporarily inconsistent reconstructions. We match reconstructed road elements and the ground truth in each frame independently with Chamfer distance, as in the standard mAP process, then remove temporarily inconsistent matches with the following check. First, for baseline methods that do not predict tracking information, we form tracks of reconstructed road elements using the same algorithm we used to get ground-truth temporal correspondences (we also extend the algorithm to re-identify a lost element by trading off the speed; see Appendix for details). Next, let an "ancestor" be a road element that belongs to the same track in a prior frame. From the beginning of the sequence, we remove a per-frame match (of reconstructed and ground-truth elements) as temporarily inconsistent if any of their ancestors was not a match. The standard mAP is then calculated with the remaining temporarily consistent matches. See Appendix for complete algorithmic details.

5 Experiments

We build our system based on the StreamMapNet codebase, while using 8 NVIDIA RTX A5000 GPUs to train our model for 72 epochs on nuScenes (18, 6, and 48 epochs for the three stages) and 35 epochs on Argoverse2 (12, 3, and 20 epochs for the three stages). The batch sizes for the three training stages are

16, 48, and 16, respectively. The training takes roughly three days, while the inference speed is roughly 10 FPS. After explaining the datasets, the metrics, and the baseline methods, the section provides the experimental results.

Datasets. We use the nuScenes [2] and Argoverse2 [37] datasets. nuScenes dataset is annotated with 2 Hz with 6 synchronized surrounding cameras. Input perspective images are of size 480×800. Argoverse2 dataset is annotated with 10 Hz, using 7 surrounding cameras. Input perspective images are of size 608×608. We follow MapTRv2 [19] and use an interval of 4 to subsample the sequences of Argoverse2. We evaluate the methods with the official dataset splits as well as the geographically non-overlapping splits proposed in StreamMapNet [41].

Metrics. We follow prior works [18,19,24,41] and use Average Precision (AP) as the main evaluation metric, where Chamfer distance is the matching criterion. The AP is averaged across three distance thresholds $\{0.5m, 1.0m, 1.5m\}$. The final mean AP (mAP) is computed by averaging the results over the three road element types: pedestrian crossing, lane-divider, and road-boundary. We provide both the original scores and the new consistency-aware augmented scores (Sect. 4.2).

Baselines. MapTRv2 [19] and StreamMapNet [41] are the main baselines due to their popularity and superior performance. We run their official codebase and train the models until complete convergence. The results of recent competing methods [7,36,46] are also included for reference by copying numbers from their corresponding papers.

Table 1. Results on nuScenes [2]. The first column shows three different ground truth used for training and testing. "Consistent" is our temporarily consistent ground truth. The standard AP scores are reported for pedestrian crossing, lane-divider, road-boundary, and their average. C-mAP is our consistency-aware metric, which requires tracking information in the ground truth and is reported only for Consistent. [+]: Numbers are from the original papers. [†]: Epochs for our multi-frame training.

G.T. data	Method	Backbone	Epoch	AP_p	AP_d	AP_b	mAP	C-mAP
MapTR	MapTR[+] [18]	R50	110	56.2	59.8	60.1	58.7	–
	PivotNet[+] [7]	SwinT	110	62.6	68.0	69.7	66.8	
	MapTRv2[+] [19]	R50	110	68.1	68.3	69.7	68.7	
	GeMap[+] [46]	R50	110	67.1	69.8	71.4	69.4	
StmMapNet	StreamMapNet [41]	R50	110	68.0	71.2	68.0	69.1	–
	SQD-MapNet[+] [36]	R50	24	63.6	66.6	64.8	65.0	
	MapTracker (Ours)	R50	72[†]	77.3	72.4	74.2	74.7	
Consistent	MapTRv2 [19]	R50	110	69.6	68.5	70.3	69.5	50.5
	StreamMapNet [41]	R50	110	70.0	72.9	68.3	70.4	56.4
	MapTracker (Ours)	R50	72[†]	80.0	74.1	74.1	76.1	69.1

5.1 Quantitative Evaluations

One of our contributions is the temporarily consistent ground truth (GT) over the two existing counterparts (i.e., MapTR [18,19] and StreamMapNet [41]). Table 1 and Table 2 show the results where a system is trained and tested on one of the three GTs (shown in the first column). Since our codebase is based on StreamMapNet, we evaluate our system on the StreamMapNet GT and our temporarily consistent GT.

nuScenes Results. Table 1 shows that both MapTRv2 and StreamMapNet achieve better mAP with our GT, which is expected as we fixed the inconsistencies in their original GT (explained in Sect. 4.1). StreamMapNet's improvement is slightly higher since it has temporal modeling (whereas MapTR does not) and exploits temporal consistency in the data. MapTracker significantly outperforms the competing methods, especially with our consistent GT by more than 8% and 22% in the original and the consistency-aware mAP scores. Note that MapTracker is the only system to produce explicit tracking information (i.e., correspondences of reconstructed elements across frames), which is required for the consistency-area mAP. A simple matching algorithm creates tracks for the baseline methods (See Appendix for details).

Table 2. Results on Argoverse2 [37]. [+]: Numbers are from the original papers. [†]: Epochs for our multi-frame training.

G.T. data	Method	Backbone	Epoch	AP_p	AP_d	AP_b	mAP	C-mAP
MapTR	MapTRv2[+] [19]	R50	6*4	62.9	72.1	67.1	67.4	–
	GeMap[+] [46]	R50	24*4	69.2	75.7	70.5	71.8	
StmMapNet	StreamMapNet[+] [41]	R50	30	62.0	59.5	63.0	61.5	–
	StreamMapNet [41]	R50	72	65.0	62.2	64.9	64.0	
	SQD-MapNet[+] [36]	R50	30	64.9	60.2	64.9	63.3	
	MapTracker (Ours)	R50	35[†]	74.5	66.4	73.4	71.4	
Consistent	MapTRv2 [19]	R50	24*4	68.3	75.6	68.9	70.9	56.1
	StreamMapNet [41]	R50	72	70.5	74.2	66.1	70.3	57.5
	MapTracker (Ours)	R50	35[†]	77.0	80.0	73.7	76.9	68.3

Argoverse2 Results. Table 2 shows that both MapTRv2 and StreamMapNet achieve better mAP scores with our consistent GT, which has higher quality GT (for pedestrian crossings and dividers) besides being temporarily consistent, benefiting all the methods. MapTracker outperforms all the other baselines by significant margins (*i.e.*, 11% or 8%, respectively) in all settings. The consistency-aware score (C-mAP) further demonstrates our superior consistency, showing an improvement of more than 18% over StreamMapNet.

Table 3. Results with geographically non-overlapping data proposed in StreamMapNet [41]. Our consistent ground truth is used. †: Epochs for our multi-frame training.

Dataset	Method	Backbone	Epoch	AP_p	AP_d	AP_b	mAP	C-mAP
nuScenes [2]	StreamMapNet [41]	R50	110	31.6	28.1	40.7	33.5	22.2
	MapTracker (Ours)	R50	72†	45.9	30.0	45.1	40.3	32.5
Argoverse2 [37]	StreamMapNet [41]	R50	72	61.8	68.2	63.2	64.4	54.4
	MapTracker (Ours)	R50	35†	70.0	75.1	68.9	71.3	63.2

5.2 Results with Geographically Non-overlapping Data

Official training/testing splits of nuScenes and Agroverse2 datasets have geographical overlaps (i.e., the same roads appear in both training/testing), which allows overfitting [20]. Table 3 compares the best baseline method StreamMapNet and MapTracker, based on geographically non-overlapping splits, proposed by StreamMapNet. MapTracker consistently outperforms with significant margins, demonstrating robust cross-scene generalization. Note that the performance for nuScenes datasets degrades for both methods. Upon careful inspection, the detection of road elements is successful but the regressed coordinates have large errors, leading to low performance.

5.3 Ablation Studies

Ablation studies in Table 4 demonstrate the contributions of key design elements in MapTracker. The first "baseline" entry is StreamMapNet without its temporal reasoning capabilities (i.e., without its BEV and vector streaming memories and modules). The second entry is StreamMapNet. Both methods are trained for 110 epochs till full convergence. The last three entries are the variants of MapTracker with or without the key design elements. The first variant drops the memory fusion components in the BEV/VEC modules. This variant utilizes the tracking formulation but relies on a single BEV/VEC memory to hold the

Table 4. Ablation studies on the key design elements of MapTracker, evaluated on the nuScenes dataset with our consistent ground truth.

Method	Task	Memory			Metrics				
		Embed.	+Fusion	+Stride	AP_p	AP_d	AP_b	mAP	C-mAP
Baseline [41]	Detection	-	-	-	69.5	71.7	68.5	69.9	56.1
StmMapNet [41]	Cond. detect.	✓	-	-	70.0	72.9	68.3	70.4	56.4
MapTracker	Tracking	✓	-	-	73.8	69.2	69.4	70.8	62.4
		✓	✓	-	78.6	73.3	72.8	74.9	68.1
		✓	✓	✓	80.0	74.1	74.1	76.1	69.1

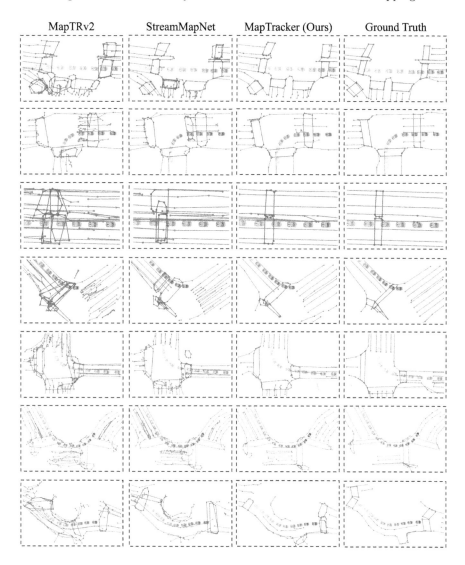

Fig. 4. Qualitative comparisons of the two representative baselines, MapTracker (Ours), and the ground truth. A simple online algorithm merges per-frame vector HD map reconstructions across a single drive-through into a global vector HD map. The top five examples are from nuScenes, while the bottom two are from Argoverse2.

past information, like the GRU embedding of StreamMapNet. The second variant adds the memory buffers and the memory fusion components but without the striding strategy, that is, using the latest 4 frames for the fusion. This variant boosts performance, demonstrating the effectiveness of our memory mechanism. The last variant adds memory striding, which makes more effective use of the memory mechanism and improves performance.

5.4 Qualitative Evaluations

Figure 4 presents qualitative comparisons of MapTracker and the baseline methods on both nuScenes and Argoverse2 datasets. For better visualization, we use a simple algorithm to merge per-frame vector HD maps into a global vector HD map. Please refer to Appendix for the details of the merging algorithm and the visualization of per-frame reconstructions. MapTracker produces much more accurate and cleaner results, demonstrating superior overall quality and temporal consistency. For scenarios where the vehicle is turning or not trivially moving forward (including the two examples in Fig. 1), StreamMapNet and MapTRv2 can produce unstable results, thus leading to broken and noisy merged results. This is mainly because the detection-based formulation has difficulties maintaining temporally coherent reconstructions under complex vehicle motions.

6 Conclusion

This paper introduces MapTracker, which formulates vector HD mapping as a tracking task and leverages a history of raster and vector latents to maintain temporal consistency. We employ a query propagation mechanism to associate tracked road elements across frames, and fuse a subset of memory entries selected with distance strides to enhance consistency. We also improve existing benchmarks by generating consistent ground truth with tracking labels and augmenting the original mAP metric with temporal consistency checks. MapTracker significantly outperforms existing methods on nuScenes and Agroverse2 datasets when evaluated with the traditional metrics and demonstrates superior temporal consistency when evaluated with our consistency-aware metrics.

Acknowledgements. This research is partially supported by NSERC Discovery Grants, NSERC Alliance Grants, and John R. Evans Leaders Fund (JELF). We thank the Digital Research Alliance of Canada and BC DRI Group for providing computational resources.

References

1. Online hd map construction challenge for autonomous driving on cvpr 2023 workshop on end-to-end autonomous driving. https://github.com/Tsinghua-MARS-Lab/Online-HD-Map-Construction-CVPR2023 (2023)
2. Caesar, H., et al.: nuscenes: a multimodal dataset for autonomous driving. In: Proceedings of the IEEE/CVF Conference on Computer Vision and Pattern Recognition, pp. 11621–11631 (2020)
3. Cai, J., et al.: Memot: multi-object tracking with memory. In: Proceedings of the IEEE/CVF Conference on Computer Vision and Pattern Recognition, pp. 8090–8100 (2022)
4. Carion, N., Massa, F., Synnaeve, G., Usunier, N., Kirillov, A., Zagoruyko, S.: End-to-end object detection with transformers. In: Vedaldi, A., Bischof, H., Brox, T., Frahm, J.-M. (eds.) ECCV 2020. LNCS, vol. 12346, pp. 213–229. Springer, Cham (2020). https://doi.org/10.1007/978-3-030-58452-8_13

5. Chen, J., Deng, R., Furukawa, Y.: Polydiffuse: Polygonal shape reconstruction via guided set diffusion models. arXiv preprint arXiv:2306.01461 (2023)
6. Deng, J., Dong, W., Socher, R., Li, L.J., Li, K., Fei-Fei, L.: Imagenet: A large-scale hierarchical image database. In: 2009 IEEE Conference on Computer Vision and Pattern Recognition, pp. 248–255. Ieee (2009)
7. Ding, W., Qiao, L., Qiu, X., Zhang, C.: Pivotnet: vectorized pivot learning for end-to-end hd map construction. In: Proceedings of the IEEE/CVF International Conference on Computer Vision, pp. 3672–3682 (2023)
8. Gao, R., Wang, L.: Memotr: long-term memory-augmented transformer for multi-object tracking. In: Proceedings of the IEEE/CVF International Conference on Computer Vision, pp. 9901–9910 (2023)
9. Gu, J., et al.: Vip3d: end-to-end visual trajectory prediction via 3d agent queries. In: Proceedings of the IEEE/CVF Conference on Computer Vision and Pattern Recognition, pp. 5496–5506 (2023)
10. Han, C., et al.: Exploring recurrent long-term temporal fusion for multi-view 3d perception. arXiv preprint arXiv:2303.05970 (2023)
11. He, K., Zhang, X., Ren, S., Sun, J.: Deep residual learning for image recognition. In: Proceedings of the IEEE Conference on Computer Vision and Pattern Recognition, pp. 770–778 (2016)
12. Huang, J., Huang, G.: Bevdet4d: Exploit temporal cues in multi-camera 3d object detection. arXiv preprint arXiv:2203.17054 (2022)
13. Li, E., Casas, S., Urtasun, R.: Memoryseg: online lidar semantic segmentation with a latent memory. In: Proceedings of the IEEE/CVF International Conference on Computer Vision (2023)
14. Li, F., Zhang, H., Liu, S., Guo, J., Ni, L.M., Zhang, L.: Dn-detr: accelerate detr training by introducing query denoising. In: Proceedings of the IEEE/CVF Conference on Computer Vision and Pattern Recognition, pp. 13619–13627 (2022)
15. Li, H., et al.: Delving into the devils of bird's-eye-view perception: A review, evaluation and recipe. IEEE Trans. Pattern Analy. Mach. Intell. (2023)
16. Li, Q., Wang, Y., Wang, Y., Zhao, H.: Hdmapnet: an online hd map construction and evaluation framework. In: 2022 International Conference on Robotics and Automation (ICRA), pp. 4628–4634. IEEE (2022)
17. Li, Z., et al.: Bevformer: learning bird's-eye-view representation from multi-camera images via spatiotemporal transformers. In: European conference on computer vision. pp. 1–18. Springer (2022). https://doi.org/10.1007/978-3-031-20077-9_1
18. Liao, B., et al.: Maptr: Structured modeling and learning for online vectorized hd map construction. arXiv preprint arXiv:2208.14437 (2022)
19. Liao, B., et al.: Maptrv2: An end-to-end framework for online vectorized hd map construction. arXiv preprint arXiv:2308.05736 (2023)
20. Lilja, A., Fu, J., Stenborg, E., Hammarstrand, L.: Localization is all you evaluate: Data leakage in online mapping datasets and how to fix it. arXiv preprint arXiv:2312.06420 (2023)
21. Lin, T.Y., Goyal, P., Girshick, R., He, K., Dollár, P.: Focal loss for dense object detection. In: Proceedings of the IEEE International Conference on Computer Vision, pp. 2980–2988 (2017)
22. Lin, X., Lin, T., Pei, Z., Huang, L., Su, Z.: Sparse4d v2: Recurrent temporal fusion with sparse model. arXiv preprint arXiv:2305.14018 (2023)
23. Lin, X., Pei, Z., Lin, T., Huang, L., Su, Z.: Sparse4d v3: Advancing end-to-end 3d detection and tracking. arXiv preprint arXiv:2311.11722 (2023)

24. Liu, Y., Yuan, T., Wang, Y., Wang, Y., Zhao, H.: Vectormapnet: end-to-end vectorized hd map learning. In: International Conference on Machine Learning, pp. 22352–22369. PMLR (2023)
25. Loshchilov, I., Hutter, F.: Fixing weight decay regularization in adam. ArXiv abs/ arXiv: 1711.05101 (2017)
26. Ma, Y., et al.: Vision-centric bev perception: A survey. arXiv preprint arXiv:2208.02797 (2022)
27. Meinhardt, T., Kirillov, A., Leal-Taixe, L., Feichtenhofer, C.: Trackformer: multi-object tracking with transformers. In: Proceedings of the IEEE/CVF Conference on Computer Vision and Pattern Recognition, pp. 8844–8854 (2022)
28. Milletari, F., Navab, N., Ahmadi, S.A.: V-net: fully convolutional neural networks for volumetric medical image segmentation. In: 2016 Fourth International Conference on 3D Vision (3DV), pp. 565–571. IEEE (2016)
29. Philion, J., Fidler, S.: Lift, splat, shoot: encoding images from arbitrary camera rigs by implicitly unprojecting to 3D. In: Vedaldi, A., Bischof, H., Brox, T., Frahm, J.-M. (eds.) ECCV 2020. LNCS, vol. 12359, pp. 194–210. Springer, Cham (2020). https://doi.org/10.1007/978-3-030-58568-6_12
30. Qiao, L., Ding, W., Qiu, X., Zhang, C.: End-to-end vectorized hd-map construction with piecewise bezier curve. In: Proceedings of the IEEE/CVF Conference on Computer Vision and Pattern Recognition, pp. 13218–13228 (2023)
31. Qiao, L., et al.: Machmap: End-to-end vectorized solution for compact hd-map construction. arXiv preprint arXiv:2306.10301 (2023)
32. Shan, T., Englot, B.: Lego-loam: lightweight and ground-optimized lidar odometry and mapping on variable terrain. In: 2018 IEEE/RSJ International Conference on Intelligent Robots and Systems (IROS), pp. 4758–4765. IEEE (2018)
33. Shan, T., Englot, B., Meyers, D., Wang, W., Ratti, C., Rus, D.: Lio-sam: tightly-coupled lidar inertial odometry via smoothing and mapping. In: 2020 IEEE/RSJ International Conference on Intelligent Robots and Systems (IROS), pp. 5135–5142. IEEE (2020)
34. Sun, P., et al.: Transtrack: Multiple object tracking with transformer. arXiv preprint arXiv:2012.15460 (2020)
35. Vaswani, A., et al.: Attention is all you need. Adv. Neural Inform. Process. Syst. **30** (2017)
36. Wang, S., et al.: Stream query denoising for vectorized hd map construction. arXiv preprint arXiv:2401.09112 (2024)
37. Wilson, B., et al.: Argoverse 2: Next generation datasets for self-driving perception and forecasting. arXiv preprint arXiv:2301.00493 (2023)
38. Xu, Z., Wong, K.K., Zhao, H.: Insightmapper: A closer look at inner-instance information for vectorized high-definition mapping. arXiv preprint arXiv:2308.08543 (2023)
39. Yang, C., et al.: Bevformer v2: Adapting modern image backbones to bird's-eye-view recognition via perspective supervision. In: Proceedings of the IEEE/CVF Conference on Computer Vision and Pattern Recognition, pp. 17830–17839 (2023)
40. Yilmaz, A., Javed, O., Shah, M.: Object tracking: a survey. ACM Comput. Surv. (CSUR) **38**(4), 13–es (2006)
41. Yuan, T., Liu, Y., Wang, Y., Wang, Y., Zhao, H.: Streammapnet: streaming mapping network for vectorized online hd map construction. In: Proceedings of the IEEE/CVF Winter Conference on Applications of Computer Vision, pp. 7356–7365 (2024)

42. Zeng, F., Dong, B., Zhang, Y., Wang, T., Zhang, X., Wei, Y.: Motr: End-to-end multiple-object tracking with transformer. In: European Conference on Computer Vision, pp. 659–675. Springer (2022). https://doi.org/10.1007/978-3-031-19812-0_38
43. Zhang, G., et al.: Online map vectorization for autonomous driving: A rasterization perspective. arXiv preprint arXiv:2306.10502 (2023)
44. Zhang, J., Singh, S.: Loam: Lidar odometry and mapping in real-time. In: Robotics: Science and Systems (2014)
45. Zhang, Y., Wang, T., Zhang, X.: Motrv2: bootstrapping end-to-end multi-object tracking by pretrained object detectors. In: Proceedings of the IEEE/CVF Conference on Computer Vision and Pattern Recognition, pp. 22056–22065 (2023)
46. Zhang, Z., Zhang, Y., Ding, X., Jin, F., Yue, X.: Online vectorized hd map construction using geometry. arXiv preprint arXiv:2312.03341 (2023)
47. Zhu, X., Su, W., Lu, L., Li, B., Wang, X., Dai, J.: Deformable detr: Deformable transformers for end-to-end object detection. arXiv preprint arXiv:2010.04159 (2020)

Image Demoiréing in RAW and sRGB Domains

Shuning Xu[1], Binbin Song[1], Xiangyu Chen[1,2], Xina Liu[1,2], and Jiantao Zhou[1(✉)]

[1] State Key Laboratory of Internet of Things for Smart City, Department of Computer and Information Science, University of Macau, Macao, China
{yc07425,yb97426,jtzhou}@um.edu.mo
[2] Shenzhen Institutes of Advanced Technology, Chinese Academy of Sciences, Beijing, China
xn.liu95@siat.ac.cn

Abstract. Moiré patterns frequently appear when capturing screens with smartphones or cameras, potentially compromising image quality. Previous studies suggest that moiré pattern elimination in the RAW domain offers greater effectiveness compared to demoiréing in the sRGB domain. Nevertheless, relying solely on RAW data for image demoiréing is insufficient in mitigating the color cast due to the absence of essential information required for the color correction by the image signal processor (ISP). In this paper, we propose to jointly utilize both RAW and sRGB data for image demoiréing (RRID), which are readily accessible in modern smartphones and DSLR cameras. We develop Skip-Connection-based Demoiréing Module (SCDM) with Gated Feedback Module (GFM) and Frequency Selection Module (FSM) embedded in skip-connections for the efficient and effective demoiréing of RAW and sRGB features, respectively. Subsequently, we design a RGB Guided ISP (RGISP) to learn a device-dependent ISP, assisting the process of color recovery. Extensive experiments demonstrate that our RRID outperforms state-of-the-art approaches, in terms of the performance in moiré pattern removal and color cast correction by 0.62 dB in PSNR and 0.003 in SSIM. Code is available at https://github.com/rebeccaeexu/RRID.

Keywords: Image demoiréing · RAW domain · Image restoration

1 Introduction

With the increasing prevalence of smartphone cameras, it is common to record on-screen content using smartphone cameras for convenience. However, this type of photography often results in the presence of moiré patterns in the captured images. The appearance of moiré patterns is caused by the spatial frequency aliasing between the camera's color filter array (CFA) and the screen's LCD subpixel [22], resulting in an unsatisfactory visual experience.

Supplementary Information The online version contains supplementary material available at https://doi.org/10.1007/978-3-031-72658-3_7.

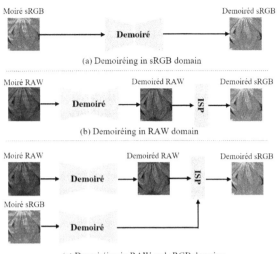

Fig. 1. Comparison of image demoiréing in different domains. (a) Image demoiréing in sRGB domain. (b) Image demoiréing in RAW domain. (c) Our proposed RRID, which performs image demoiréing in both RAW domain and sRGB domain

Moiré patterns present challenges to be eliminated due to their different scales, indistinct shapes, diverse colors, and varying frequencies. Nowadays, many learning-based image demoiréing approaches [1,5,6,14,15,18,21–23,27,32] have been proposed. Among the various methods employed, a commonly used approach is utilizing a multi-scale architecture to remove various sizes of moiré patterns [1,5,18,21,27]. Besides, some works advocate the leverage of frequency components and delicately design frequency-based demoiréing networks [6,14,22,32].

The aforementioned methods are developed to demoiré in the sRGB domain, as depicted in Fig. 1(a). While this scheme seems to be the most direct approach, its effectiveness is limited in removing moiré patterns from complex scenes. The reason for this limitation probably lies in the fact that the nonlinear operations in ISP such as demosaicing deteriorate the moiré patterns originally in the RAW domain. Many schemes consequently suggest that eliminating moiré patterns in the RAW domain is more effective than performing demoiréing in the sRGB domain [28,29]. Due to the easy accessibility of RAW data in modern smartphones and DSLR cameras, image demoiréing in RAW domain can be both feasible and advantageous. RDNet [29] is the pioneering study that investigates image demoiréing in the RAW domain. They explore the demoiréing of RAW images using an encoder-decoder and class-specific learning approach, followed by the utilization of a pre-trained ISP for the RAW-to-sRGB conversion, as depicted in Fig. 1(b). Nevertheless, relying exclusively on RAW data can lead to color cast due to the uncertainty during RAW-to-sRGB conversion.

In order to correct the color cast and make the best use of RAW and sRGB information for the moiré pattern removal, we propose to jointly utilize both RAW and sRGB data for image demoiréing (RRID), as demonstrated in Fig. 1(c). We advocate that introducing paired RAW-sRGB data is advantageous to the moiré pattern removal due to the following reasons: (1) RAW pixels provide more information compared to sRGB pixels since they are typically 12 or 14 bits. (2) Moiré patterns in the RAW domain are less apparent because they are not further affected by nonlinear operations in the ISP. (3) The paired RAW and sRGB data allows the model to learn a device-dependent ISP to aid the process of color recovery. In RRID, we adopt a multi-scale architecture in the Skip-Connection-based Demoiréing Module (SCDM), to effectively eliminate moiré patterns of different sizes. In order to effectively perform demoiréing on RAW and sRGB features, the Gated Feedback Module (GFM) and Frequency Selective Module (FSM) are integrated into skip-connections within SCDM, as opposed to brutally injecting them into the multi-scale layers. Specifically, for RAW features demoiréing, we introduce the GFM into the skip connection, enabling adaptive differentiation between texture details and moiré patterns through feature gating. Moreover, we design the FSM for the moiré sRGB features, leveraging a learnable band stop filter to mitigate moiré patterns in the frequency domain. Subsequent to the pre-demoiréing operations, we develop the RGB Guided Image Signal Processor (RGISP) to integrate color information from the demoiréd sRGB features with the demoiréd RAW features and learn a device-dependent ISP, facilitating the color recovery process. Ultimately, for the image reconstruction, we utilize multiple Residual Swin Transformer Blocks (RSTBs) [12] and convolutions to accomplish global tone mapping and detail refinement.

In summary, our contributions are listed as follows:

- We propose to exploit image demoiréing in both RAW and sRGB domains. The incorporation of RAW data enhances the moiré pattern removal process, while the inclusion of sRGB data facilitates the RAW-to-sRGB conversion and aids in color restoration of images.
- We develop SCDM with specifically designed modules, GFM and FSM, embedded in the skip-connections to perform efficient and effective demoiréing on the RAW and sRGB features.
- RGISP is designed for the RAW-to-sRGB conversion, incorporating color information from the demoiréd sRGB features with demoiréd RAW features and learn a device-dependent ISP, aiding the color deviation correction.
- Our RRID surpasses state-of-the-arts methods in both qualitative and quantitative evaluations.

2 Related Works

2.1 Image and Video Demoiréing

The purpose of image demoiréing is to restore the original clean image by removing moiré patterns and correcting color deviations from an contaminated image. Many learning-based image demoiréing [1,6,15,18,22,23,32] and video demoiréing methods [2,19,26,28] have been introduced to mitigate the moiré patterns. DMCNN [21] constructs the first real-world dataset for image demoiréing and proposes a multi-resolution neural network to handle moiré patterns across various scales. MopNet [5] devises a multi-scale aggregated network that leverages the edge guidance and pattern attributes for moiré pattern removal. WDNet [14] employs wavelet transformation to decompose images with moiré patterns into separate frequency bands and establishes a dual-branch network to restore both close-range and far-range information. FHDMi [6] formulates a two-stage approach for simultaneous removal of substantial moiré patterns while preserving image details. ESDNet [27] explores a lightweight model aimed at ultra-high-definition image demoiréing. VDMoiré [2] collects the first video demoiréing dataset and presents a baseline video demoiréing model with a relation-based temporal consistency loss. The aforementioned methods all operate in the sRGB domain. Since it has been observed that moiré patterns are less prominent in the RAW domain, some schemes advocate performing image or video demoiréing in the RAW domain. RDNet [29] introduces the first image demoiréing dataset with RAW data and conducts the demoiréing process in the RAW domain. RawVDemoiré [28] proposes a temporal alignment method for RAW video demoiréing. To remove moiré patterns in the RAW domain, the model have to simultaneously possess the capabilities of demoiréing and RAW-to-sRGB conversion. However, previous methods encounter challenges in color recovery due to the uncertainty of RAW-to-sRGB conversion. Consequently, we propose an image demoiréing network that utilizes paired RAW-sRGB data, facilitating the color recovery process in image demoiréing.

2.2 Learning-Based RAW Image Processing

RAW pixels offer the potential for leveraging additional information due to their broader bit depth, inherently containing a great wealth of data. Early research in RAW image processing primarily centers on integrating demosaicing and denoising techniques [4,9,10,13]. RTF [9] devices a machine learning approach to the demosaicing and denoising on RAW inputs. SGNet [13] introduces a self-guidance network that leverages the inherent density-map guidance and green-channel guidance for demosaicing and denoising on the RAW images. Recently, some low-level vision tasks endorse the utilization of RAW data, including low-light enhancement [3,7,8], super-resolution [16,25,30], and reflection removal [11,20]. The majority of the aforementioned RAW image processing methods only take RAW images as input and generate restored sRGB images as output, indicating that the entire model is responsible for both the restoration and ISP tasks. CR3Net [20] presents a cascaded network that harnesses both the RGB data and their corresponding RAW versions for image reflection removal. Within CR3Net, a module is formulated to emulate pointwise mappings within the ISP, converting the features from RAW to sRGB domain. To further mitigate the color casting problem, our proposed method jointly utilize RAW and sRGB data to learn a device-dependent ISP for image demoiréing.

3 Methodology

Originally caused by frequency interference between the camera's CFA and the screen's LCD subpixel, moiré patterns are further deteriorated by the nonlinear processes in ISP. Therefore, introducing RAW data for image demoiréing may offer greater effectiveness. However, relying solely on RAW data for image demoiréing is insufficient in mitigating color cast due to the uncertainty during the RAW-to-sRGB conversion. Consequently, in this work, we propose to utilize paired RAW and sRGB data, which are readily accessible in modern smartphones and DSLR cameras (e.g., iPhone 15 Pro, HUAWEI P60 Pro). The RAW-sRGB pairs essentially allows the model to learn a device-dependent ISP, aiding the color recovery process. The potential applications in real-world scenarios can be described as follows. When demoiréing function is enabled, the in-device ISP generated sRGB image will not be directly outputted. Instead, together with its corresponding RAW version, will be processed by our RRID to generate a new sRGB image with much suppressed moiré patterns. Afterwards, the RAW image could be discarded, depending on settings of the camera.

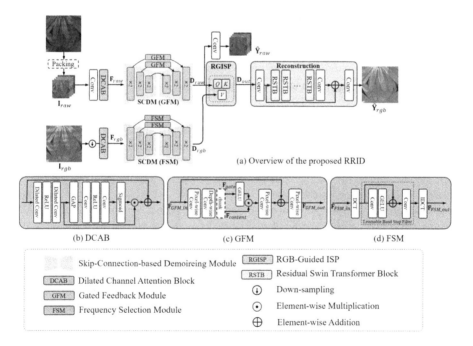

Fig. 2. The overview and detailed structures of our proposed RRID.

More specifically, Fig. 2 presents an overview of our proposed RRID for image demoiréing with the paired RAW and sRGB data. As for the RGB branch, the input is an sRGB image $\mathbf{I}_{rgb} \in \mathbb{R}^{H \times W \times 3}$, where $H \times W$ represents the spatial resolution. Given an input RAW image, it is first packed into the 4-channel RGGB format $\mathbf{I}_{raw} \in \mathbb{R}^{\frac{H}{2} \times \frac{W}{2} \times 4}$, which serves as the input for the RAW branch. RRID is designed to generate the final output sRGB image ($\hat{\mathbf{Y}}_{rgb}$).

Concretely, shallow features $\mathbf{F}_{raw}$ and $\mathbf{F}_{rgb}$ are respectively extracted from the input data $\mathbf{I}_{raw}$ and $\mathbf{I}_{rgb}$ using a convolutional layer and the Dilated Channel Attention Block (DCAB). It is worth noting that a down-sampling layer is applied to the sRGB data to ensure the uniformity of feature shapes. Then, we apply SCDM to acquire pre-demoiréd features $\mathbf{D}_{raw}$ and $\mathbf{D}_{rgb}$ in RAW branch and RGB branch, respectively. GFM is introduced in the skip-connection of SCDM for RAW features, enabling adaptive differentiation between texture details and moiré patterns through feature gating. Similarly, FSM is utilized to to mitigate moiré patterns in the frequency domain. More details will be given in Sect. 3.1. Subsequently, RGISP is designed to realize a RAW-to-sRGB conversion, generating $\mathbf{D}_{out}$. RGISP incorporates color information from coarsely demoiréd sRGB features $\mathbf{D}_{rgb}$ in a cross-attention mechanism, which will be introduced in Sect. 3.2. For the final image reconstruction, 4 RSTBs [12] are utilized to generate sRGB output $\hat{\mathbf{Y}}_{rgb}$ for their advantages in building long-range dependencies for global tone mapping and detail refinement.

3.1 Design of SCDM

We propose SCDM, a multi-scale architecture designed to remove moiré patterns from shallow RAW features F_{raw} and shallow sRGB features F_{rgb}, as shown in Fig. 2(a). Unlike the vanilla U-Net, SCDM is constructed upon DCAB, facilitating the encoding and decoding of informative features within the respective layers. The features are downsampled twice in the spatial dimension during the encoding process and then upsampled to the original resolution of the input features during decoding. To emphasize the contextual information, we enhance the learned features from the corresponding layers of the encoder with additional modules at skip connections, that is GFM for RAW and FSM for sRGB. The introduction of GFM and FSM in skip-connections is also intended to enhance model efficiency. Then, we can obtain the pre-demoiréd features $\mathbf{D}_{raw}$ and $\mathbf{D}_{rgb}$.

DCAB. DCAB is the essential building block of encoder-decoders for performing RAW demoiréing and the RGB demoiréing, as shown in Fig. 2(b). DCAB is designed using a series of dilated convolution layers followed by ReLU activation and the channel attention mechanism. The receptive field at each block is expanded with the dilated convolution kernels, facilitating the demoiréing operations. Furthermore, by incorporating the channel attention mechanism, DCAB adaptively rescales features according to interdependencies among channels. Also, shortcuts are introduced to establish connections between the input and output of DCAB, enabling the block to learn the residual features.

GFM. Embeded in the skip-connection of SCDM for RAW features, GFM is introduced to enable adaptive differentiation between the texture and moiré patterns through feature gating. The detailed structure of the GFM is shown in Fig. 2(c). For efficiency, we employ point-wise and depth-wise convolutions to aggregate channel and local content information, respectively. Then, the features are chunked along the channel dimension to generate $\mathbf{F}_{gate}$ and $\mathbf{F}_{content}$. A point-wise multiplication is applied on $\mathbf{F}_{content}$ and $\mathbf{F}_{gate}$ after a GELU activation. During feature gating, we expect that the original image content is adaptively selected and merged along both spatial and channel dimensions. The operations can be formulated as:

$$\{\mathbf{F}_{gate}, \mathbf{F}_{content}\} = \text{DConv}(\text{PConv}(\mathbf{F}_{GFM\_in})), \qquad (1)$$

$$\mathbf{F}_{GFM\_out} = \text{PConv}(\mathbf{F}_{content} \odot \text{GELU}(\mathbf{F}_{gate}) + \mathbf{F}_{GFM\_in}), \qquad (2)$$

where DConv and PConv represent depth-wise convolution and point-wise convolution, respectively. Here, $\odot$ denotes the element-wise multiplication.

FSM. For sRGB demoiréing, we leverage a learnable band stop filter [32] to mitigate moiré patterns in the frequency domain. Considering the formation of moiré patterns is caused by the spatial frequency aliasing between the camera CFA and screen's LCD subpixel, demoiréing in the frequency domain becomes a favorable approach. However, the frequency spectrum of the moiré pattern is often mixed with that of the original contents, thereby making it challenging to disentanglement. Therefore, we propose to perform the disentanglement within small patches, where the moiré patterns tend to be more pronounced in a specific range of frequency bands, facilitating the removal of moiré patterns. Inspired by Zheng et al. [32], we utilize a band stop filter to amplify certain frequencies and suppress others using Block DCT. However, obtaining the frequency prior to separate moiré patterns from normal image textures is challenging and time-consuming. To address this issue, a set of convolution layers are employed as the learnable band stop filter $B(\cdot)$ to attenuate the specific frequencies of moiré patterns while preserving the original image contents. The process of the FSM can be represented as:

$$\mathbf{F}_{FSM\_out} = \text{IDCT}(B(\text{DCT}(\mathbf{F}_{FSM\_in}))), \qquad (3)$$

where DCT and IDCT denote the process of Block DCT and the corresponding inverse process. The block size is set to 8×8 for DCT and IDCT.

3.2 Design of RGISP

RGISP is proposed to transform pre-demoiréd RAW features $\mathbf{D}_{raw}$ into sRGB domain $\mathbf{D}_{out}$. With the assistance of pre-demoiréd sRGB features $\mathbf{D}_{rgb}$, a device-dependent ISP can be learned in RGISP. Matrix transformation is commonly employed in the traditional ISP pipelines [17]. It enables the enhancement or conversion of image colors to another color space through a channel-wise matrix

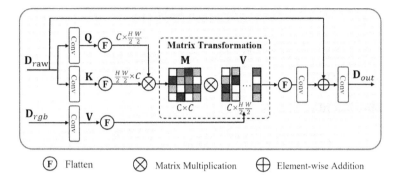

Fig. 3. The structure of RGISP.

transformation, which is facilitated by globally shared settings such as environmental illumination and color space specifications [8]. Building upon this principle, we introduce matrix transformation as a means to perform color transformation, as illustrated in Fig. 3. The design of this matrix transformation is motivated by recent advancements in transposed cross-attention. Given $\mathbf{D}_{raw} \in \mathbb{R}^{C \times \frac{H}{2} \times \frac{W}{2}}$, the vectors of query $\mathbf{Q} \in \mathbb{R}^{C \times \frac{H}{2} \frac{W}{2}}$, key $\mathbf{K} \in \mathbb{R}^{\frac{H}{2} \frac{W}{2} \times C}$ are generated through the projection with a 1×1 convolutional layer Conv and a flatten operation Flatten, where C is the intermediate channel number. Similarily, value $\mathbf{V} \in \mathbb{R}^{C \times \frac{H}{2} \frac{W}{2}}$ can be produced by operations on $\mathbf{D}_{rgb} \in \mathbb{R}^{C \times \frac{H}{2} \times \frac{W}{2}}$ as:

$$\{\mathbf{Q}, \mathbf{K}\} = \text{Flattern}(\text{Conv}(\mathbf{D}_{raw})), \quad (4)$$
$$\mathbf{V} = \text{Flattern}(\text{Conv}(\mathbf{D}_{rgb})). \quad (5)$$

Then, the transformation matrix $\mathbf{M} \in \mathbb{R}^{C \times C}$ is obtained by matrix multiplication. The procedure can be formulated as:

$$\mathbf{M} = \text{Softmax}(\mathbf{Q} \cdot \mathbf{K}^T / \lambda), \quad (6)$$

where a scaling coefficient λ is applied for numerical stability. Subsequently, the vector $\mathbf{V}$ is transformed by the matrix $\mathbf{M}$, performing color space conversion in feature-level. The output feature after color transformation can be obtained by $\mathbf{D}_{out} = \mathbf{M} \cdot \mathbf{V}$. As a complement to the global matrix transformation, we use a depth-wise convolution and a point-wise convolution to refine the local details, generating the output in sRGB domain $\mathbf{D}_{out}$.

3.3 Loss Function

To effectively remove the moiré patterns and adjust the color deviations, we introduce supervision in both RAW and sRGB domains. The overall training objective can be expressed as:

$$\mathcal{L} = \alpha \|\hat{\mathbf{Y}}_{raw} - \mathbf{Y}_{raw}\|_1 + \|\hat{\mathbf{Y}}_{rgb} - \mathbf{Y}_{rgb}\|_1, \quad (7)$$

where $\mathbf{Y}_{raw}$ and $\mathbf{Y}_{rgb}$ are the ground-truth RAW and sRGB images, respectively. We empirically set $\alpha = 0.5$.

4 Experiments

4.1 Experimental Setup

Dataset. The experiments are conducted on the RAW image demoiréing dataset published in TMM22 [29], which contains recaptured scenes of natural images, documents, and webpages. In TMM22, the dataset consists of 540 RAW and sRGB image pairs with ground-truth for training, and 408 pairs for testing. To facilitate the training and comparison process, patches of size 256×256 and 512×512 are cropped respectively for the training and testing sets. The moiré images are captured using four different smartphone cameras and three display screens of varying sizes. In addition, we utilize an image demoiréing dataset FHIMi [6] to verify the generalization ability of our method. FHDMi contains 9981 sRGB image pairs for training and 2019 for testing, featuring diverse and intricate moiré patterns. Notably, images in FHDMi have a higher resolution of 1920×1080, compared to TMM22.

Training Details. We train RRID using the AdamW optimizer with $\beta_1 = 0.9$ and $\beta_2 = 0.999$. A multistep learning rate schedule is employed, with the learning rate initialized at 2×10^{-4}. The RRID model is trained for 500 epochs using a batch size of 80 on 4 NVIDIA RTX 3090 GPUs.

4.2 Quantitative Results

We compare our approach with several image demoiréing methods in sRGB domain, including DMCNN [21], WDNet [14], ESDNet [27], and a RAW image demoiréing method RDNet [29]. We further add comparison with a low-light enhancement method DNF [8] with RAW input, a video demoiréing method RawVDmoiré [28] with RAW input, and a de-reflection method CR3Net [20] with paired RAW-sRGB data. To evaluate the performance of our proposed RRID, we adopt the following three standard metrics to assess pixel-wise accuracy and the perceptual quality: PSNR, SSIM [24], and LPIPS [31]. Additionally, the number of parameters, FLOPs, and inference time are introduced to measure the model complexity. To ensure a fair comparison, we modify the inputs of the sRGB-based models to align with the dataset requirements. Subsequently, all models are fine-tuned using the default settings as provided in their papers on TMM22 dataset. We select the better results from the pretrained and retrained models for comparison, which provides an advantage over the competing methods.

Table 1 presents the quantitative and the computational complexity comparisons. RRID outperforms the second-best method, RawVDmoiré, by 0.62dB in PSNR and 0.003 in SSIM. As for LPIPS, our methods accomplishes the second-best result at 0.079, surpassing most of the comparison methods. Even though RRID utilizes both RAW and sRGB data as inputs, our approach can still achieve

Table 1. Quantitative comparison with the state-of-the-art demoiréing approaches and RAW image restoration methods on TMM22 dataset [29] in terms of average PSNR, SSIM, LPIPS and computational complexity. The best results are highlighted with **bold**. The second-best results are highlighted with underline.

Index	Methods							
	DMCNN [21]	WDNet [14]	ESDNet [27]	RDNet [29]	DNF [8]	RawVDmoiré [28]	CR3Net [20]	RRID (Ours)
# Input type	sRGB	sRGB	sRGB	RAW	RAW	RAW	sRGB+RAW	sRGB+RAW
PSNR↑	23.54	22.33	26.77	26.16	23.55	<u>27.26</u>	23.75	**27.88**
SSIM↑	0.885	0.802	0.927	0.921	0.895	<u>0.935</u>	0.922	**0.938**
LPIPS↓	0.154	0.166	0.089	0.091	0.162	**0.075**	0.102	<u>0.079</u>
Params (M)	<u>1.55</u>	3.36	5.93	6.04	**1.25**	5.33	2.68	2.38
TFLOPs	0.102	0.055	0.141	0.161	**0.013**	<u>0.022</u>	0.883	0.093
Inference time (s)	**0.052**	0.284	0.115	1.094	0.070	0.182	<u>0.058</u>	0.089

a satisfactory demoiréing effect with a inference time of merely 0.089 s. Compared to previous methods, our RRID demonstrates superiority with acceptable parameters and FLOPs, further demonstrating the superiority of our approach.

Table 2. Quantitative comparison on FHDMi dataset [6].

Method	PSNR↑	SSIM↑
DMCNN [21]	21.54	0.773
MopNet [5]	22.76	0.796
MBCNN [32]	22.31	0.810
FHDe^{2}Net [6]	22.93	0.789
ESDNet [27]	**24.50**	**0.835**
Ours	<u>24.39</u>	<u>0.830</u>

To verify the generalization performance of our method, we conduct additional experiments on an image demoiréing dataset FHIMi [6]. In order to adapt to the FHDMi dataset with sRGB data as input, the RAW branch and RGISP are removed. We select methods tailored for image demoiréing: DMCNN [21], MopNet [5], MDCNN [32], FHDe^{2}Net [6], and ESDNet [27] for comparison, as shown in Table 2. Although RRID is initially designed for demoiréing on the RAW-sRGB image pairs, it still achieves the second-best performance by 24.39 dB in PSNR and 0.830 in SSIM. Compared to ESDNet, an image demoiréing network designed for ultra-high-definition sRGB images with more than twice of the parameters, RRID also yields competitive results. The qualitative results on the FHDMi dataset demonstrate that our method can not only perform well in demoiréing on RAW-sRGB pairs but also solve sRGB image demoiréing, exhibiting a certain degree of generalization.

4.3 Qualitative Results

We present visual comparisons among our approach and the existing methods in Fig. 4. The results demonstrate the advantage of RRID in effectively removing moiré artifacts and correcting color deviations. For the first scene demonstrating a recaptured flower with moiré patterns, previous methods tend to exhibit residual moiré patterns in their restored images. While ESDNet successfully eliminates a significant portion of these patterns, it struggles to accurately restore the original colors of the image. Apart from natural scenes, we also present a case of webpages contaminated with moiré patterns. Through the utilization of both RAW and sRGB data, our approach eliminates moiré patterns while preserving fine image details, even in scenarios characterized by severe color deviations. This ensures a more accurate restoration of the original colors and results in enhanced visual effects. Additional qualitative comparisons can be found in the supplementary file.

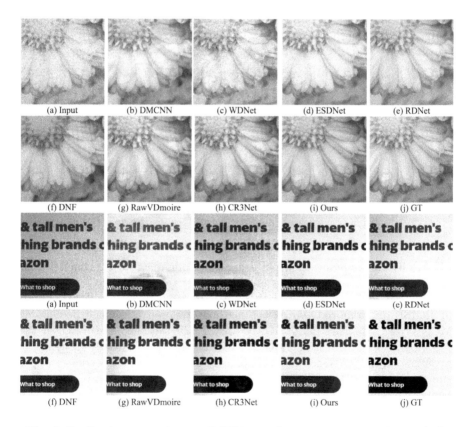

Fig. 4. Qualitative comparison on RAW image demoiréing TMM22 dataset [29].

4.4 Ablation Study

1) Ablation study on the model architecture and inputs. In Table 3, we investigate the architecture and inputs of RRID and validate the importance of different individual components in the whole RRID. We conduct this ablation study by comparing the proposed RRID and the following variants of RRID:

- B_1. Both the RAW input and the RAW demoiréing branch are removed. Due to the lack of RAW features, the entire RGISP is also deleted.
- B_2. Both the sRGB input and sRGB demoiréing branch are removed.
- B_3. The sRGB input is substituted with RAW input. The input channel number is adjusted, while the rest of the architecture is kept unchanged.
- B_4. We replace RGISP with a two convolutional layers after a concatenation of the RAW and sRGB features.
- B_5. Reconstruction module is replaced with two convolutional layers.
- B_6. The full model of RRID with both RAW and sRGB images as input.

Table 3. Comparison of baseline models for the evaluation of architecture and inputs.

Models	PSNR↑	SSIM↑
B_1 (w/o RAW in and RAW branch)	25.79	0.915
B_2 (w/o sRGB in and sRGB branch)	27.24	0.929
B_3 (w/o sRGB in)	26.51	0.923
B_4 (w/o RGISP)	27.38	0.932
B_5 (w/o Reconstruction)	26.93	0.926
B_6 (full version)	**27.88**	**0.938**

Our proposed RRID leverages paired RAW-sRGB data as input. Removing the RAW input and its corresponding RAW demoiréing module (B_1) results in a drastic decrease in PSNR of approximately 2 dB. Likewise, eliminating the sRGB input and its corresponding RGB demoiréing module (B_2) leads to a PSNR of only 27.24 dB. In addition to the decline in numerical metrics, we can also assess the visual performance from Fig. 5. When relying only on sRGB data and its corresponding demoiréing modules, moiré patterns are less effectively removed, as depicted in Fig. 5(b). Conversely, Fig. 5(c) demonstrates that using only RAW data as input generates images with more pronounced color deviations. To further validate the significance of RAW input for image demoiréing, we compare B_3 to B_1 by solely removing the RAW input data while retaining the remaining model structures. Although B_3 demonstrates some improvements in numerical metrics compared to B_1, its performance remains unsatisfactory due to the lack of information provided by the RAW input and the fact that the original RAW demoiréing module is specifically designed for RAW data. Additionally, we conducted further comparisons by removing the originally designed RGISP (B_4) and Reconstruction modules (B_5), resulting in inferior numerical metrics and performance in Fig. 5 compared to the complete RRID model (B_6).

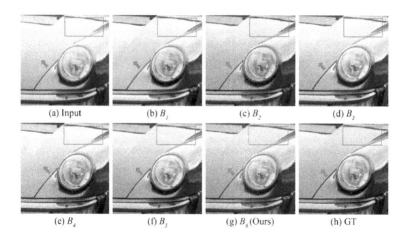

Fig. 5. Visualization of the ablation study for architecture and inputs comparison.

2) Ablation study on SCDM. SCDM is designed to perform effective demoiréing on both RAW and sRGB features. More specifically, we introduce GFM for RAW features demoiréing and FSM for sRGB features demoiréing.

Table 4. Ablation study on SCDM.

Models	PSNR↑	SSIM↑
S_1 (w/o GFM & FSM)	27.00	0.926
S_2 (w/o GFM)	27.17	0.927
S_3 (w/o FSM)	27.39	0.930
S_4 (all GFM)	27.50	0.933
S_5 (all FSM)	27.32	0.930
S_6 (w/o DCAB)	26.55	0.923
S_7 (SCDM)	**27.88**	**0.938**

Table 4 demonstrates the effectiveness and indispensability of the meticulously designed GFM and FSM integrated within the skip-connections. Removing the demoiréing modules on the skip-connections (S_1–S_3) individually result in varying degrees of performance decline of the model. For instance, removing both GFM and FSM leads to only 27.00 dB in PSNR. Furthermore, Fig. 6(b) demonstrates that the model's ability to remove moiré patterns decreases upon removing the GFM and FSM modules. Furthermore, to assess the specificity of GFM and FSM, we substitute the original modules in S_4 and S_5, respectively. Despite no reduction in the number of parameters, the PSNR decreases by approximately 0.3–0.5 dB. Since DCAB is a crucial component module in SCDM, we substitute DCAB with two convolutional layers to confirm its necessity. Figure 6(c) showcases more residual moiré patterns and significant color deviation compared to our complete SCDM presented in Fig. 6(d).

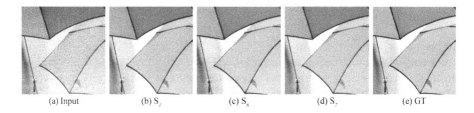

Fig. 6. Visualization of the ablation study for SCDM.

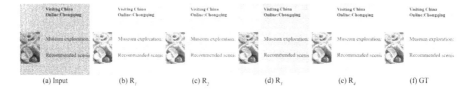

Fig. 7. Visualization of the ablation study for RGISP.

3) Ablation study on RGISP. RGISP is designed to integrate color information from the coarsely demoiréd sRGB features during the ISP stage, facilitating the color recovery process. The process of the RGISP can be recognized as a cross-attention mechanism. To assess the efficacy of the RGISP design, we compare the following structures in Table 5. For R_1, we concatenate the pre-demoiréd RAW features with the sRGB features and subsequently apply a self-attention mechanism. In R_2, the pre-demoiréd RAW features undergo a self-attention mechanism and are then concatenated with the pre-demoiréd sRGB features. Table 5 shows that our proposed RGISP surpasses the self-attention mechanisms (R_1, R_2), as well as other RAW-to-sRGB conversion methods RRM [20] (R_3). In addition, Fig. 7 depicts a webpage with globally distributed moiré patterns. R_1-R_4 are all capable of effectively removing the moiré patterns, but R_4 exhibits the best performance in correcting color deviation. With the introduction of sRGB pre-demoiréd features, RGISP is able to learn a device-dependent ISP and restore the original colors of the image effectively.

Table 5. Ablation study on RGISP.

Models	PSNR↑	SSIM↑
R_1 (self-atten([raw, rgb]))	27.54	0.933
R_2 ([self-atten(raw), rgb])	27.42	0.931
R_3 (RRM)	27.21	0.929
R_4 (RGISP)	**27.88**	**0.938**

5 Conclusion

In this paper, we propose to jointly utilize RAW and sRGB data for image demoiréing (RRID). Firstly, we introduce SCDM with GFM and FSM embedded in skip-connections to handle the demoiréing of RAW and sRGB features, respectively. Additionally, we present RGISP to learn a device-dependent ISP, aiding the color recovery process. Extensive experiments demonstrate the superior performance of our RRID method in both moiré pattern removal and color cast correction. Our approach outperforms state-of-the-art methods by 0.62 dB in PSNR and 0.003 in SSIM.

Acknowledgements. This work was supported in part by Macau Science and Technology Development Fund under SKLIOTSC-2021-2023, 0072/2020/AMJ, 0022/2022/A, and 0014/2022/AFJ; in part by Research Committee at University of Macau under MYRG-GRG2023-00058-FST-UMDF and MYRG2022-00152-FST; in part by Natural Science Foundation of Guangdong Province of China under EF2023-00116-FST.

References

1. Cheng, X., Fu, Z., Yang, J.: Multi-scale dynamic feature encoding network for image demoiréing. In: 2019 IEEE/CVF International Conference on Computer Vision Workshop (ICCVW), pp. 3486–3493. IEEE (2019)
2. Dai, P., et al.: Video demoireing with relation-based temporal consistency. In: Proceedings of the IEEE/CVF Conference on Computer Vision and Pattern Recognition, pp. 17622–17631 (2022)
3. Dong, X., et al.: Abandoning the bayer-filter to see in the dark. In: Proceedings of the IEEE/CVF Conference on Computer Vision and Pattern Recognition, pp. 17431–17440 (2022)
4. Gharbi, M., Chaurasia, G., Paris, S., Durand, F.: Deep joint demosaicking and denoising. ACM Trans. Graph. (ToG) **35**(6), 1–12 (2016)
5. He, B., Wang, C., Shi, B., Duan, L.Y.: Mop moire patterns using mopnet. In: Proceedings of the IEEE/CVF International Conference on Computer Vision, pp. 2424–2432 (2019)
6. Denninger, M., Triebel, R.: 3D scene reconstruction from a single viewport. In: Vedaldi, A., Bischof, H., Brox, T., Frahm, J.-M. (eds.) ECCV 2020. LNCS, vol. 12367, pp. 51–67. Springer, Cham (2020). https://doi.org/10.1007/978-3-030-58542-6_4
7. Huang, H., Yang, W., Hu, Y., Liu, J., Duan, L.Y.: Towards low light enhancement with raw images. IEEE Trans. Image Process. **31**, 1391–1405 (2022)
8. Jin, X., Han, L.H., Li, Z., Guo, C.L., Chai, Z., Li, C.: Dnf: decouple and feedback network for seeing in the dark. In: Proceedings of the IEEE/CVF Conference on Computer Vision and Pattern Recognition, pp. 18135–18144 (2023)
9. Khashabi, D., Nowozin, S., Jancsary, J., Fitzgibbon, A.W.: Joint demosaicing and denoising via learned nonparametric random fields. IEEE Trans. Image Process. **23**(12), 4968–4981 (2014)
10. Kokkinos, F., Lefkimmiatis, S.: Deep image demosaicking using a cascade of convolutional residual denoising networks. In: Proceedings of the European conference on computer vision (ECCV), pp. 303–319 (2018)

11. Lei, C., Chen, Q.: Robust reflection removal with reflection-free flash-only cues. In: Proceedings of the IEEE/CVF Conference on Computer Vision and Pattern Recognition, pp. 14811–14820 (2021)
12. Liang, J., Cao, J., Sun, G., Zhang, K., Van Gool, L., Timofte, R.: Swinir: image restoration using swin transformer. In: Proceedings of the IEEE/CVF international conference on computer vision, pp. 1833–1844 (2021)
13. Liu, L., Jia, X., Liu, J., Tian, Q.: Joint demosaicing and denoising with self guidance. In: Proceedings of the IEEE/CVF Conference on Computer Vision and Pattern Recognition, pp. 2240–2249 (2020)
14. Liu, L., et al.: Wavelet-based dual-branch network for image demoiréing. In: Vedaldi, A., Bischof, H., Brox, T., Frahm, J.-M. (eds.) ECCV 2020. LNCS, vol. 12358, pp. 86–102. Springer, Cham (2020). https://doi.org/10.1007/978-3-030-58601-0_6
15. Liu, S., Li, C., Nan, N., Zong, Z., Song, R.: Mmdm: multi-frame and multi-scale for image demoiréing. In: Proceedings of the IEEE/CVF Conference on Computer Vision and Pattern Recognition Workshops, pp. 434–435 (2020)
16. Luo, F., Wu, X., Guo, Y.: And: adversarial neural degradation for learning blind image super-resolution. Adv. Neural Inform. Process. Syst. **36** (2024)
17. Nakamura, J.: Image sensors and signal processing for digital still cameras. CRC press (2017)
18. Niu, Y., Lin, Z., Liu, W., Guo, W.: Progressive moire removal and texture complementation for image demoireing. IEEE Trans. Circ. Syst. Video Technol. (2023)
19. Oh, G., Gu, H., Kim, S., Kim, J.: Fpanet: Frequency-based video demoireing using frame-level post alignment. arXiv preprint arXiv:2301.07330 (2023)
20. Song, B., Zhou, J., Chen, X., Zhang, S.: Real-scene reflection removal with raw-rgb image pairs. IEEE Trans. Circ. Syst. Video Technol. (2023)
21. Sun, Y., Yu, Y., Wang, W.: Moiré photo restoration using multiresolution convolutional neural networks. IEEE Trans. Image Process. **27**(8), 4160–4172 (2018)
22. Wang, C., He, B., Wu, S., Wan, R., Shi, B., Duan, L.Y.: Coarse-to-fine disentangling demoiréing framework for recaptured screen images. IEEE Trans. Pattern Anal. Mach. Intell. (2023)
23. Wang, H., Tian, Q., Li, L., Guo, X.: Image demoiréing with a dual-domain distilling network. In: 2021 IEEE International Conference on Multimedia and Expo (ICME), pp. 1–6. IEEE (2021)
24. Wang, Z., Bovik, A.C., Sheikh, H.R., Simoncelli, E.P.: Image quality assessment: from error visibility to structural similarity. IEEE Trans. Image Process. **13**(4), 600–612 (2004)
25. Xing, W., Egiazarian, K.: End-to-end learning for joint image demosaicing, denoising and super-resolution. In: Proceedings of the IEEE/CVF Conference on Computer Vision and Pattern Recognition, pp. 3507–3516 (2021)
26. Xu, S., Song, B., Chen, X., Zhou, J.: Direction-aware video demoireing with temporal-guided bilateral learning. In: Proceedings of the AAAI Conference on Artificial Intelligence, vol. 38, pp. 6360–6368 (2024)
27. Yu, X., Dai, P., Li, W., Ma, L., Shen, J., Li, J., Qi, X.: Towards efficient and scale-robust ultra-high-definition image demoiréing. In: European Conference on Computer Vision. pp. 646–662. Springer (2022). https://doi.org/10.1007/978-3-031-19797-0_37
28. Yue, H., Cheng, Y., Liu, X., Yang, J.: Recaptured raw screen image and video demoireing via channel and spatial modulations. arXiv preprint arXiv:2310.20332 (2023)

29. Yue, H., Cheng, Y., Mao, Y., Cao, C., Yang, J.: Recaptured screen image demoiréing in raw domain. IEEE Trans. Multimedia (2022)
30. Yue, H., Zhang, Z., Yang, J.: Real-rawvsr: real-world raw video super-resolution with a benchmark dataset. In: European Conference on Computer Vision. pp. 608–624. Springer (2022). https://doi.org/10.1007/978-3-031-20068-7_35
31. Zhang, R., Isola, P., Efros, A.A., Shechtman, E., Wang, O.: The unreasonable effectiveness of deep features as a perceptual metric. In: Proceedings of the IEEE Conference on Computer Vision and Pattern Recognition, pp. 586–595 (2018)
32. Zheng, B., Yuan, S., Slabaugh, G., Leonardis, A.: Image demoireing with learnable bandpass filters. In: Proceedings of the IEEE/CVF Conference on Computer Vision and Pattern Recognition, pp. 3636–3645 (2020)

LiDAR-Event Stereo Fusion
with Hallucinations

Luca Bartolomei[✉], Matteo Poggi, Andrea Conti,
and Stefano Mattoccia

University of Bologna, Bologna, Italy
luca.bartolomei5@unibo.it
https://eventvppstereo.github.io/

Abstract. Event stereo matching is an emerging technique to estimate depth from neuromorphic cameras; however, events are unlikely to trigger in the absence of motion or the presence of large, untextured regions, making the correspondence problem extremely challenging. Purposely, we propose integrating a stereo event camera with a fixed-frequency active sensor – e.g., a LiDAR – collecting sparse depth measurements, overcoming the aforementioned limitations. Such depth hints are used by hallucinating – i.e., inserting fictitious events – the stacks or raw input streams, compensating for the lack of information in the absence of brightness changes. Our techniques are general, can be adapted to any structured representation to stack events and outperform state-of-the-art fusion methods applied to event-based stereo.

1 Introduction

Depth estimation is a fundamental task with many applications ranging from robotics, 3D reconstruction, and augmented/virtual reality to autonomous vehicles. Accurate, prompt, and high-resolution depth information is crucial for most of these tasks, but obtaining it remains an open challenge. Among the many possibilities, depth-from-stereo is one of the longest-standing approaches to deal with it, with a large literature of deep architecture [51] proposed in the last decade for processing rectified color images.

Event cameras [18] (or neuromorphic cameras) are recently emerging as an alternative to overcome the limitations of traditional imaging devices, such as their low dynamic range or the motion blur caused by fast movements. Unlike their traditional counterparts, event cameras do not capture frames at synchronous intervals. Instead, they mimic the dynamic nature of human vision by reporting pixel intensity changes, which can have *positive* or *negative* polarities, as soon as they happen. This peculiarity endows them with unparalleled features – notably microsecond temporal resolution, and an exceptionally high

Supplementary Information The online version contains supplementary material available at https://doi.org/10.1007/978-3-031-72658-3_8.

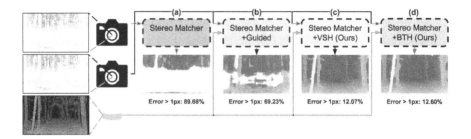

Fig. 1. LiDAR-Event Stereo Fusion with Hallucinations. In the absence of motion or brightness changes, sparse event streams lead stereo models to catastrophic failures (a). A LiDAR sensor can be used with existing strategies [49] to soften this problem, yet with limited impact (b), whereas our proposals are superior (c,d).

dynamic range – making them perfectly suited for applications where fast motion and challenging light conditions are persistent issues (*e.g.* autonomous driving). The events streams are often encoded in W×H×C tensors, thus being fully compatible with the CNNs used for classical stereo [43], capable of estimating dense disparity maps driven by data, despite the sparse nature of events.

However, as the events trigger only with brightness changes any derived data is *semi-dense* and uninformative, for instance, when facing large untextured regions or in the absence of any motion – *e.g.*, as in the example in Fig. 1. This makes the downstream stereo network struggle to match events across left and right cameras, as shown in Fig. 1 (a). According to the RGB stereo literature, fusing color information with sparse depth measurements from an active sensor [4,12,49,82] (*e.g.*, a LiDAR) considerably softens the weaknesses of passive depth sensing, despite the much lower resolution at which depth points are provided. We argue that such a strategy would counter the aforementioned issues even if applied to the event stereo paradigm, yet with a notable nodus caused by the fixed rate at which depth sensors work – usually, 10 Hz for LiDARs – being in contrast with the asynchronous acquisition rate of event cameras. This would cause to either i) use depth points only when available, harming the accuracy of most fusion strategies known from the classical stereo literature [12,49,82], or ii) limiting processing to the LiDAR pace, nullifying one of the greatest strength of event cameras – *i.e.*, microseconds resolution. Nonetheless, this track on event stereo/active sensors fusion has remained unexplored so far.

In this paper, starting from the RGB literature [4,12,49,82], we embark on a comprehensive investigation into the fusion of event-based stereo with sparse depth hints from active sensors. Inspired by [4], which projects distinctive color patterns on the images consistently with measured depth, we design a hallucination mechanism to generate fictitious events over time to densify the stream collected by the event cameras. Purposely, we propose two different strategies, respectively consisting of i) creating distinctive patterns directly at the stack level, i.e. a **Virtual Stack Hallucination** (VSH), just before the deep network processing, or ii) generating raw events directly in the stream, starting from the time instant t_d for which we aim to estimate a disparity map and performing **Back-in-Time Hallucination** (BTH). Both strategies, despite the different constraints – VSH requires explicit access to the stacked representation, whereas BTH does not – dramatically improve the accuracy of pure event-based stereo systems, overcoming some of their harshest limitations as shown in Fig. 1 (c,d). Furthermore, despite depth sensors having a fixed acquisition rate that is in contrast with the asynchronous capture rate of event cameras, VSH and BTH can leverage depth measurements not synchronized with t_d (thus collected at $t_z < t_d$) with marginal drops in accuracy compared to the case of perfectly synchronized depth and event sensors ($t_z = t_d$). This strategy allows for exploiting both VSH and BTH while preserving the microsecond resolution peculiar of event cameras. Exhaustive experiments support the following claims:

- We prove that LiDAR-stereo fusion frameworks can effectively be adapted to the event stereo domain
- Our VSH and BTH frameworks are general and work effectively with any structured representation among the eight we surveyed
- Our strategies outperform existing alternatives inherited from RGB stereo literature on DSEC [21] and M3ED [9] datasets
- VSH and BTH can exploit even outdated LiDAR data to increase the event stream distinctiveness and ease matching, preserving the microsecond resolution of event cameras and eliminating the need for synchronous processing dictated by the constant framerate of the depth sensor

2 Related Work

Stereo Matching on Color Images. It is a longstanding open problem, with a large body of literature spanning from traditional approaches grounded on handcrafted features and priors [5,24,31,36,62,68,75,76,78] to contemporary deep learning approaches that brought significant improvements over previous methods, starting with [79]. Nowadays, the most effective solutions have emerged as end-to-end deep stereo networks [51], replacing the whole stereo pipeline with a deep neural network architecture through 2D and 3D architectures. The former, inspired by the U-Net model [53], adopts an encoder-decoder design [37,42,45,50,54,59,63,64,74,77]. In contrast, the latter constructs a feature cost volume from image pair features and estimates the disparity map through 3D convolutions at the cost of substantially higher memory and runtime demands [10,11,13,16,23,27,28,57,70,73,80]. A recent trend in this field [34,38,65,71,83,84] introduced innovative deep stereo networks that embrace an iterative refinement paradigm or use Vision Transformers [22,35].

Stereo Matching with Event Cameras. This topic attracted significant attention due to the unique advantages of event sensors over traditional frame-based cameras. Similarly to conventional stereo matching, the first approaches focused on developing traditional algorithms by building structured representations, such as voxel grids [56], matched through handcrafted similarity functions

[30,56,60,85]. However, pseudo-images lose the high temporal resolution of the stream: to face this problem, [8,52] handle events without an intermediate representation using an event-to-event matching approach, where for each reference event, a set of possible matches is given. Camuñas-Mesa et al. [7] add filters to exploit orientation cues and increase matching distinctiveness. Instead, [47] revisited the cooperative network from [41]. Neural networks also showed promising results on event stereo matching with models directly processing raw events or using structured representation. The former are often inspired by [41] and typically employ Spiking Neural Networks (SNN) [1,15,44]. The latter adopts data-driven Convolutional Neural Networks (CNNs) to infer dense depth maps [43,66,67]. A detailed review of different event-based stereo techniques can be found in [17].

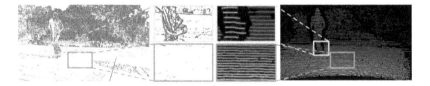

Fig. 2. Event cameras vs LiDARs – strengths and weaknesses. Event cameras provide rich cues at object boundaries where LiDARs cannot (cyan), yet LiDARs can measure depth where the lack of texture makes event cameras uninformative (green). (Color figure online)

Sensor Fusion for Stereo. Recent research has delved into the fusion of color-cameras stereo vision with active sensors, starting with handcrafted algorithms: Badino et al. [2] integrated LiDAR data directly into the stereo algorithm using dynamic programming, Gandhi et al. [19] proposed an efficient seed-growing algorithm to fuse time-of-flight (ToF) depth data with stereo pairs, while Marin et al. [40] and Poggi et al. [48] exploited confidence measures. Eventually, contemporary approaches integrated depth from sensors with modern stereo networks, either by concatenating them to images as input [12,46,69,81] or by using them to guide the cost optimization process by modulating existing cost volumes [25,49,69,82]. More recently, Bartolomei et al. [4] followed a different path with Virtual Pattern Projection (VPP). Although LiDAR sensors and event cameras have been deployed together for some applications [6,14,20,33,55,58,61], this paper represents the first attempt at combining LiDAR with an event stereo framework. We argue that the two modalities are complementary, as shown in Fig. 2 – e.g., the lack of texture and motion makes an event camera uninformative, whereas this does not affect LiDAR systems.

3 Preliminaries: Event-Based Deep Stereo

Event cameras measure brightness changes as an asynchronous stream of events. Accordingly, an event $e_k = (x_k, y_k, p_k, t_k)$ is triggered at time t_k if the intensity sensed by pixel (x_k, y_k) on the W×H sensor grid changes and surpasses a specific contrast threshold. Depending on the sign of this change, it will have polarity $p_k \in \{-1, 1\}$. Since this unstructured flow is not suitable for standard CNNs – as those proposed in the classical stereo literature [51] – converting it into W×H×C structured representations is necessary if we are interested in obtaining a dense disparity map [21]. Purposely, given a timestamp t_d at which we want to estimate a disparity map, events are sampled backward in time from the stream, either based on a time interval (SBT) or a maximum number of events (SBN), and *stacked* according to various strategies – among them:

Histogram [39]. Events of the two polarities are counted into per-pixel histograms, yielding a W×H×2 stack.

Voxel Grid [86]. The timelapse from which events are sampled is split into B uniform bins: polarities are accumulated in each bin of a W×H×B stack.

Mixed-Density Event Stack (MDES) [43]. Similar to the voxel grid strategy, the timelapse is split into bins covering $1, \frac{1}{2}, \frac{1}{4}, ..., \frac{1}{2^{N-2}}, \frac{1}{2^{N-1}}$ of the total interval. The latest event in each bin is kept, yielding a W×H×N binary stack.

Concentrated stack [43]. A shallow CNN is trained to process a pre-computed stack (*e.g.*, an MDES) and aggregate it to a W×H×1 data structure.

Time-Ordered Recent Event (TORE) [3]. It stores event timestamps into Q per-pixel queues for each polarity, yielding a W×H×2Q stack.

Time Surface [32]. A surface is derived from the timestamp distributions of the two polarities. S values are sampled for each, yielding a W×H×2S stack.

ERGO-12 [87]. An optimized representation of 12 channels, each built according to different strategies from the previous. It yields a W×H×12 stack.

Tencode [26]. A color image representation in which R and B channels encode positive and negative polarities, with G encoding the timestamp relative to the total timelapse. It produces an RGB image, *i.e.* a W×H×3 stack.

We can broadly classify stereo frameworks using these representations into three categories: i) *white boxes*, for which we have full access to the implementation of both the stereo backbone and the stacked event construction; ii) *gray boxes*, in case we do not have access to the stereo backbone; iii) *black boxes*, when the stacked event representation is not accessible neither.

4 Proposed Method

According to the sensor fusion literature for conventional cameras, the main strategies for combining stereo images with sparse depth measurements from active sensors consist of i) concatenating the two modalities and processing them as joint inputs with a stereo network [12,46,69,81], ii) modulating the internal cost volume computed by the backbone itself [25,49,69,82] or, more recently, iii) projecting distinctive patterns on images according to depth hints [4].

We follow the latter path, since it is more effective and flexible than the alternatives – which can indeed be applied to *white box* frameworks only. For this purpose, we design two alternative strategies suited even for gray and black box frameworks, respectively, as depicted in Fig. 3.

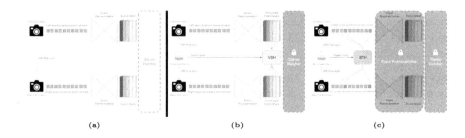

(a) (b) (c)

Fig. 3. Overview of a generic event-based stereo network and our hallucination strategies. State-of-the-art event-stereo frameworks (a) pre-process raw events to obtain event stacks fed to a deep network. In case the stacks are accessible, we define the model as a *gray box*, otherwise as a *black box*. In the former case (b), we can hallucinate patterns directly on it (VSH). When dealing with a black box (c), we can hallucinate raw events that will be processed to obtain the stacks (BTH).

4.1 Virtual Stack Hallucination – VSH

Given left and right stacks $\mathcal{S}_L, \mathcal{S}_R$ of size W×H×C and a set Z of depth measurements $z(x,y)$ by a sensor, we perform a Virtual Stack Hallucination (VSH), by augmenting each channel $c \in C$, to increase the distinctiveness of local patterns and thus ease matching. This is carried out by injecting the same virtual stack $\mathcal{A}(x,y,x',c)$ into $\mathcal{S}_L, \mathcal{S}_R$ respectively at coordinates (x,y) and (x',y).

$$\begin{aligned} \mathcal{S}_L(x,y,c) &\leftarrow \mathcal{A}(x,y,x',c) \\ \mathcal{S}_R(x',y,c) &\leftarrow \mathcal{A}(x,y,x',c) \end{aligned} \quad (1)$$

with x' obtained as $x - d(x,y)$, with disparity $d(x,y)$ triangulated back from depth $z(x,y)$ as $\frac{bf}{z(x,y)}$, according to the baseline and focal lengths b, f of the stereo system. We deploy a generalized version of the random pattern operator $\mathcal{A}$ proposed in [4], agnostic to the stacked representation:

$$\mathcal{A}(x,y,x',c) \sim \mathcal{U}(\mathcal{S}^-, \mathcal{S}^+) \quad (2)$$

with $\mathcal{S}^-$ and $\mathcal{S}^+$ the minimum and maximum values appearing across stacks $\mathcal{S}_L, \mathcal{S}_R$ and $\mathcal{U}$ a uniform random distribution. Following [4], the pattern can either cover a single pixel or a local window. This strategy alone is sufficient already to ensure distinctiveness and to dramatically ease matching across stacks, even more than with color images [4], since acting on semi-dense structures – *i.e.*, stacks are uninformative in the absence of events. It also ensures a straightforward application of the same principles used on RGB images, e.g., to combine the original content (color) with the virtual projection (pattern) employing alpha blending [4]. Nevertheless, we argue that acting at this level i) requires direct access to the stacks, i.e., a gray-box deep event-stereo network, and ii) might be sub-optimal as stacks encode only part of the information from streams.

4.2 Back-in-Time Hallucination – BTH

A higher distinctiveness to ease correspondence can be induced by hallucinating patterns directly in the continuous events domain. Specifically, we act in the so-called *event history*: given a timestamp t_d at which we want to estimate disparity, raw events are sampled from the left and right streams starting from t_d and going backward, according to either SBN or SBT stacking approaches, to obtain a pair of event histories $\mathcal{E}_L = \{e_k^L\}_{k=1}^N$ and $\mathcal{E}_R = \{e_k^R\}_{k=1}^M$, where e_k^L, e_k^R are the k-th left and right events. Events in the history are sorted according to their timestamp – i.e., inequality $t_k \leq t_{k+1}$ holds for every two adjacent e_k, e_{k+1}.

At this point, we intervene to hallucinate novel events: given a depth measurement $z(\hat{x}, \hat{y})$, triangulated back into disparity $d(\hat{x}, \hat{y})$, we inject a pair of fictitious events $\hat{e}^L = (\hat{x}, \hat{y}, \hat{p}, \hat{t})$ and $\hat{e}^R = (\hat{x}', \hat{y}, \hat{p}, \hat{t})$ respectively inside $\mathcal{E}_L$ and $\mathcal{E}_R$, producing $\hat{\mathcal{E}}_L = \{e_1^L, \ldots, \hat{e}^L, \ldots, e_N^L\}$ and $\hat{\mathcal{E}}_R = \{e_1^R, \ldots, \hat{e}^R, \ldots, e_M^R\}$. By construction, $\hat{e}^L$ and $\hat{e}^R$ adhere to i) the time ordering constraint, ii) the geometry constraint $\hat{x}' = \hat{x} - d(\hat{x}, \hat{y})$ and iii) a similarity constraint – i.e., $\hat{p}, \hat{t}$ are the same for $\hat{e}^L$ and $\hat{e}^R$. Fictitious polarity $\hat{p}$ and fictitious timestamp $\hat{t}$ are two degrees of freedom useful to ensure distinctiveness along the epipolar line and ease matching, according to which we can implement different strategies summarized in Fig. 4, and detailed in the remainder.

Fig. 4. Overview of Back-in-Time Hallucination (BTH). To estimate disparity at t_d, if LiDAR data is available – e.g., at timestamp $t_z = t_d$ (green) or $t_z = t_d - 15$ (yellow) – we can naïvely inject events of random polarities at the same timestamp t_z (a). More advanced injection strategies can be used – e.g. by hallucinating multiple events, starting from t_d, back-in-time at regular intervals (b). (Color figure online)

Single-Timestamp Injection. The simplest way to increase distinctiveness is to insert synchronized events at a fixed timestamp. Accordingly, for each depth measurement $d(\hat{x}, \hat{y})$, a total of $K_{\hat{x}, \hat{y}}$ pairs of fictitious events are inserted in $\mathcal{E}_L, \mathcal{E}_R$, having polarity $\hat{p}_k$ randomly chosen from the discrete set $\{-1, 1\}$.

Timestamp $\hat{t}$ is fixed and can be, for instance, t_z at which the sensor infers depth, that can coincide with timestamp t_d at which we want to estimate disparity – e.g., $t_z = t_d = 0$ in the case depicted in Fig. 4 (a). Inspired by [4], events might be optionally hallucinated in patches rather than single pixels. However, as depth sensors usually work at a fixed acquisition frequency – e.g., 10 Hz for LiDARs – sparse points might be unavailable at any specific timestamp. Nonetheless, since $\mathcal{E}_L, \mathcal{E}_R$ encode a time interval, we can hallucinate events even if derived from depth scans performed *in the past* – e.g., at $t_z < t_d$, – by placing them in the proper position inside $\mathcal{E}_L, \mathcal{E}_R$.

Repeated Injection. The previous strategy does not exploit one of the main advantages of events over color images, *i.e.* the temporal dimension, at its best. Purposely, we design a more advanced hallucination strategy based on *repeated* naïve injections performed along the time interval sampled by $\mathcal{E}_L, \mathcal{E}_R$. As long as we are interested in recovering depth at t_d only, we can hallucinate as many events as we want in the time interval *before t* – *i.e.*, for $t_z = t_d = 0$, over the entire interval as shown in Fig. 4 (b) – consistent with the depth measurements at t_d itself, which will increase the distinctiveness in the event histories and will ease the match by hinting the correct disparity. Inspired by the stacked representations introduced in Sect. 3, we can design a strategy for injecting multiple events along the stream. Accordingly, we define the *conservative* time range $[t^-, t^+]$ of the events histories $\mathcal{E}_L, \mathcal{E}_R$, with $t^- = \min\{t_0^L, t_0^R\}$ and $t^+ = \max\{t_N^L, t_M^R\}$ and divide it into B equal temporal bins. Then, inspired by MDES [43], we run B single-timestamp injections at $\hat{t}_b = \frac{2^b-1}{2^B}(t^+ - t^-) + t^-$, with $b \in \{1, \dots, B\}$. Additionally, each depth measurement is used only once – *i.e.*, the number of fictitious events $K_{b,\hat{x},\hat{y}}$ in the *b*-th injection is set as $K_{b,\hat{x},\hat{y}} \leftarrow K_{\hat{x},\hat{y}} \delta(b, D_{\hat{x},\hat{y}})$ where $\delta(\cdot,\cdot)$ is the Kronecker delta and $D_{\hat{x},\hat{y}} \leftarrow \text{round}(X^{\mathcal{U}}(B-1)+1)$ is a random slot assignment. We will show in our experiment how this simple strategy can improve the results of BTH, in particular increasing its robustness against misaligned LiDAR data – i.e., measurements retrieved at a timestamp $t_z < t_d$.

5 Experiments

5.1 Implementation and Experimental Settings

We implement VSH and BTH in Python, using the Numba package.

General Framework. We build our code base starting from SE-CFF [43] – state-of-the-art for event-based stereo – assuming the same stereo backbone as in their experiments, *i.e.* derived from AANet [72], and run SBN to generate the event history to be stacked. While we select a single architecture, we implement a variety of stacked representations: purposely, we implement a single instance of the stereo backbone for any stacked representations introduced in Sect. 3, taking the opportunity to evaluate their performance with the event stereo task. For Concentration representation, we use MDES as the prior stacking function following [43] and avoid considering future events during training. Furthermore, in this case, VSH is applied before the concentration network since it would interfere with gradient back-propagation during training – while this cannot occur with BTH. From [4], we adapt occlusion handling and hallucination on uniform/not uniform patches. We also implement alpha-blending, for VSH only – as it loses its purpose when acting on the raw streams. For all our methods, we inherit the same hyper-parameters from [4]; yet, we discard occluded points as the occlusion handling strategy for BTH since an equivalent strategy to deal with sparse event histories is not trivial. For VSH on Voxel Grids, we use the 5-th and 95-th percentile to calculate $\mathcal{S}^-$ and $\mathcal{S}^+$ due to the frequent presence of extreme values in the stack. For BTH, we perform 12 injections (*i.e.*, $B = 12$).

Existing Fusion Methodologies. We compare our proposal with existing methods from the RGB stereo literature, consisting of i) modulating the cost volume built by the backbone – Guided Stereo Matching [49], ii) concatenating the sparse depth values to the inputs to the stereo network – *e.g.*, as done by LidarStereoNet [12], iii) a combination of both the previous strategies – in analogy to CCVNorm [69]. Any strategy is adapted to the same common stereo backbone [43] (see **supplementary material**). Running BTH and VSH adds respectively 10ms and 2–15ms (depending on representations) on the CPU.

Training Protocol. Any model we train – either the original event stereo backbones or those implementing fusion strategies – runs for 25 epochs with a batch size of 4 and a maximum disparity set to 192. We use Adam [29] with beta (0.9, 0.999) and weight decay set to 10^{-4}. The learning rate starts at $5 \cdot 10^{-4}$ and decays with cosine annealing. We apply random crops and vertical flips to augment data during training.

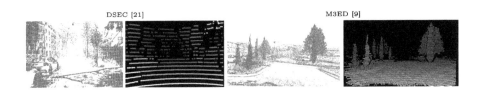

Fig. 5. Qualitative comparison – DSEC vs M3ED. DSEC features 640 × 480 event cameras and a 16-line LiDAR, M3ED has 1280 × 720 event cameras and a 64-line LiDAR. LiDAR scans have been dilated with a 7 × 7 kernel to ease visualization.

5.2 Evaluation Datasets and Protocol

We introduce datasets and metrics used in our experiments.

DSEC [21]. An outdoor event stereo dataset, captured using wide-baseline (50 cm) stereo event cameras at 640 × 480 resolution. Ground-truth disparity is obtained by accumulating 16-line LiDAR scans, for a total of 26 384 maps organized into 41 sequences. We split them into train/test sets following [43].

From the training set, we retain a further *search* split for hyper-parameters tuning and ablation experiments. Sparse LiDAR measurements are obtained by aligning the raw scans with the ground-truth – both provided by the authors – by running a LiDAR inertial odometry pipeline followed by ICP registration (see the **supplementary material** for details).

M3ED [9]. This dataset provides 57 indoor/outdoor scenes collected with a compact multi-sensor block mounted on three different vehicles – *i.e.*, a car, a UAV, and a quadruped robot. A 64-line LiDAR generates semi-dense ground-truth depth, while the event stereo camera has a shorter baseline (12 cm) and a higher resolution (1280 × 720). We use 5 sequences from this dataset for evaluation purposes only – some of which contain several frames acquired with the cameras being static – to evaluate the generalization capacity of the models both to different domains and the density of the LiDAR sensor. Similarly to DSEC, we derived sparse LiDAR depth maps from the raw scans. Thanks to the SDK made available by the authors, we could derive LiDAR measurements aligned to any desired temporal offset according to linear interpolation of the ground-truth poses (see the **supplementary material** for details). This allows us to run dedicated experiments to assess the effect of time-misaligned depth measurements. Figure 5 shows a qualitative comparison between the two datasets.

Evaluation Metrics. We compute the percentage of pixels with an error greater than 1 or 2 pixels (1PE, 2PE), and the mean absolute error (MAE). We highlight the best and second best methods per row on each metric.

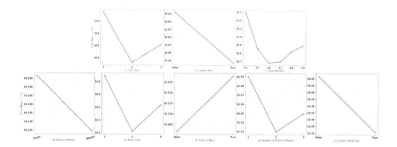

Fig. 6. Hyperparameters search. Results on DSEC search split. On top, we study the impact of (a) patch size, (b) uniform patches, and (c) alpha blending on VSH. At the bottom, we consider (d) single vs repeated injection, (e) patch size, (f) uniform patches, (g) number of fictitious events, and (h) uniform polarities on BTH.

5.3 Ablation Study

We ran hyper-parameters search and ablation experiments for VSH and BTH on the DSEC search split, reporting the 1PE error. We conducted these experiments using any representation listed in Sect. 3 – except for Concentration [43], which starts from pre-computed MDES stacks – and report the average results.

VSH. Figure 6 (top) shows the impact of different hyper-parameters on VSH strategy. In (a), we can observe how VSH is improved by using 3 × 3 patches, while 5 × 5 cannot yield further benefits. Consequently, we select it as the default configuration from now on. In (b), we show that uniform patterns are more effective than random ones, and in (c) alpha equal to 0.5 works the best.

BTH. Figure 6 (bottom) focuses on our second strategy. In (d) we show how repeated injection can improve the results; thus, we select it as the default configuration from now on. In the remainder, we will better appreciate how this setting is much more robust when dealing with misaligned LiDAR data. Next, (e) outlines how hallucinating events with 3 × 3 patches lead to the best results. Applying a uniform patch of events following (f) yields, again, better results. In (g), we tested different numbers $K_{\hat{x},\hat{y}}$ of injected fictitious events. Injecting more than one event is beneficial, yet saturating with two. Finally, (h) shows that using uniform polarities yields lower errors.

5.4 Experiments on DSEC

We now report experiments on the DSEC testing split, either when applying fusion strategies to pre-trained stereo models without retraining them or when training the networks from scratch to exploit LiDAR data.

Pre-trained Models. Table 1 reports, on each row, the results yielded by using a specific stacked representation. In the columns, we report the different fusion strategies involved in our experiments, starting with the baseline – *i.e.*, a stereo backbone processing events only. In the last row, we report the average ranking – for the three metrics – achieved by any fusion strategy over the eight representations. Starting from baseline models, we can notice how the different representations have an impact on the accuracy of the stereo backbone, with those modeling complex behaviors – *e.g.*, Time Surface [32] or ERGO-12 [87] – yielding up to 2% lower 1PE than simpler ones such as Histogram [39].

Table 1. Results on DSEC [21] pre-trained. We test the different stacked representations (rows) with several fusion strategies applied to pre-trained stereo backbones.

	Stacked representation	Baseline			Guided [49]			VSH (ours)			BTH (ours)		
		1PE	2PE	MAE	1PE	2PE	MAE	1PE	2PE	MAE	1PE	2PE	MAE
(A)	Histogram [39]	16.21	4.73	0.74	16.07	4.68	0.73	13.71	4.20	0.69	13.32	3.92	0.66
(B)	MDES [43]	15.32	4.40	0.70	15.13	4.34	0.70	12.94	3.52	0.63	12.61	3.50	0.62
(C)	Concentration [43]	15.97	4.33	0.70	15.79	4.27	0.70	13.70	3.60	0.65	14.66	3.77	0.66
(D)	Voxelgrid [86]	16.49	4.56	0.72	16.29	4.50	0.71	13.12	3.69	0.65	12.44	3.60	0.62
(E)	TORE [3]	15.91	4.57	0.71	15.72	4.50	0.71	12.53	3.65	0.63	12.27	3.68	0.62
(F)	Time Surface [32]	15.33	4.29	0.70	15.18	4.24	0.69	12.16	3.38	0.62	12.28	3.45	0.62
(G)	ERGO-12 [87]	15.02	4.20	0.68	14.87	4.14	0.68	12.02	3.40	0.61	11.98	3.42	0.61
(H)	Tencode [26]	14.46	4.17	0.68	14.29	4.11	0.67	12.12	3.37	0.61	11.86	3.45	0.61
	Avg. Rank.	-	-	-	3.00	3.00	3.00	1.75	1.38	1.50	1.25	1.63	1.13

Table 2. Results on DSEC [21] – retrained. We test different stacked representations (rows) with several fusion strategies applied during training.

	Concat [12]			Guided+Concat [69]			Guided [49]			VSH (ours)			BTH (ours)		
	1PE	2PE	MAE	1PE	2PE	MAE	1PE	2PE	MAE	1PE	2PE	MAE	1PE	2PE	MAE
(A)	12.57	3.37	0.62	12.81	3.41	0.63	15.57	4.58	0.72	9.90	3.26	0.53	10.91	3.41	0.59
(B)	12.37	3.17	0.61	12.40	3.25	0.61	14.66	4.36	0.70	9.31	3.01	0.51	9.62	3.01	0.54
(C)	12.38	3.41	0.63	12.74	3.44	0.66	15.15	4.45	0.71	9.70	3.04	0.53	9.66	2.98	0.55
(D)	12.23	3.18	0.60	11.90	3.10	0.60	14.52	4.21	0.68	10.16	3.20	0.56	9.68	2.90	0.54
(E)	12.99	3.33	0.62	12.62	3.25	0.61	16.00	4.56	0.73	9.91	3.05	0.53	9.83	2.98	0.54
(F)	12.18	3.09	0.61	12.47	3.17	0.61	14.40	4.21	0.68	9.47	2.90	0.52	9.58	2.92	0.54
(G)	12.43	3.14	0.61	12.82	3.19	0.62	13.85	3.97	0.66	9.25	2.88	0.50	9.37	2.87	0.54
(H)	11.95	3.08	0.60	11.75	3.10	0.60	14.72	4.21	0.69	9.39	3.00	0.52	9.59	2.97	0.55
	3.38	3.00	3.13	3.63	3.50	3.38	5.00	5.00	5.00	1.38	1.88	1.13	1.63	1.38	1.88

Table 3. Results on M3ED [9] – pre-trained. We test the different stacked representations (rows) with several fusion strategies applied to pre-trained stereo backbones.

	Stacked representation	Baseline			Guided [49]			VSH (ours)			BTH (ours)		
		1PE	2PE	MAE	1PE	2PE	MAE	1PE	2PE	MAE	1PE	2PE	MAE
(A)	Histogram [39]	37.70	19.49	1.76	37.18	19.29	1.75	20.19	11.19	1.19	22.32	12.37	1.27
(B)	MDES [43]	43.17	19.50	1.85	42.27	19.16	1.83	29.42	14.80	1.52	22.58	12.20	1.30
(C)	Concentration [43]	45.78	20.84	1.82	45.06	20.57	1.80	33.63	16.19	1.53	25.22	12.68	1.28
(D)	Voxelgrid [86]	37.33	17.66	1.70	36.64	17.38	1.68	20.40	11.41	1.22	20.94	11.72	1.23
(E)	TORE [3]	41.70	19.09	1.81	41.00	18.78	1.80	28.25	14.01	1.47	21.91	12.34	1.30
(F)	Time Surface [32]	38.58	18.52	1.72	37.91	18.23	1.70	24.89	13.34	1.37	22.60	12.77	1.31
(G)	ERGO-12 [87]	36.33	17.81	1.66	35.61	17.50	1.64	22.53	12.33	1.26	20.41	11.69	1.21
(H)	Tencode [26]	43.56	20.07	1.82	42.66	19.76	1.80	28.24	14.46	1.43	22.61	12.75	1.26
	Avg. Rank.	-			3.00	3.00	3.00	1.75	1.75	1.75	1.25	1.25	1.25

The Guided framework [49] can improve the results only moderately: this is caused by the very sparse measurements retrieved from the 16-line LiDAR sensor used in DSEC, as well as by the limited effect of the cost volume modulation in regions where events are not available for matching. Nonetheless, VSH and BTH consistently outperform Guided, always improving the baseline by 2-3% points on 1PE. In general, BTH achieves the best 1PE and MAE metrics in most cases; this strategy is the best when re-training the stereo backbone is not feasible.

Training from Scratch. Table 2 reports the results obtained by training the stereo backbones from scratch to perform LiDAR-event stereo fusion. This allows either the deployment of strategies that process the LiDAR data directly as input [12,69] or those not requiring it, i.e., [49] and ours. Specifically, Concat [12] and Guided+Concat [69] strategy achieve results comparable to those by VSH and BTH observed before, thus outperforming Guided [49] which, on the contrary, cannot benefit much from the training process. When deploying our solutions during training, their effectiveness dramatically increases, often dropping 1PE error below 10%. VSH often yields the best 1PE and MAE overall, nominating it as the most effective – yet intrusive – among our solutions.

Table 4. Results on M3ED [9] – retrained. We test different stacked representations (rows) with several fusion strategies applied during training.

	Concat [12]			Guided+Concat [69]			Guided [49]			VSH (ours)			BTH (ours)		
	1PE	2PE	MAE	1PE	2PE	MAE	1PE	2PE	MAE	1PE	2PE	MAE	1PE	2PE	MAE
(A)	34.67	15.21	1.92	38.65	17.00	1.94	37.45	18.98	1.76	19.34	12.93	1.46	19.83	13.20	1.39
(B)	37.72	16.91	1.85	37.32	17.16	2.14	37.00	18.66	1.76	19.24	13.17	1.44	18.70	11.79	1.24
(C)	39.88	19.01	2.33	38.45	17.76	2.47	38.14	19.62	1.80	19.84	13.68	1.90	19.46	12.29	1.35
(D)	33.89	16.21	1.89	33.54	15.85	1.75	37.85	18.81	1.74	18.56	11.76	1.32	21.02	14.30	1.80
(E)	38.83	18.38	2.27	35.80	16.63	2.05	40.51	19.96	1.95	20.03	13.97	1.86	20.04	12.65	1.39
(F)	40.26	18.44	2.19	35.48	17.74	2.15	38.77	18.41	1.75	19.61	13.01	1.55	21.91	14.33	1.72
(G)	42.43	19.31	2.31	42.24	18.42	2.34	37.95	17.83	1.76	18.45	12.31	1.55	19.12	11.60	1.24
(H)	37.46	17.87	2.15	33.69	16.47	1.95	39.78	19.42	1.82	19.49	12.21	1.38	19.28	11.68	1.33
	4.38	4.00	4.50	3.63	3.38	4.38	4.00	4.63	2.75	1.38	1.63	1.88	1.63	1.38	1.50

5.5 Experiments on M3ED

We test the effectiveness of BTH and alternative approaches on M3ED, using the backbones trained on DSEC **without** any fine-tuning on M3ED itself.

Pre-trained Models. Table 3 collects the outcome of this experiment by applying Guided, VSH, and BTH to pre-trained models. Looking at the baselines, we can appreciate how M3ED is very challenging for models trained in a different domain, with 1PE errors higher than 30%. This is caused by both the domain shift and the higher resolution of the event cameras used. Even so, complex event representations – e.g., ERGO-12 [87] – can better generalize. Guided confirms its limited impact, this time mainly because of the ineffectiveness of the cost volume modulation in the absence of any information from the events domain. On the contrary, we can appreciate even further the impact of VSH and BTH, almost halving the 1PE error. Specifically, BTH is the absolute winner with 6 out of 8 representations, and the best choice for pre-trained frameworks.

Training from Scratch. Table 4 resumes the results obtained when training on DSEC the backbones implementing LiDAR-event stereo fusion strategies. The very different distribution of depth points observed across the two datasets – sourced respectively from 16 and 64-line LiDARs – yields mixed results for existing methods [12,49,69], with rare cases for which they fail to improve the baseline model (e.g., Concat and Guided with Time Surface and ERGO-12, Guided+Concat with Histogram). On the contrary, backbones trained with VSH and BTH consistently improve over the baseline, often with larger gains compared to their use with pre-trained models. Overall, BTH is the best on 2PE and MAE, confirming it is better suited for robustness across domains and different LiDAR sensors.

Figure 7 shows qualitative results. On DSEC (top), BTH dramatically improves results over the baseline and Guided, yet cannot fully recover some details in the scene except when retraining the stereo backbone. On M3ED (bottom), both VSH and BTH with pre-trained models reduce the error by 5×.

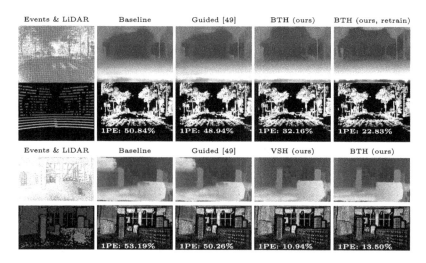

Fig. 7. Qualitative results. Results on DSEC *zurich_10_b* with Voxelgrid [86] (top) and M3ED *spot_indoor_obstacles* with Histogram [39] (bottom).

5.6 Experiments on M3ED Time-Misaligned LiDAR

We conclude by assessing the robustness of the considered strategies against the use of LiDAR not synchronized with the timestamp at which we wish to estimate disparity – occurring if we wish to maintain the microsecond resolution of the event cameras. Purposely, we extract raw LiDAR measurements collected 3, 13, 32, 61, and 100 ms in the past with the M3ED SDK.

Figure 8 shows the trend of the 1PE metric achieved by Guided (red), VSH (yellow) and BTH (black and green) on pre-trained backbones. Not surprisingly, the error rates arise at the increase of the temporal distance: while this is less evident with Guided because of its limited impact, this becomes clear with VSH and BTH. Nonetheless, both can always retain a significant gain over the baseline model (blue) – i.e., the stereo backbone processing events only – even with the farthest possible misalignment with a 10 Hz LiDAR (100ms). We can appreciate how BTH is often better than VSH (coherently with Table 3), yet only when repeated injections are performed (green). Indeed, using a single injection (black) rapidly leads BTH to an accuracy drop when increasing the misalignment, except when using Histogram representation. Overall, BTH with ERGO-12 is the most robust solution. Figure 9 shows the results achieved by VSH (yellow) and BTH (green) after retraining, against the best competitor according to average ranks in Table 4 – i.e., Guided+Concat (red). The impact of this latter is limited and sometimes fails to improve the baseline (see Histogram and ERGO-12). On the contrary, our solutions confirm their robustness and effectiveness even when dealing with time-misaligned LiDAR data.

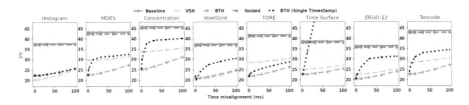

Fig. 8. Experiments with time-misaligned LiDAR on M3ED [9] – pre-trained. We measure the robustness of different fusion strategies against the use of out-of-sync LiDAR data, without retraining the stereo backbone. (Color figure online)

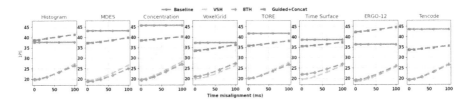

Fig. 9. Experiments with time-misaligned LiDAR on M3ED [9] – retrained. We measure the robustness of different fusion strategies against the use of out-of-sync LiDAR data when training the stereo backbone from scratch. (Color figure online)

6 Conclusion

This paper proposes a novel framework for implementing event stereo and LiDAR fusion. It works by hallucinating fictitious events either in the stacked representation processed by stereo backbones or the continuous streams sensed by event cameras, easing the matching process to the downstream stereo model estimating disparity. Our exhaustive experiments prove that our solutions, VSH and BTH, dramatically outperform alternative fusion strategies from the RGB stereo literature, retaining the microsecond resolution typical of event cameras despite the discrete frame rate of LiDARs and depth sensors in general.

Limitations. Despite the robustness shown with misaligned LiDAR data, a marginal drop in accuracy compared to the case of having LiDAR measurements at the very same timestamp at which we aim to infer disparity maps occurs. Future work will focus on studying new design mechanisms to deal with it.

Acknowledgement. This study was carried out within the MOST - Sustainable Mobility National Research Center and received funding from the European Union Next-GenerationEU - PIANO NAZIONALE DI RIPRESA E RESILIENZA (PNRR) - MISSIONE 4 COMPONENTE 2, INVESTIMENTO 1.4 - D.D. 1033 17/06/2022, CN00000023. This manuscript reflects only the authors' views and opinions, neither the European Union nor the European Commission can be considered responsible for them.

We acknowledge the CINECA award under the ISCRA initiative, for the availability of high-performance computing resources and support.

References

1. Andreopoulos, A., Kashyap, H.J., Nayak, T.K., Amir, A., Flickner, M.D.: A low power, high throughput, fully event-based stereo system. In: Proceedings of the IEEE Conference on Computer Vision and Pattern Recognition, pp. 7532–7542 (2018)
2. Badino, H., Huber, D.F., Kanade, T.: Integrating lidar into stereo for fast and improved disparity computation. In: 2011 International Conference on 3D Imaging, Modeling, Processing, Visualization and Transmission, pp. 405–412 (2011)
3. Baldwin, R.W., Liu, R., Almatrafi, M., Asari, V., Hirakawa, K.: Time-ordered recent event (tore) volumes for event cameras. IEEE Trans. Pattern Anal. Mach. Intell. **45**(2), 2519–2532 (2022)
4. Bartolomei, L., Poggi, M., Tosi, F., Conti, A., Mattoccia, S.: Active stereo without pattern projector. In: Proceedings of the IEEE/CVF International Conference on Computer Vision (ICCV), pp. 18470–18482 (October 2023)
5. Boykov, Y., Veksler, O., Zabih, R.: Fast approximate energy minimization via graph cuts. IEEE Trans. Pattern Anal. Mach. Intell. **23**(11), 1222–1239 (2001)
6. Brebion, V., Moreau, J., Davoine, F.: Learning to estimate two dense depths from lidar and event data. In: Scandinavian Conference on Image Analysis. pp. 517–533. Springer (2023). https://doi.org/10.1007/978-3-031-31438-4_34
7. Camuñas-Mesa, L.A., Serrano-Gotarredona, T., Ieng, S.H., Benosman, R.B., Linares-Barranco, B.: On the use of orientation filters for 3d reconstruction in event-driven stereo vision. Front. Neurosci. **8**, 48 (2014)
8. Carneiro, J., Ieng, S.H., Posch, C., Benosman, R.: Event-based 3d reconstruction from neuromorphic retinas. Neural Netw. **45**, 27–38 (2013)
9. Chaney, K., et al.: M3ed: Multi-robot, multi-sensor, multi-environment event dataset. In: Proceedings of the IEEE/CVF Conference on Computer Vision and Pattern Recognition (CVPR) Workshops, pp. 4015–4022 (June 2023)
10. Chang, J.R., Chen, Y.S.: Pyramid stereo matching network. In: IEEE/CVF Conference on Computer Vision and Pattern Recognition (CVPR), pp. 5410–5418 (2018)
11. Cheng, X., Wang, P., Yang, R.: Learning depth with convolutional spatial propagation network. IEEE Trans. Pattern Anal. Mach. Intell. **42**(10), 2361–2379 (2019)
12. Cheng, X., Zhong, Y., Dai, Y., Ji, P., Li, H.: Noise-aware unsupervised deep lidar-stereo fusion. In: Proceedings of the IEEE/CVF Conference on Computer Vision and Pattern Recognition, pp. 6339–6348 (2019)
13. Cheng, X., et al.: Hierarchical neural architecture search for deep stereo matching. Adv. Neural Inform. Process. Syst. **33** (2020)
14. Cui, M., Zhu, Y., Liu, Y., Liu, Y., Chen, G., Huang, K.: Dense depth-map estimation based on fusion of event camera and sparse lidar. IEEE Trans. Instrum. Meas. **71**, 1–11 (2022). https://doi.org/10.1109/TIM.2022.3144229
15. Dikov, G., Firouzi, M., Röhrbein, F., Conradt, J., Richter, C.: Spiking cooperative stereo-matching at 2 ms latency with neuromorphic hardware. In: Mangan, M., Cutkosky, M., Mura, A., Verschure, P.F.M.J., Prescott, T., Lepora, N. (eds.) Living Machines 2017. LNCS (LNAI), vol. 10384, pp. 119–137. Springer, Cham (2017). https://doi.org/10.1007/978-3-319-63537-8_11
16. Duggal, S., Wang, S., Ma, W.C., Hu, R., Urtasun, R.: Deeppruner: learning efficient stereo matching via differentiable patchmatch. In: Proceedings of the IEEE/CVF International Conference on Computer Vision, pp. 4384–4393 (2019)
17. Gallego, G., et al.: Event-based vision: a survey. IEEE Trans. Pattern Anal. Mach. Intell. **44**(1), 154–180 (2020)

18. Gallego, G., et al.: Event-based vision: a survey. IEEE Trans. Pattern Anal. Mach. Intell. **44**(1), 154–180 (2022). https://doi.org/10.1109/TPAMI.2020.3008413
19. Gandhi, V., Čech, J., Horaud, R.: High-resolution depth maps based on tof-stereo fusion. In: 2012 IEEE International Conference on Robotics and Automation, pp. 4742–4749. IEEE (2012)
20. Gao, L., et al.: Vector: a versatile event-centric benchmark for multi-sensor slam. IEEE Robot. Autom. Lett. **7**(3), 8217–8224 (2022)
21. Gehrig, M., Aarents, W., Gehrig, D., Scaramuzza, D.: Dsec: a stereo event camera dataset for driving scenarios. IEEE Robot. Autom. Lett. (2021). https://doi.org/10.1109/LRA.2021.3068942
22. Guo, W., et al.: Context-enhanced stereo transformer. In: Proceedings of the European Conference on Computer Vision (ECCV) (2022)
23. Guo, X., Yang, K., Yang, W., Wang, X., Li, H.: Group-wise correlation stereo network. In: Proceedings of the IEEE/CVF Conference on Computer Vision and Pattern Recognition, pp. 3273–3282 (2019)
24. Hirschmuller, H.: Stereo processing by semiglobal matching and mutual information. IEEE Trans. Pattern Anal. Mach. Intell. **30**(2), 328–341 (2007)
25. Huang, Y.K., et al.: S3: learnable sparse signal superdensity for guided depth estimation. In: Proceedings of the IEEE/CVF Conference on Computer Vision and Pattern Recognition, pp. 16706–16716 (2021)
26. Huang, Z., Sun, L., Zhao, C., Li, S., Su, S.: Eventpoint: self-supervised interest point detection and description for event-based camera. In: Proceedings of the IEEE/CVF Winter Conference on Applications of Computer Vision (WACV), pp. 5396–5405 (January 2023)
27. Kendall, A., et al.: End-to-end learning of geometry and context for deep stereo regression. In: The IEEE International Conference on Computer Vision (ICCV) (Oct 2017)
28. Khamis, S., Fanello, S., Rhemann, C., Kowdle, A., Valentin, J., Izadi, S.: Stereonet: guided hierarchical refinement for real-time edge-aware depth prediction. In: Proceedings of the European Conference on Computer Vision (ECCV), pp. 573–590 (2018)
29. Kingma, D.P., Ba, J.: Adam: A method for stochastic optimization. arXiv preprint arXiv:1412.6980 (2014)
30. Kogler, J., Sulzbachner, C., Humenberger, M., Eibensteiner, F.: Address-event based stereo vision with bio-inspired silicon retina imagers. Advances in theory and applications of stereo vision, pp. 165–188 (2011)
31. Kolmogorov, V., Zabin, R.: What energy functions can be minimized via graph cuts? IEEE Trans. Pattern Anal. Mach. Intell. **26**(2), 147–159 (2004)
32. Lagorce, X., Orchard, G., Galluppi, F., Shi, B.E., Benosman, R.D.: Hots: a hierarchy of event-based time-surfaces for pattern recognition. IEEE Trans. Pattern Anal. Mach. Intell. **39**(7), 1346–1359 (2016)
33. Li, B., et al.: Enhancing 3-d lidar point clouds with event-based camera. IEEE Trans. Instrum. Meas. **70**, 1–12 (2021)
34. Li, J., et al.: Practical stereo matching via cascaded recurrent network with adaptive correlation. In: Proceedings of the IEEE/CVF Conference on Computer Vision and Pattern Recognition, pp. 16263–16272 (2022)
35. Li, Z., et al.: Revisiting stereo depth estimation from a sequence-to-sequence perspective with transformers. In: Proceedings of the IEEE/CVF International Conference on Computer Vision, pp. 6197–6206 (2021)

36. Liang, C.K., Cheng, C.C., Lai, Y.C., Chen, L.G., Chen, H.H.: Hardware-efficient belief propagation. IEEE Trans. Circuits Syst. Video Technol. **21**(5), 525–537 (2011)
37. Liang, Z., et al.: Learning for disparity estimation through feature constancy. In: Proceedings of the IEEE Conference on Computer Vision and Pattern Recognition (CVPR) (June 2018)
38. Lipson, L., Teed, Z., Deng, J.: Raft-stereo: multilevel recurrent field transforms for stereo matching. In: International Conference on 3D Vision (3DV) (2021)
39. Maqueda, A.I., Loquercio, A., Gallego, G., García, N., Scaramuzza, D.: Event-based vision meets deep learning on steering prediction for self-driving cars. In: Proceedings of the IEEE Conference on Computer Vision and Pattern Recognition, pp. 5419–5427 (2018)
40. Marin, G., Zanuttigh, P., Mattoccia, S.: Reliable fusion of ToF and stereo depth driven by confidence measures. In: Leibe, B., Matas, J., Sebe, N., Welling, M. (eds.) ECCV 2016. LNCS, vol. 9911, pp. 386–401. Springer, Cham (2016). https://doi.org/10.1007/978-3-319-46478-7_24
41. Marr, D.C., Poggio, T.A.: Cooperative computation of stereo disparity. Science **194**(4262), 283–7 (1976)
42. Mayer, N., et al.: A large dataset to train convolutional networks for disparity, optical flow, and scene flow estimation. In: The IEEE Conference on Computer Vision and Pattern Recognition (CVPR) (June 2016)
43. Nam, Y., Mostafavi, M., Yoon, K.J., Choi, J.: Stereo depth from events cameras: Concentrate and focus on the future. In: Proceedings of the IEEE/CVF Conference on Computer Vision and Pattern Recognition, pp. 6114–6123 (2022)
44. Osswald, M., Ieng, S.H., Benosman, R., Indiveri, G.: A spiking neural network model of 3d perception for event-based neuromorphic stereo vision systems. Sci. Rep. **7**(1), 40703 (2017)
45. Pang, J., Sun, W., Ren, J.S., Yang, C., Yan, Q.: Cascade residual learning: A two-stage convolutional neural network for stereo matching. In: The IEEE International Conference on Computer Vision (ICCV) (Oct 2017)
46. Park, K., Kim, S., Sohn, K.: High-precision depth estimation with the 3d lidar and stereo fusion. In: 2018 IEEE International Conference on Robotics and Automation (ICRA), pp. 2156–2163. IEEE (2018)
47. Piatkowska, E., Belbachir, A., Gelautz, M.: Asynchronous stereo vision for event-driven dynamic stereo sensor using an adaptive cooperative approach. In: Proceedings of the IEEE International Conference on Computer Vision Workshops, pp. 45–50 (2013)
48. Poggi, M., Agresti, G., Tosi, F., Zanuttigh, P., Mattoccia, S.: Confidence estimation for tof and stereo sensors and its application to depth data fusion. IEEE Sens. J. **20**(3), 1411–1421 (2020). https://doi.org/10.1109/JSEN.2019.2946591
49. Poggi, M., Pallotti, D., Tosi, F., Mattoccia, S.: Guided stereo matching. In: Proceedings of the IEEE/CVF Conference on Computer Vision and Pattern Recognition, pp. 979–988 (2019)
50. Poggi, M., Tosi, F.: Federated online adaptation for deep stereo. In: CVPR (2024)
51. Poggi, M., Tosi, F., Batsos, K., Mordohai, P., Mattoccia, S.: On the synergies between machine learning and binocular stereo for depth estimation from images: a survey. IEEE Trans. Pattern Anal. Mach. Intell. **44**(9), 5314–5334 (2022)
52. Rogister, P., Benosman, R., Ieng, S.H., Lichtsteiner, P., Delbruck, T.: Asynchronous event-based binocular stereo matching. IEEE Trans. Neural Netw. Learn. Syst. **23**(2), 347–353 (2011)

53. Ronneberger, O., Fischer, P., Brox, T.: U-Net: convolutional networks for biomedical image segmentation. In: Navab, N., Hornegger, J., Wells, W.M., Frangi, A.F. (eds.) MICCAI 2015. LNCS, vol. 9351, pp. 234–241. Springer, Cham (2015). https://doi.org/10.1007/978-3-319-24574-4_28
54. Saikia, T., Marrakchi, Y., Zela, A., Hutter, F., Brox, T.: Autodispnet: improving disparity estimation with automl. In: Proceedings of the IEEE/CVF International Conference on Computer Vision, pp. 1812–1823 (2019)
55. Saucedo, M.A., et al.: Event camera and lidar based human tracking for adverse lighting conditions in subterranean environments. IFAC-PapersOnLine **56**(2), 9257–9262 (2023)
56. Schraml, S., Belbachir, A.N., Milosevic, N., Schön, P.: Dynamic stereo vision system for real-time tracking. In: Proceedings of 2010 IEEE International Symposium on Circuits and Systems, pp. 1409–1412. IEEE (2010)
57. Shen, Z., Dai, Y., Rao, Z.: Cfnet: Cascade and fused cost volume for robust stereo matching. In: Proceedings of the IEEE/CVF Conference on Computer Vision and Pattern Recognition (CVPR). pp. 13906–13915 (June 2021)
58. Song, R., Jiang, Z., Li, Y., Shan, Y., Huang, K.: Calibration of event-based camera and 3d lidar. In: 2018 WRC Symposium on Advanced Robotics and Automation (WRC SARA), pp. 289–295. IEEE (2018)
59. Song, X., Zhao, X., Hu, H., Fang, L.: Edgestereo: a context integrated residual pyramid network for stereo matching. In: ACCV (2018)
60. Sulzbachner, C., Zinner, C., Kogler, J.: An optimized silicon retina stereo matching algorithm using time-space correlation. In: CVPR 2011 WORKSHOPS, pp. 1–7. IEEE (2011)
61. Ta, K., Bruggemann, D., Brödermann, T., Sakaridis, C., Van Gool, L.: L2e: lasers to events for 6-dof extrinsic calibration of lidars and event cameras. In: 2023 IEEE International Conference on Robotics and Automation (ICRA), pp. 11425–11431. IEEE (2023)
62. Taniai, T., Matsushita, Y., Naemura, T.: Graph cut based continuous stereo matching using locally shared labels. In: Proceedings of the IEEE Conference on Computer Vision and Pattern Recognition, pp. 1613–1620 (2014)
63. Tankovich, V., Hane, C., Zhang, Y., Kowdle, A., Fanello, S., Bouaziz, S.: Hitnet: hierarchical iterative tile refinement network for real-time stereo matching. In: Proceedings of the IEEE/CVF Conference on Computer Vision and Pattern Recognition (CVPR), pp. 14362–14372 (June 2021)
64. Tonioni, A., Tosi, F., Poggi, M., Mattoccia, S., Stefano, L.D.: Real-time self-adaptive deep stereo. In: Proceedings of the IEEE/CVF Conference on Computer Vision and Pattern Recognition (CVPR) (June 2019)
65. Tosi, F., Tonioni, A., De Gregorio, D., Poggi, M.: Nerf-supervised deep stereo. In: Proceedings of the IEEE/CVF Conference on Computer Vision and Pattern Recognition (CVPR), pp. 855–866 (June 2023)
66. Tulyakov, S., Fleuret, F., Kiefel, M., Gehler, P., Hirsch, M.: Learning an event sequence embedding for dense event-based deep stereo. In: Proceedings of the IEEE/CVF International Conference on Computer Vision, pp. 1527–1537 (2019)
67. Uddin, S.N., Ahmed, S.H., Jung, Y.J.: Unsupervised deep event stereo for depth estimation. IEEE Trans. Circuits Syst. Video Technol. **32**(11), 7489–7504 (2022)
68. Veksler, O.: Stereo correspondence by dynamic programming on a tree. In: 2005 IEEE Computer Society Conference on Computer Vision and Pattern Recognition (CVPR 2005), vol. 2, pp. 384–390. IEEE (2005)

69. Wang, T.H., Hu, H.N., Lin, C.H., Tsai, Y.H., Chiu, W.C., Sun, M.: 3d lidar and stereo fusion using stereo matching network with conditional cost volume normalization. In: 2019 IEEE/RSJ International Conference on Intelligent Robots and Systems (IROS), pp. 5895–5902. IEEE (2019)
70. Wang, Y., et al.: Anytime stereo image depth estimation on mobile devices. In: 2019 International Conference on Robotics and Automation (ICRA), pp. 5893–5900 (2019)
71. Xu, G., Wang, X., Ding, X., Yang, X.: Iterative geometry encoding volume for stereo matching. In: Proceedings of the IEEE/CVF Conference on Computer Vision and Pattern Recognition, pp. 21919–21928 (2023)
72. Xu, H., Zhang, J.: Aanet: adaptive aggregation network for efficient stereo matching. In: Proceedings of the IEEE/CVF Conference on Computer Vision and Pattern Recognition, pp. 1959–1968 (2020)
73. Yang, G., Manela, J., Happold, M., Ramanan, D.: Hierarchical deep stereo matching on high-resolution images. In: Proceedings of the IEEE/CVF Conference on Computer Vision and Pattern Recognition, pp. 5515–5524 (2019)
74. Yang, G., Zhao, H., Shi, J., Deng, Z., Jia, J.: SegStereo: exploiting semantic information for disparity estimation. In: Ferrari, V., Hebert, M., Sminchisescu, C., Weiss, Y. (eds.) ECCV 2018. LNCS, vol. 11211, pp. 660–676. Springer, Cham (2018). https://doi.org/10.1007/978-3-030-01234-2_39
75. Yang, Q., Wang, L., Ahuja, N.: A constant-space belief propagation algorithm for stereo matching. In: 2010 IEEE Computer Society Conference on Computer Vision and Pattern Recognition, pp. 1458–1465. IEEE (2010)
76. Yang, Q., Wang, L., Yang, R., Stewénius, H., Nistér, D.: Stereo matching with color-weighted correlation, hierarchical belief propagation, and occlusion handling. IEEE Trans. Pattern Anal. Mach. Intell. **31**(3), 492–504 (2008)
77. Yin, Z., Darrell, T., Yu, F.: Hierarchical discrete distribution decomposition for match density estimation. In: Proceedings of the IEEE/CVF Conference on Computer Vision and Pattern Recognition, pp. 6044–6053 (2019)
78. Zabih, R., Woodfill, J.: Non-parametric local transforms for computing visual correspondence. In: Third European Conference on Computer Vision (Vol. II). pp. 151–158. 3rd European Conference on Computer Vision (ECCV), Springer-Verlag New York, Inc., Secaucus, NJ, USA (1994)
79. Zbontar, J., LeCun, Y., et al.: Stereo matching by training a convolutional neural network to compare image patches. J. Mach. Learn. Res. **17**(1), 2287–2318 (2016)
80. Zhang, F., Prisacariu, V., Yang, R., Torr, P.H.: GA-Net: guided aggregation net for end-to-end stereo matching. In: IEEE/CVF Conference on Computer Vision and Pattern Recognition (CVPR) (2019)
81. Zhang, J., Ramanagopal, M.S., Vasudevan, R., Johnson-Roberson, M.: Listereo: generate dense depth maps from lidar and stereo imagery. In: 2020 IEEE International Conference on Robotics and Automation (ICRA), pp. 7829–7836. IEEE (2020)
82. Zhang, Y., Zou, S., Liu, X., Huang, X., Wan, Y., Yao, Y.: Lidar-guided stereo matching with a spatial consistency constraint. ISPRS J. Photogramm. Remote. Sens. **183**, 164–177 (2022)
83. Zhao, H., Zhou, H., Zhang, Y., Chen, J., Yang, Y., Zhao, Y.: High-frequency stereo matching network. In: Proceedings of the IEEE/CVF Conference on Computer Vision and Pattern Recognition, pp. 1327–1336 (2023)
84. Zhao, H., Zhou, H., Zhang, Y., Zhao, Y., Yang, Y., Ouyang, T.: Eai-stereo: error aware iterative network for stereo matching. In: Proceedings of the Asian Conference on Computer Vision, pp. 315–332 (2022)

85. Zhou, Y., Gallego, G., Rebecq, H., Kneip, L., Li, H., Scaramuzza, D.: Semi-dense 3D reconstruction with a stereo event camera. In: Ferrari, V., Hebert, M., Sminchisescu, C., Weiss, Y. (eds.) ECCV 2018. LNCS, vol. 11205, pp. 242–258. Springer, Cham (2018). https://doi.org/10.1007/978-3-030-01246-5_15
86. Zhu, A.Z., Yuan, L., Chaney, K., Daniilidis, K.: Unsupervised event-based learning of optical flow, depth, and egomotion. In: Proceedings of the IEEE/CVF Conference on Computer Vision and Pattern Recognition, pp. 989–997 (2019)
87. Zubić, N., Gehrig, D., Gehrig, M., Scaramuzza, D.: From chaos comes order: Ordering event representations for object recognition and detection. In: Proceedings of the IEEE/CVF International Conference on Computer Vision (ICCV), pp. 12846–12856 (October 2023)

X-Former: Unifying Contrastive and Reconstruction Learning for MLLMs

Swetha Sirnam[1(✉)], Jinyu Yang[2], Tal Neiman[2], Mamshad Nayeem Rizve[2], Son Tran[2], Benjamin Yao[2], Trishul Chilimbi[2], and Mubarak Shah[1,2]

[1] Center for Research in Computer Vision,
University of Central Florida, Orlando, USA
swetha.sirnam@ucf.edu, shah@crcv.ucf.edu
[2] Amazon, Seattle, USA
{viyjy,taneiman,mnrizve,sontran,benjamy,trishulc}@amazon.com

Abstract. Recent advancements in Multimodal Large Language Models (MLLMs) have revolutionized the field of vision-language understanding by integrating visual perception capabilities into Large Language Models (LLMs). The prevailing trend in this field involves the utilization of a vision encoder derived from vision-language contrastive learning (CL), showing expertise in capturing overall representations while facing difficulties in capturing detailed local patterns. In this work, we focus on enhancing the visual representations for MLLMs by combining high-frequency and detailed visual representations, obtained through masked image modeling (MIM), with semantically-enriched low-frequency representations captured by CL. To achieve this goal, we introduce X-Former which is a lightweight transformer module designed to exploit the complementary strengths of CL and MIM through an innovative interaction mechanism. Specifically, X-Former first bootstraps vision-language representation learning and multimodal-to-multimodal generative learning from two frozen vision encoders, i.e., CLIP-ViT (CL-based) and MAE-ViT (MIM-based). It further bootstraps vision-to-language generative learning from a frozen LLM to ensure visual features from X-Former can be interpreted by the LLM. To demonstrate the effectiveness of our approach, we assess its performance on tasks demanding detailed visual understanding. Extensive evaluations indicate that X-Former excels in visual reasoning tasks involving both structural and semantic categories in the GQA dataset. Assessment on fine-grained visual perception benchmark further confirms its superior capabilities in visual understanding.

Keywords: Multi-Modal Learning · Masked Image Modeling · MLLMs

---

Supplementary Information The online version contains supplementary material available at https://doi.org/10.1007/978-3-031-72658-3_9.

© The Author(s), under exclusive license to Springer Nature Switzerland AG 2025
A. Leonardis et al. (Eds.): ECCV 2024, LNCS 15064, pp. 146–162, 2025.
https://doi.org/10.1007/978-3-031-72658-3_9

1 Introduction

Recently, Large Language Models (LLMs) have demonstrated remarkable success in diverse natural language tasks [3,38], prompting researchers to explore the integration of visual understanding capabilities into these models, leading to multimodal LLMs (MLLMs). MLLMs aim to leverage the vast knowledge contained within off-the-shelf LLMs and vision encoders to tackle complex visual understanding tasks, thereby opening up new possibilities in the domain of vision-language understanding. Flamingo [2] is one of the early MLLMs to align frozen visual encoders to LLMs, where it introduces a Perceiver Resampler module to extract a fixed set of features from image by optimizing image-to-text generation loss, in order to bridge the modality gap. Improving upon Flamingo, BLIP-2 [23] proposed a Querying Transformer (Q-Former) that performs vision-language alignment through cross modality fusion by employing both discriminative (contrastive & classification) and generative (image-to-text generation) losses to extract a fixed set of most useful visual features for LLM. Other concurrent works [26,43] have explored different strategies to align visual representations with LLM input space for improving vision-language understanding.

It is noteworthy that all aforementioned MLLMs employ CLIP-ViT [32] as the vision encoder, hence, inherit its limitations including: (i) poor fine-grained vision-language alignment [28], and (ii) spatially-invariant global representations [30]. As a consequence, these models struggle to encode detailed visual nuances, including object orientation, structural intricacies, spatial relationships, and multiple object instances [34], thereby hindering the ability of LLMs to comprehend local visual patterns. To alleviate this issue, there has been growing interest to learn better visual representations for MLLMs. For instance, Shikra [6] proposes to learn visual grounding for objects by adding spatial coordinates in natural language for LLM. However, this requires high-quality curated data with bounding box annotations referring to the objects in the image.

GVT [36] on the other hand distills features from pre-trained CLIP [32] via L_1 loss and uses the distilled model as the image encoder for extracting visual tokens. However, this approach relies on instruction tuning utilizing LLaVA-150k [26] dataset. Most recently, MMVP [34] proposes to leverage self-supervised pre-trained vision encoder along with CLIP-ViT to learn Mixture of Features from multiple encoders in LLaVA framework with *LLM fine-tuning*. However, they do instruction tuning with LLaVA-150k [26] dataset. Therefore, Its not clear whether such an approach can work on commonly available image-text data without relying on instruction tuning using curated datasets. An additional avenue of exploration involves constructing a self-supervised vision encoder capable of capturing both global, semantically enriched, and local, detailed visual features. The central concept involves linearly combining the training objectives of CL [32] and MIM [13]. This is motivated by the fact that MIM can effectively capture local and high-frequency representations, complementing the global and low-frequency representations captured by CL. However, this hasn't been explored for vision-laguage understanding and is also the focus of this work.

In this paper, we present X-Former, a lightweight transformer module designed to achieve effective vision-language alignment from both a global and local perspective. Particularly, X-Former adopts a two-stage training approach. The first stage involves vision-language representation learning and multimodal-to-multimodal generative learning by leveraging two frozen image encoders. Specifically, X-Former utilizes learnable query vectors to extract visual features by utilizing both CLIP-ViT [32] and MAE-ViT [13] encoders as well as employ a dual cross-attention module to dynamically fuse the extracted features. Aimed at image reconstruction and text generation, X-Former is incentivized to extract visual features covering both low frequency and high frequency.

Our main technical contributions can be summarized as:

- We propose to leverage vision encoders from CL [32] and MIM [13] to capture both global and local visual representations from frozen image encoders to improve vision-language understanding.
- We introduce X-Former with dual cross-attention to bootstrap multimodal-to-multimodal generative learning using image-text pairs, entirely without the need for curated or visual instruction data.

Empirical studies showcase the notable enhancement of our model in fine-grained visual perception tasks that demand a nuanced understanding of visual details. Specifically, in object counting tasks, X-Former demonstrates substantial improvement over BLIP-2 [23] (39.64 vs. 34.3 on COCO and 27.24 vs. 18.9 on VCR). Further, we perform fine-grained analysis comparing the image-text queries of our model and BLIP-2 to demonstrate our approach learns more diverse queries over BLIP-2 indicating the ability to capture detailed visual features. It's worth noting that BLIP-2 is pre-trained on a dataset of 129 Million image-text pairs, approximately 10× larger than the dataset used for training X-Former (14 Million). This underscores the effectiveness and efficiency of our approach.

2 Method

In this section, we first briefly recapitulate the preliminaries of Q-Former [23]. Following this, we embark on early endeavors aimed at enhancing its visual learning capabilities by leveraging off-the-shelf vision encoders, namely CLIP-ViT and MAE-ViT. Specifically, CLIP-ViT is pre-trained through vision-language contrastive learning strategies, whereas MAE-ViT is trained through masked image modeling mechanisms. Our empirical studies reveal that naively combining these two encoders fails to yield significant performance improvements, especially in tasks necessitating detailed visual comprehension. To mitigate this limitation, we introduce a lightweight transformer module, dubbed X-Former, which extends Q-Former to encapsulate both global and local information.

2.1 Preliminaries of Q-Former

Q-Former is introduced in BLIP2 [23] as a solution designed to bridge the gap between a frozen CLIP-ViT and a frozen LLM (Fig. 1(a)). Given a collection of image-text pairs $\{(I_k, T_k)\}_{k=1}^{N}$, Q-Former operates by taking a predetermined number of learnable query embeddings z, T_k, and C as input, where C indicates CLIP image features of I_k. These queries engage in mutual interaction through self-attention layers and interact with frozen image features C through cross-attention layers in every alternate layer as shown in Fig. 1(a) L1. The resulting query representation is denoted by Z', which is anticipated to encapsulate visual information derived from the frozen CLIP-ViT.

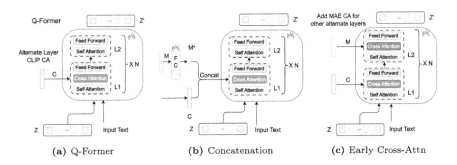

Fig. 1. (a) Vanilla Q-Former extracts a fixed number of output features Z' from the CLIP image encoder, where C and z denotes CLIP-ViT's image features and the query input, respectively; (b) Concatenated MAE-ViT (M^*) and CLIP-ViT (C) features are passed as input to Q-Former, (c) A Cross-Attention layer is added in L2 to enable MAE-ViT interaction in Q-Former.

Though Q-Former has exhibited remarkable performance on various downstream tasks like VQA and image captioning, it encounters challenges in detailed visual feature comprehension. This limitation primarily stems from the training objective of CLIP, which incentivizes ViT to prioritize low-frequency signals and global visual patterns [30]. Fortunately, MAE-ViT [13], trained to reconstruct masked image patches, excels in understanding detailed visual features. However, the integration of CLIP-ViT and MAE-ViT in multimodal understanding remains unclear, given their inherently divergent perspectives when 'viewing' images. To address this inquiry, we embark on early attempts to combine CLIP-ViT and MAE-ViT in a straightforward manner as discussed below.

2.2 Simple Combinations of CLIP-ViT and MAE-ViT

Visual Feature Concatenation. As shown in Fig. 1(b), our first attempt is to concatenate the frozen image features from CLIP-ViT and MAE-ViT, which are denoted by C and M, respectively. To accommodate the discrepancy between C and M, a linear layer is applied to align M with C, resulting in M^*, which is subsequently concatenated with C. This combined feature (C, M^*) serves as input to the Q-Former, which undergoes training in both stages following the

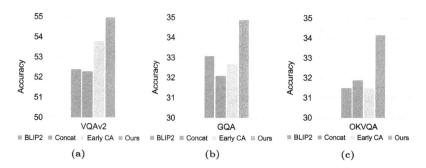

Fig. 2. Performance comparison of BLIP2, BLIP2+Concatenation, BLIP2+Early Cross-Attention, and our method on VQAv2 (a), GQA (b), and OKVQA (c) datasets.

methodology outlined in [23]. Our experiments show that the simple concatenation approach performs on par with BLIP-2, as illustrated in Fig. 2. This observation highlights the non-trivial nature of integrating C and M to leverage their complementary strengths. The distinct information provided by MAE and CLIP presents challenges for the model in simultaneously learning both global and local information while preserving visual-text coherence. Moreover, it is crucial to note that introducing additional vision encoders does not necessarily guarantee improved performance.

Early Cross-Attention. Inspired by the observations from the concatenation strategy outlined earlier, we delve into early interactions akin to CLIP-style cross-attention within Q-Former. To pursue this, we introduce early cross-attention by integrating new cross-attention layers, alternating with non-CLIP interaction layers, as depicted in Fig. 1(c). While this approach modestly improves performance compared to the concatenation strategy (see Fig. 2), it notably escalates the number of parameters in Q-Former, resulting in a total of 183M trainable parameters (approximately 75M more than BLIP-2). Importantly, increasing parameters doesn't inherently enhance performance. While enhancements are observed for the VQAv2 dataset, there's a decline in performance for the GQA dataset and comparable results for the OKVQA dataset against BLIP-2. To mitigate this and facilitate the extraction of local information from MAE, we advocate for incorporating late-interaction for the Masked Image Modeling (MIM) objective during training.

2.3 X-Former Overview

In Fig. 3, we present an overview of our method, comprising two frozen image encoders (CLIP-ViT and MAE-ViT), a frozen image decoder, and a trainable X-Former aimed at bridging the modality gap and extracting interpretable visual features for the LLM. For MAE-ViT, random masking of patches in the input image is performed. X-Former processes a set of learnable queries Z along with the input text T_k and the image features (C, M) as input. Our model extends

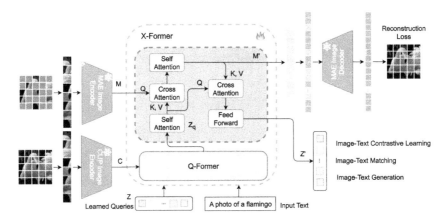

Fig. 3. An overview of X-Former which extends Q-Former by introducing a dual cross-attention module to capture both local and global visual features. First it computes CLIP visual features (C) and MAE features (M) (with random masking) from the input image-text pair. Q-Former employs C, Z, Text to generate output queries optimized for three objectives - ITC, ITM and ITG. The proposed block (purple) enriches Q-Former global representation (Z_q) with local information from MAE features (M). Initially, M is aligned and enriched by Z_q resulting in enriched MAE representation (M'), optimized for image reconstruction. Then, M' enhances Z_q with local representations through cross-attentions, optimized using VL objectives. Jointly optimizing these four objectives facilitates the learning of both global and local representations.

the framework of BLIP2 by incorporating Image-Text Matching (ITM), Image-Text Contrastive (ITC), and Image-Text Generation (ITG) losses, while also introducing a reconstruction loss for the image decoder.

X-Former. To address the limitations of Q-Former, primarily its lack of fine-grained alignment and its focus on capturing global information, we propose integrating MAE features (M) into our X-Former module, depicted as an orange block in Fig. 3. This addition facilitates the extraction of both local and global information, optimizing image reconstruction alongside the ITC, ITM, and ITG objectives, represented by the purple block in Fig. 3. The first cross-attention block employs MAE features (M) as queries and Q-Former output (Z_q) as keys and values to align and enhance M by integrating global semantic information from Q-Former, resulting in enriched MAE features (M'). Subsequently, these enriched MAE features enhance the Q-Former output (Z_q) to Z' by integrating both global and local information through cross-attention, as depicted. The enhanced queries (Z') are optimized for ITC, ITM, and ITG, along with a reconstruction objective applied to M'. Finally, M' is passed to the frozen MAE decoder to reconstruct the masked patches.

Stage 1: Pre-training. During the pre-training stage, the X-Former learns to extract both local and global representation by optimizing Reconstruction, ITC, ITM and ITG losses. The reconstruction loss together with the image-text alignment objectives enforces to align and capture local representation, while the

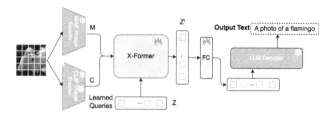

Fig. 4. LLM Alignment. X-Former queries are aligned with a *frozen* decoder-based LLM. FC layer adapts the query output(Z') to LLM embedding space.

VL objectives align it with text representation. The incorporation of MAE and CLIP features ensures that the queries extract a enhanced visual representation that is aligned with the accompanying text. We follow BLIP-2 [23] for computing ITC, ITM and ITG losses. For ITC, we compute similarity between [CLS] token of the text-embedding and each of the final output query embeddings Z', selecting the *highest* as the image-text similarity. For this objective, to prevent data leak a unimodal self-attention mask is employed, ensuring that the queries and text do not interact with each other. It maximizes the image-text similarity of positive pairs by contrasting with in-batch negatives.

For ITM, the model is asked to predict whether image-text pair match (positive) or not (negative). Here, a bi-directional self-attention mask is employed, allowing all queries and texts to attend to each other. Consequently, the output query embeddings capture multimodal information, which is then fed to a two-class linear classifier to obtain logits. These logits are averaged across all the queries to compute the final matching score. To generate negative pairs, a hard negative mining strategy [24] is employed. In the context of ITG, X-Former utilizes an input image as a condition to generate text. A multimodal causal self-attention mask is used, allowing queries to attend to each other while excluding text tokens, and enabling text tokens to attend to all queries and previous text tokens. The [CLS] token is substituted with the [DEC] token as the first text token, serving as an indicator for the decoding task.

Stage 2: LLM Alignment. During pre-training, the X-Former acquires the ability to extract information from both MAE and CLIP, resulting in queries that capture a blend of global and local information. Subsequently, we align the features of the X-Former with the frozen LLM, aiming to harness the comprehensive visual representations acquired by the X-Former module and integrate them with the robust language generation capabilities of the LLM. This integration involves connecting the pre-trained X-Former output (Z') to the LLM via a single fully-connected layer, aligning it with the LLM representation space, as depicted in Fig. 4. Specifically, we experiment with the OPT model, which is a decoder-based LLM, and train it using a language modeling loss keeping *both* image encoders and LLM *frozen*.

Table 1. Zero-shot Visual Question Answering results on the VQAv2 dataset. Note that * indicates the result is obtained using the official checkpoint.

Method	#Trainable Params	Data	VQAv2 Accuracy			
			Overall	Other	Yes/No	Number
Open-ended generation models						
FewVLM [19]	740M	9.1M	47.7	–	–	–
Frozen [35]	40M		29.5	–	–	–
VLKD [7]	406M	3.7M	42.6	–	–	–
BLIP-2 $OPT_{6.7B}$* [23]	108M	129M	55.1	47.3	72.6	34.6
BLIP-2 $OPT_{2.7B}$ [23]	107M	14M	49.9	39.3	71.5	27.3
X-Former (Ours) $OPT_{2.7B}$	129M	14M	**51.3**	**41.5**	71.2	**30.9**
BLIP-2 $OPT_{6.7B}$ [23]	108M	14M	52.4	43.6	71.5	30.8
X-Former (Ours) $OPT_{6.7B}$	130M	14M	**55.0**	**45.6**	**73.3**	**37.8**

3 Experiments

Pre-trained Models. We employ pre-trained ViT-G model from EVA-CLIP [10] as CLIP-ViT. For MAE, we utilize the pre-trained ViT-H model [13]. Our choice for the LLM involves the OPT model [42]. Our model undergoes pre-training for nine epochs in Stage-1 and one epoch in Stage-2, with OPT employed for Stage-2 alignment. See Supplementary Sect. 1 for implementation details.

Datasets and Tasks. To demonstrate the effectiveness of our approach, we leverage a standard dataset comprising 14M Image-Text pairs sourced from COCO [25], Visual Genome [21], SBU [29], CC3M [33], and CC12M [4] for model pre-training. Our evaluation spans across various benchmarks, including COCO [25], NoCaps [1], VQAv2 [12], GQA [15], OK-VQA [27], Flickr30k [31], and VCR [41]. Furthermore, we employ a fine-grained visual perception benchmark [36], featuring Object Counting (OC) and Multi-Class Identification (MCI) tasks, to assess the model's fine-grained visual understanding capabilities.

3.1 Experimental Results

Zero-Shot Visual Question Answering. First, we present the results for zero-shot visual question answering on the VQAv2-val dataset, which encompasses three question types: open-ended (other), Yes/No, and Number questions, as illustrated in Table 1. We utilize the prompt "Question: Short Answer:" for the generation process, employing beam search with a beam width of 5. We set the length-penalty to 0 to encourage short answers. Our results indicate that our approach surpasses BLIP-2 for both $OPT_{2.7B}$ and $OPT_{6.7B}$ LLMs by 1.4% and 2.6% respectively, highlighting superior visual comprehension. Particularly noteworthy are the significant enhancements observed for the Number task, which

Table 2. Zero-shot Visual Question Answering Results on GQA and OKVQA datasets. Note that * indicates the result is obtained using the official checkpoint.

Method	Data	GQA	OKVQA
FewVLM [19]	9.1M	29.3	16.5
Frozen [35]		–	5.9
VLKD [7]	3.7M	–	13.3
Flamingo3B [2]	>2B	–	41.2
Flamingo9B [2]	>2B	–	44.7
Flamingo80B [2]	>2B	–	50.6
BLIP-2 $OPT_{6.7B}$* [23]	129M	34.2	35.3
BLIP-2 $OPT_{2.7B}$ [23]	14M	33.6	24.2
X-Former (Ours) $OPT_{2.7B}$	14M	**34.1**	**27.7**
BLIP-2 $OPT_{6.7B}$ [23]	14M	33.1	31.5
X-Former (Ours) $OPT_{6.7B}$	14M	**34.9**	**34.2**

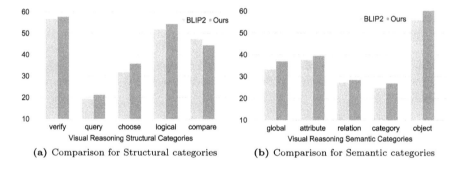

(a) Comparison for Structural categories (b) Comparison for Semantic categories

Fig. 5. Detailed Comparison for both Structural and Semantic categories in GQA.

demands precise local understanding for object counting or identification. Fine-tuning results are reported in Supplementary Sect. 2, while large-scale experimental findings are detailed in Supplementary Sect. 3.

In Table 2, we report zero-shot visual question answering results for the GQA test-dev dataset. The results demonstrate the superior performance of our method over BLIP-2. Furthermore, we conducted a comprehensive comparison to demonstrate the effectiveness of our approach across both structural and semantic categories in the GQA dataset. The structural category encompasses five question types (verify, open-ended query questions, choose from options, logical inference, and object comparison) as depicted in Fig. 5. Our results indicate that we outperform in the majority of these categories. In Fig. 5, we provide a comparison for the semantic categories, which include questions related to object existence, object attributes, object category, global scene, and object relationships. Across all these categories which includes both global and local reasoning, our approach consistently demonstrates better performance, highlighting its detailed visual understanding capabilities.

Table 3. Zero-shot Fine-Grained Visual Perception evaluation of MLLMs on Object Counting (OC) & Multi-class Identification (MCI) tasks. For fair comparison, we compare with models trained only on image-text data. *evaluated using official checkpoint.

Method	Data	OC		MCI	
		COCO	VCR	COCO	VCR
BLIP-2* [23]	129M	34.3	18.9	69.44	74.16
BLIP-2 [23]	14M	25.88	21.12	61.5	65.3
X-Former (Ours)	14M	**39.64**	**27.24**	**69.44**	**69.28**

Table 4. Zero-Shot Image Captioning Results on COCO & NoCaps without fine-tuning for captioning task. B:BLEU, C: CIDEr, S: SPICE. *evaluated using official checkpoint

Method	Data	COCO			NoCaps	
		B@4	C	S	C	S
BLIP-2* [23]	129M	39.9	134.3	24.3	113.4	15.2
BLIP-2 [23]	14M	39.2	131.0	23.7	113.1	14.9
X-Former (Ours)	14M	**39.3**	**131.1**	23.6	**113.2**	14.9

We report zero-shot visual question answering performance on OKVQA test dataset in Table 2. This dataset poses a significant challenge as it requires methods to draw upon external knowledge to answer questions effectively. Our method demonstrates a significant improvement in accuracy over BLIP-2, achieving a 2.7% and 3.5% gain with $OPT_{6.7B}$ and $OPT_{2.7B}$ LLM respectively. This signifies the robustness of our approach in accurately aligning visual information with LLM and effectively leveraging external knowledge to answer the questions.

Fine-Grained Visual Perception Evaluation. To demonstrate that our approach has better visual understanding, we evaluate perception abilities at fine-grained scale [36], we evaluate our approach for fine-grained visual perception capabilities OC and MCI task. We use the prompt "Question: {} Short Answer:". to evaluate for this task. For generation, we use beam search with a beam width of 5. We also set the length-penalty to 0 to encourage shorter answers. The questions for object counting tasks is of the form "How many {objects} are there in the image?" and for multi-class identification task it is "Does {objects} exist in the image?". For fair comparison, we compare with methods that only employ image-text datasets for training. In Table 3, we show that our model outperforms BLIP-2 on both datasets i.e., COCO and VCR. It can be seen that for Object Counting task our approach improves BLIP-2 by 13% on COCO and 6.1% on VCR datasets respectively. This indicates that X-Former is able to extract detailed visual features. Please refer to Supplementary Sects. 6, 7 for more fine-grained evaluations.

Table 5. Ablation study with early layer features from CLIP-ViT. L_i indicates i^{th} layer of CLIP-ViT.

Method	VQAv2	GQA	OKVQA
Ours	**55.0**	**34.9**	**34.2**
L_{26}	53.7	32.6	31.2
L_{28}	52.5	31.9	28.0
L_{30}	52.4	32.8	30.9

Table 6. Ablation for Effect of Reconstruction (Recon.) Loss. Reconstruction during pre-training plays an important role in aligning the MAE to extract meaningful information.

	Stage 1 Recon.	Stage 2 Recon.	VQAv2	GQA	OKVQA
			33.1	25.4	12.1
	✓	✓	52.4	32.2	29.2
Ours	✓		**55.0**	**34.9**	**34.2**

Zero-Shot Image Captioning. In addition to the visual reasoning tasks, we report results for image captioning without fine-tuning in Table 4 for COCO and NoCaps dataset. Captioning task requires image-level semantic understanding as the annotated captions briefly describe the image. We show that our approach improves on fine-grained visual reasoning tasks without impacting the captioning performance.

3.2 Qualitative Results

To effectively demonstrate the capabilities of our model, we present qualitative results that highlight its performance in the object counting task. Accurate object counting requires a deep understanding of local contexts within an image. As shown in Fig. 6(a), our method correctly counts six donuts in an image, while BLIP-2 incorrectly predicts four. This demonstrates the model's ability to distinguish individual objects even when they are closely clustered together. Figure 6(b) presents a more challenging scenario where four airplanes are flying in close proximity. BLIP-2 struggles with this task, erroneously predicting six airplanes instead of the correct number of four. Our method, on the other hand, accurately identifies the four airplanes, showcasing its robustness in handling dense object arrangements. In Fig. 6(c), we encounter two cups and a plate that share a similar color, for which BLIP-2 incorrectly predicts three cups. Our method, however, correctly identifies the two cups, demonstrating its ability to handle objects with similar visual properties. Figure 6(d) depicts a dog that blends into the background due to its similar coloring. BLIP-2 makes an incorrect prediction. Our method, in contrast, correctly identifies the dog. These qualitative results collectively demonstrate the effectiveness of our model in the object counting task, outperforming BLIP-2 in various scenarios that demand robust local understanding and the ability to handle challenging object arrangements and color similarities.

In the Multi-Class Identification task, BLIP-2's object recognition capabilities exhibit limitations when presented with Fig. 6(e). BLIP-2 mistakenly interprets the shape of a parking pole as a fire hydrant. Figure 6(f) presents a challenge where a potted plant is positioned in the background, occupying a relatively small portion of the image. BLIP-2 fails to detect the presence of the

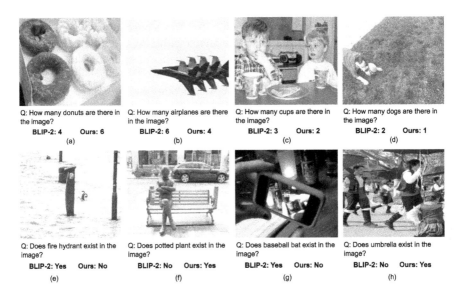

Fig. 6. Qualitative Comparison demonstrating Fine-Grained Visual Understanding in Object Counting and Multi-class Identification Tasks. Our model showcases better visual understanding by accurately counting objects and effectively identifying them without confusion based on shape or color.

potted plant, whereas our method successfully identifies it. Figure 6(g) showcases image of a bottle whose shape closely resembles that of a baseball bat, leading BLIP-2 to identify it as a baseball bat. Figure 6(h) presents a challenge due to the subtle color of an umbrella, making it difficult task. BLIP-2 fails to recognize the object, while our method accurately identifies it as an umbrella. A comprehensive qualitative analysis is provided in Supplementary Sect. 5. We also present query diversity analysis for fine-grained qualitative comparison and present them in Supplementary Sect. 6.

3.3 Ablation Analysis

We perform the following ablations to analyze the various components of our approach. Please refer to Supplementary Sect. 4 for more ablation analysis.

Leveraging Early Layer CLIP features To illustrate the efficacy of MAE embeddings in learning better local representations, we conduct experiments using features from intermediate layers of CLIP. Specifically, we explore the utilization of features from layers 26, 28, 30 as substitutes for MAE features in our proposed approach. It is important to note that for this training, there is no reconstruction loss since we are employing CLIP features. From Table 5, it can be seen that employing MAE features leads to best results. We present more results in Supplementary Sect. 4.

Impact of Image Reconstruction Loss in Pre-training. We examine the influence of the reconstruction loss during the pre-training stage. In this pre-training phase, we employ four objectives: image-text contrastive and matching loss (for discriminative vision-language alignment), reconstruction loss, and text generation loss. Our findings demonstrate that combining alignment and reconstruction objectives during pre-training, the image reconstruction loss becomes effective in extracting aligned and meaningful representations. To illustrate this, we conducted an experiment without the MAE decoder and reconstruction objective during pre-training and computed MAE features for the entire image without any masking. As shown in Table 6, row 1, the significant performance drop highlights the crucial role of the reconstruction loss in aligning and extracting useful information from MAE. This suggests that network cannot find a shortcut thus leading to drop in performance while MIM enables extracting meaningful and aligned representations leading to best performance.

Impact of Image Reconstruction Loss in LLM Alignment. We investigate the influence of reconstruction loss in stage-2. In the context of LLM alignment, the model undergoes training with language-modeling loss exclusively, without the inclusion of image-text contrastive and matching losses, and Q-Former does not receive any text-input. It is evident that combining only language modeling loss and reconstruction loss yields suboptimal results, as indicated by the performance drop shown in Table 6, row 2.

4 Related Works

Multimodal Large Language Models (MLLMs). The success of Large Language Models (LLMs) has prompted researchers to delve into the exploration of integrating visual components into these models, culminating in the development of Multimodal LLMs (MLLMs) [22,40]. MLLMs have garnered significant traction in both academic and industrial spheres due to their remarkable proficiency in comprehension and generation. The key idea is to leverage off-the-shelf pre-trained vision encoders and LLMs and keep them frozen during the training. However, the most critical challenge in utilizing a frozen LLM lies in narrowing down the gap between visual features and the text space. Existing MLLMs can be broadly divided into three categories according to the modules/components they used for bridging the modality gap: (i) Perceiver-based [16,17], (ii) Q-Former-based [23], and (iii) linear projection layer-based. In Perceiver-based methods such as Flamingo [2], they employ a Perceiver Resampler to produce a small fixed number of visual tokens per image, subsequently amalgamating them with text tokens as input for LLMs. In other words, the Perceiver Resampler relies on the image-to-text generative learning to bridge the modality gap. Q-Former shares the similar spirits with the Perceiver Resampler, except that Q-Former relies on an extra vision-language representation learning stage. Owing to its simplicity and efficiency, Q-Former is widely used in such as BLIP-2 [23], SEED [11], MiniGPT-4 [43], and InstructBLIP [8]. In linear projection layer-based methods, the common practice is to align visual features with text features through a singular linear layer before incorporating them into LLMs. The effectiveness of this simple strategy is evidenced by recent studies such as LLaVa [26] and FROMAGe [20]. Our work is inspird by Q-Former but with the following differences: (i) we extend Q-Former to handle two off-the-shelf vision encoders, i.e.,

CLIP-ViT and MAE-ViT, and (ii) we introduce multimodality-to-multimodality generative learning to further bridge the modality gap.

Self-supervised Vision Encoders. Self-supervised vision encoders (VEs) play a crucial role in MLLMs by providing visual features that are understandable by LLMs. Among them, VEs that are pre-trained by vision-language-based contrastive learning (CL) has been the most popular one, where the VE is trained to bring representations of matched image-text pair close together and push representations of unmatched pairs apart [5,9,14,18,32,39]. This encourages the VE to capture semantic similarities and differences in visual content. However, recent work reveals that CL mainly focuses on low frequency signals and longer-range global patterns inheriting from its training objective [30]. In consequence, CL-based MLLMs suffer from understanding detailed perceptions which are essential for tasks that require fine-grained visual understanding such as object counting. As a counterpart, masked image modeling (MIM) involves masking parts of an image and tasking the vision encoder with predicting the masked image patches [13]. This enhances the VE's ability to understand detailed visual features by promoting contextual understanding, encouraging the learning of spatial relationships, and facilitating the development of transferable representations. Inspired by these observations, recent work attempt to build VEs that is able to understand both global semantic and detailed local patterns [30,37]. The key idea is to leverage the strength of CL and MIM by linearly combining two training objectives with a shared VE. While simple and effective, these models are not readily applicable to MLLMs due to their sole pre-training on limited datasets and modest model sizes, significantly lagging behind their CL and MLM counterparts. Although scaling up data and model size is possible, it introduces substantial carbon emissions and fails to capitalize on the advantages offered by off-the-shelf VEs from both CL and MIM. In contrast, our approach incorporates a lightweight transformer that harnesses the benefits of pre-trained CL and MIM models, showcasing superior performance in fine-grained perception understanding without imposing a significant computational burden.

5 Conclusion

In this paper, we introduce X-Former, a novel architecture designed to enhance visual representations for Multimodal Language Models (MLLMs) by integrating pre-trained MAE and CLIP vision encoders. Our motivation stems from several observations: (i) existing MLLMs primarily rely on CLIP-ViT, which often fails to capture fine-grained visual signals; (ii) our empirical studies reveal that simply combining CLIP-ViT and MAE-ViT does not necessarily yield performance improvements; and (iii) the efficacy of MLLMs heavily depends on large-scale image-text pairs for pre-training and meticulously curated instruction tuning datasets for fine-tuning. X-Former effectively tackles these limitations by integrating CLIP-ViT and MAE-ViT through a dual cross-attention mechanism, all while keeping computational demands manageable. Our approach is plug-and-play and can be applied to other models. Our experimental results unequivocally show that X-Former surpasses BLIP-2 in a variety of visual reasoning tasks requiring robust visual comprehension. Remarkably, these superior results are achieved using only one-tenth of the image-text pair dataset, without the need for any instruction tuning datasets.

References

1. Agrawal, H., et al.: Nocaps: novel object captioning at scale. In: Proceedings of the IEEE/CVF International Conference on Computer Vision (2019)
2. Alayrac, J.B., et al.: Flamingo: a visual language model for few-shot learning. Advances in Neural Information Processing Systems (2022)
3. Brown, T., et al.: Language models are few-shot learners. Adv. Neural Inf. Processing Syst. (2020)
4. Changpinyo, S., Sharma, P., Ding, N., Soricut, R.: Conceptual 12m: pushing webscale image-text pre-training to recognize long-tail visual concepts. In: Proceedings of the IEEE/CVF Conference on Computer Vision and Pattern Recognition (2021)
5. Chen, C., et al.: Why do we need large batchsizes in contrastive learning? A gradient-bias perspective. Adv. Neural Inf. Process. Syst. (2022)
6. Chen, K., Zhang, Z., Zeng, W., Zhang, R., Zhu, F., Zhao, R.: Shikra: unleashing multimodal llm's referential dialogue magic (2023)
7. Dai, W., Hou, L., Shang, L., Jiang, X., Liu, Q., Fung, P.: Enabling multimodal generation on CLIP via vision-language knowledge distillation. In: Muresan, S., Nakov, P., Villavicencio, A. (eds.) Findings of the Association for Computational Linguistics: ACL 2022 (2022)
8. Dai, W., et al.: Instructblip: towards general-purpose vision-language models with instruction tuning (2023)
9. Duan, J., et al.: Multi-modal alignment using representation codebook. In: Proceedings of the IEEE/CVF Conference on Computer Vision and Pattern Recognition (2022)
10. Fang, Y., et al.: Eva: Exploring the limits of masked visual representation learning at scale. In: Proceedings of the IEEE/CVF Conference on Computer Vision and Pattern Recognition (2023)
11. Ge, Y., Ge, Y., Zeng, Z., Wang, X., Shan, Y.: Planting a seed of vision in large language model (2023)
12. Goyal, Y., Khot, T., Summers-Stay, D., Batra, D., Parikh, D.: Making the V in VGA matter: elevating the role of image understanding in visual question answering. In: Proceedings of the IEEE Conference on Computer Vision and Pattern Recognition (2017)
13. He, K., Chen, X., Xie, S., Li, Y., Dollár, P., Girshick, R.: Masked autoencoders are scalable vision learners. In: Proceedings of the IEEE/CVF Conference on Computer vision and Pattern Recognition (2022)
14. Hu, Z., Zhu, X., Tran, S., Vidal, R., Dhua, A.: Provla: compositional image search with progressive vision-language alignment and multimodal fusion. In: Proceedings of the IEEE/CVF International Conference on Computer Vision (2023)
15. Hudson, D.A., Manning, C.D.: GGA: a new dataset for real-world visual reasoning and compositional question answering. In: Proceedings of the IEEE/CVF Conference on Computer Vision and Pattern Recognition (2019)
16. Jaegle, A., et al.: Perceiver IO: a general architecture for structured inputs & outputs. arXiv preprint arXiv:2107.14795 (2021)
17. Jaegle, A., Gimeno, F., Brock, A., Vinyals, O., Zisserman, A., Carreira, J.: Perceiver: general perception with iterative attention. In: International Conference on Machine Learning. PMLR (2021)
18. Jiang, Q., et al.: Understanding and constructing latent modality structures in multi-modal representation learning. In: Proceedings of the IEEE/CVF Conference on Computer Vision and Pattern Recognition (2023)

19. Jin, W., Cheng, Y., Shen, Y., Chen, W., Ren, X.: A good prompt is worth millions of parameters: low-resource prompt-based learning for vision-language models. In: Proceedings of the 60th Annual Meeting of the Association for Computational Linguistics (2022)
20. Koh, J.Y., Salakhutdinov, R., Fried, D.: Grounding language models to images for multimodal inputs and outputs. International Conference on Machine Learning (2023)
21. Krishna, R., et al.: Visual genome: connecting language and vision using crowd-sourced dense image annotations. Int. J. Comput. Vision 123(1), 32–73 (2017)
22. Li, C.: Large multimodal models: notes on CVPR 2023 tutorial. arXiv preprint arXiv:2306.14895 (2023)
23. Li, J., Li, D., Savarese, S., Hoi, S.: BLIP-2: bootstrapping language-image pre-training with frozen image encoders and large language models. In: ICML (2023)
24. Li, J., Li, D., Xiong, C., Hoi, S.: Blip: bootstrapping language-image pre-training for unified vision-language understanding and generation. In: International Conference on Machine Learning. PMLR (2022)
25. Lin, T.Y., et al.: Microsoft coco: common objects in context. In: Proceedings of the Computer Vision–ECCV 2014: 13th European Conference, Zurich, 6–12 September 2014, Part V 13. Springer (2014)
26. Liu, H., Li, C., Wu, Q., Lee, Y.J.: Visual instruction tuning. In: Thirty-Seventh Conference on Neural Information Processing Systems (2023)
27. Marino, K., Rastegari, M., Farhadi, A., Mottaghi, R.: OK-VGA: a visual question answering benchmark requiring external knowledge. In: Proceedings of the IEEE/CVF Conference on Computer Vision and Pattern Recognition (2019)
28. Mukhoti, J., et al.: Open vocabulary semantic segmentation with patch aligned contrastive learning. In: Proceedings of the IEEE/CVF Conference on Computer Vision and Pattern Recognition (2023)
29. Ordonez, V., Kulkarni, G., Berg, T.: Im2text: describing images using 1 million captioned photographs. Adv. Neural Inf. Process. Syst. (2011)
30. Park, N., Kim, W., Heo, B., Kim, T., Yun, S.: What do self-supervised vision transformers learn? arXiv preprint arXiv:2305.00729 (2023)
31. Plummer, B.A., Wang, L., Cervantes, C.M., Caicedo, J.C., Hockenmaier, J., Lazebnik, S.: Flickr30k entities: collecting region-to-phrase correspondences for richer image-to-sentence models. In: Proceedings of the IEEE International Conference on Computer Vision (2015)
32. Radford, A., et al.: Learning transferable visual models from natural language supervision. In: International Conference on Machine Learning. PMLR (2021)
33. Sharma, P., Ding, N., Goodman, S., Soricut, R.: Conceptual captions: a cleaned, hypernymed, image alt-text dataset for automatic image captioning. In: Proceedings of the 56th Annual Meeting of the Association for Computational Linguistics (Volume 1: Long Papers) (2018)
34. Tong, S., Liu, Z., Zhai, Y., Ma, Y., LeCun, Y., Xie, S.: Eyes wide shut? exploring the visual shortcomings of multimodal LLMS (2024)
35. Tsimpoukelli, M., Menick, J.L., Cabi, S., Eslami, S.M.A., Vinyals, O., Hill, F.: Multimodal few-shot learning with frozen language models. In: Ranzato, M., Beygelzimer, A., Dauphin, Y., Liang, P., Vaughan, J.W. (eds.) Advances in Neural Information Processing Systems (2021)
36. Wang, G., Ge, Y., Ding, X., Kankanhalli, M., Shan, Y.: What makes for good visual tokenizers for large language models? (2023)

37. Weers, F., Shankar, V., Katharopoulos, A., Yang, Y., Gunter, T.: Masked autoencoding does not help natural language supervision at scale. In: Proceedings of the IEEE/CVF Conference on Computer Vision and Pattern Recognition (CVPR) (2023)
38. Wei, J., et al.: Emergent abilities of large language models. arXiv preprint arXiv:2206.07682 (2022)
39. Yang, J., et al.: Vision-language pre-training with triple contrastive learning. In: Proceedings of the IEEE/CVF Conference on Computer Vision and Pattern Recognition (2022)
40. Yin, S., et al.: A survey on multimodal large language models. arXiv preprint arXiv:2306.13549 (2023)
41. Zellers, R., Bisk, Y., Farhadi, A., Choi, Y.: From recognition to cognition: visual commonsense reasoning. In: The IEEE Conference on Computer Vision and Pattern Recognition (CVPR) (2019)
42. Zhang, S., et al.: Opt: open pre-trained transformer language models. arXiv preprint arXiv:2205.01068 (2022)
43. Zhu, D., Chen, J., Shen, X., Li, X., Elhoseiny, M.: Minigpt-4: enhancing vision-language understanding with advanced large language models. arXiv preprint arXiv:2304.10592 (2023)

Learning Anomalies with Normality Prior for Unsupervised Video Anomaly Detection

Haoyue Shi[1,3], Le Wang[1](✉), Sanping Zhou[1], Gang Hua[2], and Wei Tang[3]

[1] National Key Laboratory of Human-Machine Hybrid Augmented Intelligence, National Engineering Research Center for Visual Information and Applications, Institute of Artificial Intelligence and Robotics, Xi'an Jiaotong University, Xi'an, China
lewang@xjtu.edu.cn
[2] Multimodal Experiences Research Lab Dolby Laboratories, London, UK
[3] University of Illinois Chicago, Chicago, USA

Abstract. Unsupervised video anomaly detection (UVAD) aims to detect abnormal events in videos without any annotations. It remains challenging because anomalies are rare, diverse, and usually not well-defined. Existing UVAD methods are purely data-driven and perform unsupervised learning by identifying various abnormal patterns in videos. Since these methods largely rely on the feature representation and data distribution, they can only learn salient anomalies that are substantially different from normal events but ignore the less distinct ones. To address this challenge, this paper pursues a different approach that leverages data-irrelevant prior knowledge about normal and abnormal events for UVAD. We first propose a new normality prior for UVAD, suggesting that the start and end of a video are predominantly normal. We then propose normality propagation, which propagates normal knowledge based on relationships between video snippets to estimate the normal magnitudes of unlabeled snippets. Finally, unsupervised learning of abnormal detection is performed based on the propagated labels and a new loss re-weighting method. These components are complementary to normality propagation and mitigate the negative impact of incorrectly propagated labels. Extensive experiments on the ShanghaiTech and UCF-Crime benchmarks demonstrate the superior performance of our method. The code is available at https://github.com/shyern/LANP-UVAD.git.

Keywords: Video anomaly detection · Self-training · Prior Knowledge

---

H. Shi—Part of this work was done while Haoyue Shi was a visiting scholar at University of Illinois Chicago.

Supplementary Information The online version contains supplementary material available at https://doi.org/10.1007/978-3-031-72658-3_10.

1 Introduction

Video anomaly detection aims to detect abnormal events in video sequences along the temporal dimension. This task is essential for intelligent surveillance [55] and crime detection [5]. Most existing methods are trained with partial supervision that requires manual annotations. For instance, one-class methods [23,40,41] use normal videos as the training set, while weakly-supervised methods [5,33,36] rely on video-level annotations during training. In contrast, unsupervised video anomaly detection (UVAD) methods [28,37,46,47] attempt to detect anomalies without any annotations. It remains challenging because abnormal events in UVAD are rare, diverse, and usually not well-defined.

Existing UVAD methods [1,28,35,37,38,46,47] are based on self-training or reconstruction. Reconstruction-based methods [37,38,46] learn normal patterns from all training videos by gradually filtering out anomalies with large reconstruction errors. Self-training methods [1,27,35,47] first generate pseudo-labels, which are then used as supervision for training, achieving state-of-the-art performance recently.

All current UVAD methods are purely data-driven, performing unsupervised learning by identifying abnormal patterns in videos. However, these methods heavily rely on the feature representation and data distribution, often struggling to capture less salient anomalies. In reconstruction-based methods [37,38,46], where normal and abnormal patterns are easily memorized by the autoencoder, imperceptible abnormal events may be easily overlooked. Self-training methods use data-driven strategies to generate pseudo-labels. For instance, GCL identifies anomalies based on local contrast in video snippets. They focus on high-contrast anomalies, potentially leading to the problem of anomalies attenuation, *i.e.*, overlooking less salient anomalies with low contrast. Al *et al.* [1] separate normal and anomalies by clustering over the entire dataset, making it challenging to distinguish them without pre-defined normal events. This challenge is illustrated in Fig. 1, showing results from previous representative methods [1,46,47]. We can see that results from different methods vary significantly, and they can easily miss less salient anomalies.

In essence, UVAD is highly ill-posed. There still lacks a common definition of "what an anomaly is" in the community, and simply relying on data to detect anomalies is prone to failure. We tackle this problem from a different perspective. Unlike previous methods driven by data, our idea is to leverage *data-irrelevant* prior knowledge about normal and abnormal events to aid in identifying anomalies. By characterizing what normal and abnormal events look like beyond the data, this prior knowledge helps address ambiguities between normal and abnormal events that cannot be resolved from a purely data-driven perspective, leading to more effective anomaly detection.

What is Good Prior Knowledge for UVAD? It should be both informative and of high quality. We draw inspiration from the area of saliency detection [7,16,17,45]. Concretely, inspired by the boundary prior [7,45] commonly used in traditional saliency detection methods, we introduce a normality prior, that is, the start and end of a video are mostly normal. It is more robust than

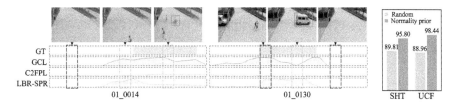

Fig. 1. Two examples of abnormal videos on ShanghaiTech [20]. The gray barcode is the ground truth. The line charts are anomaly scores of three methods [1,46,47]. The instances in the orange boxes are easily predicted incorrectly. This is because these methods are data-driven and highly rely on feature representation and data distribution, often struggling to capture less salient anomalies. (Color figure online)

Fig. 2. Accuracy of normal snippets selected by the normality prior and random.

the center prior [14,19] for anomalies because anomalies can be located far away from the temporal center but rarely touch the temporal boundary. We validate the normality prior on several datasets, as shown in Fig. 2, which shows that compared to random selection, using our normality prior to select normal snippets is significantly more accurate. A more detailed experimental discussion is presented in Sect. 4.3.

How Should the Prior Knowledge be Used? A straightforward method is to directly compare other snippets with the normal prior, *i.e.*, start and end snippets of a video. However, since normal frames vary over time, the similarities between the normal prior and distant normal frames can be very large. They can be easily mislabeled without considering the temporal and semantic consistency of video snippets. In this paper, we propose Normality Propagation, which aims to propagate the normal information based on relationships between video snippets to estimate the normal magnitudes of unlabeled snippets, which represent the normal degree they received. Different from traditional label propagation in semi-supervised learning, our normality propagation features several innovations: 1) To overcome the limitation of no labeled snippets, we propose to use the normality prior to specify normal snippets. 2) We propose a temporally-modulated feature-based similarity matrix to model pairwise similarities. 3) We apply the propagation in a more efficient way, *i.e.*, over snippets in a video instead of the whole dataset used in traditional label propagation. We then perform unsupervised learning of abnormal detection based on the propagated labels and a new loss re-weighting method. They are complementary to normality propagation and mitigate the negative impact of incorrectly propagated labels.

The main contributions are summarized as follows:

- Unlike previous UVAD methods that are purely data-driven, we propose to use the data-irrelevant normality prior to identify abnormal events. To the best of our knowledge, such prior has never been studied before in the area of UVAD.

- We introduce normality propagation to effectively propagate the normality prior to unlabeled snippets for pseudo label generation.
- We perform unsupervised learning of abnormal detection based on the propagated labels and a new loss re-weighting method. They are complementary to normality propagation and mitigate the negative impact of incorrectly propagated labels.
- Extensive experiments on ShanghaiTech [20] and UCF-Crime [33] demonstrate the effectiveness of the proposed method.

2 Related Work

The literature on video anomaly detection (VAD) is rich. We mostly restrict the discussion to approaches for unsupervised video anomaly detection, label propagation, self-training and pseudo labeling.

2.1 Unsupervised Video Anomaly Detection

Video Anomaly Detection (VAD) aims to detect abnormal events in video sequences. Full-supervised methods [18,39] tackle this task with precise annotations. Most existing methods are trained in partial supervision that needs manual annotations. For instance, one-class methods [6,9,20,22,23,29,34,40,41,54] use only normal videos to train the detection model, weakly-supervised methods [5,24,25,31–33,36,42,49,51,55] need video-level annotations during training. In contrast, unsupervised methods [28,37,38,46,47] attempt to detect anomalies without any manual annotations, which are laborious, expensive, and prone to large variations. It remains challenging because abnormal events in UVAD are rare, diverse, and usually not well-defined. In the current work, we explore unsupervised mode for video anomaly detection.

Existing UVAD methods [28,37,38,46,47] can be categorized into reconstruction based methods and self-training based methods. These methods are purely data-driven, performing unsupervised learning by identifying abnormal patterns in videos. Reconstruction-based methods [37,38,46] learn normal patterns from all training videos by gradually filtering out anomalies with large reconstruction errors. For instance, Yu *et al.* [46] design a novel self-paced refinement scheme to remove anomalies with the reconstruction model. Tur *et al.* [37] leverage the reconstruction capability of diffusion models to detect anomalies with larger reconstruction errors. They [38] further employ conditional diffusion models to improve detection performance conditioned on compact motion representations. However, as normal and abnormal patterns are easily memorized by the autoencoder, they easily overlook imperceptible abnormal events.

Self-training-based methods [1,35,47] have achieved good performance recently. They generate pseudo labels first, then use pseudo labels as self-supervision. Pseudo labels are generated relying on data-driven strategies. For instance, Zaheer [47] propose a generative cooperative learning method. It identifies anomalies based on the local contrast in video snippets *i.e.*, the difference between consecutive snippets. They may label the boundaries of abnormal events well but can attenuate interior anomalies. Al et al. [1] cluster over the entire dataset to detect anomalies belonging to a smaller cluster. Normal and abnormal videos may appear in the same scene, but it is hard to separate them without a pre-defined normal. Thakare et al. [35] use OneClassSVM and iForest to find anomalies that lie outside the constructed hypersphere. However, previous UVAD methods are driven by data. They largely rely on feature representation and data distribution. Thus they easily overlook less salient anomalies. Our idea is to leverage prior knowledge about normal events that are irrelevant to the data to help identify anomalies.

2.2 Label Propagation

Variants of label propagation [56] have been applied to various computer vision tasks, including learning with noisy labels [11,26], few-shot learning [3,21,30], and semi-supervised learning [10,57]. Method [11] for learning with noisy labels proposes a neighbour consistency regularization loss that encourages examples with similar feature representations to have similar predictions. Multi-Objective Interpolation Training (MOIT) [26] identifies and refines noisy examples using neighbours' predictions. Liu et al. [21] consider label propagation with stochastic gradient descent for episodic few-shot learning. For semi-supervised learning, Iscen et al. [10] use label propagation to obtain labels for unsupervised data based on their neighbours in the feature space. Label propagation is not directly applicable to UVAD as both the initial labels and the similarity matrix are unavailable. We propose normality propagation to address these challenges. On the one hand, we propose the data-irrelevant normality prior to specifying the initial labels. On the other hand, we design a temporally-modulated similarity matrix to effectively propagate the normality prior across the video.

2.3 Self-training and Pseudo Labeling

Self-training is a simple but effective technique used in unsupervised learning [12,13,52,53], semi-supervised learning [10,43], and weakly-supervised video anomaly detection [5,55]. It initializes pseudo labels first, then uses labels predicted by the model as self-supervision. Zhang et al. [52] first generate saliency pseudo labels by using contrast-based SOD methods [4,44], then designed a noise modelling module to deal with noises in saliency cues. The unsupervised person re-identification method [53] generates pseudo labels based on clustering results, then refines pseudo labels with clustering consensus over training epochs and temporal ensembling techniques. Lee et al. [10] uses the current network to infer pseudo labels of unlabeled data, and then re-trains the model on both labeled and unlabeled data. Zhong et al. [55] propose an alternate training framework for weakly supervised VAD, which generates frame-level pseudo labels for abnormal videos according to an action classifier [2]. In this work, we use self-training as an approach for UVAD. Our method differs from all prior work in that we propose a normality prior and introduce normality propagation, which effectively propagates the normality prior to unlabeled snippets for pseudo label generation. We also introduce a loss re-weighting strategy for robust abnormal detection.

3 Method

Problem Statement. In Unsupervised Video Anomaly Detection (UVAD), both normal and abnormal videos are included in the training set, while annotations are not provided. Assume that we have a set of N training videos, and each video is divided into a series of non-overlapping snippets. The goal of UVAD is to learn a snippet-level anomaly classifier $f_\theta(\cdot)$ that predicts the anomaly score of the snippet. A higher score indicates the snippet is more likely to be abnormal.

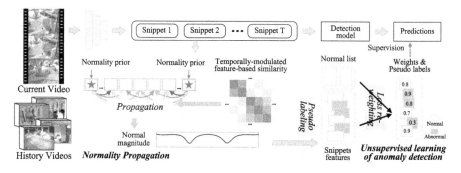

Fig. 3. Overview of our method. We first use the normality prior to specify that the start and end of a video are normal. Then we propose a new normality propagation method to propagate the normal information based on the temporally-modulated feature-based similarity for estimating normal magnitudes. After that, We perform unsupervised learning of abnormal detection based on the propagated labels and a new loss re-weighting method.

Overview. As previous methods are purely data-driven, they heavily rely on feature representation and data distribution, thereby often struggling to capture less distinct anomalies. To address this challenge, our key idea is to leverage data-irrelevant prior knowledge about normal/abnormal events to aid in identifying anomalies in videos. By characterizing what normal and abnormal events look like beyond the data, this prior knowledge helps address ambiguities between normal and abnormal events that cannot be resolved from a purely data-driven perspective, leading to more effective anomaly detection. Figure 3 illustrates an overview of our method. We first describe the normality prior in natural videos, and use it to specify normal snippets in a video. Then we propose normality propagation to propagate normal knowledge based on the temporally-modulated feature-based similarity matrix to estimate the normal magnitudes of unlabeled snippets. After propagation, unsupervised learning of abnormal detection is performed based on the propagated labels and a new re-weighting method.

Feature Extraction. Following [33,47], we utilize a fixed-weight backbone network to extract features for each snippet. Formally, we denote features in a video as $\mathbf{X} \in \mathbb{R}^{D \times L}$, where L and D are the number of video snippets of a video and the feature dimension.

3.1 Normality Prior in Our Method

We propose to use a prior about normal events in natural videos, namely normality prior. Such prior has never been studied in previous methods, but it can help address challenges faced by purely data-driven methods.

Our normality prior is that the start and end of a video are mostly normal. It is inspired by the boundary prior [7,16,17,45] widely used in traditional saliency detection methods: the image boundary is mostly background. The boundary prior is more general that the center prior [14,19] to identify anomalies because anomalies can be located far away from the temporal center, but they rarely touch the temporal boundary. This prior is validated on several anomaly detection datasets, as statistical analysis in Fig. 2 and experimental discussion in Sect. 4.3.

3.2 Normality Propagation

Normality propagation aims to effectively propagate the normality prior to the unlabeled snippets for pseudo label generation. Given features of a video with L snippets $\mathbf{X}$, we propagate normal knowledge over video snippets based on pairwise similarities for estimating their normal magnitudes $\mathbf{z}$, which represent the normal degree they received.

As normal videos contain normal snippets only, we only specify the first and end snippets in a video as normal and do not mark any abnormal ones. We first define a label vector $\mathbf{y} \in \mathbb{R}^L$ with $y_i = 1$ if $i = 1$ or $i = L$ and $y_i = 0$ otherwise. Clearly, $\mathbf{y}$ is consistent with the initial normal magnitudes of video snippets.

Considering that normal and abnormal events are temporally consistent, and they are similar by themselves but frames between them vary greatly, we propose a temporally-modulated feature-based similarity matrix to model pairwise similarities, as shown in Fig. 4. We define a matrix $\mathbf{W} \in \mathbb{R}^{L \times L}$ to describe the proximity of snippets in the feature space and temporal domain. We construct $\mathbf{W}$ by computing the snippets' feature space similarities and then modulating them by their respective temporal positions. Specifically, we define a similarity matrix $\mathbf{W}^f$ computed in the feature-space as:

$$\mathbf{W}^f_{i,j} = \begin{cases} \exp\left(-\frac{\|\mathbf{x}_i - \mathbf{x}_j\|^2}{2\sigma^2}\right), & \text{if } i \neq j \\ 0, & \text{otherwise,} \end{cases} \quad (1)$$

where $\mathbf{x}_i$ denotes the i-th snippet feature in a video, $\sigma = 0.1$ is a hyper-parameter to control the strength of the similarity. A similar matrix $\mathbf{W}^t$ is defined in the time-space, and elements are computed from the time stamps. For L snippets,

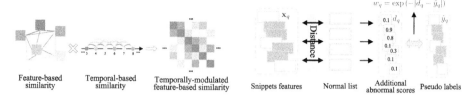

Fig. 4. Temporally-modulated feature-based similarity.

Fig. 5. Loss re-weighting strategy.

the time stamps are defined as $\mathbf{t} = \{1, 2, \cdots, L\}$, and each element in $\mathbf{W}^t$ is computed as:

$$\mathbf{W}^t_{i,j} = \begin{cases} \exp\left(-\frac{|t_i - t_j|}{L}\right), & \text{if } i \neq j \\ 0, & \text{otherwise,} \end{cases} \quad (2)$$

where t_i denotes the i-th snippet time stamp in a video. We then compute temporally-modulated feature-based similarity matrix $\mathbf{W}$ as:

$$\mathbf{W}_{i,j} = \mathbf{W}^f_{i,j} \cdot \mathbf{W}^t_{i,j}. \quad (3)$$

$\mathbf{W}_{i,j}$ therefore specifies the temporally weighted similarity between snippets i and j. Its symmetrically normalized counterpart $\mathbf{S} = \mathbf{D}^{-1/2}\mathbf{W}\mathbf{D}^{-1/2}$, in which $\mathbf{D}$ is a diagonal matrix with its (i, i)-element equal to the sum of the i-th row of $\mathbf{W}$.

Normality propagation is to let every snippet iteratively spread its normal knowledge to its neighbours until a global stable state is achieved, that is, iterate $\mathbf{z}(n+1) = \alpha\mathbf{S}\mathbf{z}(n) + (1-\alpha)\mathbf{y}$ until convergence, where α is a hyper-parameter in the range $(0, 1)$, n is the index of an iteration, and $\mathbf{z}(0) = \mathbf{y}$. The iterative process converges to a simple solution: $\mathbf{z}^* = (1-\alpha)(\mathbf{I}-\alpha\mathbf{S})^{-1}\mathbf{y}$, which is clearly equivalent to

$$\mathbf{z}^* = (\mathbf{I} - \alpha\mathbf{S})^{-1}\mathbf{y}, \quad (4)$$

where $\mathbf{I}$ is the identity matrix. $\mathbf{z}^*$ can be efficiently obtained by using the conjugate gradient from the linear system $(\mathbf{I} - \alpha\mathbf{S})\mathbf{z}^* = \mathbf{y}$. Derivations refer to the supplementary material.

It is interesting to observe that the vector $\mathbf{z}^*$ as defined by Eq.(4) is equivalent to the solver of the following objective function:

$$\mathcal{Q}(\mathbf{z}) = \alpha\mathbf{z}^T\mathbf{S}\mathbf{z} + (1-\alpha)(\mathbf{z}-\mathbf{y})^T(\mathbf{z}-\mathbf{y}), \quad (5)$$

The first term encourages similar snippets to get the same predictions, while the second term attempts to maintain predictions for the specified normal examples [56].

The optimized normal magnitudes of snippets in a video $\mathbf{z}^*$ indicate different degrees of normal knowledge that they received from the normality prior and their neighbours. A higher value means that this snippet is more likely to be normal, and vice versa.

Discussion. The normality prior specifies normal snippets effectively. By characterizing what normal events look like beyond the data, this prior knowledge helps address ambiguities between normal and abnormal events that cannot be resolved from reconstruction-based methods [37,38,46] and the global cluster-based methods [1]. Besides, normality propagation leverages the normality prior in a more effective way than directly comparing the video snippets with the normal prior. It estimates the normal magnitudes of each snippet not only based on the labeled normal snippets but also takes into account their neighbours' normal magnitudes. Thus normal snippets that are far away from specified ones will be affected by their normal neighbours and then can be labeled correctly. Besides, the previous "anomalies attenuation" problem is significantly alleviated because anomalies in a video are usually different from normal. In addition, normality propagation is a transductive method. It is easy to implement, fast and suitable for generating pseudo labels for UVDA.

3.3 Unsupervised Learning of Abnormal Detection

We perform unsupervised learning of abnormal detection based on the propagated labels and a new loss re-weighting method. They are complementary to normality propagation and mitigate the negative impact of incorrectly propagated labels.

Pseudo Labeling. We generate pseudo-labels based on normal magnitudes. Video-level pseudo labels are generated first, and then snippet-level pseudo labels are generated. Specifically, we select N^{nor} videos that have lower video scores than normal videos, where the video score is defined as the standard deviation of normal magnitudes of snippets in a video. Standard deviations of abnormal videos will be large as the abnormal video contains both normal and abnormal snippets. After that, we select snippets that have lower $r\%$ normality scores in the abnormal video as abnormal snippets. Finally, we formulate the pseudo label of the t-th snippet in the i-th video as

$$\hat{y}_t^i = \begin{cases} 1, & \text{if } i \in \mathcal{I}^{\text{abn}} \wedge t \in \mathcal{I}_i^{\text{abn}} \\ 0, & \text{otherwise,} \end{cases} \quad (6)$$

where $\mathcal{I}^{\text{abn}}$ is the set of indexes of abnormal videos, $\mathcal{I}_i^{\text{abn}}$ is the set of indexes of abnormal snippets in the i-th video.

Loss Re-weighting. Because a few videos may not follow the normality prior and the normality propagation is imperfect, we will have incorrect pseudo labels.

We propose a loss re-weighting strategy to mitigate the negative impact of the noisy pseudo labels. As shown in Fig. 5, we first select N^{hnor} high-confident normal videos and use the mean feature of each selected video to construct the normal list $\mathcal{M} = \{\mathbf{m}_0, \mathbf{m}_1, \cdots, \mathbf{m}_{N^{\text{hnor}}}\}$. We then estimate additional abnormal scores based on the global normal list. For a snippet $\mathbf{x}_q$, its additional score is defined as

$$d_q = \min_{\mathbf{m}_i \in \mathcal{M}} 1 - \langle \mathbf{x}_q, \mathbf{m}_i \rangle, \quad (7)$$

where $\langle \cdot, \cdot \rangle$ is the cosine similarity between two vectors. The additional abnormal score estimates the probability of a snippet being abnormal in a global view. It is complementary to the pseudo labels generated by normality propagation in a local view. Then, the reliability is measured by the discrepancy between pseudo labels and additional normal-based anomaly scores. We further formulate the loss weight of the snippet $\mathbf{x}_q$ as

$$w_q = \exp\left(-|d_q - \hat{y}_q|\right). \quad (8)$$

The loss re-weighting strategy penalizes pseudo-labels with high discrepancy to guide the learning through reliable pseudo-labels.

After pseudo labels and their loss weights are generated, we train a classification model with the weighted binary cross-entropy loss as

$$\mathcal{L} = \frac{1}{|\mathcal{X}|} \sum_{\mathbf{x}_i \in \mathcal{X}} -w_i \hat{y}_i \log(s_i) - w_i(1 - \hat{y}_i)\log(1 - s_i), \quad (9)$$

where $\mathcal{X}$ is the set of video snippets in a mini-batch, $s_i = f_\theta(\mathbf{x}_i)$ is the anomaly prediction of the i-th snippet.

4 Experiment

In this section, we first provide experimental details, then draw comparisons with the existing UVAD methods, and finally study different components of our method.

4.1 Experimental Details

Evaluation Datasets. We evaluate our method on two popular video anomaly detection datasets: ShanghaiTech [20] and UCF-Crime [33].

ShanghaiTech [20] is a popular dataset used in video anomaly detection. It contains 437 campus surveillance videos (330 normal videos, 107 abnormal videos) with different locations spanning and camera angles. Recent UVAD methods follow the data organization of [55] but do not use annotations in the training set. Specifically, the training set contains 63 abnormal videos and 175 normal videos, and the testing set contains 44 abnormal videos and 155 normal videos.

UCF-Crime [33] is a large-scale video anomaly detection dataset. It consists of 1900 real-world surveillance videos (950 normal videos, 950 abnormal videos) with 13 different types of realistic abnormal events. It is a complex dataset due to videos containing diverse backgrounds and durations. Recent UVAD methods follow the data organization of [33] but without using training video labels. Specifically, the training split has 810 abnormal and 800 normal videos, while the testing split has 140 abnormal and 150 normal videos.

Evaluation Metrics. Following prior work [37,47], we use the frame-level area under the ROC curve (AUC) for evaluation and comparisons. Lager AUC values indicate better performance.

Implementation Details. We use two backbones, *i.e.*, ResNext3d [8] and I3D [2], to extract features for each snippet receptively. Our detection model is composed of a temporal convolution and two linear layers. During our training procedure, each mini-batch consists of 30 randomly selected videos, and each video is sampled into 32 snippets. We train the model for 300 epochs with the RMSprop optimizer using a learning rate of 0.0001 and a momentum of 0.6. We set $\alpha = 0.99$ for normality propagation, N^{nor} as the number of normal videos in the training set, and the abnormal ratio $r\% = 40\%$. Besides, we set the number of high confident normal videos $N^{\text{hnor}} = h * N^{\text{nor}}$, and $h = 0.1$.

4.2 Comparison with State-of-the-Art Methods

In Table 1, we compare the proposed method with existing unsupervised video anomaly detection methods [37,38,47] on two different datasets, *i.e.*, ShanghaiTech [20] and UCF-Crime [33]. Selected weakly-supervised and one-class methods [36,40,48,51] are presented for reference. We reimplement LBR-SPR [46] on our dataset splits and C2FPL [1] on ResNext features for fair comparisons. As we can see, our method outperforms almost all previous methods on different features. We establish a new state-of-the-art on ShanghaiTech with 86.46% AUC and UCF-Crime with 76.64% AUC with ResNext as the backbone. We also achieve comparable performance using I3D as the backbone.

Table 1. Comparison with state-of-the-art methods in AUC (%) on ShanghaiTech and UCF-Crime. We divide the methods into weakly-supervised, one-class, and unsupervised. Best results are in **bold**. We implemented [1,46] and computed their AUC scores.

Method		Features	ShanghaiTech	UCF-Crime
Weakly-supervised	Sultani et al. [33]	C3D	–	75.41
	CLAWS [49]	C3D	89.67	83.03
	CLAWS Net+ [50]	C3D	90.12	83.37
	MIST [5]	C3D	93.13	81.40
	RTFM [36]	C3D	91.51	83.28
	MIST [5]	I3D	94.83	82.30
	RTFM [36]	I3D	97.21	84.30
	Zhang et al. [51]	I3D	–	86.22
	CLAWS [49]	ResNext	–	82.61
	CLAWS Net+ [50]	ResNext	91.46	84.16
	Zaheer et al. [47]	ResNext	86.21	79.84
One-class	Lu et al. [23]	–	68.00	65.51
	BODS [40]	I3D	–	68.26
	GODS [40]	I3D	–	70.46
	OGNet [48]	ResNext	69.90	69.47
	Zaheer et al. [47]	ResNext	79.62	74.20
Unsupervised	DyAnNet [35]	I3D	–	79.76
	C2FPL [1]	I3D	–	**80.65**
	Ours	I3D	**88.32**	80.02
	Kim et al. [15]	ResNext	56.47	52.00
	LBR-SPR [46]	ResNext	77.12	57.18
	GCL [47]	ResNext	78.93	71.04
	Tur et al. [37]	ResNext	68.88	62.91
	Tur et al. [38]	ResNext	66.36	63.52
	C2FPL [1]	ResNext	67.36	74.71
	Ours	ResNext	**86.46**	**76.64**

In addition, Unsupervised methods [1,37,47] still perform inferior to weakly-supervised ones [5,33] because no annotations are provided in UVAD. But our method achieves better performance than one-class methods [40,50], verifying the effectiveness of the unsupervised setting in video anomaly detection.

4.3 Ablation Study and Analysis

We conduct a series of ablation studies to understand better how the proposed method works, where we use ResNext [8] as the backbone to extract features.

Validation of Normality Prior. In this experiment, we validate the normality prior to the experimental performance. Table 2 shows the precision and recall of pseudo labels generated by using normality propagation with random, the data-driven prior, and the normality prior as labeled normal snippets. It also shows the corresponding testing performance. As we can see, specified normal snippets with the normality prior generate better pseudo labels and testing performance. It is more informative than data-driven prior because the latter often favors simple, similar normal snippets, e.g., assuming snippets exhibiting minimal contrast are considered normal. We also validate its robustness beyond these standard benchmarks in the supplementary material.

Table 2. Precision (%) and Recall (%) of pseudo labels generated by normality propagation with Random, Data-driven Prior, and Normality Prior as labeled normal snippets, as well as corresponding testing performance (AUC %).

	ShanghaiTech			UCF-Crime		
	Precision	Recall	TestAUC	Precision	Recall	TestAUC
Random	17.90	29.27	81.69	13.28	42.41	60.33
Data-driven Prior	19.74	24.64	83.59	15.19	48.23	68.51
Normality Prior	34.39	42.92	85.96	20.37	54.28	75.99

Table 3. Ablation study of normality propagation. T, F, and T&F mean Temporal-based, Feature-based, and Temporally-modulated Feature-based pairwise similarities. The metric is AUC (%).

	Pairwise Similarity	ShanghaiTech	UCF-Crime
Direct Comparison	–	72.33	71.22
Normality Propagation	T	79.62	57.99
	F	79.73	63.01
	T&F	85.96	75.99

Ablation Study of Normality Propagation. How to use the normality prior is very important. We compare different strategies for generating pseudo labels using the specified normal snippets, including making direct comparisons with specified normal snippets, normality propagation with temporal-based (T), feature-based (F), and temporally-modulated feature-based (T&F) pairwise similarities. The testing performances are presented in Table 3. Direct comparison has inferior performance than normality propagation with T&F. This is because direct comparison overlooks relationships between unlabeled snippets. In addition, our T&F pairwise similarity outperforms T or F pairwise similarity that

Table 4. Ablation study on the effectiveness of the loss re-weighting (LR) in our method.

	ShanghaiTech	UCF-Crime
Ours w/o LR	85.96	75.99
Ours	86.46	76.64

Table 5. Ablation study on different values of high confident normal videos, where we set different h.

	0.1	0.3	0.5
ShanghaiTech	86.46	86.18	86.24
UCF-Crime	76.64	76.63	76.42

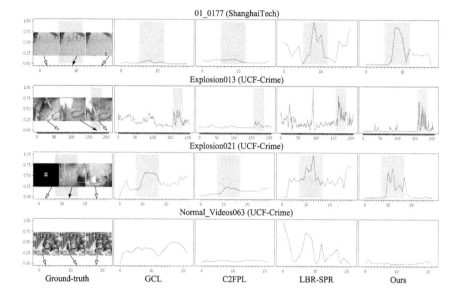

Fig. 6. Qualitative results. Curves represent the predicted anomaly scores. The grey background corresponds to the ground truth. The white and black arrows denote the locations of normal and abnormal frames displayed on the left.

only uses the similarity in the time-space or feature-space because we take into account temporal and feature similarities simultaneously. In summary, the proposed normality propagation effectively propagates the normality prior to unlabeled snippets based on the T&F pairwise similarities matrix. It mostly benefits from knowledge in labeled normal snippets as well as temporal and semantic consistency in unlabeled snippets to generate pseudo labels.

Ablation Study on Loss Re-weighting. In this experiment, we validate the effectiveness of the proposed loss re-weighting strategy. The results are shown in Table 4. As we can see, our method with normality propagation achieves good performance over two datasets. By adding the normal-based loss re-weight, we obtain better performance. This is because it provides the global normal information for re-weighting pseudo labels, which is complementary to the normality propagation that generates pseudo labels in a local view. This also verified that

Fig. 7. A failure case on UCF-Crime.

our loss re-weighting strategy can mitigate the negative effects of incorrectly propagated labels.

Ablation Study of Hyperparameters. The number of highly confident normal videos N^{hnor} is an important hyperparameter in our normal-based loss re-weighting. We set it as $h*N^{\text{nor}}$, and N^{nor} is set as the number of normal videos of the corresponding dataset. Here we test its sensitivity on two datasets. Results are presented in Table 5. Our method consistently achieves AUC higher than 86% on ShanghaiTech and 76% on UCF-Crime. Thus, our method is insensitive to N^{hnor}. More experimental results refer to the supplementary material.

4.4 Qualitative Analysis

Figure 6 shows some examples of detection results of our method and three previous representative methods, *i.e.*, one reconstruction-based method LBR-SPR [46], and two self-training methods GCL [47] and C2FPL [1]. Our method can detect the abnormal event well and predict abnormal scores of the normal frames very close to zero. In abnormal examples of *Explosion013* (UCF-Crime) and *01_0177* (ShanghaiTech), compared with other data-driven methods, our method locates anomalies more accurately. This is because our method uses the normality prior effectively. It helps address ambiguities between normal and abnormal events that cannot be resolved from a purely data-driven perspective, leading to more effective anomaly detection.

Limitation. As our method is built on the semantic consistency of normal events, inevitably, it fails when there are multiple types of normal events in a video, as shown in Fig. 7 shows a typical failure case: A normal video captured from different angles. Nevertheless, our method performs well in most videos, as shown in Fig. 6. In the future, we will explore combining with other methods that focus on detecting anomalies in multiple scenes to overcome this limitation.

5 Conclusion

Unlike existing methods driven by data, this paper leverages data-irrelevant prior knowledge for UVAD. We first propose a new normality prior, suggesting that the start and end of a video are predominantly normal. We then introduce normality propagation, which propagates normal knowledge based on relationships between video snippets to estimate normal magnitudes of unlabeled snippets. Finally, we perform unsupervised learning of abnormal detection based on the propagated labels and a new loss re-weighting method. Extensive experiments indicate that our method outperforms existing state-of-the-art methods.

Acknowledgements. This work was supported in part by National Science and Technology Major Project under Grant 2023ZD0121300, National Natural Science Foundation of China under Grants 62088102, 12326608 and 62106192, Natural Science Foundation of Shaanxi Province under Grant 2022JC-41, and Fundamental Research Funds for the Central Universities under Grant XTR042021005.

References

1. Al-Lahham, A., Tastan, N., Zaheer, M.Z., Nandakumar, K.: A coarse-to-fine pseudo-labeling (C2FPL) framework for unsupervised video anomaly detection. In: WACV, pp. 6793–6802 (2024)
2. Carreira, J., Zisserman, A.: Quo vadis, action recognition? a new model and the kinetics dataset. In: CVPR, pp. 6299–6308 (2017)
3. Chen, C., Yang, X., Xu, C., Huang, X., Ma, Z.: ECKPN: explicit class knowledge propagation network for transductive few-shot learning. In: CVPR, pp. 6596–6605 (2021)
4. Cheng, M.M., Mitra, N.J., Huang, X., Torr, P.H., Hu, S.M.: Global contrast based salient region detection. IEEE TPAMI **37**(3), 569–582 (2014)
5. Feng, J.C., Hong, F.T., Zheng, W.S.: Mist: multiple instance self-training framework for video anomaly detection. In: CVPR, pp. 14009–14018 (2021)
6. Gong, D., et al.: Memorizing normality to detect anomaly: memory-augmented deep autoencoder for unsupervised anomaly detection. In: ICCV, pp. 1705–1714 (2019)
7. Grady, L., Jolly, M.P., Seitz, A.: Segmentation from a box. In: ICCV, pp. 367–374 (2011)
8. Hara, K., Kataoka, H., Satoh, Y.: Can spatiotemporal 3D CNNs retrace the history of 2D CNNs and imagenet? In: CVPR, pp. 6546–6555 (2018)
9. Hasan, M., Choi, J., Neumann, J., Roy-Chowdhury, A.K., Davis, L.S.: Learning temporal regularity in video sequences. In: CVPR, pp. 733–742 (2016)
10. Iscen, A., Tolias, G., Avrithis, Y., Chum, O.: Label propagation for deep semi-supervised learning. In: CVPR, pp. 5070–5079 (2019)
11. Iscen, A., Valmadre, J., Arnab, A., Schmid, C.: Learning with neighbor consistency for noisy labels. In: CVPR, pp. 4672–4681 (2022)
12. Ji, H., Wang, L., Zhou, S., Tang, W., Zheng, N., Hua, G.: Meta pairwise relationship distillation for unsupervised person re-identification. In: ICCV, pp. 3661–3670 (2021)
13. Ji, W., Li, J., Bi, Q., Guo, C., Liu, J., Cheng, L.: Promoting saliency from depth: deep unsupervised RGB-D saliency detection. In: ICLR (2022)
14. Judd, T., Ehinger, K., Durand, F., Torralba, A.: Learning to predict where humans look. In: ICCV, pp. 2106–2113 (2009)
15. Kim, J.H., Kim, D.H., Yi, S., Lee, T.: Semi-orthogonal embedding for efficient unsupervised anomaly segmentation. arXiv preprint arXiv:2105.14737 (2021)
16. Lempitsky, V., Kohli, P., Rother, C., Sharp, T.: Image segmentation with a bounding box prior. In: ICCV, pp. 277–284 (2009)

17. Li, H., Lu, H., Lin, Z., Shen, X., Price, B.: Inner and inter label propagation: salient object detection in the wild. IEEE TIP **24**(10), 3176–3186 (2015)
18. Liu, K., Ma, H.: Exploring background-bias for anomaly detection in surveillance videos. In: ACM MM, pp. 1490–1499 (2019)
19. Liu, T., et al.: Learning to detect a salient object. IEEE TPAMI **33**(2), 353–367 (2010)
20. Liu, W., Luo, W., Lian, D., Gao, S.: Future frame prediction for anomaly detection–a new baseline. In: CVPR, pp. 6536–6545 (2018)
21. Liu, Y., et al.: Learning to propagate labels: transductive propagation network for few-shot learning. In: ICLR (2019)
22. Liu, Z., Nie, Y., Long, C., Zhang, Q., Li, G.: A hybrid video anomaly detection framework via memory-augmented flow reconstruction and flow-guided frame prediction. In: ICCV, pp. 13588–13597 (2021)
23. Lu, C., Shi, J., Jia, J.: Abnormal event detection at 150 fps in Matlab. In: ICCV, pp. 2720–2727 (2013)
24. Lv, H., Yue, Z., Sun, Q., Luo, B., Cui, Z., Zhang, H.: Unbiased multiple instance learning for weakly supervised video anomaly detection. In: CVPR, pp. 8022–8031 (2023)
25. Lv, H., Zhou, C., Cui, Z., Xu, C., Li, Y., Yang, J.: Localizing anomalies from weakly-labeled videos. IEEE TIP **30**, 4505–4515 (2021)
26. Ortego, D., Arazo, E., Albert, P., O'Connor, N.E., McGuinness, K.: Multi-objective interpolation training for robustness to label noise. In: CVPR, pp. 6606–6615 (2021)
27. Pang, G., Shen, C., van den Hengel, A.: Deep anomaly detection with deviation networks. In: ACM SIGKDD, pp. 353–362 (2019)
28. Pang, G., Yan, C., Shen, C., Hengel, A.v.d., Bai, X.: Self-trained deep ordinal regression for end-to-end video anomaly detection. In: CVPR, pp. 12173–12182 (2020)
29. Park, H., Noh, J., Ham, B.: Learning memory-guided normality for anomaly detection. In: CVPR, pp. 14372–14381 (2020)
30. Rodríguez, P., Laradji, I., Drouin, A., Lacoste, A.: Embedding propagation: Smoother manifold for few-shot classification. In: ECCV, pp. 121–138 (2020)
31. Sapkota, H., Yu, Q.: Bayesian nonparametric submodular video partition for robust anomaly detection. In: CVPR, pp. 3212–3221 (2022)
32. Shi, H., Wang, L., Zhou, S., Hua, G., Tang, W.: Abnormal ratios guided multi-phase self-training for weakly-supervised video anomaly detection. IEEE TMM **26**, 5575–5587 (2023)
33. Sultani, W., Chen, C., Shah, M.: Real-world anomaly detection in surveillance videos. In: CVPR, pp. 6479–6488 (2018)
34. Sun, S., Gong, X.: Hierarchical semantic contrast for scene-aware video anomaly detection. In: CVPR, pp. 22846–22856 (2023)
35. Thakare, K.V., Raghuwanshi, Y., Dogra, D.P., Choi, H., Kim, I.J.: Dyannet: a scene dynamicity guided self-trained video anomaly detection network. In: WACV, pp. 5541–5550 (2023)
36. Tian, Y., Pang, G., Chen, Y., Singh, R., Verjans, J.W., Carneiro, G.: Weakly-supervised video anomaly detection with robust temporal feature magnitude learning. In: ICCV, pp. 4975–4986 (2021)
37. Tur, A.O., Dall'Asen, N., Beyan, C., Ricci, E.: Exploring diffusion models for unsupervised video anomaly detection. In: ICIP, pp. 2540–2544 (2023)

38. Tur, A.O., Dall'Asen, N., Beyan, C., Ricci, E.: Unsupervised video anomaly detection with diffusion models conditioned on compact motion representations. In: ICIAP, pp. 49–62 (2023)
39. Wan, B., Jiang, W., Fang, Y., Luo, Z., Ding, G.: Anomaly detection in video sequences: a benchmark and computational model. IET Image Proc. **15**(14), 3454–3465 (2021)
40. Wang, J., Cherian, A.: GODS: generalized one-class discriminative subspaces for anomaly detection. In: ICCV, pp. 8201–8211 (2019)
41. Wang, L., Tian, J., Zhou, S., Shi, H., Hua, G.: Memory-augmented appearance-motion network for video anomaly detection. PR **138**, 109335 (2023)
42. Wu, P., et al.: Not only look, but also listen: learning multimodal violence detection under weak supervision. In: ECCV, pp. 322–339 (2020)
43. Xie, Q., Luong, M.T., Hovy, E., Le, Q.V.: Self-training with noisy student improves imagenet classification. In: CVPR, pp. 10687–10698 (2020)
44. Yan, Q., Xu, L., Shi, J., Jia, J.: Hierarchical saliency detection. In: CVPR, pp. 1155–1162 (2013)
45. Yang, C., Zhang, L., Lu, H., Ruan, X., Yang, M.H.: Saliency detection via graph-based manifold ranking. In: CVPR, pp. 3166–3173 (2013)
46. Yu, G., Wang, S., Cai, Z., Liu, X., Xu, C., Wu, C.: Deep anomaly discovery from unlabeled videos via normality advantage and self-paced refinement. In: CVPR, pp. 13987–13998 (2022)
47. Zaheer, M.Z., Mahmood, A., Khan, M.H., Segu, M., Yu, F., Lee, S.I.: Generative cooperative learning for unsupervised video anomaly detection. In: CVPR, pp. 14744–14754 (2022)
48. Zaheer, M.Z., Lee, J.h., Astrid, M., Lee, S.I.: Old is gold: redefining the adversarially learned one-class classifier training paradigm. In: CVPR, pp. 14183–14193 (2020)
49. Zaheer, M.Z., Mahmood, A., Astrid, M., Lee, S.I.: CLAWS: clustering assisted weakly supervised learning with normalcy suppression for anomalous event detection. In: ECCV, pp. 358–376 (2020)
50. Zaheer, M.Z., Mahmood, A., Astrid, M., Lee, S.I.: Clustering aided weakly supervised training to detect anomalous events in surveillance videos. In: IEEE TNNLS (2022)
51. Zhang, C., et al.: Exploiting completeness and uncertainty of pseudo labels for weakly supervised video anomaly detection. In: CVPR, pp. 16271–16280 (2023)
52. Zhang, J., Zhang, T., Dai, Y., Harandi, M., Hartley, R.: Deep unsupervised saliency detection: a multiple noisy labeling perspective. In: CVPR, pp. 9029–9038 (2018)
53. Zhang, X., Ge, Y., Qiao, Y., Li, H.: Refining pseudo labels with clustering consensus over generations for unsupervised object re-identification. In: CVPR, pp. 3436–3445 (2021)
54. Zhao, Y., Deng, B., Shen, C., Liu, Y., Lu, H., Hua, X.S.: Spatio-temporal autoencoder for video anomaly detection. In: ACM MM, pp. 1933–1941 (2017)
55. Zhong, J.X., Li, N., Kong, W., Liu, S., Li, T.H., Li, G.: Graph convolutional label noise cleaner: train a plug-and-play action classifier for anomaly detection. In: CVPR, pp. 1237–1246 (2019)
56. Zhou, D., Bousquet, O., Lal, T., Weston, J., Schölkopf, B.: Learning with local and global consistency. NeurIPS **16**, 321–328 (2003)
57. Zhuang, F., Moulin, P.: Deep semi-supervised metric learning with mixed label propagation. In: CVPR, pp. 3429–3438 (2023)

Revisiting Supervision for Continual Representation Learning

Daniel Marczak[1,2(✉)], Sebastian Cygert[1,3], Tomasz Trzciński[1,2,4], and Bartłomiej Twardowski[1,5,6]

[1] IDEAS NCBR, Warsaw, Poland
daniel.marczak.dokt@pw.edu.pl
[2] Warsaw University of Technology, Warsaw, Poland
[3] Gdańsk University of Technology, Gdańsk, Poland
[4] Tooploox, Wroclaw, Poland
[5] Autonomous University of Barcelona, Barcelona, Spain
[6] Computer Vision Center, Barcelona, Spain

Abstract. In the field of continual learning, models are designed to learn tasks one after the other. While most research has centered on supervised continual learning, there is a growing interest in unsupervised continual learning, which makes use of the vast amounts of unlabeled data. Recent studies have highlighted the strengths of unsupervised methods, particularly self-supervised learning, in providing robust representations. The improved transferability of those representations built with self-supervised methods is often associated with the role played by the multi-layer perceptron projector. In this work, we depart from this observation and reexamine the role of supervision in continual representation learning. We reckon that additional information, such as human annotations, should not deteriorate the quality of representations. Our findings show that supervised models when enhanced with a multi-layer perceptron head, can outperform self-supervised models in continual representation learning. This highlights the importance of the multi-layer perceptron projector in shaping feature transferability across a sequence of tasks in continual learning. The code is available on github.

Keywords: Continual Learning · Representation Learning

1 Introduction

In continual learning (CL), the goal of the model is to learn new tasks sequentially. Most of the works focus on supervised continual learning (SCL) for image classification where the learner is provided with labeled training data and the metric of interest is accuracy on all the tasks seen so far. More recently, unsupervised continual learning (UCL) gained more attention [10,11,25]. UCL considers

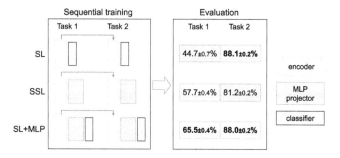

Fig. 1. In a two-task continual learning scenario, supervised learning (SL) results in representations that perform well on the second task but poorly on the first task due to high forgetting. On the other hand, representations trained with self-supervised learning (SSL) have higher first-task performance but they underperform on the second task. We show that simple modifications to supervised learning (SL+MLP) yield representations that are superior on the first task and on par with SL on the second task. We report average task-aware k-NN accuracy on 6 different 2-task combinations of CIFAR100, CIFAR10 and SVHN datasets (3 runs for each scenario).

the problem of learning robust and general representations on a sequence of tasks, without accessing the data labels. Effective UCL methods would allow the utilization of vast amounts of unlabeled data emerging on a daily basis and continually improve existing models.

A number of recent works study continual learning from a representation learning perspective and show that unsupervised approaches build more robust representations when trained continually [7,25]. More specifically, [25] shows that self-supervised learning (SSL) methods build representations that are more robust to forgetting than supervised learning (SL). [7] notice that training Sim-CLR [4] have advantageous properties for continual learning compared to supervised training. However, it is still counter-intuitive that access to more information (labels) results in worse representations in continual learning.

One of the potential reasons is the transferability gap between supervised and unsupervised learning. It was believed that the superior transferability of unsupervised learning can be attributed to a special design of contrastive loss [15, 46] or lack of annotations during training [8,36]. However, recent works [35,41] identify that a multi-layer perceptron (MLP) projector commonly used in SSL [4, 5,12,45] is a crucial component that improves transferability of SSL models. Following that finding [35,41] use an MLP projector to improve transferability of supervised learning and achieve state-of-the-art transfer learning performance, surpassing unsupervised methods.

In this work, encouraged by these advancements in improving the transferability of supervised models, we revisit supervision for continual representation learning. We argue that additional information (human annotations) should not hurt the quality of representations in continual learning, as suggested by [25]. Motivated by the latest study on transferability of representations in self-supervised and supervised learning, we aim to improve transferability between tasks in continual learning. We are the first to show that supervised models can continually learn representations of higher quality than self-supervised models when trained with a simple MLP head (see Fig. 1). We identify the crucial role of an MLP projector in representation learning through the perspective of feature transferability, forgetting, and retention for continually trained models.

The main contributions of this paper are as follows:

- We empirically show that SL equipped with a simple MLP projector can learn higher-quality representations than SSL methods in continual finetuning scenarios in both in-distribution and transfer learning scenarios
- We show that the use of the MLP projector can be coupled with several continual learning methods, further improving their performance.
- We shed light on the reasons behind the strong performance of supervised learning with MLP projector: better transferability, lower forgetting, and increasing diversity of representations.

2 Related Work

Self-supervised Learning (SSL). Learning effective visual representations without annotations is a long-standing problem that aims at leveraging large volumes of unlabeled data. Recent SSL methods show impressive performance, matching or even exceeding the performance of their supervised equivalents [3–5,12,45]. The majority of these techniques rely on image augmentation methods to produce multiple views for a given sample. They train a model to be insensitive to these augmentations by ensuring that the network generates similar representations for the views of the same image and different representations for views of other images. In this work, we use BarlowTwins [45] which considers an objective function measuring the cross-correlation matrix between the features and SimCLR [4] which uses contrastive learning based on noise-contrastive estimation. A number of studies [1,5,16,45] show that an MLP projector between the encoder and the loss function is a crucial component to prevent the collapse of the representations and improve their transferability.

Transferable Representations. [41] seeks to understand the transferability gap between unsupervised (SSL) and supervised pretraining. They found out that adding a projection network (which is commonly used in SSL) boosts the transferability of the supervised models' features. This was further explored in [35] and it was shown that it is possible to build representations that are good for both the source and the downstream tasks. In this work, we revisit those findings in the context of models learned on a sequence of tasks. Contrary to the transfer learning literature [8,35], which usually focuses on the downstream task performance, we evaluate the model on all tasks during the sequential training. This allows us to gain more insight into learned representations, i.e. representation forgetting.

Supervised Continual Learning (SCL). SCL aims to create systems that can acquire the ability to solve novel tasks using new annotated data while retaining the knowledge acquired from previously learned tasks [31]. A popular formulation of CL is class-incremental learning (CIL) [27,39] where each task introduces unseen classes that will not occur in the following tasks. In an exemplar-free setting, the model is not allowed to store any samples from previous tasks which might be important in situations where privacy concerns apply and such a setting remains a great challenge [37]. A popular strategy is *feature distillation* [23,43] which minimizes representational changes in subsequent learning stages by enforcing consistent output between the current model and the one trained in the previous task. Despite the recent progress in SCL, [19] highlight the fact that state-of-the-art SCL methods focus on eliminating bias and forgetting on a level of last classification layer. As a result they fail to improve the feature extractor during the continual training which is a main goal of this paper.

Unsupervised Continual Learning (UCL). Despite the success of SSL methods, they are designed to learn from large static datasets. UCL methods aim to overcome this issue and allow the models to learn from an ever-changing stream of data without excessive memory requirements. Recent works [10,11,25] apply SSL in the UCL setting and claim their superior results for continual representation learning. Most successful methods apply feature distillation through learnable non-linear projector: CaSSLe [10] distills features outputted by the projector while PFR [11] distills the features outputted by the backbone. UCL models are evaluated by measuring their representation strength through linear probing or k-nearest neighbors (k-NN) and this paper follows this evaluation protocol.

3 Experimental Setup

Datasets. We utilize four different datasets: CIFAR10 [22] (C10), CIFAR100 [22] (C100), SVHN [28] and ImageNet100 [38] (IN100), 100-class subset of the ILSVRC2012 dataset [34] with $\approx$ 130k images in high resolution (resized to 224×224). We consider popular settings in continual learning: CIFAR10/5, CIFAR100/5, CIFAR100/20 and ImageNet100/5 sequences, where D/N denotes that dataset D is split into N tasks with an equal number of classes in each task without overlapping ones. To gain further insight, we construct multiple two-task settings where we investigate representation strength and stability. We denote task shift with "$\rightarrow$", e.g. sequence $A \rightarrow B$ means that the model was trained on two tasks, the first one was dataset A and the second one was dataset B. We consider C10→C100 and C100→C10 scenarios as having low distribution shifts, while C10→SVHN and SVHN→C10 scenarios involve higher distribution shifts. We also perform transfer learning experiments on a set of diverse array of datasets: Food101 (Food) [2], Oxford-IIIT Pets (Pets) [32], Oxford Flowers-102 (Flowers) [29], Caltech101 (Caltech) [9], Stanford Cars (Cars) [21], FGVC-Aircraft (Aircrafts) [26], Describable Textures (DTD) [6] and Caltech-Birds-200 (Birds) [40].

Methods. We use the following supervised methods: (1) SL - the standard approach of training a model with linear classification head with a cross-entropy loss function [27]. (2) SL+MLP - SL with MLP projector added between the backbone and a linear head that is discarded at test-time (see Fig. 1), (3) t-ReX [35], and (4) SupCon [17]. Note that SL is the only method that does not utilize an additional MLP projector during training. For SSL approaches we choose BarlowTwins [45] and SimCLR [4]. Results denoted as SSL were obtained using BarlowTwins. We use ResNet-18 [13] as a feature extractor network for all the experiments. For CL strategies we use LwF [23], CaSSLe [10] and PFR [11]. Note that we do not include most of the approaches designed for class-incremental learning because most of them fail to improve their feature extractor during continual training [19].

Training. We use the code repository from CaSSLe [10] and we follow their training procedure. We train SSL models for 500 epochs per task using SGD optimizer with momentum with batch size 256 and cosine learning rate schedule. We adapt the procedure to SL by reducing the number of epochs to 100 per task and the batch size to 64. We tune the learning rate on CIFAR100/5 for each method. We use augmentations from SimCLR [4] for SSL and SupCon and augmentations proposed in [41] for SL approaches(SL, SL+MLP and t-ReX). Note that, unless stated otherwise, we investigate continual finetuning scenario and we do not employ any methods for continual learning nor replay buffer.

Evaluation. We use k-NN classifier to evaluate the quality of representations following [10,25] and Nearest Mean Classifier (NMC) as in [33,44] to evaluate the stability of representations. We use CKA [20] to measure the similarity between representations of two models. Moreover, we use forgetting (F) and forward transfer (FT) commonly used in continual learning [24]. We also measure task exclusion difference EXC [14] to evaluate the level of retention of task-specific features. We use subscripts to indicate the evaluation dataset, e.g. Acc_{C10} means "accuracy on C10 dataset". We report means and standard deviations computed across 3 runs unless stated otherwise.

4 Main Results

This section presents the experimental results of continual representation learning. In Sect. 4.1 we present our main results showing that supervised models can outperform self-supervised models in continual representation learning. In Sect. 4.2 we perform continual transfer learning evaluation that further supports this claim. In Sect. 4.3 we combine different models with CL strategies to investigate the synergy between them. Then we follow with an extensive analysis that sheds light on the reasons for improved performance. Section 5.1 investigates the quality of representations, including forgetting, task exclusion comparison, similarity, and forward transfer. Section 5.2 presents a spectral analysis of representations. In Sect. 3, we evaluate the stability of the representations learned by different methods under various distribution shifts. Finally, in Sect. 5.3 we present an ablation study on impact of different components of the projector as well as data efficiency and robustness to label noise.

4.1 Continual Representation Learning

Figure 2 presents our main finding. Namely, we show that supervised models can build stronger representations than self-supervised models under continual finetuning, contrary to previous beliefs [25]. We identify that the key component to improving the performance of supervised models is an additional MLP projector used during training and discarded afterward - without it, SL significantly underperforms compared to SSL.

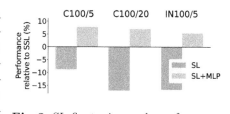

Fig. 2. SL finetuning underperforms compared to SSL. However, when equipped with the MLP projector it consistently outperforms SSL. We report the difference in k-NN accuracy (%) between supervised approaches and SSL.

Figure 3 presents the performance of SL, SSL and SL+MLP after each task. We identify two factors contributing to superior results of SL+MLP. Firstly, we observe that the performance of supervised models after the initial task is largely improved by the addition of the MLP projector, resulting in accuracy close to SSL models. In order to achieve good task-agnostic accuracy on the whole dataset (seen and unseen classes), the model trained on a single task needs to perform well on unseen data. Therefore, we attribute the advantage of SL+MLP to the increased transferability of representations induced by MLP projector, which is in line with [35,41]. Secondly, we notice that SL+MLP is the only method able to incrementally accumulate knowledge and consistently improve performance. This observation is in line with the increasing diversity of features presented in Sect. 5.2.

Table 1 presents extended results including multiple SL and SSL approaches in continual finetuning. We observe that all the supervised methods equipped with the projector significantly outperform simple SL. SL+MLP, t-ReX, and SupCon achieve much higher results than SL in all the finetuning experiments.

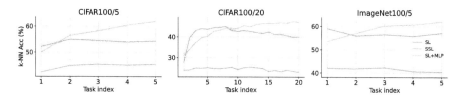

Fig. 3. SL+MLP: (1) achieves strong performance after the initial task compared to SL which indicates that it produces representations that are transferable to the unseen tasks; (2) is the only method that is able to accumulate knowledge learned on a sequence of tasks. We report task-agnostic k-NN accuracy after each task on the whole dataset (notice that yet unseen tasks are also included in the evaluation).

Table 1. Supervised methods that utilize MLP projector largely outperform vanilla SL. They also outperform SSL on most of the datasets. We report k-NN accuracy of the learned representations. The best result in **bold** and the second best underlined.

Method	C10/5	C100/5	C100/20	IN100/5
SL	59.8±1.8	45.3±0.7	23.1±0.2	40.4±1.2
SL+MLP	65.9±0.7	**61.9±0.5**	47.1 ± 0.7	**62.4±0.4**
t-ReX	69.3±1.1	59.2 ± 0.6	**50.8±0.1**	59.2 ± 0.6
SupCon	60.4±0.6	49.4±0.3	30.0±0.7	57.6±0.6
BarlowTwins	**76.2±1.2**	54.1±0.3	40.0±0.8	57.0±0.4
SimCLR	72.4 ± 1.3	48.9±0.4	33.4±0.5	54.7±0.4

What is worth noting is the fact that all these methods were trained with different supervised losses: SL+MLP uses cross-entropy, t-ReX uses cosine softmax cross-entropy and SupCon uses supervised contrastive loss. However, they all utilize the MLP projector and all outperform vanilla SL and SSL on most of the datasets.

4.2 Continual Transfer Learning

In this Section we evaluate the continually trained feature extractors on a set of diverse downstream classification tasks described in Sect. 3: Food, Pets, Flowers, Caltech, Cars, Aircrafts, DTD and Birds. We evaluate the performance on downstream datasets after each task in the continual learning sequence and present the results in Fig. 4. We observe that SL+MLP outperforms the competitors on all but one dataset. Interestingly, the pattern of results is similar to the *in-distribution* results presented in Fig. 3: (1) SL+MLP is on-par with SSL after the first task and significantly outperforms SL, and (2) improves its performance when trained continually. These results highlight the superior transferability of the representations learned by the proposed SL+MLP approach.

4.3 Synergy with CL Methods

In this Section, we investigate our findings combined with different CL approaches. More precisely, we combine both supervised and self-supervised methods with existing CL methods to verify the synergy between them.

Firstly, we evaluate the resulting combinations *in-distribution* as in Sect. 4.1 and present the results in Table 2. We observe that most of the combinations of methods and CL strategies are outperforming simple finetuning (the only exceptions are some combinations of LwF with supervised methods). Moreover, the conclusions from simple finetuning scenarios still hold – supervised methods equipped with MLP projectors outperform vanilla SL and SSL. However, with the help of CL methods, the performance gap is much smaller.

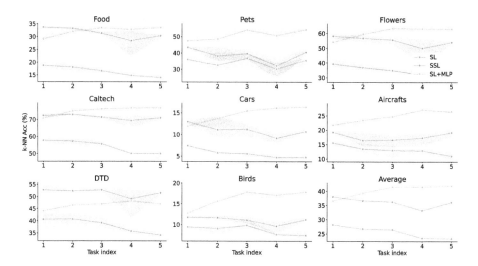

Fig. 4. Representations learned by SL+MLP are more transferable than those learned by SL and SSL. They also improve when trained on new tasks. We present the results of the models trained continually on ImageNet/5 and evaluated after each task. We report k-NN accuracy (%) on a set of 8 diverse downstream classification tasks and an average performance.

We also evaluate transfer learning performance as in Sect. 4.2 and present the results in Table 3. We observe that CL methods bring an improvement over fine-tuning in almost all experiments. Similarly to the in-distribution experiments, the best results are obtained by the supervised methods that utilize MLP projector combined with one of CL strategies.

To sum up, models that achieve the best transfer learning performance are those: (1) trained in a supervised way (2) with the use of the MLP projector and (3) coupled with CL strategy based on temporal learnable projection, namely CaSSLe or PFR. These conclusions are coherent with the previous Section where a similar set of models performed the best.

5 Analysis

5.1 Quality of Representations

Forgetting. In Table 4 we observe high representation forgetting for SL, significantly lower for SSL, and the lowest for SL equipped with MLP projector. Low forgetting is an important factor contributing to the superior performance of SL+MLP.

Task Exclusion Difference. In the two-task sequence EXC answers the question: *what is the performance gap between the model trained on B and a model trained on a sequence $A \rightarrow B$ when evaluated on A?* Results from Table 4 show

Table 2. The effect of MLP projector compounds with the positive effect of CL strategies leading to combinations that outperform SSL on most datasets. We report k-NN accuracy of the learned representations. The best result in each group in **bold** and the second best underlined.

Method	CL strategy	C10/5 Acc	Δ	C100/5 Acc	Δ	C100/20 Acc	Δ	IN100/5 Acc	Δ
\multicolumn{10}{c}{SUPERVISED CONTINUAL LEARNING}
SL	Finetune	59.8±1.8		45.3±0.7		23.1±0.2		40.4±1.2	
	LwF	69.6±1.1	+9.8	62.9±0.1	+17.6	51.5±0.2	+28.4	67.0±0.1	+26.6
	PFR	71.0±2.0	+11.2	63.3±0.2	+18.0	52.3±1.2	+29.2	65.0±0.3	+24.6
SL+MLP	Finetune	65.9±0.7		61.9±0.5		47.1±0.7		62.4±0.4	
	LwF	72.6±3.4	+6.7	58.7±0.2	-3.2	51.9±0.1	+4.8	60.4±0.2	-2.0
	PFR	76.3±1.0	+10.4	**63.6±0.2**	+1.7	**54.5±0.2**	+7.4	65.2±0.1	+2.8
t-ReX	Finetune	69.3±1.1		59.2±0.6		50.8±0.1		59.2±0.6	
	LwF	74.5±0.7	+5.2	58.3±0.4	-0.9	50.4±0.1	-0.4	58.6±1.0	-0.6
	PFR	75.9±1.2	+6.6	60.9±0.5	+1.7	53.4±0.3	+2.6	63.9±0.6	+4.7
SupCon	Finetune	60.4±0.6		49.4±0.3		30.0±0.7		57.6±0.6	
	CaSSLe	75.1±0.4	+14.7	61.1±0.2	+11.7	49.2±1.2	+19.2	**70.4±0.6**	+12.8
	PFR	**78.1±1.0**	+17.7	57.0±0.2	+7.6	51.2±0.8	+21.2	68.0±0.7	+10.4
\multicolumn{10}{c}{UNSUPERVISED CONTINUAL LEARNING}
BarlowTwins	Finetune	76.2±1.2		54.1±0.3		40.0±0.8		57.0±0.4	
	CaSSLe	**80.9±0.2**	+4.7	**58.6±0.6**	+4.5	49.3±0.1	+9.3	**64.9±0.1**	+7.9
	PFR	78.8±0.6	+2.6	57.2±0.2	+3.1	46.0±0.7	+6.0	61.1±0.2	+4.1
SimCLR	Finetune	72.4±1.3		48.9±0.4		33.4±0.5		54.7±0.4	
	CaSSLe	80.6±0.5	+8.2	55.9±0.5	+7.0	48.2±0.4	+14.8	59.3±0.5	+4.6
	PFR	79.2±0.7	+6.8	53.8±0.3	+4.9	**49.4±0.1**	+16.0	57.7±0.2	+3.0

that SL achieves small positive EXC meaning that it forgets most features specific to the initial task (but not all of them). SL+MLP achieves the highest EXC which shows that it is able to successfully retain a large portion of task-specific features. Surprisingly, SSL exhibits negative EXC. It means that it is better to train SSL model from scratch on another task than to finetune the model pretrained on the task of interest.

CKA Similarity. In Table 4 we report CKA similarity between the models trained on C10 and the rest of the models. We observe that usage of MLP head in SL increases CKA between the C10 model and other models. Moreover, in the case of SL+MLP, the models pretrained on C10 and finetuned on another task have higher similarity to C10 models than the models trained on another dataset from scratch. This is not necessarily the case for SL models. SSL models have the highest CKA scores, however, they usually underperform compared to SL+MLP. This suggests that SSL produces similar features when trained on different datasets but their discriminative power for a classification task is worse than those learned with SL+MLP.

Table 3. SL+MLP combined with PFR achieves the best average performance on transfer learning tasks. CL methods also improve the performance of SL and SSL but they achieve worse overall performance. Best results in each group (finetuning and continual learning methods) in **bold**.

Method	Food	Pets	Flowers	Caltech	Cars	Aircrafts	DTD	Birds	Avg
FINETUNING									
SL	14.0±0.6	35.6±1.6	29.6±1.0	50.0±2.1	4.8±0.2	11.0±1.1	34.2±1.0	7.4±0.8	23.3
SSL	30.4±0.4	40.5±1.6	54.0±0.8	71.1±1.0	10.7±0.4	19.1±1.1	**51.6±0.8**	11.2±0.5	36.1
SL+MLP	**33.6±0.5**	**54.3±1.3**	**63.0±1.1**	**76.9±0.9**	**16.3±0.2**	**26.4±1.4**	47.1±0.5	**17.8±0.2**	**41.9**
CONTINUAL LEARNING METHODS									
SL+LwF	30.2±0.4	57.8±0.8	57.9±0.7	72.7±1.2	13.6±0.3	22.2±0.3	46.6±0.5	17.1±0.4	39.8
SL+PFR	30.4±0.6	57.4±0.3	58.5±0.7	72.6±0.2	14.0±0.2	21.9±0.6	46.3±1.1	16.3±0.2	39.7
SSL+PFR	35.1±0.3	43.4±1.0	59.9±1.1	74.6±0.4	12.2±0.5	19.8±0.3	53.9±0.7	12.4±0.4	38.9
SSL+CaSSLe	39.1±0.2	48.5±0.7	65.1±0.8	75.5±0.4	14.2±0.6	21.1±0.9	**56.4±0.5**	13.5±0.2	41.7
SL+MLP+LwF	34.7±0.4	55.4±0.9	61.9±1.3	75.3±0.4	15.1±0.1	24.5±0.5	47.3±1.0	16.2±0.2	41.3
SL+MLP+PFR	**38.0±0.2**	**58.8±0.7**	**67.8±0.1**	**79.3±0.2**	**17.5±0.3**	**26.2±0.3**	49.2±0.4	**18.1±0.2**	**44.4**

Table 4. We observe high representation forgetting for SL, significantly lower for SSL, and the lowest for SL equipped with MLP projector. SL is able to preserve a small fraction of task-specific features while SL+MLP can retain much more, based on their EXC scores. Surprisingly, SSL achieves negative EXC meaning that pretraining on a given task hurts the performance on this task after the finetuning. We report evaluation on CIFAR10 dataset. The best value between SL, SSL, and SL-MLP in **bold**.

Training	SL				SSL				SL+MLP			
sequence	Acc ↑	F ↓	EXC ↑	CKA ↑	Acc ↑	F ↓	EXC ↑	CKA ↑	Acc ↑	F ↓	EXC ↑	CKA ↑
C10	**93.7±0.1**	-	-	-	88.8±0.1	-	-	-	93.3±0.1	-	-	-
C100	79.4±0.3	-	-	0.46	80.8±0.1	-	-	0.56	84.5±0.4	-	-	0.49
C10→C100	81.3±0.6	12.4±0.6	1.9±0.3	0.50	79.1±0.2	9.7±0.3	-1.8±0.2	0.52	88.8±0.2	4.5±0.3	4.3±0.6	0.57
SVHN	27.3±0.2	-	-	0.05	**58.6±1.2**	-	-	0.27	56.3±0.2	-	-	0.20
C10→SVHN	29.6±1.1	64.1±1.1	2.3±1.3	0.05	54.9±0.7	33.8±0.7	-3.7±1.9	0.25	62.7±0.8	30.6±0.8	6.4±1.0	0.25

Positive and Negative Forward Transfer. We present the results of the forward transfer evaluation in Table 5. All the methods benefit from pretraining on CIFAR100 which is semantically close to CIFAR10. However, pretraining on semantically distant SVHN hinders the performance of SL but it hardly influences the performance of SSL and SL+MLP.

5.2 Spectra of Representations

To gain further insight into the properties of continually trained representations, we analyze the spectrum of their covariance matrix. We follow the procedure from [16]. We gather the representations of the validation set and compute the covariance matrix of the representations, C. We perform singular value decomposition of the covariance matrix $C = USV^T$, where $S = diag(\sigma^k)$ and σ^k is k-th singular value of C. We call the representations *diverse* when a large number

Table 5. All methods benefit from pretraining on C100 which is semantically close to C10. However, pretraining on semantically distant SVHN hinders the performance of SL. Best method in **bold**.

Method	C10	C100→C10		SVHN→C10	
	$Acc_{C10}\uparrow$	$Acc_{C10}\uparrow$	$FT_{C10}\uparrow$	$Acc_{C10}\uparrow$	$FT_{C10}\uparrow$
SL	93.7±0.1	94.3±0.1	0.6±0.0	93.3±0.1	-0.4±0.1
SSL	88.8±0.1	89.2±0.1	0.5±0.2	88.5±0.1	-0.3±0.2
SL+MLP	**93.3±0.1**	**94.3±0.1**	**1.0±0.0**	**93.2±0.2**	**-0.1±0.1**

of principal directions (independent features) is needed to explain most of the variance. Figure 5 presents how singular value spectra change after each task for different training methods and different sequences of tasks.

Representation Collapse. Figure 5 reveals that supervised learning exhibits signs of neural collapse [30] - a large fraction of variance is described by a few dimensions roughly equal to the number of classes in the training set. This is an undesirable property in continual representation learning as the representations should be more versatile and useful not only for current but also for past and future tasks. SSL, on the other hand, learns a more diverse set of features resulting in a flatter singular values spectrum. In our experiments adding MLP to SL prevents neural collapse and results in features' properties more similar to SSL.

Evolution of Representations. An important property of representations learned in continual learning is the change in their diversity: the diversity that increases after each task is desired. In Fig. 5 we can observe that for SL, the diversity of the features usually decreases, except for C10→C100 where the increase is caused by a higher number of training classes [30]. For SSL, the diversity increases in the five-task scenario and remains close to constant for two-task settings. SL+MLP is able to improve the diversity of the representations consistently across all the presented scenarios suggesting its superiority in continual representation learning. It may be related to its ability to effectively accumulate knowledge when trained on a sequence of tasks, as presented in Fig. 3.

5.3 Ablation Study

Architecture of the Projector. In Table 6 we examine the impact of different projector architectures on supervised learning methods. All the methods achieve the worst performance when not using a projector. Moreover, all the methods achieve similarly high accuracy when with a suitable projector. These results highlight the importance of MLP projector for supervised continual representation learning and diminish the importance of other factors (*e.g.* loss function) on the final performance.

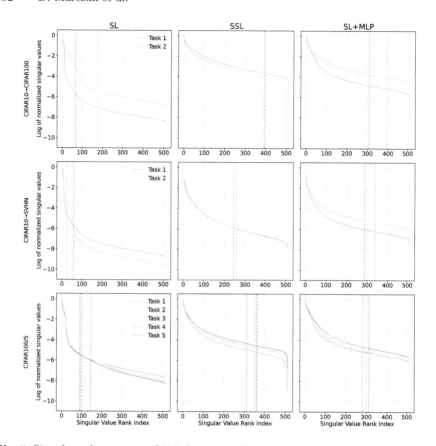

Fig. 5. Singular value spectra of 512-dimensional representation space. Representations learned with SL+MLP (right) exhibit desirable properties from the continual learning point of view: (1) they consist of a more diverse set of features (contrary to SL, left); (2) they improve feature diversity when learning new tasks consistently across all the presented settings. Singular values are ordered descending, are normalized by σ^1 (the largest singular value) and the scale is logarithmic. Vertical dashed lines denote 95% of the variance explained. Intuitively, it indicates how many relevant independent features the representation contains.

To gain a deeper insight into the impact of different components of the projector we systematically investigate the impact of the depth and width of the projector. We follow the MLPP [41] architecture for the projector, using a linear layer, batch normalization (BN), and ReLU activation, which we call a *block* with a hidden dimension d_h. This block is followed by an output linear layer with a hidden dimension d_o, and then a classification layer. The projector contains n blocks, where $n = 0$ means a linear projector. We present the results in Fig. 6. We observe that the width and depth of the projector have a very limited impact on the final performance as long as the projector has at least one block. Therefore, we further decompose the projector starting from an MLP with a

Table 6. Impact of projector architecture on different methods. Projector architectures are arranged in order of increasing number of parameters. We report k-NN accuracy (*Acc*) and average improvement over the method without projector (Δ). Best results for each method in **bold**.

Method	Projector architecture	CIFAR10/5 Acc	Δ	CIFAR100/5 Acc	Δ
SL	None	59.8±1.8		45.3±0.7	
	MLPP [41]	65.9±0.7	+6.1	**61.9±0.5**	**+16.6**
	t-ReX [35]	**67.3±0.1**	**+7.5**	58.3±0.2	+13.0
t-ReX	None	60.0±1.2		36.7±1.1	
	MLPP [41]	66.7±1.4	+6.7	58.1±0.2	+21.4
	t-ReX [35]	**69.3±1.1**	**+9.3**	**59.2±0.6**	**+22.5**
SupCon	None	58.7±0.4		23.7±0.5	
	Linear	61.0±0.7	+2.3	37.7±0.6	+14.0
	SupCon [17]	60.4±0.6	+1.7	49.4±0.3	+25.7
	MLPP [41]	**66.2±0.7**	**+7.5**	54.3±0.5	+30.6
	t-ReX [35]	63.5±0.2	+4.8	**58.1±0.1**	**+34.4**

single block ($d_h = 4096$, $d_o = 512$) and remove basic components. We present the results in Fig. 7. We observe that BN has the highest positive impact on performance. However, using BN without the linear layer and ReLU lags 4 p.p. behind the performance of the full block, highlighting the importance of all the components.

Data Efficiency. SL with MLP projector outperforms SSL in continual representation learning when having access to the same amount of data. However, in real-world scenarios, SSL is able to utilize vast amounts of unlabeled data while SL needs costly data annotations. Therefore, we examine how SL behaves when trained on a fraction of data available for SSL in each task to simulate limited access to labeled data. We restrict the data and select a fraction of it (equal for each task). We present the results in Fig. 8 (left). SL+MLP surpasses SimCLR with less than 20% of data and BarlowTwins with about 30% of data.

Robustness to Mislabelling. We evaluate the robustness of our claims against noisy labels following previous study [18] and randomly change the labels of the fraction of samples in the train dataset. Figure 8 (right) presents the results on CIFAR100/5. We observe that SL+MLP outperforms SSL even with 40% label noise. It validates the efficacy of SL+MLP in the presence of imperfect annotations.

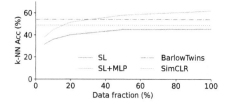

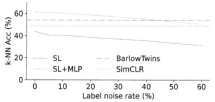

$n =$	0 (linear)	1		2		3	
$h_o =$	512 2048	512	2048	512	2048	512	2048
$d_h = 512$		60.06	60.88	60.61	60.45	60.59	60.93
$d_h = 1024$	46.74 47.45	60.61	60.50	61.04	60.61	60.84	61.25
$d_h = 2048$		60.54	60.88	61.42	61.05	60.97	61.00
$d_h = 4096$		61.46	61.10	60.95	61.46	61.09	60.91

Input Lin.	✗	✗	✓	✓
ReLU	✗	✓	✗	✓
BN ✗	46.74	49.06	47.59	53.87
✓	57.33	59.15	58.47	61.46

Fig. 6. One MLP block brings significant improvements and the next blocks have minor effect. Impact of projector depth and width on CIFAR100/5. Note that d_h does not apply to linear projectors.

Fig. 7. BatchNorm seems to have the highest positive impact on the final performance. Impact of block components on CIFAR100/5.

Fig. 8. SL+MLP outperforms SSL and SL approaches: (left) trained on full dataset even with only 30% of data available and (right) with 40% label noise. We present the accuracy (%) after the final task on CIFAR100/5 for different data fractions (left) and varying label noise level (right).

6 Discussion and Limitations

Although supervised learning with the MLP projection head seems to be more effective in continual representation learning, it comes at a price. SL requires mundane image labeling of the whole dataset which can be costly and impractical at scale. Self-supervised learning, on the other hand, is not dependent on image annotations and, therefore, can operate on a vast amount of unlabeled data.

However, SSL faces its own limitations. Firstly, most SSL approaches depend on strong image augmentations and learn representations that are invariant to them [3,4,45]. This can hinder the performance on the downstream tasks which require attention to the traits that it has been trained to be invariant to [42]. Moreover, SSL usually requires longer training which increases computational requirements in comparison to SL. It is also worth noting that both SL+MLP and SSL introduce additional costs to the model during the training, as both introduce MLP projector that requires more computational requirements. However, at test time every method operates at the same number of parameters, as we discard MLP projectors after training.

Furthermore, it's worth re-emphasizing that this work focuses on continual representation learning. While we utilize data from previous tasks to construct k-NN and nearest mean classifiers for evaluating learned representations, our primary objective is not centered on the continual approach to the downstream task (classification). We are not delving into class-incremental learning, a prevalent continual learning setting. Nonetheless, our analysis of representation strength and stability can offer valuable insights into continual learning dynamics, potentially aiding in the creation of more effective algorithms for continual downstream task solutions.

7 Conclusions

In this work, we are first to show that supervised learning can significantly outperform self-supervised learning in continual representation learning. We achieve it by equipping SL with a simple MLP projector discarded after the training, following the common practice from SSL. We show that SL+MLP can be successfully coupled with several continual learning strategies, further improving the performance. Finally, we shed some light on the reasons for improved performance when using MLP with SL: better transferability, lower forgetting, and higher diversity of learned features.

Acknowledgments. Daniel Marczak is supported by National Centre of Science (NCN, Poland) Grant No. 2021/43/O/ST6/02482. This research was partially funded by National Science Centre, Poland, grant no 2020/39/B/ST6/01511, grant no 2022/45/B/ST6/02817, and grant no 2023/51/D/ST6/02846. Bartłomiej Twardowski acknowledges the grant RYC2021-032765-I. This paper has been supported by the Horizon Europe Programme (HORIZON-CL4-2022-HUMAN-02) under the project "ELIAS: European Lighthouse of AI for Sustainability", GA no. 101120237. We gratefully acknowledge Polish high-performance computing infrastructure PLGrid (HPC Center: ACK Cyfronet AGH) for providing computer facilities and support within computational grant no. PLG/2023/016393.

References

1. Bordes, F., Balestriero, R., Garrido, Q., Bardes, A., Vincent, P.: Guillotine regularization: why removing layers is needed to improve generalization in self-supervised learning. TMLR (2023)
2. Bossard, L., Guillaumin, M., Van Gool, L.: Food-101 – mining discriminative components with random forests. In: Fleet, D., Pajdla, T., Schiele, B., Tuytelaars, T. (eds.) ECCV 2014. LNCS, vol. 8694, pp. 446–461. Springer, Cham (2014). https://doi.org/10.1007/978-3-319-10599-4_29
3. Caron, M., et al.: Emerging properties in self-supervised vision transformers. In: ICCV (2021)
4. Chen, T., Kornblith, S., Norouzi, M., Hinton, G.E.: A simple framework for contrastive learning of visual representations. In: ICML (2020)
5. Chen, X., He, K.: Exploring simple siamese representation learning. In: CVPR (2020)
6. Cimpoi, M., Maji, S., Kokkinos, I., Mohamed, S., , Vedaldi, A.: Describing textures in the wild. In: CVPR (2014)

7. Davari, M.R., Asadi, N., Mudur, S., Aljundi, R., Belilovsky, E.: Probing representation forgetting in supervised and unsupervised continual learning. In: CVPR (2022)
8. Ericsson, L., Gouk, H., Hospedales, T.M.: How well do self-supervised models transfer? In: CVPR (2020)
9. Fei-Fei, L., Fergus, R., Perona, P.: One-shot learning of object categories. IEEE TPAMI (2006)
10. Fini, E., da Costa, V.G.T., Alameda-Pineda, X., Ricci, E., Alahari, K., Mairal, J.: Self-supervised models are continual learners. In: CVPR (2022)
11. Gomez-Villa, A., Twardowski, B., Yu, L., Bagdanov, A.D., van de Weijer, J.: Continually learning self-supervised representations with projected functional regularization. In: CVPR Workshops (2021)
12. Grill, J.B., et al.: Bootstrap your own latent: A new approach to self-supervised learning. In: NeurIPS (2020)
13. He, K., Zhang, X., Ren, S., Sun, J.: Deep residual learning for image recognition. In: CVPR (2016)
14. Hess, T., Verwimp, E., van de Ven, G.M., Tuytelaars, T.: Knowledge accumulation in continually learned representations and the issue of feature forgetting. TMLR (2024)
15. Islam, A., Chen, C.F., Panda, R., Karlinsky, L., Radke, R., Feris, R.: A broad study on the transferability of visual representations with contrastive learning. In: ICCV (2021)
16. Jing, L., Vincent, P., LeCun, Y., Tian, Y.: Understanding dimensional collapse in contrastive self-supervised learning. In: ICLR (2022)
17. Khosla, P., et al.: Supervised contrastive learning. In: NeurIPS (2020)
18. Kim, C.D., Jeong, J., Moon, S., Kim, G.: Continual learning on noisy data streams via self-purified replay. In: ICCV (2021)
19. Kim, D., Han, B.: On the stability-plasticity dilemma of class-incremental learning. In: CVPR (2023)
20. Kornblith, S., Norouzi, M., Lee, H., Hinton, G.E.: Similarity of neural network representations revisited. In: ICML (2019)
21. Krause, J., Stark, M., Deng, J., Fei-Fei, L.: 3d object representations for fine-grained categorization. In: 4th International IEEE Workshop on 3D Representation and Recognition (3dRR-13) (2013)
22. Krizhevsky, A.: Learning multiple layers of features from tiny images. University of Toronto (2009)
23. Li, Z., Hoiem, D.: Learning without forgetting. IEEE TPAMI (2018)
24. Lopez-Paz, D., Ranzato, M.: Gradient episodic memory for continual learning. In: NeurIPS (2017)
25. Madaan, D., Yoon, J., Li, Y., Liu, Y., Hwang, S.J.: Representational continuity for unsupervised continual learning. In: ICLR (2022)
26. Maji, S., Rahtu, E., Kannala, J., Blaschko, M., Vedaldi, A.: Fine-grained visual classification of aircraft (2013)
27. Masana, M., Liu, X., Twardowski, B., Menta, M., Bagdanov, A.D., van de Weijer, J.: Class-incremental learning: survey and performance evaluation on image classification. IEEE TPAMI (2023)
28. Netzer, Y., Wang, T., Coates, A., Bissacco, A., Wu, B., Ng, A.Y.: Reading digits in natural images with unsupervised feature learning (2011)
29. Nilsback, M.E., Zisserman, A.: Automated flower classification over a large number of classes. In: Indian Conference on Computer Vision, Graphics and Image Processing (2008)

30. Papyan, V., Han, X.Y., Donoho, D.L.: Prevalence of neural collapse during the terminal phase of deep learning training. PNAS (2020)
31. Parisi, G.I., Kemker, R., Part, J.L., Kanan, C., Wermter, S.: Continual lifelong learning with neural networks: a review. Neural Netw. (2019)
32. Parkhi, O.M., Vedaldi, A., Zisserman, A., Jawahar, C.V.: Cats and dogs. In: CVPR (2012)
33. Rebuffi, S., Kolesnikov, A., Sperl, G., Lampert, C.H.: icarl: incremental classifier and representation learning. In: CVPR (2017)
34. Russakovsky, O., et al.: ImageNet large scale visual recognition challenge. IJCV (2015)
35. Sariyildiz, M.B., Kalantidis, Y., Alahari, K., Larlus, D.: No reason for no supervision: improved generalization in supervised models. In: ICLR (2023)
36. Sariyildiz, M.B., Kalantidis, Y., Larlus, D., Karteek, A.: Concept generalization in visual representation learning. In: ICCV (2021)
37. Smith, J.S., Tian, J., Halbe, S., Hsu, Y., Kira, Z.: A closer look at rehearsal-free continual learning. In: CVPR Workshops (2023)
38. Tian, Y., Krishnan, D., Isola, P.: Contrastive multiview coding. In: Vedaldi, A., Bischof, H., Brox, T., Frahm, J.-M. (eds.) ECCV 2020. LNCS, vol. 12356, pp. 776–794. Springer, Cham (2020). https://doi.org/10.1007/978-3-030-58621-8_45
39. van de Ven, G.M., Tuytelaars, T., Tolias, A.S.: Three types of incremental learning. Nat. Mach. Intell. (2022)
40. Wah, C., Branson, S., Welinder, P., Perona, P., Belongie, S.: The caltech-ucsd birds-200-2011 dataset (2011)
41. Wang, Y., Tang, S., Zhu, F., Bai, L., Zhao, R., Qi, D., Ouyang, W.: Revisiting the transferability of supervised pretraining: an mlp perspective. In: CVPR (2021)
42. Xiao, T., Wang, X., Efros, A.A., Darrell, T.: What should not be contrastive in contrastive learning. In: ICLR (2021)
43. Yan, S., Xie, J., He, X.: DER: dynamically expandable representation for class incremental learning. In: CVPR (2021)
44. Yu, L., et al.: Semantic drift compensation for class-incremental learning. In: CVPR (2020)
45. Zbontar, J., Jing, L., Misra, I., LeCun, Y., Deny, S.: Barlow twins: self-supervised learning via redundancy reduction. In: ICML (2021)
46. Zhao, N., Wu, Z., Lau, R.W.H., Lin, S.: What makes instance discrimination good for transfer learning? In: ICLR (2020)

FLAT: Flux-Aware Imperceptible Adversarial Attacks on 3D Point Clouds

Keke Tang[1], Lujie Huang[1], Weilong Peng[1](✉), Daizong Liu[2], Xiaofei Wang[3], Yang Ma[1], Ligang Liu[3], and Zhihong Tian[1]

[1] Guangzhou University, Guangzhou, China
{wlpeng,tianzhihong}@gzhu.edu.cn
[2] Peking University, Beijing, China
dzliu@stu.pku.edu.cn
[3] University of Science and Technology of China, Hefei, China
wxf9545@mail.ustc.edu.cn, lgliu@ustc.edu.cn

Abstract. Adversarial attacks on point clouds play a vital role in assessing and enhancing the adversarial robustness of 3D deep learning models. While employing a variety of geometric constraints, existing adversarial attack solutions often display unsatisfactory imperceptibility due to inadequate consideration of uniformity changes. In this paper, we propose FLAT, a novel framework designed to generate imperceptible adversarial point clouds by addressing the issue from a flux perspective. Specifically, during adversarial attacks, we assess the extent of uniformity alterations by calculating the flux of the local perturbation vector field. Upon identifying a high flux, which signals potential disruption in uniformity, the directions of the perturbation vectors are adjusted to minimize these alterations, thereby improving imperceptibility. Extensive experiments validate the effectiveness of FLAT in generating imperceptible adversarial point clouds, and its superiority to the state-of-the-art methods.

Keywords: Adversarial attacks · Point clouds · Imperceptibility · Flux

1 Introduction

With advancements in deep learning techniques [20,39] and the growing availability of affordable depth-sensing devices, deep neural network (DNN)-driven 3D point cloud perception has become a leading approach. Nevertheless, recent studies have highlighted the vulnerability of DNN classifiers to adversarial attacks [26,50]. Notably, subtle perturbations to the input point clouds can result in incorrect model predictions, posing significant challenges for real-world deployment. Consequently, exploring adversarial attacks on point clouds is essential for evaluating and improving the adversarial robustness of 3D deep learning models.

K. Tang and L. Huang—Joint first authors.

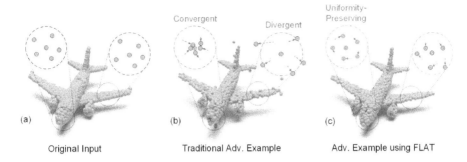

Fig. 1. Visualization of adversarial attacks on 3D point clouds: (a) the original point cloud; (b) the adversarial point cloud generated by SI-Adv; (c) the adversarial point cloud generated by FLAT. Notably, FLAT achieves greater imperceptibility by preserving the uniformity of the point cloud, as compared to SI-Adv.

In achieving imperceptibility for adversarial attacks on 3D point clouds, methods predominantly fall into two categories. The first category encompasses the incorporation of recognizable geometries, such as spheres or airplanes [50], or the application of physical-intuitive deformations [42], both designed to be hidden in the human psyche. The second, more widely-adopted category, emphasizes constraints to minimize perturbations. Traditional methods within this paradigm employ metrics such as the l_2-norm, Chamfer distance, and Hausdorff distance. Recent advancements, however, have sought to harness the intrinsic geometric properties of 3D point clouds, focusing on minimizing distortions by constraining geometric regularity [46] or guiding perturbations along established normal [25] or tangential directions [16]. While the distortion has been notably reduced, traces of adversarial attacks remain perceptible.

This raises a question: why are adversarial point clouds still perceptible under these geometric constraints? We observe that adversarial manipulation, even under these constraints, significantly alters the uniformity of the perturbed samples relative to their original configuration, see the zoomed areas of divergence and convergence in Fig. 1(b). Since the human eye can easily notice changes in uniformity, mitigating these variations during adversarial attacks could enhance their imperceptibility. However, preserving uniformity presents a challenge as it is a regional attribute, characterized by the anisotropic perturbation of points [1], rather than a property of individual points. This distinction is often overlooked by traditional methods.

In this paper, we introduce a novel **FL**ux-aware imperceptible adversarial **AT**tacks (FLAT) on 3D point clouds. By attributing the issue of significant alterations in uniformity to irregular flow of points, we propose to handle it through a novel flux perspective. Specifically, we adapt the flux concept to perturbation vectors, creating a simplified flux measure to monitor the uniformity changes, e.g., divergence within local regions, during attacks. When encountering significant flux, we suppress it by fine-tuning the directions of the perturbation vectors for preserving uniformity. This flux-aware refinement effectively results in enhanced imperceptibility, as illustrated in Fig. 1(c). We validate the effectiveness of our FLAT in attacking common DNN classifiers for 3D point clouds under various metrics. Extensive experimental results show that the adversarial point clouds generated by FLAT are significantly more imperceptible than those generated by state-of-the-art methods. Besides, we demonstrate that our flux-based approach can be readily integrated with other attack techniques to improve their imperceptibility.

Overall, our contribution is summarized as follows:

- We are the first to attribute the inadequate imperceptibility of adversarial attacks on 3D point clouds to the large deviation of uniformity.
- We develop a novel adversarial attack framework that preserves point cloud uniformity by suppressing the simplified flux of perturbation vectors.
- We show by experiments that our FLAT with preserving uniformity achieves superior performance in terms of imperceptibility under various metrics.

2 Related Work

Adversarial Attacks on Point Clouds. Adversarial attacks, designed to craft samples that mislead target models, were initially developed for 2D image classification [38] and subsequently adapted for 3D point clouds. These 3D point cloud attacks fall into three categories: addition-based attacks [50]; deletion-based attacks [47,52,54,57]; and perturbation-based attacks [18,50,56]. Our study concentrates on the perturbation-based category.

Early perturbation-based attacks [26,50] adapted C&W [5] and FGSM [11] from 2D to 3D. Zhao et al. [56] introduced isometric transformations for manipulating point clouds, and Kim et al. [18] focused on minimal point perturbations. Generative methods by Lee et al. [21] and Zhou et al. [58] explored latent space noise and GANs. Tang et al. [42] altered the 2-manifold surface. Despite high success rates, improving attack imperceptibility remains challenging.

Imperceptibility of Adversarial Attacks. To ensure the imperceptibility of attacks, common constraints include managing the l_2-norm, Chamfer and Hausdorff distance [26,50,58]. GeoA^3 [46] focuses on preserving geometric regularity. Directional perturbations guided by normal vectors [25,41] and tangential plane perturbations [16] have also been explored. Tang et al. [40] adapted constraints to follow these directions. Our framework emphasizes the often-overlooked role of uniformity deviations, introducing flux metrics to measure and regulate changes in uniformity. Unlike GeoA^3 [46], which aims for uniform distribution, we ensure consistency in uniformity before and after perturbations.

Uniformity of 3D Point Clouds. Uniform distribution in 3D point clouds is essential for accurate geometry capture, high-quality mesh generation in surface reconstruction [2,15], and effective noise removal in point cloud denoising while preserving structural integrity [13,27]. It also improves segmentation and classification accuracy [19], and is crucial for applications like architectural modeling [44] and autonomous navigation [24]. This paper enhances the imperceptibility of adversarial point clouds by preserving uniformity during attacks.

Deep Point Cloud Classification. Deep learning techniques [10] for point cloud classification have significantly advanced [3,7,12,36]. Initial approaches adapted 2D methodologies using 3D voxel grids [22,28]. The emergence of PointNet enabled direct point cloud processing [30], followed by innovations like hierarchical structures [31], point-specific convolutions [23,43,48,51], and graph-based CNNs [6,33,34,45,55]. For more comprehensive reviews, refer to survey papers [12,17]. Our research aims to attack these classifiers imperceptibly.

3 Problem Formulation

Typical Adversarial Attacks. Given a point cloud $\mathcal{P} \in \mathbb{R}^{n \times 3}$ and its label $y \in \{1, ..., K\}$, where K is the category number, perturbation-based adversarial attack aims to mislead a 3D deep classification model $\mathcal{F}$ by feeding an adversarial point cloud $\mathcal{P}^{adv}$ instead of $\mathcal{P}$ via applying an intentionally designed perturbation, such that the model $\mathcal{F}$ makes an error prediction,

$$P_i^{adv} = P_i + \sigma_{P_i} \cdot \overrightarrow{d_{P_i}}, \tag{1}$$

where σ_{P_i} is the perturbation size for the i-th point in $\mathcal{P}$, i.e., P_i, and $\overrightarrow{d_{P_i}}$ is the unit perturbation direction. Formally, the perturbation $\sigma_{P_i} \cdot \overrightarrow{d_{P_i}}$ can be obtained by solving the below general-form equation, e.g., via gradient descent,

$$\min L_{mis}(\mathcal{F}, \mathcal{P}^{adv}, y) + \lambda_1 D(\mathcal{P}, \mathcal{P}^{adv}), \tag{2}$$

where $L_{mis}(\cdot, \cdot, \cdot)$ is the loss to promote misclassification, e.g., the negation of cross-entropy loss, $\mathcal{P}^{adv}$ is the adversarial point clouds consists of $\{P_i^{adv}\}_{i=1:n}$, $D(\cdot, \cdot)$ is the constraints on distortion to facilitate imperceptibility, and λ_1 is a weighting parameter. Here, all referenced attacks are untargeted unless specified otherwise.

To achieve imperceptibility, adversarial attack solutions typically apply geometric constraints such as l_2-norm, Chamfer distance, Hausdorff distance, and curvature [46] to limit perturbations. However, these methods neglect an essential aspect: uniformity, a characteristic whose alterations are readily detected by the human eye.

Uniformity-Preserving Adversarial Attacks. To mitigate changes in uniformity, we formulate a novel type of *Uniformity-Preserving Adversarial Attacks*, building upon Eq. (2) with an additional constraint,

$$\text{Uniformity}(\Omega(P_i^{adv})) = \text{Uniformity}(\Omega(P_i)), \tag{3}$$

where $\Omega(P_i)$ denotes the local region centered at P_i, represented discretely by a set of points, and Uniformity$(\cdot)$ is the operator employed to compute uniformity.

By preserving the uniformity of local regions, this type of adversarial attack achieves better imperceptibility compared to conventional methods.

4 Method

In this section, we start by introducing the concept and mathematical definition of flux, as well as its association with point cloud uniformity. Following that, we delve into how flux can be utilized to maintain point cloud uniformity. Finally, we demonstrate the application of these principles in implementing the flux-aware imperceptible attacks.

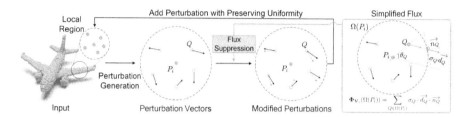

Fig. 2. Workflow of FLAT. Given an input point cloud, the model iteratively generates perturbation vectors for each local region. These vectors are then refined by measuring the simplified flux and applying suppression, thus ensuring the preservation of local shape uniformity.

4.1 Preliminary

Definition of Flux. Flux is commonly used to describe the magnitude of a vector fields flow through a surface [1]. Given a 3D vector field $\mathbf{F}$ and a simple, directed surface Ω, the flux of $\mathbf{F}$ through Ω is defined as the integral of the component of the field vector $\vec{f}$ at each point on the surface in the direction of the surface's normal vector,

$$\Phi_{\mathbf{F}}(\Omega) = \iint_{\Omega} \vec{f} \cdot \vec{n} \, \mathrm{d}S, \qquad (4)$$

where $\mathrm{d}S$ represents the integration over infinitesimal segments of the surface Ω, and $\vec{n}$ signifies the outward-pointing unit normal vector at each point on Ω.

Flux and Point Cloud Uniformity. For a closed surface, by analogizing the movement of points within the region to the flow of a field, the flux values can reflect a trend towards divergence and convergence within that region. Hence, employing the flux concept holds promise for managing uniformity alterations in point clouds during adversarial attacks.

4.2 Flux-Based Uniformity Preserving

For perturbation vectors $\mathbf{V} = \{\sigma_{P_i} \cdot \overrightarrow{d_{P_i}}\}_{i=1:n}$ calculated for an attack, our objective is to modify $\mathbf{V}$ in a way that, upon application, preserves the uniformity of the point clouds $\mathcal{P}$.

Given a subset of perturbation vectors within $\Omega(P_i)$, i.e., $\mathbf{V}_i = \{\sigma_Q \cdot \vec{d_Q} | Q \in \Omega(P_i)\}$, we employ the concept of flux to develop a metric that indicates uniformity changes in this subset. While a direct method would involve interpolating these vectors into a continuous field and integrating within $\Omega(P_i)$, as described in Eq. (4), the irregularity and sparsity of the vectors make managing this interpolation and integration challenging. Therefore, we opt for a simplified approach that concentrates simply on the discrete vectors.

Simplified Flux for Perturbation Vector Field. In the region $\Omega(P_i)$, we assume the field intensity to be constant along the perturbation vector $\vec{d_Q}$, i.e., σ_Q, originating from point Q, see the right part of Fig. 2. In this configuration, each perturbation vector intersects a point on the boundary surface. Therefore, the simplified flux can be viewed as the integration of vectors at these specific intersection points,

$$\begin{aligned}\Phi_{\mathbf{V}_i}(\Omega(P_i)) &= \sum_{Q \in \Omega(P_i)} \sigma_Q \cdot \vec{d_Q} \cdot \vec{n_Q}, \\ &= \sum_{Q \in \Omega(P_i)} A_Q \sqrt{1 + B_Q \cos^2(\theta_Q)},\end{aligned} \quad (5)$$

where

$$A_Q = \frac{\sigma_Q \cdot |\vec{d_Q}|}{r} \sqrt{r^2 - |\vec{P_iQ}|^2},$$

$$B_Q = \frac{|\vec{P_iQ}|^2}{r^2 - |\vec{P_iQ}|^2},$$

$\vec{n_Q}$ is the unit outward normal vector at the intersection point on the boundary of $\Omega(P_i)$, where $\vec{d_Q}$ meets, starting from point Q. The angle θ_Q is defined between $\vec{d_Q}$ and $\vec{P_iQ}$, with r representing the radius of the local shape.

Flux Suppression by Rotating Perturbation Vectors. To preserve the uniformity of point clouds during attacks, we suppress the simplified flux of the perturbation vector field. Specifically, we utilize gradient descent to adjust Θ,

$$\Theta' = \Theta - \alpha \frac{\partial \Phi_{\mathbf{V}_i}}{\partial \Theta},$$

$$\text{with} \quad \Theta = \{\theta_Q | Q \in \Omega(P_i)\}, \quad (6)$$

$$\frac{\partial \Phi_{\mathbf{V}_i}}{\partial \theta_Q} = -\frac{A_Q B_Q \sin(\theta_Q) \cos(\theta_Q)}{\sqrt{1 + B_Q \cos^2(\theta_Q)}},$$

where α denotes the step size for angle adjustment.

By suppressing flux in regions with significant values, point divergence is avoided. Essentially, by preventing divergence locally, convergence effects in other regions are also precluded. As a result, the uniformity of the point cloud is preserved after perturbation.

4.3 Flux-Aware Imperceptible Adversarial Attacks

Given a point cloud $\mathcal{P}$, we utilize farthest point sampling (FPS) to select m center points. Subsequently, local regions are constructed around these points, each with a radius of r. Following this, we generate initial perturbation for each point in the point cloud, refine their directions to maintain uniformity from a flux perspective, and then determine their magnitudes, see Fig. 2.

Generating Initial Perturbations. We employ IFGM [8] to generate initial perturbations with a consistent magnitude. It is noteworthy that alternative methods could also be employed to similar effect.

Refining Perturbation Directions with Flux. Within each local region, we calculate the simplified flux. If the region exhibits a substantial flux value, exceeding a threshold t, we adjust the directions of the perturbation vectors within it as described in Sect. 4.2 to preserve uniformity.

Determining Perturbation Magnitudes. With the refined perturbation directions established, we proceed to determine perturbation magnitudes following [25]. By iteratively executing the above three steps, FLAT creates highly imperceptible adversarial point clouds. Note that we initially considered optimizing both perturbation direction and magnitude simultaneously for flux suppression. However, this approach tended to favor reducing perturbation magnitude. Thus, we opt for separate optimization.

5 Experimental Results

5.1 Experimental Setup

Implementation. The FLAT framework is implemented in PyTorch [29]. We start by sampling $m = 20$ key points using the farthest point sampling (FPS) method. Around these points, we define local regions with a radius of $r = 0.1$ for simplified flux computation. We focus on regions demonstrating significant flux, particularly those whose flux values surpass the threshold t, determined as the median flux among all regions. To suppress deviations in uniformity, we adjust the perturbation directions using a step size of $\alpha = 0.02$, over a total of 8 cycles. All experiments are executed on a workstation equipped with dual 2.40 GHz CPUs, 128 GB of RAM, and four NVIDIA RTX 3090 GPUs.

Datasets. We adopt two public datasets for evaluation: ModelNet40 [49] and ShapeNet Part [53]. For ModelNet40, we designate 9,843 point clouds for training and 2,468 for testing. For ShapeNet Part, we allocate 14,007 point clouds for training and 2,874 for testing. Particularly, we uniformly sample 1,024 points from the surface of each object and rescale them into a unit cube following [30].

Victim Classifiers. We use four well-established deep point cloud classifiers as victim models, including PointNet [30], PointNet++ [31], DGCNN [45], and PointConv [48]. We train these models according to their original papers.

Table 1. Comparison of perturbation sizes needed by various methods to achieve their highest ASR in untargeted attacks.

Model	Attack	ModelNet40						ShapeNet Part					
		ASR (%)	CD (10^{-4})	HD (10^{-2})	l_2	GR	Curv (10^{-2})	EMD (10^{-2})	ASR (%)	CD (10^{-4})	HD (10^{-2})	l_2	GR	Curv (10^{-2})	EMD (10^{-2})
PointNet	PGD	100	7.155	5.025	0.981	0.302	1.624	2.315	100	13.172	17.068	1.569	0.521	3.679	3.358
	IFGM	100	4.039	5.565	0.789	0.314	0.775	0.864	100	3.328	10.269	0.785	0.408	0.619	0.556
	GeoA^3	100	4.646	0.497	1.307	0.121	0.396	2.319	100	7.531	1.444	2.655	0.146	0.465	4.104
	3d-Adv	100	6.115	4.372	0.863	0.250	1.215	1.410	100	15.659	5.495	1.787	0.279	4.006	3.693
	SI-Adv	100	2.768	2.595	0.731	0.220	0.271	0.725	100	3.435	3.692	0.881	0.233	0.441	0.825
	ITA	100	2.747	0.414	0.534	0.122	0.555	1.214	100	5.872	1.917	1.002	0.181	1.016	2.035
	Ours	100	**1.539**	**0.371**	**0.426**	**0.114**	**0.249**	**0.460**	100	**1.905**	**1.327**	**0.545**	**0.120**	**0.301**	**0.284**
PointNet++	PGD	100	5.182	0.636	0.753	0.125	1.508	2.146	100	10.090	3.257	1.342	0.215	1.508	1.584
	IFGM	100	3.558	1.162	0.640	0.146	1.149	1.454	100	4.532	3.608	0.548	0.220	1.824	1.584
	GeoA^3	100	6.579	0.461	1.615	0.114	0.762	2.919	100	7.701	0.847	2.875	0.105	1.375	4.176
	3d-Adv	100	8.915	3.564	1.535	0.141	1.288	2.784	100	9.564	3.778	2.014	0.197	3.021	3.590
	SI-Adv	100	9.399	2.377	1.422	0.185	1.061	2.684	100	9.266	3.233	1.535	0.203	**1.146**	2.811
	ITA	100	6.792	0.708	0.998	0.121	3.533	2.272	100	5.202	0.802	0.999	0.110	3.423	2.152
	Ours	100	**1.156**	**0.325**	**0.337**	**0.110**	**0.373**	**0.491**	100	**3.067**	**0.729**	**0.349**	**0.102**	1.701	**1.295**
DGCNN	PGD	100	19.968	5.098	1.933	0.267	4.924	4.785	100	63.556	27.557	5.224	0.511	7.275	9.233
	IFGM	100	15.791	12.391	1.622	0.363	2.849	3.777	100	19.623	26.040	2.069	0.504	4.954	4.387
	GeoA^3	100	7.566	0.546	1.585	0.119	0.741	3.083	100	27.612	3.748	5.798	0.199	1.695	7.502
	3d-Adv	100	10.345	3.807	3.589	0.227	5.997	6.685	100	21.553	8.531	2.258	0.282	5.119	4.628
	SI-Adv	100	7.146	1.691	1.087	0.143	0.666	2.495	100	11.685	**3.019**	1.772	0.160	2.054	3.646
	ITA	100	3.249	0.524	0.552	0.114	0.971	1.359	100	27.633	4.597	2.492	0.244	3.847	4.696
	Ours	100	**2.576**	**0.420**	**0.540**	**0.100**	**0.556**	**1.027**	100	**9.003**	4.693	**1.653**	**0.135**	**0.112**	**3.465**
PointConv	PGD	100	14.551	2.216	1.442	0.184	3.491	3.862	100	42.202	9.949	3.784	0.252	6.866	7.277
	IFGM	100	7.959	2.608	1.015	0.184	1.741	2.427	100	16.139	8.776	1.812	0.231	3.526	3.807
	GeoA^3	100	6.809	0.644	2.169	0.119	1.119	3.556	100	9.383	1.222	4.224	0.120	**1.190**	5.391
	3d-Adv	100	11.213	1.763	1.179	0.163	3.279	2.807	100	21.034	3.687	2.277	0.193	4.912	4.548
	SI-Adv	100	6.060	1.784	0.977	0.144	0.576	2.081	100	11.281	3.500	1.741	0.165	1.949	3.514
	ITA	100	5.539	0.480	0.833	0.111	1.904	1.971	100	9.082	1.452	1.375	0.146	3.645	2.925
	Ours	100	**2.139**	**0.344**	**0.478**	**0.107**	**0.301**	**0.975**	100	**5.034**	**1.194**	**0.817**	**0.117**	1.288	**1.186**

Baseline Attack Methods. We assess the effectiveness of our approach by comparing it with six state-of-the-art techniques: the gradient-based IFGM [8] and PGD [8], the direction-based SI-Adv [16] and ITA [25], and the optimization-based methods GeoA^3 [46] and 3d-Adv [50]. This diverse set of attacks provides a robust baseline to validate the effectiveness of our approach.

Evaluation Setting and Metrics. To ensure fair comparisons, we configure each attack method to attain its maximum attack success rate (ASR), which is the percentage of adversarial point clouds that successfully fool the victim model. In this maximal adversarialness condition [37,42], we assess the imperceptibility of attacks. Specifically, we employ six widely recognized metrics: Chamfer distance (CD) [9], Hausdorff distance (HD) [35], l_2-norm (l_2), curvature (Curv), geometric regularity (GR) [46], and earth mover's distance (EMD) [32]. Unless explicitly mentioned, all discussions regarding attack results are assumed to be about untargeted attacks.

5.2 Performance Comparison and Analysis

Performance of Untargeted Attacks. We evaluate the ASR and imperceptibility of various adversarial attack methods in an untargeted setting on the ModelNet40 and ShapeNet Part datasets, with results detailed in Table 1. It is observed that all these adversarial attack methods can achieve 100% ASR. However, methods like PGD, which impose larger perturbations in a single iteration, result in greater distortion and thus underperform across most metrics due to their lack of subtlety. In contrast, direction-based methods such as SI-Adv and ITA demonstrate lower distortion, showcasing their efficacy in maintaining attack stealth. Particularly, by modulating the flux of perturbation vector fields during attacks, our FLAT outperforms these state-of-the-art methods in the majority of metrics, confirming its effectiveness and superiority.

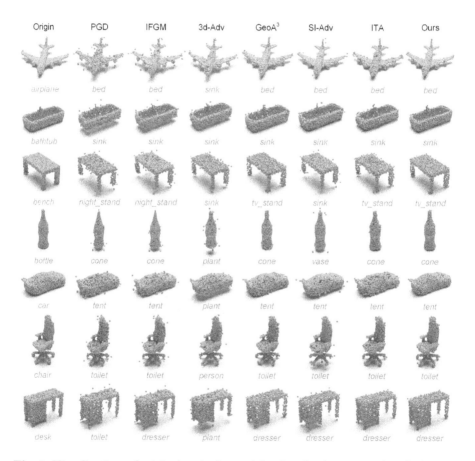

Fig. 3. Visualizations of original and adversarial point clouds generated to fool PointNet on ModelNet40 by various adversarial attack methods.

Visualization of Adversarial Point Clouds. To vividly showcase the enhanced imperceptibility afforded by our approach, we present visualizations of adversarial examples crafted using diverse attack methodologies for seven distinct classes from ModelNet40, designed to deceive PointNet, as depicted in Fig. 3. The adversarial point clouds generated by PGD and IFGM reveal pronounced outliers due to their relatively lax deformation constraints. In contrast, GeoA^3, which incorporates geometric constraints such as curvature, tends to produce samples with fewer outliers. SI-Adv and ITA, which utilize geometric characteristics for perturbation along tangential and normal vectors respectively, also result in fewer discernible outliers. Particularly, by restricting the flux of the perturbation vector field during the attack process, FLAT generates adversarial point clouds with surfaces that are noticeably more uniform and nearly devoid of outliers, thereby affirming the effectiveness and superiority of our method in terms of imperceptibility.

Table 2. Comparison of perturbation sizes needed by various methods to achieve their highest ASR in targeted attacks against PointNet and DGCNN on ModelNet40.

Attack	PointNet							DGCNN						
	ASR (%)	CD (10^{-4})	HD (10^{-2})	l_2	GR	Curv (10^{-2})	EMD (10^{-2})	ASR (%)	CD (10^{-4})	HD (10^{-2})	l_2	GR	Curv (10^{-2})	EMD (10^{-2})
PGD	100	24.428	2.042	2.630	0.180	1.787	5.548	100	65.040	16.099	9.644	0.416	2.528	14.893
IFGM	100	3.832	1.002	0.676	0.140	0.905	1.534	100	8.912	3.633	1.142	0.199	0.694	2.976
GeoA^3	100	6.501	**0.620**	4.029	0.100	0.314	4.213	100	63.106	10.464	6.272	0.318	1.769	9.234
3d-Adv	100	3.248	1.594	0.614	0.165	0.282	1.440	100	7.266	3.066	**0.902**	0.172	0.466	2.939
SI-Adv	100	4.185	1.795	1.054	0.144	0.301	1.594	100	13.974	12.761	2.141	0.184	1.309	3.511
ITA	100	38.081	1.889	1.969	0.186	2.011	4.055	100	48.321	1.623	1.542	0.193	2.351	3.553
Ours	100	**2.344**	1.044	**0.594**	**0.091**	**0.212**	**0.793**	100	**6.198**	**0.617**	1.533	**0.109**	**0.357**	**2.668**

Table 3. Comparison of uniformity changes measured by SDM [14] during attacks on four classifiers across ModelNet40 and ShapeNet Part.

Attack	ModelNet40				ShapeNet Part			
	PointNet	PointNet++	DGCNN	PointConv	PointNet	PointNet++	DGCNN	PointConv
IFGM	0.1493	0.1766	0.5049	0.2980	0.1201	0.2849	1.6553	0.2029
GeoA^3	0.1956	0.2723	0.2799	0.3079	0.3823	0.3540	2.0009	0.4162
SI-Adv	0.1211	0.4115	0.3599	0.3548	0.1072	0.3488	1.1416	0.3776
ITA	0.1207	0.1756	0.1649	0.2698	0.2995	0.1778	0.4937	0.2465
Ours	**0.0972**	**0.0785**	**0.1550**	**0.1420**	**0.0580**	**0.1480**	**0.2880**	**0.0058**

Performance of Targeted Attacks. To further corroborate the superiority of our approach, we extend our evaluation to the targeted attack setting. Specifically, we randomly select 25 instances from each of the 10 categories in the ModelNet40 test set. For each instance, we craft adversarial examples targeting the remaining nine classes, resulting in a total of 2250 targeted attack point clouds, following the methodology of [50]. The results of these targeted attacks on PointNet and DGCNN are summarized in Table 2. It is observed that targeted

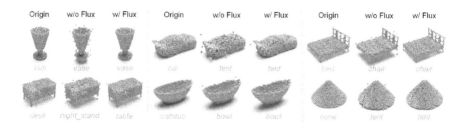

Fig. 4. Visualization of local regions of adversarial point clouds generated by FLAT with and without applying flux-based uniformity preservation in attacking PointNet.

Table 4. Comparison of our FLAT solution with a modified version, namely FLAT-SDM, that utilizes a uniformity-preserving constraint instead of flux adjustment.

Attack	ASR (%)	CD (10^{-4})	HD (10^{-2})	l_2	GR	Curv (10^{-2})	EMD (10^{-2})	SDM [14]	Iteration
FLAT-SDM	100	2.502	1.713	0.608	0.214	0.452	0.565	**0.084**	15
Ours	100	**1.539**	**0.371**	**0.426**	**0.114**	**0.249**	**0.460**	0.097	8

attacks necessitate larger perturbations compared to the untargeted attacks presented in Table 1. Although ITA performed well in the untargeted setting, its relatively fixed perturbation pattern led to larger CD values in the targeted scenario. Our method remains consistent, outperforming other techniques across the majority of metrics, which further validates its effectiveness and superiority.

5.3 Ablation Studies and Other Analysis

Importance of Flux in Uniformity Preserving. To underscore the essential role of flux in preserving the uniformity of point clouds during adversarial attacks, we compare the uniformity changes induced by our method with those brought by state-of-the-art methods. Specifically, we employ the symmetric density metric (SDM) introduced in [14] to measure differences in nearest neighbor counts and average distances between original and adversarial point clouds. The results presented in Table 3 demonstrate that our method is more effective in preserving uniformity.

To better illustrate the effectiveness of our method in preserving uniformity, we present visualizations of local regions post-attack both with and without flux suppression. As depicted in Fig. 4, the center point is highlighted in red, while the surrounding points within its local region are marked in yellow. The adversarial samples generated with flux consideration exhibit only minimal deviations, closely resembling the original point cloud structure. Conversely, those produced without considering flux demonstrate noticeable deformations, emphasizing the crucial role of flux awareness in preserving point cloud uniformity.

Flux-Based vs. SDM-Based Uniformity Preserving. To further demonstrate the importance of our flux-based solution, we compare it with a variant of FLAT, named FLAT-SDM, that incorporates an SDM constraint [14] in the initial perturbation generation phase to create initial directions with better awareness of uniformity preservation, yet does not utilize flux for perturbation direction adjustment. The results in Table 4 indicate that directly applying constraints can indeed enhance uniformity. Nevertheless, the strong nature of the constraint complicates the execution of successful attacks. As a result, it requires 15 iterations, as opposed to the original 8, with these additional iterations leading to a deterioration in other imperceptibility metrics.

Generalization of Flux-Based Uniformity Preserving. To evaluate the generalizability of our flux-based uniformity preservation approach, we incorporate it into four established iterative adversarial attack methods: PGD, IFGM [8], SI-Adv [16], and ITA [25]. As demonstrated in Table 5, these methods, when augmented with our flux-aware technique, exhibit marked improvements across most performance metrics under identical parameter settings. To illustrate the impact of flux-based uniformity preservation more vividly, we visualize the adversarial point clouds generated with and without this module in Fig. 5. It is evident that the integration of flux-based uniformity preservation significantly enhances the imperceptibility of the attack, exemplified by a substantial reduction in outliers. Consequently, we affirm the broad applicability of our flux-based uniformity preservation strategy.

Table 5. Comparison of perturbation sizes required by different methods, both with and without flux-based iformity preservation, to achieve their highest ASR in untargeted attacks against PointNet and DGCNN on ModelNet40.

Attack	PointNet							DGCNN						
	ASR (%)	CD (10^{-4})	HD (10^{-2})	l_2	GR	Curv (10^{-2})	EMD (10^{-2})	ASR (%)	CD (10^{-4})	HD (10^{-2})	l_2	GR	Curv (10^{-2})	EMD (10^{-2})
PGD	100	7.155	5.025	0.981	0.302	1.624	2.315	100	19.968	5.098	1.933	0.267	4.924	4.785
PGD+flux	100	**6.259**	**3.721**	**0.894**	**0.265**	**1.529**	**2.220**	100	**18.100**	**3.285**	**1.799**	**0.221**	**4.678**	**4.549**
IFGM	100	4.039	5.565	0.789	0.314	0.775	0.864	100	15.791	12.391	1.622	0.363	2.849	3.777
IFGM+flux	100	**2.944**	**4.036**	**0.662**	**0.276**	**0.648**	**0.712**	100	**12.270**	**9.869**	**1.167**	**0.244**	**1.982**	**2.287**
SI-Adv	100	2.768	2.595	0.731	0.220	0.271	0.725	100	7.146	1.691	1.087	0.143	0.666	2.495
SI-Adv+flux	100	**2.165**	**1.914**	**0.580**	**0.187**	**0.202**	**0.609**	100	**7.055**	**1.676**	**1.037**	**0.126**	**0.658**	**2.309**
ITA	100	2.747	0.414	0.534	0.122	0.555	1.214	100	2.940	0.524	0.552	0.114	0.971	1.359
ITA+flux	100	**2.253**	**0.579**	**0.506**	**0.119**	**0.534**	**1.022**	100	**2.845**	**0.422**	**0.431**	**0.103**	**0.871**	**1.207**

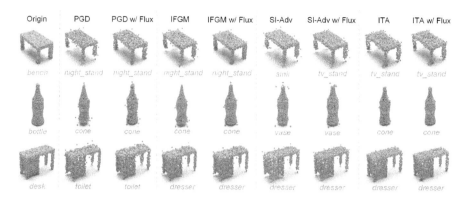

Fig. 5. Visualization of adversarial point clouds generated by various attack methods in attacking PointNet, with and without integrating flux-based uniformity preservation.

Impact of Initial Perturbation Directions. We assess the influence of initial perturbation directions on the effectiveness of adversarial attacks by considering three alternatives: (1) gradient descent direction as implemented in IFGM [8]; (2) tangent plane direction as utilized in SI-Adv [16]; (3) normal vector direction as adopted in ITA [25]. The comparative analysis, depicted in Fig. 6, reveals only slight variations in the resulting distortions across these initial directions, underscoring our method's robustness. Notably, the gradient descent direction, as derived from IFGM, yield the best results; hence, we select it in our solution.

Impact of Local Region Number and Size. We evaluate the impact of local region number and size on the performance of our flux-based uniformity preservation. As illustrated in Fig. 7(a), augmenting the number of local regions corresponds to reduced l_2 and EMD metrics, indicating enhanced flux suppression. This trend stabilizes upon reaching 20 regions. As for the simplified flux values, we do not observe any significant variation. Furthermore, we analyze the influence of the radius in Fig. 7(b). It reveals an initial reduction in l_2 and EMD metrics with radius expansion, suggesting a wider influence on point suppression.

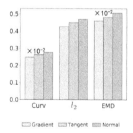
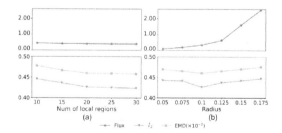

Fig. 6. Comparison of FLAT perturbation sizes with various initial directions.

Fig. 7. Comparison of flux values and perturbation sizes for FLAT across varying numbers of local regions and different radii.

Table 6. Comparison of perturbation size and running time for FLAT in attacking PointNet using both formal and simplified flux methods.

Flux	ASR (%)	CD (10^{-4})	HD (10^{-2})	l_2	GR	Curv (10^{-2})	EMD (10^{-2})	Time (s)
Formal flux	100	1.73	1.03	0.41	0.20	0.27	0.34	150.07
Ours	100	1.54	0.37	0.43	0.11	0.25	0.46	2.02

This trend, however, inverts at a radius exceeding 0.1, where metrics begin to climb again. Notably, flux magnitude continuously grows with the inclusion of more points. Hence, we adopt 20 local regions with radius of 0.1 in our solution.

Formal Flux Based on Continuous Field. To assess the performance impact of our simplified flux approximation, we employ a linear radial basis function to interpolate a continuous field within each local region. This interpolation is based on the Euclidean distances between virtual points within the region and the fixed discretized points. As reported in Table 6, the results from attacking PointNet indicate that our simplified version performs comparably to the formal flux based on a continuous field. Moreover, our method incurs significantly lower computational costs than the formal flux. These findings collectively validate the efficiency and effectiveness of our simplified flux.

Analysis on Undefendability and Transferability. To evaluate the resilience of our method against different defense mechanisms, we compare it with other adversarial attack techniques targeting PointNet, under two defense strategies: statistical outlier removal (SOR) and DUP-Net [59]. The results, depicted in Fig. 8(a,b), show that all attack methods, FLAT included, experience reduced ASR when countered with these defenses, with DUP-Net causing a more significant decrease. While methods like IFGM and PGD achieve higher ASRs, they do so at the cost of greater distortion. Conversely, our approach results in lower distortion while still achieving comparatively high ASR, showcasing its superior ability to withstand defenses.

Table 7. The average time required by different methods to generate an adversarial example to attack PointNet on ModelNet40

Attack	PGD	IFGM	GeoA^3	3d-Adv	SI-Adv	ITA	Ours
Time (s)	0.440	0.322	66.093	26.815	1.059	44.529	2.020

Additionally, we assess the transferability of FLAT and other methods by launching attacks on one classifier and testing the generated adversarial point clouds on other classifiers. The outcomes, showcased in Fig. 8(c,d), highlight a notable decline in ASR after transfer, with all methods dropping below 50%. Our method not only shows a competitive ASR but also the least amount of

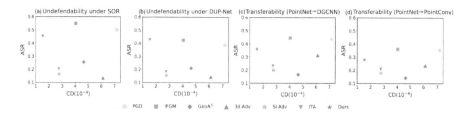

Fig. 8. Visualization of (a-b) the undefendability of different methods under the SOR and DUP-Net defense [59], and (c-d) the transferability of different methods from PointNet to DGCNN and from PointNet to PointConv.

distortion, validating its capability to strike a balance between successful attacks and maintaining imperceptibility.

Analysis on Time Complexity. To assess the time efficiency of our method, we report the average minimal time required by different methods to successfully generate an adversarial example to attack PointNet on ModelNet40, as shown in Table 7. The results demonstrate that our method's time complexity is competitive compared to existing approaches.

6 Conclusion

In this paper, we have proposed FLAT, a novel flux-aware attack framework for generating imperceptible adversarial point clouds. The rationale involves modeling adversarial perturbations as a vector field and subsequently suppressing their flux within localized regions to preserve point cloud uniformity. Extensive experiments validate that FLAT generates adversarial point clouds with enhanced imperceptibility. We hope our work can inspire further research into enhancing the imperceptibility of 3D adversarial attacks. In the future, we plan to delve deeper into field-based factors to continue advancing the imperceptibility.

Acknowledgements. This work was supported in part by National Natural Science Foundation of China (62102105, U20B2046, 62025207), Guangdong Basic and Applied Basic Research Foundation (2022A1515011501, 2022A1515010138, 2024A1515012064), Guangzhou Sci-Tech Program (SL2022A04J01112), Guangzhou University Program (YJ2023048), and Academician Binxing Fang's Specialized Class.

References

1. Aris, R.: Vectors, tensors and the basic equations of fluid mechanics. Courier Corporation (2012)
2. Berger, M., et al.: A survey of surface reconstruction from point clouds. In: Computer Graphics Forum, vol. 36, pp. 301–329 (2017)
3. Bronstein, M.M., Bruna, J., LeCun, Y., Szlam, A., Vandergheynst, P.: Geometric deep learning: going beyond euclidean data. IEEE Signal Process. Mag. **34**(4), 18–42 (2017)

4. Cai, G., He, L., Zhou, M., Alhumade, H., Hu, D.: Learning smooth representation for unsupervised domain adaptation. IEEE TNNLS (2021)
5. Carlini, N., Wagner, D.: Towards evaluating the robustness of neural networks. In: IEEE Symposium on Security and Privacy, pp. 39–57 (2017)
6. Chen, L., Zhang, Q.: Ddgcn: graph convolution network based on direction and distance for point cloud learning. Vis. Comput. **39**(3), 863–873 (2023)
7. Chen, Y., Peng, W., Tang, K., Khan, A., Wei, G., Fang, M.: Pyrapvconv: efficient 3d point cloud perception with pyramid voxel convolution and sharable attention. Comput. Intell. Neurosci. **2022**(1), 2286818 (2022)
8. Dong, X., Chen, D., Zhou, H., Hua, G., Zhang, W., Yu, N.: Self-robust 3d point recognition via gather-vector guidance, In: CVPR. pp. 11513–11521 (2020)
9. Fan, H., Su, H., Guibas, L.J.: A point set generation network for 3d object reconstruction from a single image. In: CVPR, pp. 605–613 (2017)
10. Fang, X., Liu, D., Zhou, P., Hu, Y.: Multi-modal cross-domain alignment network for video moment retrieval. IEEE TMM **25**, 7517–7532 (2022)
11. Goodfellow, I.J., Shlens, J., Szegedy, C.: Explaining and harnessing adversarial examples. In: ICLR (2015)
12. Guo, Y., Wang, H., Hu, Q., Liu, H., Liu, L., Bennamoun, M.: Deep learning for 3d point clouds: a survey. IEEE TPAMI **43**(12), 4338–4364 (2020)
13. Han, X.F., Jin, J.S., Wang, M.J., Jiang, W., Gao, L., Xiao, L.: A review of algorithms for filtering the 3d point cloud. Signal Process. Image Commun. **57**, 103–112 (2017)
14. He, Y., Ren, X., Tang, D., Zhang, Y., Xue, X., Fu, Y.: Density-preserving deep point cloud compression. In: CVPR, pp. 2333–2342 (2022)
15. Huang, H., Li, D., Zhang, H., Ascher, U., Cohen-Or, D.: Consolidation of unorganized point clouds for surface reconstruction. ACM TOG **28**(5), 1–7 (2009)
16. Huang, Q., Dong, X., Chen, D., Zhou, H., Zhang, W., Yu, N.: Shape-invariant 3d adversarial point clouds. In: CVPR, pp. 15335–15344 (2022)
17. Ioannidou, A., Chatzilari, E., Nikolopoulos, S., Kompatsiaris, I.: Deep learning advances in computer vision with 3d data: a survey. ACM Comput. Surv. **50**(2), 1–38 (2017)
18. Kim, J., Hua, B.S., Nguyen, T., Yeung, S.K.: Minimal adversarial examples for deep learning on 3d point clouds. In: ICCV, pp. 7797–7806 (2021)
19. Landrieu, L., Simonovsky, M.: Large-scale point cloud semantic segmentation with superpoint graphs. In: CVPR, pp. 4558–4567 (2018)
20. LeCun, Y., Bengio, Y., Hinton, G.: Deep learning. Nature **521**(7553), 436–444 (2015)
21. Lee, K., Chen, Z., Yan, X., Urtasun, R., Yumer, E.: Shapeadv: Generating shape-aware adversarial 3d point clouds. arXiv preprint arXiv:2005.11626 (2020)
22. Li, B.: 3d fully convolutional network for vehicle detection in point cloud. In: IROS, pp. 1513–1518 (2017)
23. Li, Y., Bu, R., Sun, M., Wu, W., Di, X., Chen, B.: Pointcnn: convolution on χ-transformed points. In: NeurIPS, pp. 820–830 (2018)
24. Li, Y., et al.: Deep learning for lidar point clouds in autonomous driving: A review. IEEE TNNLS **32**(8), 3412–3432 (2020)
25. Liu, D., Hu, W.: Imperceptible transfer attack and defense on 3d point cloud classification. IEEE TPAMI **45**(4), 4727–4746 (2022)
26. Liu, D., Yu, R., Su, H.: Extending adversarial attacks and defenses to deep 3d point cloud classifiers. In: ICIP, pp. 2279–2283 (2019)
27. Luo, S., Hu, W.: Score-based point cloud denoising. In: ICCV, pp. 4583–4592 (2021)

28. Maturana, D., Scherer, S.: Voxnet: a 3d convolutional neural network for real-time object recognition. In: IROS, pp. 922–928 (2015)
29. Paszke, A., et al.: Pytorch: an imperative style, high-performance deep learning library. In: NeurIPS, vol. 32 (2019)
30. Qi, C.R., Su, H., Mo, K., Guibas, L.J.: Pointnet: deep learning on point sets for 3d classification and segmentation. In: CVPR, pp. 652–660 (2017)
31. Qi, C.R., Yi, L., Su, H., Guibas, L.J.: Pointnet++: deep hierarchical feature learning on point sets in a metric space. In: NeurIPS, pp. 5105–5114 (2017)
32. Rubner, Y., Tomasi, C., Guibas, L.J.: The earth mover's distance as a metric for image retrieval. IJCV **40**, 99–121 (2000)
33. Shi, W., Rajkumar, R.: Point-gnn: graph neural network for 3d object detection in a point cloud. In: CVPR, pp. 1711–1719 (2020)
34. Sun, Y., Miao, Y., Chen, J., Pajarola, R.: Pgcnet: patch graph convolutional network for point cloud segmentation of indoor scenes. Vis. Comput. **36**(10), 2407–2418 (2020)
35. Taha, A.A., Hanbury, A.: Metrics for evaluating 3d medical image segmentation: analysis, selection, and tool. BMC Med. Imaging **15**(1), 1–28 (2015)
36. Tang, K., et al.: Reppvconv: attentively fusing reparameterized voxel features for efficient 3d point cloud perception. Vis. Comput. **39**(11), 5577–5588 (2023)
37. Tang, K., et al.: Manifold constraints for imperceptible adversarial attacks on point clouds. In: AAAI, vol. 38, pp. 5127–5135 (2024)
38. Tang, K., Lou, T., Peng, W., Chen, N., Shi, Y., Wang, W.: Effective single-step adversarial training with energy-based models. IEEE TETCI (2024). https://doi.org/10.1109/TETCI.2024.3378652
39. Tang, K., et al.: Decision fusion networks for image classification. IEEE TNNLS (2022). https://doi.org/10.1109/TNNLS.2022.3196129
40. Tang, K., et al.: Rethinking perturbation directions for imperceptible adversarial attacks on point clouds. IEEE Internet Things J. **10**(6), 5158–5169 (2023)
41. Tang, K., et al.: Normalattack: curvature-aware shape deformation along normals for imperceptible point cloud attack. Sec. Commun. Netw. **2022**(1), 1186633 (2022)
42. Tang, K., et al.: Deep manifold attack on point clouds via parameter plane stretching. In: AAAI, vol. 37, pp. 2420–2428 (2023)
43. Thomas, H., Qi, C.R., Deschaud, J.E., Marcotegui, B., Goulette, F., Guibas, L.J.: Kpconv: flexible and deformable convolution for point clouds. In: ICCV, pp. 6411–6420 (2019)
44. Wang, R., Peethambaran, J., Chen, D.: Lidar point clouds to 3-d urban models: a review. IEEE J. Selected Topics Appli. Earth Observat. Remote Sensing **11**(2), 606–627 (2018)
45. Wang, Y., Sun, Y., Liu, Z., Sarma, S.E., Bronstein, M.M., Solomon, J.M.: Dynamic graph cnn for learning on point clouds. ACM Trans. Graph. **38**(5), 1–12 (2019)
46. Wen, Y., Lin, J., Chen, K., Chen, C.P., Jia, K.: Geometry-aware generation of adversarial point clouds. IEEE TPAMI **44**(6), 2984–2999 (2020)
47. Wicker, M., Kwiatkowska, M.: Robustness of 3d deep learning in an adversarial setting. In: CVPR, pp. 11767–11775 (2019)
48. Wu, W., Qi, Z., Fuxin, L.: Pointconv: deep convolutional networks on 3d point clouds. In: CVPR, pp. 9621–9630 (2019)
49. Wu, Z., et al.: 3d shapenets: a deep representation for volumetric shapes. In: CVPR, pp. 1912–1920 (2015)
50. Xiang, C., Qi, C.R., Li, B.: Generating 3d adversarial point clouds. In: CVPR, pp. 9136–9144 (2019)

51. Xu, M., Ding, R., Zhao, H., Qi, X.: Paconv: position adaptive convolution with dynamic kernel assembling on point clouds. In: CVPR (2021)
52. Yang, J., Zhang, Q., Fang, R., Ni, B., Liu, J., Tian, Q.: Adversarial attack and defense on point sets. arXiv preprint arXiv:1902.10899 (2019)
53. Yi, L., et al.: A scalable active framework for region annotation in 3d shape collections. ACM TOG **35**(6), 1–12 (2016)
54. Zhang, J., Jiang, C., Wang, X., Cai, M.: Td-net: topology destruction network for generating adversarial point cloud. In: ICIP, pp. 3098–3102 (2021)
55. Zhao, H., Jiang, L., Fu, C.W., Jia, J.: Pointweb: enhancing local neighborhood features for point cloud processing. In: CVPR, pp. 5565–5573 (2019)
56. Zhao, Y., Wu, Y., Chen, C., Lim, A.: On isometry robustness of deep 3d point cloud models under adversarial attacks. In: CVPR, pp. 1201–1210 (2020)
57. Zheng, T., Chen, C., Yuan, J., Li, B., Ren, K.: Pointcloud saliency maps. In: ICCV, pp. 1598–1606 (2019)
58. Zhou, H., et al.: Lg-gan: Label guided adversarial network for flexible targeted attack of point cloud based deep networks. In: CVPR, pp. 10356–10365 (2020)
59. Zhou, H., Chen, K., Zhang, W., Fang, H., Zhou, W., Yu, N.: Dup-net: denoiser and upsampler network for 3d adversarial point clouds defense. In: ICCV, pp. 1961–1970 (2019)

MMBench: Is Your Multi-modal Model an All-Around Player?

Yuan Liu[1], Haodong Duan[1], Yuanhan Zhang[3], Bo Li[3], Songyang Zhang[1], Wangbo Zhao[4], Yike Yuan[5], Jiaqi Wang[1], Conghui He[1], Ziwei Liu[3(✉)], Kai Chen[1(✉)], and Dahua Lin[1,2(✉)]

[1] Shanghai AI Laboratory, Shanghai, China
[2] The Chinese University of Hong Kong, Hong Kong, China
[3] S-Lab, Nanyang Technological University, Singapore, Singapore
ziwei.liu@ntu.edu.sg
[4] National University of Singapore, Singapore, Singapore
[5] Zhejiang University, Hangzhou, China
https://liuziwei7.github.io/

Abstract. Large vision-language models (VLMs) have recently achieved remarkable progress, exhibiting impressive multimodal perception and reasoning abilities. However, effectively evaluating these large VLMs remains a major challenge, hindering future development in this domain. Traditional benchmarks like VQAv2 or COCO Caption provide quantitative performance measurements but lack fine-grained ability assessment and robust evaluation metrics. Meanwhile, subjective benchmarks, such as OwlEval, offer comprehensive evaluations of a model's abilities by incorporating human labor, which is not scalable and may display significant bias. In response to these challenges, we propose MMBench, a bilingual benchmark for assessing the multi-modal capabilities of VLMs. MMBench methodically develops a comprehensive evaluation pipeline, primarily comprised of the following key features: 1. MMBench is meticulously curated with well-designed quality control schemes, surpassing existing similar benchmarks in terms of the number and variety of evaluation questions and abilities; 2. MMBench introduces a rigorous CircularEval strategy and incorporates large language models to convert free-form predictions into pre-defined choices, which helps to yield accurate evaluation results for models with limited instruction-following capabilities. 3. MMBench incorporates multiple-choice questions in both English and Chinese versions, enabling an apples-to-apples comparison of VLMs' performance under a bilingual context. To summarize, MMBench is a systematically designed **objective** benchmark for a **robust** and **holistic** evaluation of vision-language models. We hope MMBench will assist

Y. Liu, H. Duan, Y. Zhang, B. Li and S. Zhang—Equal Contribution.
H. Duan—Project Lead.

Supplementary Information The online version contains supplementary material available at https://doi.org/10.1007/978-3-031-72658-3_13.

the research community in better evaluating their models and facilitate future progress in this area. MMBench has been supported in VLMEvalKit (https://github.com/open-compass/VLMEvalKit).

1 Introduction

Recently, notable progress has been achieved within the realm of large language models (LLMs). For instance, the latest LLMs, such as OpenAI's ChatGPT and GPT-4 [26], have demonstrated remarkable reasoning capabilities that are comparable to, and in some cases, even surpass human capabilities. Drawing inspiration from these promising advancements in LLMs, large vision-language models (LVLMs) have also experienced a revolutionary transformation. Notable works, such as GPT-4v [26], Gemini-Pro-V [31] and LLaVA [22], have demonstrated enhanced capabilities in image content recognition and reasoning within the domain of vision-language models, exhibiting superior performance compared to earlier works. Nevertheless, a large proportion of the early studies [14, 22, 41] tend to emphasize showcasing qualitative examples rather than undertaking comprehensive and quantitative experiments to thoroughly assess their model performance. The lack of quantitative assessment poses a considerable challenge for comparing various models. Recent studies have primarily explored two approaches to conduct quantitative evaluations. The first approach involves utilizing existing public datasets [7, 15] for objective evaluation, while the second approach employs human annotators [35, 36] to perform subjective evaluations. However, it is worth noting that both approaches exhibit some inherent limitations.

A multitude of public datasets, such as VQAv2 [15], COCO Caption [7], GQA [18], and OK-VQA [24], have long served as valuable resources for the quantitative evaluation of VLMs. These datasets offer **objective** metrics, including accuracy, BLEU, CIDEr, *etc*. However, when employed to evaluate more advanced LVLMs, these benchmarks encounter the following challenges. 1. **False Negative Issues**: Most existing evaluation metrics require an exact match between the prediction and the reference target, leading to potential limitations. For instance, in the VQA task, even if the prediction is "bicycle" while the reference answer is "bike", the existing metric would assign a negative score to the prediction, resulting in a considerable number of false-negative samples. 2. **Lacking Finegrained Analysis**: Current public datasets predominantly focus on evaluating a model's performance on specific tasks, offering limited insights into the fine-grained capabilities of these models. Thus, they provide insufficient feedback regarding potential directions for future improvements.

Given the aforementioned challenges, recent studies, such as OwlEval [36] and LVLM-eHub [35] propose human-involved **subjective** evaluation strategies, aiming to address existing methods' limitations by incorporating human judgment and perception in the evaluation process. OwlEval artificially constructs 82 open-ended questions based on images from public datasets and employs

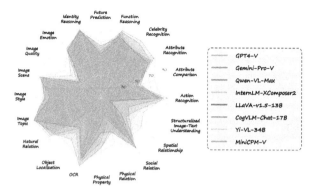

Fig. 1. Results of eight representative large vision-language models (VLMs) across the 20 ability dimensions defined in MMBench-test.

human annotators to assess the quality of VLM predictions. Similarly, inspired by FastChat [39], LVLM-eHub develops an online platform where two models are prompted to answer the same question related to an image. A participant then compares the answers provided by two models. Subjective evaluation strategies offer numerous benefits. These include **accurate matching**, where humans can precisely correlate a prediction with the target, even when expressed in different words, and **comprehensive assessment**, where humans are inclined to juxtapose two predictions considering multiple facets. The ultimate score is computed as the mean score across diverse abilities, facilitating a holistic evaluation of the model's capabilities.

While subjective evaluation allows for a more comprehensive assessment, it also introduces new challenges. Firstly, human evaluations are inherently biased. Consequently, it becomes challenging to reproduce the results presented in a work with a different group of annotators. Also, existing subjective evaluation strategies face scalability issues. Employing annotators for model evaluation after each experiment is an expensive endeavor. Moreover, small evaluation datasets can result in statistical instability. To ensure a robust evaluation, collecting more data is necessary, which in turn demands a significant amount of human labor.

In light of the challenges faced by conventional objective and subjective benchmarks, we propose **MMBench**, a systematically designed objective benchmark to robustly evaluate different abilities of LVLMs. MMBench contains over 3000 multiple-choice questions covering 20 ability dimensions, such as object localization and social reasoning, for evaluating VLMs. Each ability dimension encompasses over 125 questions, with the quantity of questions per ability maintained at a roughly equal level. The distribution facilitates a balanced and thorough assessment. Since some existing VLMs have limited instruction-following capability and cannot directly output choice labels (A, B, C, *etc.*) for multi-choice questions, the evaluation based on exact matching may not yield accurate and reasonable conclusions. To reduce the number of false-negative samples during answer matching, we employ LLMs to match a model's prediction to candidate choices and then output the label for the matched choice. We conduct a comparison between LLM-based choice matching and human evaluations, and discovered that GPT-4 can accurately match human assessments in 91.5% of cases, demonstrating its good alignment and robustness as a choice extractor. To make the evaluation more robust, we propose a novel evaluation strategy, named **CircularEval** (details in Sect. 4.3). We comprehensively evaluate 21 well-known vision-language models (across different model architectures and scales) on MMBench and report their performance on different ability dimensions. The performance ranking offers a direct comparison between various models and provides valuable feedback for future optimization. In summary, our main contributions are three-fold:

- **Systematically-constructed Dataset**: To thoroughly evaluate the capacity of a VLM, we carefully curated a dataset comprising a total of 3,217 meticulously selected questions, covering a diverse spectrum of 20 fine-grained skills.
- **Robust Evaluation**: We introduce a novel circular evaluation strategy (CircularEval) to improve the robustness of our evaluation process. After that, GPT-4 is employed to match the model's prediction with given choices, which can successfully extract choices even from predictions of a VLM with poor instruction-following capability.
- **Analysis and Observations**: We perform a comprehensive evaluation of a series of well-known vision-language models using MMBench, and the evaluation results can provide insights to the research community for future improvement.

2 Related Work

2.1 Multimodal Datasets

Large-scale VLMs have shown promising potential in multimodal tasks such as complex scene understanding and visual question answering. Though qualitative results so far are encouraging, quantitative evaluation is of great necessity to systematically evaluate and compare the abilities of different VLMs. Recent works have evaluated their models on numerous existing public multimodality datasets. COCO Caption [7], Nocaps [2], and Flickr30k [38] provide human-generated image captions and the corresponding task is to describe the image content in the form of text. Visual question answering datasets, such as GQA [18], OK-VQA [24], VQAv2 [15], and Vizwiz [16], contain question-answer pairs related to the given image, used to measure the model's ability on visual perception and reasoning. Some datasets provide more challenging question-answering scenarios by incorporating additional tasks. For example, TextVQA [30] proposes questions about text shown in the image, thus involving the OCR task in question-answering. ScienceQA [23] focuses on scientific topics, requiring the model to integrate commonsense into reasoning. Youcook2 [40] replaces images with video clips, introducing additional temporal information. However, the aforementioned datasets are designed on specific domains, and can only evaluate the model's performance on one or several tasks. Besides, different data formats and evaluation metrics across datasets make it more difficult to comprehensively assess a model's capability. Ye et al. [36] constructed OwlEval, an evaluation set encompassing a variety of visual-related tasks, albeit of a limited size. Fu et al. [13] introduced MME, which assesses a VLM's capabilities from various perspectives at a small scale. Diverging from prior works, in this paper, we present a novel multimodal benchmark, MMBench. We also devise a suite of evaluation standards aimed at ensuring the stability and accuracy of the evaluation results.

2.2 Multimodal Models

Building upon the success of Large Language Models (LLMs) such as GPTs [5, 28,29], LLaMA [33], and Vicuna [39], recent advancements have been made in multimodal models. Flamingo [3], an early attempt at integrating LLMs into vision-language pretraining, has made significant strides. To condition effectively on visual features, it incorporates several gated cross-attention dense blocks within pretrained language encoder layers. OpenFlamingo [3] offers an open-source version of this model. BLIP-2 [20] introduces a Querying Transformer (Q-former) to bridge the modality gap between the frozen image encoder and the large language encoder. Subsequently, InstructBLIP [9] extends BLIP-2 [20] with vision-language instruction tuning, achieving superior performance. MiniGPT-4 [41] attributes the prowess of GPT-4 [26] to advanced LLMs and proposes the use of a single projection layer to align the visual representation with the language model. LLaVA [22] also utilizes GPT-4 to generate instruction-following data for vision-language tuning. The learning paradigm and the multimodal instruction tuning corpus proposed by LLaVA are widely adopted by subsequent works [1,6,8,21]. During the instruction tuning, Low-Rank Adaptation (LoRA [17]) has been adopted by recent works [8,10,36] on language models to achieve better performance on multimodal understanding. In the realm of proprietary models, the APIs of multiple powerful VLMs have also been made publicly available to prosper downstream applications, including GPT-4v [26], Gemini-Pro-V [31], and Qwen-VL-Max [4]. After conducting a thorough evaluation of these models on the proposed MMBench, we offer insights for future multimodal research.

3 The Construction of MMBench

Three characteristics differentiate MMBench from existing benchmarks for multi-modality understanding: i) MMBench adopts images/problems from various sources to evaluate diversified abilities in a hierarchical taxonomy; ii) MMBench performs rigorous quality control to ensure the correctness and validity of testing samples; iii) MMBench is a bilingual multi-modal benchmark and enables an apple-to-apple comparison of VLM performance under English and Chinese contexts. Below we will delve into more details of the construction of MMBench.

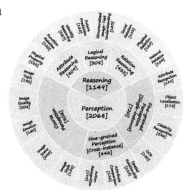

Fig. 2. Ability dimensions in MMBench. Currently, MMBench incorporates three levels of ability dimensions, encompassing 20 distinct leaf abilities.

3.1 The Hierachical Ability Taxonomy of MMBench

Human possess remarkable perception and reasoning capabilities. These abilities have been crucial in human evolution and serve as a foundation for complex cognitive processes. Perception refers to gathering information from sensory inputs, while reasoning involves drawing conclusions based on this information. Together, they form the basis of most tasks in the real world, including recognizing objects, solving problems, and making decisions [12,25]. In pursuit of genuine general artificial intelligence (AGI), vision-language models (VLMs) are also expected to exhibit strong perception and reasoning abilities. Therefore, we adopt **Perception** and **Reasoning** as level-1 (**L-1**) abilities in our taxonomy. After that, we incorporate more fine-grained ability dimensions into the taxonomy, and categorize them into six **L-2** and twenty **L-3** ability dimensions. We display the ability taxonomy in Fig. 2 and you can find detailed definitions of each fine-grained ability in the Appendix.

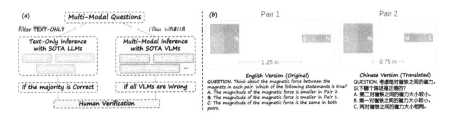

Fig. 3. The construction of MMBench. (a). The quality control strategies adopted in MMBench; (b) An illustration of questions in MMBench-CN.

3.2 Data Collection and Quality Control

Question Collection. In MMBench, we collect vision-language QAs in the format of multiple-choice problems for each L-3 ability. A problem P_i corresponds to a quadruple (Q_i, C_i, I_i, A_i). Q_i denotes the question, C_i represents a set with n ($2 \leq n \leq 4$) choices $c_1, c_2, ..., c_n$, I_i corresponds to the image associated with the question, and A_i is the correct answer. The data—including images, choices, and questions—are manually collected from multiple sources by a group of volunteers. For each **L-3** ability, we first set an example by compiling 10 ∼ 20 multiple-choice questions. Then we enlist the volunteers, all of whom are undergraduate or graduate students from various disciplines, to expand the problem set. The expansion is based on the ability definition and potential data sources, which include both public datasets and the Internet. According to the statistics, more than 80% of questions in MMBench are collected from the Internet. For the remaining 20% samples, the images are gathered from the validation set of public datasets (if they exist) while the questions are self-constructed, which is not supposed to be used for training. In the Appendix, we list data sources used in collection and provide visualization of samples corresponding to each **L-3** ability.

Quality Control. Raw data collected from volunteers may include wrong or unqualified samples. During investigation, we find that there exist two major patterns for such samples: i) the answer to the question can be inferred with **text-only** inputs, which makes it inappropriate for evaluating the multimodal understanding capability of VLMs; ii) the sample is simply **wrong**, either with a flawed question, choices, or an incorrect answer. We design two strategies to filter those low-quality samples, which is visualized in Fig. 3(a). We adopt 'majority voting' to detect **text-only** samples: data samples are inferred with state-of-the-art LLMs (GPT-4 [26], Gemini-Pro [31], *etc.*). If more than half of the LLMs can answer the question correctly with text-only inputs, the question will be manually verified and then removed if it is unqualified. To detect **wrong** samples, we also implement an automatic filtering mechanism. We select several state-of-the-art VLMs (including both open-source and proprietary ones), to answer all questions in MMBench . If all VLMs fail to answer the question correctly, we consider this question potentially problematic. Such questions will be manually checked and excluded if they are actually wrong. The quality control paradigm helps us to construct high-quality datasets and can also be used to clean other existing benchmarks.

MMBench-CN. We further convert the curated MMBench into a Chinese version. During the process, all content in questions and choices are translated to Chinese based on GPT-4, except for proper nouns, symbols, and code. All those translations are verified by humans to ensure the validity. MMBench-CN enables an apple-to-apple comparison of VLM performance under English and Chinese contexts. An example in MMBench-CN is illustrated in Fig. 3(b).

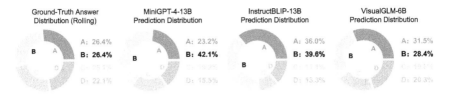

Fig. 4. The choice distribution of ground-truth answers and predictions of sample VLMs (all CircularEval records). Since there exist questions with only 2/3 choices in MMBench, the choice distribution of ground-truth is not exactly even.

3.3 MMBench Statistics

Data Statistics. In the present study, we have gathered a total of 3,217 data samples spanning across 20 distinct **L-3** abilities. We depict the problem counts of all the 3 levels of abilities in Fig. 2. To ensure a balanced and comprehensive evaluation for each ability, we try to maintain an even distribution among problems associated with different abilities during data collection, with at least 125 samples for each **L-3** category.

Data Splits. We follow the standard practice in previous works [24] to split MMBench into dev and test subsets at a ratio of 4:6. For the dev subset, we make all data samples publicly available along with the ground truth answers for all questions. For the test subset, only the data samples are released, while the ground truth answers remain confidential. To obtain the test subset evaluation results, one needs to submit the predictions to MMBench evaluation server.

4 Evaluation Strategy

In MMBench, we propose a new strategy that yields robust evaluation results with affordable costs. To deal with the free-form outputs of VLMs, we propose utilizing state-of-the-art LLMs as a helper for choice extraction. We conduct extensive experiments to study the LLM-involved evaluation procedure. The results well support the effectiveness of GPT-4 as a choice extractor. We further adopt a new evaluation strategy named **CircularEval**, which feeds a question to a VLM multiple times (with shuffled choices) and checks if a VLM succeeds in all attempts. With **CircularEval**, we deliver a rigorous evaluation and more effectively display the performance gap between VLMs.

4.1 LLM-Involved Choice Extraction

In our initial attempts to evaluate on MMBench questions, we observed that the instruction-following capabilities of VLMs can vary significantly. Though problems are presented as clear multiple-choice questions with well-formatted options, many VLMs still output the answers in free-form text[1], especially for VLMs that have not been trained with multiple-choice questions or proprietary VLMs for general purposes (GPT-4v, Qwen-VL-Max, *etc.*). Extracting choices from free-form predictions is straight-forward for human beings, but might be difficult with rule-based matching. To this end, we design a universal evaluation strategy for all VLMs with different instruction-following capabilities:

Step 1. Matching Prediction. Initially, we attempt to extract choices from VLM predictions using heuristic matching. We aim to extract the choice label (e.g., A, B, C, D) from the VLM's output. If successful, we use this as the prediction. If not, we attempt to extract the choice label using an LLM.

Step 2. Matching LLM's Output. If step 1 fails, we try to extract the choice with LLMs (**gpt-4-0125** by default). We first provide ChatGPT with the question, choices, and model prediction. Then, we request it to align the prediction with one of the given choices, and subsequently produce the label of the corresponding option. If the LLM finds that the model prediction is significantly different from all choices, we ask it to return a pseudo choice 'Z'. In experiments, we find that for almost all cases we encountered, the LLM can output a valid choice according to the instruction.

For each sample, we compare the model's label prediction (after GPT's similarity readout) with the actual ground truth label. If the prediction matches the label, the test sample is considered correct.

4.2 LLM as the Choice Extractor: A Feasibility Analysis

Instruction following (IF) Capabilities of VLMs Vary a Lot. We conduct pilot experiments to study the effectiveness of LLMs as the choice extractor. As a first step, we perform single-pass inference on all MMBench questions with VLMs in our evaluation core set (defined in Sect. 5.2). While there exist VLMs that perfectly follow the multiple-choice format and achieve high success rates (>99%) in heuristic matching, all proprietary models and a significant proportion of open-source VLMs failed to generate well-formatted outputs. In Table 1, we list the success rates of different VLMs in heuristic matching[2]. Among all VLMs, VisualGLM achieves the lowest matching success rate, which is merely 65%. For those VLMs, incorporating LLMs as the choice extractor leads to significant change in the final accuracy. Another noteworthy thing is that the IF capability and the overall multimodal understanding capability is not necessarily correlated. For example, OpenFlamingo v2 [3] demonstrates top IF capability among all VLMs, while also achieving one of the worst performances on MMBench (Table 3).

---

[1] For example, the model output can be the meaning of choice "A" rather than "A".
[2] VLMs that achieve > 99% matching rates are not listed, including LLaVA series, Yi-VL series, mPLUG-Owl2, OpenFlamingo v2, and CogVLM-Chat.

Table 1. Statistics of IF capabilities of VLMs.
We report the heuristic matching success rate of
VLMs, and the accuracy before and after LLM-
based choice extraction. In 'X+Y', X denotes the
matching-based accuracy, Y indicates the gain of
using LLM as the choice extractor.

Model Name	Match Rate	DEV Acc	Model Name	Match Rate	DEV Acc
MiniGPT4-7B	85.7	47.9 +8.8	MiniGPT4-13B	84.8	52.1 +8.7
InstructBLIP-7B	93.6	57.1 +1.3	InstuctBLIP-13B	93.7	58.4 +5.6
IDEFICS-9B-Instruct	96.6	58.4 +1.3	Qwen-VL-Chat	93.8	73.3 +3.6
MiniCPM-V	95.2	70.9 +1.5	VisualGLM-6B	64.8	39.9 +23.2
GPT-4v	91.8	81.5 +3.0	GeminiProVision	97.5	81.8 +0.8
Qwen-VL-Plus	77.4	64.5 +15.0	Qwen-VL-Max	96.0	82.0 +3.2

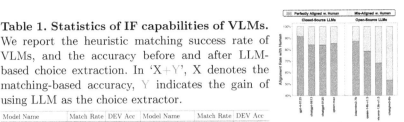

Fig. 5. Alignment rates between human and different LLMs. 'chatgpt' is 'gpt-3.5-turbo'. Open-source LLMs are 'chat' variants.

Fig. 6. CircularEval strategy. In **CircularEval**, a problem is tested multiple times with circular shifted choices and the VLM needs to succeed in all testing passes. In this example, the VLM failed in pass 3 and thus considered failed the problem.

Quality and Stability of LLM Choice Extractors. For VLM predictions that cannot be parsed by heuristic matching, we adopt GPT-4 as the choice extractor. To validate its efficacy, we first build a subset of the inference records. Each item in the set is a pair of questions and VLM predictions, which cannot be parsed by step-1 matching. We sample 10% of those hard examples ($\sim$ 420 samples), and ask volunteers to perform manual choice extraction on these data samples. Such annotations enable us to validate the choice extraction of LLMs, by measuring their alignment rates with humans.

Figure 5 reports the alignment rates (extracted choices are exactly the same) between LLMs and humans. We find that a great number of LLMs can complete the task well and achieve decent alignment rate with human. Among proprietary LLMs, GPT-4 achieves the highest level of alignment rate, which is 91.5%, while GPT-3.5-Turbo and Qwen-Max achieve around 85%. Open-source LLMs achieve more diversified performance on the choice matching task. InternLM2-7B [32] achieves an 87% alignment rate and significantly outperforms other open-source LLMs and GPT-3.5-Turbo. In the following experiments, we adopt **gpt-4-0125** as the choice extractor due to its superior alignment capability. Meanwhile, we also note that the slight difference in top-performing LLMs' alignment rates has little effect on the quantitative performance of VLMs.

4.3 CircularEval Strategy

In MMBench, the problems are presented as multiple-choice questions. Such formulation poses an evaluation challenge: random guessing will lead to ∼25% Top-1 accuracy for 4-choice questions, potentially reducing the discernible performance differences among VLMs. Besides, we noticed that VLMs may prefer to predict a certain choice among all given choices (Fig. 4), which further amplifies the bias in evaluation. To this end, we introduce a more robust evaluation strategy termed **Circular Evaluation** (or **CircularEval**). Under this setting, each question is fed to a VLM N times (N is the number of choices). Each time, circular shifting is applied to the choices and the answer to generate a new prompt for VLMs (example in Fig. 6). A VLM is considered successful in solving a question only if it correctly predicts the answer in all circular passes. In practice, once a VLM fails on a circular passes, there is no need to infer the remaining passes, which makes the actual cost of CircularEval less than $N\times$ under practical scenarios. CircularEval can achieve a good trade-off between robustness and cost.

5 Evaluation Results

5.1 Experimental Setup

For the main results, we evaluate various models belonging to three major categories on MMBench: (a) *Text-Only* GPT-4 [26]; (b) *Open-Source VLMs* including model variants of OpenFlamingo [3], MiniGPT4 [41], InstructBLIP [9], LLaVA [21], IDEFICS [19], CogVLM [34], Qwen-VL [4], Yi-VL [1], mPLUG-Owl [37], InternLM-XComposer [10], and MiniCPM-V [27]; (c) *Proprietary VLMs* including Qwen-VL-[Plus/Max] [4], Gemini-Pro-V [31], and GPT-4v [26]. For a fair comparison, we adopt the zero-shot setting to infer MMBench questions with all VLMs, based on the same prompt. For all VLMs, open-ended generation is adopted to obtain the prediction, and '**gpt-4-0125**' is used as the choice extractor. In the Appendix, we provide detailed information regarding the architecture and the parameter size for all Open-Source VLMs evaluated in this paper, as well as additional results for more VLMs under various settings.

5.2 Main Results

CircularEval vs. VanillaEval. We first compare our **CircularEval** (infer a question over multiple passes, consistency as a must) with **VanillaEval** (infer a question only once). In Table 2, we present the results with two evaluation strategies on MMBench-**dev**. For most VLMs, switching from VanillaEval to CircularEval leads to a significant drop in model accuracy. In general, comparisons under CircularEval can reveal a more significant performance gap between different VLMs. LLaVA-v1.5-13B outperforms its 7B counterpart by 2.1% Top-1 accuracy under VanillaEval, while a much larger performance gap (4.7% Top-1)

Table 2. CircularEval vs. VanillaEval. We report the **CircularEval** Top-1 accuracy and accuracy drop (compared to **VanillaEval**) of all VLMs on MMBench-dev.

VLM	Circular	Acc. Change	VLM	Circular	Acc. Change	VLM	Circular	Acc. Change
MiniGPT4-7B	32.7%	-24.1%	MiniGPT4-13B	37.5%	-23.2%	Yi-VL-6B	65.6%	-9.8%
InstructBLIP-7B	37.4%	-24.0%	InstructBLIP-13B	40.9%	-23.0%	Yi-VL-34B	68.2%	-9.5%
LLaVA-v1.5-7B	62.5%	-11.2%	LLaVA-v1.5-13B	67.2%	-8.6%	MiniCPM-V	64.8%	-10.6%
IDEFICS-9B-Instruct	37.2%	-22.6%	LLaVA-InternLM2-20B	72.8%	-7.0%	Qwen-VL-Plus	62.9%	-16.6%
VisualGLM-6B	36.1%	-27.0%	CogVLM-Chat-17B	62.4%	-15.6%	Qwen-VL-Max	76.4%	-8.7%
Qwen-VL-Chat	59.5%	-17.4%	mPLUG-Owl2	63.5%	-8.7%	Gemini-Pro-V	70.9%	-11.7%
OpenFlamingo v2	2.6%	-34.1%	InternLM-XComposer2	79.1%	-4.7%	GPT-4v	74.3%	-10.8%

is observed under CircularEval. As a special case, the performance of OpenFlamingo v2 drops from 36.7% to only 2.6% when we move from VanillaEval to CircularEval. CircularEval is such a challenging setting that it even makes state-of-the-art proprietary VLMs (GPT-4v, Qwen-VL-Max, *etc.*) suffer from ~10% Top-1 accuracy drops. In the following experiments, we adopt the more rigorous and well-defined **CircularEval** as our default evaluation paradigm.

We exhaustively evaluate all VLMs on all existing leaf abilities of MMBench. In Table 3, we report the models' overall performance and the performance in six **L-2** abilities on the test split, namely Coarse Perception (**CP**), Fine-grained Perception (single-instance, **FP-S**; cross-instance, **FP-C**), Attribute Reasoning (**AR**), Logic Reasoning (**LR**), and Relation Reasoning (**RR**).[3] The results offer valuable insights into the individual strengths and limitations of each VLM in multi-modal understanding.

Performance on MMBench-test. We first conduct a sanity check by inferring MMBench questions with GPT-4, using text-only inputs. After conducting the rigorous quality control paradigm in Sect. 3.2, GPT-4 demonstrates a random-level overall accuracy. Among open-source VLMs, InternLM-XComposer2 [10] achieves the best performance and surpass other open-source or proprietary models by a large margin, w.r.t. the overall score, demonstrating its superior ability in multimodal understanding. After that, models adopting the architecture of LLaVA [22] (LLaVA series and Yi-VL series) also showcase strong overall performance, which is just inferior to the state-of-the-art closed-source GPT-4v and Qwen-VL-Max. With a small parameter size ($\leq$ 3B), MiniCPM-V achieves over 60% Top-1 accuracy, highlighting the potential of small-scale VLMs. Models including MiniGPT, IDEFICS, VisualGLM, and InstructBLIP demonstrate significantly inferior performance compared to other VLMs, while OpenFlamingo v2 shows random-level performance due to the lack of instruction tuning.

LLM Plays a Vital Role. From the evaluation results, we find that the large language model (LLM) adopted plays a vital role in the VLM performance. For instance, all LLaVA series VLMs (v1.5-7B, v1.5-13B, InternLM2-20B) adopt

---
[3] Please refer to appendix for more fine-grained results and MMBench-dev split results.

Table 3. CircularEval results on MMBench test set (L-2 abilities). Abbreviations adopted: Logical Reasoning (LR), Attribute Reasoning (AR), Relation Reasoning (RR), Fine-grained Perception, X-Instance (FP-C), Fine-grained Perception, Single Instance (FP-S), Coarse Perception (CP). Open-source models tagged with * incorporate in-house data in model training.

Model	Overall	CP	FP-S	FP-C	AR	LR	RR
Large Language Models							
GPT-4-Turbo (0125) [26]	2.9%	0.6%	1.2%	4.1%	3.7%	4.9%	7.4%
OpenSource VLMs							
OpenFlamingo v2 [3]	2.3%	1.1%	3.5%	1.5%	5.3%	0.0%	2.7%
MiniGPT4-7B [41]	30.5%	37.0%	31.8%	17.2%	49.8%	9.2%	25.6%
IDEFICS-9B-Instruct [19]	35.2%	48.3%	31.3%	29.6%	47.8%	11.4%	25.2%
VisualGLM-6B [11]	35.4%	40.2%	38.5%	26.2%	47.8%	19.6%	29.5%
InstructBLIP-7B [9]	38.3%	46.7%	39.0%	31.8%	55.5%	8.7%	31.0%
MiniGPT4-13B [41]	38.8%	44.6%	42.9%	23.2%	64.9%	8.2%	32.9%
InstructBLIP-13B [9]	39.8%	47.2%	42.9%	21.0%	60.4%	12.5%	38.8%
Qwen-VL-Chat* [4]	60.9%	68.5%	67.7%	50.2%	78.0%	37.0%	45.7%
MiniCPM-V [27]	61.4%	65.6%	69.4%	51.3%	70.6%	35.3%	59.7%
LLaVA-v1.5-7B [21]	63.4%	70.0%	68.0%	57.7%	76.6%	33.2%	56.2%
mPLUG-Owl2 [37]	63.5%	68.1%	69.1%	55.8%	78.4%	37.0%	57.0%
CogVLM-Chat-17B [34]	63.6%	72.8%	66.6%	55.4%	71.4%	33.7%	62.0%
Yi-VL-6B* [1]	65.5%	72.8%	72.9%	56.2%	75.5%	41.3%	55.4%
LLaVA-v1.5-13B [21]	66.9%	73.1%	72.4%	60.3%	75.5%	35.9%	65.5%
Yi-VL-34B* [1]	68.4%	72.0%	78.0%	54.7%	81.2%	38.6%	68.2%
LLaVA-InternLM2-20B [8]	72.3%	78.3%	76.6%	68.2%	78.4%	46.2%	69.4%
InternLM-XComposer2* [10]	78.1%	80.4%	83.5%	73.0%	83.7%	63.6%	74.4%
Proprietary VLMs							
Qwen-VL-Plus [4]	64.6%	66.5%	79.1%	50.2%	73.9%	42.9%	57.8%
Gemini-Pro-V [31]	70.2%	70.0%	78.9%	65.9%	82.9%	46.2%	65.9%
GPT-4v [26]	74.3%	77.6%	73.8%	71.5%	85.3%	63.6%	68.6%
Qwen-VL-Max [4]	75.4%	74.8%	87.2%	67.0%	85.3%	54.9%	70.5%

the same vision backbone and are trained with the same multimodal corpus, while switching the LLM from Vicuna-v1.5 [39] to the more powerful InternLM2-20B [32] leads to steady improvement across all L-2 capabilities (especially significant for reasoning tasks). The scaling also holds for variants with different sizes from the same LLM family. By adopting the 13B variant of Vicuna rather than the 7B variant, VLMs in the MiniGPT, InstructBLIP, and LLaVA v1.5 series outperform their 7B counterparts by 8.3%, 1.5%, and 3.5% overall Top-1 accuracies on the MMBench-test split, respectively.

Performance on MMBench-CN. Figure 7 compares the performance of different VLMs on MMBench and MMBench-CN. Most VLMs display a lower performance on MMBench-CN compared to the results on MMBench, except OpenFlamingo v2, VisualGLM, and Qwen-VL-Plus. The difference may be attributed to the unbalanced English and Chinese corpora used in the pre-training and instruction-tuning of VLMs and their corresponding LLMs. We notice that most top-

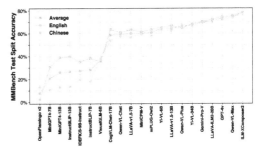

Fig. 7. The performance on the test split of MMBench and MMBench-CN. Models are sorted with the ascending order of average performance. ILM stands for InternLM.

performing VLMs on MMBench also display outstanding performance under the bilingual context. The largest EN-CN performance gap for models that achieve 70+% Top-1 accuracy on MMBench is a mere 2%, For InternLM-XComposer2, the accuracy only drops by less than 1% when evaluated on MMBench-CN. The advantage can be attributed to utilizing LLMs with better bilingual capabilities or tuning the VLM with more balanced cross-language multimodal corpora.

Table 4. 'Upper-bound' Acc Estimation for Proprietary VLMs.

Model	MMBench-test	Upper Bound
GPT-4v	74.3	76.2
Gemini-Pro-V	70.2	72.6
Qwen-VL-Max	75.4	75.5

Fig. 8. Content Moderation Cases of Proprietary VLMs.

5.3 Fine-Grained Analysis

Content Moderation of Proprietary VLMs. Taking an in-depth look at predictions of proprietary VLMs, we notice that all of them apply explicit content moderation. GPT-4v, Gemini-Pro-V, and Qwen-VL-Max reject answering in 1.8%, 1.6%, and 0.1% of cases across all CircularEval passes in MMBench, respectively. 74% of questions rejected by GPT-4v are related to celebrity recognition (Fig. 8), while no obvious rejection pattern is observed for Gemini. Such moderation has a negative impact on the evaluated accuracy. To estimate an **upper-bound** performance, we assume that VLMs can perfectly answer all rejected questions and re-calculate the accuracy. Table 4 shows that the content moderation policy affects the MMBench-test accuracy by up to 2.4%, which is not significant.

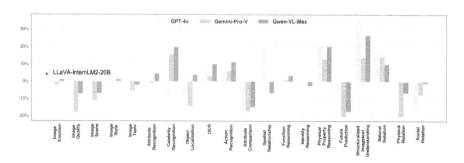

Fig. 9. Proprietary VLMs vs. Open-Source ones at a fine-grained level.

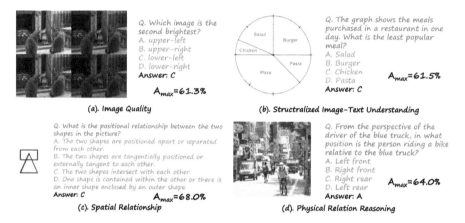

Fig. 10. Hard examples that belong to the 4 L-3 abilities with lowest A_{max}. All VLMs have made the wrong prediction for the visualized examples under CircularEval.

Proprietary vs. Open-Source: What is the Gap?. Compared to the varied performance of open-source VLMs, most proprietary models demonstrate competitive performance on MMBench. This raises a question we care about: are proprietary models generally more powerful, or do each kind of model display unique strengths and weaknesses across different types of ability? To answer this question, we perform a fine-grained comparison of three proprietary VLMs and LLaVA-InternLM2-20B, the top-performing model trained on open-source datasets only, and visualize the result in Fig. 9. We observe that proprietary models significantly outperform the open-source ones under two major scenarios: i) **Structuralized image-text understanding**, which requires VLMs to understand complex codes, tables, diagrams, or layouts. ii) **Tasks requiring external knowledge to solve**, which correspond to abilities including celebrity recognition, physical property reasoning, natural relation reasoning, *etc.* Meanwhile, proprietary VLMs do not display advantages on tasks corresponding to other perception or reasoning capabilities.

Hard Cases in MMBench. For most VLMs, the fine-grained accuracies vary a lot across different ability categories. To provide insights for future VLM optimization, we find the maximum accuracy (A_{max}) across all evaluated VLMs on each L-3 capability. Samples belonging to L-3 capabilities with the lowest A_{max} are visualized in Fig. 10. Generally, we find that all existing VLMs have the following limitations: 1. Poor at recognizing the low-level features on visual inputs, *i.e.*, they cannot accurately recognize and compare the brightness, sharpness, contrast ratio, or artifacts of images. 2. Difficulty in understanding structuralized visual inputs like tables, diagrams, or layouts, even for relatively simple cases like Fig. 10(b); 3. Perform badly on recognizing or reasoning about the inter-object spatial relationships, either in 2D or 3D space.

6 Conclusion

We introduce MMBench, a multi-modality benchmark that performs objective evaluation for VLMs with over 3,000 multiple-choice questions covering 20 ability dimensions. To produce robust and reliable evaluation results, we introduce a new evaluation strategy named **CircularEval**. The strategy is much stricter than the vanilla 1-pass evaluation and can yield reliable evaluation results at an affordable cost. Considering the limited instruction following ability of some VLMs, to yield more accurate evaluation results, we additionally adopt LLMs to extract choices from the model's predictions. We comprehensively evaluate over 20 mainstream VLMs on MMBench, covering different architectures and parameter sizes. The evaluation results provide valuable insights for future improvements.

Acknowledgement. This project is supported by the National Key R&D Program of China (No. 2022ZD 0161600), the Shanghai Postdoctoral Excellence Program (No. 2023023), China Postdoctoral Science Fund (No. 2024M751559), and Shanghai Artificial lntelligence Laboratory. This project is also supported under the RIE2020 Industry Alignment Fund - Industry Collaboration Projects (IAF-ICP) Funding Initiative, as well as cash and in-kind contribution from the industry partner(s).

References

1. 01-ai: Yi-vl (2023). https://huggingface.co/01-ai/Yi-VL-34B
2. Agrawal, H., et al.: Nocaps: novel object captioning at scale. In: Proceedings of the IEEE/CVF International Conference on Computer Vision, pp. 8948–8957 (2019)
3. Alayrac, J.B., Donahue, J., Luc, P., Miech, A., Barr, I., Hasson, Y., Lenc, K., Mensch, A., Millican, K., Reynolds, M., et al.: Flamingo: a visual language model for few-shot learning. Adv. Neural. Inf. Process. Syst. **35**, 23716–23736 (2022)
4. Bai, J., et al.: Qwen-vl: a frontier large vision-language model with versatile abilities. arXiv preprint arXiv:2308.12966 (2023)
5. Brown, T., et al.: Language models are few-shot learners. Adv. Neural. Inf. Process. Syst. **33**, 1877–1901 (2020)

6. Chen, L., et al.: Sharegpt4v: Improving large multi-modal models with better captions. arXiv preprint arXiv:2311.12793 (2023)
7. Chen, X., et al.: Microsoft coco captions: Data collection and evaluation server. arXiv preprint arXiv:1504.00325 (2015)
8. Contributors, X.: Xtuner: A toolkit for efficiently fine-tuning llm (2023). https://github.com/InternLM/xtuner
9. Dai, W., et al.: Instructblip: Towards general-purpose vision-language models with instruction tuning. arXiv preprint arXiv:2305.06500 (2023)
10. Dong, X., et al.: Internlm-xcomposer2: Mastering free-form text-image composition and comprehension in vision-language large model. arXiv preprint arXiv:2401.16420 (2024)
11. Du, Z., et al.: Glm: general language model pretraining with autoregressive blank infilling. In: Proceedings of the 60th Annual Meeting of the Association for Computational Linguistics (Volume 1: Long Papers), pp. 320–335 (2022)
12. Fodor, J.A.: The modularity of mind. MIT press (1983)
13. Fu, C., et al.: Mme: A comprehensive evaluation benchmark for multimodal large language models. ArXiv **abs/2306.13394** (2023). https://api.semanticscholar.org/CorpusID:259243928
14. Gong, T., et al.: Multimodal-gpt: a vision and language model for dialogue with humans (2023)
15. Goyal, Y., Khot, T., Summers-Stay, D., Batra, D., Parikh, D.: Making the v in vqa matter: Elevating the role of image understanding in visual question answering. In: Proceedings of the IEEE Conference on Computer Vision and Pattern Recognition, pp. 6904–6913 (2017)
16. Gurari, D., et al.: Vizwiz grand challenge: Answering visual questions from blind people. In: Proceedings of the IEEE Conference on Computer Vision and Pattern Recognition, pp. 3608–3617 (2018)
17. Hu, E.J., et al.: Lora: Low-rank adaptation of large language models. arXiv preprint arXiv:2106.09685 (2021)
18. Hudson, D.A., Manning, C.D.: Gqa: a new dataset for real-world visual reasoning and compositional question answering. In: Proceedings of the IEEE/CVF Conference on Computer Vision And Pattern Recognition, pp. 6700–6709 (2019)
19. Laurençon, H., et al.: Obelics: An open web-scale filtered dataset of interleaved image-text documents (2023)
20. Li, J., Li, D., Savarese, S., Hoi, S.: Blip-2: Bootstrapping language-image pre-training with frozen image encoders and large language models. arXiv preprint arXiv:2301.12597 (2023)
21. Liu, H., Li, C., Li, Y., Lee, Y.J.: Improved baselines with visual instruction tuning. arXiv preprint arXiv:2310.03744 (2023)
22. Liu, H., Li, C., Wu, Q., Lee, Y.J.: Visual instruction tuning. arXiv preprint arXiv:2304.08485 (2023)
23. Lu, P., et al.: Learn to explain: multimodal reasoning via thought chains for science question answering. Adv. Neural. Inf. Process. Syst. **35**, 2507–2521 (2022)
24. Marino, K., Rastegari, M., Farhadi, A., Mottaghi, R.: Ok-vqa: a visual question answering benchmark requiring external knowledge. In: Proceedings of the IEEE/CVF Conference on Computer Vision and Pattern Recognition, pp. 3195–3204 (2019)
25. Oaksford, M., Chater, N.: Bayesian rationality: The probabilistic approach to human reasoning. Oxford University Press (2007)
26. OpenAI: Gpt-4 technical report. ArXivabs/arXiv: 2303.08774 (2023)

27. OpenBMB: Minicpm: Unveiling the potential of end-side large language models (2024)
28. Ouyang, L., et al.: Training language models to follow instructions with human feedback. Adv. Neural. Inf. Process. Syst. **35**, 27730–27744 (2022)
29. Radford, A., Wu, J., Child, R., Luan, D., Amodei, D., Sutskever, I., et al.: Language models are unsupervised multitask learners. OpenAI Blog **1**(8), 9 (2019)
30. Singh, A., et al.: Towards vqa models that can read. In: Proceedings of the IEEE/CVF Conference on Computer Vision and Pattern Recognition, pp. 8317–8326 (2019)
31. Team, G., et al.: Gemini: a family of highly capable multimodal models. arXiv preprint arXiv:2312.11805 (2023)
32. Team, I.: Internlm: a multilingual language model with progressively enhanced capabilities (2023). https://github.com/InternLM/InternLM-techreport
33. Touvron, H., et al.: Llama: Open and efficient foundation language models. arXiv preprint arXiv:2302.13971 (2023)
34. Wang, W., et al.: Cogvlm: Visual expert for pretrained language models. ArXiv abs/ arXiv: 2311.03079 (2023), https://api.semanticscholar.org/CorpusID:265034288
35. Xu, P., et al.: Lvlm-ehub: a comprehensive evaluation benchmark for large vision-language models (2023)
36. Ye, Q., et al.: mplug-owl: Modularization empowers large language models with multimodality. arXiv preprint arXiv:2304.14178 (2023)
37. Ye, Q., et al.: mplug-owl2: Revolutionizing multi-modal large language model with modality collaboration. ArXiv arXiv: 2311.04257 (2023). https://api.semanticscholar.org/CorpusID:265050943
38. Young, P., Lai, A., Hodosh, M., Hockenmaier, J.: From image descriptions to visual denotations: new similarity metrics for semantic inference over event descriptions. Trans. Associat. Comput. Linguist. **2**, 67–78 (2014)
39. Zheng, L., et al.: Judging llm-as-a-judge with mt-bench and chatbot arena (2023)
40. Zhou, L., Xu, C., Corso, J.: Towards automatic learning of procedures from web instructional videos. In: Proceedings of the AAAI Conference on Artificial Intelligence, vol. 32 (2018)
41. Zhu, D., Chen, J., Shen, X., Li, X., Elhoseiny, M.: Minigpt-4: Enhancing vision-language understanding with advanced large language models. arXiv preprint arXiv:2304.10592 (2023)

Implicit Filtering for Learning Neural Signed Distance Functions from 3D Point Clouds

Shengtao Li[1,2], Ge Gao[1,2(✉)], Yudong Liu[1,2], Ming Gu[1,2], and Yu-Shen Liu[2]

[1] Beijing National Research Center for Information Science and Technology (BNRist), Tsinghua University, Beijing, China
{list21,liuyd23}@mails.tsinghua.edu.cn, guming@tsinghua.edu.cn
[2] School of Software, Tsinghua University, Beijing, China
{gaoge,liuyushen}@tsinghua.edu.cn

Abstract. Neural signed distance functions (SDFs) have shown powerful ability in fitting the shape geometry. However, inferring continuous signed distance fields from discrete unoriented point clouds still remains a challenge. The neural network typically fits the shape with a rough surface and omits fine-grained geometric details such as shape edges and corners. In this paper, we propose a novel non-linear implicit filter to smooth the implicit field while preserving high-frequency geometry details. Our novelty lies in that we can filter the surface (zero level set) by the neighbor input points with gradients of the signed distance field. By moving the input raw point clouds along the gradient, our proposed implicit filtering can be extended to non-zero level sets to keep the promise consistency between different level sets, which consequently results in a better regularization of the zero level set. We conduct comprehensive experiments in surface reconstruction from objects and complex scene point clouds, the numerical and visual comparisons demonstrate our improvements over the state-of-the-art methods under the widely used benchmarks. Project page: https://list17.github.io/ImplicitFilter.

Keywords: Implicit filtering · Signed distance functions · Point cloud reconstruction

1 Introduction

Reconstructing surfaces from 3D point clouds is an important task in 3D computer vision. Recently signed distance functions (SDFs) learned by neural networks have been a widely used strategy for representing high-fidelity 3D geometry. These methods train the neural networks to predict the signed distance for every position in the space by signed distances from ground truth or inferred from the raw 3D point cloud. With the learned signed distance field, we can

Supplementary Information The online version contains supplementary material available at https://doi.org/10.1007/978-3-031-72658-3_14.

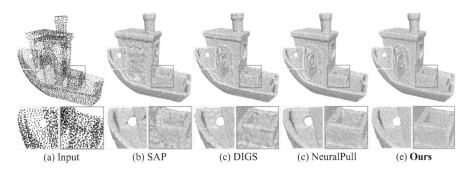

(a) Input (b) SAP (c) DIGS (c) NeuralPull (e) **Ours**

Fig. 1. Visualization of the comparisons on FAMOUS dataset [12]. Our implicit filter can improve the reconstruction by removing the noise and keeping the geometric details compared with other methods.

obtain the surface by running the marching cubes algorithm [27] to extract the zero level set.

Without signed distance ground truth, inferring the correct gradient and distance for each query point could be hard. Since the gradient of the neural network also indicates the direction in which the signed distance field changes, recent works [1,2,4,14,29,38] typically add constraints on the network gradient to learn a stable field. In terms of the rate at which the field is changing, the eikonal term [1,2,5,38] is widely used to ensure the norm of the gradient to be one everywhere. For the gradient direction constraint, some methods [4,10] use the direction from the query point to the nearest point on the surface as guidance. Leveraging the continuity of the neural network and the gradient constraint, all these methods could reconstruct discrete points. However, the continuity cannot guarantee the prediction is correct everywhere. Therefore, reconstructed surfaces of previous methods usually contain noise and ignore geometry details when there are not enough points to guide the reconstruction, as shown in Fig. 1.

The above issue arises from the fact that these methods overlook the geometric information within the neighborhood but only focus on adding constraints on individual points to optimize the network. To resolve this issue, we introduce the bilateral filter for implicit fields that reduces surface noise while preserving the high-frequency geometric characteristics of the shape. Our designed implicit filter takes into account both the position of point clouds and the gradient of learned implicit fields. Based on the assumption of all input points lying on the surface, we can filter noise points on the zero level set by minimizing the weighted projection distance to gradients of the neighbor input points. Moreover, by moving the input points along the gradient of the field to other level sets, we can easily extend the filter to the whole field. This helps constrain the signed distance field near the surface and achieve better consistency through different level sets. To evaluate the effectiveness of our proposed implicit filtering, we validate it under widely used benchmarks including object and scene reconstructions. Our contributions are listed below.

- We introduce the implicit filter on SDFs to smooth the surface while preserving geometry details for learning better neural networks to represent shapes or scenes.
- We improve the implicit filter by extending it to non-zero level sets of signed distance fields. This regularization of the field aligns different level sets and provides better consistency within the whole SDF field.
- Both object and scene reconstruction experiments validate our implicit filter, demonstrating its effectiveness and ability to produce high-fidelity reconstruction results, surpassing the previous state-of-the-art methods.

2 Related Work

With the rapid development of deep learning, neural networks have shown great potential in surface reconstruction from 3D point clouds. In the following, we briefly review methods related to implicit learning for 3D shapes and reconstructions from point clouds.

Implicit Learning from 3D Supervision. The most commonly used strategy to train the neural network is to learn priors in a data-driven manner. These methods require signed distances or occupancy labels as 3D supervision to learn global priors [6,12,26,31,32] or local priors [7,17,18,22,35,40,41,44,45]. With large-scale training datasets, the neural network can perform well with similar shapes, but may not generalize well to unseen cases with large geometric variations. These models often have limited inputs that can be difficult to scale for varying sizes of point clouds.

Implicit Learning from Raw Point Clouds. Different from the supervised methods, we can learn implicit functions by overfitting neural networks on single point clouds globally or locally to learn SDFs [1–4,10,21,28,30,34]. These unsupervised methods rely on neural networks to infer implicit functions without learning any priors. Therefore, apart from the guidance of original input point clouds, we also need constraints on the direction [3,4,10,21] or the norm [1,2,30] of the gradients, specially designed priors [3,28], or differentiable poisson solver [34] to infer SDFs. This unsupervised approach heavily depends on the fitting capability and continuity of neural networks. However, these SDFs lack accuracy because there is no reliable guidance available for each query point across the entire space when working with discrete point clouds. Therefore, deducing the correct geometry for free space becomes particularly crucial. Our implicit filtering enhances SDFs by inferring the geometric details through the implicit field information of neighbor points.

Feature Preserving Point Cloud Reconstruction. Early works [16,23,33] reconstruct point clouds with sharp features usually by point cloud consolidation. The key idea of these methods is to enhance the quality of point clouds with sharp features. One popular category is the local projection operation (LOP) [25] and its variants [15,16,23,36]. The projection operator provides a stable and easily generalizable method for point cloud filtering, which is also the foundation of our implicit filter. The difference lies in that we do not need any normal or other priors and our filtering can be directly applied to implicit fields to extract high-fidelity meshes. Some other learning-based methods [47,48] try to consolidate point clouds with edge points in a data-driven manner. Although capable of generating high-quality point clouds, these methods still require a proper reconstruction method [13] to inherit the details in meshes.

With the advancement of deep learning in point cloud reconstruction, some approaches [5,24,38,42] also explored employing neural networks to reconstruct high-precision models. FFN [39], SIREN [38], and IDF [43] introduce high-frequency features into the neural network in different ways to preserve the geometric details of the reconstructed shape. DIGS [5] and EPI [42] smooth the surface by using the divergence as guidance to alleviate the implicit surface roughness. Compared with these methods, we first introduce local geometric features through filtering to optimize the implicit field, so that we can achieve higher accuracy.

3 Method

Neural SDFs Overview. This section will briefly describe the concepts we used in our implicit filtering. We focus on the SDF $f : \boldsymbol{R}^3 \to \boldsymbol{R}$ inferred from the point cloud $\boldsymbol{P} = \{\boldsymbol{p}_i | \boldsymbol{p}_i \in \boldsymbol{R}^3\}_{i=1}^N$ without ground truth signed distances and normals. f predicts a signed distance $s \in \boldsymbol{R}$ for an arbitrary query point $\boldsymbol{q}$, as formulated by $s = f_\theta(\boldsymbol{q})$, where θ denotes the parameters of the neural network.

The level set $\mathcal{S}_d$ of SDF is defined as a set of continuous query points with the same signed distance d, formulated as $\mathcal{S}_d = \{\boldsymbol{q} | f_\theta(\boldsymbol{q}) = d\}$. The goal of our implicit filtering is to smooth each level set with geometry details. Then we can extract the zero level set as a mesh by running the marching cubes algorithm [27].

Level Set Bilateral Filtering. Filtering for 2D images replaces the intensity of each pixel with the weighted intensity values from nearby pixels. Different from images, the resolution of implicit fields is infinite and we need to find the neighborhood on each level set for filtering. By minimizing the following loss function,

$$L_{dist} = \frac{1}{N} \sum_{i=1}^{N} |f_\theta(\boldsymbol{p}_i)|, \quad (1)$$

we can approximate that all points in $\boldsymbol{P}$ are located on level set $\mathcal{S}_0$, which makes it feasible to find neighbor points on $\mathcal{S}_0$. For a given point $\bar{\boldsymbol{p}}$ on $\mathcal{S}_0$, one simple strategy of filtering is to average positions of neighbor points $\mathcal{N}(\bar{\boldsymbol{p}}, \mathcal{S}_0) \subset \boldsymbol{P}$ on $\mathcal{S}_0$ by a Gaussian filter based on relative positions as follows:

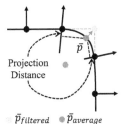

Fig. 2. By minimizing the weighted projection distance, our filter can preserve the sharp feature but the average method leads to a wrong result.

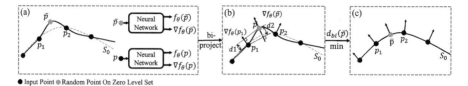

Fig. 3. Overview of filtering the zero level set. (a) We assume all input points lying on the surface and compute gradients as normals. (b) Calculating bidirectional projection distances $d1 = |\mathbf{n}_{p_j}^T (\bar{\mathbf{p}} - \mathbf{p}_j)|$, $d2 = |\mathbf{n}_{\bar{p}}^T (\bar{\mathbf{p}} - \mathbf{p}_j)|$ and the weights in Eq. (4). (c) By minimizing Eq. (4), we can remove the noise on the zero level set. The gradient ∇f_θ in this figure defaults to be regularized.

$$\bar{\mathbf{p}}_{average} = \frac{\sum_{\mathbf{p}_j \in \mathcal{N}(\bar{\mathbf{p}}, \mathcal{S}_0)} \mathbf{p}_j \phi(\|\bar{\mathbf{p}} - \mathbf{p}_j\|)}{\sum_{\mathbf{p}_j \in \mathcal{N}(\bar{\mathbf{p}}, \mathcal{S}_0)} \phi(\|\bar{\mathbf{p}} - \mathbf{p}_j\|)}, \quad (2)$$

where the Gaussian function ϕ is defined as $\phi(\|\bar{\mathbf{p}} - \mathbf{p}_j\|) = \exp\left(-\frac{\|\bar{\mathbf{p}} - \mathbf{p}_j\|^2}{\sigma_p^2}\right)$.

However, as depicted in Fig. 2, it is evident that this weighted mean position yields excessively smooth surfaces, causing sharp features and details to be further obscured. To keep the geometric details, our filtering operator suggests measuring the projection distance to the gradient of neighbor points as shown in Fig. 2 and Fig. 3(b). When calculating weights, it is vital to account for both the impact of relative positions and the gradient similarity. Following the principles of bilateral filtering, to compute the filtered point for $\bar{\mathbf{p}}$, we simply need to minimize the following distance equation:

$$d(\bar{\mathbf{p}}) = \frac{\sum_{\mathbf{p}_j \in \mathcal{N}(\bar{\mathbf{p}}, \mathcal{S}_0)} |\mathbf{n}_{p_j}^T (\bar{\mathbf{p}} - \mathbf{p}_j)| \phi(\|\bar{\mathbf{p}} - \mathbf{p}_j\|) \psi(\mathbf{n}_{\bar{p}}, \mathbf{n}_{p_j})}{\sum_{\mathbf{p}_j \in \mathcal{N}(\bar{\mathbf{p}}, \mathcal{S}_0)} \phi(\|\bar{\mathbf{p}} - \mathbf{p}_j\|) \psi(\mathbf{n}_{\bar{p}}, \mathbf{n}_{p_j})}, \quad (3)$$

where the gradient $\mathbf{n}_{\bar{p}}$, $\mathbf{n}_{p_j}$ and the Gaussian function ψ are defined as $\mathbf{n}_{\bar{p}} = \frac{\nabla f_\theta(\bar{\mathbf{p}})}{\|\nabla f_\theta(\bar{\mathbf{p}})\|}$, $\mathbf{n}_{p_j} = \frac{\nabla f_\theta(\mathbf{p}_j)}{\|\nabla f_\theta(\mathbf{p}_j)\|}$, $\psi(\mathbf{n}_{\bar{p}}, \mathbf{n}_{p_j}) = \exp\left(-\frac{1 - \mathbf{n}_{\bar{p}}^T \mathbf{n}_{p_j}}{1 - \cos(\sigma_n)}\right)$.

In addition to projection to the gradient $\mathbf{n}_{p_j}$, we observe that the projection distance to $\mathbf{n}_{\bar{p}}$ can assist in learning a more stable gradient for point $\bar{\mathbf{p}}$ which is also adopted in EAR [16]. Taking into account the bidirectional projection, our final bilateral filtering operator can be formulated as follows:

$$d_{bi}(\bar{\mathbf{p}}) = \frac{\sum_{\mathbf{p}_j \in \mathcal{N}(\bar{\mathbf{p}}, \mathcal{S}_0)} \left(|\mathbf{n}_{p_j}^T (\bar{\mathbf{p}} - \mathbf{p}_j)| + |\mathbf{n}_{\bar{p}}^T (\bar{\mathbf{p}} - \mathbf{p}_j)|\right) \phi(\|\bar{\mathbf{p}} - \mathbf{p}_j\|) \psi(\mathbf{n}_{\bar{p}}, \mathbf{n}_{p_j})}{\sum_{\mathbf{p}_j \in \mathcal{N}(\bar{\mathbf{p}}, \mathcal{S}_0)} \phi(\|\bar{\mathbf{p}} - \mathbf{p}_j\|) \psi(\mathbf{n}_{\bar{p}}, \mathbf{n}_{p_j})}.$$

(4)

Although similar filtering methods have been widely studied in applications such as point cloud denoising and resampling [16,48], there are two critical problems when applying these methods in implicit fields:

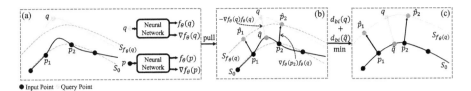

Fig. 4. Overview of sampling points. (a) Sampling query points near the surface. (b) Pulling the query point to the zero level set and input points to the level set where the query point is located. (c) Applying the filter on each level set. The gradient ∇f_θ in this figure defaults to be regularized.

1. Filtering the zero level set needs to sample points on the level set $\mathcal{S}_0$, which necessitates the resolution of the equation $f_\theta = 0$, or the utilization of the marching cubes algorithm [27]. Both methods pose challenges in achieving fast and uniform point sampling. For the randomly sampled point q on non-zero level set $\mathcal{S}_{f_\theta(q)}$, we can also not filter this level set since there are no neighbor points on $\mathcal{S}_{f_\theta(q)}$.
2. The normals utilized in our filtering are derived from the gradients of the neural network f_θ. While the network typically offers reliable gradients, we may find that $\nabla f_\theta = 0$ is also the optimal solution to the minimum value of Eqs. (3) and (4). This degenerate solution is unexpected, as it implies a scenario where there is no surface when the gradient is zero everywhere.

We will focus on addressing the two issues in the subsequent sections.

Sampling Points for Filtering. Inspired by NeuralPull [4], we can pull a query point to the zero level set by the gradient of the neural network f_θ. For a given query point q as input, the pulled location $\hat{q}$ can be formulated as follows:

$$\hat{q} = q - f_\theta(q)\nabla f_\theta(q)/\|\nabla f_\theta(q)\|. \tag{5}$$

The point q and $\hat{q}$ lie respectively on level set $\mathcal{S}_{f_\theta(q)}$ and $\mathcal{S}_0$ as illustrate in Fig. 4(b). By adopting the sampling strategy in NeuralPull, we can generate samples $Q = \{q_i | q_i \in \mathbf{R}^3\}_{i=1}^M$ on different level sets near the surface and pull them to $\mathcal{S}_0$ by Eq. (5), to obtain $\hat{Q} = \{\hat{q}_i | \hat{q}_i = q_i - f_\theta(q_i)\nabla f_\theta(q_i)/\|\nabla f_\theta(q_i)\|, q_i \in Q\}_{i=1}^M$. Hence, we can filter the zero level set by minimizing Eq. (4) across all pulled query points $\hat{Q}$, which is equivalent to optimizing the following loss:

$$L_{zero} = \sum\nolimits_{\hat{q} \in \hat{Q}} d_{bi}(\hat{q}), \tag{6}$$

where for each $\hat{q} \in \hat{Q}$, $\mathcal{N}(\hat{q}, \mathcal{S}_0)$ denotes finding the neighbors of $\hat{q}$ within the input points P, since P is assumed to be located on $\mathcal{S}_0$.

This filtering mechanism can be easily extended to non-zero level sets in a similar inverse manner. To be more specific, as for level set $\mathcal{S}_{f_\theta(q)}$, the neighbor points for query point $q \in Q$ are required. These points should lie on the level set $S_{f_\theta(q)}$ same as q, allowing us to filter the level set $S_{f_\theta(q)}$ using the same filter as described in Eq. (4).

However, obtaining $\mathcal{N}(q, S_{f_\theta(q)})$ in P is not feasible, since all input points P are situated on the zero level set instead of the $S_{f_\theta(q)}$ level set. To address this issue, we propose a technique for identifying neighbors of q on level set $S_{f_\theta(q)}$, by projecting the input points P inversely onto the specific level set $S_{f_\theta(q)}$ based on the gradient, as depicted in Fig. 4(b). The projected neighbor points can be represented as in Eq. (7). Filtering across multiple level sets helps to enhance the performance of our method by optimizing the consistency between different level sets within the SDF field, We further showcase this evidence in the ablation study detailed in Section Sect. 4.4.

$$\mathcal{N}(q, S_{f_\theta(q)}) = \{\hat{p} | \hat{p} = p + f_\theta(q) \frac{\nabla f_\theta(p)}{||\nabla f_\theta(p)||}, p \in \mathcal{N}(\hat{q}, S_0))\}. \tag{7}$$

Based on the above analysis, we can filter the level sets $S_{f_\theta(q)}$ by minimizing Eq. (4) over all sample points Q through Eq. (7), equivalent to optimizing the following loss:

$$L_{field} = \sum_{q \in Q} d_{bi}(q). \tag{8}$$

It is worth noting that for a fixed query point q, the pulled query point $\hat{q}$ dynamically changes when training the neural network, which results in a time-consuming process to repeatedly conduct neighbor searching for $\hat{q}$. To handle this matter, we substitute the $\mathcal{N}(\hat{q}, S_0)$ with $\mathcal{N}(NN(q), S_0)$, where $NN(q)$ denotes the nearest point of q within the point cloud P as shown in Fig. 5. While this substitution may introduce a slight bias for training, it also ensures the neighbor points are close to $\hat{q}$, therefore this trade-off between efficiency and accuracy is reasonable.

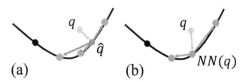

Fig. 5. (a) Searching neighbors directly for $\hat{q}$. (b) Searching neighbors for $NN(q)$ instead of $\hat{q}$.

Gradient Constraint. The other problem of implicit filtering is gradient degeneration. Overfitting the neural network requires the SDF to be geometrically initialized. We can consider the initialized implicit field as the noisy field and apply our filter directly to train the network from the beginning to fit the raw point cloud by removing the 'noise'. However, if the denoise target is too complex, gradient degeneration will occur during the training process. Therefore, we need to add a constraint to the gradient of the SDF.

There are two ways for training the neural network to pull query points onto the surface based on NeuralPull [4] and CAP-UDF [49]. One is minimizing the distance between the pulled point $\hat{q}$ and the nearest point $NN(q)$ as formulated below:

$$L_{pull} = \frac{1}{M} \sum_{i \in [1,M]} ||\hat{q}_i - NN(q_i)||_2. \tag{9}$$

The other is minimizing the Chamfer distance between moved query points and the raw point cloud:

$$L_{CD} = \frac{1}{M} \sum_{i \in [1,M]} \min_{j \in [1,N]} ||\hat{q}_i - p_j||_2 + \frac{1}{N} \sum_{j \in [1,N]} \min_{i \in [1,M]} ||p_j - \hat{q}_i||_2. \tag{10}$$

A stable SDF can be trained by the losses above since they are trying to move the query points to be in the same distribution with the point cloud, which can provide the constraint for our implicit filter. Here we choose L_{CD} since the filtered points are likely not the nearest points and L_{CD} is a more relaxed constraint.

Loss Function. Finally, our loss function is formulated as:

$$L = L_{zero} + \alpha_1 L_{field} + \alpha_2 L_{dist} + \alpha_3 L_{CD}, \tag{11}$$

where α_1, α_2, and α_3 is the balance weights for our implicit filtering loss.

Implementation Details. We employ a neural network similar to OccNet [31] and the geometric network initialization proposed in SAL [1] with a smaller radius the same as GridPull [10] to learn the SDF. We use the strategy in NeuralPull [4] to sample queries around each point p in P. We set the weight α_3 to 10 to constrain the learned SDF and α_1 and α_2 to 1. The parameters σ_n, σ_p are set to $15°, \max_{p_j \in \mathcal{N}(\bar{p}, S_{f_\theta(\bar{p})})}(||\bar{p} - p_j||)$ respectively.

4 Experiments

We conducted experiments to assess the performance of our implicit filter for surface reconstruction from raw point clouds. The results are presented for general shapes in Sect. 4.1, real scanned raw data including 3D objects in Sect. 4.2, and complex scenes in Sect. 4.3. Additionally, ablation experiments were carried out to validate the theory and explore the impact of various parameters in Sect. 4.4.

4.1 Surface Reconstruction for Shapes

Datasets and Metrics. For surface reconstruction of general shapes from raw point clouds, we conduct evaluations on three widely used datasets including a subset of ShapeNet [8], ABC [20], and FAMOUS [12]. We use the same setting with NeuralPull [4] for the dataset ShapeNet.

Table 1. Comparisons on ABC and Famous datasets. The threshold of F-score (F-S.) is 0.01.

Methods	ABC			FAMOUS		
	CD_{L2}	CD_{L1}	F-S.	CD_{L2}	CD_{L1}	F-S.
P2S [12]	0.298	0.015	0.598	0.012	0.008	0.752
IGR [14]	2.675	0.063	0.448	1.474	0.044	0.573
NP [4]	0.095	0.011	0.673	0.100	0.012	0.746
PCP [3]	0.252	0.023	0.373	0.037	0.014	0.435
SIREN [38]	0.022	0.012	0.493	0.025	0.012	0.561
DIGS [5]	0.021	0.010	0.667	0.015	0.008	0.772
Ours	0.011	0.009	0.691	0.008	0.007	0.778

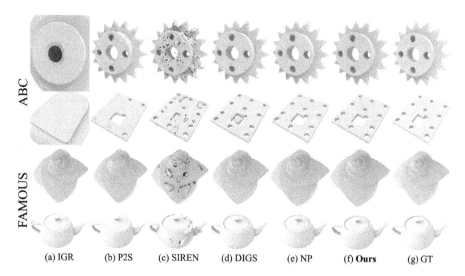

(a) IGR (b) P2S (c) SIREN (d) DIGS (e) NP (f) **Ours** (g) GT

Fig. 6. Visual comparisons of surface reconstruction on ABC and FAMOUS datasets. Our method can reconstruct objects with sharp edges and less noise compared with other methods.

For datasets ABC and FAMOUS, we use the train/test splitting released by Points2Surf [12] and we sample points directly from the mesh in the ABC dataset without other mesh preprocessing to keep the sharp features.

For evaluating the performance, we follow NeuralPull to sample 1×10^5 points from the reconstructed surfaces and the ground truth meshes on the ShapeNet dataset and sample 1×10^4 on the ABC and FAMOUS datasets. For the evaluation metrics, we use L1 and L2 Chamfer distance (CD_{L1} and CD_{L2}) to measure the error. Moreover, we adopt normal consistency (NC) and F-score to evaluate the accuracy of the reconstructed surface, the threshold is the same with NeuralPull.

Comparisons. To evaluate the validity of our implicit filter, we compare our method with a variety of methods including SPSR [19], Points2Surf (P2S) [12], IGR [14], NeuralPull (NP) [4], LPI [9], PCP [3], GridPull (GP) [10], SIREN [38], DIGS [5]. The quantitative results on ABC and FAMOUS datasets are shown in Tab. 1, and selectively visualized in Fig. 6. Our model reaches state-of-the-art performance on both datasets, accomplishing the goal of eliminating noise on each level set while preserving the geometric details. To more intu-

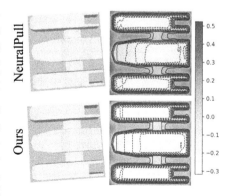

Fig. 7. Visualization of level sets on a cross section.

Table 2. Comparisons on ShapeNet dataset.

	SPSR [19]	NP [4]	LPI [9]	PCP [3]	GP [10]	Ours
$CD_{L2} \times 100$	0.286	0.038	0.0171	0.0136	0.0086	**0.0032**
NC	0.866	0.939	0.9596	0.9590	0.9723	**0.9779**
F-Score (0.002)	0.407	0.961	0.9912	0.9871	0.9896	**0.9976**
F-Score (0.004)	0.618	0.976	0.9957	0.9899	0.9923	**0.9985**

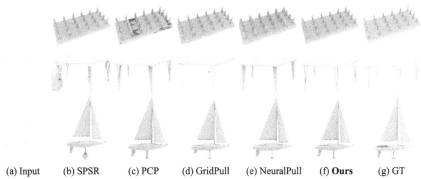

(a) Input   (b) SPSR   (c) PCP   (d) GridPull   (e) NeuralPull   (f) **Ours**   (g) GT

Fig. 8. Visual comparisons of surface reconstruction on ShapeNet dataset.

itively validate the efficacy of our filtering, we visualize the level sets on a cross section in Fig. 7. We also report the results on ShapeNet which contains over 3000 objects in terms of CD_{L2}, NC, and F-Score with thresholds of 0.002 and 0.004 in Table 2. The detailed comparison for each class of ShapeNet can be found in the supplementary material. Our method outperforms previous methods over most classes. The visualization comparisons in Fig. 8 show that our method can reconstruct a smoother surface with fine details.

To validate the effect of our filter on sharp geometric features. We evaluate the edge points by the edge Chamfer distance metric used in [11]. We sample 100k points uniformly on the surface of both the reconstructed mesh and ground truth. The edge point p is calculated by finding whether there exists a point $q \in \mathcal{N}_\epsilon(p)$ satisfied $|n_q n_p| < \sigma$, where $\mathcal{N}_\epsilon(p)$ represents the neighbor points within distance ϵ from p. The results are shown in Table 3 and visualized in Fig. 9. We set $\epsilon = 0.01$ and $\sigma = 0.1$.

4.2 Surface Reconstruction for Real Scans

Dataset and Metrics. For surface reconstruction of real point cloud scans, we follow VisCo [37] to evaluate our method under the Surface Reconstruction Benchmarks (SRB) [46]. We use Chamfer and Hausdorff distances (CD_{L1} and HD) between the reconstruction meshes and the ground truth. Furthermore,

Table 3. Edge Chamfer distance comparisons on ABC dataset, $ECD_{L2} \times 100$.

Methods	P2S [12]	IGR [14]	NP [4]	PCP [3]	SIREN [38]	DIGS [5]	Ours
ECD_{L1}	0.0496	0.0835	0.0501	0.0628	0.0695	0.0786	**0.0256**
ECD_{L2}	1.055	2.365	1.255	1.265	1.407	2.493	**0.399**

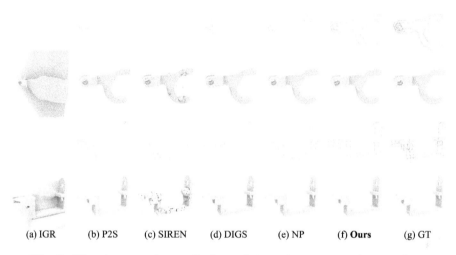

(a) IGR   (b) P2S   (c) SIREN   (d) DIGS   (e) NP   (f) **Ours**   (g) GT

Fig. 9. Visual comparisons of edge points and reconstruction results.

we report their corresponding one-sided distances ($d_{\overrightarrow{C}}$ and $d_{\overrightarrow{H}}$) between the reconstructed meshes and the input noisy point cloud.

Comparisons. We compare our method with state-of-the-art methods under the real scanned SRB dataset, including IGR [14], SPSR [19], Shape As Points (SAP) [34], NeuralPull (NP) [4], and GridPull (GP) [10]. The numerical comparisons are

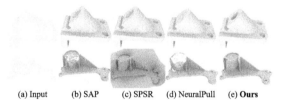

(a) Input   (b) SAP   (c) SPSR   (d) NeuralPull   (e) **Ours**

Fig. 10. Visual comparisons on SRB dataset.

shown in Table 4, where we achieve the best accuracy in most cases. The visual comparisons in Fig. 10 demonstrate that our method can reconstruct a continuous and smooth surface with geometry details.

4.3 Surface Reconstruction for Scenes

Dataset and Metrics. To further demonstrate the advantage of our method in the surface reconstruction of real scene scans, we conduct experiments using the 3D Scene dataset. The 3D Scene dataset is a challenging real-world dataset with

Table 4. Comparisons on SRB dataset.

		SPSR [19]	IGR [14]	SIREN [38]	VisCo [37]	SAP [34]	NP [4]	GP [10]	DIGS [5]	Ours
Anchor	CD_{L1}	0.60	0.22	0.32	0.21	0.12	0.122	0.093	0.063	**0.052**
	HD	14.89	4.71	8.19	3.00	2.38	3.243	1.804	1.447	**1.232**
	$d_{\vec{C}}$	0.60	0.12	0.10	0.15	0.08	0.061	0.066	0.030	**0.025**
	$d_{\vec{H}}$	14.89	1.32	2.432	1.07	0.83	3.208	0.460	0.270	**0.265**
Daratech	CD_{L1}	0.44	0.25	0.21	0.21	0.26	0.375	0.062	**0.049**	0.051
	HD	7.24	4.01	4.30	4.06	0.87	3.127	**0.648**	0.858	0.751
	$d_{\vec{C}}$	0.44	0.08	0.09	0.14	0.04	0.746	0.039	0.025	**0.028**
	$d_{\vec{H}}$	7.24	1.59	1.77	1.76	0.41	3.267	**0.293**	0.441	0.423
DC	CD_{L1}	0.27	0.17	0.15	0.15	0.07	0.157	0.066	0.042	**0.041**
	HD	3.10	2.22	2.18	2.22	1.17	3.541	1.103	**0.667**	0.815
	$d_{\vec{C}}$	0.27	0.09	0.06	0.09	0.04	0.242	0.036	0.022	**0.019**
	$d_{\vec{H}}$	3.10	2.61	2.76	2.76	**0.53**	3.523	0.539	0.729	0.724
Gargoyle	CD_{L1}	0.26	0.16	0.17	0.17	0.07	0.080	0.063	0.047	**0.044**
	HD	6.80	3.52	4.64	4.40	1.49	1.376	1.129	**0.971**	1.089
	$d_{\vec{C}}$	0.26	0.06	0.08	0.11	0.05	0.063	0.045	0.028	**0.022**
	$d_{\vec{H}}$	6.80	0.81	0.91	0.96	0.78	0.475	0.700	0.271	**0.246**
Lord Quas	CD_{L1}	0.20	0.12	0.17	0.12	0.05	0.064	0.047	0.031	**0.030**
	HD	4.61	1.17	0.82	1.06	0.98	0.822	0.569	**0.496**	0.554
	$d_{\vec{C}}$	0.20	0.07	0.12	0.07	0.04	0.053	0.031	0.017	**0.014**
	$d_{\vec{H}}$	4.61	0.98	0.76	0.64	0.51	0.508	0.370	**0.181**	0.230

complex topology and noisy open surfaces. We uniformly sample 1000 points per m^2 of each scene as the input and follow PCP [3] to sample 1M points on both the reconstructed and the ground truth surfaces. We leverage L1 and L2 Chamfer distance (CD_{L1}, CD_{L2}) and normal consistency (NC) to evaluate the reconstruction quality.

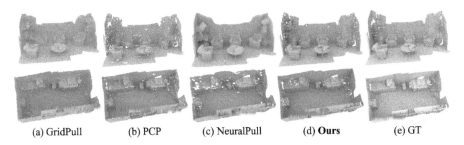

(a) GridPull (b) PCP (c) NeuralPull (d) **Ours** (e) GT

Fig. 11. Visual comparisons of surface reconstruction on 3D Scene dataset.

Comparisons. We compare our method with the state-of-the-art methods ConvONet [35], LIG [18], DeepLS [7], NeuralPull (NP) [4], PCP [3], GridPull (GP) [10]. The numerical comparisons in Table 5 demonstrate our superior performance in all scenes even compared with the local-based methods. We further present visual comparisons in Fig. 11. The visualization further shows that our method can achieve smoother with high-fidelity surfaces in complex scenes. It should be noted that the surface we extract here is not the zero level set but the 0.001 level set since the scene is not watertight. For NeuralPull we use the threshold of 0.005 instead of 0.001 to extract the complete surface therefore the mesh looks thicker.

Table 5. Comparisons on 3D Scene dataset, $CD_{L2} \times 1000$.

	Burghers			Lounge			Copyroom			Stonewall			Totempole		
	CD_{L2}	CD_{L1}	NC	CD_{L2}	CD_{L1}	NC	CD_{L2}	CD_{L1}	NC	CD_{L2}	CD_{L1}	NC	CD_{L2}	CD_{L1}	NC
ConvONet [35]	27.46	0.079	0.907	9.54	0.046	0.894	10.97	0.045	0.892	20.46	0.069	0.905	2.054	0.021	0.943
LIG [18]	3.055	0.045	0.835	9.672	0.056	0.833	3.61	0.036	0.810	5.032	0.042	0.879	9.58	0.062	0.887
DeepLS [7]	0.401	0.017	0.920	6.103	0.053	0.848	0.609	0.021	0.901	0.320	0.015	0.954	0.601	0.017	0.950
GP [10]	1.367	0.028	0.873	4.684	0.053	0.827	2.327	0.030	0.857	2.234	0.024	0.913	2.278	0.034	0.878
PCP [3]	1.339	0.031	0.929	0.432	0.014	**0.934**	0.405	0.014	**0.914**	0.266	0.014	**0.957**	1.089	0.029	**0.954**
NP [4]	0.897	0.025	0.883	0.855	0.022	0.887	0.479	0.018	0.862	0.434	0.018	0.929	1.604	0.032	0.923
Ours	**0.133**	**0.011**	**0.934**	**0.120**	**0.008**	0.926	**0.111**	**0.009**	0.913	**0.082**	**0.009**	**0.957**	**0.203**	**0.013**	0.944

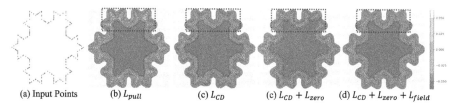

(a) Input Points (b) L_{pull} (c) L_{CD} (c) $L_{CD} + L_{zero}$ (d) $L_{CD} + L_{zero} + L_{field}$

Fig. 12. The 2D level sets show the distance field learned by different losses. The red lines represent the learned zero level set. (Color figure online)

4.4 Ablation Studies

We conduct ablation studies on the FAMOUS dataset to demonstrate the effectiveness of our proposed implicit filter and explore the effect of some important hyperparameters. We report the performance in terms of L1 and L2 Chamfer distance ($CD_{L1}, CD_{L2} \times 10^3$), normal consistency (NC), and F-Score (F-S.).

Effect of Eikonal Loss. We select the L_{CD} to prevent the degeneration of the gradient since it both constrains the value and the gradient of the SDF.

Table 6. Effect of the Eikonal term.

Loss	CD_{L1}	CD_{L2}	F-S.	NC
w/Eikonal, w/o CD	0.009	0.021	0.738	0.899
w/Eikonal, w/ CD	0.008	0.009	0.774	0.910
w/o Eikonal, w/ CD	**0.007**	**0.008**	**0.778**	**0.911**

It also guides how to pull the query point onto the surface. Therefore we omit the Eikonal term used in previous methods like the IGR [14], SIREN [38], and DIGS [5] which have no other direct supervision for the gradient. To verify this selection, we conduct the following experiments by trade-off these two functions.

Table 7. Effect of different losses.

Loss	CD_{L1}	CD_{L2}	F-S.	NC
L_{pull}	0.012	0.083	0.742	0.884
L_{CD}	0.010	0.031	0.757	0.891
$L_{CD} + L_{zero}$	0.008	0.018	0.772	0.905
$L_{CD} + L_{zero} + L_{field}$	0.008	0.011	0.769	0.908
Ours	**0.007**	**0.008**	**0.778**	**0.911**

Table 8. Effect of bidirectional projection.

	$d(\bar{p})$	$d_{bi}(\bar{p})$
CD_{L1}	0.010	**0.007**
CD_{L2}	0.024	**0.008**
F-S.	0.726	**0.778**
NC	0.890	**0.911**

With the experimental results in Table 6, we find that only applying the Eikonal term is not as effective as CD alone. At the same time combining the Eikonal term with CD does not further enhance the experiment results, but the difference is small.

Effect of Level Set Filtering. To justify the effectiveness of each term in our loss function. We report the results trained by different combinations in Table 7. The L_{CD} is more applicable for training SDF from raw point clouds. The zero-level filter can help remove the noise and keep the geometric features. Filtering across non-zero level sets can improve the overall consistency of the entire signed distance field. Since we assume all input points lie on the surface, the function L_{dist} is also necessary. Figure 12 shows a 2D comparison of these losses, showing that our filter loss functions can reconstruct a field that is aligned at all level sets and maintains geometric characteristics.

Effect of the Bidirectional Projection. To validate our bidirectional projection distance, we report the results in Table 8. The numerical comparisons show that projecting the distance to both normals can improve the reconstruction quality. Note that only using $d(\bar{p})$ can also improve the results.

Weight of Level Set Projection Loss. We explore the effect of the L_{CD} loss function by adjusting the weight α_3 in Eq. (11). We report our results with different candidates $\{0, 1, 10\}$ in Table 9, where 0 means we do not use the L_{CD} to constrain the gradient. The comparisons in Table 9 show that although our implicit filter can directly learn SDFs, it is better to adopt the L_{CD} for a more stable field. However, if the weight is too large, the filtering effect will decrease. It is recommended to select weights ranging from 1 to 10, which is usually adequate. For the weights α_1 and α_2, setting them to 1 is always necessary.

Effect of Filter Parameters. We compare the effect of different parameters σ_n, σ_p in Table 10. The diagonal weight for σ_p means the length of the diagonal of the bounding box for the local patch mentioned in [48]. The results indicate that the method is relatively robust to parameter variation in a certain range.

Table 9. Effect of weight α_3.

α_3	CD_{L1}	CD_{L2}	F-S.	NC
0	0.008	0.013	0.758	0.903
1	**0.007**	0.011	0.772	0.910
10	**0.007**	**0.008**	**0.778**	**0.911**
100	0.008	0.009	0.774	0.909

Table 10. Effect of filter parameters σ_n and σ_p.

		CD_{L1}	CD_{L2}	F-S.	NC
σ_n	15°	**0.007**	**0.008**	**0.778**	**0.911**
	30°	0.007	0.011	0.771	0.907
	45°	0.008	0.012	0.764	0.903
	60°	0.008	0.010	0.767	0.901
σ_p	max	**0.007**	**0.008**	**0.778**	**0.911**
	diagonal	0.008	0.011	0.763	0.904

5 Conclusion

We introduce implicit filtering on SDFs to reduce the noise of the signed distance field while preserving geometry features. We filter the distance field by minimizing the weighted bidirectional projection distance, where we can generate sampling points on the zero level set and neighbor points on non-zero level sets by the pulling procedure. By leveraging the Chamfer distance, we address the issue of gradient degeneration problem. The visual and numerical comparisons demonstrate our effectiveness and superiority over state-of-the-art methods.

Acknowledgements. This work was supported by Beijing Science and Technology Program (Z231100001723014).

References

1. Atzmon, M., Lipman, Y.: Sal: Sign agnostic learning of shapes from raw data. In: IEEE/CVF Conference on Computer Vision and Pattern Recognition (CVPR), June 2020
2. Atzmon, M., Lipman, Y.: SALD: sign agnostic learning with derivatives. In: 9th International Conference on Learning Representations, ICLR 2021 (2021)
3. Baorui, M., Yu-Shen, L., Matthias, Z., Zhizhong, H.: Surface reconstruction from point clouds by learning predictive context priors. In: Proceedings of the IEEE/CVF Conference on Computer Vision and Pattern Recognition (CVPR) (2022)
4. Baorui, M., Zhizhong, H., Yu-Shen, L., Matthias, Z.: Neural-pull: Learning signed distance functions from point clouds by learning to pull space onto surfaces. In: International Conference on Machine Learning (ICML) (2021)
5. Ben-Shabat, Y., Hewa Koneputugodage, C., Gould, S.: Digs: divergence guided shape implicit neural representation for unoriented point clouds. In: Proceedings of the IEEE/CVF Conference on Computer Vision and Pattern Recognition, pp. 19323–19332 (2022)
6. Boulch, A., Marlet, R.: Poco: Point convolution for surface reconstruction. In: Proceedings of the IEEE/CVF Conference on Computer Vision and Pattern Recognition (CVPR). pp. 6302–6314 (June 2022)

7. Chabra, R., et al.: Deep local shapes: Learning local sdf priors for detailed 3d reconstruction. In: Computer Vision–ECCV 2020: 16th European Conference, Glasgow, UK, August 23–28, 2020, Proceedings, Part XXIX 16, pp. 608–625. Springer (2020)
8. Chang, A.X., et al.: ShapeNet: an information-rich 3D model repository. Tech. Rep. arXiv:1512.03012 [cs.GR], Stanford University — Princeton University — Toyota Technological Institute at Chicago (2015)
9. Chao, C., Yu-shen, L., Zhizhong, H.: Latent partition implicit with surface codes for 3d representation. In: European Conference on Computer Vision (ECCV) (2022)
10. Chen, C., Liu, Y.S., Han, Z.: Gridpull: towards scalability in learning implicit representations from 3d point clouds. In: Proceedings of the IEEE International Conference on Computer Vision (ICCV) (2023)
11. Chen, Z., Tagliasacchi, A., Zhang, H.: Bsp-net: generating compact meshes via binary space partitioning. Proceedings of IEEE Conference on Computer Vision and Pattern Recognition (CVPR) (2020)
12. Erler, P., Guerrero, P., Ohrhallinger, S., Mitra, N.J., Wimmer, M.: POINTS2SURF learning implicit surfaces from point clouds. In: Vedaldi, A., Bischof, H., Brox, T., Frahm, J.-M. (eds.) ECCV 2020. LNCS, vol. 12350, pp. 108–124. Springer, Cham (2020). https://doi.org/10.1007/978-3-030-58558-7_7
13. Fleishman, S., Cohen-Or, D., Silva, C.T.: Robust moving least-squares fitting with sharp features. ACM Trans. Graph. (TOG) **24**(3), 544–552 (2005)
14. Gropp, A., Yariv, L., Haim, N., Atzmon, M., Lipman, Y.: Implicit geometric regularization for learning shapes. In: Proceedings of Machine Learning and Systems 2020, pp. 3569–3579 (2020)
15. Huang, H., Li, D., Zhang, H., Ascher, U., Cohen-Or, D.: Consolidation of unorganized point clouds for surface reconstruction. ACM Trans. Graph. (TOG) **28**(5), 1–7 (2009)
16. Huang, H., Wu, S., Gong, M., Cohen-Or, D., Ascher, U., Zhang, H.: Edge-aware point set resampling. ACM Trans. Graph. (TOG) **32**(1), 1–12 (2013)
17. Huang, J., Gojcic, Z., Atzmon, M., Litany, O., Fidler, S., Williams, F.: Neural kernel surface reconstruction. In: Proceedings of the IEEE/CVF Conference on Computer Vision and Pattern Recognition, pp. 4369–4379 (2023)
18. Jiang, C.M., Sud, A., Makadia, A., Huang, J., Nießner, M., Funkhouser, T.: Local implicit grid representations for 3d scenes. In: Proceedings IEEE Conf. on Computer Vision and Pattern Recognition (CVPR) (2020)
19. Kazhdan, M., Hoppe, H.: Screened poisson surface reconstruction. ACM Trans. Graph. (ToG) **32**(3), 1–13 (2013)
20. Koch, S., et al.: Abc: a big cad model dataset for geometric deep learning. In: Proceedings of the IEEE/CVF Conference on Computer Vision and Pattern Recognition, pp. 9601–9611 (2019)
21. Koneputugodage, C.H., Ben-Shabat, Y., Campbell, D., Gould, S.: Small steps and level sets: fitting neural surface models with point guidance. In: Proceedings of the IEEE/CVF Conference on Computer Vision and Pattern Recognition, pp. 21456–21465 (2024)
22. Li, S., Gao, G., Liu, Y., Liu, Y.S., Gu, M.: Gridformer: point-grid transformer for surface reconstruction. In: Proceedings of the AAAI Conference on Artificial Intelligence (2024)
23. Liao, B., Xiao, C., Jin, L., Fu, H.: Efficient feature-preserving local projection operator for geometry reconstruction. Comput. Aided Des. **45**(5), 861–874 (2013)
24. Lindell, D.B., Van Veen, D., Park, J.J., Wetzstein, G.: Bacon: band-limited coordinate networks for multiscale scene representation. In: Proceedings of the

IEEE/CVF Conference on Computer Vision and Pattern Recognition, pp. 16252–16262 (2022)
25. Lipman, Y., Cohen-Or, D., Levin, D., Tal-Ezer, H.: Parameterization-free projection for geometry reconstruction. ACM Trans. Graph. (TOG) **26**(3), 22–es (2007)
26. Liu, S.L., Guo, H.X., Pan, H., Wang, P.S., Tong, X., Liu, Y.: Deep implicit moving least-squares functions for 3d reconstruction. In: Proceedings of the IEEE/CVF Conference on Computer Vision and Pattern Recognition, pp. 1788–1797 (2021)
27. Lorensen, W.E., Cline, H.E.: Marching cubes: A high resolution 3d surface construction algorithm. In: Seminal graphics: pioneering efforts that shaped the field, pp. 347–353 (1998)
28. Ma, B., Liu, Y.S., Han, Z.: Reconstructing surfaces for sparse point clouds with on-surface priors. In: Proceedings of the IEEE/CVF Conference on Computer Vision and Pattern Recognition, pp. 6315–6325 (2022)
29. Ma, B., Zhou, J., Liu, Y.S., Han, Z.: Towards better gradient consistency for neural signed distance functions via level set alignment. In: Conference on Computer Vision and Pattern Recognition (CVPR) (2023)
30. Marschner, Z., Sellán, S., Liu, H.T.D., Jacobson, A.: Constructive solid geometry on neural signed distance fields. In: SIGGRAPH Asia 2023 Conference Papers, pp. 1–12 (2023)
31. Mescheder, L., Oechsle, M., Niemeyer, M., Nowozin, S., Geiger, A.: Occupancy networks: Learning 3d reconstruction in function space. In: Proceedings IEEE Conf. on Computer Vision and Pattern Recognition (CVPR) (2019)
32. Mi, Z., Luo, Y., Tao, W.: Ssrnet: scalable 3d surface reconstruction network. In: Proceedings of the IEEE/CVF Conference on Computer Vision and Pattern Recognition, pp. 970–979 (2020)
33. Öztireli, A.C., Guennebaud, G., Gross, M.: Feature preserving point set surfaces based on non-linear kernel regression. In: Computer graphics forum. vol. 28, pp. 493–501. Wiley Online Library (2009)
34. Peng, S., Jiang, C., Liao, Y., Niemeyer, M., Pollefeys, M., Geiger, A.: Shape as points: a differentiable poisson solver. Adv. Neural. Inf. Process. Syst. **34**, 13032–13044 (2021)
35. Peng, S., Niemeyer, M., Mescheder, L., Pollefeys, M., Geiger, A.: Convolutional occupancy networks. In: European Conference on Computer Vision (ECCV) (2020)
36. Preiner, R., Mattausch, O., Arikan, M., Pajarola, R., Wimmer, M.: Continuous projection for fast l1 reconstruction. ACM Trans. Graph. **33**(4), 47–1 (2014)
37. Pumarola, A., Sanakoyeu, A., Yariv, L., Thabet, A., Lipman, Y.: Visco grids: surface reconstruction with viscosity and coarea grids. Adv. Neural. Inf. Process. Syst. **35**, 18060–18071 (2022)
38. Sitzmann, V., Martel, J., Bergman, A., Lindell, D., Wetzstein, G.: Implicit neural representations with periodic activation functions. Neural Information Processing Systems, Neural Information Processing Systems (Jun (2020)
39. Tancik, M., et al.: Fourier features let networks learn high frequency functions in low dimensional domains. Adv. Neural. Inf. Process. Syst. **33**, 7537–7547 (2020)
40. Tang, J., Lei, J., Xu, D., Ma, F., Jia, K., Zhang, L.: Sa-convonet: Sign-agnostic optimization of convolutional occupancy networks. In: Proceedings of the IEEE/CVF International Conference on Computer Vision (2021)
41. Tretschk, E., Tewari, A., Golyanik, V., Zollhöfer, M., Stoll, C., Theobalt, C.: PatchNets: patch-based generalizable deep implicit 3D shape representations. In: Vedaldi, A., Bischof, H., Brox, T., Frahm, J.-M. (eds) ECCV 2020. LNCS, vol. 12361, pp. 293–309. Springer, Cham (2020). https://doi.org/10.1007/978-3-030-58517-4_18

42. Wang, X., Cheng, Y., Wang, L., Lu, J., Xu, K., Xiao, G.: Edge preserving implicit surface representation of point clouds. arXiv preprint arXiv:2301.04860 (2023)
43. Wang, Y., Rahmann, L., Sorkine-Hornung, O.: Geometry-consistent neural shape representation with implicit displacement fields. In: The Tenth International Conference on Learning Representations. OpenReview (2022)
44. Wang, Z., et al.: Alto: alternating latent topologies for implicit 3d reconstruction. In: Proceedings of the IEEE/CVF Conference on Computer Vision and Pattern Recognition, pp. 259–270 (2023)
45. Williams, F., et al.: Neural fields as learnable kernels for 3d reconstruction. In: Proceedings of the IEEE/CVF Conference on Computer Vision and Pattern Recognition, pp. 18500–18510 (2022)
46. Williams, F., Schneider, T., Silva, C., Zorin, D., Bruna, J., Panozzo, D.: Deep geometric prior for surface reconstruction. In: Proceedings of the IEEE/CVF Conference on Computer Vision and Pattern Recognition, pp. 10130–10139 (2019)
47. Yu, L., Li, X., Fu, C.-W., Cohen-Or, D., Heng, P.-A.: EC-net: an edge-aware point set consolidation network. In: Ferrari, V., Hebert, M., Sminchisescu, C., Weiss, Y. (eds.) ECCV 2018. LNCS, vol. 11211, pp. 398–414. Springer, Cham (2018). https://doi.org/10.1007/978-3-030-01234-2_24
48. Zhang, D., Lu, X., Qin, H., He, Y.: Pointfilter: point cloud filtering via encoder-decoder modeling. IEEE Trans. Visual Comput. Graphics **27**(3), 2015–2027 (2020)
49. Zhou, J., Ma, B., Liu, Y.S., Fang, Y., Han, Z.: Learning consistency-aware unsigned distance functions progressively from raw point clouds. In: Advances in Neural Information Processing Systems (NeurIPS) (2022)

Unsupervised Exposure Correction

Ruodai Cui[1], Li Niu[2(✉)], and Guosheng Hu[3]

[1] Qualcomm Technologies, Inc., Shanghai, China
ruodcui@qti.qualcomm.com
[2] Department of Computer Science and Engineering, MoE Key Lab of Artificial Intelligence, Shanghai Jiao Tong University, Shanghai, China
ustcnewly@sjtu.edu.cn
[3] University of Bristol, Bristol, UK

Abstract. Current exposure correction methods have three challenges, labor-intensive paired data annotation, limited generalizability, and performance degradation in low-level computer vision tasks. In this work, we introduce an innovative Unsupervised Exposure Correction (UEC) method that eliminates the need for manual annotations, offers improved generalizability, and enhances performance in low-level downstream tasks. Our model is trained using *freely* available paired data from an emulated Image Signal Processing (ISP) pipeline. This approach does not need expensive manual annotations, thereby minimizing individual style biases from the annotation and consequently improving its generalizability. Furthermore, we present a large-scale Radiometry Correction Dataset, specifically designed to emphasize exposure variations, to facilitate unsupervised learning. In addition, we develop a transformation function that preserves image details and outperforms state-of-the-art supervised methods [12], while utilizing only 0.01% of their parameters. Our work further investigates the broader impact of exposure correction on downstream tasks, including edge detection, demonstrating its effectiveness in mitigating the adverse effects of poor exposure on low-level features. The source code and dataset are publicly available at https://github.com/BeyondHeaven/uec_code.

Keywords: Exposure correction · Unsupervised learning

1 Introduction

Exposure, a pivotal factor in photography, significantly impacts image quality by influencing visual clarity. In radiometry, the exposure is defined as scene irradiance, the amount of light that reaches the image sensor. This can be controlled by exposure value (EV), which combines essential factors such as aperture, shutter speed, and ISO settings. Despite advancements in Image Signal Processor (ISP), which facilitate automatic EV adjustments during image capture, challenges persist under non-ideal lighting conditions. As a result, post-processing of

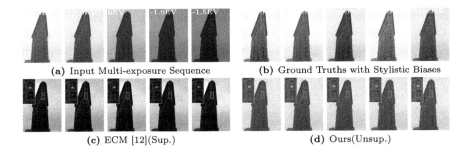

Fig. 1. Visual Comparison: ECM [12] vs. Our UEC method.

sRGB images remains crucial. The advent of deep learning has spurred extensive research in this field, with numerous studies proposing models for exposure correction [1,4,14,16,19,24,25,27,32,34,42,44], demonstrating significant achievements. Nevertheless, these methods still face three challenges.

(1) A primary challenge arises from the dependence on the paired data, where the ground truth is from proficient photographers. This process is inherently complex and labor-intensive, as it involves detailed editing and refinement of each image, demanding more manual work than just labeling as in classification tasks. (2) Previous methods often suffer from limited generalizability. To be specific, (a) as noted, the efficiency of manual adjustments is low, resulting in small datasets available in academia. (b) the manual adjustments inevitably introduce different stylistic biases. Personal preferences vary significantly between individuals, suggesting that such ground truths are intrinsically noisy. (3) Previous methodologies have predominantly focused on generating aesthetically pleasing images. However, they often yield images with notable degradations in low-level features. Such degradations render these images less suitable for various downstream computer vision tasks, like edge detection and segmentation, where preserving these features is important.

Inspired by Afifi et al.'s work of exposure dataset [2], we realize that acquiring paired data does not necessarily require extensive manual intervention. They apply an emulated Image Signal Processing (ISP) pipeline on RAW data, thereby creating multi-exposure sequences, as shown in Fig. 1a. The work [2] learns the mapping from the inputs (Fig. 1a) to the ground truth (ExpertC in Fig. 1b), which is a typical supervised learning. However, this approach introduces ambiguities due to inconsistencies in the ground truths, e.g. five experts in Fig. 1b. Unlike existing supervised learning methods [2,12,14], in this work, we introduce an innovative Unsupervised Exposure Correction (UEC) method. (1) To achieve this, we creatively ask the images in the same multi-exposure sequence, which can be generated freely, to mutually serve as a ground truth for learning exposure adjustments across a diverse range. Specifically, we simply generate multi-exposure sequences and EV labels via an emulated ISP pipeline. In this way, we do not need the expensive manual annotations to generate ground truth. (2) The existing methods have inconsistent ground truth which might have different standards on colors, leading to degraded performance. In contrast, in our UEC method, the images in multi-exposure sequence do not have such "standards" because they mutually work as the ground truth. (3) We employ a pixel-wise exposure transformation in sRGB color space for preserving visual details as shown in Fig. 1d which greatly outperforms the SOTA method in Fig. 1c. This detail-preserving property is crucial for downstream computer vision tasks that rely on low-level features, such as edge detection and segmentation. To further facilitate the analysis without individual stylistic biases in exposure correction, we contribute a new dataset, Radiometry Correction Dataset, that is large and encompasses broad exposures. This dataset maintains a consistent style with only radiometry variations.

Our contributions can be summarized as follows: (1) By modeling radiometry, we introduce an innovative unsupervised learning methodology that fundamentally addresses the problem of expensive annotations. To our knowledge, this is the first unsupervised learning solution for exposure correction, which achieves competitive performance as SOTA supervised models [12] with only 0.01% of their parameters, and minimizes individual stylistic biases. (2) We propose a pixel-wise exposure transformation that can well preserve the details of images and flexibly accept different resolutions of input images. By this way, the enhanced images can improve the performance of many downstream tasks, e.g. edge detection. (3) We contribute a new dataset focusing on radiometry for better generalizability. Input images of this dataset are synthesized from ground truth data, ensuring a uniform style across the dataset.

2 Related Work

Exposure correction has evolved from traditional techniques to modern deep learning methods. Traditional methods, like histogram equalization [17], and Retinex-based algorithms [22,33], laid the groundwork. Recent deep learning studies, including Ren et al. [34], Zhang et al. [44], Guo et al. [14], and Loh et al. [29], focus on low-light enhancement. However, these methods exposed a gap in addressing both underexposed and overexposed images. Afifi et al. [2] introduced a dataset featuring varying exposure attributes, enabling comprehensive methodologies.

Exposure correction methods fall into two categories: image-to-image translation and color transformation. Image-to-image translation aims to create dense translations between input and output pairs, leveraging deep learning as seen in works by Eyiokur et al. [12], Afifi et al. [2], and others [4,8,19]. Color transformation methods, on the other hand, predict mapping curve parameters to enhance images, employing techniques like quadratic transforms [7,26,37,41], local affine transforms [13], curve-based transforms [5,14,20,23,31], filters [11,30], lookup tables [38,43], and MLP models [15,28,39]. These techniques define specific functions or models for the enhancement process, offering diverse tools for image exposure challenges.

3 Method

Traditionally, exposure correction learns a mapping from one input image I_1 with one (usually weaker) exposure to another image I_2 with a different (usually stronger and better) exposure. To learn this mapping, conventionally, people *manually* generate the well-exposed image I_2 as the ground-truth of the existing I_1, which is quite expensive and is hard to scale up.

To address this, we creatively propose a way to exploit the information from the *freely* generated multi-exposure sequence from RAW data. In the same sequence, the images with different exposures can mutually work as the ground-truth to learn the mapping for exposure adjustment. In Sect. 3.1, we formulate our Unsupervised Exposure Correction (UEC). Then, the neural architectures in Sect. 3.2 and loss functions in Sect. 3.3 are introduced to achieve UEC. Following that, Sect. 3.4 outlines the testing methodology. Moreover, we propose our Radiometry Correction Dataset where the paired images are freely generated from RAW data in Sect. 4.

3.1 Unsupervised Exposure Correction Modeling

The emulated ISP enables us to generate extensive multi-exposure sequences. To make full use of the large data, we introduce a task with self-supervision. A reference image guides exposure adjustments, serving as the calibration target, while a transformation curve shifts the exposure level. This method involves varying the exposure over a wide range, rather than directly aligning it to an optimal value, thereby categorizing it as a specific form of style transfer.

We initiate the process by employing a style encoder to extract the exposure feature from the image, denoted as $E = e(I)$. Diverging from conventional style transfer methods, which typically rely on a one-hot vector as a prior [9] or leverage features refined through triplet loss [21], we have developed an alternative optimization approach that better accommodates exposure dynamics. We discovered a sequential relationship within varying exposure images, allowing us to order them by their exposures. Consequently, to quantify the difference between any two images, we can employ a single scalar.

$$\Delta E = d(E_1, E_2) = d(e(I_1), e(I_2)). \tag{1}$$

Then, we can perform the style transformation from ΔE:

$$I'_1 = f(\Delta E, I_1). \tag{2}$$

Having established the transformation function, , we now focus on developing optimization strategies. We identify two principles to guide the optimization process.

Restoration Supervision for Pretext Task. Consider a scenario where we sample two images, I_1 and I_2, from the same sequence I, leading to the equation $I'_1 = I_2$. We call this "Restoration Supervision", which can train $f(\cdot)$ under the restoration guidance of I_2, as illustrated in Fig. 2a.

However, it is important to note that both I_1 and I_2 originate from the same sequence I. For the real task, we need to handle image pairs from different scenes. Since our method learn from data which varies solely in exposure, the trained $f(\cdot)$ can only modify exposures. Moreover, we calculate the exposure difference in a latent space, not in pixel space, which omits low-level features. This approach enables the difference function $d(\cdot,\cdot)$ to adapt images from different scenes to a certain degree. Therefore, when we select reference images from a different sequence, the alterations remain confined to exposure adjustment, even though the ΔE computed may not precisely indicating the optimal exposure variation.

Monopoly Principle for Real Task. In order to compute this value accurately, we proceed to train the network with images from different scenes. However, comparing exposures in images from different scenes poses a challenge. Instead of directly contrasting the input and output images, we compare two outputs from the same input, ensuring scene consistency and enabling pixel-to-pixel optimization. Notably, if an over-exposed image is selected as a reference, the resulting image will be brighter; conversely, choosing an under-exposed image as a reference will lead to a darker result. Therefore, by selecting reference images with varying EVs, we can create a bright-dark image pair. We call this "Monopoly Principle" to indicate that the output's brightness should be monopoly with the exposure of reference image.

We select two reference images, J_1 and J_2, from a sequence J, distinct from I, where we sample input images I_1. We assume the EV of J_1 exceeds that of J_2, which means every pixel value of J_1 should be equal or over that of J_2. This condition can be expressed as:

$$\forall (x,y), \quad J_1(x,y) \geq J_2(x,y) \quad \text{with EV}(J_1) > \text{EV}(J_2). \tag{3}$$

Employing J_1 and J_2 as reference images, we apply $d(\cdot,\cdot)$ and $f(\cdot)$ to the same image I_1, resulting in two transformed images, I'_{J_1} and I'_{J_2}, respectively. In this case, I'_{J_1} is expected to exhibit a brighter appearance compared to I'_{J_2}. This can be described as

$$I'_{J_1} = f(d(e(I_1), e(J_1)), I_1), \quad I'_{J_2} = f(d(e(I_1), e(J_2)), I_1). \tag{4}$$

Combine Eq. (3) and Eq. (4), we can get

$$\forall (x,y), \quad I'_{J_1}(x,y) \geq I'_{J_2}(x,y) \quad \text{with EV}(J_1) > \text{EV}(J_2). \tag{5}$$

As demonstrated in Fig. 2b, this learning criterion enables the network to align exposures across diverse scenes, thereby enhancing the applicability of our method in real-world situations.

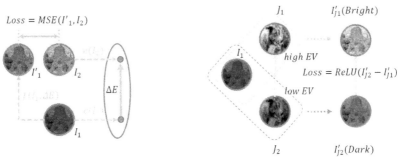

(a) Pretext Task training. Notations are defined in Eq. (1) and Eq. (2)

(b) Real Task training. Notations are defined in Eq. (4)

Fig. 2. Schematic diagrams of training UEC.

Summarization. We design a transformation to adjust exposures, which can be optimized by two principles. Firstly, we select images from a single multi-exposure sequence, establishing a baseline for exposure manipulation. Secondly, we aim to adapt to varied scenes. This involves selecting two reference images from different exposures, but differing from the sequence where we sample the input image. By preserving the relative brightness in the outputs, we facilitate exposure style transformations across different scenes. We concurrently train the network towards using two specific loss functions for each goal.

3.2 Unsupervised Exposure Correction Architecture

We have introduced three crucial functions: (1) $e(\cdot)$ for the extraction of exposure-related features, (2) $d(\cdot, \cdot)$ for the computation of exposure difference between two images in a latent space, and (3) $f(\cdot)$ for the exposure correction of images given the exposure difference. Our UEC network is structured around these functions, as illustrated in the overview shown in Fig. 3a. Specifically, our UEC framework incorporates three distinct neural networks: (1) Exposure Feature Encoder, detailed in Fig. 3b; (2) Parameter Predictor based on the difference, detailed in Fig. 3c; and (3) Exposure Corrector, detailed in Fig. 3d.

Firstly, a decoder comprises two convolutional layers followed by a global pooling layer, producing a 96D feature representation through the combination of maximum, average, and standard deviation across channels. These statistical metrics links to global attributes of images, e.g., contrast, histogram distribution. Secondly, Parameter Predictor evaluates the exposure difference between image pairs, and then computes the parameter λ for exposure correction. Thirdly, the exposure correction is modeled as an interpolation of direct scaling and a non-linear adjustment through a network:

$$I_{out}(x,y) = \lambda \times I_{in}(x,y) + (1-\lambda) \times h(I_{in}(x,y)), \qquad (6)$$

where λ modulates the transformed between the original input image I_{in} and its corresponding output I_{out}. $h(\cdot)$ is the non-linear transformation function imple-

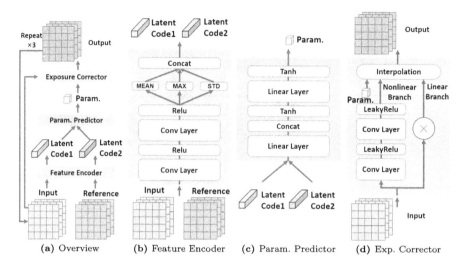

Fig. 3. The architecture of our UEC method.

mented by 1 × 1 convolution layers. This iterative process is repeated three times to enhance the correction effectiveness. For more details, please refer to supplementary material.

3.3 Loss Function

We jointly train the Exposure Feature Encoder, Parameter Predictor, and Exposure Corrector in an end-to-end fashion. We employ three types of losses and compute their weighted sum as the final loss.

Restoration Loss. Based on the Restoration Supervision discussed in Sect. 3.1, when we sample the input and reference from the same multi-exposure sequence, the synthesized exposure-adjusted image should be identical to the reference image. This is illustrated with I_1 and I_2 in Fig. 2a. We utilize the L2 loss for the restoration:

$$L_{\text{restoration}} = \frac{1}{CHW} \left\| I^{\text{out}} - I^{\text{ref}} \right\|_2. \tag{7}$$

Here, C, H, and W denote the image's channels, height, and width, respectively. I^{out} is the neural network output, and I^{ref} the reference image.

Monopoly Loss. Based on the Monopoly Principle discussed in Sect. 3.1, when we sample the two references from the same multi-exposure sequence, which is different from the input's sampling source, the relative brightness relation should also be transferred to the output image pair. This is shown with I'_{J1} and I'_{J2} in Fig. 2b. We utilize the ReLU function for the monopoly loss:

$$L_{\text{monopoly}} = \frac{1}{CHW}\text{ReLU}\left(I^{\text{out2}} - I^{\text{out1}}\right) \quad \text{with } \text{EV}(I^{\text{ref1}}) > \text{EV}(I^{\text{ref2}}). \quad (8)$$

In this equation, I^{out1} and I^{out2} represent the neural network outputs corresponding to the references I^{ref1} and I^{ref2}, respectively. It is important to note that the EV of I^{ref1} is greater than that of I^{ref2}. The ReLU function ensures that the loss calculation emphasizes the positive differences, aligning with the expected brightness relation between the two outputs.

Semantic-Preserving Loss. For the preservation of semantics, we use total variation loss [3], a proven regularization technique to maintain spatial coherence and semantic integrity. This approach smooths transitions between adjacent pixels:

$$L_{\text{semantic}} = \frac{1}{CHW}\left\|\nabla I^{\text{out}}\right\|_2. \quad (9)$$

Here, $\nabla(\cdot)$ denotes the gradient operator, computing the image's spatial derivative.

Summarization. The final loss function is expressed as:

$$L = \alpha_1 \cdot L_{\text{restoration}} + \alpha_2 \cdot L_{\text{monopoly}} + \alpha_3 \cdot L_{\text{semantic}}, \quad (10)$$

where α_1, α_2, and α_3 are the balancing weights. Since our framework (Exposure Feature Encoder, Parameter Predictor, Exposure Corrector) works in an end-to-end way, the total loss function, Eq. (10), is added on the end of exposure corrector to update the weights of all the modules during training.

3.4 Testing Details

We have outlined the training procedure for our UEC model, which employs reference images of diverse quality to address the full spectrum of exposure adjustments. For the testing phase, it is crucial to identify the optimal value for the final calibration. The testing inference largely replicates the training process as depicted in Fig. 3a, with the primary distinction being the use of a single reference image for all testing inputs. Specifically, we hard code the exposure features derived from this image across all test cases.

The efficiency of this method is underscored by its minimal data requirements. Mastering exposure adjustments is complex, but identifying an image's optimal exposure level is considerably more straightforward. Our approach emphasizes radiometry to ensure consistency, which differs from colorimetry, thereby minimizing the data needed to such an extent that a single ground truth image is sufficient. After training, the UEC model is adept at generating a sequence of images with varying exposures for any given input. The goal during testing is to select the most suitable image from this sequence. To enhance efficiency, our method bypasses the need to generate and then choose from multiple exposures, instead directly yielding the image with the best exposure.

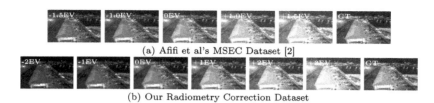

(a) Afifi et al's MSEC Dataset [2]

(b) Our Radiometry Correction Dataset

Fig. 4. Visualization of Datasets: (a) Afifi et al.'s [2] vs. (b) Ours.

4 Our Radiometry Correction Dataset

In Sect. 1, we have explored the benefits of disentanglement in radiometry adjustment for enhancing the generalizability of exposure correction. For this purpose, a specialized dataset is essential. Hence, we developed Radiometry Correction Dataset using MIT-Adobe FiveK Dataset [5], which includes 5,000 RAW photographs and their expertly modified sRGB images. These adjustments were performed by five specialists who worked directly with the RAW images. Drawing from their expertise, we adopted a reverse engineering approach to generate ill-exposed images. This process entails adjusting the exposure of the reference images to generate synthetic versions that are either underexposed or overexposed, while while freezing other post-processing ISP procedures to keep the individual style constant. Following the established convention, we selected the versions edited by ExpertC as our reference images. Based on this, we adjusted the exposures relative to the original images, spanning -2EV, -1EV, 0EV, +1EV, +2EV, and +3EV. Considering that input images are often underexposed, this range ensures a balanced exposure spectrum in relation to the reference images.

Figure 4 provides a comparison between Afifi's MSEC Dataset [2] and our Radiometry Correction Dataset. Evidently, in this example, the ground exhibits a blue hue as ExpertC's individual style. For additional comparisons, please consult the supplementary material.

5 Experiments

5.1 Experiment Settings

Datasets. The prevailing benchmark primarily utilizes the MSEC Dataset from the study by Afifi et al. [2]. To mitigate bias stemming from individual styles on colorimetry, we introduce our Radiometry Correction Dataset. Both datasets are employed for training and testing various methods. To assess the models' generalizability, we conduct training on the Exposure Dataset and perform testing on LOL dataset [6], which is specifically designed for low-light image enhancement.

Implementation Details. The weights in Eq. (10) are empirically set $\alpha_1 = \alpha_2 = 1$ and $\alpha_3 = 0.1$. We use only one well-exposed reference image during testing. We select the second image from left to right in Fig. 9 as our reference in our evaluations. More details are described in our supplementary material.

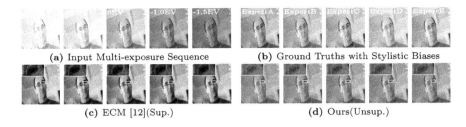

Fig. 5. Visual Comparison: ECM [12] vs. Our UEC method.

Evaluation Metrics. In Sect. 1, we highlighted the importance of evaluating exposure correction in terms of both overall aesthetics and low-level features. To assess overall aesthetics, we compare the output to the ground truth images using metrics such as PSNR and SSIM [35]. For low-level feature evaluation, we analyze the output through edge detection and make comparisons using PSNR and F1 scores, applying a threshold based on median pixel values.

5.2 Exposure Correction Results

Results on MSEC Dataset. We compare our results with the ground truth, as presented in Table 1a. Even though our Unsupervised Exposure Correction (UEC) model has not been trained on manually calibrated data and our adjustments are confined to radiometry, the proficiency of our UEC model remains competitive when compared to supervised SOTA models like ECM [12] and Afifi et al. [2].

In Fig. 6, we present a comparison of test images using various methods, including our approach, ground truth images, and results from supervised models. Our method demonstrates competitive performance in radiometry correction, matching the SOTA supervised techniques. Furthermore, in Fig. 1 and Fig. 5, we compare our UEC method with ECM [12]. In Fig. 1, it is important to note that the ground truth colorimetry, as determined by five experts, exhibits variability, indicating that learning from human-adjusted images may introduce biases reflective of their individual styles. In Fig. 5, the ECM's results [12] (Fig. 5c) manifest a color cast in the background, where the background turns green. Our method addresses this issue by applying radiometry correction. Furthermore, to highlight the superior detail resolution achieved by our UEC method, we have magnified a segment of the image located in the upper left corner. Further comparisons are elaborated in the supplementary material.

Results on Generalizability. To assess the models' generalizability, we utilized pretrained models and evaluated their performance on LOL dataset [40]. LOL dataset comprises 485 pairs of training images, with each pair featuring one low-light image and one corresponding normal-light image. For our assessment, we exclusively employed the training image pairs from LOL dataset to serve as our evaluation set.

Table 1. Quantitative comparison of performance and generalizability.

Method	PSNR	SSIM
HDRCNN w/PS [10]	17.032	0.687
DPED (iPhone) [18]	16.274	0.629
DPED (BlackBerry) [18]	17.890	0.671
DPE(HDR) [8]	16.206	0.623
DPE(S-FiveK) [8]	17.510	0.677
Zero-DCE [14]	12.597	0.549
Afifi et al. [2]	19.483	0.739
ECM [12]	**20.874**	**0.877**
Ours	18.756	0.812

(a) Testing Result

Method	PSNR	SSIM
Afifi et al.(E) [2]	14.268	0.638
ECM(E) [12]	15.439	0.650
ECM(R) [12]	17.537	0.725
Ours(E)	**18.571**	**0.728**

(b) generalizability Results. "E" denotes pretraining on MSEC Dataset [2]; "R" signifies pretraining on our Radiometry Correction Dataset. We use LOL dataset [40] for evaluation.

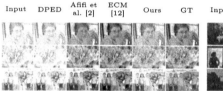
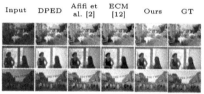

Fig. 6. Results on MSEC Dataset [2]. We take images from [12] and compare with our model.

For evaluation, we selected one ground truth image from MSEC Dataset to serve as our reference image. As illustrated in Table 1b, while the performance of supervised models significantly decreases, our results demonstrate stability and outperform ECM [12]. It is noteworthy that ECM [12], when trained on our Radiometry Correction Dataset, displayed superior generalizability compared to its performance on MSEC Dataset. The visualization in Fig. 7 reveals that our approach ensures better color consistency without stylistic biases. These results underscores the robustness of radiometry correction.

Results on Radiometry Correction Dataset. To solely assess radiometric performance, we trained and evaluated exposure correction methods using our Radiometry Correction Dataset. We benchmarked our UEC method against the SOTA supervised model, ECM [12]. Both methods demonstrated similar performance in terms of PSNR; however, our UEC method significantly outperformed in terms of SSIM. The results are detailed in Table 2. To ensure an fair comparison, we utilized the official implementations and adhered to the default hyperparameter configurations.

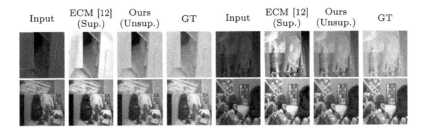

Fig. 7. Generalizability performance by ECM [12] and our UEC. We pretrain models on MSEC Dataset [2] and assess their performance on LOL Dataset [40].

Table 2. Quantitative comparison of performance and generalizability.

	ECM [12]		Ours			ECM [12]		Ours
EV	PSNR	SSIM	PSNR	SSIM	EV	PSNR	F1-Score	PSNR	F1-Score
−2	20.122	0.718	19.475	0.812	−2	15.510	0.912	20.059	0.959
−1	20.301	0.735	20.144	0.847	−1	15.983	0.918	21.616	0.966
0	20.546	0.751	**20.968**	0.884	0	16.449	0.923	**23.746**	0.974
+1	20.667	0.758	**21.614**	0.907	+1	16.730	0.925	**25.550**	0.978
+2	20.588	0.758	**21.187**	0.897	+2	16.758	0.926	**24.090**	0.975
+3	**20.445**	0.743	19.901	0.862	+3	16.443	0.925	**20.929**	0.964
Avg.	20.445	0.744	**20.548**	0.868	Avg.	16.312	0.922	**22.665**	0.969

(a) Results on Radiometry Correction Dataset. (b) Edge detection results on Radiometry Correction Dataset.

Results of Varying Exposure Manipulation. Compared to traditional supervised methods that learn the mapping from a poorly exposed image to a well-exposed one, our method can adjust the exposure within a certain range. To demonstrate the effectiveness of our method at varying exposure levels, we generate images with different exposures by controlling the exposure of the reference image. The results are shown in Table 3. Our method performs better on the MSEC dataset compared to our dataset. This discrepancy might be attributed to the wider range of exposure values in our dataset, which varies from −2EV to +3EV, presenting a greater challenge. In contrast, the MSEC dataset has a narrower exposure range of −1.5EV to +1.5EV. In our dataset, underexposure and overexposure lead to a loss of texture, making it difficult to achieve satisfactory results solely by adjusting brightness. Additionally, in our dataset, darker images yield better results, possibly because they are closer to completely black, making conversion easier. However, overexposed images are not entirely white. The large exposure range from underexposure to overexposure further increases the challenge. We provide the visualized results in our supplementary material.

Results on Edge Detection. Exposure correction not only enhances image aesthetics but also supports downstream tasks in computer vision. Since many algorithms rely on images with standard exposure, variations in real-world exposure can challenge their effectiveness. By improving visibility and detail clarity, exposure correction enhances image features crucial for accurate object recog-

Table 3. Performance of exposure control on both datasets.

EV	−1.5	−1.0	0	+1.0	+1.5
PSNR	27.764	27.733	27.823	27.959	28.097

(a) Exposure control on MSEC dataset.

EV	−2	−1	0	+1	+2	+3
PSNR	22.577	20.528	18.336	17.820	15.752	15.138

(b) Exposure control on Our dataset.

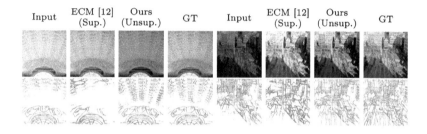

Fig. 8. Edge detection performance by ECM [12] and our UEC.

nition. However, it is important to maintain feature integrity to avoid compromising details with artifacts. To assess its effect on crucial features, we focused on edge detection—a task highly sensitive to exposure changes-using the LDC network [36] as a benchmark to compare performance across different exposures.

Using Radiometry Correction Dataset, we compared ECM's [12] method against our UEC approach. The findings, detailed in Table 2b and illustrated in Fig. 8, highlight ECM's underperformance, even below that of unmodified inputs. This shortfall is attributed to ECM's tendency to lose fine details, negating the benefits of exposure correction. Conversely, our UEC method outperforms, demonstrated by higher PSNR values indicating superior image quality. Notably, our unsupervised approach excels in adapting to varied exposures for diverse computer vision tasks, requiring only the preservation of RAW files during data collection.

Impact of Reference Images on UEC Results. Figure 9 demonstrates the impact of utilizing different reference images on the outcomes of our UEC model. Notably, the results show minor variations within a certain range, while maintaining their overall performance. This observation emphasizes the UEC model's robust generalizability, its proficiency in extracting exposure features from diverse images, and its competence in achieving exposure alignment across a range of scenes.

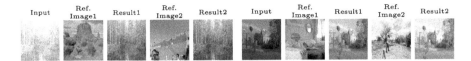

Fig. 9. Impact of Reference Images on UEC Results.

Table 4. Comparison at a resolution of 256×256 between our UEC model and ECM [12].

Method	Parameters	Model Size	Time(GPU)	Time(CPU)
ECM. [12]	182,443,267	695 MB	7.08 ms	149.39 ms
ours	19,388	0.079 MB	1.46 ms	6.38 ms

5.3 Time and Memory Consumption

Our approach achieves comparable performance to ECM [12] with only 0.01% of their parameters. It supports real-time operation at 4K resolution, processing high-definition images directly and avoiding low-level feature loss from downsampling. As shown in Table 4, our method is 4.85 times faster on GPU and 23.41 times faster on CPU compared to ECM [12]. Additional results can be found in the supplementary material.

6 Conclusion

In this study, we introduce an unsupervised methodology for correcting exposure, overcoming three primary limitations observed in prior methods: the dependence on labor-intensive paired datasets, constrained generalizability, and the deterioration of low-level features. To address these issues, our approach unfolds in three stages. Initially, we establish an unsupervised learning framework grounded in the inherent principles of exposure, thereby obviating the requirement for ground truth data. Subsequently, we eschew the traditional reliance on ground truth, which often incorporates subjective stylistic biases, and instead, draw insights from a diverse array of images differing solely in their radiometric properties. Finally, we employ color transformation techniques to maintain the intrinsic pixel relationships, effectively preserving low-level features and enhancing the robustness of our method.

References

1. Afifi, M., Derpanis, K.G., Ommer, B., Brown, M.S.: Learning to correct overexposed and underexposed photos. arXiv preprint arXiv:2003.11596 **13** (2020)
2. Afifi, M., Derpanis, K.G., Ommer, B., Brown, M.S.: Learning multi-scale photo exposure correction. In: Proceedings of the IEEE Computer Society Conference on Computer Vision and Pattern Recognition, pp. 9153–9163 (2021)

3. Aly, H.A., Dubois, E.: Image up-sampling using total-variation regularization with a new observation model. IEEE Trans. Image Process. **14**(10), 1647–1659 (2005)
4. Bhattacharya, J., Modi, S., Gregorat, L., Ramponi, G.: D2bgan: a dark to bright image conversion model for quality enhancement and analysis tasks without paired supervision. IEEE Access **10**, 57942–57961 (2022)
5. Bychkovsky, V., Paris, S., Chan, E., Durand, F.: Learning photographic global tonal adjustment with a database of input/output image pairs. In: Proceedings of the IEEE Conference on Computer Vision and Pattern Recognition (CVPR), pp. 97–104. IEEE (2011)
6. Cai, J., Gu, S., Zhang, L.: Learning a deep single image contrast enhancer from multi-exposure images. IEEE Trans. Image Process. **27**(4), 2049–2062 (2018)
7. Chai, Y., Giryes, R., Wolf, L.: Supervised and unsupervised learning of parameterized color enhancement. In: The IEEE Winter Conference on Applications of Computer Vision (WACV), pp. 992–1000 (2020)
8. Chen, Y.S., Wang, Y.C., Kao, M.H., Chuang, Y.Y.: Deep photo enhancer: Unpaired learning for image enhancement from photographs with gans. In: Proceedings of the IEEE Conference on Computer Vision and Pattern Recognition, pp. 6306–6314 (2018)
9. Choi, Y., Choi, M., Kim, M., Ha, J.W., Kim, S., Choo, J.: Stargan: unified generative adversarial networks for multi-domain image-to-image translation. In: Proceedings of the IEEE Conference on Computer Vision and Pattern Recognition, pp. 8789–8797 (2018)
10. Dayley, L.D., Dayley, B.: Photoshop CS5 Bible. John Wiley & Sons (2010)
11. Deng, Y., Loy, C., Tang, X.: Aesthetic-driven image enhancement by adversarial learning. In: 2018 ACM Multimedia Conference on Multimedia Conference (MM), pp. 870–878. ACM (2018)
12. Eyiokur, F.I., Yaman, D., Ekenel, H.K., Waibel, A.: exposure correction model to enhance image quality. In: IEEE Computer Society Conference on Computer Vision and Pattern Recognition Workshops **2022-June**, pp. 675–685 (2022)
13. Gharbi, M., Chen, J., Barron, J.T., Hasinoff, S.W., Durand, F.: Deep bilateral learning for real-time image enhancement. ACM Trans. Graph. (TOG) **36**(4), 118 (2017)
14. Guo, C., Li, C., Guo, J., Loy, C., Hou, J., Kwong, S., Cong, R.: Zero-reference deep curve estimation for low-light image enhancement. In: Proceedings of the IEEE/CVF Conference on Computer Vision and Pattern Recognition (CVPR), pp. 1780–1789 (2020)
15. He, J., Liu, Y., Qiao, Y., Dong, C.: Conditional sequential modulation for efficient global image retouching. In: European Conference on Computer Vision (ECCV), pp. 679–695. Springer (2020)
16. Hu, S., Yan, J., Deng, D.: Contextual information aided generative adversarial network for low-light image enhancement. Electronics **11**(1), 32 (2021)
17. Ibrahim, H., Kong, N.S.P.: Brightness preserving dynamic histogram equalization for image contrast enhancement. IEEE Trans. Consum. Electron. **53**(4), 1752–1758 (2007)
18. Ignatov, A., Kobyshev, N., Timofte, R., Vanhoey, K., Van Gool, L.: Dslr-quality photos on mobile devices with deep convolutional networks. In: Proceedings of the IEEE International Conference on Computer Vision, pp. 3277–3285 (2017)
19. Jiang, Y., Gong, X., Liu, D., Cheng, Y., Fang, C., Shen, X., Yang, J., Zhou, P., Wang, Z.: Enlightengan: Deep light enhancement without paired supervision. IEEE Trans. Image Process. **30**, 2340–2349 (2021)

20. Kim, H.-U., Koh, Y.J., Kim, C.-S.: Global and local enhancement networks for paired and unpaired image enhancement. In: Vedaldi, A., Bischof, H., Brox, T., Frahm, J.-M. (eds.) ECCV 2020. LNCS, vol. 12370, pp. 339–354. Springer, Cham (2020). https://doi.org/10.1007/978-3-030-58595-2_21
21. Kotovenko, D., Sanakoyeu, A., Lang, S., Ommer, B.: Content and style disentanglement for artistic style transfer. In: Proceedings of the IEEE/CVF International Conference on Computer Vision, pp. 4422–4431 (2019)
22. Land, E.H., McCann, J.J.: Lightness and retinex theory. Josa **61**(1), 1–11 (1971)
23. Li, C., Guo, C., Ai, Q., Zhou, S., Loy, C.C.: Flexible piecewise curves estimation for photo enhancement. arXiv preprint arXiv:2010.13412 (2020)
24. Li, C., Guo, C., Feng, R., Zhou, S., Loy, C.C.: Cudi: curve distillation for efficient and controllable exposure adjustment. arXiv preprint arXiv:2207.14273 (2022)
25. Li, C., Guo, C., Loy, C.C.: Learning to enhance low-light image via zero-reference deep curve estimation. IEEE Trans. Pattern Anal. Mach. Intell. **44**(8), 4225–4238 (2021)
26. Liu, E., Li, S., Liu, S.: Color enhancement using global parameters and local features learning. In: Proceedings of the Asian Conference on Computer Vision (2020)
27. Liu, R., Ma, L., Zhang, J., Fan, X., Luo, Z.: Retinex-inspired unrolling with cooperative prior architecture search for low-light image enhancement. In: Proceedings of the IEEE/CVF Conference on Computer Vision and Pattern Recognition, pp. 10561–10570 (2021)
28. Liu, Y., et al.: Very lightweight photo retouching network with conditional sequential modulation. CoRR abs/2104.06279 (2021)
29. Loh, Y.P., Chan, C.S.: Getting to know low-light images with the exclusively dark dataset. Comput. Vis. Image Underst. **178**, 30–42 (2019)
30. Moran, S., Marza, P., McDonagh, S., Parisot, S., Slabaugh, G.: Deeplpf: Deep local parametric filters for image enhancement. In: IEEE Conference on Computer Vision and Pattern Recognition (CVPR). pp. 12826–12835 (2020)
31. Moran, S., McDonagh, S., Slabaugh, G.: Curl: neural curve layers for global image enhancement. In: 2020 25th International Conference on Pattern Recognition (ICPR), pp. 9796–9803. IEEE (2021)
32. Nsampi, N.E., Hu, Z., Wang, Q.: Learning exposure correction via consistency modeling. In: Proc. Brit. Mach. Vision Conf. (2021)
33. Rahman, Z.u., Jobson, D.J., Woodell, G.A.: Retinex processing for automatic image enhancement. J. Electron. imaging **13**(1), 100–110 (2004)
34. Ren, W., et al.: Low-light image enhancement via a deep hybrid network. IEEE Trans. Image Process. **28**(9), 4364–4375 (2019)
35. Sara, U., Akter, M., Uddin, M.S.: Image quality assessment through fsim, ssim, mse and psnr-a comparative study. J. Comput. Commun. **7**(3), 8–18 (2019)
36. Soria, X., Pomboza-Junez, G., Sappa, A.D.: Ldc: lightweight dense cnn for edge detection. IEEE Access **10**, 68281–68290 (2022)
37. Wang, B., Yu, Y., Xu, Y.: Example-based image color and tone style enhancement. ACM Trans. Graph. (TOG) **30**(4), 1–12 (2011)
38. Wang, T., et al.: Real-time image enhancer via learnable spatial-aware 3d lookup tables. In: Proceedings of the IEEE/CVF International Conference on Computer Vision (ICCV), pp. 2471–2480, October 2021
39. Wang, Y., et al.: Neural color operators for sequential image retouching. In: European Conference on Computer Vision, pp. 38–55. Springer (2022)
40. Wei, C., Wang, W., Yang, W., Liu, J.: Deep retinex decomposition for low-light enhancement. arXiv preprint arXiv:1808.04560 (2018)

41. Yan, Z., Zhang, H., Wang, B., Paris, S., Yu, Y.: Automatic photo adjustment using deep neural networks. ACM Transactions on Graphics (TOG) **35**(2), 11 (2016)
42. Yang, K.F., Cheng, C., Zhao, S.X., Yan, H.M., Zhang, X.S., Li, Y.J.: Learning to adapt to light. Int. J. Comput. Vision **131**(4), 1022–1041 (2023)
43. Zeng, H., Cai, J., Li, L., Cao, Z., Zhang, L.: Learning image-adaptive 3d lookup tables for high-performance photo enhancement in real-time. IEEE Trans. Pattern Anal. Mach. Intelli. (TPAMI) (2020)
44. Zhang, Y., Zhang, J., Guo, X.: Kindling the darkness: a practical low-light image enhancer. In: Proceedings of the 27th ACM International Conference on Multimedia, pp. 1632–1640 (2019)

Anytime Continual Learning for Open Vocabulary Classification

Zhen Zhu[✉][iD], Yiming Gong[iD], and Derek Hoiem[iD]

University of Illinois at Urbana-Champaign, Champaign, USA
{zhenzhu4,yimingg8,dhoiem}@illinois.edu

Abstract. We propose an approach for anytime continual learning (AnytimeCL) for open vocabulary image classification. The AnytimeCL problem aims to break away from batch training and rigid models by requiring that a system can predict any set of labels at any time and efficiently update and improve when receiving one or more training samples at any time. Despite the challenging goal, we achieve substantial improvements over recent methods. We propose a dynamic weighting between predictions of a partially fine-tuned model and a fixed open vocabulary model that enables continual improvement when training samples are available for a subset of a task's labels. We also propose an attention-weighted PCA compression of training features that reduces storage and computation with little impact to model accuracy. Our methods are validated with experiments that test flexibility of learning and inference.

Keywords: Continual learning · Anytime learning · Open-vocabulary classification

1 Introduction

Continual learning aims to improve a system's capability as it incrementally receives new data and labels, which gains increasing importance with the expanding scale and applications of visual learning. Continual learning for classification is traditionally approached with discrete label spaces. Adding labels or tasks over time changes the problem landscape, thereby increasing the difficulty of learning. The open vocabulary setting, however, frames classification as comparing continuous feature and label embeddings. Any images and textual labels can be embedded, so learning in this setting involves only improving on the existing problem definition, which may be more amenable to continuous improvement. Although open vocabulary models can predict over arbitrary label sets, even models like CLIP [37] that are trained on internet-scale data have unsatisfactory performance on many tasks [47,48].

Our work aims to continually improve open vocabulary image classifiers as new labeled data is received. We call this "**anytime continual learning**"

Supplementary Information The online version contains supplementary material available at https://doi.org/10.1007/978-3-031-72658-3_16.

because the goal is to improve efficiently *any time* new examples are received and to maintain the ability to predict over arbitrary label sets at *any time*.

Recent work by Zhu et al. [62] extends CLIP by training linear classifiers and combining their predictions with those of label vectors from the original text embedding. To enable efficient and distributed learning, linear classifiers are trained for each partition of the feature space defined by online hierarchical clustering. This approach requires storing all training examples, but a new example is efficiently incorporated by training only on the data from the example's nearest cluster. In experiments on their own benchmark designed for open vocabulary continual learning and existing benchmarks, Zhu et al. [62] outperform other recent approaches (Fig. 1).

Our approach is to train online, fine-tuning the last transformer block while keeping the label embedding fixed. When a new training sample is received, we fill a class-balanced batch with stored samples and update the last transformer block in a single step. Our experiments show this performs better than retraining a classifier layer, and each new sample can be incorporated in milliseconds. Additionally, we introduce a modified loss, enabling the prediction of "none of the above" when the true label is absent from the candidate set, which improves overall performance.

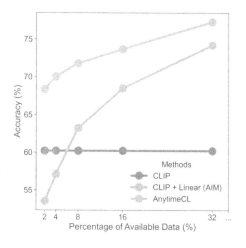

Whenever a new training example arrives, our method updates an estimate of the tuned and original model's accuracy for the given label. The tuned and original model's predictions are then weighted in proportion to their expected accuracy for each label. This weighting accounts for the growing effectiveness of the tuned model, and greatly outperforms Zhu et al. [62]'s AIM weighting based on whether an example's label is likely to have been seen in training.

Fig. 1. Our AnytimeCL algorithm can be efficiently updated with each new example and continuously improve. By dynamically weighting predictions between a tunable model and a frozen open vocabulary model, our method can predict over any label set while gaining expertise. This figure shows our method outperforms previous SotA Zhu et al. [62] in every stage in the data incremental setting. Our method also outperforms in other settings like task-incremental and class-incremental.

The partial fine-tuning approach requires storing either full images that need to be re-encoded during training, or data-consumptive patch-level features at intermediate layers. We apply a per-image weighted PCA, which provides a 30x reduction of data with little impact on prediction accuracy. Such compression could be especially beneficial in a large-scale federated learning setting.

In experiments on the open vocabulary continual learning benchmark proposed by Zhu et al. [62], our approach outperforms under all settings and stages, including data-incremental, class-incremental, and task-incremental learning, and zero shot prediction. Our ablations evaluate the effects of all key design parameters, including the partial fine-tuning, model prediction weighting, batch sampling, and the regularization loss term. We also show that our method can be applied to combine non-open vocabulary models like DINOv2 with CLIP to improve further while maintaining open vocabulary ability.

In summary, our **main contribution** is an open vocabulary continual learning method that can quickly incorporate new training examples received in any order and continually improve while maintaining open vocabulary performance. Our experiments demonstrate that our system's accuracy and efficiency stem from multiple proposed innovations:

- Partial fine-tuning of features with a fixed label embedding;
- Online training with each batch composed of the new training sample and class-balanced stored samples;
- Online learning of per-label accuracy for effective combination of original and tuned model predictions;
- Loss modification to enable "none of the above" prediction, which also stabilizes open vocabulary training;
- Intermediate layer feature compression that reduces storage of training samples and improves speed without much loss to accuracy.

2 Related Work

The goal of continual learning (CL) is to continually improve the performance when seeing more data, while online continual learning [1,3,8,24,34,36] aims to achieve a good trade-off between performance and learning efficiency. Our problem setup and goals are distinguished by these important characteristics:

- *Flexibly learn*, receiving labeled examples in data-incremental (random order), class incremental (grouped by category), or task incremental (grouped by dataset);
- *Efficiently update* from a single example or a batch of examples;
- *Retain training data*, though potentially in compact, privacy-preserving forms;
- *Cumulatively improve* from new data without forgetting other classes or tasks;
- *Flexibly infer*, predicting over label sets defined at inference time, without rigid task boundaries.

Continual learning approaches can be broadly categorized into regularization [19,23,59], replay/rehearsal methods [24,38,42,56], and parameter isolation or expansion [2,27,35,36,40,41,57,60]. Regularization techniques generally impose constraints on the learning process to alleviate forgetting. Rehearsal methods involve storing and replaying past data samples during training [5,35,

36,42]. So far, when training data can be stored, simple replay methods tend to outperform others, whether limiting the number of stored examples [36] or training computation [35]. Parameter isolation methods maintain learning stability by fixing subsets of parameters [27,41] or extending model with new parameters [40,57,60]. Recently, several works [16–18,45,51–53] adopt prompt tuning for continual learning, which can also fall into the parameter expansion category. These methods can be used for stage-wise incremental continual learning, but none demonstrates the efficacy on online incremental learning that is our focus. Prompt tuning provides an alternative to weight-based fine-tuning with less forgetting. We consider weight-tuning approaches here, but preliminary experiments indicate the same strategy is applicable to prompt tuning, with similar accuracy but slower training.

We now focus on those most relevant and influential to our work; see Wang et al. [50] for a comprehensive survey of continual learning.

2.1 CL for Open-Vocabulary Classification

WiSE-FT [54] fine-tunes CLIP encoders on target tasks and averages fine-tuned weights with original weights for robustness to distribution shifts. PAINT [15] shares a similar idea as WiSE-FT but the weight mixing coefficient is found via a held-out set. With such approaches, generality degrades with each increment of continual learning. CLS-ER [4] exponentially averages model weights in different paces for its plastic and stable model to balance learning and forgetting. Closer to our work, ZSCL [61] fine-tunes CLIP encoders using a weight ensemble idea, similar to WiSE-FT, and applies a distillation loss with a large unsupervised dataset to reduce forgetting. Unlike these, we partially fine-tune CLIP with fixed label embeddings and use a dynamically weighted combination of the predictions from the tuned and original models, which enables improvement on target tasks without sacrificing the generality of the original model.

Another class of methods [13,44] use attribute-based recognition [12,22], continually learning attributes and attribute-class relations as a way to generalize from seen to unseen categories. We use CLIP, which provides zero-shot ability by training to match image-text pairs on a large training corpus, instead of the attribute-based approach. In reported results [13,37,44] and our own tests, CLIP has much higher zero-shot accuracy (e.g. for the SUN and aPY datasets) than the continual attribute-based methods achieve even after training, supporting the idea of building on vision-language models for open vocabulary image classification.

2.2 CL with Constrained Computation

Prabhu et al. [35] evaluate a wide variety of continual learning techniques under a setting of retaining data but limiting compute. Compared to all other techniques, they find that simple replay strategies, such as fine-tuning with uniformly random or class-balanced batches, are most effective in this setting. They also provide some evidence that partial-fine tuning of a well-initialized network may be an efficient and effective approach. In this direction, we find that fine-tuning only the last transformer block of CLIP by performing one mini-batch update per incoming sample works surprisingly well.

2.3 Biologically Inspired CL

The complimentary learning systems theory [31] (CLS theory) posits that humans achieve continual learning through the interplay of sparse retrieval-based and dense consolidated memory systems. This has inspired many works in continual learning, e.g. [4,33,62]. The most closely related is Zhu et al. [62], as detailed in the introduction. Many works are also inspired by the idea of wake/sleep phases of learning, where faster updates are required during the wake stage. Recent studies of mammalian memory and learning indicate that consolidation occurs in similar ways via replay during wakefulness and sleep [7,43,49]. Our online learning method mixes use of individual examples and consolidated networks for continuous improvement for continuous improvement, reflecting the idea of wakeful consolidation through replay.

2.4 Memory Compression

Other work has compressed training data [9], or distilled training data into a network [58], or learned prompts [16–18,45,51–53] instead of maintaining original examples. AQM [9] uses an online VQ-VAE [29] to compress training samples to reduce storage for continual learning. We instead compress intermediate features to improve both storage and computation of partial fine-tuning.

3 Method

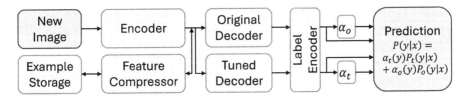

Fig. 2. Overview. On receiving a new training sample, the batch is completed with stored samples, the prediction is made using a label embedding, the tuned decoder and confidence weights (α_o, α_t) are updated in one step, and the new sample is stored. To save space and time, stored examples are encoded and compressed. In testing, the probability of each candidate label is determined by predictions from both decoders and their class-wise confidence weights. Our method enables constant-time updates from new samples while continually improving and maintaining open vocabulary performance. Green blocks are updated during training; blue blocks are not updated. (Color figure online)

Our system receives training examples $(x, y, \mathcal{Y})$ one by one to update its models, in order to continually improve in prediction of y given $(x, \mathcal{Y})$, where x is an input image, y is its target label, and $\mathcal{Y}$ is its set of candidate labels.

Figure 2 offers a system overview aimed at enhancing open vocabulary image classifiers by integrating a tuned model for learning target tasks with a frozen original model. The tuned model uses the same encoder as the original but incorporates a trainable decoder. For an image x, both the tuned model and the original model produce the probabilities for all candidate labels, denoted as $P_t(y|x)$ and $P_o(y|x)$. The final probability for the image is weighted through our Online Class-wise Weighting (OCW, Sect. 3.2):

$$P(y|x) = \alpha_o(y) P_t(y|x) + \alpha_t(y) P_o(y|x), \qquad (1)$$

where $\alpha_o(y)$ and $\alpha_t(y)$ are the relative expected accuracies for the original and tuned models for label y. During training, new samples are encoded to intermediate features (feature vectors for patches plus a CLS token), optionally compressed (Sect. 3.3), and stored, for reuse in future steps.

3.1 Models

Original Model. In our experiments, the original model is the publicly available CLIP [37] ViT model that has been trained contrastively on image-text pairs. In our experiments, the shared encoder is all but the last transformer block, and the original decoder is the last block. The CLIP model produces a probability for label y for image x given a set of candidate text labels $\mathcal{Y}$ based on the dot product of image embedding e_x (CLS token) and text embedding e_y:

$$P_o(y|x) = \frac{\exp(100 \cdot \cos(e_x, e_y))}{\sum_{y_k \in \mathcal{Y}} \exp(100 \cdot \cos(e_x, e_{y_k}))}. \qquad (2)$$

Tuned Model. With each new sample, our goal is to efficiently update the tuned model to improve accuracy in received labels while maintaining general features that are effective for future learning. We tune only the last image transformer block while keeping the label embedding fixed, which helps the features to stay correlated with the text modality and reduces overfitting to received labels. Note that more blocks can be tuned, which could enable better performance but with increased computation. The tuned decoder is initialized with the weights of the original decoder.

Given a new sample, we form a batch of that sample and a class-balanced sampling of stored training samples. Specifically, we first determine the class count for selection: for a batch size of B and all seen labels $\mathcal{Y}_t$, we uniformly select $\min(B-1, |\mathcal{Y}_t|)$ classes from $\mathcal{Y}_t$, where $|\cdot|$ indicates the length of a set. Then we uniformly sample an equal number of instances from each chosen class to form the batch. This class-balanced sampling ensures continual retention and improvement of prediction of training labels. Then, the tuned model is updated with one iteration based on a cross-entropy loss.

Additionally, we develop a regularization loss that helps with performance. The idea is that, if the true label is not in the label candidates, a low score should be predicted for every candidate label. We implement this with an "other" option within the candidate set, but since "other" does not have an appearance, we model it with just a learnable bias term. The combined loss for training the tuned model is therefore:

$$\mathcal{L}(x, y, \mathcal{Y}) = \mathcal{L}_{\text{ce}}(x, y, \mathcal{Y} \cup \text{other}) + \beta \mathcal{L}_{\text{ce}}(x, \text{other}, (\mathcal{Y} \cup \text{other}) \setminus y), \qquad (3)$$

where β is a hyperparameter balancing the two loss terms, set to 0.1 based on validation experiments.

3.2 Online Class-Wise Weighting

We estimate the probability of each candidate label by weighted voting of the tuned and original models (Eq. 1). We would like to assign more weight to the model that is more likely to be correct for a given label. There are two big problems: 1) training samples would provide a highly biased estimate of correctness for the tuned model; 2) maintaining a held out set or performing cross-validation would either reduce available training samples or be impractically slow. Our solution is to use each training sample, before update, to update our estimate of the likelihood of correctness for its label based on the tuned and original predictions.

Let $c_t(y)$ and $c_o(y)$ be the estimated accuracy of the tuned and original model for label y. We apply an exponential moving average (EMA) updating method to estimate them online, ensuring the estimations are reliable when evaluated at any time, concurring with our anytime continual learning goals. Assuming the EMA decay is set to η (=0.99 in our experiments), the estimated accuracy of the tuned model at the current step is:

$$c_t(y) = \eta \hat{c}_t(y) + (1 - \eta) \mathbb{1}[y_t(x) = y]. \qquad (4)$$

Here, $\hat{c}_t(y)$ is the estimated accuracy of label y in the previous step; $y_t(x)$ denotes the predicted label of the tuned model for x. Since the exponential moving average depends on past values, we compute $c_t(y)$ as the average accuracy for the first $\lfloor \frac{1}{1-\eta} \rfloor$ samples. $c_o(y)$ is updated in the same way.

After getting $c_t(y)$ and $c_o(y)$, the weights of the two models are:

$$\alpha_t(y) = \frac{c_t(y)}{c_t(y) + c_o(y) + \epsilon}, \qquad \alpha_o(y) = 1 - \alpha_t(y). \qquad (5)$$

Here, ϵ is a very small number (1e-8) to prevent division by zero. For labels not seen by the tuned model, we set $\alpha_t(y) = 0$, so $\alpha_o(y) = 1$.

3.3 Storage Efficiency and Privacy

Partial tuning of our model requires either storing each image or storing the features (or tokens) that feed into the tuned portion of the model. Storing images has disadvantages of lack of privacy and inefficiency in space and computation, due to need to re-encode in training. Storing features alleviates some of these problems, but still uses much memory or storage. Therefore, we investigate how to compress the training features, aiming to increase storage and computational efficiency while maintaining training effectiveness. The compressed features also provide more privacy than storing original images, though we do not investigate how well the images can be recovered from compressed features.

Well-trained networks learn data-efficient representations that are difficult to compress. Indeed, if we try to compress feature vectors with VQ-VAE [29] or PCA (principal component analysis) trained on a dataset, we are not able to achieve any meaningful compression without great loss in training performance. The features within *each image*, however, contain many redundancies. We, therefore, compute PCA vectors on the features in each image and store those vectors along with the coefficients of each feature vector. Further, not all tokens are equally important for prediction. Hence, we train a per-image attention-weighted PCA, weighted by attention between each token and the CLS token. Finally, we can compress further by storing min/max floating point values for each vector and for the coefficients and quantizing them to 8-bit or 16-bit unsigned integers. See the supplemental material for details. By storing only five PCA vectors and their coefficients this way, we can reduce the storage of fifty 768-dim tokens (7×7 patch tokens + CLS token) from 153K bytes to 5K bytes with less than 1% difference in prediction accuracy.

4 Experiments

Using the setup from [62], we sequentially receive training samples for a target task in a **data-incremental** (random ordering) or **class-incremental** (class-sorted ordering). Or, in a **task-incremental** setting, we receive all examples for each task sequentially. Within each set, we incorporate new examples **one by one** in an online fashion. After receiving each set, the model is evaluated on the entire task (including classes not yet seen in training) for data- or class-incremental, or on all tasks (including tasks not yet seen in training) for task-incremental.

Our main experiments use the same setup as Zhu et al. [62], which are designed to test flexible learning and flexible inference. A **target task** contains a subset of seen labels while a **novel task** contains none of these labels. Target tasks are CIFAR100 [21], SUN397 [55], FGVCAircraft [26], EuroSAT [14], OxfordIIITPets [32], StanfordCars [20], Food101 [6], and Flowers102 [28]. Novel tasks are ImageNet [39], UCF101 [46], and DTD [10]. Training examples are received for target tasks, but not for novel tasks. Under this setup, we have 226,080 training samples and 1,034 classes in total.

- **Task Incremental Learning** updates one model with each of the eight target tasks sequentially.
- **Class Incremental Learning** updates a per-task model with examples from a subset (one-fifth) of classes for a stage.
- **Data Incremental Learning** updates a per-task model with a random subset of the data in eight stages: 2%, 4%, 8%, 16%, 32%, 64%, 100%.

After each stage, accuracy is averaged across all tasks (including tasks/classes not yet seen).

Additionally, we provide a **union data incremental** scenario where we mix all target tasks together and union all target labels, mainly for hyperparameter selection and ablation experiments. **Flexible inference** is evaluated after the task-incremental learning for prediction over novel tasks and sets of candidate labels $\mathcal{Y}$: zero-shot, union of all target tasks/labels and zero-shot tasks/labels, and a mix of some target and zero-shot tasks/labels. See [62] for details. The union and mixed settings are most challenging because they require effective combined use of both the original and tuned models, without any task identifiers or indication whether an example's label is in the tuned model's domain.

Implementation Details. We employ the ViT-B/32 model from CLIP as our network backbone. We use AdamW [25] as the optimizer with a weight decay of 0.05, and we adhere to CLIP's standard preprocessing without additional data augmentations. For offline training, following [11], we set the learning rate at 6e−4 and the batch size at 2048, employing a cosine annealing scheduler for the learning rate with a minimum rate of 1e−6. For online training, we use a batch size of 32 and a learning rate of $9.375e = \frac{32 \times 6e-4}{2048}$, based on adjustments from the offline training hyperparameters. More method and training details are in the supplemental.

4.1 Main Result

In this evaluation, we compare our method with the previous SotA Zhu et al. [62] (CLIP+LinProbe (AIM-Emb)) under the task-, class-, data-incremental scenarios, as well as flexible inference. From all plots in Fig. 3, our AnytimeCL method consistently outperforms CLIP+LinProbe (AIM). For task incremental, the improvement is mainly due to the partial finetuning with fixed label embeddings. To provide a more direct comparison with Zhu et al. [62], we train our approach only after receiving all data of a task (*e.g.*, 100% CIFAR100) and term this as AnytimeCL (Offline), which also shows steady improvements over Zhu et al. [62]. For early stages in class incremental and data incremental, the performance improvement mainly comes from the advantage of our online class-wise weighting method over AIM, which only considers whether the testing sample is likely to be from a known class.

4.2 Comparison Under MTIL Task Incremental Learning [61]

We compare the performance of AnytimeCL with Zhu et al. [62] and ZSCL [61] under the task incremental learning setting as introduced in ZSCL. Results are summarized in Table 1, using the Transfer, Last, and Average metrics from [61] for comparison. Improvements over CLIP are denoted in green under the Δ columns, while declines are marked in red. Our method consistently achieves zero loss in Transfer in both experiments, indicating an absence of forgetting. On this benchmark, our method performs comparably to LinProbe (AIM) and outperforms other methods. This experimental setup does not fully capture our

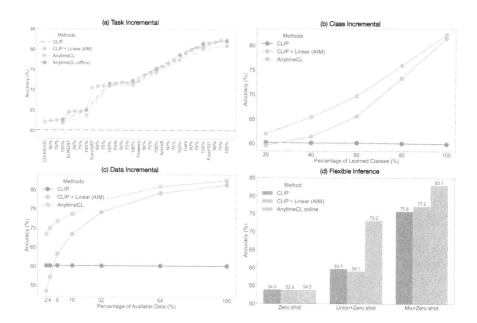

Fig. 3. Performance comparison of CLIP, CLIP + Linear (AIM), and AnytimeCL variants in task incremental (a); class incremental (b); data incremental (c); and flexible inference (d) settings. In (a), the online approach is represented by a solid line, while offline methods are depicted with dashed lines, assessed when the online algorithms receive 25%, 50%, 75%, and 100% data of a task, labeling tasks at the 25% point.

Table 1. Comparison of different methods on MTIL in Order I from ZSCL [61]. CLIP is our tested zero-shot performance. All other results are taken Zhu et al. [62] and ZSCL [61]'s papers. "Transfer" evaluates the model's performance on zero-shot tasks; "Last" is the averaged accuracy on all target tasks after finishing the final task while "Avg." computes the average task performance on all training stages.

Method	Transfer	Δ	Avg.	Δ	Last	Δ
CLIP	69.4	0.0	65.3	0.0	65.3	0.0
LwF [23]	56.9	-12.5	64.7	-0.6	74.6	+9.3
iCaRL [38]	50.4	-19.0	65.7	+0.4	80.1	+14.8
WiSE-FT [54]	52.3	-17.1	60.7	-4.6	77.7	+12.4
ZSCL [61]	68.1	-1.3	75.4	+10.1	83.6	+18.3
TreeProbe (AIM) (50k) [62]	69.3	-0.1	75.9	+10.6	85.5	+20.2
LinProbe (AIM) [62]	69.3	-0.1	**77.1**	**+11.8**	86.0	+20.7
AnytimeCL	**69.4**	0	77.0	+11.7	85.8	+20.5
AnytimeCL (Offline)	**69.4**	0	77.0	+11.7	**86.2**	+20.9

method's key advantages of online training, ability for data and class incremental learning, and open vocabulary prediction. Also, in this setting, a simple weighting method is effective—to use the tuned model's prediction for tasks with all candidate labels in the training set, or the original model's prediction otherwise.

4.3 Training Features Compression

We test our per-image attention-weighted PCA compresssion for training features (Table 2). We compare the data size, average full time to process one 32-sample batch (including data loading, uncompressing, and forward/backward training passes), and final accuracy after completing fine-tuning. This test is on CIFAR100. Compared to processing the full image or full features, using our compressed features saves 30x the storage while achieving nearly the same accuracy. We were not able to achieve good accuracy when compressing with VQ-VAE, even when using large vocabularies, nor when using PCA vectors computed over the entire dataset. Computing PCA vectors per instance (over the 50 tokens) and weighting the tokens by their CLS-attention each improve the tuned accuracy. Quantizing the PCA vectors and coefficients further saves space at a small loss to accuracy. Using the compressed features also gives a nearly 2x speedup vs. using the full features (because the compressed features fit entirely in memory) and 6x vs. using the full image, though all times are fast. We provide more method details in the supplemental along with a comparison of using different per-instance PCA compression methods under the union data incremental scenario, to show how these compression methods influence the result at various timesteps.

Table 2. Comparison of different compression methods in terms of memory usage, processing speed, and test accuracy on CIFAR100. See supplementary material for details on how to get ms/batch and FT Accuracy. We use the same seed for all methods.

Compression	KB/example	ms/batch	FT Accuracy
Full image	150.5	43.9	77.8
Full features	153.6	25.6	77.8
VQ	0.2	4.5	64.8
Dataset-wide PCA (200 components)	40.2	7.6	76.4
Per-instance PCA (5 components)	19.4	7.7	74.6
+ CLS-weight (5 components)	19.4	8.9	77.9
+ int-quantization (5 components)	5.3	13.9	77.5

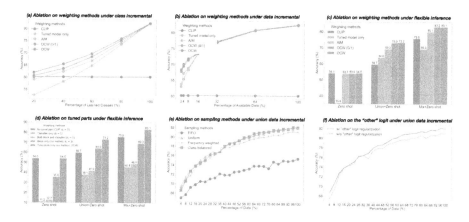

Fig. 4. Ablation results. Best viewed in color and zoomed-in.

4.4 Ablations

In the following, we ablate several important factors that contribute to our method. Additional ablations and results are provided in the supplemental.

Weighting Strategies. Weighting is vital for our dual decoder approach. We compare several ways to compute the weights: 1) CLIP: namely $\alpha_t(y) = 0$ for any images; 2) Tuned model only: $\alpha_t(y) = 1$ for any images; 3) AIM [62]; 4) OCW (0/1): a variant of OCW where we round $\alpha_t(y)$ to 0 or 1 to use either the original or tuned model; 5) Our proposed OCW. We partially finetune the decoder with fixed label embeddings and combine the tuned model with the original model using different weighting strategies. The results are shown in Fig. 4 (a), (b) and (c), under data-, class-incremental and flexible inference. OCW performs the best of all methods in every setting and stage. Notably, OCW enables improvements even in the early stages when limited data is available for a class or task and better handles tasks that mix novel and target labels.

Tuned Parts. Our proposed method tunes the last transformer block while keeping the label encoder fixed. In Fig. 4 (d), we compare this approach to alternatives of tuning only the label encoder, both the block and the label encoder, or neither of them, under the flexible inference test. When only using the tuned model for comparison ($\alpha_t = 1$), fine tuning only the last transformer best retains predictive ability for novel labels. Further analysis under the task incremental scenario is provided in the supplemental material.

Sampling Methods. We compare different methods for sampling from the stored samples. FIFO cycles through samples in order of first appearance, "uniform" randomly draws samples for each batch, class-balanced (which we use in all other experiments) samples classes, and frequency weighted sampling (FWS) samples based on how many times a given sample has been batched in training. Class-balanced and uniform are similar in practice, and perform best (Fig. 4 (e)). This result is consistent with Prabhu et al. [35]'s finding that simple sampling methods outperform complex ones under a computational budget. See supplemental for more details.

The "Other" Logit Regularization. The ablation experiments depicted in Fig. 4 (f) assess the impact of "other" logit regularization in the union data incremental scenario. The results demonstrate consistent enhancements when this regularization is applied, compared to its absence.

4.5 Scalability

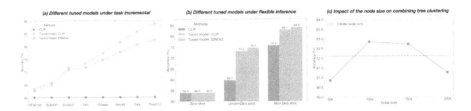

Fig. 5. (a) & (b): The "CLIP" method refers to the original model. For Tuned model: DINOv2, we use the ViT-B/14 checkpoint; (c): Infinite node size indicates only one tuned model regardless of the number of samples received.

DINOv2 as the Tuned Model. Using self-supervised DINOv2 [30] as our tuned model showcases our method's adaptability beyond the original CLIP model [62]. Our approach allows for tuning the feature embedding space while keeping label embeddings fixed, enabling the incorporation of diverse encoders through an additional linear layer for dimension matching. This layer is trained in joint with the tuned decoder. As illustrated in Fig. 5 (a), replacing the tuned model with DINOv2 results in consistent performance improvements at every stage, with a notably steeper improvement curve in later stages. Furthermore, Fig. 5 (b) demonstrates that this modification preserves Zero-shot performance while significantly boosting Union+Zero-shot and Mix+Zero-shot outcomes, attributed to DINOv2's enhanced performance on target tasks.

Tree Clustering Models. Zhu et al. [62] propose tree-based clustering as a way to limit the training time needed to incorporate new examples. Our method can already incorporate an example much faster in a way that does not depend on the number of training samples, but the clustering idea is still intriguing as a path to scalability in storage, memory, and model capacity. In Fig. 5 (b), we evaluate different node capacities and find that moderate-sized partitions may slightly increase performance. In these experiments, each node stores a copy of the last CLIP transformer block and updates it based on data that it receives.

5 Discussion

Our AnytimeCL approach shows promise of breaking free of staged train and deploy processes, with many benefits to applications that constantly or sporadically receive new data and have customizable or evolving label spaces.

The *anytime* has three main parts: (1) being able to do a task even when training data do not fully cover the label space for that task; (2) being able to incorporate a new training example quickly; and (3) not losing the benefit of older training data. Our key to the first part is using complimentary models. For that we are motivated by CLS theory [31] and more specifically adopt the framework of Zhu et al. [62], improving it with our prediction weighting method and by partially fine-tuning the encoder. Our key to the second part is that we do not need to incorporate one million examples quickly, just one. This reframes the computationally budgeted CL problem of Prabhu et al. [35], who aim to incorporate a new large batch of examples in time that allows visiting only a portion of the existing data. For some settings, our goal is more practical, as it enables an opportunistic approach that interleaves pure online learning during busy times and offline learning when time allows. While humans are often held up as excellent continual learners, they also learn one thing fast but many things slowly over time and make use of restful wake and sleep to continue learning. The effectiveness of our online learning is also due, in part, to leveraging a strong vision-language model, just recently possible. The key to the third goal is simple—do not throw out the data. If one cares about privacy and storage, then finding ways of compressing and preserving privacy, like our PCA-based approach, may be better than trying to retain the benefits of a large dataset without one.

Potential directions for future work:

- Beyond classification: Our partial fine-tuning approach is applicable to other tasks, such as semantic segmentation, visual question answering, and object detection. Our weighting method is applicable to open vocabulary detection and segmentation.
- Multi-model inference: Extending our approach to multiple models, including task-specific models, would further extend flexible learning and model re-use opportunities.
- Scalability: We expect that tree-based data clustering and training one model per cluster, as in [62], provides a good mechanism for scalability. Larger scale experiments, involving dozens of datasets and many millions of examples, are needed to fully test this.
- Federated learning: Combining tree-based clustering and feature compression, we can encode training examples in the client or central server and transmit to a node that stores and updates the tunable block and its data. This enables fully distributed training on inexpensive nodes.

6 Conclusion

We propose an effective approach for anytime continual learning of open vocabulary image classification. The main innovation is dynamic class-senstive weighting for combining predictions of open-vocabulary and example-based tuned models, and we also find that partial fine-tuning with sufficiently small learning rates enables a surprisingly effective online learner. We offer a per-image attention-weighted PCA approach to compress features, with benefits to storage, computation, and privacy. Our experiments show that our approach is very promising, and further exploration is merited to more fully understand its effective scope and limitations.

Acknowledgement. This work is supported in part by ONR award N00014-21-1-2705, ONR award N00014-23-1-2383, and U.S. DARPA ECOLE Program No. #HR00112390060. The views and conclusions contained herein are those of the authors and should not be interpreted as necessarily representing the official policies, either expressed or implied, of DARPA, ONR, or the U.S. Government.

References

1. Aljundi, R., et al.: Online continual learning with maximally interfered retrieval. CoRR (2019)
2. Aljundi, R., Chakravarty, P., Tuytelaars, T.: Expert gate: Lifelong learning with a network of experts. In: CVPR (2017)
3. Aljundi, R., Lin, M., Goujaud, B., Bengio, Y.: Gradient based sample selection for online continual learning. In: NeurIPS (2019)
4. Arani, E., Sarfraz, F., Zonooz, B.: Learning fast, learning slow: A general continual learning method based on complementary learning system. In: ICLR (2022)
5. Bang, J., Kim, H., Yoo, Y., Ha, J., Choi, J.: Rainbow memory: continual learning with a memory of diverse samples. In: CVPR (2021)
6. Bossard, L., Guillaumin, M., Van Gool, L.: Food-101 – mining discriminative components with random forests. In: Fleet, D., Pajdla, T., Schiele, B., Tuytelaars, T. (eds.) ECCV 2014. LNCS, vol. 8694, pp. 446–461. Springer, Cham (2014). https://doi.org/10.1007/978-3-319-10599-4_29
7. Brodt, S., Inostroza, M., Niethard, N., Born, J.: Sleep-a brain-state serving systems memory consolidation. In: Neuron (2023)
8. Caccia, L., Aljundi, R., Asadi, N., Tuytelaars, T., Pineau, J., Belilovsky, E.: New insights on reducing abrupt representation change in online continual learning. In: ICLR (2022)
9. Caccia, L., Belilovsky, E., Caccia, M., Pineau, J.: Online learned continual compression with adaptive quantization modules. In: ICML (2020)
10. Cimpoi, M., Maji, S., Kokkinos, I., Mohamed, S., , Vedaldi, A.: Describing textures in the wild. In: CVPR (2014)
11. Dong, X., et al.: CLIP itoolf is a strong fine-tuner; achieving 85.7% and 88.0% top-1 accuracy with VIT-B and VIT-L on imagenet. CoRR (2022)
12. Farhadi, A., Endres, I., Hoiem, D., Forsyth, D.: Describing objects by their attributes. In: CVPR. IEEE (2009)
13. Gautam, C., Parameswaran, S., Mishra, A., Sundaram, S.: Online lifelong generalized zero-shot learning. Neural networks (2021)
14. Helber, P., Bischke, B., Dengel, A., Borth, D.: Eurosat: a novel dataset and deep learning benchmark for land use and land cover classification. IEEE J. Sel. Topics Appl. Earth Observ. Remote Sens. (2019)
15. Ilharco, G., et al.: Patching open-vocabulary models by interpolating weights. In: NeurIPS (2022)

16. Jin, W., Cheng, Y., Shen, Y., Chen, W., Ren, X.: A good prompt is worth millions of parameters: low-resource prompt-based learning for vision-language models. In: Proceedings of the ACL (2022)
17. Khattak, M.U., Rasheed, H.A., Maaz, M., Khan, S.H., Khan, F.S.: Maple: Multimodal prompt learning. In: CVPR (2023)
18. Khattak, M.U., Wasim, S.T., Naseer, M., Khan, S., Yang, M., Khan, F.S.: Self-regulating prompts: foundational model adaptation without forgetting. In: ICCV (2023)
19. Kirkpatrick, J., et al.: Overcoming catastrophic forgetting in neural networks. CoRR (2016)
20. Krause, J., Stark, M., Deng, J., Fei-Fei, L.: 3D object representations for fine-grained categorization. In: 4th International IEEE Workshop on 3D Representation and Recognition (3dRR-13) (2013)
21. Krizhevsky, A., Hinton, G.: Learning multiple layers of features from tiny images. Master's thesis, Department of Computer Science, University of Toronto (2009)
22. Lampert, C.H., Nickisch, H., Harmeling, S.: Attribute-based classification for zero-shot visual object categorization. IEEE TPAMI (2014)
23. Li, Z., Hoiem, D.: Learning without forgetting. In: Leibe, B., Matas, J., Sebe, N., Welling, M. (eds.) ECCV 2016. LNCS, vol. 9908, pp. 614–629. Springer, Cham (2016). https://doi.org/10.1007/978-3-319-46493-0_37
24. Lopez-Paz, D., Ranzato, M.: Gradient episodic memory for continual learning. In: NeurIPS (2017)
25. Loshchilov, I., Hutter, F.: Decoupled weight decay regularization. In: ICLR (2019)
26. Maji, S., Rahtu, E., Kannala, J., Blaschko, M.B., Vedaldi, A.: Fine-grained visual classification of aircraft. CoRR (2013)
27. Mallya, A., Lazebnik, S.: PackNet: adding multiple tasks to a single network by iterative pruning. In: CVPR (2018)
28. Nilsback, M.E., Zisserman, A.: Automated flower classification over a large number of classes. In: 2008 Sixth Indian Conference on Computer Vision, Graphics & Image Processing (2008)
29. van den Oord, A., Vinyals, O., kavukcuoglu, k.: Neural discrete representation learning. In: NeurIPS (2017)
30. Oquab, M., et al.: Dinov2: learning robust visual features without supervision. CoRR (2023)
31. O'Reilly, R.C., Bhattacharyya, R., Howard, M.D., Ketz, N.: Complementary learning systems. Cogn. Sci. (2014)
32. Parkhi, O.M., Vedaldi, A., Zisserman, A., Jawahar, C.V.: Cats and dogs. In: CVPR (2012)
33. Pham, Q., Liu, C., Hoi, S.: DualNet: continual learning, fast and slow. In: NeurIPS (2021)
34. Prabhu, A., Cai, Z., Dokania, P.K., Torr, P.H.S., Koltun, V., Sener, O.: Online continual learning without the storage constraint. CoRR (2023)
35. Prabhu, A., et al.: Computationally budgeted continual learning: What does matter? CVPR (2023)
36. Prabhu, A., Torr, P.H.S., Dokania, P.K.: GDumb: a simple approach that questions our progress in continual learning. In: Vedaldi, A., Bischof, H., Brox, T., Frahm, J.-M. (eds.) ECCV 2020. LNCS, vol. 12347, pp. 524–540. Springer, Cham (2020). https://doi.org/10.1007/978-3-030-58536-5_31
37. Radford, A., et al.: Learning transferable visual models from natural language supervision. In: ICML (2021)

38. Rebuffi, S., Kolesnikov, A., Sperl, G., Lampert, C.H.: ICARL: incremental classifier and representation learning. In: CVPR (2017)
39. Russakovsky, O., et al.: ImageNet large scale visual recognition challenge. IJCV **115**, 211–252 (2015)
40. Rusu, A.A., et al.: Progressive neural networks. arXiv:1606.04671 (2016)
41. Serrà, J., Suris, D., Miron, M., Karatzoglou, A.: Overcoming catastrophic forgetting with hard attention to the task. In: ICML (2018)
42. Shin, H., Lee, J.K., Kim, J., Kim, J.: Continual learning with deep generative replay. In: NeurIPS (2017)
43. Siegel, J.M.: Memory consolidation is similar in waking and sleep. Curr. Sleep Med. Rep. **7**, 15–18 (2021)
44. Skorokhodov, I., Elhoseiny, M.: Class normalization for (continual)? generalized zero-shot learning. In: ICLR (2021)
45. Smith, J.S., et al.: Coda-prompt: continual decomposed attention-based prompting for rehearsal-free continual learning. In: CVPR (2023)
46. Soomro, K., Zamir, A.R., Shah, M.: UCF101: a dataset of 101 human actions classes from videos in the wild. CoRR (2012)
47. Tong, S., Jones, E., Steinhardt, J.: Mass-producing failures of multimodal systems with language models. In: NeurIPS (2023)
48. Tong, S., Liu, Z., Zhai, Y., Ma, Y., LeCun, Y., Xie, S.: Eyes wide shut? exploring the visual shortcomings of multimodal LLMS. CoRR (2024)
49. Wamsley, E.J.: Offline memory consolidation during waking rest. Nat. Rev. Psychol. **23**, 171–173 (2022)
50. Wang, L., Zhang, X., Su, H., Zhu, J.: A comprehensive survey of continual learning: theory, method and application. CoRR (2023)
51. Wang, Y., Huang, Z., Hong, X.: S-prompts learning with pre-trained transformers: an OCCAM's razor for domain incremental learning. In: NeurIPS (2022)
52. Wang, Z., et al.: Learning to prompt for continual learning. In: CVPR (2022)
53. Wang, Z., et al.: Dualprompt: complementary prompting for rehearsal-free continual learning. In: Avidan, S., Brostow, G., Cissé, M., Farinella, G.M., Hassner, T. (eds.) ECCV 2022. LNCS, vol. 13686, pp. 631–648. Springer, Cham (2022)
54. Wortsman, M., et al.: Robust fine-tuning of zero-shot models. In: CVPR (2022)
55. Xiao, J., Hays, J., Ehinger, K.A., Oliva, A., Torralba, A.: Sun database: large-scale scene recognition from abbey to zoo. In: CVPR (2010)
56. Yan, S., Xie, J., He, X.: DER: dynamically expandable representation for class incremental learning. In: CVPR (2021)
57. Yoon, J., Yang, E., Lee, J., Hwang, S.J.: Lifelong learning with dynamically expandable networks. In: ICLR (2018)
58. Yu, R., Liu, S., Wang, X.: Dataset distillation: a comprehensive review (2023)
59. Zenke, F., Poole, B., Ganguli, S.: Continual learning through synaptic intelligence. In: ICML, pp. 3987–3995 (2017)
60. Zhang, J.O., Sax, A., Zamir, A., Guibas, L., Malik, J.: Side-tuning: a baseline for network adaptation via additive side networks. In: Vedaldi, A., Bischof, H., Brox, T., Frahm, J.-M. (eds.) ECCV 2020. LNCS, vol. 12348, pp. 698–714. Springer, Cham (2020). https://doi.org/10.1007/978-3-030-58580-8_41
61. Zheng, Z., Ma, M., Wang, K., Qin, Z., Yue, X., You, Y.: Preventing zero-shot transfer degradation in continual learning of vision-language models. CoRR (2023)
62. Zhu, Z., Lyu, W., Xiao, Y., Hoiem, D.: Continual learning in open-vocabulary classification with complementary memory systems. CoRR (2023)

External Knowledge Enhanced 3D Scene Generation from Sketch

Zijie Wu[1], Mingtao Feng[2(✉)], Yaonan Wang[1], He Xie[1],
Weisheng Dong[2], Bo Miao[3], and Ajmal Mian[3]

[1] Hunan University, Changsha, China
[2] Xidian University, Xi'an, China
mintfeng@hnu.edu.cn
[3] University of Western Australia, Claremont, USA

Abstract. Generating realistic 3D scenes is challenging due to the complexity of room layouts and object geometries. We propose a sketch based knowledge enhanced diffusion architecture (SEK) for generating customized, diverse, and plausible 3D scenes. SEK conditions the denoising process with a hand-drawn sketch of the target scene and cues from an object relationship knowledge base. We first construct an external knowledge base containing object relationships and then leverage knowledge enhanced graph reasoning to assist our model in understanding hand-drawn sketches. A scene is represented as a combination of 3D objects and their relationships, and then incrementally diffused to reach a Gaussian distribution. We propose a 3D denoising scene transformer that learns to reverse the diffusion process, conditioned by a hand-drawn sketch along with knowledge cues, to regressively generate the scene including the 3D object instances as well as their layout. Experiments on the 3D-FRONT dataset show that our model improves FID, CKL by 17.41%, 37.18% in 3D scene generation and FID, KID by 19.12%, 20.06% in 3D scene completion compared to the nearest competitor DiffuScene.

Keywords: Scene Generation · Knowledge Enhanced System · Diffusion

1 Introduction

There is an increasing demand for tools that automate the creation of artificial 3D environments for applications in game development, movies, augmented/virtual reality, and interior design. Sketch based 3D scene generation allows users to control the generated scene entities through a rough hand-drawn

---

Z. Wu—Work performed during visit at the University of Western Australia.

Supplementary Information The online version contains supplementary material available at https://doi.org/10.1007/978-3-031-72658-3_17.

sketch. Several methods for 3D scene generation rely on an input image [34,40] to guide the generation process for alignment with the input. However, such methods focus on leveraging 2D-3D consistency for supervision, which restricts diversity in the generated scene. Moreover, obtaining an image that serves as a 2D rendering of the intended 3D scene is not always straightforward.

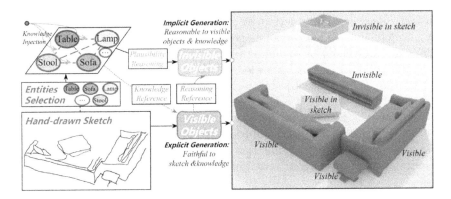

Fig. 1. Our method generates a 3D scene from an input sketch and entities, enhanced by external knowledge. It follows explicit visual cues in the sketch for *visible* objects along with their relationships and employs plausibility reasoning to add objects that are not explicitly depicted (*invisible*) in the sketch, to generate a coherent scene.

Recently, sketch based methods [32,43,51,58] have been proposed for user-specified 3D modeling. However, these methods focus primarily on generating single 3D objects. While much progress has been made in generating high-quality 3D objects, generation of complete 3D scenes is still challenging given the complex scene layouts, diverse object geometries and strong coherence between objects. For example, a chair can be placed underneath a table or around it, or may even be to the side of a bed. Various arrangements are possible for a chair and its neighboring objects, and each one must follow some rules including object-object relationships and space occupancy. To improve the representation and comprehension of 3D scenes, external knowledge has been introduced for multiple primary tasks, e.g., scene graph generation [11], robotic grounding [13], visual question answering [36], and semantic segmentation [18]. This involves reusing ontologies and integrating existing knowledge for improved outcomes. External knowledge has been a prevailing technique in transferring implicit representations between scenes for improved performance in various 3D vision tasks. In this paper, we leverage external knowledge to provide auxiliary information for completing implicit scene patterns, that are not obvious in the sparse ambiguous hand-drawn sketch, and guide our proposed indoor 3D scene generation model.

Existing methods use simple hand-crafted object relationships for generating 3D scenes. For instance, GRAINS [25] organize the scene objects into simple scene graph hierarchies that are manually defined. Furthermore, numerous works generate indoor scene layouts [29,38,40] in the form of object identities and bounding boxes and then *retrieve* existing furniture shapes from a repository for placement inside those bounding boxes. Hence, the generated layouts as well as the object shapes both lack diversity.

We propose a 3D scene generation method (see Fig. 1) that creates custom, diverse and plausible 3D scenes from hand-drawn sketches and entities, enhanced by external knowledge of object relationships. Our method takes a sketch as the main scene description and leverages external knowledge cues to reduce ambiguity in inferring *visible* objects (shapes and layout) in the sketch and enhance the generation diversity by including *invisible* objects that are not drawn in the sketch. We build an external knowledge base to contain rich knowledge priors of relationships. *Invisible* objects are inferred from the knowledge base across the *invisible* and *visible* objects to maintain diversity, plausibility and alignment with user specifications. Based on the sketch and knowledge reasoning, the proposed conditional scene diffusion *simultaneously* generates a 3D scene layout with detailed object geometries (see Fig. 2) with plausible structure and coherence among objects. Our contributions are summarized below:

- We propose an end-to-end generative model (SEK) to *simultaneously* generate realistic 3D room layouts and object shapes based on hand-drawn sketches and object entities, enhanced by external knowledge.
- We construct an external knowledge base that defines various object relationships, and serves as a foundational entity-relationship prior to provide additional guidance to the inference process. This improves the plausibility of the generated scenes, including layout and object shapes.
- We learn novel reasoning from external knowledge cues and hand-drawn sketches to extract a relationship subgraph of the specified entities during inference and integrate it with sketch features to form the diffusion condition.
- We propose a 3D denoising scene transformer that operates in the latent space and converts the denoising scene representation into the frequency domain to alleviate the influence of constants corresponding to invalid objects (padded zeros in scene representation) that are added to make the number of objects per scene constant.

2 Related Works

Sketch Based 3D Object Generation: Sketches have been used as a sparse representation of natural images and 3D shapes [44,58] as they are quite illustrative, despite their simplicity and abstract nature. Some works [28] estimate depth and 3D normals from a set of viewpoints for an input sketch, which are then integrated to form a 3D point cloud. Others [20] represent the 3D shape and its occluding contours in a joint VAE latent space during training, enabling them to retrieve a sketch during inference to generate a 3D shape. Recently, Kong *et al.* [23] trained a diffusion model conditioned on sketches using multi-stage training and fine-tuning. Sanghi *et al.* [43] used local semantic features from a frozen large pre-trained image encoder, such as CLIP, to map the sketch into a latent space for diverse shape generation. These methods mainly focus on object level generation, which prefers simple shape information instead of hierarchical relationship generation in scenes. However, 3D scenes contain rich information, including different furniture types, object geometries, room layouts, etc., presenting significant challenges to the generation process.

Knowledge Graphs in 3D Scenes: Prior knowledge has proven to be an effective source of information to enhance object and relation recognition [10]. Pioneering works, including ConceptNet [45], VisualGenome [24], DBPedia [2], and WordNet [33], have extensively studied the acquisition of label-pairs frequency as a primary source of relations. These methods have achieved great success in many applications such as image generation [21,54], visual question answering [8,47], camera localization [1], and robotic grounding [13,19,42]. Nevertheless, the integrated knowledge is not very useful in isolation since it is hard-coded in the form of intrinsic parameters. Hence, some methods bring external knowledge bases into 3D tasks related inference. Gu *et al.* [15] extracted knowledge triplets from the ConceptNet knowledge bases to help scene graph generation. GBNet [57] adopted auxiliary edges as bridges that facilitates message passing between knowledge graph and scene graph. Li *et al.* [25] introduced hand-crafted relationships for 3D scene generation in a recursive manner. The complete scene is encoded as multiple properties, including geometry and relationship, and is then recovered in a suggested pattern. Although previous studies have taken notice of knowledge in the 3D area, they only implicitly mine the extra knowledge base or define the relationship pairs to strengthen the iterative scene recovery between relationships and objects while ignoring the intrinsic properties of the data for specific 3D scene knowledge representation.

3D Scene Generation: Early methods for 3D scene generation are based on GANs [26,55], VAEs [5,26,39], and Autoregressive models [14,38,50]. They are renowned for their ability to generate high-quality results quickly, yet they often face challenges of limited diversity and difficulties in producing samples that align well with user specifications. Numerous methods learn to produce faithful results under different input conditions, such as images [34,35,48], text [30,46], sketches [52], and wall layouts [14,22,38]. Another approach in 3D scene generation is based on graph conditioning [7,29]. Graph-to-3D [7] jointly optimizes models to learn both scene layouts and shapes conditioned on a scene graph. However, the scene graph does not directly reveal the relationships among objects. This necessitates a complete graph description, impacting their realism and applicability. Conditional 3D scene synthesis methods offer faithful scene recovery tailored to user specifications, while generative methods strike a balance between diversity and alignment with user specifications.

3 Diffusion Model for Scene Generation

In the diffusion process, the data distribution is gradually destroyed into Gaussian noise following the Markov forward chain. A denoising process then recovers

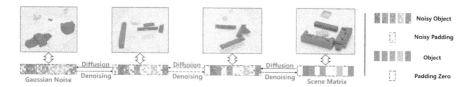

Fig. 2. Demonstration of the scene diffusion/denoising process of the matrix field and the spatial field. The denoising process samples from a Gaussian distribution and progressively denoises the sample for plausible and realistic scene generation. Note how the layout and 3D shapes are both *simultaneously* denoised.

data from the Gaussian distribution with an iterative reverse chain. To devise our scene diffusion model for generating 3D scenes, we introduce matrix conversion to represent an indoor scene in the form of a matrix. All processes operate on the matrix field. Figure 2 illustrates how the diffusion and denoising processes mutually transform the Gaussian and target data distributions.

Matrix conversion is proposed to encode the scene objects into parameters that specify their locations and shape attributes. Given an indoor scene $\mathcal{O}$ containing a set of objects $\{\mathbf{o}_i\}|_{i=1}^{K}$, each object o_i is characterized by a 1-D vector which concatenates its spatial location and latent shape attributes, i.e. $\mathbf{o}_i = [\mathcal{G}_i, \mathcal{F}_i] \in \mathbb{R}^{D \times 1}$. Every 3D scene is normalized by relocating it in a world coordinate system where the floor center is the origin.

The placement location of each object $\mathcal{G}_i = [\alpha_i, \mathbf{s}_i, \mathbf{t}_i]$ is defined by its axis-aligned 3D bounding box size $\mathbf{s} \in \mathbb{R}^{3 \times 1}$, translation $\mathbf{t} \in \mathbb{R}^{3 \times 1}$ and yaw angle $\alpha \in \mathbb{R}^{2 \times 1}$. Following [56], the yaw angle is parameterized as a 2D vector of sine and cosine values. All objects are normalized to $[-1, 1]^3$ first, and subsequently encoded into latent space. The shape latent $\mathcal{F} \in \mathbb{R}^{\mathcal{F}}$ is trained on DeepSDF [37] to obtain a unique code per object.

Since the number of objects can vary across scenes, we pad zero-vectors $\{\mathbf{z}_i | i \in (K+1, M)\} \in \mathbb{R}^{D \times 1}$ so that all scenes have a fixed number of M objects. Here, $K < M$. All objects are concatenated to form a full scene representation: $\mathcal{O} \in \mathbb{R}^{D \times M}$. A scene is encoded as a unique matrix, where each row corresponds to an object with shape, location and size attributes. By synthesizing various combinations of object parameters, we can generate diverse scenes.

Diffusion Process: In the forward chain of scene diffusion, the original scene matrix $\mathcal{O}_0 \sim q(\mathcal{O}_0)$ is gradually corrupted into a pre-defined T-step noised scene distribution following the Markov chain assumption until the Gaussian distribution is reached. Based on the Markov property, the joint distribution $\mathcal{O}_{1:T}$ is straight derived from the original scene matrix $\mathcal{O}_0$:

$$q\left(\mathcal{O}_{0:T}\right) = q\left(\mathcal{O}_0\right) \prod_{t=1}^{T} q\left(\mathcal{O}_t \mid \mathcal{O}_{t-1}\right), \quad q\left(\mathcal{O}_t \mid \mathcal{O}_{t-1}\right) = \mathcal{N}\left(\sqrt{1-\beta_t}\mathcal{O}_{t-1}, \beta_t \mathbf{I}\right), \tag{1}$$

where $\mathcal{N}(\mu, \sigma^2)$ denotes a Gaussian distribution and β_t is the known variance defined during the diffusion process.

Deniosing Process: Since the forward chain concludes with a Gaussian distribution, we apply the reverse chain starting from a standard Gaussian prior and ending with the desired scene representation $\mathcal{O}_0$:

$$p_\theta(\mathcal{O}_{0:T}) = p(\mathcal{O}_T) \prod_{t=1}^T p_\theta(\mathcal{O}_{t-1} \mid \mathcal{O}_t), \quad p_\theta(\mathcal{O}_{t-1} \mid \mathcal{O}_t) = \mathcal{N}\left(\mu_\theta(\mathcal{O}_t, t), \sigma_t^2 \mathbf{I}\right), \tag{2}$$

where p_θ is the inference step using parameterized network of the proposed scene denoiser. SEK is trained by minimizing the cross-entropy loss between two diffusion chains in relation to the sketch and knowledge enhanced conditional feature c, and by learning the scene denoiser parameters θ:

$$\min_\theta \mathbb{E}_{\mathcal{O}_0 \sim q(\mathcal{O}_0), \mathcal{O}_{1:T} \sim q(\mathcal{O}_{1:T})} \left[\sum_{t=1}^T \log p_\theta(\mathcal{O}_{t-1} \mid \mathcal{O}_t, c)\right]. \tag{3}$$

Following [17] to minimize Eq. 3, SEK learns to match each $q(\mathcal{O}_{t-1} \mid \mathcal{O}_t, \mathcal{O}_0)$ and $p_\theta(\mathcal{O}_{t-1} \mid \mathcal{O}_t)$ by estimating the noise ϵ_θ under the condition $(c, t, \mathcal{O}_t)$ to match the added noise ϵ in the diffusion process:

$$\mathcal{L}_{\text{sce}} = \mathbb{E}_{c, t, \epsilon, \mathcal{O}_0}\left[\|\epsilon - \epsilon_\theta(c, t, \mathcal{O}_t)\|^2\right], \epsilon \sim \mathcal{N}(0, \mathbf{I}). \tag{4}$$

Scene diffusion progressively generates the 3D scene using the reverse chain:

$$\mathcal{O}_{t-1} = 1/\sqrt{\alpha_t}\left(\mathcal{O}_t - 1 - \alpha_t/\sqrt{1 - \bar{\alpha}_t}\epsilon_\theta(c, t, \mathcal{O}_t)\right) + \sqrt{\beta_t}\epsilon, \tag{5}$$

where $\alpha_t = 1 - \beta_t$, $\tilde{\alpha} = \prod_{s=1}^t \alpha_s$, and ϵ is the standard Gaussian noise. Building upon a well-defined scene diffusion, the proposed model is theoretically capable of generating high-quality and diverse 3D scenes.

4 Knowledge Enhanced Sketch Based Guidance

Equipped with the scene diffusion model, the problem we need to address is how to ensure that the generated scene aligns with user description while preserving diversity, quality and plausibility. The input modality requires capturing the essence of the scene, providing a comprehensive description of the backbone while allowing flexibility and ambiguity, without confining to specific details, to foster diversity in generation. To this end, we deploy a model with *sketch* conditioning, offering strong flexibility, user-friendliness and diversity. Since all desired objects are not necessarily drawn in the sketch, *object entities* are also provided as input to complement the sketch and extract cues from the knowledge base. As depicted in Fig. 3(a), the knowledge-enhanced sketch is integrated via multi-head attention. We employ a spectrum-filter (SF) to enhance meaningful object features. The conditional denoising process iteratively predicts noise for scene matrix generation (Fig. 3(c)). Once the scene matrix is generated, the complete scene is decoded into a spatial field through data pop-out decoding along the object number dimension (Fig. 3(b)). Meaningless padding zeros are discarded.

Sketch serves as our primary medium for convenient user interaction, however, it lacks details to provide precise instructions for the scene generation. This aligns well with our objective of allowing diversity in generation while remaining faithful to user specifications. To enhance the instructions contained in a sketch, we integrate external knowledge that clarifies vague information. For example, when faced with a sparse sketch depicting either a *"table-aligned-sofa"* o a *"stool-aligned-sofa,"* querying knowledge can assist in determining which scenario is more likely to occur in indoor scenes.

Knowledge serves as a complement or extension to the sketch in our framework. For *visible* objects in the sketch, knowledge facilitates bidirectional validation to complement sketch descriptions. When a user provides object entities that are not visible in the sketch, knowledge helps the model to accommodate plausible object shapes based on the visible side of the relationships. In summary, knowledge enhancement provides several advantages: 1) Visible sketch enhancement: When object relations are depicted, knowledge can complement any ambiguous object descriptions in the hand-drawn sketch to enhance the plausibility of the generation. 2) Invisible description complement: If the depicted relations in the sketch do not contain the desired object entities, knowledge can accommodate the *invisible* objects by relating them to the depicted ones. For example, placing an appropriate "table" alongside a "sofa".

4.1 Knowledge Base

Generating a complete scene from a sketch often encounters challenges due to vague description and insufficient details. The proposed knowledge base helps in keeping the generated scenes realistic. We view knowledge as an essential complement to the scene sketch, effectively addressing its inherent sparsity and ambiguity. This section introduces our external knowledge base designed to retain extensive relationship priors for injection into the inference process. Knowledge of object relationships is first extracted from this external knowledge base and then dynamically learned to correspond with the sketch, facilitating interaction between the desired object entities.

We define knowledge base $KB = (\mathcal{V}, \mathcal{R}, p)$ as a repository containing a set of triplet relations where $\mathcal{V}$, $\mathcal{R}$, and p denote object nodes, their inner relation edges, and their edge probability respectively. The nodes $\mathcal{V}$ consist of a set of object types $f = \{f_1, f_2, \cdots, f_N\}$ and the relation edges $\mathcal{R}$ contain multiple predefined relations. Each triplet indicates the probability of the given relation existing between objects. This information is considered as external knowledge to reveal object relationships and then enhance the sketch descriptions in our framework. The predefined edge relationships $\mathcal{R}$, which include knowledge cues, are initially extracted from all indoor scenes in the given dataset. These relationships are

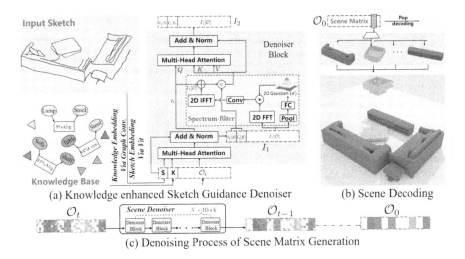

Fig. 3. Proposed SEK framework. (a) Sketch features are extracted by ViT and integrated with knowledge-enhanced reasoning features to form the denoising condition. The proposed 3D scene denoiser simultaneously generates plausible layouts and realistic 3D shapes in the matrix field. (b) The generated scene matrix is decoded to form the complete scene. (c) The denoising process: The scene denoiser starts from random noise and iteratively generates the scene matrix.

then normalized and stored in a unique knowledge graph, forming the designed external knowledge base. Here, we introduce our scheme details to extract the priors of object relations, which comprise the external knowledge base KB.

Object Relationship Construction: External knowledge integration aids in generating reasonable semantic entities and their relationships. Hence, we extract multiple knowledge relations $\mathcal{R}$ from the indoor scenes dateset. For generating this knowledge relation, each scene is divided into smaller functional groups using the density-based clustering algorithm, DBSCAN [3]. Initially, each object in the scene is voxelized, and then clustered into groups.

Overall, for a 3D scene S, we cluster it as $S = \{\langle \mathbf{g}_1, \mathbf{g}_2, \ldots, \mathbf{g}_K \rangle \| \mathbf{g}_i = (o_i, \ldots, o_j)\}$, where $\mathbf{g}_i$ represents the ith clustered group comprising variable number of objects. Inspired by [14,25,49], the extracted relationships between paired objects are categorized as adjacent relations within the same group $\mathbf{g}_i, \mathbf{g}_j, i = j$:

$$\mathcal{R}_a = \{\text{``Attachment''}, \text{``Alignment''}, \text{``Dependent''}\} \quad (6)$$

and distant relations cross different groups $\mathbf{g}_i, \mathbf{g}_j, i \neq j$:

$$\mathcal{R}_d = \{\text{``Co-occurrence''}, \text{``ParallelCollinearity''}\} \quad (7)$$

Unlike previous works that focus on object-wise relationships, we extract multiple relations to depict how entities are organized. These relations are defined as follows: (a) *Attachment*: The minimum distance between two adjacent object voxels is smaller than voxel length. (b) *Alignment*: Any plane of bounding boxes from two adjacent objects is coplanar. (c) *Dependent*: Adjacent pairwise relations in the same group that do not belong to attachment and alignment. (d) *Parallel Collinearity*: Horizontal axes of bounding boxes of the objects of different groups are parallel. (e) *Co-occurrence*: Identifying two objects that co-occur in the same scene of different groups.

Relationship Probability Counting: We count the number n_{ij} of these relationships using a clustered scene structure derived from the 3D-FRONT Dataset [12] to perform knowledge initialization. i, j denote indices of the entities, such as *chair*, *desk*. Each probability p is normalized by:

$$p_{ij} = 1/1 + e^{-10 \cdot n_{ij}^{\mathcal{R}}/max(n)^{\mathcal{R}}}. \tag{8}$$

The defined relationship edges $\mathcal{R}$ and their corresponding probability p are then integrated into the built-in knowledge base KB. The proposed SEK learns the object correlations among the given objects types, enabling it to dynamically interact with the knowledge base and assist in the diffusion module for scene generation. We build the knowledge base from 5,754 scenes in the dataset. More details are in the supplementary material.

4.2 Knowledge-Enhanced Graph Reasoning

The constructed knowledge base effectively reveals the potential spatial relationships among various pieces of objects. Based on the desired object types (*chair, table, desk, etc.*), we propose a knowledge-enhanced graph reasoning module (KeGR) to incorporate external knowledge from the initialized knowledge base for comprehensive room generation reasoning. For object types $\{f_i\}_{i=1}^n$ that the scene demands, we initialize each object representation h_i of f_i via GloVe so that $h_i \in \mathbb{R}^{1 \times D_w}$. Next, we obtain a subset of $\{f_i\}_{i=1}^n$ to construct a fully connected subgraph $G_i^E = (h_i^E, \mathcal{E}_i^E, \mathcal{P}_i^E) \in \{KB\}$. $h_i^E \in \mathbb{R}^{n \times D_w}$ denoting its node feature matrix. We represent edge and probability $\mathcal{E}_i^E, \mathcal{P}_i^E$ as the initial adjacency matrix $A_i^E \in \mathbb{R}^{n \times n}$. Knowledge-enhanced graph reasoning is achieved via multistep graph convolutions:

$$H_i^{E(j)} = \delta\left(A_i^E H_i^{E(j-1)} W^{E(j)}\right), \tag{9}$$

where j denotes the jth step of graph reasoning and δ is the activation function. $W^{E(j)}$ is a learnable parameter, and $H^{E(j)}$ is the node feature matrix of G^E at j_{th} step. After J iterations, we term $H_i^{E(J)}$ as the final node feature matrix of the graph reasoning of the current relationship. With all relation feature update, we get a final feature matrix group $\{H_i^{E(J)}\}|_{i=0}^r$, where r is the index related to relationships (6, 7) of knowledge base. We perform a 1×1 convolution across the relationship dimension to get the graph feature that is used to condition the scene diffusion:

$$H^G = \delta(conv_{1 \times 1}(H_0, H_1, \dots, H_r)). \tag{10}$$

4.3 Knowledge Enhanced Sketch Guided Denoiser

The denoiser serves as the key module of the scene diffusion model. It predicts the noise ϵ_θ for denoising process, thereby enabling the iterative generation of the 3D scene using Eq. 5. With the external knowledge, our KeGR module produces a rich feature representation for guiding scene generation. Given the specified entities, very diverse scenes can possibly be generated. Hence, to guide the generation process w.r.t. user alignment, we include a sketch as a complementary description. Sketches are highly expressive, inherently capturing subjective and fine-grained visual cues. Furthermore, its advantage lies in the combination of easy access and vivid description. Conditioned on sketch based knowledge reasoning features, the proposed SEK denoises the 3D scene from a random point in Gaussian distribution. We employ ViT [9] as our sketch embedding backbone to obtain the sketch condition H^S maintaining the details. The conditional feature c is formed as the concatenation $[H^S, H^G] \in \mathbb{R}^{D \times 2}$ of sketch and graph features.

In the forward chain of scene diffusion, the scene representation matrix $\mathcal{O}^{D \times M}$ (along with the padding) is diffused by adding Gaussian noise. As depicted in Fig. 2, padding occupies a significant portion of the scene matrix, potentially overwhelming the object information as the noise level increases. Padding is introduced in scene representation to make its dimension fixed during training. However, during inference, there is no mask available to filter out any padded values that are generated but have no specific meaning and overwhelm the desired shapes. The absence of a padding mask and the unavailability of the number of generated objects make it difficult to filter out disturbance components effectively. To address this problem, we propose component enhancement through a spectrum-filter with the intention to filter out the padding, ensuring that the prediction receives sufficient information from the valid object components. Compared to the valid object representations, we observe that the padding zeros have a low-frequency variance distribution. Note that this low-frequency distribution vanishes as noise is systematically added to $\mathcal{O}_0$ step-by-step, following the Markov chain assumption, and finally reaches the single-kernel Gaussian distribution. We apply a high-pass filter to suppress the low-frequency padding in the spectral domain. Let $\mathcal{O}_I$ denote the output of the attention blocks; the proposed spectrum-filter is computed as

$$\text{EF}(\mathcal{O}_I, B) = \mathcal{O}_I + e^{-t} \Theta_{IFFT} \left(\text{Conv} \left(\sigma(\mathcal{O}_I, B) \circledast \Theta_{FFT}(\mathcal{O}_I) \right) \right),$$

where t is the time step and $\circledast$ denotes high-pass filtering with adaptive Gaussian smoothed filters $\sigma(\mathcal{O}_I, B)$ (with bandwidth B), which has the same spatial size as $\mathcal{O}_I$. Θ denotes the spectrum operation using Fourier transform. Following [31], we create an initial 2D Gaussian map based on bandwidth B and apply the predefined weight parameter associated with time step t to scale the filter.

Our spectrum-filter block enhances meaningful object features in the encoded scene representation. As shown in Fig. 3(a), the scene transformer performs feature embedding first to get the embedding context initialization $\mathcal{I} = [c, t, \mathcal{O}_t] \in \mathbb{R}^{D \times (M+3)}$, where $t \in \mathbb{R}^{D \times 1}$ is the time step embedding and $\mathcal{O}_t \in \mathbb{R}^{D \times M}$ is the scene representation at time t. We begin by applying multihead attention at dimension M to capture the relevance of each element to every other element in the sequence. For example, we explore the guidance correlation between sketches and knowledge, the relevance of guidance among different conditions and every piece of object, as well as the interactions among different pieces of objects:

$$I_1 \in \mathbb{R}^{D \times (M+3)} = Atten_1(Q_1 = K_1 = V_1 = \mathcal{I}). \tag{11}$$

Next, we encode $\mathcal{I}_1$ using a transformer encoder to enrich its semantic information for each instance and then follow it by the proposed spectrum-filter block:

$$I_2 \in \mathbb{R}^{D \times (M+3)} = Atten_2(Q_2 = I_1, K_2 = V_3 = I_1[S] © I_1[K] © EF(I_1[\mathcal{O}])), \tag{12}$$

where, © denotes concatenation. Finally, we sample ϵ_{t-1} with dimensions congruent to those of scene $\mathcal{O}$ for prediction, using a set of regressive steps, where ϵ_{t-1} is the current predicted noise using the scene denoiser.

Overall, the 3D scene denoiser takes the context embedding $\mathcal{I}$ as input to perform spatial self-attention using the multi-head attention block. It is then fed to the spectrum-filter to enhance the features of valid objects and suppress invalid padding. Finally we take the output with dimension congruent to $\mathcal{O}_t$ as the predicted noise ϵ_{t-1} for supervision.

5 Experiments

Datasets: We train and test our method on three downstream tasks: 3D scene generation, 3D scene completion, and knowledge transfer validation. For the generation task, we use three types of indoor rooms from the 3D-FRONT dataset [12], including 4041 *Bedrooms*, 900 *Diningrooms*, 813 *Livingrooms*. We randomly split the data into training-test sets at 80–20% ratio. To acquire sketch, we first render images from 21 views using BlenderProc from each 3D scene uniformly when the viewpoint axis is $z_{vp} > 0$. We then apply Canny edge detection [27] to the rendered scene images to acquire their edge sketches. We manually remove the walls so that each scene contains hand-drawn looking sketches of the object. In scene completion, we randomly mask 30–80% of the objects in the scene and render it to acquire the sketches with viewpoint the same as in generation task following the above rendering process. Finally, we test the effectiveness of knowledge transfer by transferring knowledge from the 3D-FRONT dataset to the ScanNet dataset [6]. ScanNet is a real indoor scene dataset with 1,513 rooms of 21 different types. Common categories between ScanNet and 3D-FRONT dataset are selected for our knowledge transfer experiment. We retrieve objects of ScanNet from ShapeNet [4] to acquire consistent objects across scenes to maintain the same setting as in 3D-FRONT.

Baselines: We compare with state-of-the-art scene generation methods which can be categorised into retrieval-based and generation-based methods. In the

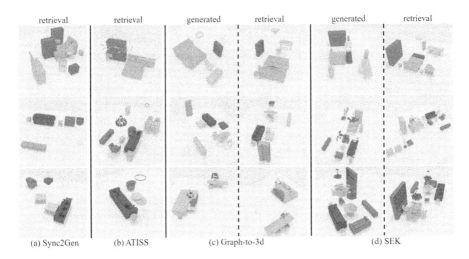

Fig. 4. Qualitative comparison. Syn2Gen and ATISS perform retrieval using 3D bounding boxes. Graph-to-3d and our method perform generation but we also show the corresponding retrieval results by searching nearest neighbor of shape code for comparison. Our method performs higher quality generation with detailed shapes and better plausibility of relationships.

former category, current works focus on the 3D scene layout generation followed by object placement using shape retrieval to form the complete scenes. We select the major floor plan [38,40,50], room size [55], and graph [29] based scene generation methods. Besides, some unconditional generation methods [46,53] are also included for comparison. In the latter category, methods generate both shape and layouts to directly form the 3D scenes. We select the graph [7] based and unconditional [34] generation methods for comparison. For a fair comparison, we ensure the training data of baselines is the same and that each model has its required modality. Furthermore, we also compare with the most relevant sketch based method, Sketch2Scene [52]. The ATISS and Sceneformer are adopted to accept module plugin of our sketch condition for comparison.

Evaluation Metrics: Following previous works [38,50,53], we use Frechet Inception Distance (FID), Kernel Inception Distance (KID × 0.001), Scene Classification Accuracy (SCA), and Category KL Divergence (CKL × 0.01) to measure the plausibility and diversity of 1,000 generated scenes. Additional information regarding evaluation metrics can be found in the supplementary materials.

5.1 Comparisons with State-of-the-Art Methods

Generative Quality Evaluation: Table 1 compares the indoor scene generation quality of our method with existing state-of-the-art. Only our method performs (single view) sketch and knowledge guided 3D scene generation. Among the unconditional methods in Table 1, the diffusion based DiffuScene [46] achieves better performance than Sync2Gen [53]. Although graph-based methods perform well on individual object generation (Fig. 4(c)), these methods require a complete scene graph description that still does not specify the relative locations of objects. Hence, graph-based methods do not perform well in complete scene generation and deviate significantly from the target scene. Graph-to-Box, a variant of Graph-to-3D, focuses only on learning the object layout. Image based methods synthesize scenes under strict 2D-3D consistency. ScenePrior [34] achieves better CKL, indicating the accuracy of object classes. Layout-based methods, such as ATISS [38], start from a given layout, often a top-down wall rendering image, and perform better in terms of generation diversity and quality. However, they do not always generate reasonable scene results. Our SEK outperforms current state-of-the-art methods in quality evaluation and achieves 17.41% FID, 3.63% SCA, and 37.18% CKL better than the nearest competitor DiffuScene [46] in the Dining Room category. Figure 4 shows a qualitative comparison between Sync2Gen, ATISS, Graph-to-3D, and SEK.

Sketch based Generation Evaluation: Table 2 compares the quality of scene generation from sketches. As there are no previous generative methods specifically designed for sketch-based generation, for a fair comparison, we adopt the current state-of-the-art methods, ATISS and Sceneformer, that allow the plugin of additional conditions. We append the additional sketch to the original attention modules of ATISS and Sceneformer and refer to them as ATISS-S and Sceneformer-S, where S indicates the addition of the sketch condition. We further augment these methods by concatenating the knowledge-enhanced sketch condition, resulting in the baseline models ATISS-SK and Sceneformer-SK respectively. Additionally, we compare with the most relevant prior work in sketch-based scene synthesis, Sketch2Scene [52]. Sketch2Scene optimizes scenes

Table 1. Comparative results on the 3D-FRONT dataset. For Scene Classification Accuracy (SCA), a score closer to 50% is better as it means that the generated distribution is closer to target distribution.

Method	Bedroom			Dining room			Living room		
	FID ↓	SCA %	CKL ↓	KID ↓	SCA %	CKL ↓	KID ↓	SCA %	CKL ↓
DepthGAN [55]	40.15	96.04	5.04	81.13	98.59	9.72	88.10	97.85	7.95
Sync2Gen [53]	31.07	82.97	2.24	46.05	88.02	4.96	48.45	84.57	7.52
ATISS [38]	18.60	61.71	0.78	38.66	71.34	0.64	40.83	72.66	0.69
DiffuScene [46]	18.29	53.52	0.35	32.60	55.50	0.22	36.18	57.81	0.21
Graph-to-3D [7]	61.24	74.03	1.79	54.11	76.18	1.68	41.13	79.37	2.04
Graph-to-Box [7]	55.28	69.48	1.02	50.29	73.25	1.42	48.77	78.41	1.81
3D-SLN [29]	58.17	71.27	1.38	49.67	75.39	1.44	47.29	76.29	1.77
ScenePrior [34]	24.88	83.26	0.43	46.25	89.27	0.58	44.28	88.07	0.31
FastSynth [40]	31.89	83.61	2.40	51.26	90.12	5.26	57.22	88.21	6.27
Sceneformer [50]	33.61	85.38	1.86	61.08	85.94	5.18	63.54	90.20	3.13
Ours	**15.21**	**51.24**	**0.18**	**25.46**	**51.78**	**0.16**	**31.24**	**52.91**	**0.15**

Table 2. Results for 3D scene generation based on a given sketch (and knowledge).

Method	Bedroom			Dining room			Living room		
	FID ↓	KID ↓	SCA %	FID ↓	KID ↓	SCA %	FID ↓	KID ↓	SCA %
Sceneformer-S [46]	37.21	13.04	88.37	64.38	14.21	87.16	65.78	15.03	91.14
ATISS-S [46]	21.33	3.87	64.37	42.53	6.97	74.28	44.24	7.29	77.61
Sceneformer-SK [46]	24.68	8.46	81.66	52.07	8.78	80.33	59.43	10.48	86.91
ATISS-SK [46]	18.47	1.58	58.37	37.24	2.47	63.05	39.10	3.08	63.35
Sketch2Scene [38]	22.47	7.18	55.78	41.35	6.91	72.38	58.79	7.27	84.81
Ours	**15.21**	**1.12**	**51.24**	**25.46**	**0.49**	**51.78**	**31.24**	**0.71**	**52.91**

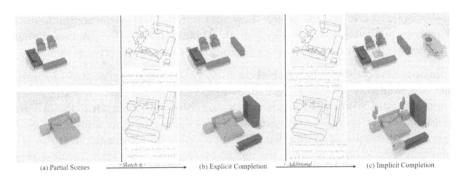

Fig. 5. Demonstration of sketch & knowledge guided scene completion. In explicit completion, the sketch and user-specified entities complement each other. Beyond explicit instructions, the additional invisible entities are inferred based on knowledge and the current visible objects to generate plausible extra objects in the scene.

to closely resemble examples in a repository while adhering to constraints from input sketches. This is done through sketch-based co-retrieval and co-placement of 3D models, ensuring similarity to reference scenes while maintaining originality. In the implementation of Sketch2Scene, since our sketch is included as a whole and lacks related pixel class information, we employ DBSCAN to manually cluster the object sketches as required in the inference stage. Note that while Sketch2Scene is the most relevant prior work in sketch-based scene generation, it necessitates an additional 3D model repository. In contrast, our method generates scenes in an end-to-end manner, without the need for such repositories.

Table 3. Results for 3D scene completion from random initial 3–5 objects.

Method	Bedroom			Dining room			Living room		
	FID ↓	KID ↓	SCA %	FID ↓	KID ↓	SCA %	FID ↓	KID ↓	SCA %
ATISS [38]	30.54	2.38	26.73	42.65	8.32	43.99	45.39	8.08	41.26
DiffuScene [46]	27.32	1.92	40.30	40.99	6.31	49.06	43.72	8.37	46.48
Ours	**21.84**	**1.58**	**45.47**	**33.03**	**5.18**	**49.45**	**35.74**	**6.51**	**48.76**

Scene Completion: We compare against ATISS [38] and DiffuScene [46] for scene completion. For our SEK and DiffuScene, we apply the DDIM inversion process, akin to image inpainting [41], to the scene representation of the known furniture. A partial scene is obtained by the learned reverse chain following Eq. 2, and is then combined with the known scene to form the completed scene. More specifically, we retrain the model by randomly masking furniture in the sketch and test the scene completion performance. Results are given in Table 3 which show that our method performs the best on all metrics achieving average 19.12%FID, 20.06%KID, 2.61%SCA better than the nearest competitor DiffuScene [46]. Figure 5 shows qualitative results.

Table 4. Module ablation study

Sketch/Knowledge	SF	FID ↓	SCA %	CKL ↓
✗/✓	✓	32.26	58.31	0.61
ResNet50/✗	✓	34.76	58.94	1.07
ViT/✗	✓	33.29	56.81	0.85
ResNet50/✓	✓	24.68	52.70	0.18
ViT/✓	✗	25.83	54.19	0.37
ViT/✓	✓	**23.97**	**51.97**	**0.16**

Ablation Study: We perform ablation study on the condition modules to verify their contributions in Table 4. We employ ResNet50 [16] and ViT [9] with 8 attention blocks and 8 attention heads as our sketch encoder. Sketch alone can not achieve good performance and performs worse than using only knowledge guidance. In the absence of knowledge, ViT shows some improvement over ResNet, but when knowledge is present, the enhancement from the sketch encoder type (ViT vs ResNet) becomes minimal. We also drop SF module for comparison (6th row). We show the More importantly, sketch and knowledge base complement each other really well and significantly improve performance when working together to jointly promote the overall quality of generation.

5.2 Knowledge Transfer to ScanNet

Knowledge transfer study is conduct to evaluate the effectiveness of the knowledge base across datasets. In our architecture, the sketch guides the spatial distribution of the objects, while the knowledge helps establish their relationships and resolves ambiguities in the sketch to generate plausible scenes. As shown in Table 5, we compare three baseline implementations of knowledge base: 1) Empty: We use an empty relationship knowledge base (parameters set to zero). 2) 3DFRONT: We directly use the external knowledge base constructed on 3D-FRONT for generation on ScanNet. 3) ScanNet: We re-train the knowledge base on ScanNet and then use it for generation on ScanNet. As expected, without relationship knowledge base, the results are much worse than when knowledge base is used. Interestingly, the knowledge extracted on 3DFRONT generates ScanNet scenes (row 2) as good as when knowledge is extracted from ScanNet itself to generate ScanNet scenes (row 3) with a very minor drop in performance on all metrics i.e. 0.34 FID, 0.02 KID, 0.95%SCA, and 0.01 CKL. This shows that our constructed knowledge base effectively transfers across datasets.

6 Conclusion

We proposed a novel sketch based knowledge-enhanced diffusion method for generating customized, diverse, and plausible 3D scenes. Our method conditions the denoising process with a hand-drawn sketch of the required scene and cues from object relationship knowledge. Given the strong generative ability of the base diffusion model, our method can take a hand-drawn sketch along with entity information to generate diverse scenes that align well with user specifications. We introduced a new condition for generation that incorporates external knowledge graphs, consisting of a set of well-defined relationship tuples. External knowledge helps resolve ambiguities for visible objects and their relationships in the hand-drawn sketches as well as introduce additional objects that are specified entities but not drawn in the sketch. Experimental results demonstrate that our model achieves state-of-the-art performance in 3D scene generation and shows promising results for the task of 3D scene completion as well.

Table 5. Knowledge transfer to ScanNet dataset.

KB Source	FID ↓	KID ↓	SCA %	CKL ↓
Empty	44.27	3.08	75.30	1.54
3DFRONT	33.81	0.83	55.18	0.18
ScanNet	**33.47**	**0.81**	**54.23**	**0.17**

Acknowledgement. This research was supported by National Key R & D Program of China under Grant 2023YFB4704800, National Natural Science Foundation of China under Grant 62293512, 62373293, 62293515, 62203160, and by ARC Discovery Project DP240101926. Ajmal Mian is the recipient of an ARC Future Fellowship Award (project number FT210100268) funded by the Australian Government.

References

1. Armeni, I., et al.: 3D scene graph: a structure for unified semantics, 3d space, and camera. In: Proceedings of the IEEE/CVF International Conference on Computer Vision, pp. 5664–5673 (2019)
2. Auer, S., Bizer, C., Kobilarov, G., Lehmann, J., Cyganiak, R., Ives, Z.: DBpedia: a nucleus for a web of open data. In: Aberer, K., et al. (eds.) ASWC/ISWC -2007. LNCS, vol. 4825, pp. 722–735. Springer, Heidelberg (2007). https://doi.org/10.1007/978-3-540-76298-0_52
3. Campello, R.J., Kröger, P., Sander, J., Zimek, A.: Density-based clustering. Wiley Interdiscip. Rev. Data Min. Knowl. Disc. **10**(2), e1343 (2020)
4. Chang, A.X., et al.: Shapenet: an information-rich 3d model repository. arXiv preprint arXiv:1512.03012 (2015)
5. Chattopadhyay, A., Zhang, X., Wipf, D.P., Arora, H., Vidal, R.: Learning graph variational autoencoders with constraints and structured priors for conditional indoor 3D scene generation. In: Proceedings of the IEEE/CVF Winter Conference on Applications of Computer Vision (WACV), pp. 785–794 (2023)
6. Dai, A., Chang, A.X., Savva, M., Halber, M., Funkhouser, T., Nießner, M.: Scannet: Richly-annotated 3d reconstructions of indoor scenes. In: Proceedings of the IEEE Conference on Computer Vision and Pattern Recognition, pp. 5828–5839 (2017)

7. Dhamo, H., Manhardt, F., Navab, N., Tombari, F.: Graph-to-3D: end-to-end generation and manipulation of 3d scenes using scene graphs. In: Proceedings of the IEEE/CVF International Conference on Computer Vision, pp. 16352–16361 (2021)
8. Ding, Y., Yu, J., Liu, B., Hu, Y., Cui, M., Wu, Q.: Mukea: multimodal knowledge extraction and accumulation for knowledge-based visual question answering. In: Proceedings of the IEEE/CVF Conference on Computer Vision and Pattern Recognition, pp. 5089–5098 (2022)
9. Dosovitskiy, A., et al.: An image is worth 16x16 words: transformers for image recognition at scale. ICLR (2021)
10. Feng, M., et al.: Exploring hierarchical spatial layout cues for 3D point cloud based scene graph prediction. IEEE Trans. Multimedia (2023)
11. Feng, M., Hou, H., Zhang, L., Wu, Z., Guo, Y., Mian, A.: 3D spatial multimodal knowledge accumulation for scene graph prediction in point cloud. In: Proceedings of the IEEE/CVF Conference on Computer Vision and Pattern Recognition, pp. 9182–9191 (2023)
12. Fu, H., et al.: 3D-front: 3D furnished rooms with layouts and semantics. In: Proceedings of the IEEE/CVF International Conference on Computer Vision, pp. 10933–10942 (2021)
13. Gao, C., Chen, J., Liu, S., Wang, L., Zhang, Q., Wu, Q.: Room-and-object aware knowledge reasoning for remote embodied referring expression. In: Proceedings of the IEEE/CVF Conference on Computer Vision and Pattern Recognition, pp. 3064–3073 (2021)
14. Gao, L., Sun, J.M., Mo, K., Lai, Y.K., Guibas, L.J., Yang, J.: SceneHGN: hierarchical graph networks for 3d indoor scene generation with fine-grained geometry. IEEE Trans. Pattern Anal. Mach. Intell. (2023)
15. Gu, J., Zhao, H., Lin, Z., Li, S., Cai, J., Ling, M.: Scene graph generation with external knowledge and image reconstruction. In: Proceedings of the IEEE/CVF Conference on Computer Vision and Pattern Recognition, pp. 1969–1978 (2019)
16. He, K., Zhang, X., Ren, S., Sun, J.: Deep residual learning for image recognition. In: Proceedings of the IEEE Conference on Computer Vision and Pattern Recognition, pp. 770–778 (2016)
17. Ho, J., Jain, A., Abbeel, P.: Denoising diffusion probabilistic models. Adv. Neural. Inf. Process. Syst. **33**, 6840–6851 (2020)
18. Hou, Y., Zhu, X., Ma, Y., Loy, C.C., Li, Y.: Point-to-voxel knowledge distillation for lidar semantic segmentation. In: Proceedings of the IEEE/CVF Conference on Computer Vision and Pattern Recognition, pp. 8479–8488 (2022)
19. Hughes, N., Chang, Y., Carlone, L.: Hydra: a real-time spatial perception system for 3d scene graph construction and optimization. arXiv preprint arXiv:2201.13360 (2022)
20. Jin, A., Fu, Q., Deng, Z.: Contour-based 3D modeling through joint embedding of shapes and contours. In: Symposium on Interactive 3D Graphics and Games, pp. 1–10 (2020)
21. Johnson, J., Gupta, A., Fei-Fei, L.: Image generation from scene graphs. In: Proceedings of the IEEE Conference on Computer Vision and Pattern Recognition, pp. 1219–1228 (2018)
22. Jyothi, A.A., Durand, T., He, J., Sigal, L., Mori, G.: LayoutVAE: stochastic scene layout generation from a label set. In: Proceedings of the IEEE/CVF International Conference on Computer Vision, pp. 9895–9904 (2019)
23. Kong, D., Wang, Q., Qi, Y.: A diffusion-refinement model for sketch-to-point modeling. In: Proceedings of the Asian Conference on Computer Vision, pp. 1522–1538 (2022)

24. Krishna, R., et al.: Visual genome: Connecting language and vision using crowd-sourced dense image annotations. Int. J. Comput. Vision **123**, 32–73 (2017)
25. Li, M., et al.: Grains: generative recursive autoencoders for indoor scenes. ACM Trans. Graphics (TOG) **38**(2), 1–16 (2019)
26. Li, S., Li, H., et al.: Deep generative modeling based on VAE-GAN for 3D indoor scene synthesis. Int. J. Comput. Games Technol. **2023** (2023)
27. Li, Y., Liu, B.: Improved edge detection algorithm for canny operator. In: 2022 IEEE 10th Joint International Information Technology and Artificial Intelligence Conference (ITAIC), vol. 10, pp. 1–5. IEEE (2022)
28. Lun, Z., Gadelha, M., Kalogerakis, E., Maji, S., Wang, R.: 3D shape reconstruction from sketches via multi-view convolutional networks. In: 2017 International Conference on 3D Vision (3DV), pp. 67–77. IEEE (2017)
29. Luo, A., Zhang, Z., Wu, J., Tenenbaum, J.B.: End-to-end optimization of scene layout. In: Proceedings of the IEEE/CVF Conference on Computer Vision and Pattern Recognition (CVPR) (2020)
30. Ma, R., et al.: Language-driven synthesis of 3D scenes from scene databases. ACM Trans. Graphics (TOG) **37**(6), 1–16 (2018)
31. Miao, B., Bennamoun, M., Gao, Y., Mian, A.: Spectrum-guided multi-granularity referring video object segmentation. arXiv preprint arXiv:2307.13537 (2023)
32. Mikaeili, A., Perel, O., Safaee, M., Cohen-Or, D., Mahdavi-Amiri, A.: Sked: sketch-guided text-based 3D editing. In: Proceedings of the IEEE/CVF International Conference on Computer Vision, pp. 14607–14619 (2023)
33. Miller, G.A.: Wordnet: a lexical database for English. Commun. ACM **38**(11), 39–41 (1995)
34. Nie, Y., Dai, A., Han, X., Nießner, M.: Learning 3D scene priors with 2D supervision. In: Proceedings of the IEEE/CVF Conference on Computer Vision and Pattern Recognition, pp. 792–802 (2023)
35. Nie, Y., Han, X., Guo, S., Zheng, Y., Chang, J., Zhang, J.J.: Total3dunderstanding: joint layout, object pose and mesh reconstruction for indoor scenes from a single image. In: Proceedings of the IEEE/CVF Conference on Computer Vision and Pattern Recognition, pp. 55–64 (2020)
36. Parelli, M., et al.: Clip-guided vision-language pre-training for question answering in 3d scenes. In: Proceedings of the IEEE/CVF Conference on Computer Vision and Pattern Recognition (CVPR) Workshops, pp. 5606–5611 (2023)
37. Park, J.J., Florence, P., Straub, J., Newcombe, R., Lovegrove, S.: DeepSDF: learning continuous signed distance functions for shape representation. In: Proceedings of the IEEE/CVF Conference on Computer Vision and Pattern Recognition, pp. 165–174 (2019)
38. Paschalidou, D., Kar, A., Shugrina, M., Kreis, K., Geiger, A., Fidler, S.: ATISS: autoregressive transformers for indoor scene synthesis. Adv. Neural. Inf. Process. Syst. **34**, 12013–12026 (2021)
39. Purkait, P., Zach, C., Reid, I.: SG-VAE: scene grammar variational autoencoder to generate new indoor scenes. In: Vedaldi, A., Bischof, H., Brox, T., Frahm, J.-M. (eds.) ECCV 2020. LNCS, vol. 12369, pp. 155–171. Springer, Cham (2020). https://doi.org/10.1007/978-3-030-58586-0_10
40. Ritchie, D., Wang, K., Lin, Y.A.: Fast and flexible indoor scene synthesis via deep convolutional generative models. In: Proceedings of the IEEE/CVF Conference on Computer Vision and Pattern Recognition (CVPR) (2019)
41. Rombach, R., Blattmann, A., Lorenz, D., Esser, P., Ommer, B.: High-resolution image synthesis with latent diffusion models. In: Proceedings of the IEEE/CVF Conference on Computer Vision and Pattern Recognition, pp. 10684–10695 (2022)

42. Rosinol, A., et al.: KIMERA: from slam to spatial perception with 3D dynamic scene graphs. Int. J. Robot. Res. **40**(12–14), 1510–1546 (2021)
43. Sanghi, A., et al.: Sketch-a-shape: zero-shot sketch-to-3D shape generation. arXiv preprint arXiv:2307.03869 (2023)
44. Shen, Y., Zhang, C., Fu, H., Zhou, K., Zheng, Y.: Deepsketchhair: deep sketch-based 3d hair modeling. IEEE Trans. Visual Comput. Graphics **27**(7), 3250–3263 (2021). https://doi.org/10.1109/TVCG.2020.2968433
45. Speer, R., Chin, J., Havasi, C.: Conceptnet 5.5: an open multilingual graph of general knowledge. In: Proceedings of the AAAI Conference on Artificial Intelligence, vol. 31 (2017)
46. Tang, J., Nie, Y., Markhasin, L., Dai, A., Thies, J., Nießner, M.: Diffuscene: scene graph denoising diffusion probabilistic model for generative indoor scene synthesis. arXiv preprint arXiv:2303.14207 (2023)
47. Teney, D., Liu, L., van Den Hengel, A.: Graph-structured representations for visual question answering. In: Proceedings of the IEEE Conference on Computer Vision and Pattern Recognition, pp. 1–9 (2017)
48. Tulsiani, S., Gupta, S., Fouhey, D.F., Efros, A.A., Malik, J.: Factoring shape, pose, and layout from the 2D image of a 3D scene. In: Proceedings of the IEEE Conference on Computer Vision and Pattern Recognition, pp. 302–310 (2018)
49. Wald, J., Dhamo, H., Navab, N., Tombari, F.: Learning 3D semantic scene graphs from 3D indoor reconstructions. In: Proceedings of the IEEE/CVF Conference on Computer Vision and Pattern Recognition, pp. 3961–3970 (2020)
50. Wang, X., Yeshwanth, C., Nießner, M.: Sceneformer: indoor scene generation with transformers. In: 2021 International Conference on 3D Vision (3DV), pp. 106–115. IEEE (2021)
51. Wu, Z., Wang, Y., Feng, M., Xie, H., Mian, A.: Sketch and text guided diffusion model for colored point cloud generation. In: Proceedings of the IEEE/CVF International Conference on Computer Vision, pp. 8929–8939 (2023)
52. Xu, K., Chen, K., Fu, H., Sun, W.L., Hu, S.M.: Sketch2scene: Sketch-based co-retrieval and co-placement of 3D models. ACM Trans. Graphics (TOG) **32**(4), 1–15 (2013)
53. Yang, H., et al.: Scene synthesis via uncertainty-driven attribute synchronization. In: Proceedings of the IEEE/CVF International Conference on Computer Vision, pp. 5630–5640 (2021)
54. Yang, L., et al.: Diffusion-based scene graph to image generation with masked contrastive pre-training. arXiv preprint arXiv:2211.11138 (2022)
55. Yang, M.J., Guo, Y.X., Zhou, B., Tong, X.: Indoor scene generation from a collection of semantic-segmented depth images. In: Proceedings of the IEEE/CVF International Conference on Computer Vision, pp. 15203–15212 (2021)
56. Yin, T., Zhou, X., Krahenbuhl, P.: Center-based 3D object detection and tracking. In: Proceedings of the IEEE/CVF Conference on Computer Vision and Pattern Recognition, pp. 11784–11793 (2021)
57. Zareian, A., Karaman, S., Chang, S.-F.: Bridging knowledge graphs to generate scene graphs. In: Vedaldi, A., Bischof, H., Brox, T., Frahm, J.-M. (eds.) ECCV 2020. LNCS, vol. 12368, pp. 606–623. Springer, Cham (2020). https://doi.org/10.1007/978-3-030-58592-1_36
58. Zhang, S.H., Guo, Y.C., Gu, Q.W.: Sketch2model: view-aware 3D modeling from single free-hand sketches. In: Proceedings of the IEEE/CVF Conference on Computer Vision and Pattern Recognition, pp. 6012–6021 (2021)

G3R: Gradient Guided Generalizable Reconstruction

Yun Chen[1,2](✉), Jingkang Wang[1,2], Ze Yang[1,2], Sivabalan Manivasagam[1,2], and Raquel Urtasun[1,2]

[1] Waabi, Toronto, Canada
{ychen,jwang,zyang,smanivasagam,urtasun}@waabi.ai
[2] University of Toronto, Toronto, Canada

Abstract. Large scale 3D scene reconstruction is important for applications such as virtual reality and simulation. Existing neural rendering approaches (*e.g.*, NeRF, 3DGS) have achieved realistic reconstructions on large scenes, but optimize per scene, which is expensive and slow, and exhibit noticeable artifacts under large view changes due to overfitting. Generalizable approaches, or large reconstruction models, are fast, but primarily work for small scenes/objects and often produce lower quality rendering results. In this work, we introduce G3R, a generalizable reconstruction approach that can efficiently predict high-quality 3D scene representations for large scenes. We propose to learn a reconstruction network that takes the gradient feedback signals from differentiable rendering to iteratively update a 3D scene representation, combining the benefits of high photorealism from per-scene optimization with data-driven priors from fast feed-forward prediction methods. Experiments on urban-driving and drone datasets show that G3R generalizes across diverse large scenes and accelerates the reconstruction process by at least 10× while achieving comparable or better realism compared to 3DGS, and also being more robust to large view changes. Please visit our project page for more results: https://waabi.ai/g3r.

Keywords: Generalizable Reconstruction · Neural Rendering · Learned Optimization · 3DGS · Large Reconstruction Models

1 Introduction

Reconstruction of large real world scenes from sensor data, such as urban traffic scenarios, is a long-standing problem in computer vision and computer graphics. Scene reconstruction enables applications such as virtual reality and high-fidelity

---

Y. Chen and J. Wang—Equal contributions.

Supplementary Information The online version contains supplementary material available at https://doi.org/10.1007/978-3-031-72658-3_18.

camera simulation, where robots such as autonomous vehicles can learn and be evaluated safely at scale. To be effective, the 3D reconstructions must have high photorealism at novel views, be efficient to generate, enable scene manipulation, and enable real-time image rendering.

Recently, neural rendering approaches such as NeRF [36] and 3D Gaussian Splatting (3DGS) [18] have achieved realistic reconstructions for large scenes using camera and optionally LiDAR data. However, they require a costly per-scene optimization process to reconstruct the scene by recreating the input sensor data via differentiable rendering, which may take several hours to achieve high-quality. Moreover, they typically focus on the novel view synthesis (NVS) setting where the target view is close to the source views and often exhibit artifacts when the viewpoint changes are large (*e.g.*, meter-scale shifts), as it can overfit to the input images while not learning the true underlying 3D representation.

To enable faster reconstruction and better performance at novel views, recent works aim to synthesize a *generalizable* representation with a single pre-trained network, which can be used for NVS on unseen scenes in a zero-shot manner. These methods utilize an encoder to predict the intermediate scene representation by aggregating image features extracted from multiple source views according to camera and geometry priors, and then decode the representation for NVS via volume rendering or a transformer [7,26,65,66,82]. The encoder and decoder networks are trained across many scenes to reconstruction priors. Most recently, large reconstruction models (LRMs) are proposed to learn reconstruction priors by training on large-scale synthetic datasets for generalizable single-step 2D to 3D reconstruction [14,22,30,68,83]. However, both generalizable NVS and LRMs are primarily applied to objects or small scenes due to the complexity of large scenes, which are difficult to predict accurately from a single step network prediction. Furthermore, the computation resources and memory needed to utilize many input scene images (>100) with existing techniques that aggregate ray features [65], build cost volumes [7] or perform image-based rendering [66] are prohibitive.

In this paper, we present Gradient Guided Generalizable Reconstruction (G3R), the first method that enables fast and generalizable reconstruction of large scenes. Given a sequence of images and an approximate geometry scaffold (*e.g.*, points from LiDAR or multi-view stereo), G3R can produce a modifiable digital twin as a set of 3D Gaussian primitives in two minutes or less for large scenes ($> 10,000\,\text{m}^2$). This representation can be directly used for high-fidelity novel-view rendering at interactive frame rates (> 90 FPS). Our key idea is to learn a single reconstruction network that iteratively updates the 3D scene representation, combining the benefits of data-driven priors from fast prediction methods with the iterative gradient feedback signal from per-scene optimization methods. G3R can be viewed as a "learned optimizer" [4,69] for scene reconstruction. Towards this goal, we first initialize a neural scene representation which we call *3D Neural Gaussians* from the geometry scaffold that can be differentiably rendered. Rather than select a few close-by source views for unprojection like existing generalizable works, we propose a novel way of lifting 2D

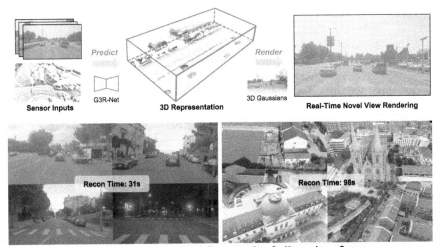

Fig. 1. Gradient Guided Generalizable Reconstruction (G3R): Our method learns a single reconstruction network that takes multi-view camera images and an initial point set to predict the 3D representation for large scenes ($> 10,000 \, \text{m}^2$) in two minutes or less, enabling realistic and real-time camera simulation.

images to 3D space by rendering and backpropagating to obtain gradients w.r.t the current 3D representation. These 3D gradients can be seen as 2D images unprojected to 3D with the current representation as the 3D proxy, which takes the rendering procedure into account, and is thus naturally occlusion aware and contains a useful feedback signal. Moreover, it provides a unified representation that can efficiently aggregate as many 2D images as needed by just aggregating the gradients. Then, our reconstruction network (G3R-Net) takes the 3D gradients and current 3D representation as inputs and iteratively predicts updates to refine the representation. Since the G3R-Net incorporates the rendering feedback signal at each step and is trained across multiple scenes, it can significantly accelerate the convergence compared to standard gradient descent algorithms (*i.e.*, 24 iterations v.s. 1000s of iterations). G3R-Net is trained across multiple scenes, enabling high quality reconstruction and improving robustness for NVS.

Experiments on two outdoor datasets with large-scale scenes demonstrate the generalizability of G3R. With as little as 24 iterations, G3R reconstructs large scenes with comparable or better realism at novel views than the per-scene optimization approaches while being at least 10× faster. To the best of our knowledge, this is the first generalizable reconstruction approach that can reconstruct a faithful 3D representation for such large-scale scenes ($> 10,000 \, \text{m}^2$) in high-resolution (> 100 source images at 1080×1920), showing the potential to build digital twins for the metaverse and simulation at large scale (Fig. 1).

2 Related Work

Optimization-based scene reconstruction: The current state-of-the-art in scene reconstruction is optimizing differentiable radiance fields, such as NeRF [36] or 3DGS [18], which model the 3D scene either as neural networks or as Gaussian primitives, and then alpha-composite along the ray via either ray-marching or rasterization, respectively. To extend to city-scale scenes, some works decompose the scene into sub-components and represent each with a network to increase model capacity [27,59,63,85]. To enable realistic and controllable sensor simulation, another line of work decomposes dynamic scenes (*e.g.*, urban driving scenes) into static background and moving objects [16,29,40,61,72,74,76–79,87] or conduct inverse rendering for geometry, material, lighting and semantics decomposition [28,41,64,67]. These works require time-consuming (hours or days) per-scene optimization for large scenes and often exhibit artifacts at large view changes due to overfitting. In contrast, G3R predicts a high-quality and robust 3D representation for large scenes in a few minutes or less.

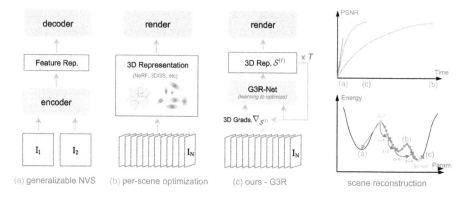

Fig. 2. Three paradigms for scene reconstruction and novel view synthesis (NVS). (a) Existing generalizable approaches select a few reference images (usually ≤ 5) for feed-forward prediction of intermediate representation and then decode/render the feature representation to produce the rendered images. (b) Per-scene optimization approaches take all source images (*e.g.*, > 100 for large scenes) and reconstructs a 3D representation via energy minimization and differentiable rendering. (c) G3R conducts iterative prediction to refine the 3D representation with the 3D gradient guidance (*i.e*, learned optimization) taking all source images. Compared to the other two paradigms, G3R leverages the benefits of both worlds (data-driven priors, gradient feedback) and achieves the best trade-off between the reconstruction quality and time (rightmost).

Generalizable reconstruction: To generalize to novel scenes, researchers train neural networks across diverse scenes and incorporate proxy geometry like depth maps for image-based rendering [3,20,44,45,70]. However, it is usually challenging or expensive to obtain high-quality geometry for real-world large scenes. To address this issue, recent works adopt transformers to either directly map the source images and camera embedding to the target view without any physical constraints [21,47–50] or aggregate points from source images along the epipolar lines for rendering [9,11,39,42,51,56,57,62,65,66,75]. Another popular approach is to lift 2D images to 3D cost volumes with geometry priors [7,9,17,26,31] but struggles with large camera movement. These methods do not produce a unified 3D representation, suffer from noticeable artifacts under large view changes, and are slow to render. On the other hand, some works that directly predict 3D representations such as multi-plane images (MPI) [12,55,86] or implicit representations [7,38,39,52,80] only work well on objects or small scenes. Concurrent work [6] predicts 3D Gaussians for generalizable reconstruction, but is limited to low-resoluation image pairs. In contrast, G3R take all available source images and predicts a unified representation for large-scale scenes including dynamics, enabling scalable and realistic simulation. Most recently, large reconstruction models [14,22,30,68,83] (LRMs) achieve strong generalizability across small objects by training on large synthetic dataset such as Objaverse. To our best knowledge, G3R is the first LRM that generalizes across diverse large scenes and handles large view changes by training on large-scale real-world datasets.

Iterative Networks for 3D: Our method falls under the "iterative network" framework, which conduct iterative updates to gradually refine the output. Prior works have studied iterative approaches on low-dimensional inverse problems [2,5,24,34,35] such 6-DOF pose and illumination estimation. In contrast, G3R solves a challenging high-dimensional inverse problem (*i.e*, scene reconstruction) using a learned optimizer [4,23,69]. Specifically, we train a neural network that exploits spatial correlation to expedite the reconstruction process. Similar to G3R, DeepView [12] also employs an iterative network with gradient guidance to reconstruct a 3D representation (MPI), but for small baselines only. Moreover, it unfolds the optimization through a series of distinct CNN networks and loss-agnostic gradient components at each stage for each source image, limiting the number of input images, and leading to large memory usage and slow speed.

3 Gradient Guided Generalizable Reconstruction (G3R)

Given a set of source camera images $\mathbf{I}^{\mathrm{src}} = \{\mathbf{I}_i\}_{1 \leq i \leq N}$ and an approximate geometry scaffold $\mathcal{M}$ (*e.g.*, obtained from either LiDAR or points from multi-view stereo) captured in-the-wild by a sensor platform moving through a large dynamic scene, our goal is to efficiently reconstruct a realistic and editable 3D representation $\mathcal{S}$ for accurate real-time camera simulation. In this paper, we introduce Gradient Guided Generalizable Reconstruction (G3R), the first method that can create modifiable digital clones of large real world scenes ($> 10,000\,\mathrm{m}^2$) in two minutes or less, and that renders novel views with high photorealism at > 90 FPS. Our method overview is shown in Fig. 3. G3R combines data-driven priors from fast prediction methods with the iterative gradient feedback signal from per-scene optimization methods by learning to optimize for large scene reconstruction (Fig. 2-left). G3R iteratively updates a representation we call 3D neural Gaussians, initialized from the scaffold $\mathcal{M}$, with a single neural network. The network takes the gradient feedback signals from differentiably rendering the representation to reconstruct the source images $\mathbf{I}^{\text{src}}$. G3R achieves the best trade-off between realism and reconstruction speed, achieving performance and scalability (see Fig. 2-right).

In what follows, we first introduce our scene representation (3D neural Gaussians) designed for handling dynamic and unbounded large scenes (Sect. 3.1). Then we show how to lift 2D images to 3D space by propagating the gradients (Sect. 3.2), followed by iterative refinements in Sect. 3.3. We describe training the network across multiple scenes in Sect. 3.4.

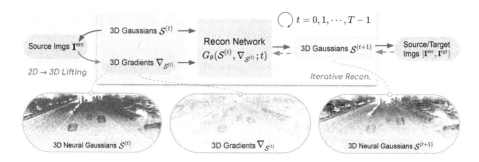

Fig. 3. Method overview. We model the generalizable reconstruction as an iterative process, where the 3D neural Gaussians $\mathcal{S}^{(t)}$ are iteratively refined with reconstruction network G_θ. We first lift the source 2D images $\mathbf{I}^{\text{src}}$ to 3D space by backpropagating the rendering procedure to get the gradients w.r.t the representation $\nabla_{\mathcal{S}^{(t)}}$ (blue arrow). Then the reconstruction network G_θ takes the 3D representation $\mathcal{S}^{(t)}$, the gradient $\nabla_{\mathcal{S}^{(t)}}$ and the iteration step t as input, and predicts an updated 3D representation $\mathcal{S}^{(t+1)}$. To train the network, we render $\mathcal{S}^{(t+1)}$ at source and novel views, and compute loss. The backward gradient flow for training G_θ is highlighted with dashed blue arrows. (Color figure online)

3.1 G3R's Scene Representation

3D Gaussian Splatting [18] (3DGS) is a differentiable rasterization technique that allows real-time rendering of photorealistic scenes learned from posed images and an intitial set of points from SfM [53]. 3DGS represents the scene with a set of 3D Gaussians (i.e, points) $\mathcal{G} = \{g_i\}_{1 \leq i \leq M}$, where $g_i \in \mathbb{R}^{14}$ consists of position ($\mathbb{R}^3$), scale ($\mathbb{R}^3$), orientation ($\mathbb{R}^4$), color ($\mathbb{R}^3$) and opacity ($\mathbb{R}^1$).

These gaussian points $\mathcal{G}$ can be rendered to 2D images with camera poses Π using a differentiable tile rasterizer $f_{\text{rast}}(\mathcal{G}, \Pi)$, where each point is projected and splatted to the image plane based on the scale and orientation, then the color is blended with other points based on the opacity and depth to camera. However, 3DGS's explicit representation lacks modelling capacity useful for learning-based optimization. Furthermore, 3DGS [18] focuses on small static scenes or individual objects, and has challenges modeling large-scale dynamic scenes, such as self-driving scenarios. In this paper, we make two enhancements to 3DGS's representation. First, we augment its representation with a latent feature vector, which we call *3D neural Gaussians*, providing additional capacity for generalizable reconstruction and learning-based optimization. Second, we decompose the scene into the nearby static scene, dynamic actors, and a distant region to enable modelling of large unbounded dynamic scenes. We now describe these two enhancements and then detail the rendering process.

3D Neural Gaussians: We define our scene representation $\mathcal{S}$ as a set of *3D Neural Gaussians*, $\mathcal{S} = \{h_i\}_{1 \leq i \leq M}$, where each point is represented by a feature vector $h_i \in \mathbb{R}^C$. This latent representation helps encode information about the scene during the iterative updates in the learning-based optimization described in Sect. 3.3. To render, we convert the 3D neural Gaussians to a set of explicit color 3D Gaussians $\mathcal{G} = \{g_i\}_{1 \leq i \leq M}$, using a Multi-Layer Perceptron (MLP) network $g_i = f_{\text{mlp}}(h_i)$. To encode geometry and additional physical information about the scene into h_i and ensure stable optimization, we designate the first 14 channels as the 3D Gaussian attributes and add a skip connection in f_{mlp} such that it updates these channels to generate g_i.

Representing Rigid Dynamic Objects and Unbounded Scenes: We decompose the dynamic scene and its set of 3D neural Gaussians $\mathcal{S}$ into a static background $\mathcal{S}^\mathcal{B}$, a set of dynamic actors $\mathcal{S}^\mathcal{A}$ and a distant region $\mathcal{S}^\mathcal{Y}$ (*e.g.*, far-away buildings and sky). We assume rigid motion $\mathcal{T}(\mathcal{S}^\mathcal{A}, \boldsymbol{\xi}^\mathcal{A})$ for dynamic actors, where $\mathcal{T}$ is the rigid transformation and $\boldsymbol{\xi}^\mathcal{A}$ are the actor extrinsics. The dynamic points $\mathcal{S}^\mathcal{A}$ are moved across different frames using 3D bounding boxes that specify each foreground actor's size and location. We initialize the 3D neural Gaussians for the static background and dynamic actors using the provided approximate geometry scaffolds $\mathcal{M}$ (*e.g.*, aggregated LiDAR points or multi-view stereo points). We further position a fixed number of points at a large distance to model the distant region. See Section 4 and supp. for details.

Rendering: Given $\mathcal{S}$ and camera poses $\Pi = \{\mathbf{K}_i, \boldsymbol{\xi}_i\}$, where $\mathbf{K}_i$ and $\boldsymbol{\xi}_i$ are the camera intrinsics and extrinsics for view i, we convert $\mathcal{S}$ to 3D Gaussians $\mathcal{G}$ and then leverage the differentiable tile rasterizer [18] to render the images $\hat{\mathbf{I}}$:

$$f_{\text{render}}(\mathcal{S}; \Pi) := f_{\text{rast}}(\mathcal{G}; \Pi) = f_{\text{rast}}(f_{\text{mlp}}(\mathcal{S}); \Pi) \quad (1)$$
$$= f_{\text{rast}}(f_{\text{mlp}}(\mathcal{S}^\mathcal{B}, \mathcal{S}^\mathcal{Y}, \mathcal{T}(\mathcal{S}^\mathcal{A}, \boldsymbol{\xi}^\mathcal{A})); \Pi) \quad (?)$$

3.2 Lift 2D Images to 3D as Gradients

Previous generalizable works [7,26,65] lift a few 2D images (*e.g.*, ≤ 5) to 3D by aggregating image features extracted from source views according to camera and geometry priors (*e.g.*, epipolar geometry or multi-view stereo). Since each image is processed by a neural network separately, it cannot take many source images due to the high memory usage in both training and inference, limiting its applicability to small objects under small viewpoint changes. This is because large scenes usually have complex topology/geometry and cannot be reconstructed accurately with only a small set of source images. Moreover, it can be challenging to select and merge source views and also ensure spatial consistency.

Instead, we propose to lift 2D images to 3D space by "rendering and backpropagating" to obtain gradients w.r.t the 3D representation. Compared to leveraging networks to process images independently, 3D gradients provide a unified representation that can efficiently aggregate as many images as needed. Moreover, 3D gradients take the rendering procedure into account, naturally handling occlusions. It also enables adjustment of the 3D representation, which is not done in traditional depth rendering for view warping. Finally, the 3D gradients are fast to compute with modern differentiable rasterization engines.

Specifically, given the 3D representation $\mathcal{S}$, we first render the scene to source input views $\hat{\mathbf{I}}^{\text{src}} = f_{\text{render}}(\mathcal{S}; \Pi^{\text{src}})$ using Eq. 2. Then, we compare the rendered images with the inputs $\mathbf{I}^{\text{src}}$, compute the reconstruction loss L, and backpropagate the difference to 3D representation $\mathcal{S}$ to get accumulated gradients $\nabla_{\mathcal{S}} := \nabla_{\mathcal{S}} L(\mathcal{S}, \mathbf{I}^{\text{src}}; \Pi^{\text{src}})$ as shown in Fig. 3, with

$$L(\mathcal{S}, \mathbf{I}^{\text{src}}; \Pi^{\text{src}}) = \sum_i \left\| \mathbf{I}_i^{\text{src}} - \hat{\mathbf{I}}_i^{\text{src}} \right\|_2 = \sum_i \left\| \mathbf{I}_i^{\text{src}} - f_{\text{render}}(\mathcal{S}; \Pi_i^{\text{src}}) \right\|_2, \quad (3)$$

$$\nabla_{\mathcal{S}} L(\mathcal{S}, \mathbf{I}^{\text{src}}; \Pi^{\text{src}}) = \frac{\partial L(\mathcal{S}, \mathbf{I}^{\text{src}}; \Pi_i^{\text{src}})}{\partial \mathcal{S}} = \sum_i \frac{\partial \left\| \mathbf{I}_i^{\text{src}} - f_{\text{render}}(\mathcal{S}; \Pi_i^{\text{src}}) \right\|_2}{\partial \mathcal{S}}. \quad (4)$$

The differentiable function f_{render} builds a connection between 2D and 3D, and the gradient $\nabla_{\mathcal{S}}$ encodes the 2D images in 3D using $\mathcal{S}$ as the proxy.

3.3 Iterative Reconstruction with a Neural Network

We now describe how we iteratively refine the scene representation $\mathcal{S}$ given the source images $\mathbf{I}^{\text{src}}$. At each step t, we take the current 3D representation $\mathcal{S}^{(t)}$ as a proxy to compute the gradient $\nabla_{\mathcal{S}^{(t)}}$ via differentiable rendering, thereby unprojecting 2D source images $\mathbf{I}^{\text{src}}$ to 3D, and then feed $\nabla_{\mathcal{S}^{(t)}}$ into the network G_θ to predict the updated 3D representation $\mathcal{S}^{(t+1)}$:

$$\mathcal{S}^{(t+1)} = \mathcal{S}^{(t)} + \gamma(t) \cdot G_\theta(\mathcal{S}^{(t)}, \nabla_{\mathcal{S}^{(t)}} L(\mathcal{S}^{(t)}, \mathbf{I}^{\text{src}}; \Pi^{\text{src}}); t), \quad t = 0, 1, \ldots, T-1. \quad (5)$$

$\gamma(t)$ defines the update scale at different step t. Intuitively, similar to gradient descent, we desire a decaying schedule $\gamma(t)$ and a small T so that the network can predict an initial coarse representation and then quickly refine it. We use the cosine scheduler from DDIM [54] for $\gamma(t)$. We use a 3D UNet [10] with sparse convolution [60] as G_θ to process the neural Gaussians $\mathcal{S}$. The iterative process allows us to refine the 3D representation to achieve better quality and use a smaller network that is more efficient and easier to learn.

3.4 Training and Inference

We now describe the training process to train the learned optimizer G_θ and neural decoding MLP f_{mlp}. For each scene, we initialize the scene representation

$\mathcal{S}^{(0)}$ from the geometry scaffold $\mathcal{M}$. We iteratively refine $\mathcal{S}$ with the network prediction for T steps. To enhance the generalizability of reconstruction network, we render the updated representation to both source views $\mathbf{I}^{\text{src}}$ and novel views $\mathbf{I}^{\text{tgt}}$ during training ($\mathbf{I} = [\mathbf{I}^{\text{src}}, \mathbf{I}^{\text{tgt}}]$), and backpropagate the gradients to the parameters of the reconstruction network G_θ and the f_{mlp}. Note that in Eq. 3 only the gradients from source views are used as input to G_θ for the next iteration, as the target views will not be available at test time. G_θ is trained to minimize final rendering loss for every iteration step t. We train the networks across many large outdoor scenes. The total loss $\mathcal{L}$ is:

$$\mathcal{L} = \mathcal{L}_{\text{mse}}(\hat{\mathbf{I}}, \mathbf{I}) + \lambda_{\text{lpips}} \mathcal{L}_{\text{lpips}}(\hat{\mathbf{I}}, \mathbf{I}) + \lambda_{\text{reg}} \mathcal{L}_{\text{reg}}(\mathcal{G}), \tag{6}$$

Table 1. Comparison to reconstruction methods on PandaSet. The methods with best photorealism are marked using gold ●, silver ●, and bronze ● medals. † denotes the method needs to reconstruct the scene again with different source images when rendering each new view.

Models		Novel View Synthesis			Inference Time	
		PSNR↑	SSIM↑	LPIPS↓	Recon Time	Render FPS
Generalizable	MVSNeRF$_{ft}$ [7]	23.68	0.659	0.482	35min 31s	0.0392
	ENeRF [26]	24.43	0.736 ●	0.306 ●	0.057s†	6.93
	GNT [65]	23.99	0.693	0.408	0.32s†	0.00498
	PixelSplat [6]	23.21	0.653	0.490	0.74s†	147
Per-scene Opt.	Instant-NGP [37]	24.34	0.729	0.436	7min 16s	3.24
	3DGS [18]	25.14	0.747 ●	0.372 ●	50min 14s	121
Ours	G3R (turbo)	24.76 ●	0.720	0.438	31s	121
	G3R	25.22 ●	0.742 ●	0.371 ●	123s	121

where $\hat{\mathbf{I}}$ is the rendered images, $\mathcal{L}_{\text{mse}}$ is the photometric loss, $\mathcal{L}_{\text{lpips}}$ is the perceptual loss [84], and $\mathcal{L}_{\text{reg}}$ is the regularization term applied on the shape of the transformed Gaussians $\mathcal{G}$ to be flat for better alignment with the surface.

$$\mathcal{L}_{\text{reg}}(\mathcal{G}) = \sum_i \max(0, d_i^{\min} - \epsilon), \tag{7}$$

where $d_i^{\min}$ is the minimal value of the 3-channel scale for each Gaussian g_i. We encourage it to be smaller than a threshold ϵ.

Inference: Given the pre-trained reconstruction network G_θ and neural Gaussian decoder MLP f_{mlp}, we can now reconstruct novel scenes not seen during training. Specifically, we take all input images $\mathbf{I}^{\text{src}}$ for the novel scene and the 3D neural Gaussian initialization $\mathcal{S}^{(0)}$ to iteratively compute the gradients $\nabla_\mathcal{S}$ and refine the 3D representation. Finally, we export $\mathcal{S}^{(T)}$ to standard 3D Gaussians $\mathcal{G}^{(T)}$ for real-time rerasterization.

4 Experiments

We compare G3R against state-of-the-art (SoTA) generalizable and per-scene optimization approaches, ablate our design choices, and demonstrate the capability of generalization across datasets. Finally, we show that G3R-predicted representation is editable and we can generate realistic multi-camera videos.

4.1 Experimental Setup

Datasets: We conduct experiments on two public datasets with large real-world scenes: PandaSet [73], which contains dynamic actors in driving scenes and BlendedMVS [81], which contains large static infrastructure. We select 7 diverse scenes for testing with each covering around $200 \times 80\,\mathrm{m}^2$, and the rest (96 scenes) for training. BlendedMVS-large is a collection of 29 real-world scenes captured by a drone, ranging in size from $10,000\,\mathrm{m}^2$ to over $100,000\,\mathrm{m}^2$, and also includes reconstructed meshes from multi-view stereo [1]. We select 25 scenes for training and 4 for testing. For both datasets, we use every other frame as source and the remaining for test. BlendedMVS has more challenging novel views, as the distance between two nearby views can be large (See supp. for more anlaysis).

Implementation Details: We initialize the 3D neural Gaussians' $\mathcal{S}^{(0)}$ positions using downsampled 3D points from LiDAR points in PandaSet or mesh faces in BlendedMVS. To ensure geometry coverage, the scale for each Gaussian is initialized isotropically as the distance to its third nearest point. The rotation is set to identity and the opacity to 0.7. The other feature channels are randomly initialized. We disable view-dependent spherical harmonics from the original 3DGS [18] for simplicity and improved memory usage. We normalize the 3D gradients $\nabla_{\mathcal{S}^{(t)}} L(\mathcal{S}^{(t)})$ by channel across all the points before feeding to the network. For dynamic scenes, we adopt 3 separate networks for the background, actors, and the distant region. tanh activation is applied in the output layer. The per-scene reconstruction step T is set as 24 during training. We train for 1000 scene iterations in total using Adam optimizer [19] with learning rate 1e−4. This takes roughly 30 h on 2 RTX 3090 GPUs. We adopt a warm-up strategy during training that gradually increases the scene reconstruction steps in the first few scene iterations. The network is updated at each reconstruction step. We provide two variants during evaluation, where the faster model, G3R (turbo), uses fewer iterations and fewer 3D neural Gaussians. See supp. for more details.

Baselines: We compare G3R against both generalizable NVS (Fig. 2a) and per-scene optimization approaches (Fig. 2b). For generalizable NVS, we compare against MVSNeRF [7], ENeRF [26], GNT [65] and concurrent work PixelSplat [6]. MVSNeRF warps 2D image features onto a plane sweep and then applies a 3D CNN to reconstruct a NeRF which can be finetuned further. Similarly, ENeRF also warps multi-view source images and leverages depth-guided sampling for efficient reconstruction and rendering. GNT samples points along

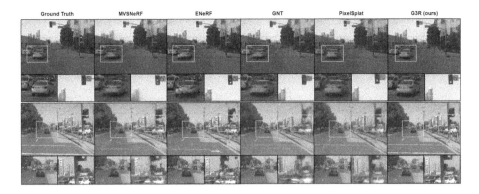

Fig. 4. Qualitative comparison to generalizable approaches on PandaSet.

each target ray and predicts the pixel color by learning the aggregation of viewwise features from the epipolar lines using transformers. PixelSplat predicts 3D Gaussians with a 2-view epipolar transformer to extract features and then predict the depth distribution and pixel-aligned Gaussians. Except for MVSNeRF, which finetunes the predicted representation on new scenes, all generalizable methods need to reconstruct the scene again with different nearest neighboring source images when rendering a new view. Unless stated otherwise, we train and evaluate all generalizable models using the same data as G3R. For per-scene optimization approaches, we compare against Instant-NGP [37] and 3DGS [18]. Instant-NGP is an efficient NeRF framework with multi-hash grid encoding and tiny MLP for fast reconstruction. We enhance Instant-NGP with depth supervision for better performance. 3DGS models the scene with 3D Gaussians and uses a differentiable rasterizer for fast scene reconstruction and real-time rendering. We enhance 3DGS to support dynamic actors and unbounded scenes with the same implementation as G3R. We optimize each test scene separately using all source frames. Please see supp. for additional details.

4.2 Generalizable Reconstruction on Large Scenes

Scene Reconstruction on PandaSet: We report scene reconstruction results on PandaSet in Table 1 and Fig. 4. Compared to SoTA generalizable approaches, G3R achieves significantly better photorealism and real-time rendering with an affordable reconstruction cost (2 min or less). In contrast, baselines conduct image-based rendering and result in noticeable artifacts for dynamic actors due to the lack of explicit 3D representation that can model dynamics. Moreover, they often produce blurry rendering results, especially in nearby regions where there are large view changes, due to flawed representation prediction and poor geometry estimation for view warping. We note that ENeRF achieves good LPIPS with image warping, but has severe visual artifacts and low PSNR. We also compare G3R with SoTA per-scene optimization approaches including Instant-NGP and 3DGS. Our approach achieves on par or better photorealism while shortening the reconstruciton time to 2 min. We note that PixelSplat leads to a higher FPS since it can only process low-resolution images and predicts a smaller number of 3D Gaussian points compared to G3R due to memory limitations.

Scene Reconstruction on BlendedMVS: We further consider BlendedMVS to evaluate the robustness of different methods to handle many source inputs and large view changes. As shown in Fig. 5 and Table 2, existing generalizable approaches including ENeRF, GNT and PixelSplat cannot handle large view changes and produce bad rendering results with significant visual artifacts due to bad geometry estimation (*e.g.*, blurry appearance, unnatural discontinuity, wrong color palette, etc.). To address this issue, we adapt PixelSplat, named PixelSplat++, to leverage the 3D scaffold to reduce ambiguity and take all available source images for good coverage. Please see supp. for details. While achieving signficiant performance boost over existing generalizable methods, PixelSplat++ is still far from per-scene optimization approaches due to the challenge of one-step prediction with limited network capacity. Our method results in the best photorealism, minimal reconstruction time and enables real-time rendering speed, which again verifies the effectiveness of our proposed paradigm. Moreover, G3R outperforms per-scene optimization methods especially in perceptual quality. We hypothesize this is because the learned data-driven prior helps handle large view changes better.

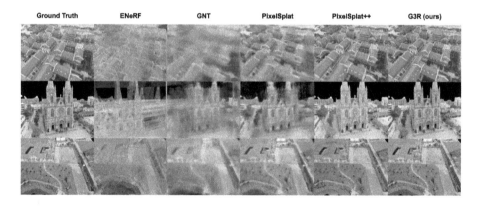

Fig. 5. Qualitative comparison to generalizable approaches on BlendedMVS.

Robust 3D Gaussian Prediction: We compare with the rendering performance of 3DGS at novel views in Fig. 6. We observe that while 3DGS has sufficient capacity to memorize the source frames, it suffers a significant performance drop when rendering at novel views due to poor underlying geometry [8,13]. In contrast, G3R predicts 3D gaussians in a more robust way because G3R is trained with novel view supervision across many scenes (Eq. 6) and this supervision helps regularize the 3D neural Gaussians to generalize rather than merely memorize the

Table 2. Comparison on BlendedMVS. The methods with best photorealism are marked using gold ◉, silver ◉, and bronze● medals. † denotes the method needs to reconstruct the scene again with different source images when rendering each new view.

	Models	Novel View Synthesis			Inference Time	
		PSNR↑	SSIM↑	LPIPS↓	Recon Time	Render FPS
Generalizable	ENeRF [26]	15.21	0.270	0.660	0.11 s†	2.65
	GNT [65]	16.42	0.366	0.707	0.35 s†	0.00249
	PixelSplat [6]	16.24	0.344	0.781	1.14 s†	176
	PixelSplat++	19.60	0.404	0.601	69 s	158
Per-scene Opt.	Instant-NGP [37]	24.86●	0.639	0.459●	26min 48 s	1.65
	3DGS [18]	25.12	0.668●	0.462	39.5min	97.0
Ours	G3R (turbo)	24.56	0.674	0.421	98 s	97.0
	G3R	25.22	0.707	0.390	210 s	97.0

Fig. 6. Robustness of G3R vs. 3DGS. 3DGS is sharper on interpolation views (*Interp.*), but has artifacts on extraopolation views (*Extrap.*).

source views. We also consider a more challenging *extrapolation* setting where we select 20 consecutive frames as source views and simulate the future 3 frames (*e.g.*, 3 - 6 m of shift) to evaluate the robustness when rendering at extrapolated views. As shown in Fig. 6, G3R results in more realistic rendering performance. In contrast, 3DGS has severe visual artifacts highlighted by pink arrows (*e.g.*, black holes or wrong colors in road, sky and actor regions). Please refer to supp. for more analysis.

Ablation study: In Table 3, we ablate the key components proposed in G3R on PandaSet, including replacing the 3D neural Gaussians with the standard 3D Gaussian representation, conducting one-step prediction in both training and inference, training the network only with source view supervision, and switching decaying schedule $\gamma(t)$ to a constant update scale (0.3) at each step. As shown in Table 3, our proposed neural Gaussian representation is more expressive, thus easing the network prediction. The iterative refinement is critical in the proposed paradigm and single-step prediction fails to generate high-quality reconstruction results. We notice that single-step G3R is worse than PixelSplat as we enforce smaller updates per step for stable convergence. Moreover, we show training the network with novel views on many scenes is necessary to enhance the robustness

Table 3. Ablation study on PandaSet.

Models	PSNR	SSIM	LPIPS
Ours	**25.22**	**0.742**	**0.371**
− 3D neural representation	24.72	0.718	0.420
− iterative reconstruction	20.03	0.510	0.623
− training with novel views	24.59	0.715	0.419
− update schedule $\gamma(t)$	25.03	0.732	0.400

Table 4. Cross-dataset Generalization. Pandaset-pretrained model outperforms baselines trained on BlendedMVS (see Table 2).

	PSNR	PSNR	PSNR
Zero-short transfer	24.11	0.653	0.448
Finetune on 2 scenes	24.99	0.676	0.428

Fig. 7. PandaSet-pretrained model generalizes to Waymo Open Dataset.

of 3D representation for realistic novel view rendering. Finally, a proper update schedule further improves performance.

Generalization Study: We further evaluate the PandaSet-trained G3R model (static background module) on BlendedMVS (self-driving → drone). The results in Table 4 show that G3R trained only on PandaSet achieves significantly better performance in BlendedMVS than generalizable baselines trained on BlendedMVS directly. We further finetune the G3R model with only 2 BlendedMVS scenes, achieving comparable results as directly training on full BlendedMVS. We also showcase applying a Pandaset-pretrained G3R model to Waymo Open Dataset (WOD) [58] scenes in Fig. 7, unveiling the potential for scalable real-world sensor simulation. See supp. for more analysis.

Realistic and Controllable Camera Simulation: We now showcase applying G3R for high-fidelity multi-camera simulation in large-scale driving scenarios. Compared to previous generalizable approaches, our method can reconstruct a standalone representation, which allows us to control, edit and interactively render the scene for various applications.

In Fig. 8, we show G3R-reconstructed scene can synthesize consistent and high-fidelity multi-camera videos from one single driving pass (top row). Moreover, we can manipulate the scene by freezing the sensors and changing the positions of dynamic actors, and render corresponding multi-camera (second row) or panorama images (bottom row).

Fig. 8. Realistic and controllable multi-camera simulation on PandaSet. G3R reconstructs a manipulable 3D scene representation.

Limitations: Our approach has artifacts in large extrapolations, which may require scene completion. Better surface regularization [8,13] and adversarial training [46,77] may mitigate these issues. G3R's performance suffers when initialized with sparse points, but can leverage LiDAR or fast MVS techniques [71] to mitigate this. We also do not model non-rigid deformations [33] and emissive lighting. See supp. for details.

5 Conclusion

In this paper, we introduce G3R, a novel approach for efficient generalizable large-scale 3D scene reconstruction. By leveraging gradient feedback signals from differentiable rendering, G3R achieves acceleration of at least 10× over state-of-the-art per-scene optimization methods, with comparable or superior photorealism. Importantly, our method predicts a standalone 3D representation that exhibits robustness to large view changes and enables real-time rendering, making it well-suited for VR and simulation. Experiments on urban-driving and drone datasets showcase the efficacy of G3R for in-the-wild 3D scene reconstruction. Our learning-to-optimize paradigm with gradient signal can apply to other 3D representations such as triplanes with NeRF rendering, or other inverse problems such as generalizable surface reconstruction [15,25,32,43].

Acknowledgement. We sincerely thank the anonymous reviewers for their insightful comments and suggestions. We thank the Waabi team for their valuable assistance and support.

References

1. Altizure: Mapping the world in 3D. https://www.altizure.com
2. Adler, J., Öktem, O.: Solving ill-posed inverse problems using iterative deep neural networks. arXiv (2017)
3. Aliev, K.-A., Sevastopolsky, A., Kolos, M., Ulyanov, D., Lempitsky, V.: Neural point-based graphics. In: Vedaldi, A., Bischof, H., Brox, T., Frahm, J.-M. (eds.) ECCV 2020. LNCS, vol. 12367, pp. 696–712. Springer, Cham (2020). https://doi.org/10.1007/978-3-030-58542-6_42

4. Andrychowicz, M., et al.: Learning to learn by gradient descent by gradient descent. In: NeurIPS (2016)
5. Carreira, J., Agrawal, P., Fragkiadaki, K., Malik, J.: Human pose estimation with iterative error feedback. In: CVPR (2015)
6. Charatan, D., Li, S., Tagliasacchi, A., Sitzmann, V.: pixelsplat: 3D gaussian splats from image pairs for scalable generalizable 3D reconstruction. arXiv (2023)
7. Chen, A., et al.: MVSNeRF: fast generalizable radiance field reconstruction from multi-view stereo. In: ICCV (2021)
8. Cheng, K., et al.: GaussianPro: 3D gaussian splatting with progressive propagation. arXiv (2024)
9. Chibane, J., Bansal, A., Lazova, V., Pons-Moll, G.: Stereo radiance fields (SRF): Learning view synthesis for sparse views of novel scenes. In: CVPR (2021)
10. Çiçek, Ö., Abdulkadir, A., Lienkamp, S.S., Brox, T., Ronneberger, O.: 3D U-Net: learning dense volumetric segmentation from sparse annotation. In: Ourselin, S., Joskowicz, L., Sabuncu, M.R., Unal, G., Wells, W. (eds.) MICCAI 2016. LNCS, vol. 9901, pp. 424–432. Springer, Cham (2016). https://doi.org/10.1007/978-3-319-46723-8_49
11. Cong, W., et al.: Enhancing nerf akin to enhancing LLMS: generalizable nerf transformer with mixture-of-view-experts. In: ICCV (2023)
12. Flynn, J., et al.: DeepView: view synthesis with learned gradient descent. In: CVPR (2019)
13. Guédon, A., Lepetit, V.: Sugar: surface-aligned gaussian splatting for efficient 3D mesh reconstruction and high-quality mesh rendering. arXiv (2023)
14. Hong, Y., et al.: LRM: large reconstruction model for single image to 3D. In: The Twelfth International Conference on Learning Representations (2024). https://openreview.net/forum?id=sllU8vvsFF
15. Huang, J., Gojcic, Z., Atzmon, M., Litany, O., Fidler, S., Williams, F.: Neural kernel surface reconstruction. In: CVPR (2023)
16. Huang, S., et al.: Neural lidar fields for novel view synthesis. arXiv (2023)
17. Johari, M.M., Lepoittevin, Y., Fleuret, F.: GeoNeRF: generalizing nerf with geometry priors. In: CVPR (2022)
18. Kerbl, B., Kopanas, G., Leimkühler, T., Drettakis, G.: 3D gaussian splatting for real-time radiance field rendering. TOG (2023)
19. Kingma, D.P., Ba, J.: Adam: a method for stochastic optimization. In: ICLR (2015)
20. Kopanas, G., Philip, J., Leimkühler, T., Drettakis, G.: Point-based neural rendering with per-view optimization. Computer graphics forum (Print) (2021)
21. Kulh'anek, J., Derner, E., Sattler, T., Babuvska, R.: ViewFormer: NeRF-free neural rendering from few images using transformers. In: Avidan, S., Brostow, G., Cissé, M., Farinella, G.M., Hassner, T. (eds.) ECCV 2022. LNCS, vol. 13675, pp. 198–216. Springer, Cham (2022). https://doi.org/10.1007/978-3-031-19784-0_12
22. Li, J., et al.: Instant3D: fast text-to-3D with sparse-view generation and large reconstruction model. In: The Twelfth International Conference on Learning Representations (2024). https://openreview.net/forum?id=2lDQLiH1W4
23. Li, K., Malik, J.: Learning to optimize. In: ICLR (2016)
24. Li, Y., Wang, G., Ji, X., Xiang, Y., Fox, D.: DeepIM: deep iterative matching for 6D pose estimation. IJCV (2018)
25. Liang, Y., He, H., Chen, Y.: RETR: modeling rendering via transformer for generalizable neural surface reconstruction. In: NeurIPS (2023)
26. Lin, H., et al.: Efficient neural radiance fields for interactive free-viewpoint video. In: SIGGRAPH Asia 2022 Conference Papers (2022)

27. Lin, J., et al.: VastGaussian: vast 3D gaussians for large scene reconstruction. arXiv (2024)
28. Lin, Z.H.,et al.: UrbaNIR: large-scale urban scene inverse rendering from a single video. arXiv (2023)
29. Liu, J.Y., Chen, Y., Yang, Z., Wang, J., Manivasagam, S., Urtasun, R.: Real-time neural rasterization for large scenes. In: ICCV (2023)
30. Liu, R., Wu, R., Van Hoorick, B., Tokmakov, P., Zakharov, S., Vondrick, C.: Zero-1-to-3: zero-shot one image to 3D object. In: Proceedings of the IEEE/CVF International Conference on Computer Vision, pp. 9298–9309 (2023)
31. Liu, Y., et al.: Neural rays for occlusion-aware image-based rendering. In: CVPR (2022)
32. Long, X., Lin, C., Wang, P., Komura, T., Wang, W.: Sparseneus: fast generalizable neural surface reconstruction from sparse views. In: Avidan, S., Brostow, G., Cissé, M., Farinella, G.M., Hassner, T. (eds.) ECCV 2022. LNCS, vol. 13692, pp. 210–227. Springer, Cham (2022)
33. Luiten, J., Kopanas, G., Leibe, B., Ramanan, D.: Dynamic 3D gaussians: tracking by persistent dynamic view synthesis. arXiv (2023)
34. Ma, W.-C., Wang, S., Gu, J., Manivasagam, S., Torralba, A., Urtasun, R.: Deep feedback inverse problem solver. In: Vedaldi, A., Bischof, H., Brox, T., Frahm, J.-M. (eds.) ECCV 2020. LNCS, vol. 12350, pp. 229–246. Springer, Cham (2020). https://doi.org/10.1007/978-3-030-58558-7_14
35. Manhardt, F., Kehl, W., Navab, N., Tombari, F.: Deep model-based 6D pose refinement in RGB. In: Ferrari, V., Hebert, M., Sminchisescu, C., Weiss, Y. (eds.) Computer Vision – ECCV 2018. LNCS, vol. 11218, pp. 833–849. Springer, Cham (2018). https://doi.org/10.1007/978-3-030-01264-9_49
36. Mildenhall, B., Srinivasan, P.P., Tancik, M., Barron, J.T., Ramamoorthi, R., Ng, R.: NeRF: Representing Scenes as Neural Radiance Fields for View Synthesis. In: Vedaldi, A., Bischof, H., Brox, T., Frahm, J.-M. (eds.) ECCV 2020. LNCS, vol. 12346, pp. 405–421. Springer, Cham (2020). https://doi.org/10.1007/978-3-030-58452-8_24
37. Müller, T., Evans, A., Schied, C., Keller, A.: Instant neural graphics primitives with a multiresolution hash encoding (2022)
38. Müller, N., et al.: AutoRF: learning 3D object radiance fields from single view observations. In: CVPR (2022)
39. Niemeyer, M., Mescheder, L., Oechsle, M., Geiger, A.: Differentiable volumetric rendering: learning implicit 3D representations without 3D supervision. In: CVPR (2020)
40. Ost, J., Mannan, F., Thuerey, N., Knodt, J., Heide, F.: Neural scene graphs for dynamic scenes. In: CVPR (2021)
41. Pun, A., et al.: Neural lighting simulation for urban scenes. In: NeurIPS (2023)
42. Reizenstein, J., Shapovalov, R., Henzler, P., Sbordone, L., Labatut, P., Novotný, D.: Common objects in 3D: large-scale learning and evaluation of real-life 3D category reconstruction. In: ICCV (2021)
43. Ren, Y., Wang, F., Zhang, T., Pollefeys, M., Susstrunk, S.E.: VolRecon: volume rendering of signed ray distance functions for generalizable multi-view reconstruction. In: CVPR (2022)
44. Riegler, G., Koltun, V.: Free view synthesis. In: Vedaldi, A., Bischof, H., Brox, T., Frahm, J.-M. (eds.) ECCV 2020. LNCS, vol. 12364, pp. 623–640. Springer, Cham (2020). https://doi.org/10.1007/978-3-030-58529-7_37
45. Riegler, G., Koltun, V.: Stable view synthesis. In: CVPR (2021)

46. Roessle, B., Müller, N., Porzi, L., Bulò, S.R., Kontschieder, P., Nießner, M.: GANERF: leveraging discriminators to optimize neural radiance fields. ACM Trans. Graph (2023)
47. Rombach, R., Esser, P., Ommer, B.: Geometry-free view synthesis: transformers and no 3D priors. In: ICCV (2021)
48. Sajjadi, M.S.M., et al.: Rust: latent neural scene representations from unposed imagery. In: CVPR (2022)
49. Sajjadi, M.S.M., et al.: Scene representation transformer: geometry-free novel view synthesis through set-latent scene representations. In: CVPR (2022)
50. Seitzer, M., van Steenkiste, S., Kipf, T., Greff, K., Sajjadi, M.S.M.: DYST: towards dynamic neural scene representations on real-world videos. arXiv (2023)
51. Sitzmann, V., Rezchikov, S., Freeman, W., Tenenbaum, J., Durand, F.: Light field networks: neural scene representations with single-evaluation rendering. NeurIPS (2021)
52. Sitzmann, V., Zollhöfer, M., Wetzstein, G.: Scene representation networks: continuous 3D-structure-aware neural scene representations. In: NeurIPS (2019)
53. Snavely, N., Seitz, S.M., Szeliski, R.: Photo tourism: exploring photo collections in 3D. In:SIGGRAPH (2006)
54. Song, J., Meng, C., Ermon, S.: Denoising diffusion implicit models. In: ICLR (2020)
55. Srinivasan, P.P., Tucker, R., Barron, J.T., Ramamoorthi, R., Ng, R., Snavely, N.: Pushing the boundaries of view extrapolation with multiplane images. arXiv (2019)
56. Suhail, M., Esteves, C., Sigal, L., Makadia, A.: Light field neural rendering. In: CVPR (2021)
57. Suhail, M., Esteves, C., Sigal, L., Makadia, A.: Generalizable patch-based neural rendering. In: Avidan, S., Brostow, G., Cissé, M., Farinella, G.M., Hassner, T. (eds.) ECCV 2022. LNCS, vol. 13692, pp. 156–174. Springer, Cham (2022)
58. Sun, P., et al.: Scalability in perception for autonomous driving: waymo open dataset. In: CVPR (2020)
59. Tancik, M., et al.: Block-nerf: scalable large scene neural view synthesis. In: CVPR (2022)
60. Tang, H., et al.: Torchsparse++: efficient training and inference framework for sparse convolution on GPUs. In: IEEE/ACM International Symposium on Microarchitecture (MICRO) (2023)
61. Tonderski, A., Lindström, C., Hess, G., Ljungbergh, W., Svensson, L., Petersson, C.: Neurad: neural rendering for autonomous driving. arXiv (2023)
62. Trevithick, A., Yang, B.: GRF: learning a general radiance field for 3D representation and rendering. In: ICCV (2021)
63. Turki, H., Ramanan, D., Satyanarayanan, M.: Mega-nerf: scalable construction of large-scale nerfs for virtual fly-throughs. In: CVPR (2022)
64. Wang, J., et al.: CADSim: robust and scalable in-the-wild 3D reconstruction for controllable sensor simulation. In: 6th Annual Conference on Robot Learning (2022)
65. Wang, P., Chen, X., Chen, T., Venugopalan, S., Wang, Z., et al.: Is attention all nerf needs? arXiv (2022)
66. Wang, Q., et al.: IBRNet: learning multi-view image-based rendering. In: CVPR (2021)
67. Wang, Z., et al.: Neural fields meet explicit geometric representations for inverse rendering of urban scenes. In: CVPR (2023)
68. Wei, X., et al.: MesHLRM: large reconstruction model for high-quality mesh. arXiv preprint arXiv:2404.12385 (2024)

69. Wichrowska, O., et al.: Learned optimizers that scale and generalize. In: ICML (2017)
70. Wiles, O., Gkioxari, G., Szeliski, R., Johnson, J.: Synsin: End-to-end view synthesis from a single image. arXiv (2019)
71. Wu, J., et al.: GOMVS: geometrically consistent cost aggregation for multi-view stereo. In: CVPR (2024)
72. Wu, Z., et al.: Mars: An instance-aware, modular and realistic simulator for autonomous driving. arXiv (2023)
73. Xiao, P., et al.: Pandaset: advanced sensor suite dataset for autonomous driving. In: ITSC (2021)
74. Yan, Y., et al.: Street gaussians for modeling dynamic urban scenes. arXiv (2024)
75. Yang, H., Hong, L., Li, A., Hu, T., Li, Z., Lee, G.H., Wang, L.: Contranerf: Generalizable neural radiance fields for synthetic-to-real novel view synthesis via contrastive learning. In: CVPR (2023)
76. Yang, J., et al.: EmerNeRF: emergent spatial-temporal scene decomposition via self-supervision. arXiv (2023)
77. Yang, Z., et al.: UNISIM: a neural closed-loop sensor simulator. In: CVPR (2023)
78. Yang, Z., Manivasagam, S., Chen, Y., Wang, J., Hu, R., Urtasun, R.: Reconstructing objects in-the-wild for realistic sensor simulation. In: ICRA (2023)
79. Yang, Z., Manivasagam, S., Liang, M., Yang, B., Ma, W.C., Urtasun, R.: Recovering and simulating pedestrians in the wild. In: Conference on Robot Learning, pp. 419–431. PMLR (2021)
80. Yang, Z., et al.: S3: neural shape, skeleton, and skinning fields for 3D human modeling. In: CVPR, pp. 13284–13293 (2021)
81. Yao, Y., et al.: BlendedMVS: a large-scale dataset for generalized multi-view stereo networks. CVPR (2020)
82. Yu, A., Ye, V., Tancik, M., Kanazawa, A.: PixelNeRF: neural radiance fields from one or few images. In: CVPR (2021)
83. Zhang, K., et al.: GS-LRM: large reconstruction model for 3D gaussian splatting. arXiv preprint arXiv:2404.19702 (2024)
84. Zhang, R., Isola, P., Efros, A.A., Shechtman, E., Wang, O.: The unreasonable effectiveness of deep features as a perceptual metric. In: CVPR (2018)
85. Zhenxing, M., Xu, D.: Switch-NeRF: learning scene decomposition with mixture of experts for large-scale neural radiance fields. In: ICLR (2022)
86. Zhou, T., Tucker, R., Flynn, J., Fyffe, G., Snavely, N.: Stereo magnification: learning view synthesis using multiplane images. In: SIGGRAPH (2018)
87. Zhou, X., Lin, Z., Shan, X., Wang, Y., Sun, D., Yang, M.H.: DrivingGaussian: composite gaussian splatting for surrounding dynamic autonomous driving scenes. arXiv (2023)

DreamScene360: Unconstrained Text-to-3D Scene Generation with Panoramic Gaussian Splatting

Shijie Zhou[1](✉), Zhiwen Fan[2], Dejia Xu[2], Haoran Chang[1], Pradyumna Chari[1], Tejas Bharadwaj[1], Suya You[3], Zhangyang Wang[2], and Achuta Kadambi[1]

[1] University of California, Los Angeles, USA
shijiezhou@ucla.edu
[2] University of Texas at Austin, Austin, USA
[3] DEVCOM Army Research Laboratory, Adelphi, USA
http://dreamscene360.github.io/

Abstract. The increasing demand for virtual reality applications has highlighted the significance of crafting immersive 3D assets. We present a text-to-3D 360° scene generation pipeline that facilitates the creation of comprehensive 360° scenes for in-the-wild environments in a matter of minutes. Our approach utilizes the generative power of a 2D diffusion model and prompt self-refinement to create a high-quality and globally coherent panoramic image. This image acts as a preliminary "flat" (2D) scene representation. Subsequently, it is lifted into 3D Gaussians, employing splatting techniques to enable real-time exploration. To produce consistent 3D geometry, our pipeline constructs a spatially coherent structure by aligning the 2D monocular depth into a globally optimized point cloud. This point cloud serves as the initial state for the centroids of 3D Gaussians. In order to address invisible issues inherent in single-view inputs, we impose semantic and geometric constraints on both synthesized and input camera views as regularizations. These guide the optimization of Gaussians, aiding in the reconstruction of unseen regions. In summary, our method offers a globally consistent 3D scene within a 360° perspective, providing an enhanced immersive experience over existing techniques. Project website at: http://dreamscene360.github.io/.

1 Introduction

The vast potential applications of text-to-3D to VR/MR platforms, industrial design, and gaming sectors have significantly propelled research efforts aimed

---

S. Zhou, Z. Fan and D. Xu—Equal contribution.

at developing a reliable method for immersive scene content creation at scale. Recent developments in the 2D domain have seen the successful generation or editing of high-quality and adaptable images/videos using large-scale pre-trained diffusion models [48,51] on large-scale datasets, allowing users to generate customized content on demand.

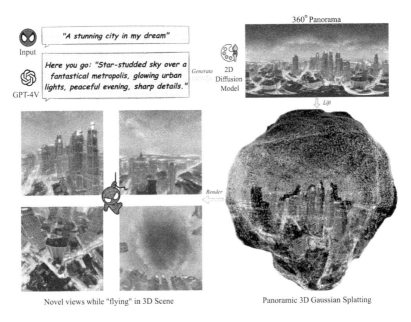

Fig. 1. DreamScene360. We introduce a 3D scene generation pipeline that creates immersive scenes with full 360° coverage from text prompts of any level of specificity.

Moving beyond 2D, the generation of 3D content, particularly 3D scenes, is constrained by the limited availability of annotated 3D image-text data pairs. Consequently, efforts in 3D content creation often rely on leveraging large-scale 2D models. This line of approach facilitates the creation of 3D scenes through a time-consuming distillation process. An example of this is DreamFusion [46], which seeks to distill the object-wise 2D priors from diffusion models into a 3D neural radiance field (NeRF) [34]. However, these approaches often suffer from low rendering quality, primarily due to the multi-view inconsistency of 2D models, and struggle to extend to scene-scale 3D structure with fine details texture creation, particularly for outdoor scenes [46] with outward-facing viewpoints and unbounded scene scale. Another avenue of 3D generation draws insights from explicit representations, such as point clouds and meshes, as demonstrated in LucidDreamer [7] and Text2Room [20]. These methods attempt to bridge the gap between 2D and 3D generation by initializing with an explicit 3D representation, and then progressively expanding the learned 3D representation to encompass a broader field-of-view. However, the progressive optimization frameworks leveraged by these methods struggle to inpaint substantial missing areas, especially when targeting 360° scenes under unconstrained conditions, resulting in notably distorted and disjointed structures. Moreover, the issue of prompt engineering in text-to-image generation [51,52], becomes more pronounced in text-to-3D generation frameworks [1,7,46] that rely on either time-consuming score distillation or complex, multi-step progressive inpainting during the scene generation process, leading to a considerable trial-and-error effort to achieve the desired 3D scene.

To address the above challenges in creating a holistic 360° text-to-3D scene generation pipeline, we introduce **DreamScene360**. Our method initially leverages the generative capabilities of text-to-panorama diffusion models [62] to produce omnidirectional 360° panoramas providing a comprehensive representation of the scene. A self-refining mechanism is used to enhance the image to alleviate prompt engineering, where GPT-4V is integrated to improve the visual quality and the text-image alignment through iterative quality assessment and prompt revision. While the generated panorama images overcome the view consistency issue across different viewpoints, they still lack depth information and any layout priors in unconstrained settings, and contains partial observations due to their single-view nature. To address this, our approach involves initializing scale-consistent scene geometry by employing a pretrained monocular depth estimator alongside an optimizable geometric field, facilitating deformable alignment for each perspective-projected pixel. The gaps, stemming from single-view observations, can be filled by deforming the Gaussians to the unseen regions by creating a set of pseudo-views with a synthesized multi-view effect and the distillation of pseudo geometric and semantic constraints from 2D models (DPT [49] and DINOv2 [43]) to guide the deformation process to alleviate artifacts.

Collectively, our framework, DreamScene360, enables the creation of immersive and realistic 3D environments from a simple user command, offering a novel solution to the pressing demand for high-quality 3D scenes (see the workflow in Fig. 1). Our work also paves the way for more accessible and user-friendly 3D scene generation by reducing the reliance on extensive manual effort.

2 Related Works

2D Assets Generation. The generation of 2D assets allows for incredible creative liberty, and the use of large-scale learning based priors for content generation. Generative Adversarial Networks [15] were the original state of the art for image generation. Variants such as StyleGAN [24] showed the ability for fine-grained control of attributes such as expression through manipulation of the latent representations as well. More recently, denoising diffusion models [9,18,56] have been the new state of the art for generative models. Text-guided image diffusion models [53] have shown the capability of generating high-quality images that are faithful to the conditioning text prompts. Subsequent work has made the generation process more efficient by performing denoising in the latent space [51], and by speeding up the denoising process [14,33,54]. Techniques such as classifier free guidance [19] have significantly improved faithfulness to text prompts. Additional control of generation has also been shown to be possible, through auxiliary inputs such as layout [74], pose [73] and depth maps [5]. More recently, text-to-image diffusion models have been finetuned to generate structured images, such as panoramas [62]. In this work, we utilize text-to-panorama generation as structured guidance for text to 360° scene generation.

Text-to-3D Scene Generation. In recent times, text-to-3D scene generation has been reinvigorated with the advent of 2D diffusion models and guidance techniques. Text-to-image diffusion models, along with techniques such as classifier free guidance, have been found to provide strong priors to guide 3D generation methods [46,57,63,64,68]. These methods are generally more frequently used than direct text-to-3D diffusion models [22,23]. More recent works [13,30,60] successfully generate multi-object compositional 3D scenes. On the other hand, a second class of methods use auxiliary inputs such as layouts [45]. The least constrained text-to-3D methods do not rely on any auxiliary input with the only input being the text prompt describing the 3D scene [7,11,20,32,44,70,72]. Our work requires a text prompt input; however, unlike prior work, we propose using panoramic images as an intermediate input for globally consistent scenes.

Efficient 3D Scene Representation. 3D scene representations include a wide variety of techniques, including point clouds [4], volumetric representations [31,41], and meshes [71]. More recently, however, learning-based scene representations have gained prominence. Implicit representations such as neural radiance fields [35] have shown high quality rendering and novel view synthesis capability. Additional learning-based representations, that span both explicit [12] and mixed [39] have been proposed to speed up learning. Most recently, 3D Gaussian splatting [25] has enabled fast learning and rendering of high quality radiance fields using a Gaussian kernel-based explicit 3D representation. Since then, several works have emerged enabling sparse view [66,76] and compressed [10,27,38,40,42] 3D Gaussian representations, as well as representation of multidimensional feature fields [75]. A variation involves learning a Gaussian-based radiance field representation using panoramic images as input [2]. In this work, we propose a method for text to 360° 3D scene generation, by using panorama images as an intermediate representation.

3 Methods

In this section, we detail the proposed DreamScene360 architecture (Fig. 2). DreamScene360 initially generates a 360° panorama utilizing a self-refinement process, which ensures robust generation and aligns the image with the text semantically (Sect. 3.1). The transformation from a flat 2D image to a 3D model begins with the initialization from a panoramic geometric field (Sect. 3.2), followed by employing semantic alignment and geometric correspondences as regularizations for the deformation of the 3D Gaussians (Sect. 3.3).

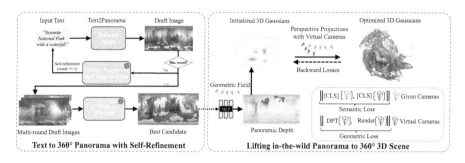

Fig. 2. Overall Architecture. Beginning with a concise text prompt, we employ a diffusion model to generate a 360° panoramic image. A self-refinement process is employed to produce the optimal 2D candidate panorama. Subsequently, a 3D geometric field is utilized to initialize the Panoramic 3D Gaussians. Throughout this process, both semantic and geometric correspondences are employed as guiding principles for the optimization of the Gaussians, aiming to address and fill the gaps resulting from the single-view input.

3.1 Text to 360° Panoramas with Self-refinement

Panoramic images provide an overview of the entire scene in one image, which is essential for generating 360° 3D scenes with global consistency.

360° Panoramic Image Generation. A crucial requirement for the generated panorama is ensuring continuity between the leftmost and rightmost sides of the image. We utilize the diffusion process from MultiDiffuser [3] to generate a panoramic image I_0 of resolution $H \times 2H$ based on a pre-trained diffusion model Φ. Starting from a noisy image I_T, we iteratively denoise the image by solving an optimization problem for each of several patches of the image, selected via a sliding window mechanism. For each patch $P_i(I_t)$, we ensure the distance against their denoised version $\Phi(P_i(I_t))$ is minimized. Though each patch may pull the denoising process in a different direction, the fused result is the weighted average of the update of each sample, whose closed form is written as follows,

$$\Phi(I_{t-1}) = \sum_{i=1}^{n} \frac{P_i^{-1}(W_i)}{\sum_{j=1}^{n} P_j^{-1}(W_j)} \otimes P_i^{-1}(\Phi(P_i(I_t))), \tag{1}$$

where W_i refers to per pixel weight, set to 1 in our experiments.

We use StitchDiffusion [62] as the pretrained 2D diffusion model, where a stitch method is employed in the generation process for synthesizing seamless 360° panoramic images. Trained on a curated paired image-text dataset containing 360° panoramas, a customized LoRA [21] module is incorporated into the MultiDiffusion [3] process. At each denoising timestamp, we not only diffuse at the original resolution $H \times (2H + 2W)$ via MultiDiffuser Φ, but also stitch the leftmost and rightmost image regions ($H \times W$ each) and diffuse the concatenated patch as well to ensure consistency at the border regions. Finally, we consider the center cropped region of $H \times 2H$ as the final 360° panoramic image. It is worth noting that Dreamscene360 is versatile in practice and can also adapt to other text-to-panorama diffusion models.

Multi-round Self-refinement. Unlike previous works that generate 3D scenes through a time-consuming score distillation process [64] or progressive inpainting [7], our work uses panorama to achieve user-friendly "one-click" 3D scene generation. We integrate GPT-4V to facilitate iterative self-refinement during the generation process, a feature that was challenging to implement in previous baselines due to the absence of global 2D image representations. Here, we draw inspiration from perspective image generation, Idea2Img [69], and implement a self-improvement framework aiming for better text-image aligned panoramic image generation. Starting from a user-provided rough prompt, we leverage GPT-4V to provide feedback and prompt revision suggestions according to the results generated by the previous step. During each round, GPT-4V is judges the generated image quality in terms of object counts, attributes, entities, relationships, sizes, appearance, and overall similarity with the original user-specified prompt. A score from 0–10 is assigned to each draft image and the one image with highest score is provided to GPT-4V for additional improvement. GPT-4V will then produce an improved text prompt based on the issues observed in the current generation results, and the new prompt will be used for another round of panorama generation. After a number of rounds, we collect the image with the highest visual quality score judged by GPT-4V in the whole generation process as our final panorama image.

Our 3D scene generation framework, therefore, enjoys a user-friendly self-improvement process, without the need for troublesome prompt engineering for users as in previous methods [7,46,64], but is able to obtain a high-quality, visually pleasing, text-aligned, and 360° consistent panoramic images that can be later converted into an immersive 3D scene via Panoramic Gaussian Splatting, which is detailed in the next sections.

3.2 Lifting in-the-Wild Panorama to 360 Scene

Transforming a single image, specifically an in-the-wild 360° panoramic image, into a 3D model poses significant challenges due to inadequate observational data to regularize the optimization process, such as those required in 3D Gaussian Splatting (see Fig. 7). Rather than beginning with a sparse point cloud (3DGS), we initialize with a dense point cloud utilizing pixel-wise depth information from the panoramic image of resolution H × W, which are then refined towards more precise spatial configurations, ensuring the creation of globally consistent 3D representations, robust to viewpoint changes.

Monocular Geometric Initialization. Given a single panoramic image $\mathbf{P}$, we project it onto N perspective tangent images with overlaps $\{(\mathbf{I}_i \in \mathbb{R}^{H \times W \times 3}, \mathbf{P}_i \in \mathbb{R}^{3 \times 4})\}_{i=1}^{N}$, following the literature's suggestion that 20 tangent images adequately cover the sphere's surface as projected by an icosahedron [50]. Unlike indoor panoramas, which benefit from structural layout priors in optimization, we employ a monocular depth estimator, DPT [49], to generate a monocular depth map $\boldsymbol{D}_i^{\text{Mono}}$ providing a robust geometric relationship. Nevertheless, these estimators still possess affine ambiguity, lacking a known scale and shift relative to metric depth. Addressing this efficiently is crucial for precise geometric initialization.

Global Structure Alignment. Previous studies in deformable depth alignment [16,50,61] and pose-free novel view synthesis [6] have explored aligning scales across multiple view depth maps derived from monocular depth estimations. In this context, we utilize a learnable global geometric field (MLPs), inspired by [50,61], supplemented by per-view scale and per-pixel shift parameters: $\{(\alpha_i \in \mathbb{R}, \boldsymbol{\beta}_i \in \mathbb{R}^{H \times W}\}_{i=1}^N$. We define the view direction ($\boldsymbol{v}$) for all pixels on the perspective images deterministically. The parameters of MLPs Θ are initialized with an input dimension of three and an output dimension of one. The parameters $\{(\alpha_i \in \mathbb{R}\}_{i=1}^N$ are initialized to ones, and $\{(\boldsymbol{\beta}_i \in \mathbb{R}^{H \times W}\}_{i=1}^N$ to zeros. With these optimizable parameters, we define our optimization goal as follows:

$$\min_{\alpha, \beta, \Theta} \left\{ ||\alpha \cdot \boldsymbol{D}^{\text{Mono}} + \boldsymbol{\beta} - \text{MLPs}(\boldsymbol{v}; \Theta)||_2^2 + \lambda_{\text{TV}} \mathcal{L}_{\text{TV}}(\boldsymbol{\beta}) + \lambda_\alpha ||\gamma(\alpha) - 1||^2 \right\} \quad (2)$$

where $\mathcal{L}_{\text{TV}}(\boldsymbol{\beta}) = \sum_{i,j} \left((\beta_{i,j+1} - \beta_{i,j})^2 + (\beta_{i+1,j} - \beta_{i,j})^2 \right)$

Here, λ_{TV} and λ_α are regularization coefficients to balance the loss weight, $\gamma(\cdot)$ is the softplus function. We set FoV as 80° during the optimization and use the predicted depth from the MLPs for the subsequent Gaussian optimization, as it can provide the depth of any view direction.

3.3 Optimizing Monocular Panoramic 3D Gaussians

While 3D Gaussians initialized with geometric priors from monocular depth maps provide a foundational structure, they are inherently limited by the lack of parallax inherent to single-view panoramas. This absence of parallax - critical for depth perception through binocular disparity - along with the lack of multiple observational cues typically provided by a baseline, poses substantial challenges in accurately determining spatial relationships and depth consistency. Thus, it is imperative to employ more sophisticated and efficient generative priors that function well in unconstrained settings and enable the scaling up and extend to any text prompts.

3D Gaussian Splatting. 3D Gaussian Splatting (3DGS) utilizes the multi-view calibrated images using Structure-from-Motion [55], and optimizes a set of Gaussians with center $\boldsymbol{x} \in \mathbb{R}^3$, an opacity value $\alpha \in \mathbb{R}$, spherical harmonics (SH) coefficients $\boldsymbol{c} \in \mathbb{R}^C$, a scaling vector $\boldsymbol{s} \in \mathbb{R}^3$ and a rotation vector $\boldsymbol{q} \in \mathbb{R}^4$ represented by a quaternion. Upon projecting the 3D Gaussians into a 2D space, the color C of a pixel is computed by volumetric rendering, which is performed using front-to-back depth order [26]:

$$C = \sum_{i \in \mathcal{N}} c_i \alpha_i T_i, \quad (3)$$

where $T_i = \prod_{j=1}^{i-1}(1-\alpha_j)$, $\mathcal{N}$ is the set of sorted Gaussians overlapping with the given pixel, T_i is the transmittance, defined as the product of opacity values of previous Gaussians overlapping the same pixel.

Synthesize Parallax with Virtual Cameras. We emulate parallax by synthesizing virtual cameras that are unseen in training but are close to the input panoramic viewpoint. We methodically create these cameras to simulate larger movements. This procedure is quantitatively described by incrementally introducing perturbation to the panoramic viewpoint coordinates (x, y, z) formalized as:

$$(x', y', z') = (x, y, z) + \delta(d_x, d_y, d_z) \quad (4)$$

where (x', y', z') denotes the new virtual camera positions, and $\delta(d_x, d_y, d_z)$ denotes the progressively increasing perturbations in each coordinate direction, under uniform distribution over $[-0.05, +0.05] \times \gamma$, where $\gamma \in \{1, 2, 4\}$ stands for 3-stage progressive perturbations, emulating the camera's movement from the original point.

Distilling Semantic Similarities. Previous research has underscored the significance of capturing the appearance of objects/scenes to establish robust feature correspondences across different views [17,29], which is crucial for tasks like co-segmentation and 2D pre-training. Inspired by these studies, we aim to establish a connection in visual feature correspondence between training views and synthetically generated virtual views. This is achieved by enforcing feature-level similarity, guiding the 3D Gaussians to fill the geometric gaps in the invisible regions effectively. To elaborate, we generate a perspective image I_i from the panoramic 3D Gaussians at a specific training camera viewpoint P_i and create another synthesized image I'_i using Eq 4 that is proximate to P_i. We employ the [CLS] token from the pre-trained DINOv2 [43] model to encapsulate compact semantic features through the equation:

$$\mathcal{L}_{sem} = 1 - \text{Cos}([\text{CLS}](I_i), [\text{CLS}](I'_i)), \quad (5)$$

where Cos denotes the cosine similarity. This approach is inspired by [59,67], assuming that the [CLS] token from a self-supervised, pre-trained Vision Transformer (ViT) can capture the high-level semantic attributes of an image. Such a mechanism is instrumental in identifying and leveraging similarities between two adjacent rendered images, facilitating our end-to-end 3DGS optimization process.

Regularizing Geometric Correspondences. While leveraging appearance cues is valuable, relying solely on them can lead to spatial inconsistencies.

Fig. 3. Diverse Generation. We demonstrate that our generated 3D scenes are diverse in style, consistent in geometry, and highly matched with the simple text inputs.

This is because models like DINOv2 might prioritize capturing semantic features over maintaining geometric coherence, potentially resulting in artifacts such as floaters. To mitigate this, we introduce a geometric regularization strategy designed to penalize discontinuities between pixels that exhibit inaccurate depth relationships. We employ a monocular depth estimator DPT [49] for this purpose [8]. Although it may not provide globally accurate depth, it allows us to ascertain the relative spatial relationships between pixels:

$$\mathcal{L}_{geo}(I_i, D_i) = 1 - \frac{\text{Cov}(D_i, \text{DPT}(I_i))}{\sqrt{\text{Var}(D_i)\text{Var}(\text{DPT}(I_i))}}, \tag{6}$$

Here, I_i represents the rendered image at the i-th camera, and D_i signifies the rendered depth. Additionally, we incorporate an unsupervised local smoothness prior (TV loss) on the rendered depth at virtual views.

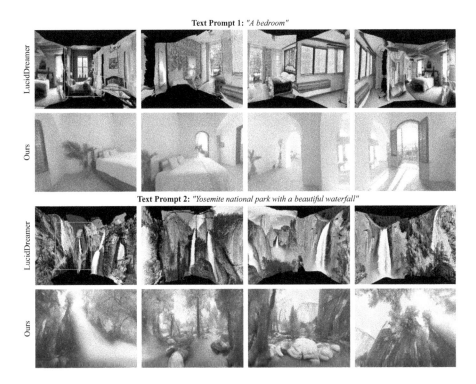

Fig. 4. Visual Comparisons. We showcase 360° 3D scene generation. In each row, from left to right, displays novel views as the camera undergoes clockwise rotation in yaw, accompanied by slight random rotations in pitch and random translations. LucidDreamer [7] hallucinates novel views from a conditioned image (indicated by a red bounding box) but lacks global semantic, stylized, and geometric consistency. In contrast, our method provides complete 360° coverage without any blind spots (black areas in baseline results), and shows globally consistent semantics.

Table 1. Quantitative comparisons between LucidDreamer and ours.

	CLIP Distance↓	Q-Align↑	NIQE↓	BRISQUE↓	Runtime
LucidDreamer [7]	0.8900	3.0566	6.2305	51.9764	6 min 15 s
Ours	**0.8732**	**3.1094**	**4.9165**	**38.3911**	7 min 20 s

3.4 Optimization

The entire pipeline for transforming a panorama into 3D can be supervised end-to-end through a composite of all the loss functions.

$$\mathcal{L} = \mathcal{L}_{RGB} + \lambda_1 \cdot \mathcal{L}_{sem} + \lambda_2 \cdot \mathcal{L}_{geo} \tag{7}$$

Here, $\mathcal{L}_{RGB}$ represents the photometric loss on projected perspective images, which consists of $\mathcal{L}_1$ and a D-SSIM term as described in [25]. $\mathcal{L}_{sem}$ denotes semantic regularization, and $\mathcal{L}_{geo}$ signifies geometric regularization. The weights λ_1 and λ_2 are set to 0.05 each, respectively.

4 Experiments

4.1 Experiment Setting

Implementation. Given the panorama generated from text, we resize it to 1024 × 2048 in order to produce a dense enough point cloud by projecting per-pixel panoramic depth values into 3D space along the per-pixel camera ray directions, where the directions can be obtained according to the coordinate transformation in spherical panoramic imaging [50]. Since the panoramic depth map is predicted from the optimized geometric field, the values in local patches are guaranteed to be consistent and smooth, thereby resulting in point cloud in 3D that accurately captures the smooth geometry at the objects' surfaces. This setup enables effective 3D Gaussians initialization within a controllable, bounded area, eliminating redundant points in empty spaces. To circumvent the common floater issue in 3DGS rendering, we disable the densification process, enhancing the overall quality and consistency of the rendered scenes.

Baseline Methods. Our work tackles the challenging problem of unconstrained 360° generation, including both indoor and outdoor scenarios. However, the works utilizing a bounded NeRF representation using score distillation do not work very well in this case. Thus, the comparisons are conducted between Dream-Scene360 (ours) and the state-of-the-art LucidDreamer [7]. We use the open-source codebase of LucidDreamer, which starts from a single image and a text prompt. The framework constructs a global point cloud by progressive inpainting into 360° views and then distills a set of 3D Gaussians. In our experiments, we set the input image for the LucidDreamer to be generated from Stable Diffusion [51] v1.5 using the input text for a fair comparison.

Metrics. Since there is no ground truth in the generated 3D scenes, we utilize CLIP [47] embedding distance, following previous works [58,68], to measure the text-image alignment as a mechanism to quantify the novel view rendering quality. Specifically, for each method, we render images with camera rotations and translations to mimic the immersive trajectory inside the 3D scenes. Additionally, we utilize multiple non-reference image quality assessment metrics. NIQE [37] and BRISQUE [36] are widely adopted non-reference quality assessment methods in measuring in-the-wild image quality. QAlign [65] is the state-of-the-art method in quality assessment benchmarks, which adopts a large multi-modal model fine-tuned on available image quality assessment datasets. We utilize the "quality" mode of QAlign, which focuses on the perceptual quality of image contents.

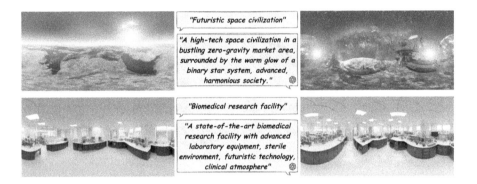

Fig. 5. Ablation of Self-refinement. We demonstrate that the self-refinement process greatly enhances the image quality by improving the text prompt. As shown in each row, the image on the left is generated using a simple user prompt, while a prompt augmented by GPT-4V obtains the image on the right. We observe that after the multi-round self-refinement, GPT-4V selects the one panorama with better visual quality, which provides solid support for the immersive 3D scene we further generate.

4.2 Main Results

360° Scene Generation. As shown in Fig. 3, our method can generate diverse 3D scenes in different styles with distinct contents, while preserving high-fidelity novel-view rendering ability and realistic scene geometry. These results showcase our method's generalization ability to diverse use cases, providing a user-friendly experience in realizing users' imaginations.

Comparisons with Baseline Methods. We show visual comparisons against LucidDreamer [7] in Fig. 4. LucidDreamer involves progressive inpainting to find agreement between multiple synthesized images and tries to fuse a point cloud that can be later distilled into 3D Gaussians. Their pipeline, which inpaints each patch separately based on the same text prompt, tends to produce repetitive results especially when generating complex scenes. In comparison, our method delivers consistent results thanks to the intermediate panorama as a global 2D representation. We provide quantitative comparisons in Table 1, where the camera undergoes clockwise rotation in yaw, accompanied by slight random rotations in pitch and random translations to capture 4 novel views roughly representing front, back, left, and right, along with two views of up and down by 90° rotation in pitch, to mimic the exploration in an immersive 360° panoramic 3D scene.

For the *bedroom* text prompt, LucidDreamer begins with a view featuring a modern style bedroom with one bed but hallucinates the novel views into multiple bedrooms, while the style of the imagined bedrooms diverges from that of the starting view. In the case of the *Yosemite* text prompt, LucidDreamer merely replicates the waterfall seen in the initial view throughout. In conclusion, our results demonstrate global semantic, stylized, and geometric consistency, offering complete 360° coverage without any blind spots.

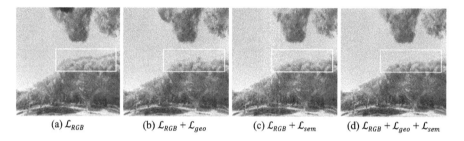

(a) $\mathcal{L}_{RGB}$ (b) $\mathcal{L}_{RGB} + \mathcal{L}_{geo}$ (c) $\mathcal{L}_{RGB} + \mathcal{L}_{sem}$ (d) $\mathcal{L}_{RGB} + \mathcal{L}_{geo} + \mathcal{L}_{sem}$

Fig. 6. Ablation of Optimization Loss. We demonstrate the impact of Semantic and Geometric losses on the synthesized virtual cameras. (a) Utilizing photometric loss on camera views from a rendered panorama induces artifacts when altering rendered camera positions. (b) Implementing Geometric and (c) Semantic regularizations effectively reduces the artifacts originating from invisible views. (d) Integrating both regularizations yields the most optimal outcomes.

4.3 Ablation Study

Self-refinement Process. We further evaluate the importance of the self-refinement process. In this part, we mainly focus on the generated panorama since panorama provides a holistic view of the 3D scene and provides an upper bound for the visual quality. As can be seen in Fig. 5, we observe that using a simple text prompt usually delivers minimalist results with fewer details. With the help of GPT-4V, our self-refinement process enjoys prompt revision and quality assessment. This ability finally facilitates the selection of a more realistic and detailed panorama image among the draft image candidates. These functionalities are otherwise hard to achieve in previous baselines that do not have global 2D representations, and as a result, our results provide a much better visual appearance than baselines as shown in Fig. 4 and Table 1.

Loss Function. We investigate the importance of multiple loss functions we adopted in Fig. 6. As shown in the figure, especially in the highlighted region, our full configuration of loss functions delivers the best visual quality when rendered from a novel viewpoint. In comparison, removing the geometric regularizations or semantic constraints will result in unpleasant artifacts.

Initialization. We showcase the importance of proper initialization in Fig. 7. As can be seen in the image, adopting our curated point initialization largely benefits the rendering quality of our 3D Panoramic Gaussian Splatting. An alternate solution involves adopting random initialization of the point cloud locations. This option, however, generates blurry results, mainly due to the challenge of learning proper point locations on the fly without ground truth geometry as supervision. Note that since our full model incorporates this high-quality point initialization, at the training stage, we can safely disable the adaptive density control of 3D Gaussians and, therefore, speed up convergence. Using random initialization of 3D Gaussians, on the other hand, has to rely on adaptive density control introduced in 3D Gaussian Splatting paper [25] (*e.g.* densify, clone, prune) to move the 3D Gaussians to desired locations via backpropagation.

(a) Random initialized geometry for optimizing 3D Gaussians.

(b) Initialized from globally aligned monocular geometry.

Fig. 7. Ablation Study on 3D Initialization. We present a comparative visualization of various initialization methods for 3D Panoramic Gaussian Splatting. In the absence of geometric priors of the scene (a), the optimized 3D Gaussian rendering yields plausible results in panoramic camera views. However, artifacts become evident when the rendered camera position is altered. To address this, we employ a monocular depth approach combined with learnable alignment factors and a geometric field (b), which ensures consistent alignment across multiple views.

5 Conclusion

In this work, we present DreamScene360, a novel framework that unrestrictedly generates high-quality, immersive 3D scenes with complete 360° coverage in various styles from text inputs, regardless of their specificity level. Our pipeline leverages panorama as a middle ground, which provides us with a self-refinement opportunity by asking GPT-4V for quality assessment and prompt revision. Subsequently, we construct a geometric field that initializes 3D Gaussians. With the help of unsupervised geometric and semantic loss on virtual views, we lift the 2D panorama into panoramic 3D Gaussians. Experiments demonstrate our superiority against baseline methods in terms of global consistency and visual quality. However, our generation results are still limited by the default resolution (512 × 1024) of the pretrained text-to-panorama diffusion model. Moving forward, we will explore generating 3D scenes at higher resolutions and even 4D dynamic scenes [28] for a better, seamless, and immersive user experience.

Acknowledgements. We thank the Visual Machines Group (VMG) at UCLA, Visual Informatics Group at UT Austin (VITA), and ARL Intelligent Perception Branch for feedback and support. This project was supported by LUCI program under the Basic Research Office and partially supported by ARL grants W911NF20-2-0158 and W911NF-21-2-0104 under the cooperative A2I2 program. Z.W. is supported by the U.S. DEVCOM Army Research Laboratory and LUCI program, and an Army Young Investigator Award. A.K. is supported by a DARPA Young Faculty Award, NSF CAREER Award IIS-2046737, and Army Young Investigator Award.

References

1. Armandpour, M., Zheng, H., Sadeghian, A., Sadeghian, A., Zhou, M.: Re-imagine the negative prompt algorithm: transform 2d diffusion into 3d, alleviate janus problem and beyond. arXiv preprint arXiv:2304.04968 (2023)
2. Bai, J., Huang, L., Guo, J., Gong, W., Li, Y., Guo, Y.: 360-gs: layout-guided panoramic gaussian splatting for indoor roaming. arXiv preprint arXiv:2402.00763 (2024)
3. Bar-Tal, O., Yariv, L., Lipman, Y., Dekel, T.: Multidiffusion: fusing diffusion paths for controlled image generation (2023)
4. Berger, M., et al.: State of the art in surface reconstruction from point clouds. In: 35th Annual Conference of the European Association for Computer Graphics, Eurographics 2014-State of the Art Reports. No. CONF, The Eurographics Association (2014)
5. Bhat, S.F., Mitra, N.J., Wonka, P.: Loosecontrol: lifting controlnet for generalized depth conditioning. arXiv preprint arXiv:2312.03079 (2023)
6. Bian, W., Wang, Z., Li, K., Bian, J.W., Prisacariu, V.A.: Nope-nerf: optimising neural radiance field with no pose prior. In: Proceedings of the IEEE/CVF Conference on Computer Vision and Pattern Recognition, pp. 4160–4169 (2023)
7. Chung, J., Lee, S., Nam, H., Lee, J., Lee, K.M.: Luciddreamer: domain-free generation of 3d gaussian splatting scenes. arXiv preprint arXiv:2311.13384 (2023)
8. Deng, C., et al.: Nerdi: single-view nerf synthesis with language-guided diffusion as general image priors. In: Proceedings of the IEEE/CVF Conference on Computer Vision and Pattern Recognition, pp. 20637–20647 (2023)

9. Dhariwal, P., Nichol, A.: Diffusion models beat gans on image synthesis. Adv. Neural. Inf. Process. Syst. **34**, 8780–8794 (2021)
10. Fan, Z., Wang, K., Wen, K., Zhu, Z., Xu, D., Wang, Z.: Lightgaussian: unbounded 3d gaussian compression with 15x reduction and 200+ fps. arXiv preprint arXiv:2311.17245 (2023)
11. Fang, C., Hu, X., Luo, K., Tan, P.: Ctrl-room: controllable text-to-3d room meshes generation with layout constraints. arXiv preprint arXiv:2310.03602 (2023)
12. Fridovich-Keil, S., Yu, A., Tancik, M., Chen, Q., Recht, B., Kanazawa, A.: Plenoxels: radiance fields without neural networks. In: Proceedings of the IEEE/CVF Conference on Computer Vision and Pattern Recognition, pp. 5501–5510 (2022)
13. Gao, G., Liu, W., Chen, A., Geiger, A., Schölkopf, B.: Graphdreamer: compositional 3d scene synthesis from scene graphs. arXiv preprint arXiv:2312.00093 (2023)
14. Geng, Z., Pokle, A., Kolter, J.Z.: One-step diffusion distillation via deep equilibrium models. Adv. Neural Inf. Process. Syst. **36** (2024)
15. Goodfellow, I., et al.: Generative adversarial networks. Commun. ACM **63**(11), 139–144 (2020)
16. Hedman, P., Kopf, J.: Instant 3d photography. ACM Trans. Graph. (TOG) **37**(4), 1–12 (2018)
17. Hénaff, O.J., et al.: Object discovery and representation networks. arXiv preprint arXiv:2203.08777 (2022)
18. Ho, J., Jain, A., Abbeel, P.: Denoising diffusion probabilistic models. Adv. Neural. Inf. Process. Syst. **33**, 6840–6851 (2020)
19. Ho, J., Salimans, T.: Classifier-free diffusion guidance. arXiv preprint arXiv:2207.12598 (2022)
20. Höllein, L., Cao, A., Owens, A., Johnson, J., Nießner, M.: Text2room: extracting textured 3d meshes from 2d text-to-image models. arXiv preprint arXiv:2303.11989 (2023)
21. Hu, E.J., et al.: Lora: low-rank adaptation of large language models. arXiv preprint arXiv:2106.09685 (2021)
22. Jun, H., Nichol, A.: Shap-e: generating conditional 3d implicit functions. arXiv preprint arXiv:2305.02463 (2023)
23. Karnewar, A., Vedaldi, A., Novotny, D., Mitra, N.J.: Holodiffusion: training a 3d diffusion model using 2d images. In: Proceedings of the IEEE/CVF Conference on Computer Vision and Pattern Recognition, pp. 18423–18433 (2023)
24. Karras, T., Laine, S., Aila, T.: A style-based generator architecture for generative adversarial networks. In: Proceedings of the IEEE/CVF Conference on Computer Vision and Pattern Recognition, pp. 4401–4410 (2019)
25. Kerbl, B., Kopanas, G., Leimkühler, T., Drettakis, G.: 3d gaussian splatting for real-time radiance field rendering. ACM Trans. Graph. (TOG) **42**(4), 1–14 (2023)
26. Kopanas, G., Philip, J., Leimkühler, T., Drettakis, G.: Point-based neural rendering with per-view optimization. In: Computer Graphics Forum, vol. 40, pp. 29–43. Wiley Online Library (2021)
27. Lee, J.C., Rho, D., Sun, X., Ko, J.H., Park, E.: Compact 3d gaussian representation for radiance field. arXiv preprint arXiv:2311.13681 (2023)
28. Li, R., et al.: 4k4dgen: panoramic 4d generation at 4k resolution. arXiv preprint arXiv:2406.13527 (2024)
29. Li, W., Hosseini Jafari, O., Rother, C.: Deep object co-segmentation. In: Jawahar, C.V., Li, H., Mori, G., Schindler, K. (eds.) ACCV 2018. LNCS, vol. 11363, pp. 638–653. Springer, Cham (2019). https://doi.org/10.1007/978-3-030-20893-6_40

30. Lin, Y., et al.: Componerf: text-guided multi-object compositional nerf with editable 3d scene layout. arXiv preprint arXiv:2303.13843 (2023)
31. Lombardi, S., Simon, T., Saragih, J., Schwartz, G., Lehrmann, A., Sheikh, Y.: Neural volumes: learning dynamic renderable volumes from images. arXiv preprint arXiv:1906.07751 (2019)
32. Mao, W., Cao, Y.P., Liu, J.W., Xu, Z., Shou, M.Z.: Showroom3d: text to high-quality 3d room generation using 3d priors. arXiv preprint arXiv:2312.13324 (2023)
33. Meng, C., et al.: On distillation of guided diffusion models. In: Proceedings of the IEEE/CVF Conference on Computer Vision and Pattern Recognition, pp. 14297–14306 (2023)
34. Mildenhall, B., Srinivasan, P.P., Tancik, M., Barron, J.T., Ramamoorthi, R., Ng, R.: NeRF: representing scenes as neural radiance fields for view synthesis. In: Vedaldi, A., Bischof, H., Brox, T., Frahm, J.-M. (eds.) ECCV 2020. LNCS, vol. 12346, pp. 405–421. Springer, Cham (2020). https://doi.org/10.1007/978-3-030-58452-8_24
35. Mildenhall, B., Srinivasan, P.P., Tancik, M., Barron, J.T., Ramamoorthi, R., Ng, R.: Nerf: representing scenes as neural radiance fields for view synthesis. Commun. ACM **65**(1), 99–106 (2021)
36. Mittal, A., Moorthy, A.K., Bovik, A.C.: No-reference image quality assessment in the spatial domain. IEEE Trans. Image Process. **21**(12), 4695–4708 (2012)
37. Mittal, A., Soundararajan, R., Bovik, A.C.: Making a "completely blind" image quality analyzer. IEEE Signal Process. Lett. **20**(3), 209–212 (2012)
38. Morgenstern, W., Barthel, F., Hilsmann, A., Eisert, P.: Compact 3d scene representation via self-organizing gaussian grids. arXiv preprint arXiv:2312.13299 (2023)
39. Müller, T., Evans, A., Schied, C., Keller, A.: Instant neural graphics primitives with a multiresolution hash encoding. arXiv preprint arXiv:2201.05989 (2022)
40. Navaneet, K., Meibodi, K.P., Koohpayegani, S.A., Pirsiavash, H.: Compact3d: compressing gaussian splat radiance field models with vector quantization. arXiv preprint arXiv:2311.18159 (2023)
41. Nguyen-Phuoc, T., Li, C., Theis, L., Richardt, C., Yang, Y.L.: Hologan: unsupervised learning of 3d representations from natural images. In: Proceedings of the IEEE/CVF International Conference on Computer Vision, pp. 7588–7597 (2019)
42. Niedermayr, S., Stumpfegger, J., Westermann, R.: Compressed 3d gaussian splatting for accelerated novel view synthesis. arXiv preprint arXiv:2401.02436 (2023)
43. Oquab, M., et al.: Dinov2: learning robust visual features without supervision. arXiv preprint arXiv:2304.07193 (2023)
44. Ouyang, H., Heal, K., Lombardi, S., Sun, T.: Text2immersion: generative immersive scene with 3d gaussians. arXiv preprint arXiv:2312.09242 (2023)
45. Po, R., Wetzstein, G.: Compositional 3d scene generation using locally conditioned diffusion. arXiv preprint arXiv:2303.12218 (2023)
46. Poole, B., Jain, A., Barron, J.T., Mildenhall, B.: Dreamfusion: text-to-3d using 2d diffusion. arXiv preprint arXiv:2209.14988 (2022)
47. Radford, A., et al.: Learning transferable visual models from natural language supervision. In: International Conference on Machine Learning, pp. 8748–8763. PMLR (2021)
48. Ramesh, A., Dhariwal, P., Nichol, A., Chu, C., Chen, M.: Hierarchical text-conditional image generation with clip latents, **1**(2), 3. arXiv preprint arXiv:2204.06125 (2022)
49. Ranftl, R., Bochkovskiy, A., Koltun, V.: Vision transformers for dense prediction. In: Proceedings of the IEEE/CVF International Conference on Computer Vision, pp. 12179–12188 (2021)

50. Rey-Area, M., Yuan, M., Richardt, C.: 360monodepth: high-resolution 360deg monocular depth estimation. In: Proceedings of the IEEE/CVF Conference on Computer Vision and Pattern Recognition, pp. 3762–3772 (2022)
51. Rombach, R., Blattmann, A., Lorenz, D., Esser, P., Ommer, B.: High-resolution image synthesis with latent diffusion models. In: Proceedings of the IEEE/CVF Conference on Computer Vision and Pattern Recognition, pp. 10684–10695 (2022)
52. Saharia, C., et al.: Photorealistic text-to-image diffusion models with deep language understanding. In: Oh, A.H., Agarwal, A., Belgrave, D., Cho, K. (eds.) Advances in Neural Information Processing Systems (2022). https://openreview.net/forum?id=08Yk-n5l2Al
53. Saharia, C., et al.: Photorealistic text-to-image diffusion models with deep language understanding. Adv. Neural. Inf. Process. Syst. **35**, 36479–36494 (2022)
54. Salimans, T., Ho, J.: Progressive distillation for fast sampling of diffusion models. arXiv preprint arXiv:2202.00512 (2022)
55. Schonberger, J.L., Frahm, J.M.: Structure-from-motion revisited. In: Proceedings of the IEEE Conference on Computer Vision and Pattern Recognition, pp. 4104–4113 (2016)
56. Song, J., Meng, C., Ermon, S.: Denoising diffusion implicit models. arXiv preprint arXiv:2010.02502 (2020)
57. Song, L., et al.: Roomdreamer: text-driven 3d indoor scene synthesis with coherent geometry and texture. arXiv preprint arXiv:2305.11337 (2023)
58. Tang, J., Ren, J., Zhou, H., Liu, Z., Zeng, G.: Dreamgaussian: generative gaussian splatting for efficient 3d content creation. arXiv preprint arXiv:2309.16653 (2023)
59. Tumanyan, N., Bar-Tal, O., Bagon, S., Dekel, T.: Splicing vit features for semantic appearance transfer. In: Proceedings of the IEEE/CVF Conference on Computer Vision and Pattern Recognition, pp. 10748–10757 (2022)
60. Vilesov, A., Chari, P., Kadambi, A.: Cg3d: compositional generation for text-to-3d via gaussian splatting. arXiv preprint arXiv:2311.17907 (2023)
61. Wang, G., Wang, P., Chen, Z., Wang, W., Loy, C.C., Liu, Z.: Perf: panoramic neural radiance field from a single panorama. arXiv preprint arXiv:2310.16831 (2023)
62. Wang, H., Xiang, X., Fan, Y., Xue, J.H.: Customizing 360-degree panoramas through text-to-image diffusion models. In: Proceedings of the IEEE/CVF Winter Conference on Applications of Computer Vision, pp. 4933–4943 (2024)
63. Wang, H., Du, X., Li, J., Yeh, R.A., Shakhnarovich, G.: Score jacobian chaining: Lifting pretrained 2d diffusion models for 3d generation. In: Proceedings of the IEEE/CVF Conference on Computer Vision and Pattern Recognition, pp. 12619–12629 (2023)
64. Wang, Z., et al.: Prolificdreamer: high-fidelity and diverse text-to-3d generation with variational score distillation. arXiv preprint arXiv:2305.16213 (2023)
65. Wu, H., et al.: Q-align: teaching lmms for visual scoring via discrete text-defined levels. arXiv preprint arXiv:2312.17090 (2023)
66. Xiong, H., Muttukuru, S., Upadhyay, R., Chari, P., Kadambi, A.: Sparsegs: real-time 360{deg} sparse view synthesis using gaussian splatting. arXiv preprint arXiv:2312.00206 (2023)
67. Xu, D., Jiang, Y., Wang, P., Fan, Z., Shi, H., Wang, Z.: Sinnerf: training neural radiance fields on complex scenes from a single image. In: Computer Vision–ECCV 2022: 17th European Conference, Tel Aviv, Israel, 23–27 October 2022, Proceedings, Part XXII, pp. 736–753. Springer, Heidelberg (2022). https://doi.org/10.1007/978-3-031-20047-2_42

68. Xu, D., Jiang, Y., Wang, P., Fan, Z., Wang, Y., Wang, Z.: Neurallift-360: lifting an in-the-wild 2d photo to a 3d object with 360deg views. arXiv preprint arXiv:2211.16431 (2022)
69. Yang, Z., et al.: Idea2img: iterative self-refinement with gpt-4v (ision) for automatic image design and generation. arXiv preprint arXiv:2310.08541 (2023)
70. Yu, H.X., et al.: Wonderjourney: going from anywhere to everywhere. arXiv preprint arXiv:2312.03884 (2023)
71. Zhang, C., Chen, T.: Efficient feature extraction for 2d/3d objects in mesh representation. In: Proceedings 2001 International Conference on Image Processing (Cat. No. 01CH37205), vol. 3, pp. 935–938. IEEE (2001)
72. Zhang, J., Li, X., Wan, Z., Wang, C., Liao, J.: Text2nerf: text-driven 3d scene generation with neural radiance fields. IEEE Trans. Visualization Comput. Graph. (2024)
73. Zhang, L., Rao, A., Agrawala, M.: Adding conditional control to text-to-image diffusion models. In: Proceedings of the IEEE/CVF International Conference on Computer Vision, pp. 3836–3847 (2023)
74. Zheng, G., Zhou, X., Li, X., Qi, Z., Shan, Y., Li, X.: Layoutdiffusion: controllable diffusion model for layout-to-image generation. In: Proceedings of the IEEE/CVF Conference on Computer Vision and Pattern Recognition, pp. 22490–22499 (2023)
75. Zhou, S., et al.: Feature 3dgs: supercharging 3d gaussian splatting to enable distilled feature fields. In: Proceedings of the IEEE/CVF Conference on Computer Vision and Pattern Recognition, pp. 21676–21685 (2024)
76. Zhu, Z., Fan, Z., Jiang, Y., Wang, Z.: FSGS: real-time few-shot view synthesis using gaussian splatting. arXiv preprint arXiv:2312.00451 (2023)

Frequency-Spatial Entanglement Learning for Camouflaged Object Detection

Yanguang Sun[1](✉), Chunyan Xu[1], Jian Yang[1], Hanyu Xuan[2](✉), and Lei Luo[1](✉)

[1] PCA Lab, Nanjing University of Science and Technology, Nanjing, China
Sunyg@njust.edu.cn
[2] School of Big Data and Statistics, Anhui University, Heifei, China

Abstract. Camouflaged object detection has attracted a lot of attention in computer vision. The main challenge lies in the high degree of similarity between camouflaged objects and their surroundings in the spatial domain, making identification difficult. Existing methods attempt to reduce the impact of pixel similarity by maximizing the distinguishing ability of spatial features with complicated design, but often ignore the sensitivity and locality of features in the spatial domain, leading to sub-optimal results. In this paper, we propose a new approach to address this issue by jointly exploring the representation in the frequency and spatial domains, introducing the Frequency-Spatial Entanglement Learning (FSEL) method. This method consists of a series of well-designed Entanglement Transformer Blocks (ETB) for representation learning, a Joint Domain Perception Module for semantic enhancement, and a Dual-domain Reverse Parser for feature integration in the frequency and spatial domains. Specifically, the ETB utilizes frequency self-attention to effectively characterize the relationship between different frequency bands, while the entanglement feed-forward network facilitates information interaction between features of different domains through entanglement learning. Our extensive experiments demonstrate the superiority of our FSEL over 21 state-of-the-art methods, through comprehensive quantitative and qualitative comparisons in three widely-used datasets. The source code is available at: https://github.com/CSYSI/FSEL.

Keywords: Camouflaged object detection · Computer vision · Frequency-Spatial entanglement learning

1 Introduction

"Camouflage" is a natural defense mechanism used by certain animals, such as chameleons, grasshoppers, and caterpillars, to blend into their surroundings and protect themselves. The study of camouflaged object detection (COD) focuses on identifying concealed targets in real-world situations. This research is crucial in developing robust visual perception models in computer vision. COD has a wide range of applications, including medical image analysis [8,9], species conservation [33], and industrial defect detection [10].

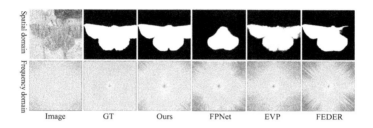

Fig. 1. The visual comparison results of the proposed FSEL and current COD methods (*i.e.*, FPNet [4], EVP [27], and FEDER [13]) in the spatial and frequency domain.

In the early stages, some COD methods [11,32,55] relied on manually crafted features to detect camouflaged objects. However, due to the extremely challenging appearance of these objects, the results obtained were often unsatisfactory. Later, with the advancement of deep learning and the availability of large-scale datasets [7,20], many COD methods [18,21,30,34,52] based on deep learning have been proposed. With a large amount of data available for training, these methods have the potential to automatically extract features and detect camouflaged objects, resulting in impressive performance. More recently, researchers have introduced various techniques for exploring useful information of input features from the spatial domain using boundary-guided [39,57,58], multi-scale strategy [6,34,38], uncertainty-aware [21,47], distraction mining [31], and *etc*.

We have observed that the COD methods [7,18,21,52,58] mentioned above primarily focus on single spatial features. While these spatial features are advantageous for COD tasks, they are often susceptible to interference from complex backgrounds. This vulnerability arises from their reliance on pixel-level information, with a primary emphasis on the local intensity and spatial position of individual pixels. Furthermore, spatial features possess local properties, meaning that pixels within a feature may only exhibit certain correlations with surrounding pixels. That being said, relying solely on spatial features can make it challenging to distinguish subtle variations within concealed objects and backgrounds. Therefore, it is crucial to find ways to overcome the limitations of spatial features to achieve accurate COD results. Recently, frequency features generated through Fourier Transform have been shown to have global characteristics and have been proven to be beneficial for understanding image contents [3,35,42]. This can help break the bottleneck of spatial features.

Some recent COD methods [4,13,22,27,56] have begun to incorporate frequency clues in their approach. These methods can be divided into two categories based on their objectives. The first category (*e.g.*, FDNet [56] and EVP

[27]) is to act directly on input images through different frequency transforms to extract frequency features, which are then combined with spatial features. However, camouflaged images often contain a lot of background noise, making the frequency features obtained from the image unreliable. When aggregated with spatial features, this may introduce some unnecessary background noises, resulting in under-segmented results (as depicted EVP [27] in Fig. 1). The second category [4,13,22] focuses on initial features from the encoder. For example, Cong et al. [4] designed a frequency-perception module to improve the detection of camouflaged objects by utilizing both high-frequency features and low-frequency features. And He et al. [13] proposed frequency attention modules that obtain important parts of corresponding features by considering both high-frequency and low-frequency components. Although these methods have shown promising results, they only focus on high-frequency and low-frequency features, overlooking some information that falls between these two frequencies. This can be seen in FPNet [4] and FEDER [13] in Fig. 1, where significant information within the frequency domain may be missed.

Based on the above discussion, we propose a novel method called Frequency-Spatial Entanglement Learning for accurate camouflaged object detection. Our method combines global frequency features and local spatial features to optimize the initial input features and enhance their discriminative ability. Specifically, we first establish a Frequency Self-attention to obtain discriminative global frequency features, which models the correlation between each frequency band and learns the dependency relationships between different frequencies in input bands. Moreover, we introduce entanglement learning between the frequency and spatial features in the Entanglement Transformer Block, allowing them to mutually learn and collaborate for optimization. Furthermore, we extend the applicability of global frequency features by utilizing the Joint Domain Perception Module and the Dual-domain Reverse Parser to optimize the input features and generate powerful representations that incorporate both frequency and spatial information. Extensive experiments on three widely-used benchmark datasets (*i.e.*, CAMO [20], COD10K [7], and NC4K [30]) demonstrate that FSEL consistently outperforms 21 state-of-the-art COD methods across different backbones.

The main contributions can be summarized as follows:

(1) We propose a Frequency-Spatial Entanglement Learning (FSEL) framework that utilizes both global frequency and local spatial features to enhance the detection of camouflaged objects.
(2) To improve the representation capability of frequency and spatial features, we have designed an Entanglement Transformer Block (ETB). This block allows for entanglement learning of frequency-spatial features, resulting in a more comprehensive understanding of the data.
(3) To reduce the sensitivity and locality limitation of spatial features, we have incorporated frequency domain transformations into both the Joint Domain Perception Module (JDPM) and the Dual-domain Reverse Parser (DRP).

2 Related Work

Camouflaged Object Detection. Recently, with the public availability of datasets (*i.e.*, CAMO [20], COD10K [7], and NC4K [30]), deep learning-based COD methods have started to surface in large numbers, which can be broadly categorized into, including multi-scale strategies [34,38,59], edge-guidance [13,39,57,58], uncertainty-aware [21,47], multi-graph learning [52], iterative manner [15,18,54], transformer [27,37], and so on. Besides, other COD methods [4,22,27] considered frequency clues to help with reasoning camouflaged objects. Particularly, Zhong *et al.* [56] processed directly the camouflaged image through discrete cosine transform to obtain frequency information. He *et al.* [13] proposed frequency attention modules to filter out the noteworthy parts of corresponding features. After that, Cong *et al.* [4] designed a frequency-perception module by learning different frequency features to achieve coarse localization of camouflaged objects. However, these methods often focus on the high- and low-frequency information, ignoring the relationship between all bands in the frequency domain. Therefore, we conduct frequency analysis on different spectral features to achieve all frequency band interactions and importance allocation. Furthermore, we perform entanglement learning on global frequency and local spatial features, which is beneficial for obtaining powerful representations.

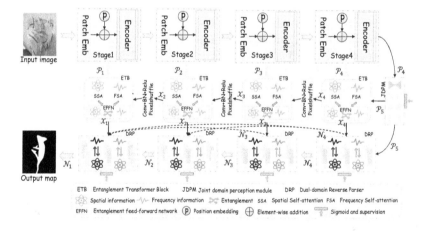

Fig. 2. Overview of the proposed FSEL model framework for camouflaged object detection. The proposed FSEL method generates predicted results through a Joint Domain Perception Module (JDPM), a series of stacked Entanglement Transformer Block (ETB), and a Dual-domain Reverse Parser (DRP).

Vision Transformer. Transformer [40] utilized self-attention to model global semantic and long-range dependencies, which helps to understand the correlation between different regions in an image, and therefore it has been widely used in some computer vision tasks, including object detection [1,45], image classification [43,60], semantic segmentation [5,17,46], *etc.* For example, Yuan *et al.* [49] obtained long-range relationships from the sequence of image patches to perform the image classification. Next, Liu *et al.* [29] split input maps into non-overlapping local windows, and then transferred the information through shift operations between the windows to improve the efficiency of the model. In addition, other transformer models have been successful in computer visions, such as Restormer [51], CrossFormer [44], EfficientViT [28], MPFormer [53], and among others. Unlike these methods, which always model relationships based on the spatial domain, we transform spatial features into the frequency domain and combine them to perform dual-domain feature optimization.

Frequency Learning. The frequency domain is very important for signal analysis, and recently it has been gradually applied in computer vision tasks. Particularly, Qin *et al.* [35] assumed channel attention as a compression problem and introduced frequency transformation in the channel attention. Yun *et al.* [50] handled the balancing problem of different frequency components of visual features. Wang *et al.* [42] proposed a frequency shortcut perspective in image classification. In addition, some frequency domain-based methods [3,4,13,23,27,41] have achieved great performance. In this paper, we extend global frequency features to different applications, involving multi-receptive fields perception, transformer, and reverse attention.

3 Method

3.1 Framework Architecture

Camouflaged objects exhibit a high level of visual similarity to their backgrounds, achieved through adaptive changes in color, texture, and shape. This creates challenges in distinguishing between object and background pixels in the spatial domain. Additionally, the locality of features in the spatial domain is limited in understanding camouflaged objects. To address this issue, we have implemented several strategies: 1) We have expanded beyond the spatial domain and utilized Fourier transformation to map features to the frequency domain, allowing for a more global perspective; 2) We have analyzed the relationships between all frequency bands to combine global frequency features with local spatial features; 3) We have extended frequency features to multiple components to fully utilize the global understanding of the object.

The complete architecture of our FSEL model is shown in Fig. 2. Given an input image $I_c \in \mathbb{R}^{H \times W \times 3}$, we first use the basic encoder (*i.e.*, PVTv2 [43]/ ResNet [14]/ Res2Net [12]) to extract initial feature $\mathcal{P}=\{\mathcal{P}_i\}_{i=1}^{4}$ with the resolution of $\frac{W}{2^{i+1}} \times \frac{H}{2^{i+1}}$. Then the JDPM (Sect. 3.2) captures a higher-level semantic feature $\mathcal{P}_5$ to guide location by integrating multi-receptive field information from frequency-spatial domains. After that, the ETB (Sect. 3.3) models the cross-long-range relationships and performs entanglement learning on the frequency and spatial domains from initial features to generate discriminative feature $\mathcal{X}=\{\mathcal{X}_i\}_{i=1}^{4}$. To ensure the quality of predicted map $\mathcal{N}=\{\mathcal{N}_i\}_{i=1}^{4}$, we design a DRP (Sect. 3.4) to aggregate feature flows through auxiliary optimization in the frequency and spatial domains.

3.2 Joint Domain Perception Module

Multi-scale information is beneficial for contextual understanding in different regions. We observe that these methods [2,26,48] often generate multi-scale features through different convolutions with multiple receptive fields in the spatial domain. However, the receptive field of convolution operations in the spatial domain is limited, and in the process of data processing, tiny fluctuations may be overlooked, resulting in sub-optimized outcomes.

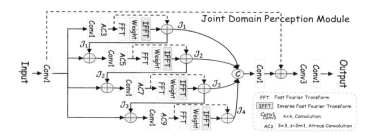

Fig. 3. Details of the joint domain perception module.

Therefore, we propose a Joint Domain Perception Module (JDPM) that reconstructs multi-receptive field information by introducing frequency transformation in multi-scale features. As depicted in Fig. 3, our JDRM uses the hierarchical structure to extract frequency-spatial information of different receptive fields. Technically, we use feature $\mathcal{P}_4$ as input and first reduce its channel numbers using 1×1 convolution (C_1), i.e., $\mathcal{P}_4^{128} = C_1 \mathcal{P}_4$. Then, we construct a set of 3×3 atrous convolutions ($\mathcal{AC}_z$) with filling rate z to capture local multi-scale spatial feature $\{\mathcal{J}_n^s\}_{n=1}^4$, i.e., $\mathcal{J}_n^s = C_1 \mathcal{AC}_z(\mathcal{P}_4^{128} + \mathcal{J}_{n-1})$, where $z = 2n+1$ and $n - 1 \geq 1$. Next, we transform local spatial features into the frequency domain using the Fast Fourier Transform ($fft(\cdot)$) and perform redundancy filtering. We then use the Inverse Fast Fourier Transform ($ifft(\cdot)$) and the modulus of complex features to obtain global frequency features $\{\mathcal{J}_n^f\}_{n=1}^4$, which is defined as:

$$\mathcal{J}_n^f = \Phi \| ifft(\sigma(fft(\mathcal{J}_n^s)) * fft(\mathcal{J}_n^s)) \|, n = 1, 2, 3, 4,$$

$$fft(u,v) = \sum_{x=0}^{W-1} \sum_{y=0}^{H-1} \mathcal{J}_n^s(x,y) e^{-2\pi i(\frac{ux}{W} + \frac{vy}{H})}, \quad (1)$$

$$ifft(x,y) = \frac{1}{WH} \sum_{u=0}^{W-1} \sum_{v=0}^{H-1} fft(u,v) e^{2\pi i(\frac{ux}{W} + \frac{vy}{H})},$$

where $\Phi\|\cdot\|$ and "$*$" denote the modulus operation and the element-wise multiplication. $\sigma(\cdot)$ presents a set of weight coefficients, which sequentially contains a convolution, a batch normalization, a ReLU, a convolution, and a sigmoid function. (u, v) and (x, y) denote frequency domain coordinates and spatial domain coordinates, i represents the imaginary part. After that, we aggregate global frequency features with local spatial features to generate intermediate multi-scale features $\{\mathcal{J}_n\}_{n=1}^4$, that is, $\mathcal{J}_n = \mathcal{J}_n^s + \mathcal{J}_n^f, n = 1, 2, 3, 4$.

Finally, we concatenate all multi-scale features and introduce residual connections to generate a coarse feature map $\mathcal{P}_5$ with 1-channel through a 3×3 and a 1×1 convolutions, which can be expressed as:

$$\mathcal{P}_5 = C_3 C_1(C_1 Cat(\mathcal{J}_1, \mathcal{J}_2, \mathcal{J}_3, \mathcal{J}_4) + \mathcal{P}_4^{128}), \tag{2}$$

where C_k presents $k \times k$ convolution. $Cat(\cdot, \cdot, \cdot, \cdot)$ and "$+$" denote concatenation and element-wise addition.

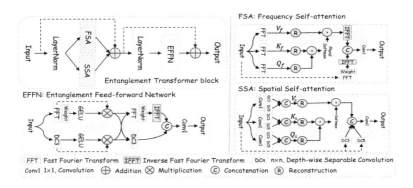

Fig. 4. Details of the entanglement transformer block.

3.3 Entanglement Transformer Block

Unlike previous methods [4,16,29], which only model long-range dependencies based on local features in the spatial domain, our ETB incorporates different relationships from the frequency and spatial domains. In addition, we propose entanglement learning for different domain features in the ETB, allowing for the integration of information such as color, texture, edge, spectral, amplitude, and energy. This approach is beneficial for learning discriminative representations by considering various types of information. As depicted in Fig. 4, our ETB consists of three key components: frequency self-attention (FSA), spatial self-attention (SSA), and entanglement feed-forward network (EFFN).

Frequency Self-attention. To better analyze frequency signals, we introduce a self-attention structure, which allows the model to obtain the relationships and interactions among different frequency bands, learn importance weights, and perform adaptive fusion. Technically, the feature $\mathcal{P}_\psi^{128}$ with 128-channel is taken as an input, which comes from the initial features at the same level and the optimized high-level features (as depicted in Fig. 2), and a layer normalization to generate feature $\hat{\mathcal{P}}(\hat{\mathcal{P}}=\mathcal{LN}(\mathcal{P}_\psi^{128}))$. After that, we use a Fast Fourier Transform to obtain $query\ Q_f=fft^Q(\hat{\mathcal{P}})$, $key\ K_f=fft^K(\hat{\mathcal{P}})$, $value\ V_f=fft^V(\hat{\mathcal{P}})$ in the frequency domain, where $fft^{(\cdot)}(\cdot)$ denotes the Fast Fourier Transform, and then we reshape $query\ (\widetilde{Q}_f \in \mathbb{R}^{C \times HW})$ and $key\ (\widetilde{K}_f \in \mathbb{R}^{HW \times C})$ projections that their dot-product generates transpose-attention map $(\Lambda_f, \Lambda_f = \widetilde{Q}_f \odot \widetilde{K}_f)$, where $\odot$ presents the matrix multiplication. Different from these models [15,16] that directly utilize the Softmax function to activate the attention map with the real type, the frequency attention map (Λ_f) is a complex type that cannot be activated directly. Therefore, we extract the real part $(\Lambda_f^{re}, \Lambda_f^{re}=(\Lambda_f+conj(\Lambda_f))/2)$ and imaginary part $(\Lambda_f^{im}, \Lambda_f^{im}=(\Lambda_f-conj(\Lambda_f))/2i)$, where $conj(\cdot)$ denotes conjugate complex number, and then activate and merge the real and imaginary parts to obtain the activated attention maps $(a\Lambda_f)$, as follows:

$$a\Lambda_f = \Theta(Sof(\Lambda_f^{re}), Sof(\Lambda_f^{im})), \tag{3}$$

where $\Theta(\cdot, \cdot)$ presents a combination function that combines the imaginary and real parts into a complex number. $Sof(\cdot)$ denotes a Softmax function. Subsequently, we use the attention map $a\Lambda_f$ to optimize the weights on the frequency feature V_f and then use the Inverse Fast Fourier Transform $(ifft(\cdot))$ to convert it to an original domain and employ the modulus operation to obtain the frequency attention feature. In addition, we introduce a frequency residual connection to increase frequency information $(\hat{\mathcal{P}}_f^r, \hat{\mathcal{P}}_f^r=\Phi||ifft(\sigma(fft(\hat{\mathcal{P}}))*fft(\hat{\mathcal{P}}))||)$, and finally fuse features to produce the frequency feature $\mathcal{X}_f^1$, which is formulated as:

$$\mathcal{X}_f^1 = C_1 Cat(\Phi||ifft(a\Lambda_f \odot \widetilde{V}_f)||, \hat{\mathcal{P}}_f^r), \tag{4}$$

where $Cat(\cdot, \cdot)$ and $\odot$ are concatenation and matrix multiplication. $\Phi||\cdot||$ presents the modulus operation. $\widetilde{V}_f$ is the reshaped V_f.

Spatial Self-attention. Considering the unfixed size of camouflaged objects, we embed abundant contextual information into spatial self-attention. As shown in the bottom right of Fig. 4, similar to the FSA, we take the feature $\hat{\mathcal{P}}$ as the input and encode the position information using a 1×1 convolution (C_1), and then we obtain the $query\ Q_s$, $key\ K_s$, and $value\ V_s$ required by the self-attention by utilizing two depth-wise separable convolution with 3×3 $(\mathcal{DC}_3)$ and 5×5 $(\mathcal{DC}_5)$. After that, we generate the attention map $(a\Lambda_s, a\Lambda_s = Sof(\widetilde{Q}_s \odot \widetilde{K}_s)\ )$ through the reconstructed $\widetilde{Q}_s$ and $\widetilde{K}_s$ and activate it using

Softmax function. Subsequently, the activated attention map $a\Lambda_s$ is used to correct the weights of V_s. Besides, to increase the spatial local information ($\hat{\mathcal{P}}_s^r$, $\hat{\mathcal{P}}_s^r = Cat(\mathcal{DC}_3\hat{\mathcal{P}}, \mathcal{DC}_5\hat{\mathcal{P}})$), we perform a residual connection to generate spatial feature $\mathcal{X}_s^r$, as shown in:

$$\mathcal{X}_s^r = C_1 Cat(a\Lambda_s \odot \widetilde{V}_s, \hat{\mathcal{P}}_s^r), \tag{5}$$

where C_1, $Cat(\cdot, \cdot)$, and $\odot$ are the same as in Eq. (4). $\widetilde{V}_s$ is the reshaped V_s.

Entanglement Feed-Forward Network. Frequency and spatial features usually contain different information. The frequency domain focuses on the global energy distribution and variation of signals, while spatial information acts on local pixel-level details and spatial structures, all of which are crucial for comprehending camouflaged objects. In our EFFN, these features are considered as two kinds of states that can perform entanglement learning to obtain more robust and powerful representations during the entanglement process.

Specifically, we first entangle the global frequency feature $\mathcal{X}_f^1$ and the local spatial feature $\mathcal{X}_s^1$ to adapt them to each other, followed by the residual connection to acquire the comprehensive feature $\mathcal{X}_c^1$, that is, $\mathcal{X}_c^1 = \mathcal{X}_s^1 + \mathcal{X}_f^1 + \mathcal{P}_\psi^{128}$, which performs the layer normalization to improve the stability, and then the normalized feature $\hat{\mathcal{X}}_c^1$ ($\hat{\mathcal{X}}_c^1 = \mathcal{LN}(\mathcal{X}_c^1)$) is subjected to non-linearity entanglement learning in the EFFN. Technically, the EFFN consists of two phases, the first stage projects the feature $\hat{\mathcal{X}}_c^1$ to the frequency and spatial domains, and utilizes the GELU function for nonlinear activation and a gate mechanism to obtain global frequency feature $\hat{\mathcal{X}}_f^2$ and local spatial feature $\hat{\mathcal{X}}_s^2$, which can be written as follows:

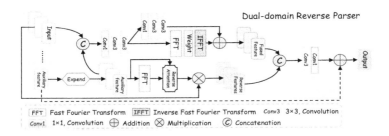

Fig. 5. Details of the dual-domain reverse parser.

$$\begin{aligned}\hat{\mathcal{X}}_f^2 &= GE(\varPhi||\sigma(fft(\hat{\mathcal{X}}_c^1))*fft(\hat{\mathcal{X}}_c^1)||)*\varPhi||\sigma(fft(\hat{\mathcal{X}}_c^1))*fft(\hat{\mathcal{X}}_c^1)||, \\ \hat{\mathcal{X}}_s^2 &= GE(\mathcal{DC}_3\hat{\mathcal{X}}_c^1)*\mathcal{DC}_3\hat{\mathcal{X}}_c^1,\end{aligned} \tag{6}$$

where $GE(\cdot)$ denotes the GELU function. Subsequently, in the second stage, the frequency and spatial features from the first stage are again entangled by interacting with each other by transferring information from different domains, and the entangled frequency-spatial features are optimized independently. They are then aggregated and reduced channels to generate comprehensive feature $\hat{\mathcal{X}}_c^3$, which can be formulated as follows:

$$\hat{\mathcal{X}}_c^3 = \mathcal{C}_1 Cat(\hat{\mathcal{X}}_f^3, \hat{\mathcal{X}}_s^3) + \mathcal{X}_c^1,$$
$$\hat{\mathcal{X}}_f^3 = \Phi\|ifft(\sigma(fft(Cat(\hat{\mathcal{X}}_f^2, \hat{\mathcal{X}}_s^2)))*fft(Cat(\hat{\mathcal{X}}_f^2, \hat{\mathcal{X}}_s^2)))\|, \quad (7)$$
$$\hat{\mathcal{X}}_s^3 = \mathcal{DC}_3 Cat(\hat{\mathcal{X}}_f^2, \hat{\mathcal{X}}_s^2),$$

where $\mathcal{C}_1$, $Cat(\cdot,\cdot)$ and $\Phi\|\cdot\|$ are the same as in Eq. (4). $fft(\cdot)$ and $ifft(\cdot)$ present the Fast Fourier Transform and the Inverse Fast Fourier Transform. $\mathcal{DC}_3$ denotes the depth-wise separable convolution with 3×3 kernel. Finally, we introduce residual connections to obtain the final feature $\mathcal{X}$ with 128-channel in the ETB, i.e., $\mathcal{X} = \mathcal{C}_1 Cat(\hat{\mathcal{X}}_c^3, \mathcal{P}_\psi^{128}) + \mathcal{P}_\psi^{128}$. Through multiple aggregation interactions, global frequency and local spatial features interact and depend on each other, leading to the entanglement of features from different states, forming rich and comprehensive representations.

3.4 Dual-Domain Reverse Parser

Different from these methods [6,16,58] that integrate multi-level features based on the spatial domain, we propose the dual-domain reverse parser (DRP), which optimizes and aggregates diverse information from multi-level feature $\mathcal{X}$ in both frequency and spatial domains. As depicted in Fig. 2, we first take the feature $\mathcal{X}_4$ from the ETB as the optimization objective and use the higher-level semantic feature $\mathcal{P}_5$ as the auxiliary objective.

The DRP consists of two branches (as shown in Fig. 5), in the first branch, we first expand the channel of auxiliary feature $\mathcal{P}_5$ to match the dimension of the optimization objective and aggregate these feature to obtain feature $\mathcal{I}_4$, i.e., $\mathcal{I}_4 = \mathcal{C}on(Cat(Ex(\mathcal{P}_5), \mathcal{X}_4))$, where $Ex(\cdot)$ denotes to expand the channel to 128, $\mathcal{C}on(\cdot)$ presents a 1×1 convolution and two 3×3 convolutions. And then feature $\mathcal{I}_4$ is separated into the spatial and frequency domains. We perform the Fast Fourier Transform ($fft(\cdot)$) and Inverse Fast Fourier Transform ($ifft(\cdot)$) in the frequency domain and adopt a series of convolution operations ($\mathcal{C}on(\cdot)$) to optimize features in the spatial domain. Subsequently, they are aggregated to obtain the fused feature $\mathcal{N}_4^1$, that is,

$$\mathcal{N}_4^1 = \Phi\|ifft(\sigma(fft(\mathcal{I}_4))*fft(\mathcal{I}_4))\| + \mathcal{C}on(\mathcal{I}_4), \quad (8)$$

where $\Phi\|\cdot\|$ denotes the modulus operation. "+" presents element-wise addition. In the second branch, we produce the hybrid reverse attention map ($\mathcal{A}_r$, $\mathcal{A}_r = (1 - Sig(\mathcal{P}_5)) + (1 - Sig(\Phi\|fft(\mathcal{P}_5)\|)))$) using the auxiliary feature, where $Sig(\cdot)$ denotes the Sigmoid function. Unlike other methods [6,13], our the reverse attention map ($\mathcal{A}_r$) contains abundant the frequency-spatial information to efficiently obtain the reverse feature $\mathcal{N}_4^2$, i.e., $\mathcal{N}_4^2 = \mathcal{A}_r * \mathcal{X}_4$. Next, we integrate the features $\mathcal{N}_4^1$ and $\mathcal{N}_4^2$ to generate final feature $\mathcal{N}_4$, that is, $\mathcal{N}_4 = C_3Cat(\mathcal{N}_4^1, \mathcal{N}_4^2) + \mathcal{P}_5$. Subsequently, $\mathcal{N}_{i+1}$ will continue to optimize features $\mathcal{X}_i(i=1,2,3)$ as an auxiliary objective in the proposed DRP. Note that there must be at least one auxiliary feature used for optimizing feature $\mathcal{X}_i$ to generate feature $\mathcal{N}_i$, and auxiliary features are input in the dense connection manner.

3.5 Loss Function

In the proposed FSEL method, we supervise multi-level feature $\mathcal{N}$ to produce an accurately predicted map. Specifically, we adopt the weighted binary cross-entropy (BCE) and the weighted intersection over union (IoU) [36] as the overall loss function to optimize the model based on ground truth (G). The loss function can be defined as:

$$\mathcal{L}_{all} = \sum_{i=1}^{5} \frac{1}{2^{i-1}} (\mathcal{L}_{bce}^w(\mathcal{N}_i, G) + \mathcal{L}_{iou}^w(\mathcal{N}_i, G)), \tag{9}$$

where $\mathcal{L}_{bce}^w$ and $\mathcal{L}_{iou}^w$ denote the weighted BCE and IoU functions. $\mathcal{N}_5$ is the feature $\mathcal{P}_5$ from the JDPM.

Table 1. Quantitative results on three COD datasets. The best result is shown in **blod**. "Ours-R50", "Ours-R2N", and "Ours-Pvt" present ResNet50 [14]/Res2Net [12]/PVTv2 [43] as backbone. "Ours-R50[†]" denotes using the same input strategy as ZoomNet [34].

Method	Year, Pub.	CAMO (250 images)						COD10K (2026 images)						NC4K (4121 images)					
		$\mathcal{M}\downarrow$	$F_\varphi^m\uparrow$	$F_\varphi^a\uparrow$	$F_\varphi^w\uparrow$	$S_m\uparrow$	$E_m\uparrow$	$\mathcal{M}\downarrow$	$F_\varphi^m\uparrow$	$F_\varphi^a\uparrow$	$F_\varphi^w\uparrow$	$S_m\uparrow$	$E_m\uparrow$	$\mathcal{M}\downarrow$	$F_\varphi^m\uparrow$	$F_\varphi^a\uparrow$	$F_\varphi^w\uparrow$	$S_m\uparrow$	$E_m\uparrow$
SINet [7]	2020, CVPR	0.100	0.762	0.709	0.606	0.751	0.835	0.051	0.708	0.593	0.551	0.770	0.797	0.058	0.805	0.768	0.723	0.807	0.883
UGTR [47]	2021, ICCV	0.086	0.800	0.748	0.684	0.784	0.858	0.036	0.772	0.671	0.666	0.815	0.850	0.052	0.833	0.778	0.747	0.839	0.888
JSOCOD [21]	2021, CVPR	0.073	0.812	0.779	0.728	0.800	0.872	0.035	0.762	0.705	0.684	0.807	0.882	0.047	0.838	0.803	0.771	0.841	0.906
MGL-S [52]	2021, CVPR	0.089	0.791	0.733	0.664	0.772	0.850	0.037	0.765	0.667	0.655	0.808	0.851	0.055	0.826	0.771	0.731	0.828	0.885
LSR [30]	2021, CVPR	0.080	0.791	0.756	0.696	0.787	0.859	0.037	0.756	0.699	0.673	0.802	0.883	0.048	0.836	0.802	0.766	0.839	0.904
PFNet [31]	2021, CVPR	0.085	0.795	0.751	0.695	0.782	0.855	0.040	0.748	0.676	0.660	0.798	0.868	0.053	0.821	0.779	0.745	0.828	0.894
SegMaR$_1$ [18]	2022, CVPR	0.072	0.821	0.772	0.728	0.808	0.870	0.035	0.765	0.699	0.682	0.811	0.881	-	-	-	-	-	-
PreyNet [54]	2022, MM	0.077	0.803	0.764	0.708	0.789	0.856	0.034	0.775	0.731	0.697	0.810	0.894	-	-	-	-	-	-
FEDER [13]	2023, CVPR	0.071	0.824	0.786	0.738	0.802	0.877	0.032	0.788	0.740	0.716	0.820	0.892	0.044	0.852	**0.822**	0.789	0.846	0.913
Ours-R50	-	**0.067**	**0.833**	**0.799**	**0.758**	**0.821**	**0.893**	0.031	**0.802**	**0.743**	**0.728**	**0.830**	**0.898**	**0.042**	**0.855**	0.818	**0.792**	**0.854**	**0.914**
ZoomNet [34]	2022, CVPR	**0.066**	0.832	0.792	0.752	0.820	0.883	**0.029**	0.810	0.741	0.729	0.835	0.893	0.043	0.851	0.815	0.784	0.852	0.907
Ours-R50[†]	-	0.068	**0.837**	**0.799**	**0.765**	**0.826**	**0.890**	**0.029**	**0.814**	**0.754**	**0.743**	**0.839**	**0.903**	**0.040**	**0.863**	**0.828**	**0.802**	**0.861**	**0.917**
C2FNet [38]	2021, IJCAI	0.080	0.803	0.764	0.719	0.796	0.865	0.036	0.764	0.703	0.686	0.811	0.886	0.049	0.832	0.788	0.762	0.838	0.901
FAPNet [57]	2022, TIP	0.076	0.823	0.776	0.734	0.815	0.877	0.036	0.781	0.707	0.694	0.820	0.875	0.047	0.846	0.804	0.775	0.850	0.903
SINet_v2 [6]	2022, TPAMI	0.071	0.800	0.770	0.748	0.820	0.884	0.037	0.770	0.682	0.680	0.813	0.864	0.048	0.842	0.792	0.770	0.847	0.901
BSANet [58]	2022, AAAI	0.079	0.804	0.768	0.717	0.794	0.866	0.034	0.776	0.724	0.699	0.815	0.894	0.048	0.839	0.805	0.771	0.841	0.906
BGNet [39]	2022, IJCAI	0.073	0.825	0.786	0.749	0.811	0.878	0.033	0.795	**0.739**	0.722	0.828	**0.902**	0.044	0.851	0.813	0.788	0.850	0.911
Ours-R2N	-	**0.065**	**0.844**	**0.803**	**0.771**	**0.831**	**0.895**	**0.030**	**0.803**	**0.739**	**0.731**	**0.837**	0.899	**0.042**	**0.858**	**0.821**	**0.798**	**0.860**	**0.915**
VST [25]	2021, ICCV	0.081	0.812	0.753	0.713	0.808	0.853	0.037	0.779	0.721	0.698	0.817	0.882	0.048	0.840	0.801	0.768	0.844	0.899
EVP [27]	2023, CVPR	0.067	0.836	0.800	0.762	0.831	0.896	0.032	0.802	0.708	0.726	0.835	0.877	-	-	-	-	-	-
FPNet [4]	2023, MM	0.056	0.863	0.838	0.802	0.851	0.912	0.029	0.817	0.765	0.755	0.847	0.909	-	-	-	-	-	-
HiNet [15]	2023, AAAI	0.055	0.857	0.833	0.809	0.849	0.910	0.023	0.850	**0.818**	**0.806**	0.868	**0.936**	0.037	0.879	0.854	0.834	0.874	0.928
FSPNet [16]	2023, CVPR	0.050	0.869	0.829	0.799	0.855	0.919	0.026	0.816	0.736	0.735	0.847	0.900	0.035	0.878	0.826	0.816	0.878	0.923
SAM [19]	2023, ICCV	-	-	-	-	-	-	0.050	0.844	0.758	0.701	0.778	0.800	0.078	0.852	0.754	0.696	0.765	0.778
Ours-Pvt	-	**0.040**	**0.891**	**0.864**	**0.851**	**0.885**	**0.942**	**0.021**	**0.853**	0.796	0.800	**0.873**	0.928	**0.030**	**0.895**	**0.864**	**0.853**	**0.892**	**0.941**

Table 2. Efficiency analysis of our FSEL and multiple COD methods.

	SINet [7]	PFNet [31]	MGL [52]	UGTR [47]	JSOCOD [21]	C2FNet [38]	ZoomNet [34]	SegMaR [18]	FSPNet [16]	HitNet [15]	FEDER [13]	Ours-R50	Ours-R2N	Ours-Pvt
Parameters (M)	48.95	46.50	63.60	48.87	217.98	28.41	32.38	55.62	273.79	**25.73**	37.37	29.15	29.31	67.13
FLOPs (G)	38.75	53.22	553.94	1000.01	112.34	26.17	203.50	33.65	283.31	56.55	**23.98**	35.64	37.07	54.73

4 Experiment

4.1 Experimental Setups

Datasets. We evaluate our FSEL model on three benchmark datasets: CAMO [20], COD10K [7], and NC4K [30]. CAMO [20] is an early dataset that contains 1,250 camouflaged images with 1,000 training images and 250 testing images. COD10K [7] is a currently large dataset of camouflaged objects, consisting of 3,040 training images and 2,026 testing images. NC4K [30] is the largest COD dataset for testing, containing 4,121 images of camouflaged objects. We use 4,040 images from CAMO [20] and COD10K [7] as training samples to train the FSEL.

Implementation Details. The proposed FSEL model is implemented in the PyTorch framework on four NVIDIA GTX 4090 GPUs with 24 GB. We utilize the pre-trained PVTv2 [43]/ResNet50 [14]/Res2Net [12] as the encoder to extract initial features. Following [6,13], we also employ data augmentation techniques such as random flipping and random clipping to enhance training data. We use the Adam optimizer with an initial learning rate of 1e-4 and decay the rates by 10 every 60 epochs. All input images are resized to 416×416, and the batch size is set to 40 for 180 epochs of training progressing.

Evaluation Metrics. We use six well-known evaluation metrics, including Mean Absolute Error ($\mathcal{M}$), Maximum F-measure (F_φ^m), Average F-measure (F_φ^a), Weighted F-measure (F_φ^w), S-measure (S_m), and E-measure (E_m).

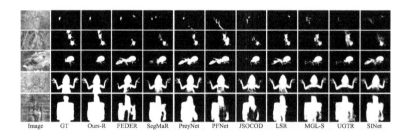

Fig. 6. Qualitative comparisons of the proposed FSEL and nine COD methods.

4.2 Comparisons with the SOTAs

We conduct a comparison of our FSEL with twenty-one COD methods, including SINet [7], C2FNet [38], UGTR [47], JSOCOD [21], MGL-S [52], LSR [30], PFNet [31], VST [25], FAPNet [57], SINet$_{v2}$ [6], BSANet [58], SegMaR [18], ZoomNet [34], BGNet [39], PreyNet [54], FEDER [13], EVP [27], FPNet [4], HitNet [15], FSPNet [16], and SAM [19]. Note that the predicted maps from all methods are provided by the authors or obtained from open-source codes.

Quantitative Evaluation. Table 1 summarizes the quantitative result of our FSEL and other 21 SOTA models. From Table 1, we can observe that the FSEL model achieves excellent performance across different backbone networks. Particularly, compared to the recently proposed FEDER [13] method, our FSEL with ResNet50 [14] backbone overall surpasses 5.97%, 3.23%, and 4.76% on three public datasets under the $\mathcal{M}$ metric. Besides, in the Res2Net [12] backbone, our FSEL achieves average performance gains of 9.23%, 2.30%, 2.16%, 2.94%, 1.34%, and 1.24% over the second-best method in terms of six public evaluation metrics on CAMO [20] dataset. Moreover, compared to the frequency-based FPNet [4] and EVP [27] methods, FSEL method with PVTv2 [43] backbone achieves average performance gains of 38.10%, 4.41%, 4.05%, 5.96%, 3.07%, and 2.09% over FPNet [4] and 52.38%, 6.36%, 12.43%, 10.19%, 4.55%, and 5.82% over EVP [27] in terms of $\mathcal{M}$, F_φ^m, F_φ^a, F_φ^w, S_m, and E_m on the COD10K [7] dataset. Furthermore, FSEL achieves excellent performance when adopting the same input strategy with ZoomNet [34]. The superiority in performance benefits from the joint optimization of the ETB, JDPM, and DRP for input features in the frequency and spatial domains. In addition, we provide the parameters and FLOPs in Table 2. It can be seen that the proposed FSEL method parameters and FLOPs are at a medium to high level, however, our performance far exceeds that of methods with similar parameters and FLOPs.

Qualitative Evaluation. Figure 6 gives the visual comparisons between our FSEL and several COD method in different scenarios. As depicted in Fig. 6, the proposed FSEL method exhibits accurate and complete segmentation for camouflaged objects with different sizes compared to current COD methods (*i.e.*, HitNet [15], FSPNet [16], and FPNet [4]). These visual results demonstrate the superiority of the FSEL method for detecting camouflaged objects through the frequency-spatial domain optimization strategy.

4.3 Ablation Study

Effectiveness of Proposed Each Component. We provide the quantitative results of different components in the proposed FSEL model, shown in Table 3. Specifically, we first adopt "ResNet50 [14] - FPN [24]" as "Baseline" (Table 3(a)) to detect camouflaged objects. And then we independently validate the effectiveness of "ETB" (Table 3(b)), "DRP" (Table 3(c)) and "JDPM" (Table 3(d)),

Table 3. Ablation analysis of our FSEL structure.

No.	Baseline	ETB	DRP	JDPM	$\mathcal{M}\downarrow$	$F_\varphi^m\uparrow$	$F_\varphi^a\uparrow$	$F_\varphi^w\uparrow$	$S_m\uparrow$	$E_m\uparrow$	$\mathcal{M}\downarrow$	$F_\varphi^m\uparrow$	$F_\varphi^a\uparrow$	$F_\varphi^w\uparrow$	$S_m\uparrow$	$E_m\uparrow$
					CAMO(250 images)						COD10K(2026 images)					
(a)	✓				0.093	0.784	0.712	0.663	0.767	0.847	0.046	0.749	0.617	0.610	0.778	0.818
(b)	✓	✓			0.076	0.811	0.763	0.723	0.801	0.873	0.034	0.789	0.712	0.702	0.821	0.886
(c)	✓		✓		0.074	0.820	0.778	0.735	0.810	0.876	0.034	0.794	0.722	0.713	0.826	0.887
(d)	✓			✓	0.081	0.797	0.743	0.697	0.787	0.863	0.039	0.769	0.676	0.668	0.804	0.863
(e)	✓	✓	✓		0.074	0.823	0.782	0.731	0.807	0.875	0.032	0.798	0.723	0.714	0.828	0.888
(f)	✓	✓		✓	0.071	0.807	0.767	0.728	0.802	0.878	0.033	0.781	0.719	0.702	0.815	0.895
(g)	✓		✓	✓	0.071	0.830	0.789	0.742	0.810	0.879	**0.031**	0.796	0.730	0.718	0.827	0.893
(h)	✓	✓	✓	✓	**0.067**	**0.833**	**0.799**	**0.758**	**0.821**	**0.893**	**0.031**	**0.802**	**0.743**	**0.728**	**0.830**	**0.898**

and it can be seen that the performance of the predicted map increases significantly when the proposed component is embedded in the "Baseline" (Table 3(a)). Additionally, we validate the compatibility among all modules. From Table 3(e), Table 3(f), and Table 3(g), it can be observed that the three components are compatible with each other. Subsequently, all components are integrated, and the performance of the model is improved once again, as shown in Table 3(h). Additionally, in Fig. 7, we show the visual results obtained by progressively adding the proposed components (*i.e.*, ETB, JDPM, and DRP), generating that the predicted map gradually approaches the ground truth (GT). The above results demonstrate the effectiveness of our proposed modules in detecting camouflaged objects.

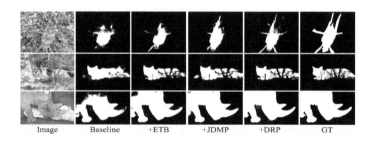

Fig. 7. Visual results of the effectiveness of our modules.

Effectiveness of Frequency-Spatial Information Within the ETB. Do we really need frequency information? To answer this question, we perform a series of experiments in the internal part of the ETB. Specifically, the ETB is first divided into two parts, with "ETB-S" (Table 4(a)) containing only spatial information, and "ETB-F" (Table 4(b)) presenting that it includes only frequency information. Based on Table 4, the performance of the separate frequency and spatial domains exhibits certain differences compared to the complete ETB (Table 4(g)). Besides, we investigate the entanglement learning of frequency-spatial information in the proposed ETB. In Table 4(c)–(f), it can be seen that the frequency and spatial features interact fusion to achieve entanglement between two states, enhancing the model's reasoning ability of camouflaged objects.

Table 4. Ablation analysis within the ETB structure.

No.	ETB				CAMO(250 images)						COD10K(2026 images)					
	FSA	SFA	FFFN	SFFA	$\mathcal{M}\downarrow$	$F^m_\varphi\uparrow$	$F^a_\varphi\uparrow$	$F^w_\varphi\uparrow$	$S_m\uparrow$	$E_m\uparrow$	$\mathcal{M}\downarrow$	$F^m_\varphi\uparrow$	$F^a_\varphi\uparrow$	$F^w_\varphi\uparrow$	$S_m\uparrow$	$E_m\uparrow$
(a)	✓		✓		0.079	0.798	0.752	0.706	**0.802**	0.864	0.037	0.770	0.697	0.679	0.816	0.874
(b)	✓		✓		**0.075**	0.810	**0.770**	**0.725**	0.798	**0.880**	**0.034**	0.778	**0.713**	0.695	0.812	**0.890**
(c)	✓		✓	✓	**0.075**	0.807	0.759	0.720	0.795	0.876	0.035	0.778	0.692	0.687	0.811	0.873
(d)		✓	✓	✓	0.078	**0.811**	0.764	0.723	0.800	0.875	**0.034**	0.777	0.703	0.692	0.814	0.882
(e)	✓	✓	✓		0.076	0.808	0.764	0.712	0.795	0.872	**0.034**	0.782	0.702	0.692	0.819	0.878
(f)	✓	✓		✓	0.077	0.801	0.757	0.712	0.794	0.873	0.036	0.778	0.697	0.686	0.814	0.875
(g)	✓	✓	✓	✓	0.076	**0.811**	0.763	0.723	0.801	0.873	**0.034**	**0.789**	0.712	**0.702**	**0.821**	0.886

4.4 Expanded Application

To demonstrate the generalization ability of our FSEL model, we extend the FSEL model to salient object detection and polyp segmentation tasks. As shown in Fig. 8, the proposed FSEL method achieves highly accurate segmentation for both salient objects and polyps, benefiting from the complementary utilization of frequency domain and spatial information. More details and data are presented in the **supplementary materials**.

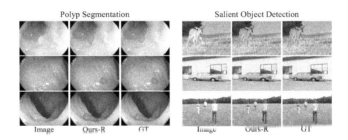

Fig. 8. Visual results of the expanded application.

5 Conclusion

In this paper, we introduce a new approach for detecting camouflaged objects called Frequency-Spatial Entanglement Learning (FSEL). The key to FSEL is to extract important information from both the frequency and spatial domains. To achieve this, we have developed a Joint Domain Perception Module that combines multi-scale information from frequency-spatial features to accurately localize regions. Additionally, we have created an Entanglement Transformer Block that can be easily integrated into existing methods to improve their performance by modeling long-range dependencies in the hybrid domain. Furthermore, we have designed a Dual-Domain Reverse Parser that interacts with diverse information in multi-layer features to achieve more precise segmentation. Our extensive comparison experiments demonstrate that FSEL outperforms 21 state-of-the-art COD methods on three popular benchmark datasets.

Acknowledgments. This work was supported in part by the National Science Fund of China (No. 62276135, 62361166670, 62372238, and 62302006).

References

1. Carion, N., Massa, F., Synnaeve, G., Usunier, N., Kirillov, A., Zagoruyko, S.: End-to-End object detection with transformers. In: Vedaldi, A., Bischof, H., Brox, T., Frahm, J.-M. (eds.) ECCV 2020. LNCS, vol. 12346, pp. 213–229. Springer, Cham (2020). https://doi.org/10.1007/978-3-030-58452-8_13
2. Chen, L.C., Papandreou, G., Kokkinos, I., Murphy, K., Yuille, A.L.: Deeplab: semantic image segmentation with deep convolutional nets, atrous convolution, and fully connected crfs. TPAMI **40**(4), 834–848 (2017)
3. Chi, L., Jiang, B., Mu, Y.: Fast fourier convolution. In: Larochelle, H., Ranzato, M., Hadsell, R., Balcan, M., Lin, H. (eds.) NeurIPS, vol. 33, pp. 4479–4488 (2020)
4. Cong, R., Sun, M., Zhang, S., Zhou, X., Zhang, W., Zhao, Y.: Frequency perception network for camouflaged object detection. In: ACM MM (2023)
5. Ding, J., Xue, N., Xia, G.S., Schiele, B., Dai, D.: Hgformer: hierarchical grouping transformer for domain generalized semantic segmentation. In: CVPR, pp. 15413–15423 (2023)
6. Fan, D.P., Ji, G.P., Cheng, M.M., Shao, L.: Concealed object detection. TPAMI **44**(10), 6024–6042 (2022)
7. Fan, D.P., Ji, G.P., Sun, G., Cheng, M.M., Shen, J., Shao, L.: Camouflaged object detection. In: CVPR, pp. 2777–2787 (2020)
8. Fan, D.-P., et al.: PraNet: parallel reverse attention network for polyp segmentation. In: Martel, A.L., et al. (eds.) MICCAI 2020. LNCS, vol. 12266, pp. 263–273. Springer, Cham (2020). https://doi.org/10.1007/978-3-030-59725-2_26
9. Fan, D.P., et al.: Inf-net: automatic covid-19 lung infection segmentation from ct images. TMI **39**(8), 2626–2637 (2020)
10. Fang, F., Li, L., Gu, Y., Zhu, H., Lim, J.H.: A novel hybrid approach for crack detection. PR **107**, 107474 (2020)
11. Galun, S., Basri, B.: Texture segmentation by multiscale aggregation of filter responses and shape elements. In: ICCV, pp. 716–723 (2003)
12. Gao, S.H., Cheng, M.M., Zhao, K., Zhang, X.Y., Yang, M.H., Torr, P.: Res2net: a new multi-scale backbone architecture. TPAMI **43**(2), 652–662 (2019)
13. He, C., et al.: Camouflaged object detection with feature decomposition and edge reconstruction. In: CVPR, pp. 22046–22055 (2023)
14. He, K., Zhang, X., Ren, S., Sun, J.: Deep residual learning for image recognition. In: CVPR, pp. 770–778 (2016)

15. Hu, X., et al.: High-resolution iterative feedback network for camouflaged object detection. In: AAAI, vol. 37, pp. 881–889 (2023)
16. Huang, Z., et al.: Feature shrinkage pyramid for camouflaged object detection with transformers. In: CVPR, pp. 5557–5566 (2023)
17. Jain, J., Li, J., Chiu, M.T., Hassani, A., Orlov, N., Shi, H.: Oneformer: one transformer to rule universal image segmentation. In: CVPR, pp. 2989–2998 (2023)
18. Jia, Q., Yao, S., Liu, Y., Fan, X., Liu, R., Luo, Z.: Segment, magnify and reiterate: detecting camouflaged objects the hard way. In: CVPR, pp. 4713–4722 (2022)
19. Kirillov, A., et al.: Segment anything. In: ICCV, pp. 4015–4026 (2023)
20. Le, T.N., Nguyen, T.V., Nie, Z., Tran, M.T., Sugimoto, A.: Anabranch network for camouflaged object segmentation. Comput. Vis. Image Underst. **184**, 45–56 (2019)
21. Li, A., Zhang, J., Lv, Y., Liu, B., Zhang, T., Dai, Y.: Uncertainty-aware joint salient object and camouflaged object detection. In: CVPR, pp. 10071–10081 (2021)
22. Lin, J., Tan, X., Xu, K., Ma, L., Lau, R.W.: Frequency-aware camouflaged object detection. ACM TMCCA **19**(2), 1–16 (2023)
23. Lin, S., et al.: Deep frequency filtering for domain generalization. In: CVPR, pp. 11797–11807 (2023)
24. Lin, T.Y., Dollár, P., Girshick, R., He, K., Hariharan, B., Belongie, S.: Feature pyramid networks for object detection. In: ICCV, pp. 2117–2125 (2017)
25. Liu, N., Zhang, N., Wan, K., Shao, L., Han, J.: Visual saliency transformer. In: ICCV, pp. 4722–4732 (2021)
26. Liu, S., Huang, D., et al.: Receptive field block net for accurate and fast object detection. In: ECCV, pp. 385–400 (2018)
27. Liu, W., Shen, X., Pun, C.M., Cun, X.: Explicit visual prompting for low-level structure segmentations. In: CVPR, pp. 19434–19445 (2023)
28. Liu, X., Peng, H., Zheng, N., Yang, Y., Hu, H., Yuan, Y.: Efficientvit: memory efficient vision transformer with cascaded group attention. In: CVPR, pp. 14420–14430 (2023)
29. Liu, Z., et al.: Swin transformer: hierarchical vision transformer using shifted windows. In: ICCV, pp. 10012–10022 (2021)
30. Lv, Y., et al.: Simultaneously localize, segment and rank the camouflaged objects. In: CVPR, pp. 11591–11601 (2021)
31. Mei, H., Ji, G.P., Wei, Z., Yang, X., Wei, X., Fan, D.P.: Camouflaged object segmentation with distraction mining. In: CVPR, pp. 8772–8781 (2021)
32. Mondal, A., Ghosh, S., Ghosh, A.: Partially camouflaged object tracking using modified probabilistic neural network and fuzzy energy based active contour. IJCV **122**, 116–148 (2017)
33. Nafus, M.G., Germano, J.M., Perry, J.A., Todd, B.D., Walsh, A., Swaisgood, R.R.: Hiding in plain sight: a study on camouflage and habitat selection in a slow-moving desert herbivore. BE **26**(5), 1389–1394 (2015)
34. Pang, Y., Zhao, X., Xiang, T.Z., Zhang, L., Lu, H.: Zoom in and out: a mixed-scale triplet network for camouflaged object detection. In: CVPR, pp. 2160–2170 (2022)
35. Qin, Z., Zhang, P., Wu, F., Li, X.: Fcanet: frequency channel attention networks. In: ICCV, pp. 783–792 (2021)
36. Rahman, M.A., Wang, Y.: Optimizing intersection-over-union in deep neural networks for image segmentation, pp. 234–244 (2016)
37. Song, Z., Kang, X., Wei, X., Liu, H., Dian, R., Li, S.: Fsnet: focus scanning network for camouflaged object detection. TIP **32**, 2267–2278 (2023)
38. Sun, Y., Chen, G., Zhou, T., Zhang, Y., Liu, N.: Context-aware cross-level fusion network for camouflaged object detection. In: IJCAI, pp. 1025–1031 (2021)

39. Sun, Y., Wang, S., Chen, C., Xiang, T.Z.: Boundary-guided camouflaged object detection. In: IJCAI (2022)
40. Vaswani, A., et al.: Attention is all you need. NeurIPS **30** (2017)
41. Wang, C., Jiang, J., Zhong, Z., Liu, X.: Spatial-frequency mutual learning for face super-resolution. In: CVPR, pp. 22356–22366 (2023)
42. Wang, S., Veldhuis, R., Brune, C., Strisciuglio, N.: What do neural networks learn in image classification? a frequency shortcut perspective. In: ICCV, pp. 1433–1442 (2023)
43. Wang, W., et al.: Pvt v2: Improved baselines with pyramid vision transformer. CVM **8**(3), 415–424 (2022)
44. Wu, H., et al.: Cvt: introducing convolutions to vision transformers. In: CVPR, pp. 22–31 (2021)
45. Wu, Y., Liang, L., Zhao, Y., Zhang, K.: Object-aware calibrated depth-guided transformer for rgb-d co-salient object detection. In: ICME, pp. 1121–1126 (2023)
46. Wu, Y., Song, H., Liu, B., Zhang, K., Liu, D.: Co-salient object detection with uncertainty-aware group exchange-masking. In: CVPR, pp. 19639–19648 (2023)
47. Yang, F., et al.: Uncertainty-guided transformer reasoning for camouflaged object detection. In: ICCV, pp. 4146–4155 (2021)
48. Yang, M., Yu, K., Zhang, C., Li, Z., Yang, K.: Denseaspp for semantic segmentation in street scenes. In: CVPR, pp. 3684–3692 (2018)
49. Yuan, L., et al.: Tokens-to-token vit: training vision transformers from scratch on imagenet. In: ICCV, pp. 558–567 (2021)
50. Yun, G., Yoo, J., Kim, K., Lee, J., Kim, D.H.: Spanet: frequency-balancing token mixer using spectral pooling aggregation modulation. In: ICCV, pp. 6113–6124 (2023)
51. Zamir, S.W., Arora, A., Khan, S., Hayat, M., Khan, F.S., Yang, M.H.: Restormer: efficient transformer for high-resolution image restoration. In: CVPR, pp. 5728–5739 (2022)
52. Zhai, Q., Li, X., Yang, F., Chen, C., Cheng, H., Fan, D.P.: Mutual graph learning for camouflaged object detection. In: CVPR, pp. 12997–13007 (2021)
53. Zhang, H., et al.: Mp-former: mask-piloted transformer for image segmentation. In: CVPR, pp. 18074–18083 (2023)
54. Zhang, M., Xu, S., Piao, Y., Shi, D., Lin, S., Lu, H.: Preynet: preying on camouflaged objects. In: ACM MM, pp. 5323–5332 (2022)
55. Zhang, X., Zhu, C., Wang, S., Liu, Y., Ye, M.: A bayesian approach to camouflaged moving object detection. TCSVT **27**(9), 2001–2013 (2016)
56. Zhong, Y., Li, B., Tang, L., Kuang, S., Wu, S., Ding, S.: Detecting camouflaged object in frequency domain. In: CVPR, pp. 4504–4513 (2022)
57. Zhou, T., Zhou, Y., Gong, C., Yang, J., Zhang, Y.: Feature aggregation and propagation network for camouflaged object detection. TIP **31**, 7036–7047 (2022)
58. Zhu, H., et al.: I can find you! boundary-guided separated attention network for camouflaged object detection. In: AAAI, vol. 36, pp. 3608–3616 (2022)
59. Zhu, J., Zhang, X., Zhang, S., Liu, J.: Inferring camouflaged objects by texture-aware interactive guidance network. In: AAAI, vol. 35, pp. 3599–3607 (2021)
60. Zhu, L., Wang, X., Ke, Z., Zhang, W., Lau, R.W.: Biformer: vision transformer with bi-level routing attention. In: CVPR, pp. 10323–10333 (2023)

VisionTrap: Vision-Augmented Trajectory Prediction Guided by Textual Descriptions

Seokha Moon[1], Hyun Woo[1], Hongbeen Park[1], Haeji Jung[1], Reza Mahjourian[2], Hyung-gun Chi[3], Hyerin Lim[4], Sangpil Kim[1], and Jinkyu Kim[1(✉)]

[1] Korea University, Seoul 02841, Republic of Korea
jinkyukim@korea.ac.kr
[2] The University of Texas at Austin, Austin, TX 78712, USA
[3] Perdue University, West Lafayette 95008, USA
[4] Hyundai Motor Company, Seongnam 13529, Republic of Korea

Abstract. Predicting future trajectories for other road agents is an essential task for autonomous vehicles. Established trajectory prediction methods primarily use agent tracks generated by a detection and tracking system and HD map as inputs. In this work, we propose a novel method that also incorporates visual input from surround-view cameras, allowing the model to utilize visual cues such as human gazes and gestures, road conditions, vehicle turn signals, etc., which are typically hidden from the model in prior methods. Furthermore, we use textual descriptions generated by a Vision-Language Model (VLM) and refined by a Large Language Model (LLM) as supervision during training to guide the model on what to learn from the input data. Despite using these extra inputs, our method achieves a latency of 53 ms, making it feasible for real-time processing, which is significantly faster than that of previous single-agent prediction methods with similar performance. Our experiments show that both the visual inputs and the textual descriptions contribute to improvements in trajectory prediction performance, and our qualitative analysis highlights how the model is able to exploit these additional inputs. Lastly, in this work we create and release the nuScenes-Text dataset, which augments the established nuScenes dataset with rich textual annotations for every scene, demonstrating the positive impact of utilizing VLM on trajectory prediction. Our project page is at https://moonseokha.github.io/VisionTrap.

Keywords: Motion Forecasting · Trajectory Prediction · Autonomous Driving · nuScenes-Text Dataset

---

Supplementary Information The online version contains supplementary material available at https://doi.org/10.1007/978-3-031-72658-3_21.

1 Introduction

Predicting agents' future poses (or trajectories) is crucial for safe navigation in dense and complex urban environments. To achieve such task successfully, it is required to model the following aspects: (i) understanding individual's behavioral contexts (*e.g.*, actions and intentions), (ii) agent-agent interactions, and (iii) agent-environment interactions (*e.g.*, pedestrians on the crosswalk). Recent works [5,12,13,24,25,31,49,50] have achieved remarkable progress, but their inputs are often limited – they mainly use a high-definition (HD) map and agents' past trajectories from a detection and tracking system as inputs.

HD map is inherently static, and only provide pre-defined information that limits their adaptability to changing environmental conditions like traffic near construction areas or weather conditions. They also cannot provide visual data for understanding agents' behavioral context, such as pedestrians' gazes, orientations, actions, gestures, and vehicle turn signals, all of which can significantly influence agents' behavior. Therefore, scenarios requiring visual context understanding may necessitate more than non-visual input for better and more reliable performance.

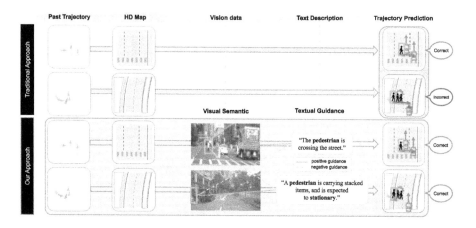

Fig. 1. Existing approaches are often conditioned only on agents' past trajectories and HD map to predict future trajectories. Here, we want to explore leveraging camera images and textual descriptions obtained from images to better learn the agent's behavioral context and agent-environment interactions by incorporating high-level semantic information into the prediction process, such as "a pedestrian is carrying stacked items, and is expected to stationary."

In this paper, we advocate for leveraging visual semantics in the trajectory prediction task. We argue that visual inputs can provide useful semantics, which non-visual inputs may not provide, for accurately predicting agents' future trajectories. Despite its potential advantages, only a few works [10,23,27,36–39] have used vision data to improve the performance of trajectory prediction in autonomous driving domain. Existing approaches often utilize images of the area where the agent is located or the entire image without explicit instructions on what information to extract. As a result, these methods tend to focus only on salient features, leading to sub-optimal performance. Additionally, because they typically rely solely on frontal-view images, it becomes challenging to fully recognize the surrounding driving environment.

To address these limitations and harness the potential of visual semantics, we propose **VisionTrap**, a vision-augmented trajectory prediction model that efficiently incorporates visual semantic information. To leverage visual semantics obtained from surround-view camera images, we first encode them into a composite Bird's Eye View (BEV) feature along with map data. Given this vision-aware BEV scene feature, we use a deformable attention mechanism to extract scene information from relevant areas (using predicted agents' future positions), and augment them into per-agent state embedding, producing scene-augmented state embedding. In addition, recent works [4,15,23,27,36] have shown that classifying intentions can improve model performance by helping predict agents' instantaneous movements. Learning with supervision of each agent's intention helps avoid training restrictions and oversimplified learning that may not yield optimal performance. However, annotating agents' intentions by dividing them into action categories involves inevitable ambiguity, which can be costly and hinder efficient scalability. Moreover, creating models that rely on these small sets can limit the model's expressiveness. Thus, as shown in Fig. 1, we leverage textual guidance as supervision to guide the model in leveraging richer visual semantics by aligning visual features (*e.g.*, an image of a pedestrian nearby a parked vehicle) with textual descriptions (*e.g.*, "a pedestrian is carrying stacked items, and is expected to stationary."). While we use additional input data, real-time processing is crucial in autonomous driving. Therefore, we designed VisionTrap based on a real-time capable model proposed in this paper. VisionTrap efficiently utilizes visual semantic information and employs textual guidance only during training. This allows it to achieve performance comparable to high-accuracy, non-real-time single-agent prediction methods [7,29] while maintaining real-time operation.

Since currently published autonomous driving datasets do not include textual descriptions, we created the nuScenes-Text dataset based on the large-scale nuScenes dataset [3], which includes vision data and 3D coordinates of each agent. The nuScenes-Text dataset collects textual descriptions that encompass high-level semantic information, as shown in Fig. 8. "A man wearing a blue shirt is talking to another man, expecting to cross the street when the signal changes." Automating this annotation process, we utilize both a Vision-Language Model (VLM) and a Large-Language Model (LLM).

Our extensive experiments on the nuScenes dataset show that our proposed text-guided image augmentation is effective in guiding our trajectory prediction model successfully to learn individuals' behavior and environmental contexts, producing a significant gain in trajectory prediction performance.

2 Related Work

Encoding Behavioral Contexts for Trajectory Prediction. Recent works in trajectory prediction utilize past trajectory observations and HD map to provide static environmental context. Traditional methods use rasterized Bird's Eye View (BEV) maps with ConvNet blocks [5,12,34,41,44], while recent approaches employ vectorized maps with graph-based attention or convolution layers for better understanding complex topologies [11,13,14,21,40,41]. However, HD maps are static and cannot adapt to changes, like construction zones affecting agent behavior. To address this, some works [10,27,37–39] aim to address these issues by utilizing images. To obtain meaningful visual semantic information about the situations an agent faces in a driving scene, it is necessary to utilize environmental information containing details from the objects themselves and from the environments they interact with. However, [27,37,39] focus solely on extracting information about agents' behavior using images near the agents, while [10,38] process the entire image at once and focus only on information about the scene without considering the parts that agents need to interact with. Therefore, in this paper, we propose an effective way to identify relevant parts of the image that each agent should focus on and efficiently learn semantic information from those parts.

Scene-Centric vs. Agent-Centric. Two primary approaches to predicting road agents' future trajectories are scene-centric and agent-centric. Scene-centric methods [32,42,47] encode each agent within a shared scene coordinate system, ensuring rapid inference speed but may exhibit slightly lower performance than agent-centric methods. Agent-centric approaches [2,8,24,25,50] standardize environmental elements and separately predict agents' future trajectories, offering improved predictive accuracy. However, their inference time and memory requirements are linearly scaled with the number of agents in the scene, posing a scalability challenge in dense urban environments with hundreds of pedestrians and vehicles. Thus, in this paper, we focus on scene-centric approaches.

Multimodal Contrastive Learning. With the increasing diversity of data sources, multimodal learning has become popular as it aims to effectively integrate information from various modalities. One of the common and effective approaches for multimodal learning is to align the modalities in a joint embedding space, using contrastive learning [18,35,45]. Contrastive Learning (CL) pulls together the positive pairs and pushes away the negative pairs, constructing an embedding space that effectively accommodates the semantic relations among the representations. Although CL is renowned for its ability to create a robust embedding space, its typical training mechanism introduces sampling bias, unintentionally incorporating similar pairs as negative pairs [6]. Debiasing strategies [6,16,17,30,48] have been introduced to mitigate such false-negatives, and it is particularly crucial in autonomous driving scenarios where multiple agents within a scene might have similar intentions in their behaviors. In our work, we carefully design our contrastive loss by filtering out the negative samples that are considered to be false-negatives. Inspired by [30,48], we do this by utilizing the sentence representations and their similarities, and finally achieve debiased contrastive learning in multimodal setting.

3 Method

This paper explores leveraging high-level visual semantics to improve the trajectory prediction quality. In addition to conventionally using agents' past trajectories and their types as inputs, we advocate for using visual data as an additional input to utilize agents' visual semantics. As shown in Fig. 2, our model consists of four main modules: (i) Per-agent State Encoder, (ii) Visual Semantic Encoder, (iii) Text-driven Guidance module, and (iv) Trajectory Decoder. Our *Per-agent State Encoder* takes as an input a sequence of state observations (which are often provided by a detection and tracking system), producing per-agent context features (Sect. 3.1). In our *Visual Semantic Encoder*, we encode multi-view images (capturing the surrounding view around the ego vehicle) into a unified Bird's Eye View (BEV) feature, followed by concatenation with a dense feature map of road segments. Given this BEV feature, the per-agent state embedding is updated in the Scene-Agent Interaction module (Sect. 3.2). We utilize *Text-driven Guidance module* to supervise the model to understand or reason about detailed visual semantics, producing richer semantics (Sect. 3.3). Lastly, given per-agent features with rich visual semantics, our *Trajectory Decoder* predicts the future positions for all agents in the scene in a fixed time horizon (Sect. 3.4).

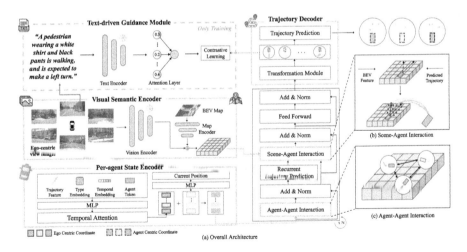

Fig. 2. An overview of VisionTrap, which consists of four main steps: (i) Per-agent State Embedding, which produces per-agent context features given agents' state observations; (ii) Visual Semantic Encoder, which transforms multi-view images with an HD map into a unified BEV feature, updating agents' state embedding via a deformable attention layer; (iii) Text-driven Guidance Module, which supervises the model to reason about detailed visual semantics and (iv) Trajectory Decoder, which predicts agents' the future poses in a fixed time horizon.

3.1 Per-Agent State Encoder

Encoding Agent State Observations. Following recent trajectory prediction approaches [31,50], we first encode per-agent state observations (*e.g.*, agent's observed trajectory and semantic attributes) provided by object detection and tracking systems. We utilize the geometric attributes with relative positions (instead of absolute positions) by representing the observed trajectory of agent i as $\{p_i^t - p_i^{t-1}\}_{t=1}^T$ where $p_i^t = (x_i^t, y_i^t)$ is the location of agent i in an ego-centric coordinate system at time step $t \in \{1, 2, \ldots, T\}$. T denotes the observation time horizon. Note that we use an ego-centric (scene-centric) coordinate system where a scene is centered and rotated around the current ego-agent's location and orientation. Given these geometric attributes and their semantic attributes a_i (*i.e.*, agent types, such as cars, pedestrians, and cyclists), per-agent state embedding $s_i^t \in \mathbb{R}^{d_s}$ for agent i at time step t is obtained as follows:

$$s_i^t = f_{\text{geometric}}(p_i^t - p_i^{t-1}) + f_{\text{type}}(a_i) + f_{\text{PE}}(e^t), \qquad (1)$$

where $f_{\text{geometric}} : \mathbb{R}^2 \to \mathbb{R}^{d_s}$, $f_{\text{type}} : \mathbb{R}^1 \to \mathbb{R}^{d_s}$, and $f_{\text{PE}} : \mathbb{R}^{d_{pe}} \to \mathbb{R}^{d_s}$ are MLP blocks. Note that we use the learned positional embeddings $e^t \in \mathbb{R}^{d_{pe}}$, guiding the model to learn (and utilize) the temporal ordering of state embeddings.

Encoding Temporal Information. Following existing approaches [46,50], we utilize a temporal Transformer encoder to learn the agent's temporal information over the observation time horizon. Given the sequence of per-agent state embeddings $\{s_i^t\}_{t=1}^T$ and an additional learnable token $s^{T+1} \in \mathbb{R}^{d_s}$ stacked into the end of the sequence, we feed these input into the temporal (self-attention) attention block, producing per-agent spatio-temporal representations $s_i' \in \mathbb{R}^{d_s}$.

LNCSsubsubsectionEncoding Interaction between Agents. We further use the cross-attention-based agent-agent interaction module to learn the relationship between agents. Further, as our model depends on the geometric attributes with relative positions, we add embeddings of the agents' current position p_i^T to make the embeddings spatially aware, producing per-agent representation $z_i = s_i' + f_{loc}(p_i^T)$ where $f_{loc} : \mathbb{R}^2 \to \mathbb{R}^{d_s}$ is another MLP block. This process is performed at once within the ego-centric coordinate system to eliminate the cost of recalculating correlation distances with other agents for each individual agent. The agent state embedding z_i is used as the query vector, and those of its neighboring agents are converted to the key and the value vectors as follows:

$$q_i^{\text{Interact}} = W_Q^{\text{Interact}} z_i, \quad k_j^{\text{Interact}} = W_K^{\text{Interact}} z_j, \quad v_j^{\text{Interact}} = W_V^{\text{Interact}} z_j, \qquad (2)$$

where $W_Q^{\text{Interact}}, W_K^{\text{Interact}}, W_V^{\text{Interact}} \in \mathbb{R}^{d_{\text{Interact}} \times d_s}$ are learnable matrices.

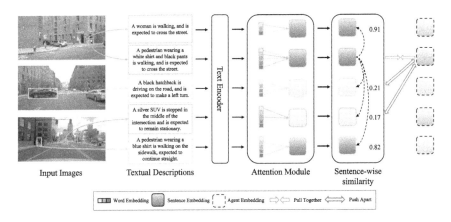

Fig. 3. An overview of our Text-driven Guidance Module. We extract word-level embeddings using pretrained BERT [9] as a text encoder, and then we use an attention module to aggregate these per-word embeddings into a composite sentence-level embedding. Based on the cosine similarity between these embeddings, we apply contrastive learning loss to ground textual descriptions into the agent's state embedding.

3.2 Visual Semantic Encoder

Vision-Augmented Scene Feature Generation. Given ego-centric multi-view images $\mathcal{I} = \{\mathcal{I}_j\}_{j=1}^{n_I}$, we feed them into Vision Encoder using the same architecture from BEVDepth [20], to produce the BEV image feature as $B_I \in \mathbb{R}^{h \times w \times d_{\text{bev}}}$. Then, we incorporate the rasterized map information into the BEV embeddings to align B_I. We utilize CNN blocks with Feature Pyramid Network (FPN) [22] to produce another BEV feature $B_{\text{map}} \in \mathbb{R}^{h \times w \times d_{\text{map}}}$. Lastly, we concatenate all generated BEV features into a composite BEV scene feature $B = [B_I; B_{\text{map}}] \in \mathbb{R}^{h \times w \times (d_{\text{bev}} + d_{\text{map}})}$. In this process, we compute map aligned around the current location and direction of the ego vehicle only once, even in the presence of n agents, as we adopt an ego-centric approach. This significantly reduces computational costs compared to agent-centric approaches, which require reconstructing and encoding map for each of the n agents.

Augmenting Visual Semantics into Agent State Embedding. When given the vision-aware BEV scene feature B, we use deformable cross-attention [51] module to augment map-aware visual scene semantics into the per-agent state embedding z_i, as illustrated in Fig. 2(b). This allows for the augmentation of agent state embedding z_i. Compared to commonly used ConvNet-based architectures [5,12,34], our approach leverages a wide receptive field and can selectively focus on scene feature, explicitly extracting multiple areas where each agent needs to focus and gather information. Additionally, as the agent state embedding is updated for each block, the focal points for the agent also require repeated refinement. To achieve this, we employ a Recurrent Trajectory Prediction module, which utilizes the same architecture as the main trajectory decoder(explained in Sect. 3.4). This module refines the agent's future trajectory $u^{\text{aux}} = \{u_i^{\text{aux}}\}_{i=1}^{T_f}$ by recurrently improving the predicted trajectories. These refined trajectories serve as reference points for agents to focus on in the Scene-Agent Interaction module, integrating surrounding information around the reference points into the agent's function. Our module is defined as follows:

$$z_i^{\text{scene}} = z_i^{\text{interact}} + \sum_{h=1}^{H} W_h \left[\sum_{o=1}^{O} \left(\alpha_{hio} W_h' \mathbf{B}_{\left(u_i^{\text{aux}} + \triangle u_{hio}^{\text{aux}}\right)} \right) \right], \quad (3)$$

where H denotes the number of attention heads and O represents the number of offset points for every reference point u_i^{aux} where we use an auxiliary trajectory predictor and use the agent's predicted future positions as reference points. Note that W_h and W_h' are learnable matrices, and α_{hio} is the attention weight for each learnable offset $\triangle u_{hio}^{\text{aux}}$ in each head. The number of attention points is typically set fewer than the number of surrounding road elements, reducing computational costs.

3.3 Text-Driven Guidance Module

We observe that our visual semantic encoder simplifies visual reasoning about a scene to focus on salient visible features, resulting in sub-optimal performance in trajectory prediction. For instance, the model may primarily focus on the vehicle itself, disregarding other semantic details, such as "a vehicle waiting in front of the intersection with turn signals on, expected to turn left." Therefore, we introduce the Text-driven Guidance Module to supervise the model, allowing the model to understand the context of the agents using detailed visual semantics. For this purpose, we employ multi-modal contrastive learning where positive pair is pulled together and negative pairs are pushed farther. However, the textual descriptions for prediction tasks in the driving domain are diverse in expression, posing an ambiguity in forming negative pairs between descriptions.

To address this, as shown in Fig. 3, we extract word-level embeddings using BERT [9], and then we use a attention module to aggregate these per-word embeddings into a composite sentence-level embedding $\mathcal{T}_i$ for agent i. Given $\mathcal{T}_i$, we measure cosine similarity with other agents' sentence-level embeddings $\mathcal{T}_j$ for $j \neq i$, and we treat as negative pairs if $\text{sim}_{\cos}(\mathcal{T}_i, \mathcal{T}_j) < \theta_{\text{th}}$ where θ_{th} is a threshold value (we set $\theta_{\text{th}} = 0.8$ in our experiments). Further, we limit the number of negative pairs within a batch for stable optimization, which is particularly important as the number of agents in a given scene varies. Specifically, given an agent i, we choose top-k sentence-level embeddings from $\{\mathcal{T}_j\}$ sorted in ascending order for $j \neq i$. Subsequently, we form a positive pair between the agent's state embedding z_i^{scene} and corresponding textual embedding $\mathcal{T}_i$, while negative pairs as z_i^{scene} and $\{\mathcal{T}_j\}_{j=1}^{k}$. Ultimately, we use the following InfoNCE loss [33] to guide agent's state embedding with textual descriptions:

$$\mathcal{L}_{\text{cl}} = -\log \frac{e^{\text{sim}_{\cos}(z_i^{\text{scene}}, \mathcal{T}_i)/\tau}}{\sum_{j=1}^{k} e^{\text{sim}_{\cos}(z_i^{\text{scene}}, \mathcal{T}_j)/\tau}}, \quad (4)$$

where τ is a temperature parameter used in the attention layer, enabling biasing the distribution of attention scores.

3.4 Trajectory Decoder

Transformation Module. For fast inference speed and compatibility with ego-centric images, we adopt ego-centric approach in the State Encoder and Scene Semantic Interaction. However, as noted by Su et al. [42], ego-centric approaches typically underperform compared to agent-centric approaches due to the need to learn invariance for transformations and rotations between scene

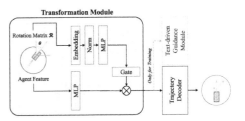

Fig. 4. An overview of transformation module, which standardizes agents' orientation.

elements. This implies that the features of agents with similar future movements are not standardized. Thus, prior to utilizing the Text-driven Guidance Module and predicting each agent's future trajectory, we employ the Transformation Module to standardize each agent's orientation, aiming to mitigate the complexity associated with learning rotation invariance. This allows us to effectively apply the Text-driven Guidance Module, as we can make the features of agents in similar situations similar. As depicted in Fig. 4, the Transformation Module takes the agent's feature and rotation matrix $\mathcal{R}$ as input and propagates the rotation matrix to the agent's feature using a Multi-Layer Perceptron (MLP). This transformation enables the determination of which situations the agent's features face along the y-axis.

Trajectory Decoder. Similar to [5,31,34,43], we use a parametric distribution over the agent's future trajectories $u = \{u_i\}_{i=1}^{T_f}$ for $u_i \in \mathbb{R}^2$ as Gaussian Mixture Model (GMM). We represent a mode at each time step t as a 2D Gaussian distribution over a certain position with a mean $\mu_t \in \mathbb{R}^2$ and covariance $\Sigma_t \in \mathbb{R}^{2\times 2}$. Our decoder optimizes a weighted set of a possible future trajectory for the agent, producing full output distribution as

$$p(u) = \sum_{m=1}^{M} \rho_m \prod_{t=1}^{T_f} \mathcal{N}(u_t - \mu_m^t, \Sigma_m^t), \tag{5}$$

where our decoder produces a softmax probability ρ over mixture components and Gaussian parameters μ and Σ for M modes and T_f time steps.

Loss Functions. We optimize trajectory predictions and their associated confidence levels by minimizing $\mathcal{L}_{\text{traj}}$ to train our model in an end-to-end manner. We compute $\mathcal{L}_{\text{traj}}$ by minimizing the negative log-likelihood function between actual and predicted trajectories and the corresponding confidence score, and it can be formulated as follows:

$$\mathcal{L}_{\text{traj}} = -\frac{1}{N} \sum_{i=1}^{N} \log \left(\sum_{m=1}^{M} \frac{\rho_{i,m}}{\sqrt{2b^2}} \exp \left(-\frac{(\mathbf{Y}_i - \hat{\mathbf{Y}}_{i,m})^2}{2} \right) \right). \tag{6}$$

Here, b and $\mathbf{Y}$ represent the scale parameters and the real future trajectory, respectively. We denote predicted future positions as $\hat{\mathbf{Y}}_{i,m}$ and the corresponding confidence scores as $\rho_{i,m}$ for agent i at future time step t across different modes $m \in M$. Furthermore, we minimize an auxiliary loss function $\mathcal{L}_{\text{traj}}^{\text{aux}}$ similar to $\mathcal{L}_{\text{traj}}$ to train the trajectory decoder used by the Recurrent Trajectory Prediction module. Ultimately, our model is trained by minimizing the following loss $\mathcal{L}$, with $\lambda_{\text{traj}}^{\text{aux}}$ and λ_{cl} controlling the strength of each loss term:

$$\mathcal{L} = \mathcal{L}_{\text{traj}} + \lambda_{\text{traj}}^{\text{aux}} \mathcal{L}_{\text{traj}}^{\text{aux}} + \lambda_{\text{cl}} \mathcal{L}_{\text{cl}}. \tag{7}$$

4 nuScenes-Text Dataset

To our best knowledge, currently available driving datasets for prediction tasks lack textual descriptions of the actions of road users during various driving events. While the DRAMA dataset [26] offers textual descriptions for agents in driving scenes, it only provides a single caption for one agent in each scene alongside the corresponding bounding box. This setup suits detection and captioning tasks but not prediction tasks. To address this gap, we collect the textual descriptions for the nuScenes dataset [3], which provides surround-view camera images, trajectories of road agents, and map data. With its diverse range of typical road agents activities, nuScenes is widely used in prediction tasks.

Textual Description Generation. We employ a three-step process for generating textual descriptions of agents from images, as illustrated in Fig. 5. Initially, we employ a pre-trained Vision-Language Model (VLM) BLIP-2 [19]. However, it often underperforms in driving-related image-to-text tasks. To address this, we fine-tune the VLM with the DRAMA dataset [26], containing textual descriptions of agents in driving scenes. We isolate the bounding box region representing the agent of interest, concatenate it with the original image (Fig. 5), and leverage the fine-tuned VLM to generate descriptions for each agent separately in the nuScenes dataset [3] as an image captioning task. However, the generated descriptions often lack correct action-related details, providing unnecessary information for prediction. To address shortcomings, we refine generated texts using GPT [1], a well-known Large Language Model (LLM). Inputs include the generated text, agent type, and maneuvering. Rule-based logic determines the agent's

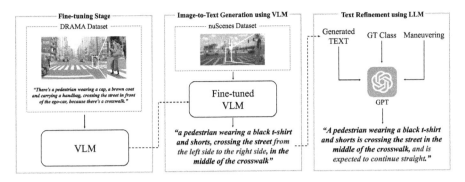

Fig. 5. To create the nuScenes-Text Dataset, three main steps are involved: (i) Fine-tuning stage using DRAMA Dataset [26], (ii) Image-to-Text Generation stage applying the fine-tuned VLM to the nuScenes Dataset [3], and (iii) Text Refinement process using ground truth information (*e.g.* GT class, Maneuvering) along with generated text and GPT [1]. The red color indicates that needs to be filtered out, while the cyan color indicates additional content related to the intention. (Color figure online)

maneuvering (*e.g.*, stationary, lane change, turn right). We use prompts to correct inappropriate descriptions, aiming to generate texts that provide prediction-related information on agent type, actions, and rationale. Examples are provided in Fig. 6, with additional details (*e.g.*, rule-based logic, GPT prompt) in the supplemental material.

Coverage of nuScenes-Text Dataset. In this section, we demonstrate how well our created nuScene-Text Dataset encapsulates the context of the agent, as depicted in Fig. 6, and discuss the coverage and benefits of this dataset. Figure 6a represents the contextual information of the agent changing over time in text form. This attribute assists in accurately predicting object trajectories under behavioral context changes. We also demonstrate in Fig. 6b that distinctive characteristics of each object can be captured (*e.g.*, "A pedestrian waiting to cross the street.", "A construction worker sitting on the lawn.") and generate three unique textual descriptions for each object, showcasing diverse perspectives. Additionally, to enhance text descriptions when the VLM generates incorrect agent types, behavior predictions, or harmful information, such as "from the left side to the right side", which can be misleading due to the directional variation in BEV depending on the camera's orientation, we refine the text using an LLM. This refinement process aims to improve text quality for identifying driving scenes through surround images. Figure 6c illustrates this improvement process, ensuring the relevance and accuracy of text by removing irrelevant details (indicated by red) and adding pertinent information (indicated by cyan).

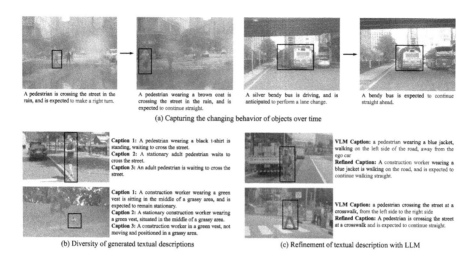

Fig. 6. Examples of our generated textual descriptions

Dataset Statistics. Our created dataset contains 1,216,206 textual descriptions for 391,732 objects (three for each object), averaging 13 words per description. In Fig. 7, we visualize frequently used words, highlighting the dataset's rich vocabulary and diversity. Further, we conduct a human evaluation using Amazon Mechanical Turk (Mturk) to quantitatively evaluate image-text alignments. 5 human evaluators are recruited, and it is performed on a subset of 1,000 randomly selected samples. Each evaluator is presented with the full image, cropped object image, and corresponding text and asked the question: "Is the image well-aligned with the text, considering the reference image?". The results show that 94.8% of the respondents chose 'yes', indicating a high level of accuracy in aligning images with texts. All results are aggregated through a majority vote. Further details on the nuScenes-Text Dataset are provided in the supplemental material.

Fig. 7. Frequency of words

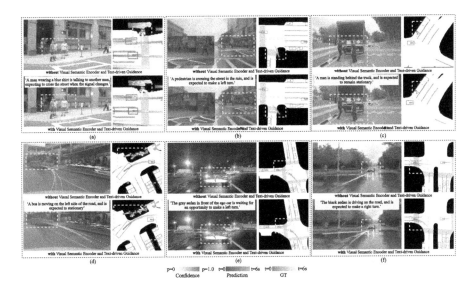

Fig. 8. Examples of trajectory prediction outputs in six different scenarios. The examples on the top row represent scenarios with pedestrians, while those on the bottom row have vehicles. We also provide ground truth textual descriptions about an object in a red box, which were not seen during inference. (Color figure online)

5 Experiments

Dataset. We conduct experiments using the nuScenes dataset [3], which offers two versions: (i) a dataset dedicated to a trajectory prediction task and (ii) a whole dataset. While the former focuses solely on single-agent prediction tasks, the latter is more suitable for our purposes. Therefore, we provide scores for both datasets in our experiments. Further implementation, evaluation, and dataset details can be found in the supplemental material.

Qualitative Analysis. Figure 8 presents the results of VisionTrap on nuScenes dataset [3], demonstrating the impact of Visual Semantic Encoder and Text-driven Guidance Module on agent trajectory prediction.

The top row shows improved results of pedestrians. For (a), while the result without visual information predicts the man will cross the crosswalk, the prediction with visual information indicates the man will remain stationary due to red traffic light and people talking to each other rather than trying to cross the road. (b) presents how gaze and body orientation help in predicting the pedestrian's intention to walk towards the crosswalk, and (c) provides visual context of the man getting on a stationary vehicle, implying the trajectory of the man would remain stationary as well. The following row exhibits the improved prediction results of vehicles. In (d), understanding that the people are standing at a bus stop enables the model to make a reasonable prediction for the bus. (e) gives a

Table 1. Trajectory prediction performance comparison on nuScenes [3] dataset regarding ADE_{10}, MR_{10}, and FDE_1. Inference times are reported in milliseconds (msec), measured based on 12 agents using a single RTX 3090 Ti GPU.

Model	Prediction Method	Using Map Data	Time ↓ (msec)	ADE_{10} ↓	MR_{10} ↓	FDE_1 ↓	
Multipath [5]	single	✔	87	1.50	0.74	–	
MHA-JAM [29]	single	✔	–	1.24	0.46	8.57	
P2T [7]	single	✔	116	1.16	0.46	10.5	
PGP [8]	single	✔	215	1.00	0.37	7.17	
LAformer [24]	single	✔	115	0.93	0.33	–	
Trajectron++ [41]	multi	✔	38	1.51	0.57	9.52	
AgentFormer [46]	multi	✔	107	1.45	–	–	Average
VisionTrap baseline	multi		13	1.48	0.56	10.75	improvement:
+ Map Encoder	multi	✔	21	1.40	0.53	10.41	4.65%
+ Visual Semantic Encoder	multi	✔	53	1.23	0.36	9.32	21.97%
+ Text-driven Guidance (Ours)	multi	✔	53	1.17	0.32	8.72	27.56%

visual cue of turn signal, indicating the vehicle's intention of turning left. Lastly, visual context in (f) leads to a more stable prediction of the vehicle turning right, as the image clearly shows the vehicle is directed towards its right.

These examples highlight the crucial role of visual data in improving trajectory prediction accuracy, offering insights that cannot obtained from non-visual data. Further qualitative analysis details are available in the supplemental material.

Quantitative Analysis. Table 1 compares our model with other methods for single and multi-agent prediction. Our query-based prediction model designed to effectively utilize visual semantic information and Text-driven Guidance Module, which we use as baseline, achieves the fastest inference speed. We also demonstrate that the Visual Semantic Encoder significantly improves performance, especially when combined with the Text-driven Guidance Module, yielding comparable results to existing single-agent prediction methods with better miss rate performance, while still maintaining real-time operation. These results suggest that vision data provides additional information inaccessible to non-vision data, and textual descriptions derived from vision data effectively guide the model.

Since our method employs egocentric surround-view images, it is feasible to effectively predict for all observed agents in the scene. We utilize the nuScenes dataset covering all scenes, enabling comprehensive evaluation of all observed agents (refer to Table 2). This demonstrates the contributions of all proposed components to predicting all agents in the scene.

Table 2. Ablation study of variant models on nuScenes [3] whole dataset.

Method	ADE_{10} ↓	FDE_{10} ↓	MR_{10} ↓
VisionTrap baseline	0.425	0.641	0.081
+ Map Encoder	0.407	0.601	0.075
+ Visual Semantic Encoder	0.382	0.551	0.056
+ Text-driven Guidance (Ours)	**0.368**	**0.535**	**0.051**

Finally, we emphasize that the purpose of this study is not to achieve state-of-the-art performance. Instead, our aim is to demonstrate that vision information, often overlooked in trajectory prediction tasks, can provide additional insights.

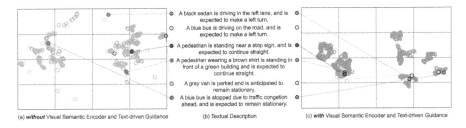

Fig. 9. UMAP [28] visualizations for per-agent state embeddings from models (a) without and (c) with leveraging visual and textual semantics. (b) We also provide corresponding ground truth textual descriptions.

These insights are inaccessible from non-vision data, thereby enhancing performance in trajectory prediction tasks. This is our original motivation for this task, and the results in Fig. 8, Table 1 and Table 2 provide justification for our method.

UMAP Visualization. We observe an overall improvement in clustering of agent state embeddings when leveraging visual and textual semantics in Fig. 9. Furthermore, extracting textual descriptions of agents within the same cluster group is shown to exhibit similar situations. This indicates that state embeddings for agents in similar situations are located in a similar embedding space.

Analyzing the Text-driven Guidance Module. To analyze the effect of each component of the proposed Text-Based Guidance Module, we removed each factor to see how the model performs, as shown in Tab. 3. In the case of A, we use simple symmetric contrastive loss that is used in [35]. However, our loss adopts asymmetric

Table 3. Performance comparison to analyze the effect of each component of Text-Based Guidance Module on the nuScenes [3] all dataset.

Method	ADE_6 ↓	FDE_6 ↓	MR_6 ↓
A. CLIP loss	0.51	0.79	0.10
B. Ours w/ symmetric loss	0.50	0.76	0.10
C. Ours w/o refining negative pair	0.49	0.72	0.09
D. Ours w/o top-k algorithm	0.46	0.67	0.08
E. Ours	**0.44**	**0.66**	**0.07**

form of contrastive loss that only calculates softmax probabilities in one direction. B gives the result of incorporating symmetric loss in our loss design. C shows the result of removing the stage of negative pair refinement, allowing potential false-negatives. In D, we skip the process of ascending sorting and limiting the number of negative pairs. Removing these steps causes variance in number of agents considered each scene, leading to different scales of loss. In the end, our asymmetric contrastive loss with negative pairs refined and its number constrained demonstrated the best performance across all metrics.

6 Conclusion

In this paper, we introduced an novel approach called VisionTrap to trajectory prediction by incorporating visual input from surround-view cameras. This enables the model to leverage visual semantic cues, which were previously inaccessible to traditional trajectory prediction methods. Additionally, we utilize text descriptions produced by a VLM and refined by a LLM to provide supervision, guiding the model in learning from the input data. Our thorough experiments demonstrate that both visual inputs and textual descriptions contribute to enhancing trajectory prediction performance. Furthermore, our qualitative analysis shows how the model effectively utilizes these additional inputs.

Acknowledgment. This work was supported by Autonomous Driving Center, Hyundai Motor Company R&D Division. This work was partly supported by IITP under the Leading Generative AI Human Resources Development(IITP-2024-RS-2024-00397085, 10%) grant, IITP grant (RS-2022-II220043, Adaptive Personality for Intelligent Agents, 10% and IITP-2024-2020-0-01819, ICT Creative Consilience program, 5%). This work was also partly supported by Basic Science Research Program through the NRF funded by the Ministry of Education(NRF-2021R1A6A1A13044830, 10%). This work also supported by Culture, Sports and Tourism R&D Program through the Korea Creative Content Agency grant funded by the Ministry of Culture, Sports and Tourism in 2024((International Collaborative Research and Global Talent Development for the Development of Copyright Management and Protection Technologies for Generative AI, RS-2024-00345025, 4%),(Research on neural watermark technology for copyright protection of generative AI 3D content, RS-2024-00348469, 25%)), Institute of Information & communications Technology Planning & Evaluation (IITP) grant funded by the Korea government(MSIT)(RS-2019-II190079, 1%). We also thank Yujin Jeong and Daewon Chae for their helpful discussions and feedback.

References

1. Brown, T., et al.: Language models are few-shot learners. Adv. Neural. Inf. Process. Syst. **33**, 1877–1901 (2020)
2. Buhet, T., Wirbel, E., Bursuc, A., Perrotton, X.: Plop: probabilistic polynomial objects trajectory planning for autonomous driving. arXiv preprint arXiv:2003.08744 (2020)
3. Caesar, H., et al.: nuscenes: a multimodal dataset for autonomous driving. In: Proceedings of the IEEE/CVF Conference on Computer Vision and Pattern Recognition, pp. 11621–11631 (2020)
4. Casas, S., Luo, W., Urtasun, R.: Intentnet: learning to predict intention from raw sensor data. In: Conference on Robot Learning, pp. 947–956. PMLR (2018)
5. Chai, Y., Sapp, B., Bansal, M., Anguelov, D.: Multipath: multiple probabilistic anchor trajectory hypotheses for behavior prediction. arXiv preprint arXiv:1910.05449 (2019)
6. Chuang, C.Y., Robinson, J., Lin, Y.C., Torralba, A., Jegelka, S.: Debiased contrastive learning. Adv. Neural. Inf. Process. Syst. **33**, 8765–8775 (2020)
7. Deo, N., Trivedi, M.M.: Trajectory forecasts in unknown environments conditioned on grid-based plans. arXiv preprint arXiv:2001.00735 (2020)
8. Deo, N., Wolff, E., Beijbom, O.: Multimodal trajectory prediction conditioned on lane-graph traversals. In: Conference on Robot Learning, pp. 203–212. PMLR (2022)

9. Devlin, J., Chang, M.W., Lee, K., Toutanova, K.: Bert: pre-training of deep bidirectional transformers for language understanding. arXiv preprint arXiv:1810.04805 (2018)
10. Fang, L., Jiang, Q., Shi, J., Zhou, B.: Tpnet: trajectory proposal network for motion prediction. In: Proceedings of the IEEE/CVF Conference on Computer Vision and Pattern Recognition, pp. 6797–6806 (2020)
11. Gao, J., et al.: Vectornet: encoding hd maps and agent dynamics from vectorized representation. In: Proceedings of the IEEE/CVF Conference on Computer Vision and Pattern Recognition, pp. 11525–11533 (2020)
12. Gilles, T., Sabatini, S., Tsishkou, D., Stanciulescu, B., Moutarde, F.: Home: heatmap output for future motion estimation. In: 2021 IEEE International Intelligent Transportation Systems Conference (ITSC), pp. 500–507. IEEE (2021)
13. Gilles, T., Sabatini, S., Tsishkou, D., Stanciulescu, B., Moutarde, F.: Thomas: trajectory heatmap output with learned multi-agent sampling. arXiv preprint arXiv:2110.06607 (2021)
14. Gilles, T., Sabatini, S., Tsishkou, D., Stanciulescu, B., Moutarde, F.: Gohome: graph-oriented heatmap output for future motion estimation. In: 2022 International Conference on Robotics and Automation (ICRA), pp. 9107–9114. IEEE (2022)
15. Girase, H., et al.: Loki: long term and key intentions for trajectory prediction. In: Proceedings of the IEEE/CVF International Conference on Computer Vision, pp. 9803–9812 (2021)
16. Hwang, I., et al.: Selecmix: debiased learning by contradicting-pair sampling. Adv. Neural. Inf. Process. Syst. **35**, 14345–14357 (2022)
17. Jang, T., Wang, X.: Difficulty-based sampling for debiased contrastive representation learning. In: Proceedings of the IEEE/CVF Conference on Computer Vision and Pattern Recognition (CVPR), pp. 24039–24048 (2023)
18. Jia, C., et al.: Scaling up visual and vision-language representation learning with noisy text supervision. In: International Conference on Machine Learning, pp. 4904–4916. PMLR (2021)
19. Li, J., Li, D., Savarese, S., Hoi, S.: Blip-2: bootstrapping language-image pretraining with frozen image encoders and large language models. arXiv preprint arXiv:2301.12597 (2023)
20. Li, Y., et al.: Bevdepth: acquisition of reliable depth for multi-view 3d object detection. In: Proceedings of the AAAI Conference on Artificial Intelligence, vol. 37, pp. 1477–1485 (2023)
21. Liang, M., et al.: Learning lane graph representations for motion forecasting. In: Vedaldi, A., Bischof, H., Brox, T., Frahm, J.-M. (eds.) ECCV 2020. LNCS, vol. 12347, pp. 541–556. Springer, Cham (2020). https://doi.org/10.1007/978-3-030-58536-5_32
22. Lin, T.Y., Dollár, P., Girshick, R., He, K., Hariharan, B., Belongie, S.: Feature pyramid networks for object detection. In: Proceedings of the IEEE Conference on Computer Vision and Pattern Recognition, pp. 2117–2125 (2017)
23. Liu, B., et al.: Spatiotemporal relationship reasoning for pedestrian intent prediction. IEEE Rob. Autom. Lett. **5**(2), 3485–3492 (2020)
24. Liu, M., et al.: Laformer: trajectory prediction for autonomous driving with lane-aware scene constraints. arXiv preprint arXiv:2302.13933 (2023)
25. Liu, Y., Zhang, J., Fang, L., Jiang, Q., Zhou, B.: Multimodal motion prediction with stacked transformers. In: Proceedings of the IEEE/CVF Conference on Computer Vision and Pattern Recognition, pp. 7577–7586 (2021)

26. Malla, S., Choi, C., Dwivedi, I., Choi, J.H., Li, J.: Drama: Joint risk localization and captioning in driving. In: Proceedings of the IEEE/CVF Winter Conference on Applications of Computer Vision, pp. 1043–1052 (2023)
27. Malla, S., Dariush, B., Choi, C.: Titan: future forecast using action priors. In: Proceedings of the IEEE/CVF Conference on Computer Vision and Pattern Recognition, pp. 11186–11196 (2020)
28. McInnes, L., Healy, J., Melville, J.: Umap: uniform manifold approximation and projection for dimension reduction. arXiv preprint arXiv:1802.03426 (2018)
29. Messaoud, K., Deo, N., Trivedi, M.M., Nashashibi, F.: Trajectory prediction for autonomous driving based on multi-head attention with joint agent-map representation (2020)
30. Miao, P., Du, Z., Zhang, J.: Debcse: rethinking unsupervised contrastive sentence embedding learning in the debiasing perspective. In: Proceedings of the 32nd ACM International Conference on Information and Knowledge Management, pp. 1847–1856 (2023)
31. Nayakanti, N., Al-Rfou, R., Zhou, A., Goel, K., Refaat, K.S., Sapp, B.: Wayformer: motion forecasting via simple & efficient attention networks. In: 2023 IEEE International Conference on Robotics and Automation (ICRA), pp. 2980–2987. IEEE (2023)
32. Ngiam, J., et al.: Scene transformer: a unified architecture for predicting multiple agent trajectories. arXiv preprint arXiv:2106.08417 (2021)
33. Oord, A.V.D., Li, Y., Vinyals, O.: Representation learning with contrastive predictive coding. arXiv preprint arXiv:1807.03748 (2018)
34. Phan-Minh, T., Grigore, E.C., Boulton, F.A., Beijbom, O., Wolff, E.M.: Covernet: multimodal behavior prediction using trajectory sets. In: Proceedings of the IEEE/CVF Conference on Computer Vision and Pattern Recognition, pp. 14074–14083 (2020)
35. Radford, A., et al.: Learning transferable visual models from natural language supervision. In: International Conference on Machine Learning, pp. 8748–8763. PMLR (2021)
36. Rasouli, A., Kotseruba, I., Kunic, T., Tsotsos, J.K.: Pie: a large-scale dataset and models for pedestrian intention estimation and trajectory prediction. In: ICCV (2019)
37. Rasouli, A., Rohani, M., Luo, J.: Bifold and semantic reasoning for pedestrian behavior prediction. In: Proceedings of the IEEE/CVF International Conference on Computer Vision (ICCV), pp. 15600–15610 (October 2021)
38. Rasouli, A., Yau, T., Lakner, P., Malekmohammadi, S., Rohani, M., Luo, J.: Pepscenes: a novel dataset and baseline for pedestrian action prediction in 3d. arXiv preprint arXiv:2012.07773 (2020)
39. Rasouli, A., Yau, T., Rohani, M., Luo, J.: Multi-modal hybrid architecture for pedestrian action prediction. In: 2022 IEEE Intelligent Vehicles Symposium (IV), pp. 91–97. IEEE (2022)
40. Rowe, L., Ethier, M., Dykhne, E.H., Czarnecki, K.: Fjmp: factorized joint multi-agent motion prediction over learned directed acyclic interaction graphs. In: Proceedings of the IEEE/CVF Conference on Computer Vision and Pattern Recognition, pp. 13745–13755 (2023)
41. Salzmann, T., Ivanovic, B., Chakravarty, P., Pavone, M.: Trajectron++: dynamically-feasible trajectory forecasting with heterogeneous data. In: Vedaldi, A., Bischof, H., Brox, T., Frahm, J.-M. (eds.) ECCV 2020. LNCS, vol. 12363, pp. 683–700. Springer, Cham (2020). https://doi.org/10.1007/978-3-030-58523-5_40

42. Su, D.A., Douillard, B., Al-Rfou, R., Park, C., Sapp, B.: Narrowing the coordinate-frame gap in behavior prediction models: Distillation for efficient and accurate scene-centric motion forecasting. In: 2022 International Conference on Robotics and Automation (ICRA), pp. 653–659. IEEE (2022)
43. Varadarajan, B., et al.: Multipath++: efficient information fusion and trajectory aggregation for behavior prediction. In: 2022 International Conference on Robotics and Automation (ICRA), pp. 7814–7821. IEEE (2022)
44. Wu, D., Wu, Y.: Air^2 for interaction prediction. arXiv preprint arXiv:2111.08184 (2021)
45. Yuan, X., et al.: Multimodal contrastive training for visual representation learning. In: Proceedings of the IEEE/CVF Conference on Computer Vision and Pattern Recognition, pp. 6995–7004 (2021)
46. Yuan, Y., Weng, X., Ou, Y., Kitani, K.: Agentformer: agent-aware transformers for socio-temporal multi-agent forecasting. In: Proceedings of the IEEE/CVF International Conference on Computer Vision (ICCV) (2021)
47. Zeng, W., Liang, M., Liao, R., Urtasun, R.: Lanercnn: distributed representations for graph-centric motion forecasting. In: 2021 IEEE/RSJ International Conference on Intelligent Robots and Systems (IROS), pp. 532–539. IEEE (2021)
48. Zhou, K., Zhang, B., Zhao, X., Wen, J.R.: Debiased contrastive learning of unsupervised sentence representations (2022)
49. Zhou, Z., Wang, J., Li, Y.H., Huang, Y.K.: Query-centric trajectory prediction. In: Proceedings of the IEEE/CVF Conference on Computer Vision and Pattern Recognition, pp. 17863–17873 (2023)
50. Zhou, Z., Ye, L., Wang, J., Wu, K., Lu, K.: Hivt: hierarchical vector transformer for multi-agent motion prediction. In: Proceedings of the IEEE/CVF Conference on Computer Vision and Pattern Recognition, pp. 8823–8833 (2022)
51. Zhu, X., Su, W., Lu, L., Li, B., Wang, X., Dai, J.: Deformable detr: deformable transformers for end-to-end object detection. arXiv preprint arXiv:2010.04159 (2020)

Occluded Gait Recognition with Mixture of Experts: An Action Detection Perspective

Panjian Huang[1], Yunjie Peng[2,4], Saihui Hou[1,3(✉)], Chunshui Cao[3], Xu Liu[3], Zhiqiang He[2,4], and Yongzhen Huang[1,3(✉)]

[1] School of Artificial Intelligence, Beijing Normal University, Beijing, China
202231081025@mail.bnu.edu.cn, {housaihui,huangyongzhen}@bnu.edu.cn
[2] School of Computer Science and Technology, Beihang University, Beijing, China
[3] WATRIX.AI, Beijing, China
[4] AI Lab, Lenovo Research, Beijing, China

Abstract. Extensive occlusions in real-world scenarios pose challenges to gait recognition due to missing and noisy information, as well as body misalignment in position and scale. We argue that rich dynamic contextual information within a gait sequence inherently possesses occlusion-solving traits: 1) Adjacent frames with gait continuity allow holistic body regions to infer occluded body regions; 2) Gait cycles allow information integration between holistic actions and occluded actions. Therefore, we introduce an action detection perspective where a gait sequence is regarded as a composition of actions. To detect accurate actions under complex occlusion scenarios, we propose an Action Detection Based Mixture of Experts (GaitMoE), consisting of Mixture of Temporal Experts (MTE) and Mixture of Action Experts (MAE). MTE adaptively constructs action anchors by temporal experts and MAE adaptively constructs action proposals from action anchors by action experts. Especially, action detection as a proxy task with gait recognition is an end-to-end joint training only with ID labels. In addition, due to the lack of a unified occluded benchmark, we construct a pioneering Occluded Gait database (OccGait), containing rich occlusion scenarios and annotations of occlusion types. Extensive experiments on OccGait, OccCASIA-B, Gait3D and GREW demonstrate the superior performance of GaitMoE. OccGait is available at https://github.com/BNU-IVC/OccGait.

Keywords: Occluded Gait Recognition · Dynamic Contextual Information · Action Detection · Mixture of Experts

1 Introduction

Gait recognition has attracted increasing attention and gained broad applications in crime prevention, forensic identification, and social security [44] due to

---
P. Huang and Y. Peng—Equal contribution.

Supplementary Information The online version contains supplementary material available at https://doi.org/10.1007/978-3-031-72658-3_22.

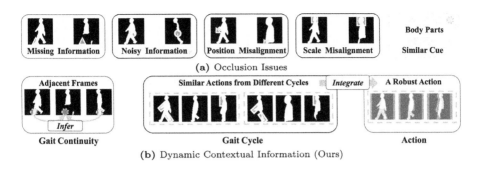

Fig. 1. (a) The main four occlusion issues in gait recognition. (b) The dynamic contextual information within a gait sequence can infer and integrate occlusion information.

its ability to accurately identify walking patterns of pedestrians from a distance in complex surveillance scenarios, e.g., viewing angles, cloth-changing and illumination conditions [40]. Applying upstream tasks such as tracking, segmentation, size normalization, and alignment to preprocess raw videos, the obtained gait representations (e.g., silhouettes or skeletons) make existing methods [3,4,10,12,23–25,31,32,34,52] achieve accurate identification. However, current studies overlook occlusions largely existing in practical scenarios, e.g., occluded by carrying, obstacles, the crowd, or moving out of camera view. As shown in Fig. 1(a), body regions occluded by obstacles or the crowd lead to missing and noisy information, while partial visual body regions with size normalization cause position and scale misalignment, which significantly degrading fine-grained feature matching [56,58]. Direct solutions, e.g., simply discarding or persisting occluded frames, do not adequately address occlusion issues since partially visual body regions in occluded frames may still contain key discriminative regions while allowing them to persist will cause erroneous feature extraction and matching. Therefore, occlusion issues have become one of the biggest bottlenecks in gait recognition.

To address occlusion issues in gait recognition, we rethink the dynamic contextual information within a gait sequence: **(i) Gait Continuity.** Adjacent frames with continual motion enable body regions in holistic frames to infer the same body regions in occluded frames. As shown in Fig. 1(b), for the current frame missing lower body information, the preceding and subsequent holistic frames can still infer approximate motion in the occluded regions. **(ii) Gait Cycle.** As shown in Fig. 1(b), we regard the current misaligned frame and adjacent frames as an action. Combined with the gait cycle, i.e., a gait sequence is formed by the repetition of a series of actions, to discover actions with similar cues, holistic and occluded actions can be integrated into a robust action.

Driven by the above analysis, we introduce a new perspective for occluded gait recognition, "Action Detection". The paradigm of action detection aims to predict the action boundaries and categories from an untrimmed video, where predefined consecutive frames as action anchors represent potential actions and further action anchors with high actionness scores generate action propos-

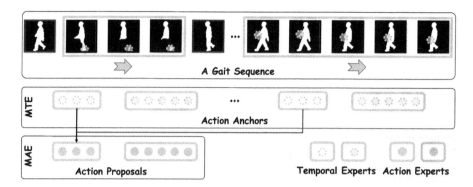

Fig. 2. Action composition. Each temporal expert focuses on one body region with individual temporal size, constructing action anchors. Each action expert integrates similar action anchors from different gait cycles, constructing one action proposal.

als [33,43,59]. Specific to gait recognition, we associate action anchors to represent adjacent frames with gait continuity and further consider the gait cycle to construct action proposals from similar action anchors. Therefore, a gait sequence can be regarded as a composition of actions.

Considering a single model that struggles to capture holistic and diverse actions under complex occlusion scenarios, *e.g.*, the uncertainty of occluded body regions and duration, we propose an Action Detection Based Mixture of Experts (GaitMoE). Mixture of Experts (MoE) follows a divide-and-conquer philosophy, breaking down complex problems into simple sub-problems. Each sub-problem is handled by a dedicated expert, contributing collectively to solve the overall complexity. As shown in Fig. 2, each temporal expert focuses on the corresponding body region and temporal granularity, sliding at the entire gait sequence to construct action anchors. Subsequently, each action expert integrates similar action anchors with one action type, constructing action proposals. Finally, instead of localization and classification in action detection, action proposals are collectively as discriminative features for identification.

Additionally, the absence of publicly available gait databases with quantifiable occlusion metrics poses an extreme challenge for occluded gait recognition. To this end, we establish a pioneering Occluded Gait database (OccGait) with two characteristics: **(i) Diverse Occlusion Scenarios.** Each subject has 4 different types of occlusion situations, including None of Occlusion, Carrying Occlusion, Crowd Occlusion, and Static Occlusion. **(ii) Explicit Occlusion Types.** OccGait provides explicit occlusion types for each gait sequence, which enables to qualify and quantify occlusion issues.

Our main contributions can be summarized as follows:

- To address occlusion challenges, we introduce an action detection perspective where an Action Detection Based Mixture of Experts (GaitMoE) structures a gait sequence as a composition of action.
- To qualify and quantify occlusion issues, we build a novel Occluded Gait recognition benchmark (OccGait), including diverse occlusion scenarios and explicit annotations of occlusion types.

- To evaluate effectiveness and robustness, extensive experimental results on OccGait, OccCASIA-B, Gait3D, and GREW demonstrate that our method significantly outperforms other state-of-the-art methods.

2 Related Work

2.1 Gait Recognition

Gait Recognition is mainly categorized into appearance-based and model-based approaches. Appearance-based methods [3,4,10,12,23–25,31,32,34,52] usually uses templates of compressing a sequence of gait silhouettes (*e.g.*, Gait Energy Image), set of frames and sequence of frames as inputs, extracting fine-grained features (*e.g.*, spatial-temporal and part-level representations). Model-based methods [14,47,48,61–63] explicitly model human body structure, *e.g.*, 2D or 3D skeletons and meshes. Additionally, some researches [2,30,42] take other data types as inputs, such as RGB frames, optical flow and point clouds. However, most of these methods usually neglect the fact that real-world scenarios introduce a significant amount of occlusion.

2.2 Occluded Gait Recognition

We introduce occluded gait recognition from two aspects: (i) **DataBases.** For synthesis-based databases, Chen *et al.* [6] simulate occluded scenarios based on CMU Mobo [15] through adding horizontal or vertical black bars. Uddin *et al.* [50] synthesize relative static and dynamic occlusions based on OUMVLP [45] by a background rectangle mask in a fixed position or gradually changed position. Delgado-Escano *et al.* [9] generate crowd occlusions based on CASIA-B [60] and TUM-GAID [20] by augmenting persons in raw videos. Xu *et al.* [56,58] synthesize occluded scenarios by simulating cropping and size-normalized silhouettes. For real-world databases, Hofmann *et al.* [21] collect TUM-IITKGP, including static occlusions (*e.g.*, standing people, a backpack, gown and hands in pocket), dynamic occlusions (*e.g.*, two walking people). Chattopadhyay *et al.* [5] construct a frontal and occluded gait database by estimating Kinect depth. Li *et al.* [29] present an OG RGB+D dataset captured by Azure Kinect DK sensor, containing occlusions with carrying, clothing and the crowd. However, these databases have some limitations, *e.g.*, the single occlusion scenario, small occlusion regions, or not publicly available yet. (ii) **Architectures.** For reconstruction-based approaches, Xu *et al.* [58] re-normalize and register silhouettes from learned holistic information (*e.g.*, scales) before the following feature extraction and matching process. Xu *et al.* [56] estimate SMPL with pose and shape features from RGB occluded videos. Uddin *et al.* [50] reconstruct a sequence from the occluded sequence by a conditional deep generative adversarial network. Peng *et al.* [37] register and recover occluded silhouettes with a self-supervised alignment module and temporal recovery transformer. Guo *et al.* [16] propose a Physics-Augmented Autoencoder (PAA) that generates physically intermediate representations through a graph-convolution-based encoder and a physics-based decoder,

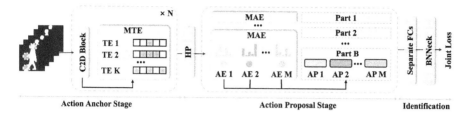

Fig. 3. The overview of GaitMoE. Best viewed in colors, the input sequence is firstly extracted to frame-level features by C2D Block (*e.g.*, 2D CNNs or a residual block), and Mixture of Temporal Experts (MTE) dispatch temporal experts (TE) with different temporal bounding boxes to the corresponding channel segments, forming action anchors. After horizontal partitioning (HP), Mixture of Action Experts (MAE) is independent for each part-level feature. For each channel segment, one action expert (AE) integrates similar action anchors along the temporal dimension by weighted sum operation, forming one action proposal. Finally, The concatenated action proposals as part features for identification.

which enhances the ability to reconstruct input skeleton sequences even when partially occluded. For reconstruction-free approaches, Gupta *et al.* [17] propose a occlusion-aware module by synthetic occlusions to detect occlusion type information to guide gait recognition training. Zhu *et al.* [64] use SMPLify-X that provides the body shape feature decoupled from its pose and strong prior, which enables to generate the complete shape even with mild occlusions.

2.3 Mixture of Experts

Mixture of Experts (MoE) is a sparsely-activated architecture where a router network output weights for aggregating multiple experts [41]. The philosophy of divide and conquer allows MoE to reduce computational cost and increase model capacity, and it has been widely extended to Vision Transformer [28,38,55]. Each expert in MoE is dispatched with one sub-data (*e.g.*, image patches, data from one domain) and maintains specialization, which is named "Expert". This work extends MoE to detect fine-grained actions in occluded gait recognition and makes each expert concentrate on one representative action.

3 Methodology

GaitMoE mainly consists of Mixture of Temporal Experts (MTE) and Mixture of Action Experts (MAE). We give a brief overview of the full process in Fig. 3.

3.1 Mixture of Temporal Experts

Although a gait sequence has filtered texture information and performed coarse-aligned registration, the complex occlusions in real scenarios cause the uncertainty of occluded body regions and duration. To this end, we adopt multi-scale mechanisms in temporal and channel dimensions to extract fine-grained features.

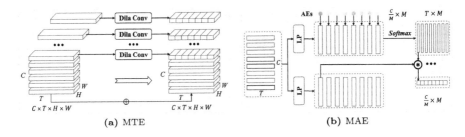

Fig. 4. (a) Mixture of Temporal Experts. Dila Conv (DC) represents Dilated Convolution, predefining action anchors with different dilated ratios. (b) Mixture of Action Experts. LP denotes the linear projection. Similar action anchors adaptively integrate into action proposals.

Action Anchors. As we know, a gait sequence possesses continuity with significant mutual information between each frame and adjacent frames. For example, when we observe a person lifting their leg, it is likely to be followed by a swinging leg. Some existing approaches employ temporal modeling to capture such relationships, *e.g.*, 3D CNNs and LSTMs. However, uncertain and complex occlusions interfere with the dynamic information. To alleviate these issues, Fig. 4(a) shows that MTE predefines various sizes of temporal experts, which are dilated convolutions with different dilated ratios for corresponding channel segments. At each temporal position, each temporal expert independently constructs action anchors from adjacent temporal positions, which is why we name it "Temporal Expert". Considering adjacent occluded frames may introduce noise information to current holistic frames, MTE preserves partial channel segments for adaptively selecting clean regions. Let $\mathcal{X} \in \mathbb{R}^{\mathcal{C} \times \mathcal{T} \times \mathcal{H} \times \mathcal{W}}$ denote frame-level features extracted from silhouettes by C2D Block, where $\mathcal{C}, \mathcal{T}, \mathcal{H}, \mathcal{W}$ represent channel, consecutive $\mathcal{T}$ frames, height and width dimensions. The process of MTE is formulated as follows:

$$\mathcal{Y} = Concat(DC_i(\mathcal{X}_i), i=1,2,\cdots,\mathcal{K}, \mathcal{X}_{[\mathcal{S}-\mathcal{K}+1,\mathcal{S}]}) \tag{1}$$

where $\mathcal{X}_i \in \mathbb{R}^{\frac{\mathcal{C}}{\mathcal{S}} \times \mathcal{T} \times \mathcal{H} \times \mathcal{W}}$, $\mathcal{Y} \in \mathbb{R}^{\mathcal{C} \times \mathcal{T} \times \mathcal{H} \times \mathcal{W}}$, $\mathcal{S}$ is the number of channel segments, $\mathcal{K}$ is the number of temporal experts, DC is dilated convolution, and i is the segment index. In our work, we set $\mathcal{S}=8, \mathcal{K}=4$, and DC_i is 3D CNN with kernel size $(3,1,1)$, stride $(1,1,1)$, padding $(i,0,0)$, and dilated ratio i. In addition, residual learning is embedded within MTE for easing training.

3.2 Mixture of Action Experts

Although action anchors contain rich dynamic information, they may have a large amount of redundant and invalid actions. Therefore, we adopt prototype-based architecture to adaptively select and integrate discriminative and similar action anchors as action proposals. Since a gait sequence can be regarded as the composition of actions, when the sequence is occluded resulting in many invalid actions, the action prototype needs to detect the most discriminative action type and aggregate this action type from occluded and holistic actions. In addition, GaitMoE adopts horizontal pooling (HP) for fine-grained part features $\mathcal{P} \in \mathbb{R}^{C \times T}$, and MAE is independent for each part. Here, we omit the part index for simplicity.

Action Proposals. To capture discriminative actions only with ID labels, MAE shown in Fig. 4(b) predefines a set of learnable action prototypes where an action prototype adaptively learns a type of action for recognition, that is why we name it "Action Expert". For fine-grained action extraction, we dispatch each action expert to the corresponding channel segment (e.g., action anchors), and action experts "watch" contextual action anchors in the entire temporal dimension for filtering action anchors with occlusions, and select and integrate similar action anchors as action proposals. Different to MoEs [13,27,36,41] where the routers generally adopt Top-K selection and sparsely memorize information, recent MoE works has shown remarkable performance with a fixed hash router [39], or convolutional experts [7]. In this work, we introduce a soft selection for balancing the training. Let $\mathcal{A} \in \mathbb{R}^{\frac{C}{M} \times M}$ represent M action experts with $\frac{C}{M}$ dimension, and we obtain action queries $\mathcal{Q}$ by identify mapping on $\mathcal{A}$, action keys $\mathcal{K}$ and values $\mathcal{V}$ by different linear projections on $\mathcal{P}$. Then, we dispatch each action query to the corresponding channel segment of $\mathcal{K}$ to evaluate action anchors by scores where the higher the score, the more reliable the action anchor, and vice versa. To make one action expert concentrate on one most representative action, we use the softmax function to calculate the scores within the corresponding channel segment of $\mathcal{K}$ and weighted sum operation with the corresponding channel segment of $\mathcal{V}$ as an action proposal. The formulation is as follows:

$$\mathcal{Q}_i = \mathcal{A}_i, \quad \mathcal{K}_i = \mathcal{P}_i \mathcal{W}^{\mathcal{K}}, \quad \mathcal{V}_i = \mathcal{P}_i \mathcal{W}^{\mathcal{V}} \qquad (2)$$

$$\mathcal{F}_i = \sum_{j=1}^{T} \mathcal{O}_{i,j} \otimes \mathcal{V}_{i,j}, \quad \mathcal{O}_{i,j} = \frac{exp(\mathcal{Q}_{i,j} \mathcal{K}_{i,j}^T)}{\sum_{j=1}^{T} exp(\mathcal{Q}_{i,j} \mathcal{K}_{i,j}^T)} \qquad (3)$$

where i is the channel segment index, $i \in 1, 2, \cdots, M$, j is the action anchor index along the temporal dimension of the i segment, $j \in 1, 2, \cdots, T$, $\mathcal{W}^{\mathcal{K}}, \mathcal{W}^{\mathcal{V}} \in \mathbb{R}^{\frac{C}{M} \times \frac{C}{M}}$, $\mathcal{P}_i \in \mathbb{R}^{T \times \frac{C}{M}}$. $\mathcal{Q}_{i,j}, \mathcal{K}_{i,j}, \mathcal{V}_{i,j} \in \mathbb{R}^{1 \times \frac{C}{M}}$. Finally, we obtain $\mathcal{F}$ as one part feature by concatenating action proposals $[\mathcal{F}_1, \mathcal{F}_2, \cdots, \mathcal{F}_M]$ along the channel dimension. $\mathcal{F}$ is fed into Separate FCs and BNNeck for identification.

3.3 Joint Loss

GaitMoE is an end-to-end joint learning framework only with ID labels, introducing action detection as a proxy task with gait recognition. The joint loss includes two types: Triplet Loss [19] $\mathcal{L}_{tp}$ and Cross Entropy Loss $\mathcal{L}_{ce}$, which constrains each part independently. This formulation is as follows:

$$\mathcal{L} = \mathcal{L}_{tp} + \beta \mathcal{L}_{ce} \qquad (4)$$

where the hyper-parameter β is for balancing the two terms.

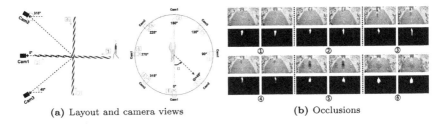

Fig. 5. (a) The layout and camera views during collection. (b) The 3 types of occlusion scenarios where the number ①② are for Carrying Occlusion, ③④ are for Crowd Occlusion and ⑤⑥ are for Static Occlusion, and None Occlusion is omitted for simplicity.

4 The OccGait Benchmark

Due to the absence of a comprehensive gait database for quantitatively and qualitatively analyzing the impact of various occlusion types, we collect the Occluded Gait Database (OccGait), including 101 subjects with 4 types of occluded scenarios, 8 camera views, and over $80k$ sequences. *It is worth noting that we hope the OccGait can serve as a starting point to promote robust gait recognition for practical applications, similar to the transition from the indoor databases, e.g., CASIA-X Series* [44,46,53,60] *and OU-X Series* [1,22,26,35,45,49,51,57] *to the wild databases, e.g., Gait3D* [62] *and GREW* [65].

4.1 Data Collection and Pre-Processing

The OccGait is collected in an indoor gait recognition laboratory. During OccGait collection, we obtain authorization from all subjects who are informed for academic data collection in advance. Privacy is also the highest priority in our research. As the left in Fig. 5(a), we place 3 cameras (Cam1 of $0°$, Cam2 of $45°$, Cam3 of $315°$) with 1920×1080 resolution in the square area. During the data collection process, the subjects follow this walk route (*i.e.*, 1-2-3-4). We filter out data with overlapping camera views caused by the combination of 3 cameras and walking directions, and obtain gait sequences with 8 camera views on the right in Fig. 5(a). To qualify and quantify realistic and complex occluded scenarios, we set 4 types of occluded situations with diverse occlusions shown in Fig. 5(b), which are None of Occlusion (*i.e.*, Normal Walking as NM), Carrying Occlusion (CA, ①②), Crowd Occlusion (CR, ③④), and Static Occlusion (ST, ⑤⑥).

All subjects walk the route four times for NM and two times for CA, CR and ST, respectively, which denotes NM01, NM02, NM03, NM04, CA01, CA02, CR01, CR02, ST01 and ST02.

None of Occlusion. To simulate walking status in real scenarios, subjects in their clothing, walk with their walking speed. As shown in Fig. 5(a), there are no obstacles in this situation, and the full body of each subject is fully visible.

Carrying Occlusion. As shown in Fig. 5(b)(①②), we establish two common carrying scenarios for daily life: umbrellas and luggage. People with an umbrella on a rainy day partially obstruct the upper of the body, while luggage occludes both the torso and the lower of the body.

Crowd Occlusion. Gait recognition often requires retrieval in crowded scenes, and human body occlusion can lead to significant interference, such as occlusion of body edges and incorrect dynamic information, as shown in Fig. 5(b)(③④). We design two types of crowd occlusion, the subject walking with another person in different directions (opposite and parallel), which results in partial occlusion of gait sequences at certain moments and complete occlusion of gait sequences at all times.

Static Occlusion. Gait recognition will be deployed in a wide range of scenarios, such as in squares, malls, and other locations with numerous static obstructions. Figure 5(b)(⑤) demonstrates our placement of plants and chairs in the walking route. Figure 5(b)(⑥) shows that the complex and irregular static obstructions hinder the lower body region.

Data Pre-processing. We adopt MaskFormer [8], a segmentation algorithm pre-trained on large datasets (including numerous occluded scenarios), to extract silhouettes from the original RGB data as input.

4.2 Evaluation Protocol

OccGait is divided into training and testing sets. The 51 individuals with odd numbers as the training set while the remaining 50 with even numbers as the test set. Our gait evaluation follows the protocols of the previous gait evaluation [60]. Given a query sequence, we measure its distance to each sequence in the gallery, retrieving the subject with the closest distance from the gallery. To quantify occluded gait analysis, we use the NM01 and NM02 of each subject in the testing set as the gallery, evaluating their Rank 1 performance under different occlusions and viewing angles.

5 Experiments

5.1 Datasets

We first conduct extensive qualitative and quantitative occlusion analyses on OccGait and OccCASIA-B [37]. Subsequently, we further validate the generalizability and practicality of our method on Gait3D [62] and GREW [65].

OccGait is for real-scenario occlusion evaluation built by this work. The details have been discussed in Sect. 4.

OccCASIA-B is a synthetic occluded gait database [37] from CASIA-B and has similar basic statistics, containing 124 subjects, 3 different walking situations, *etc..*, Walking in Normal (NM), Walking with a Bag (BG) and Walking with Different Clothes (CL), 11 camera views from uniform interval of 18° in

Table 1. The Rank-1 accuracy (%) on OccGait for different probe views excluding the identical-view cases. For evaluation, the sequences of NM01 and NM02 for each subject are taken as the gallery. The benchmark adopts None Occlusion (NO), Carrying Occlusion (CA), Crowd Occlusion (CR) and Static Occlusion (ST).

	Method	Probe View								Average
		0°	45°	90°	135°	180°	225°	270°	315°	
NM	GaitSet [4]	65.7	91.7	89.1	90.7	66.7	91.9	89.4	92.1	84.7
	GaitPart [12]	62.9	92.0	88.9	89.6	59.6	89.9	87.3	90.7	82.6
	GaitGL [32]	73.9	94.3	92.6	93.4	68.1	93.1	91.7	92.6	87.5
	STOR [37]	73.7	94.6	92.0	93.3	73.3	94.1	92.6	92.6	88.3
	GaitBase [11]	68.4	91.6	88.7	91.1	74.9	93.6	88.1	91.7	86.0
	GaitMoE(ours)	**81.0**	**96.0**	**94.0**	**95.1**	**81.3**	**94.7**	**94.1**	**94.7**	**91.4**
CA	GaitSet [4]	50.1	74.3	79.7	79.1	47.4	73.6	74.0	76.0	69.3
	GaitPart [12]	42.0	69.9	74.9	78.6	37.6	66.4	66.0	63.7	62.4
	GaitGL [32]	48.6	79.7	84.6	87.9	38.7	76.7	78.1	70.4	70.6
	STOR [37]	55.9	83.6	83.6	86.3	54.4	81.7	83.1	81.7	76.3
	GaitBase [11]	58.1	82.0	84.3	85.6	54.3	79.9	80.1	79.3	75.4
	GaitMoE(ours)	**68.3**	**87.7**	**88.9**	**89.6**	**61.7**	**86.4**	**86.6**	**87.3**	**82.1**
CR	GaitSet [4]	58.3	84.7	80.1	77.4	52.3	77.9	77.6	85.6	74.2
	GaitPart [12]	48.1	81.9	76.4	69.6	39.0	67.6	68.1	79.3	66.3
	GaitGL [32]	47.4	89.0	81.3	77.1	41.0	75.0	77.1	87.6	71.9
	STOR [37]	55.6	88.4	86.0	81.7	52.0	81.7	83.0	88.4	77.1
	GaitBase [11]	62.4	86.9	83.1	82.1	60.0	85.4	80.7	87.6	78.5
	GaitMoE(ours)	**63.1**	**90.6**	**86.6**	**84.4**	**57.6**	**81.6**	**84.9**	**90.3**	**79.9**
ST	GaitSet [4]	54.1	86.3	86.7	82.3	54.1	86.7	86.4	83.6	74.2
	GaitPart [12]	44.6	83.7	85.7	77.3	39.1	83.4	84.1	77.4	71.9
	GaitGL [32]	36.7	87.7	90.7	80.0	37.6	86.6	90.4	82.3	74.0
	STOR [37]	50.3	90.0	91.3	87.4	54.9	90.1	91.3	89.1	80.6
	GaitBase [11]	57.7	89.9	87.4	85.9	59.9	88.9	87.4	85.4	80.3
	GaitMoE(ours)	**62.9**	**93.3**	**92.1**	**89.4**	**63.6**	**91.1**	**93.6**	**91.4**	**84.7**

[0°, 180°]. To simulate occlusion situations, OccCASIA-B sets 4 types of occlusions: None Occlusion (NO), Crowd Occlusion (CO), Static Occlusion (SO) and Detect Occlusion (DO), which denotes a walking person without occlusions, occluded by another one in a crowded area, occluded by static occlusions, e.g., benches, bicycles and fire hydrants, and losing body regions in the up, down, left, or right direction. The OccCASIA-B takes the first 74 subjects as the training set where each gait sequence with 0.6 of occlusion probability generates one of the 4 types of occluded scenarios. The remaining 50 subjects are used for occlusion evaluation. For each occlusion benchmark, except for the first 4 NM sequences used as the holistic gallery set, the remaining sequences are generated with the corresponding occlusion scenario.

Gait3D samples two segments of continuous two-hour video clips from each of seven-day raw videos in a supermarket, including complex covariates (e.g., occlusions, view angles) for practical gait recognition. It contains 3000 subjects with 25309 sequences, taking 2000 subjects as the training dataset and 1000 subjects as the testing dataset.

GREW is a large-scale wild gait database, containing 26345 subjects with 128671 sequences captured by 882 cameras. It provides 4 types of silhouettes,

Table 2. The Rank-1 accuracy (%) on OccCAISA-B across different views, excluding the identical-view cases. The NO, CO, SO, and DO denote the testing sets of Non-Occlusion, Crowd Occlusion, Static Occlusion, and Detection Occlusion accordingly. Based on the walking condition, probe sequences are grouped into Normal Walking (NM), Carrying Bags (BG), and Cloth-changing Condition (CL).

Methods	NO				CO				SO				DO			
	NM	BG	CL	Mean	NM	BG	CL	Mean	NM	BG	CL	Mean	NM	BG	CL	Mean
GaitSet [4]	92.4	83.0	65.2	80.2	80.7	69.6	50.4	66.9	86.0	76.6	57.8	73.5	85.7	72.7	52.7	70.4
GaitPart [12]	92.3	84.9	68.9	82.0	80.2	70.6	53.3	68.1	85.0	77.3	60.5	74.3	80.7	68.4	52.4	67.2
GaitGL [32]	94.5	89.2	75.3	86.4	84.4	74.8	57.7	72.3	87.4	81.8	66.9	78.7	86.2	76.2	61.5	74.6
STOR [37]	95.9	90.9	77.1	88.0	88.8	80.7	64.3	77.9	91.2	85.6	69.9	82.3	93.2	86.3	70.1	83.2
GaitBase [11]	94.4	88.5	68.7	83.9	87.6	78.0	57.5	74.4	90.1	82.7	62.0	78.3	89.9	80.4	58.5	76.3
GaitMoE(ours)	**96.2**	**91.5**	**80.7**	**89.5**	**90.0**	**83.3**	**68.3**	**80.5**	**91.9**	**86.0**	**73.9**	**83.9**	**93.4**	**87.3**	**75.3**	**85.3**

Table 3. Comparisons on Gait3D and GREW.

Method	Venue	Gait3D		GREW	
		Rank-1	mAP	Rank-1	Rank-5
GaitSet [4]	AAAI19	36.7	30.0	46.3	63.6
GaitPart [12]	CVPR20	28.2	47.6	44.0	60.7
GaitGL [32]	ICCV21	29.7	22.3	47.3	63.6
SMPLGait [62]	CVPR22	46.3	37.2	–	–
MTSGait [62]	MM22	48.7	37.6	55.3	71.3
GaitBase [11]	CVPR23	64.6	–	60.1	–
DANet [34]	CVPR23	48.0	–	–	–
GaitGCI [10]	CVPR23	50.3	39.5	68.5	80.8
DyGait [54]	ICCV23	66.3	56.4	71.4	83.2
HSTL [52]	ICCV23	61.3	55.5	62.7	76.6
GaitMoE-T(ours)	–	71.3	**62.5**	74.4	84.9
GaitMoE-B(ours)	–	**73.7**	66.2	**79.6**	**89.1**

optical flow, and 2D/3D human poses. The benchmark takes 20000 subjects as the training dataset and 6000 subjects as the testing dataset, and each subject provides two sequences for the gallery set and two sequences for the probe set.

5.2 Implementation Details

We provide details about the training process. **Inputs.** All datasets are resized to 64×44. In addition, we employ spatial alignment module [37] as pre-processing to re-align input silhouettes for OccGait and OccCASIA-B. We adopt batch size [P, K] and the number of iterations, [8, 16], $40K$ for OccGait, OccCASIA-B, [32, 4], $60K$ for Gait3D, and [32, 4], $180K$ for GREW. We sample 30 frames of each gait sequence in the training stage and all frames are used for inference. **Network.** For OccGait and OccCASIA-B, we stack three 2D convolution blocks as our Baseline with the number of channels (64, 128, 256). Each 2D convolution block is followed by an MTE. After Horizontal Pooling with the part parameter of 64, we set individual MAE for each part as in Fig. 4(b) and $\mathcal{M}$ is set to 16.

Table 4. Impact of MTE and MAE on OccGait, OccCASIA-B and Gait3D.

		OccGait				OccCASIA-B				Gait3D	
MTE	MAE	NM	CA	CR	ST	NO	CO	SO	DO	Rank-1	mAP
		87.7	69.0	75.1	77.2	85.8	75.9	79.6	80.0	59.2	48.6
✓		89.7	78.3	77.3	81.2	88.2	76.9	82.1	83.6	66.2	56.6
	✓	89.4	78.4	77.4	81.8	87.6	78.3	81.7	82.5	67.2	57.3
✓	✓	**91.4**	**82.1**	**79.9**	**84.7**	**89.5**	**80.5**	**83.9**	**85.3**	**71.3**	**62.5**

Table 5. The number of Action Experts in MAE on OccGait.

MTE	MAE	NM	CA	CR	ST	Mean
✓	1	89.5	78.3	78.0	82.1	82.0
✓	8	90.6	80.7	79.1	83.3	83.4
✓	16	**91.4**	**82.1**	**79.9**	**84.7**	**84.5**
✓	32	90.8	81.2	79.9	84.6	84.1

For Gait3D and GREW, we replace our Baseline with GaitBase-like architecture (4 residual blocks or 10 residual blocks) [11,18], setting channels to (64, 128, 256, 512). More Details are shown in *Supplementary Materials*. **Optimization.** GaitMoE is an end-to-end joint training framework only with ID labels. We use the optimizer of SGD with an initial learning rate of 0.1, which is decreasing by a factor of 0.1 per $[10K, 20K, 30K]$, $[10K, 20K, 30K]$, $[20K, 40K, 50K]$, $[80K, 120K, 150K]$ for OccGait, OccCASIA-B, Gait3D and GREW, respectively. For joint loss, we set $\beta = 0.1$ for OccGait, OccCASIA-B, and $\beta = 1.0$ for Gait3D and GREW. All the models are trained on NVIDIA 8×3090 GPUs.

5.3 Main Results

To evaluate the effectiveness of GaitMoE, we first compare with other state-of-the-art methods on OccGait and OccCASIA-B with controlled occlusions, and then further make comparisons on Gait3D and GREW with complex covariates (*e.g.*, uncertain occlusions).

Comparison on OccGait. To qualitatively and quantitatively validate Gait-MoE, we build the real-scenario gait database, OccGait, containing a wide range of occlusions and explicit annotations. Meanwhile, spatial and scale misalignment may occur in all occlusion scenarios. Table 1 shows that GaitMoE outperforms other state-of-the-art methods under all types of occlusions. Notably, CR causes interference from other pedestrians to overshadow the main subject information at certain angles, causing the model to overly focus on the obstructer, not the target. However, the Average results still show SoTA in the main manuscript, which validates the occlusion-solving traits.

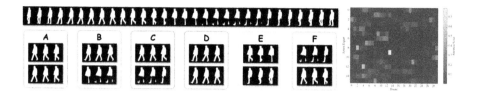

Fig. 6. The visualization of action composition. Here are an occluded gait sequence, action proposals and the selection map of action anchors. Alphabet denotes the action proposal. Each rectangle denotes one action proposal, and each row within one action proposal denotes an action anchor.

Comparison on OccCASIA-B. Table 2 shows the overall results where set-based and temporal-based gait recognition methods (*i.e.*, GaitSet, GaitPart, GaitGL, GaitBase) suffer from severe degradation under the occlusion scenario, comparing to their performances in their paper on holistic CASIA-B. Although STOR with silhouette registration alleviates the misalignment issue, the complex walking patterns under occlusions make the feature extraction difficult. Our proposed method outperforms all methods by a large margin, especially in the extremely challenging cloth-changing (CL) condition, *e.g.*, exceeds GaitBase by 12.0% in **NO**, and 16.8% in **DO**. The experimental results demonstrate that GaitMoE effectively filters invalid occluded actions and extracts robust actions.

Comparisons on Gait3D and GREW. We have validated the effectiveness of our method on the controlled occlusion environment (*i.e.*, OccCASIA-B and OccGait). The large-scale wild gait databases also provide complex and uncertain occlusion scenarios. As shown in Table 3, GaitMoE-T (4 residual blocks) and GaitMoE-B (10 residual blocks) also achieve the highest performance among state-of-the-art methods, which further proves the generalizability and practicality of our method.

5.4 Ablation Study

In this section, we mainly qualitatively and quantitatively validate the MTE and MAE in OccGait, OccCASIA-B and Gait3D. Besides, we provide the visualization of action detection on OccCASIA-B.

The Effectiveness of MTE and MAE. Table 4 shows that each of MTE and MAE can extract better dynamic information. Especially, the combination of MTE and MAE *i.e.*, GaitMoE, achieves a significant increase over Baseline, which proves that our proposed method discovers the representative actions by first coarse action extraction (*i.e.*, action anchors), and then fine-grained action extraction (*i.e.*, action proposals). Besides, we select (8, 4) for ($\mathcal{S}$, $\mathcal{K}$) on temporal experts since a smaller $\mathcal{S}$ or larger $\mathcal{K}$ degrades original information, and a larger $\mathcal{S}$ or smaller $\mathcal{K}$ restricts dynamic information. More Details are

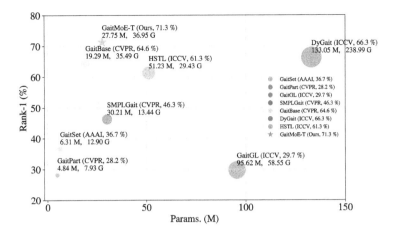

Fig. 7. The model comparisons on accuracy and efficiency. Rank-1 (%), Param. (M) and FLOPs. (G) on Gait3D.

shown in *Supplementary Materials*. We also quantify the impact on the number of action experts. Table 5 illustrates that more experts may result in redundant actions, while fewer experts may not be able to capture the diverse actions. We select 16 as the parameter of GaitMoE.

The Visualization of Action Composition. To better understand and interpret the action detection in gait recognition, we visualize the key module MAE of action detection in Fig. 6, we select part index 60, action anchor with 2 dilated ratios and 6 action proposals as the example. MTE enables to infer the occluded region information by consecutive frames. For an example in E, the right occluded frame can hallucinate the missing region through the middle frame. MAE enables to select and integrate the most discriminative similar action anchors from different gait cycles. For an example in F, although missing information occurs, MAE combines multiple similar action anchors to further confirm this action type (*i.e.*, swinging legs), integrating occluded information.

Trade-off between Accuracy and Efficiency. As Fig. 7 shows, we compare all parameters of these models. For FLOPs, models input a 30-frame gait sequence without Separate FCs and BNNeck for significant comparisons. GaitMoE makes a trade-off and achieves SoTA without substantially increasing computational cost. In contrast, DyGait demands significant computation due to 3D convolutions and GaitBase offers better efficiency with 2D convolutions but lower accuracy.

5.5 Conclusion and Limitations

In this paper, we introduce an action detection perspective where a gait sequence is regarded as a composition of actions, allowing holistic body regions to infer occluded body regions and information integration between holistic and occluded actions. To detect accurate actions under complex occlusion scenarios, we propose an Action Detection Based Mixture of Experts (GaitMoE) to leverage dynamic contextual information *i.e.*, gait continuity and gait cycle, to construct action anchors and action proposals. To obtain qualitative and quantitative occlusion analysis, we propose a novel Occluded Gait Recognition benchmark (OccGait) as a pioneering database with a wide range of occlusion scenarios and explicit annotations. Extensive experimental results have demonstrated that GaitMoE effectively captures accurate and robust actions for occluded gait recognition. In addition. we provide some limitations where the selection map of action anchors shown in Fig. 6 shows that some of the experts in GaitMoE present redundancy and similarity. We will explore the optimization and design for expert selection in the future.

Acknowledgment. This work is jointly supported by National Natural Science Foundation of China (62276025, 62206022), Beijing Municipal Science & Technology Commission (Z231100007423015) and Shenzhen Technology Plan Program (KQTD20170331093217368).

References

1. An, W., et al.: Performance evaluation of model-based gait on multi-view very large population database with pose sequences. IEEE Trans. Biometrics Behav. Identity Sci. **2**(4), 421–430 (2020)
2. Bashir, K., Xiang, T., Gong, S., Mary, Q., et al.: Gait representation using flow fields. In: BMVC, pp. 1–11 (2009)
3. Chai, T., Li, A., Zhang, S., Li, Z., Wang, Y.: Lagrange motion analysis and view embeddings for improved gait recognition. In: Proceedings of the IEEE/CVF Conference on Computer Vision and Pattern Recognition, pp. 20249–20258 (2022)
4. Chao, H., He, Y., Zhang, J., Feng, J.: Gaitset: regarding gait as a set for cross-view gait recognition. In: Proceedings of the AAAI Conference on Artificial Intelligence, vol. 33, pp. 8126–8133 (2019)
5. Chattopadhyay, P., Sural, S., Mukherjee, J.: Frontal gait recognition from occluded scenes. Pattern Recogn. Lett. **63**, 9–15 (2015)
6. Chen, C., Liang, J., Zhao, H., Hu, H., Tian, J.: Frame difference energy image for gait recognition with incomplete silhouettes. Pattern Recogn. Lett. **30**(11), 977–984 (2009)
7. Chen, X., Li, H., Li, M., Pan, J.: Learning a sparse transformer network for effective image deraining. In: Proceedings of the IEEE/CVF Conference on Computer Vision and Pattern Recognition, pp. 5896–5905 (2023)
8. Cheng, B., Misra, I., Schwing, A.G., Kirillov, A., Girdhar, R.: Masked-attention mask transformer for universal image segmentation. In: Proceedings of the IEEE/CVF Conference on Computer Vision and Pattern Recognition, pp. 1290–1299 (2022)
9. Delgado-Escano, R., Castro, F.M., Cózar, J.R., Marin-Jimenez, M.J., Guil, N.: Mupeg-the multiple person gait framework. Sensors **20**(5), 1358 (2020)
10. Dou, H., Zhang, P., Su, W., Yu, Y., Lin, Y., Li, X.: Gaitgci: generative counterfactual intervention for gait recognition. In: Proceedings of the IEEE/CVF Conference on Computer Vision and Pattern Recognition, pp. 5578–5588 (2023)

11. Fan, C., Liang, J., Shen, C., Hou, S., Huang, Y., Yu, S.: Opengait: revisiting gait recognition towards better practicality. In: Proceedings of the IEEE/CVF Conference on Computer Vision and Pattern Recognition, pp. 9707–9716 (2023)
12. Fan, C., et al.: Gaitpart: temporal part-based model for gait recognition. In: Proceedings of the IEEE/CVF Conference on Computer Vision and Pattern Recognition, pp. 14225–14233 (2020)
13. Fedus, W., Zoph, B., Shazeer, N.: Switch transformers: scaling to trillion parameter models with simple and efficient sparsity. J. Mach. Learn. Res. **23**(1), 5232–5270 (2022)
14. Fu, Y., Meng, S., Hou, S., Hu, X., Huang, Y.: Gpgait: generalized pose-based gait recognition. arXiv preprint arXiv:2303.05234 (2023)
15. Gross, R.: The cmu motion of body (mobo) database. Carnegie Mellon University. The Robotics Institute (2001)
16. Guo, H., Ji, Q.: Physics-augmented autoencoder for 3d skeleton-based gait recognition. In: Proceedings of the IEEE/CVF International Conference on Computer Vision, pp. 19627–19638 (2023)
17. Gupta, A., Chellappa, R.: You can run but not hide: improving gait recognition with intrinsic occlusion type awareness. In: Proceedings of the IEEE/CVF Winter Conference on Applications of Computer Vision, pp. 5893–5902 (2024)
18. He, K., Zhang, X., Ren, S., Sun, J.: Deep residual learning for image recognition. In: Proceedings of the IEEE Conference on Computer Vision and Pattern Recognition, pp. 770–778 (2016)
19. Hermans, A., Beyer, L., Leibe, B.: In defense of the triplet loss for person re-identification. arXiv preprint arXiv:1703.07737 (2017)
20. Hofmann, M., Geiger, J., Bachmann, S., Schuller, B., Rigoll, G.: The tum gait from audio, image and depth (gaid) database: multimodal recognition of subjects and traits. J. Vis. Commun. Image Represent. **25**(1), 195–206 (2014)
21. Hofmann, M., Wolf, D., Rigoll, G.: Identification and reconstruction of complete gait cycles for person identification in crowded scenes. In: Proceedings of International Conference on Computer Vision Theory and Applications (VISAPP), Algarve, Portugal (2011)
22. Hossain, M.A., Makihara, Y., Wang, J., Yagi, Y.: Clothing-invariant gait identification using part-based clothing categorization and adaptive weight control. Pattern Recogn. **43**(6), 2281–2291 (2010)
23. Hou, S., Cao, C., Liu, X., Huang, Y.: Gait lateral network: learning discriminative and compact representations for gait recognition. In: Vedaldi, A., Bischof, H., Brox, T., Frahm, J.-M. (eds.) ECCV 2020. LNCS, vol. 12354, pp. 382–398. Springer, Cham (2020). https://doi.org/10.1007/978-3-030-58545-7_22
24. Hou, S., Liu, X., Cao, C., Huang, Y.: Set residual network for silhouette-based gait recognition. IEEE Trans. Biometrics Behav. Identity Sci. **3**(3), 384–393 (2021)
25. Huang, X., et al.: Context-sensitive temporal feature learning for gait recognition. In: Proceedings of the IEEE/CVF International Conference on Computer Vision, pp. 12909–12918 (2021)
26. Iwama, H., Okumura, M., Makihara, Y., Yagi, Y.: The ou-isir gait database comprising the large population dataset and performance evaluation of gait recognition. IEEE Trans. Inf. Forensics Secur. **7**(5), 1511–1521 (2012)
27. Lepikhin, D., et al.: Gshard: scaling giant models with conditional computation and automatic sharding. arXiv preprint arXiv:2006.16668 (2020)
28. Li, B., Yang, J., Ren, J., Wang, Y., Liu, Z.: Sparse fusion mixture-of-experts are domain generalizable learners. arXiv e-prints pp. arXiv–2206 (2022)

29. Li, N., Zhao, X.: A multi-modal dataset for gait recognition under occlusion. Appl. Intell. **53**(2), 1517–1534 (2023)
30. Liang, J., Fan, C., Hou, S., Shen, C., Huang, Y., Yu, S.: Gaitedge: beyond plain end-to-end gait recognition for better practicality. arXiv preprint arXiv:2203.03972 (2022)
31. Lin, B., Zhang, S., Bao, F.: Gait recognition with multiple-temporal-scale 3d convolutional neural network. In: Proceedings of the 28th ACM International Conference on Multimedia, pp. 3054–3062 (2020)
32. Lin, B., Zhang, S., Yu, X.: Gait recognition via effective global-local feature representation and local temporal aggregation. In: Proceedings of the IEEE/CVF International Conference on Computer Vision, pp. 14648–14656 (2021)
33. Lin, C., et al.: Learning salient boundary feature for anchor-free temporal action localization. In: Proceedings of the IEEE/CVF Conference on Computer Vision and Pattern Recognition, pp. 3320–3329 (2021)
34. Ma, K., Fu, Y., Zheng, D., Cao, C., Hu, X., Huang, Y.: Dynamic aggregated network for gait recognition. In: Proceedings of the IEEE/CVF Conference on Computer Vision and Pattern Recognition, pp. 22076–22085 (2023)
35. Makihara, Y., Mannami, H., Yagi, Y.: Gait analysis of gender and age using a large-scale multi-view gait database. In: Kimmel, R., Klette, R., Sugimoto, A. (eds.) ACCV 2010. LNCS, vol. 6493, pp. 440–451. Springer, Heidelberg (2011). https://doi.org/10.1007/978-3-642-19309-5_34
36. Mustafa, B., Riquelme, C., Puigcerver, J., Jenatton, R., Houlsby, N.: Multimodal contrastive learning with limoe: the language-image mixture of experts. Adv. Neural. Inf. Process. Syst. **35**, 9564–9576 (2022)
37. Peng, Y., Cao, C., He, Z.: Occluded gait recognition. In: 2023 International Joint Conference on Neural Networks (IJCNN), pp. 1–8. IEEE (2023)
38. Riquelme, C., et al.: Scaling vision with sparse mixture of experts. Adv. Neural. Inf. Process. Syst. **34**, 8583–8595 (2021)
39. Roller, S., Sukhbaatar, S., Weston, J., et al.: Hash layers for large sparse models. Adv. Neural. Inf. Process. Syst. **34**, 17555–17566 (2021)
40. Sepas-Moghaddam, A., Etemad, A.: Deep gait recognition: a survey. IEEE Trans. Pattern Anal. Mach. Intell. **45**, 264–284 (2022)
41. Shazeer, N., et al.: Outrageously large neural networks: the sparsely-gated mixture-of-experts layer. arXiv preprint arXiv:1701.06538 (2017)
42. Shen, C., Fan, C., Wu, W., Wang, R., Huang, G.Q., Yu, S.: Lidargait: benchmarking 3d gait recognition with point clouds. In: Proceedings of the IEEE/CVF Conference on Computer Vision and Pattern Recognition, pp. 1054–1063 (2023)
43. Shi, D., Zhong, Y., Cao, Q., Ma, L., Li, J., Tao, D.: Tridet: temporal action detection with relative boundary modeling. In: Proceedings of the IEEE/CVF Conference on Computer Vision and Pattern Recognition, pp. 18857–18866 (2023)
44. Song, C., Huang, Y., Wang, W., Wang, L.: Casia-e: a large comprehensive dataset for gait recognition. IEEE Trans. Pattern Anal. Mach. Intell. **45**(3), 2801–2815 (2022)
45. Takemura, N., Makihara, Y., Muramatsu, D., Echigo, T., Yagi, Y.: Multi-view large population gait dataset and its performance evaluation for cross-view gait recognition. IPSJ Trans. Comput. Vision Appl. **10**, 1–14 (2018)
46. Tan, D., Huang, K., Yu, S., Tan, T.: Efficient night gait recognition based on template matching. In: 18th International Conference on Pattern Recognition (ICPR 2006), vol. 3, pp. 1000–1003. IEEE (2006)

47. Teepe, T., Gilg, J., Herzog, F., Hörmann, S., Rigoll, G.: Towards a deeper understanding of skeleton-based gait recognition. In: Proceedings of the IEEE/CVF Conference on Computer Vision and Pattern Recognition, pp. 1569–1577 (2022)
48. Teepe, T., Khan, A., Gilg, J., Herzog, F., Hörmann, S., Rigoll, G.: Gaitgraph: graph convolutional network for skeleton-based gait recognition. In: 2021 IEEE International Conference on Image Processing (ICIP), pp. 2314–2318. IEEE (2021)
49. Tsuji, A., Makihara, Y., Yagi, Y.: Silhouette transformation based on walking speed for gait identification. In: 2010 IEEE Computer Society Conference on Computer Vision and Pattern Recognition, pp. 717–722. IEEE (2010)
50. Uddin, M.Z., Muramatsu, D., Takemura, N., Ahad, M.A.R., Yagi, Y.: Spatio-temporal silhouette sequence reconstruction for gait recognition against occlusion. IPSJ Trans. Comput. Vision Appl. **11**(1), 1–18 (2019)
51. Uddin, M.Z., et al.: The ou-isir large population gait database with real-life carried object and its performance evaluation. IPSJ Trans. Comput. Vision Appl. **10**(1), 1–11 (2018)
52. Wang, L., Liu, B., Liang, F., Wang, B.: Hierarchical spatio-temporal representation learning for gait recognition. arXiv preprint arXiv:2307.09856 (2023)
53. Wang, L., Tan, T., Ning, H., Hu, W.: Silhouette analysis-based gait recognition for human identification. IEEE Trans. Pattern Anal. Mach. Intell. **25**(12), 1505–1518 (2003)
54. Wang, M., et al.: Dygait: exploiting dynamic representations for high-performance gait recognition. arXiv preprint arXiv:2303.14953 (2023)
55. Wang, W., et al.: Image as a foreign language: beit pretraining for all vision and vision-language tasks. arXiv preprint arXiv:2208.10442 (2022)
56. Xu, C., Makihara, Y., Li, X., Yagi, Y.: Occlusion-aware human mesh model-based gait recognition. IEEE Trans. Inf. Forensics Secur. **18**, 1309–1321 (2023)
57. Xu, C., Makihara, Y., Ogi, G., Li, X., Yagi, Y., Lu, J.: The ou-isir gait database comprising the large population dataset with age and performance evaluation of age estimation. IPSJ Trans. Comput. Vision Appl. **9**(1), 1–14 (2017)
58. Xu, C., Tsuji, S., Makihara, Y., Li, X., Yagi, Y.: Occluded gait recognition via silhouette registration guided by automated occlusion degree estimation. In: Proceedings of the IEEE/CVF International Conference on Computer Vision, pp. 3199–3209 (2023)
59. Yang, L., Peng, H., Zhang, D., Fu, J., Han, J.: Revisiting anchor mechanisms for temporal action localization. IEEE Trans. Image Process. **29**, 8535–8548 (2020)
60. Yu, S., Tan, D., Tan, T.: A framework for evaluating the effect of view angle, clothing and carrying condition on gait recognition. In: 18th International Conference on Pattern Recognition (ICPR 2006), vol. 4, pp. 441–444. IEEE (2006)
61. Zhang, C., Chen, X.P., Han, G.Q., Liu, X.J.: Spatial transformer network on skeleton-based gait recognition. Expert Syst. e13244 (2023)
62. Zheng, J., Liu, X., Liu, W., He, L., Yan, C., Mei, T.: Gait recognition in the wild with dense 3d representations and a benchmark. In: Proceedings of the IEEE/CVF Conference on Computer Vision and Pattern Recognition, pp. 20228–20237 (2022)
63. Zhu, H., Zheng, W., Zheng, Z., Nevatia, R.: Gaitref: gait recognition with refined sequential skeletons. arXiv preprint arXiv:2304.07916 (2023)
64. Zhu, H., Zheng, Z., Nevatia, R.: Gait recognition using 3-d human body shape inference. In: Proceedings of the IEEE/CVF Winter Conference on Applications of Computer Vision, pp. 909–918 (2023)
65. Zhu, Z., et al.: Gait recognition in the wild: a benchmark. In: Proceedings of the IEEE/CVF International Conference on Computer Vision, pp. 14789–14799 (2021)

EDTalk: Efficient Disentanglement for Emotional Talking Head Synthesis

Shuai Tan[1], Bin Ji[1], Mengxiao Bi[2], and Ye Pan[1](✉)

[1] Shanghai Jiao Tong University, Shanghai, China
{tanshuai0219,bin.ji,whitneypanye}@sjtu.edu.cn
[2] NetEase Fuxi AI Lab, Beijing, China
bimengxiao@corp.netease.com

Abstract. Achieving disentangled control over multiple facial motions and accommodating diverse input modalities greatly enhances the application and entertainment of the talking head generation. This necessitates a deep exploration of the decoupling space for facial features, ensuring that they **a)** operate independently without mutual interference and **b)** can be preserved to share with different modal inputs-both aspects often neglected in existing methods. To address this gap, this paper proposes a novel **E**fficient **D**isentanglement framework for **Talk**ing head generation (**EDTalk**). Our framework enables individual manipulation of mouth shape, head pose, and emotional expression, conditioned on video or audio inputs. Specifically, we employ three **lightweight** modules to decompose the facial dynamics into three distinct latent spaces representing mouth, pose, and expression, respectively. Each space is characterized by a set of learnable bases whose linear combinations define specific motions. To ensure independence and accelerate training, we enforce orthogonality among bases and devise an **efficient** training strategy to allocate motion responsibilities to each space without relying on external knowledge. The learned bases are then stored in corresponding banks, enabling shared visual priors with audio input. Furthermore, considering the properties of each space, we propose an Audio-to-Motion module for audio-driven talking head synthesis. Experiments are conducted to demonstrate the effectiveness of EDTalk. The code and pretrained models are released at: https://tanshuai0219.github.io/EDTalk/

Keywords: Talking head generation · Facial disentanglement

1 Introduction

Talking head animation has garnered significant research attention owing to its wide-ranging applications in education, filmmaking, virtual digital humans, and the entertainment industry [35]. While previous methods [23,39,56,57] have achieved notable advancements, most of them generate talking head videos in a

Supplementary Information The online version contains supplementary material available at https://doi.org/10.1007/978-3-031-72658-3_23.

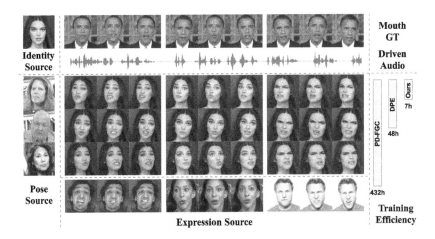

Fig. 1. Illustrative animations produced by EDTalk. Given an identity source, EDTalk synthesizes talking face videos characterized by mouth shapes, head poses, and expressions consistent with mouth GT, pose source and expression source. These facial dynamics can also be inferred directly from driven audio. Importantly, EDTalk demonstrates superior efficiency in disentanglement training compared to other methods.

holistic manner, lacking fine-grained individual control. Consequently, attaining precise and disentangled manipulation over various facial motions such as mouth shapes, head poses, and emotional expressions remains a challenge, crucial for crafting lifelike avatars [47]. Moreover, existing approaches typically cater to only one driving source: either audio [28,46] or video [17,41], thereby limiting their applicability in the multimodal context. There is a pressing need for a unified framework capable of simultaneously achieving individual facial control and handling both audio-driven and video-driven talking face generation (Fig. 1).

To tackle the challenges, an intuition is to disentangle the entirety of facial dynamics into distinct facial latent spaces dedicated to individual components. However, it is non-trivial due to the intricate interplay among facial movements [47]. For instance, mouth shapes profoundly impact emotional expressions, where one speaks happily with upper lip corners but sadly with the depressed ones [13,14]. Despite the extensive efforts made in facial disentanglement by previous studies [27,33,47,58,65], we argue there exist three key limitations. **(1)** Overreliance on external and prior information increases the demand for data and complicates the data pre-processing: One popular line [27,47,58] relies heavily on external audio data to decouple the mouth space via contrastive learning [24]. Subsequently, they further disentangle the pose space using predefined 6D pose coefficients extracted from 3D face reconstruction models [11]. However, such external and prior information escalates dataset demands and any inaccuracies therein can lead to the trained model errors. **(2)** Disentangling latent spaces without internal constraints leads to incomplete decoupling. Previous works [27,65] simply constrain each space externally with a prior during the decoupling process, overlooking inter-space constraints. This oversight fails to ensure that each space exclusively handles its designated component without interference from others, leading to training complexities, reduced efficiency, and performance degradation. **(3)** Inefficient training strategy escalates the training time and computational cost. When disentangling a new sub-space, some methods [33,47] require training the entire heavyweight network from scratch, which significantly incurs high time and computational costs [14]. It can be costly and unaffordable for many researchers. Furthermore, most methods are unable to utilize audio and video inputs simultaneously.

To cope with such issues, this paper proposes an **E**fficient **D**isentanglement framework, tailored for one-shot talking head generation with precise control over mouth shape, head pose, and emotional expression, conditioned on video or audio inputs. Our key insight lies in our requirements for decoupled space: **(a)** The decoupled spaces should be disjoint, which means each space captures solely the motion of its corresponding component without the interference from others. This also ensures that decoupling a new space will not affect the trained models, thereby avoiding the necessity of training from scratch. **(b)** Once the spaces are disentangled from video data to support video-driven paradigm, they should be stored to share with the audio inputs for further audio-driven setting.

To this end, drawing inspiration from the observation that the entire motion space can be represented by a set of directions [53], we innovatively disentangle the whole motion space into three distinct component-aware latent spaces. Each space is characterized by a set of learnable bases. To ensure that different latent spaces do not interfere with each other, we constrain bases orthogonal to each other not only *intra*-space [53] but also *inter*-space. To accomplish the disentanglement without prior information, we introduce a progressive training strategy comprising cross-reconstruction mouth-pose disentanglement and self-reconstruction complementary learning for expression decoupling. Despite comprising two stages, our decoupling process involves training only the proposed lightweight Latent Navigation modules, keeping the weights of other heavier modules fixed for efficient training.

To explicitly preserve the disentangled latent spaces, we store the base sets of disentangled spaces in the corresponding banks. These banks serve as repositories of prior bases essential for audio-driven talking head generation. Consequently, we introduce an Audio-to-Motion module designed to predict the weights of the mouth, pose, and expression banks, respectively. Specifically, we employ an audio encoder to synchronize lip motions with the audio input. Given the non-deterministic nature of head motions [61], we utilize normalizing flows [37] to generate probabilistic and realistic poses by sampling from a Gaussian distribution, guided by the rhythm of audio. Regarding expression, we aim to extract emotional cues from the audio [21] and transcripts. It ensures that the generated talking head video aligns with the tone and context of audio, eliminating the need for additional expression references. In this way, our EDTalk enables talking face generation directly from the sole audio input.

Our contributions are outlined as follows: **1)** We present EDTalk, an efficient disentanglement framework enabling precise control over talking head synthesis concerning mouth shape, head pose, and emotional expression. **2)** By introducing orthogonal bases and an efficient training strategy, we successfully achieve complete decoupling of these three spaces. Leveraging the properties of each space, we implement Audio-to-Motion modules to facilitate audio-driven talking face generation. **3)** Extensive experiments demonstrate that our EDTalk surpasses the competing methods in both quantitative and qualitative evaluation.

2 Related Work

2.1 Disentanglement on the Face

Facial dynamics typically involve coordinated movements such as head poses, mouth shapes, and emotional expressions in a global manner [45], making their separate control challenging. Several works have been developed to address this issue. PC-AVS [65] employs contrastive learning to isolate the mouth space related to audio. Yet since similar pronunciations tend to correspond to the same mouth shape [26], the constructed negative pairs in a mini-batch often include positive pairs and the number of negative pairs in the mini-batch is too small [16], both of which results in subpar results. Similarly, PD-FGC [47] and TH-PAD [58] face analogous challenges in obtaining content-related mouth spaces. Although TH-PAD incorporates lip motion decorrelation loss to extract non-lip space, it still retains a coupled space where expressions and head poses are intertwined. This coupling results in randomly generated expressions co-occurring with head poses, compromising user-friendliness and content relevance. Despite the achievement of PD-FGC in decoupling facial details, its laborious coarse-to-fine disentanglement process consumes substantial computational resources and time. DPE [33] introduces a bidirectional cyclic training strategy to disentangle head pose and expression from talking head videos. However, it necessitates two generators to independently edit expression and pose sequentially, escalating computational resource consumption and runtime. In contrast, we propose an efficient decoupling approach to segregate faces into mouth, head pose, and expression components, readily controllable by different sources. Moreover, our method requires only a unified generator, and minimal additional resources are needed when exploring a new disentangled space.

2.2 Audio-Driven Talking Head Generation

Audio-driven talking head generation [3,29] endeavors to animate images with accurate lip movements synchronized with input audio clips. Research in this area is predominantly categorized into two groups: intermediate representation based methods and reconstruction-based methods. Intermediate representation based methods [4,6,12,51,52,56,59,63,66] typically consist of two sub-modules: one predicts intermediate representations from audio, and the other synthesizes photorealistic images from these representations. For instance, Das et al.

[12] employ landmarks as an intermediate representation, utilizing an audio-to-landmark module and a landmark-to-image module to connect audio inputs and video outputs. Yin et al. [57] extract 3DMM parameters [2] to warp source images using predicted flow fields. However, obtaining such intermediate representations, like landmarks and 3D models, is laborious and time-consuming. Moreover, they often offer limited facial dynamics details, and training the two sub-modules separately can accumulate errors, leading to suboptimal performance. In contrast, our approach operates within a reconstruction-based framework [5,7,39,40,44,46,49,64]. It integrates features extracted by encoders from various modalities to reconstruct talking head videos in an end-to-end manner, alleviating the aforementioned issues. A notable example is Wav2Lip [36], which employs an audio encoder, an identity encoder, and an image decoder to generate precise lip movements. Similarly, Zhou et al. [65] incorporate an additional pose encoder for free pose control, yet disregard the nondeterministic nature of natural movement. To address this, we propose employing a probabilistic model to establish a distribution of non-verbal head motions. Additionally, none of the existing methods consider facial expressions, crucial for authentic talking head generation. Our approach aims to integrate facial expressions into the model to enhance the realism and authenticity of the generated talking heads.

2.3 Emotional Talking Head Generation

Emotional talking head generation is gaining traction due to its wide-ranging applications and heightened entertainment potential. On the one hand, some studies [14,21,43,50] identify emotions using discrete emotion labels, albeit facing a challenge to generate controllable and fine-grained expressions. On the other hand, recent methodologies [20,27,31,47] incorporate emotional images or videos as references to indicate desired expressions. Ji et al. [20], for instance, mask the mouth region of an emotional video and utilize the remaining upper face as an expression reference for emotional talking face generation. However, as mouth shape plays a crucial role in conveying emotion [45], they struggle to synthesize vivid expressions due to their failure to decouple expressions from the entire face. Thanks to our orthogonal base and efficient training strategy, we are capable of fully disentangling different motion spaces like mouth shape and emotional expression, thus achieving finely controlled talking head synthesis. Moreover, we also incorporate emotion contained within audio and transcripts. To the best of our knowledge, we are the first to achieve this goal-automatically inferring suitable expressions from audio tone and text, thereby generating consistent emotional talking face videos without relying on explicit image/video references.

3 Methodology

As illustrated in Fig. 2(a), given an identity image I^i, we aim to synthesize emotional talking face image $\hat{I}^g$ that maintains consistency in identity information,

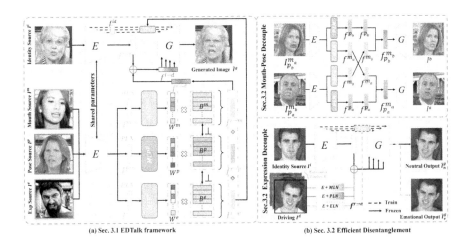

Fig. 2. Illustration of our proposed EDTalk. (a) EDTalk framework. Given an identity source I^i and various driving images I^* ($* \in \{m, p, e\}$) for controlling corresponding facial components, EDTalk animates the identity image I^i to mimic the mouth shape, head pose, and expression of I^m, I^p and I^e with the assistance of three Component-aware Latent Navigation modules: MLN, PLN and ELN. (b) Efficient Disentanglement. The disentanglement process consists of two parts: Mouth-Pose decouple and Expression Decouple. For the former, we introduce the cross-reconstruction training strategy aimed at separating mouth shape and head pose. For the latter, we achieve expression disentanglement using self-reconstruction complementary learning.

mouth shape, head pose, and emotional expression with various driving sources I^i, I^m, I^p and I^e. Our intuition is to disentangle different facial components from the overall facial dynamics. To this end, we propose EDTalk (Sect. 3.1) with learnable orthogonal bases stored in banks B^* ($*$ refers to the mouth source m, pose source p and expression source e for simplicity), each representing a distinct direction of facial movements. To ensure the bases are component-aware, we propose an efficient disentanglement strategy (Sect. 3.2), comprising Mouth-Pose Decoupling and Expression Decoupling, which decompose the overall facial motion into mouth, pose, and expression spaces. Leveraging these disentangled spaces, we further explore an Audio-to-Motion module (Sect. 3.3, Fig. 3) to produce audio-driven emotional talking face videos featuring probabilistic poses, audio-synchronized lip motions, and semantically-aware expressions.

3.1 EDTalk Framework

Figure 2(a) illustrates the structure of EDTalk, which is based on an autoencoder architecture consisting of an Encoder E, three Component-aware Latent Navigation modules (CLNs) and a Generator G. The encoder E maps the identity image I^i and various driving source I^* into the latent features $f^{i \to r} = E(I^i)$ and $f^{* \to r} = E(I^*)$. The process in inspired by FOMM [41] and LIA [53]. Instead of directly modeling motion transformation $f^{i \to *}$ from identity image I^i to driving image I^* in the latent space, we posit the existence of a canonical feature f^r, that facilitates motion transfer between identity features and driving ones, expressed as $f^{i \to *} = f^{i \to r} + f^{r \to *}$.

Thus, upon acquiring the latent features $f^{* \to r}$ extracted by E from driving images I^*, we devise three Component-aware Latent Navigation modules to transform them into $f^{r \to *} = CLN(f^{* \to r})$. For clarity, we use pose as an example, denoted as $* = p$. Within the Pose-aware Latent Navigation (PLN) module, we establish a pose bank $B^p = \{b_1^p, ..., b_n^p\}$ to store n learnable base b_i^p. To ensure each base represents a distinct pose motion direction, we enforce orthogonality between every pair of bases by imposing a constraint of $\langle b_i^p, b_j^p \rangle = 0$ $(i \neq j)$, where $\langle \cdot, \cdot \rangle$ signifies the dot product operation. It allows us to depict various head pose movements as linear combinations of the bases. Consequently, we design a Multi-Layer Perceptron layer MLP^p to predict the weights $W^p = \{w_1^p, ..., w_n^p\}$ of the pose bases from the latent feature $f^{p \to r}$:

$$W^p = \{w_1^p, ..., w_n^p\} = MLP^p(f^{p \to r}), \qquad f^{r \to p} = \sum_{i=1}^{n} w_i^p b_i^p, \qquad (1)$$

Mouth and Expression-aware Latent Navigation module share the same architecture with PLM but have different parameters, where we can also derive $f^{r \to m} = \sum_{i=1}^{n} w_i^m b_i^m, W^m = MLP^m(f^{m \to r})$ and $f^{r \to e} = \sum_{i=1}^{n} w_i^e b_i^e, W^e = MLP^e(f^{e \to r})$ in the similar manner. It's worth noting that to achieve complete disentanglement of facial components and prevent changes in one component from affecting others, we ensure orthogonality between the three banks (B^m, B^p, B^e). This also allows us to directly combine the three features to obtain the driving feature $f^{r \to d} = f^{r \to m} + f^{r \to p} + f^{r \to e}$. We further get $f^{i \to d} = f^{i \to r} + f^{r \to d}$, which is subsequently fed into the Generator G to synthesize the final result $\hat{I}^g$. To maintain identity information, G incorporates the identity features f^{id} of the identity image via skip connections. Additionally, to enhance emotional expressiveness with the assistance of the emotion feature $f^{r \to e}$, we introduce a lightweight plug-and-play Emotion Enhancement Module (EEM), which will be discussed in the subsequent subsection. In summary, the generation process can be formulated as follows:

$$\hat{I}^g = G(f^{i \to d}, f^{id}, EEM(f^{r \to e})), \qquad (2)$$

where EEM is exclusively utilized during emotional talking face generation. For brevity, we omit f^{id} in the subsequent equations.

3.2 Efficient Disentanglement

Based on the outlined framework, the crux lies in training each Component-aware Latent Navigation module to store only the bases corresponding to the motion of its respective components and to ensure no interference between different components. To achieve this, we propose an efficient disentanglement strategy comprising Mouth-Pose Decoupling and Expression Decoupling, thereby separating the overall facial dynamics into mouth, pose, and expression components.

Mouth-Pose Decouple. As depicted at the top of Fig. 2(b), we introduce cross-reconstruction technical, which involves synthesized images of switched mouths: $I_{p_b}^{m_a}$ and $I_{p_a}^{m_b}$. Here, we superimpose the mouth region of I^a onto I^b and vice versa. Subsequently, the encoder E encodes them into canonical features, which are processed through PLN and MLN to obtain corresponding features:

$$f^{p_b}, f^{m_a} = PLN(E(I_{p_b}^{m_a})), MLN(E(I_{p_b}^{m_a})) \tag{3}$$

$$f^{p_a}, f^{m_b} = PLN(E(I_{p_a}^{m_b})), MLN(E(I_{p_a}^{m_b})) \tag{4}$$

Next, we substitute the extracted mouth features and feed them into the generator G to perform cross reconstruction of the original images: $\hat{I}^b = G(f^{p_b}, f^{m_b})$ and $\hat{I}^a = G(f^{p_a}, f^{m_a})$. Additionally, we include identity features f^{id} extracted from another frame of the same identity as input to the generator G. Afterward, we supervise the Mouth-Pose Decouple module by adopting reconstruction loss $\mathcal{L}_{\text{rec}}$, perceptual loss $\mathcal{L}_{\text{per}}$ [22,60] and adversarial loss $\mathcal{L}_{\text{adv}}$:

$$\mathcal{L}_{\text{rec}} = \sum_{\#=a,b} \|I^\# - \hat{I}^\#\|_1; \quad \mathcal{L}_{\text{per}} = \sum_{\#=a,b} \|\Phi(I^\#) - \Phi(\hat{I}^\#)\|_2^2; \tag{5}$$

$$\mathcal{L}_{\text{adv}} = \sum_{\#=a,b} (\log D(I^\#) + \log(1 - D(\hat{I}^\#))), \tag{6}$$

where Φ denotes the feature extractor of VGG19 [42] and D is a discriminator tasked with distinguishing between reconstructed images and ground truth (GT). In addition, self-reconstruction of the Ground Truth (GT) is crucial, where mouth features and pose features are extracted from the same image and then input into G to reconstruct itself using $\mathcal{L}_{\text{self}}$. Furthermore, we impose feature-level constraints on the network:

$$\mathcal{L}_{\text{fea}} = \sum_{\#=a,b} (exp(-\mathcal{S}(f^{p\#}, PLN(E(I^\#)))) + exp(-\mathcal{S}(f^{m\#}, MLN(E(I^\#))))), \tag{7}$$

where we extract mouth features and pose features from I^a and I^b, aiming to minimize their disparity with those extracted from synthesized images of switched mouths using cosine similarity $\mathcal{S}(\cdot, \cdot)$. Once the losses have converged, the parameters are no longer updated for the remainder of training, significantly reducing training time and resource consumption for subsequent stages.

Expression Decouple. As illustrated in the bottom of Fig. 2(b), to decouple expression information from driving image I^d, we introduce Expression-aware Latent Navigation module (ELN) and a lightweight plug-and-play Emotion Enhancement Module (EEM), both trained via self-reconstruction complementary learning. Specifically, given an identity source I^i and a driving image I^d sharing the same identity as I^i but differing in mouth shapes, head poses and emotional expressions, our pre-trained modules (i.e., E, MLN, PLN, and G) from previous stage effectively disentangle mouth shape and head pose from I^d

and drive I^i to generate $\hat{I}_n^g$ with matching mouth shape and head pose as I^d but with the same expression with I^i. Therefore, to faithfully reconstruct I^d with the same expression, ELN is compelled to learn complementary information not disentangled by MLN, PLN, precisely the expression information. Motivated by the observation [47] that expression variation in a video sequence is typically less frequent than changes in other motions, we define a window of size K around I^d and average K extracted expression features to obtain a clean expression feature $f^{r \to e}$. $f^{r \to e}$ is then combined with extracted mouth and pose features as input to the generator G. Additionally, EEM takes $f^{r \to e}$ as input and utilizes affine transformations to produce $f^e = (f_s^e, f_b^e)$ that control adaptive instance normalization (AdaIN) [19] operations. The AdaIN operations further adapt identity feature f^{id} as emotion-conditioned features f_e^{id} by:

$$f_e^{id} := EEM(f^{id}) = f_s^e \frac{f^{id} - \mu(f^{id})}{\sigma(f^{id})} + f_b^e, \tag{8}$$

where $\mu(\cdot)$ and $\sigma(\cdot)$ represent the average and variance operations. Subsequently, we generate output $\hat{I}_e^g$ with the expression of I^d via Eq. 2. We enforce a motion reconstruction loss [47] $\mathcal{L}_{\text{mot}}$ in addition to the same reconstruction loss $\mathcal{L}_{\text{rec}}$, perceptual loss $\mathcal{L}_{\text{per}}$ and adversarial loss $\mathcal{L}_{\text{adv}}$ as Eq. 5 and Eq. 6:

$$\mathcal{L}_{\text{mot}} = \|\phi(I^d) - \phi(\hat{I}_e^g)\|_2 + \|\psi(I^d) - \psi(\hat{I}_e^g)\|_2, \tag{9}$$

where $\phi(\cdot)$ and $\psi(\cdot)$ denote features extracted by the 3D face reconstruction network and the emotion network of [11]. Moreover, to ensure that the synthesized image accurately mimics the mouth shape of the driving frame, we further introduce a mouth consistency loss $\mathcal{L}_{\text{m-c}}$:

$$\mathcal{L}_{\text{m-c}} = e^{-\mathcal{S}(MLN(E(\hat{I}_e^g)), MLN(E(I^d)))}, \tag{10}$$

where MLN and E are pretrained in the previous stage. During training, we only need to train lightweight ENL and EEM, resulting in fast training.

After successfully training the two-stage Efficient Disentanglement module, we acquire three disentangled spaces, enabling one-shot video-driven talking face generation with separate control of identity, mouth shape, pose, and expression, given different driving sources, as illustrated in Fig. 2(a).

3.3 Audio-To-Motion

Integrating the disentangled spaces, we aim to address a more appealing but challenging task: audio-driven talking face generation. In this

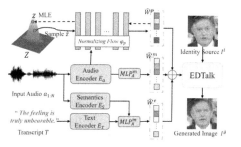

Fig. 3. The overview of Audio-to-Motion, for mouth, pose, expression prediction.

section, depicted in Fig. 3, we introduce three modules to predict the weights of pose, mouth, and expression from audio. These modules replace the driving video input, facilitating audio-driven talking face generation.

Audio-Driven Lip Generation. Prior works [31,45] generate facial dynamics, encompassing lip motions and expressions, in a holistic manner, which proves challenging for two main reasons: 1) Expressions, being acoustic-irrelevant motions, can impede lip synchronization [61]. 2) The absence of lip visual information hinders fine details synthesis at the phoneme level [34]. Thanks to the disentangled mouth space obtained in the previous stage, we naturally mitigate the influence of expression without necessitating special training strategies or loss functions like [61]. Additionally, since the decoupled space is trained during video-driven talking face generation using video as input, which offers ample visual information in the form of mouth bases b_i^m stored in the bank B^m, we eliminate the need for extra visual memory like [34]. Instead, we only need to predict the weight w_i^m of each base b_i^m, which generates the fine-grained lip motion. To achieve this, we design an Audio Encoder E_a, which embeds the audio feature into a latent space $f^a = E_a(a_{1:N})$. Subsequently, a linear layer MLP_A^m is added to decode the mouth weight $\hat{W}^m$. During training, we fix the weights of all modules and only update E_a and MLP_A^m using the weighted sum of feature loss $\mathcal{L}_{fea}^m$, reconstruction loss $\mathcal{L}_{rec}^m$ and sync loss $\mathcal{L}_{sync}^m$ [36]:

$$\mathcal{L}_{fea}^m = \|W^m - \hat{W}^m\|_2, \qquad \mathcal{L}_{rec}^m = \|I - \hat{I}\|_2, \qquad (11)$$

$$\mathcal{L}_{sync}^m = -\log(\frac{v \cdot s}{max(\|v\|_2 \cdot \|s\|_2, \epsilon)}), \qquad (12)$$

where $W^m = MLN(E(I))$ is the GT mouth weight extracted from GT image I and $\hat{I}$ is generated image using Eq. 2. $\mathcal{L}_{sync}^m$ is introduced from [36], where v and s are extracted by the speech encoder and image encoder in SyncNet [10].

Flow-Based Probabilistic Pose Generation. Due to the nature of one-to-many mapping from the input audio to head poses, learning a deterministic mapping like previous works [51,52,66] output the same results, which bring ambiguity and inferior visual results. To generate probabilistic and realistic head motions, we predict the pose weights $\hat{W}^p$ using Normalizing Flow φ_p [37], as illustrated in Fig. 3. During training (indicated by dash lines), we extract pose weights W^p from videos as the ground truth and feed them into our φ_p. By incorporating Maximum Likelihood Estimation (MLE) in Eq. 13, we embed it into a Gaussian distribution p_Z conditioned on audio feature $f^a = E_a(a_{1:N})$:

$$z_t = \varphi_p^{-1}(w_t^p, f_t^a), \qquad \mathcal{L}_{\text{MLE}} = -\sum_{t=0}^{N-1} \log p_{\mathcal{Z}}(z_t) \qquad (13)$$

As the normalizing flow φ_p is bijective, we reconstruct the pose weight $\hat{W}^p = \varphi_p(z, f_t^a)$ and utilize a pose reconstruction loss $\mathcal{L}_{rec}^p$ along with a temporal loss $\mathcal{L}_{tem}^p$ to constrain φ_p:

$$\mathcal{L}_{\text{rec}}^p = \|W^p - \hat{W}^p\|_2, \qquad \mathcal{L}_{\text{tem}} = \frac{1}{N-1} \sum_{t=1}^{N-1} \|(w_t^p - w_{t-1}^p) - (\hat{w}_t^p - \hat{w}_{t-1}^p)\|_2 \quad (14)$$

During inference, we randomly sample $\hat{z}$ from the constructed distribution p_Z and then generate pose weights $\hat{W}^p = \varphi_p(z, f_t^a)$. This process ensures the diversity of head motions while maintaining consistency with the audio rhythm.

Semantically-Aware Expression Generation. As finding videos with a desired expression may not always be feasible, potentially limiting their application [30], we aim to explore the emotion contained in audio and transcript with the aid of the introduced Semantics Encoder E_S and Text Encoder E_T. Inspired by [54], our Semantics Encoder E_S is constructed upon the pretrained HuBERT model [18], which consists of a CNN-based feature encoder and a transformer-based encoder. We freeze the CNN-based feature encoder and only fine-tuned the transformer blocks. Text Encoder E_T is inherited from the pretrained Emoberta [25], which encodes the overarching emotional context embedded within textual descriptions. We concatenate the embeddings generated by E_S and E_T and feed them into a MLP_A^e to generate the expression weights $\hat{W}^e$. Since audio or text may not inherently contain emotion during inference, such as in TTS-generated speech, in order to support the prediction of emotion from a single modality, we randomly mask ($\mathcal{M}$) a modality with probability p during training, inspired by HuBERT:

$$\hat{W}^e = \begin{cases} MLP_a^e(E_S(a), E_T(T)), & 0.5 \leq p \leq 1, \\ MLP_a^e(\mathcal{M}(E_S(a)), E_T(T)), & 0.25 \leq p < 0.5, \\ MLP_a^e(E_S(a), \mathcal{M}(E_T(T))), & 0 \leq p < 0.25. \end{cases} \quad (15)$$

We employ $\mathcal{L}_{\text{exp}} = \|W^e - \hat{W}^e\|_1$ to encourage $\hat{W}^e$ close to weight W^e generated by pretrained ELN from emotional frames. Until now, we are able to generate **probabilistic semantically-aware** talking head videos solely from an identity image and the driving audio.

4 Experiments

4.1 Experimental Settings

Implement Details. Our model is trained and evaluated on the datasets MEAD [50] and HDTF [62]. Additionally, we report results on additional datasets, including LRW [9] and Voxceleb2 [8], for further assessment of our method in the supplementary. All video frames are cropped following FOMM [41] and resized to 256×256. Our method is implemented using PyTorch and trained using the Adam optimizer on 2 NVIDIA GeForce GTX 3090 GPUs. The dimension of the latent code $f^{*\rightarrow r}$ and bases b^* is set to 512, and the number of bases of B^m, B^p and B^e are set to 20, 6 and 10, respectively. The weight for $\mathcal{L}_{\text{mot}}$ is set to 10 and the remaining weights are set to 1.

Table 1. Quantitative comparisons with state-of-the-art methods.

Method	MEAD [50]						HDTF [62]				
	PSNR↑	SSIM↑	M/F-LMD↓	FID↓	$Sync_{conf}$↑	Acc_{emo}↑	PSNR↑	SSIM↑	M/F-LMD↓	FID↓	$Sync_{conf}$↑
MakeItTalk [66]	19.442	0.614	2.541/2.309	37.917	5.176	14.64	21.985	0.709	2.395/2.182	18.730	4.753
Wav2Lip [36]	19.875	0.633	1.438/2.138	44.510	**8.774**	13.69	22.323	0.727	1.759/2.002	22.397	**9.032**
Audio2Head [51]	18.764	0.586	2.053/2.293	27.236	6.494	16.35	21.608	0.702	1.983/2.060	29.385	7.076
PC-AVS [65]	16.120	0.458	2.649/4.350	38.679	7.337	12.12	22.995	0.705	2.019/1.785	26.042	8.482
AVCT [52]	17.848	0.556	2.870/3.160	37.248	4.895	13.13	20.484	0.663	2.360/2.679	19.066	5.661
SadTalker [61]	19.042	0.606	2.038/2.335	39.308	7.065	14.25	21.701	0.702	1.995/2.147	14.261	7.414
IP-LAP [63]	19.832	0.627	2.140/2.116	46.502	4.156	17.34	22.615	0.731	1.951/1.938	19.281	3.456
TalkLip [48]	19.492	0.623	1.951/2.204	41.066	5.724	14.00	22.241	0.730	1.976/1.937	23.850	1.076
EAMM [20]	18.867	0.610	2.543/2.413	31.268	1.762	31.08	19.866	0.626	2.910/2.937	41.200	4.445
StyleTalk [31]	21.601	0.714	1.800/1.422	24.774	3.553	63.49	21.319	0.692	2.324/2.330	17.053	2.629
PD-FGC [47]	21.520	0.686	1.571/1.318	30.240	6.239	44.86	23.142	0.710	**1.626**/1.497	25.340	7.171
EAT [14]	20.007	0.652	1.750/1.668	21.465	7.984	64.40	22.076	0.719	2.176/1.781	28.759	7.493
EDTalk-A	**21.628**	**0.722**	**1.537/1.290**	**17.698**	8.115	**67.32**	**25.156**	**0.811**	1.676/**1.315**	**13.785**	7.642
EDTalk-V	**22.771**	**0.769**	**1.102/1.060**	**15.548**	6.889	**68.85**	**26.504**	**0.845**	**1.197/1.111**	**13.172**	6.732
GT	1.000	1.000	0.000/0.000	0.000	7.364	79.65	1.000	1.000	0.000/0.000	0.000	7.721

Comparison Setting. We compare our method with: (a) emotion-agnostic talking face generation methods: MakeItTalk [66], Wav2Lip [36], Audio2Head [51], PC-AVS [65], AVCT [52], SadTalker [61], IP-LAP [63], TalkLip [48]. (b) Emotional talking face generation methods: EAMM [20], StyleTalk [31], PD-FGC [47], EMMN [45], EAT [14], EmoGen [15]. Different from previous work, EDTalk encapsulates the entire face generation process without any other sources (e.g. poses [14,20], 3DMM [31,57], phoneme [31,52]) and pre-processing operations during inference, which facilitates the application. We evaluate our model in both audio-driven setting (EDTalk-A) and video-driven setting (EDTalk-V) w.r.t. (i) generated video quality using PSNR, SSIM [55] and FID [38]. (ii) audio-visual synchronization using Landmarks Distances on the Mouth (M-LMD) [6] and the confidence score of SyncNet [10]. (ii) emotional accuracy using Acc_{emo} calculated by pretrained Emotion-Fan [32] and Landmarks Distances on the Face (F-LMD). Partial results are moved to supplementary material due to limited space.

4.2 Experimental Results

Quantitative Results. The quantitative results are presented in Table 1, where our EDTalk-A and EDTalk-V achieve the best performance across most metrics, except $Sync_{conf}$. Wav2Lip pretrains their SyncNet discriminator on a large dataset [1], which might lead the model to prioritize achieving a higher $Sync_{conf}$ over optimizing visual performance. It is evident in the blur mouths generated by Wav2Lip and inferior M-LMD score to our method.

Qualitative Results. Figure 4 demonstrates comparison of visual results. TalkLip and IP-LAP struggle to generate accurate lip motions. Despite elevated lip-synchronization of SadTalker, they can only produce slight lip motions with

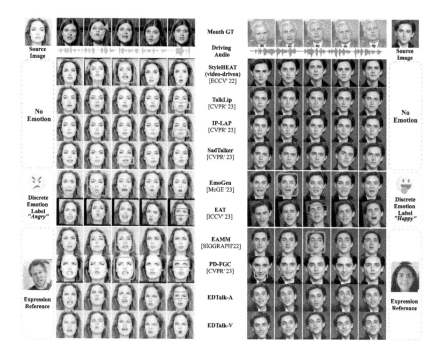

Fig. 4. Qualitative comparisons with state-of-the-art methods. See full comparison in supplementary material.

closed mouth and are also bothered by jitter between frames. StyleHEAT generates accurate mouth shape driven by Mouth GT video instead of audio but suffers from incorrect head pose and identity loss. This issue also plagues EmoGen, EAMM and PD-FGC. Besides, EmoGen and EAMM fail to perform the desired expression. Due to discrete emotion input, EAT cannot synthesize fine-grained expression like the narrowed eyes performed by expression reference. In the case of "happy", unexpected closed eyes and weird teeth are observed in EAT and PD-FGC, respectively. In contrast, both EDTalk-A and EDTalk-V excel in producing realistic expressions, precise lip synchronization and correct head poses.

Efficiency Analysis. Our approach is highly efficient in terms of training time, required data and computational resources in decoupling spaces. In the mouth-pose decoupling stage, we solely utilize the HDTF dataset, containing **15.8 h** of videos, for the decoupling. Training with a batch size of 4 on **two 3090 GPUs** for **4k** iterations achieves state-of-the-art perfor-

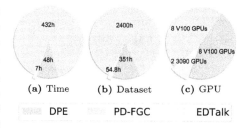

(a) Time (b) Dataset (c) GPU

DPE   PD-FGC   EDTalk

Fig. 5. Resources for training.

mance, which takes about one hour. In contrast, DPE is trained on the VoxCeleb dataset, which comprises **351 h** of video, for **100K** iterations initially, then an additional **50K** iterations with a batch size of 32 on 8 V100 GPUs, which takes over 2 days. Besides, they need to train two task-specific generators for expression and pose. Similarly, PD-FGC takes **2 days** on **4 T V100 GPUs** for lip, and another **2 days on 4 T V100 GPUs** for pose decoupling. It significantly exceeds our computational resources and training time. In the expression decouple stage, we train our model on MEAD and HDTF dataset (total **54.8 h** of videos) for 6 h. On the other hand, PD-FGC decouples expression space on Voxceleb2 dataset (**2400 h**) by discorelation loss for 2 weeks. The visualization in Fig. 5 allows for a more intuitive comparison of the differences between the different methods concerning required training time, training data, and computational arithmetic.

4.3 Ablation Study

Latent Space. To analyze the contributions of our key designs on obtaining the disentangled latent spaces, we conduct an ablation study with two variants:

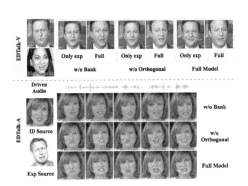

Fig. 6. Ablation results.

(1) remove base banks (**w/o Bank**). (2) remove orthogonal constraint (**w/o Orthogonal**). Fig. 6 presents our ablation study results on video-driven and audio-driven settings, respectively. Since **w/o Bank** struggles to decouple different latent spaces, *only exp* fails to extract the emotional expression. Additionally, without the visual information stored in banks, the quality of the generated full frame is poor. Although **w/o Orthogonal** improves the image quality through vision-rich banks, due to the lack of orthogonality constraints on the base, it interferes with different spaces, resulting in less obvious generated emotions. The Full Model achieves the best performance in both aspects.

5 Conclusion

This paper introduces EDTalk, a novel system designed to efficiently disentangle facial components into latent spaces, enabling fine-grained control for talking head synthesis. The core insight is to represent each space with orthogonal bases stored in dedicated banks. We propose an efficient training strategy that autonomously allocates spatial information to each space, eliminating the necessity for external or prior structures. By integrating these spaces, we enable audio-driven talking head generation through a lightweight Audio-to-Motion module. Experiments showcase the superiority of our method in achieving disentangled and precise control over diverse facial motions. We provide more discussion about the limitations and ethical considerations in the supplementary material.

Acknowledgements. This work was supported by National Natural Science Foundation of China (NSFC, NO. 62102255), NetEase Fuxi Lab Industry-University Collaboration Research Funding.

References

1. Afouras, T., Chung, J.S., Senior, A., Vinyals, O., Zisserman, A.: Deep audio-visual speech recognition. IEEE Trans. Pattern Anal. Mach. Intell. **44**(12), 8717–8727 (2018)
2. Blanz, V., Vetter, T.: A morphable model for the synthesis of 3d faces. In: Proceedings of the 26th Annual Conference on Computer Graphics and Interactive Techniques, pp. 187–194 (1999)
3. Bregler, C., Covell, M., Slaney, M.: Video rewrite: driving visual speech with audio. In: Seminal Graphics Papers: Pushing the Boundaries, vol. 2, pp. 715–722 (2023)
4. Chen, L., et al.: Talking-head generation with rhythmic head motion. In: Vedaldi, A., Bischof, H., Brox, T., Frahm, J.-M. (eds.) ECCV 2020. LNCS, vol. 12354, pp. 35–51. Springer, Cham (2020). https://doi.org/10.1007/978-3-030-58545-7_3
5. Chen, L., Li, Z., Maddox, R.K., Duan, Z., Xu, C.: Lip movements generation at a glance. In: Proceedings of the European Conference on Computer Vision (ECCV), pp. 520–535 (2018)
6. Chen, L., Maddox, R.K., Duan, Z., Xu, C.: Hierarchical cross-modal talking face generation with dynamic pixel-wise loss. In: Proceedings of the IEEE/CVF Conference on Computer Vision and Pattern Recognition, pp. 7832–7841 (2019)
7. Chen, L., et al.: Vast: vivify your talking avatar via zero-shot expressive facial style transfer. In: Proceedings of the IEEE/CVF International Conference on Computer Vision, pp. 2977–2987 (2023)
8. Chung, J.S., Nagrani, A., Zisserman, A.: Voxceleb2: deep speaker recognition. arXiv preprint arXiv:1806.05622 (2018)
9. Chung, J.S., Zisserman, A.: Lip reading in the wild. In: Lai, S.-H., Lepetit, V., Nishino, K., Sato, Y. (eds.) ACCV 2016. LNCS, vol. 10112, pp. 87–103. Springer, Cham (2017). https://doi.org/10.1007/978-3-319-54184-6_6
10. Chung, J.S., Zisserman, A.: Out of time: automated lip sync in the wild. In: Chen, C.-S., Lu, J., Ma, K.-K. (eds.) ACCV 2016. LNCS, vol. 10117, pp. 251–263. Springer, Cham (2017). https://doi.org/10.1007/978-3-319-54427-4_19
11. Daněček, R., Black, M.J., Bolkart, T.: Emoca: emotion driven monocular face capture and animation. In: Proceedings of the IEEE/CVF Conference on Computer Vision and Pattern Recognition, pp. 20311–20322 (2022)
12. Das, D., Biswas, S., Sinha, S., Bhowmick, B.: Speech-driven facial animation using cascaded GANs for learning of motion and texture. In: Vedaldi, A., Bischof, H., Brox, T., Frahm, J.-M. (eds.) ECCV 2020. LNCS, vol. 12375, pp. 408–424. Springer, Cham (2020). https://doi.org/10.1007/978-3-030-58577-8_25

13. Ekman, P., Friesen, W.V.: Facial action coding system. Environ. Psychol. Nonverbal Behav. (1978)
14. Gan, Y., Yang, Z., Yue, X., Sun, L., Yang, Y.: Efficient emotional adaptation for audio-driven talking-head generation. In: Proceedings of the IEEE/CVF International Conference on Computer Vision, pp. 22634–22645 (2023)
15. Goyal, S., et al.: Emotionally enhanced talking face generation. In: Proceedings of the 1st International Workshop on Multimedia Content Generation and Evaluation: New Methods and Practice, pp. 81–90 (2023)
16. He, K., Fan, H., Wu, Y., Xie, S., Girshick, R.: Momentum contrast for unsupervised visual representation learning. In: Proceedings of the IEEE/CVF Conference on Computer Vision and Pattern Recognition, pp. 9729–9738 (2020)
17. Hong, F.T., Zhang, L., Shen, L., Xu, D.: Depth-aware generative adversarial network for talking head video generation. In: Proceedings of the IEEE/CVF Conference on Computer Vision and Pattern Recognition, pp. 3397–3406 (2022)
18. Hsu, W.N., Bolte, B., Tsai, Y.H.H., Lakhotia, K., Salakhutdinov, R., Mohamed, A.: Hubert: Self-supervised speech representation learning by masked prediction of hidden units. IEEE/ACM Trans. Audio Speech Lang. Process. **29**, 3451–3460 (2021)
19. Huang, X., Belongie, S.: Arbitrary style transfer in real-time with adaptive instance normalization. In: Proceedings of the IEEE International Conference on Computer Vision, pp. 1501–1510 (2017)
20. Ji, X., et al.: Eamm: one-shot emotional talking face via audio-based emotion-aware motion model. In: ACM SIGGRAPH 2022 Conference Proceedings, pp. 1–10 (2022)
21. Ji, X., Zhou, H., Wang, K., Wu, W., Loy, C.C., Cao, X., Xu, F.: Audio-driven emotional video portraits. In: Proceedings of the IEEE/CVF Conference on Computer Vision and Pattern Recognition, pp. 14080–14089 (2021)
22. Johnson, J., Alahi, A., Fei-Fei, L.: Perceptual losses for real-time style transfer and super-resolution. In: Leibe, B., Matas, J., Sebe, N., Welling, M. (eds.) ECCV 2016. LNCS, vol. 9906, pp. 694–711. Springer, Cham (2016). https://doi.org/10.1007/978-3-319-46475-6_43
23. Khakhulin, T., Sklyarova, V., Lempitsky, V., Zakharov, E.: Realistic one-shot mesh-based head avatars. In: European Conference on Computer Vision, pp. 345–362. Springer, Heidelberg (2022). https://doi.org/10.1007/978-3-031-20086-1_20
24. Khosla, P., et al.: Supervised contrastive learning. Adv. Neural. Inf. Process. Syst. **33**, 18661–18673 (2020)
25. Kim, T., Vossen, P.: Emoberta: speaker-aware emotion recognition in conversation with roberta. arXiv preprint arXiv:2108.12009 (2021)
26. Li, D., et al.: Ae-nerf: audio enhanced neural radiance field for few shot talking head synthesis. arXiv preprint arXiv:2312.10921 (2023)
27. Liang, B., et al.: Expressive talking head generation with granular audio-visual control. In: Proceedings of the IEEE/CVF Conference on Computer Vision and Pattern Recognition, pp. 3387–3396 (2022)
28. Liu, T., et al.: Anitalker: animate vivid and diverse talking faces through identity-decoupled facial motion encoding. arXiv preprint arXiv:2405.03121 (2024)
29. Liu, X., Xu, Y., Wu, Q., Zhou, H., Wu, W., Zhou, B.: Semantic-aware implicit neural audio-driven video portrait generation. In: European Conference on Computer Vision, pp. 106–125. Springer, Heidelberg (2022). https://doi.org/10.1007/978-3-031-19836-6_7
30. Ma, Y., et al.: Talkclip: talking head generation with text-guided expressive speaking styles. arXiv preprint arXiv:2304.00334 (2023)

31. Ma, Y., et al.: Styletalk: one-shot talking head generation with controllable speaking styles. arXiv preprint arXiv:2301.01081 (2023)
32. Meng, D., Peng, X., Wang, K., Qiao, Y.: Frame attention networks for facial expression recognition in videos. In: 2019 IEEE International Conference on Image Processing (ICIP), pp. 3866–3870. IEEE (2019)
33. Pang, Y., et al.: DPE: disentanglement of pose and expression for general video portrait editing. In: Proceedings of the IEEE/CVF Conference on Computer Vision and Pattern Recognition, pp. 427–436 (2023)
34. Park, S.J., Kim, M., Hong, J., Choi, J., Ro, Y.M.: Synctalkface: talking face generation with precise lip-syncing via audio-lip memory. In: Proceedings of the AAAI Conference on Artificial Intelligence, vol. 36, pp. 2062–2070 (2022)
35. Pataranutaporn, P., et al.: Ai-generated characters for supporting personalized learning and well-being. Nat. Mach. Intell. **3**(12), 1013–1022 (2021)
36. Prajwal, K., Mukhopadhyay, R., Namboodiri, V.P., Jawahar, C.: A lip sync expert is all you need for speech to lip generation in the wild. In: Proceedings of the 28th ACM International Conference on Multimedia, pp. 484–492 (2020)
37. Rezende, D., Mohamed, S.: Variational inference with normalizing flows. In: International Conference on Machine Learning, pp. 1530–1538. PMLR (2015)
38. Seitzer, M.: pytorch-fid: FID Score for PyTorch (2020). https://github.com/mseitzer/pytorch-fid. version 0.3.0
39. Shen, S., Li, W., Zhu, Z., Duan, Y., Zhou, J., Lu, J.: Learning dynamic facial radiance fields for few-shot talking head synthesis. In: European Conference on Computer Vision, pp. 666–682. Springer, Heidelberg (2022). https://doi.org/10.1007/978-3-031-19775-8_39
40. Shen, S., et al.: Difftalk: crafting diffusion models for generalized audio-driven portraits animation. In: Proceedings of the IEEE/CVF Conference on Computer Vision and Pattern Recognition, pp. 1982–1991 (2023)
41. Siarohin, A., Lathuilière, S., Tulyakov, S., Ricci, E., Sebe, N.: First order motion model for image animation. Adv. Neural Inf. Process. Syst. **32** (2019)
42. Simonyan, K., Zisserman, A.: Very deep convolutional networks for large-scale image recognition. arXiv preprint arXiv:1409.1556 (2014)
43. Sinha, S., Biswas, S., Yadav, R., Bhowmick, B.: Emotion-controllable generalized talking face generation. In: International Joint Conference on Artificial Intelligence. IJCAI (2021)
44. Song, Y., Zhu, J., Li, D., Wang, A., Qi, H.: Talking face generation by conditional recurrent adversarial network. In: Proceedings of the Twenty-Eighth International Joint Conference on Artificial Intelligence (2019). https://doi.org/10.24963/ijcai.2019/129
45. Tan, S., Ji, B., Pan, Y.: Emmn: emotional motion memory network for audio-driven emotional talking face generation. In: Proceedings of the IEEE/CVF International Conference on Computer Vision, pp. 22146–22156 (2023)
46. Thies, J., Elgharib, M., Tewari, A., Theobalt, C., Nießner, M.: Neural voice puppetry: audio-driven facial reenactment. In: Vedaldi, A., Bischof, H., Brox, T., Frahm, J.-M. (eds.) ECCV 2020. LNCS, vol. 12361, pp. 716–731. Springer, Cham (2020). https://doi.org/10.1007/978-3-030-58517-4_42
47. Wang, D., Deng, Y., Yin, Z., Shum, H.Y., Wang, B.: Progressive disentangled representation learning for fine-grained controllable talking head synthesis. In: Proceedings of the IEEE/CVF Conference on Computer Vision and Pattern Recognition, pp. 17979–17989 (2023)

48. Wang, J., Qian, X., Zhang, M., Tan, R.T., Li, H.: Seeing what you said: talking face generation guided by a lip reading expert. In: Proceedings of the IEEE/CVF Conference on Computer Vision and Pattern Recognition, pp. 14653–14662 (2023)
49. Wang, J., et al.: Lipformer: high-fidelity and generalizable talking face generation with a pre-learned facial codebook. In: Proceedings of the IEEE/CVF Conference on Computer Vision and Pattern Recognition, pp. 13844–13853 (2023)
50. Wang, K., et al.: MEAD: a large-scale audio-visual dataset for emotional talking-face generation. In: Vedaldi, A., Bischof, H., Brox, T., Frahm, J.-M. (eds.) ECCV 2020. LNCS, vol. 12366, pp. 700–717. Springer, Cham (2020). https://doi.org/10.1007/978-3-030-58589-1_42
51. Wang, S., Li, L., Ding, Y., Fan, C., Yu, X.: Audio2head: audio-driven one-shot talking-head generation with natural head motion. In: International Joint Conference on Artificial Intelligence. IJCAI (2021)
52. Wang, S., Li, L., Ding, Y., Yu, X.: One-shot talking face generation from single-speaker audio-visual correlation learning. In: Proceedings of the AAAI Conference on Artificial Intelligence, vol. 36, pp. 2531–2539 (2022)
53. Wang, Y., Yang, D., Bremond, F., Dantcheva, A.: Latent image animator: learning to animate images via latent space navigation. In: International Conference on Learning Representations (2021)
54. Wang, Y., Boumadane, A., Heba, A.: A fine-tuned wav2vec 2.0/hubert benchmark for speech emotion recognition, speaker verification and spoken language understanding. arXiv preprint arXiv:2111.02735 (2021)
55. Wang, Z., Bovik, A.C., Sheikh, H.R., Simoncelli, E.P.: Image quality assessment: from error visibility to structural similarity. IEEE Trans. Image Process. **13**, 600–612 (2004)
56. Yang, K., Chen, K., Guo, D., Zhang, S.H., Guo, Y.C., Zhang, W.: Face2face ρ: Real-time high-resolution one-shot face reenactment. In: European Conference on Computer Vision, pp. 55–71. Springer, Heidelberg (2022). https://doi.org/10.1007/978-3-031-19778-9_4
57. Yin, F., et al.: Styleheat: One-shot high-resolution editable talking face generation via pre-trained stylegan. In: European Conference on Computer Vision, pp. 85–101. Springer, Heidelberg (2022). https://doi.org/10.1007/978-3-031-19790-1_6
58. Yu, Z., Yin, Z., Zhou, D., Wang, D., Wong, F., Wang, B.: Talking head generation with probabilistic audio-to-visual diffusion priors. In: Proceedings of the IEEE/CVF International Conference on Computer Vision, pp. 7645–7655 (2023)
59. Zakharov, E., Shysheya, A., Burkov, E., Lempitsky, V.: Few-shot adversarial learning of realistic neural talking head models. In: Proceedings of the IEEE/CVF International Conference on Computer Vision, pp. 9459–9468 (2019)
60. Zhang, R., Isola, P., Efros, A.A., Shechtman, E., Wang, O.: The unreasonable effectiveness of deep features as a perceptual metric. In: Proceedings of the IEEE Conference on Computer Vision and Pattern Recognition, pp. 586–595 (2018)
61. Zhang, W., et al.: Sadtalker: learning realistic 3d motion coefficients for stylized audio-driven single image talking face animation. In: Proceedings of the IEEE/CVF Conference on Computer Vision and Pattern Recognition, pp. 8652–8661 (2023)
62. Zhang, Z., Li, L., Ding, Y., Fan, C.: Flow-guided one-shot talking face generation with a high-resolution audio-visual dataset. In: Proceedings of the IEEE/CVF Conference on Computer Vision and Pattern Recognition, pp. 3661–3670 (2021)
63. Zhong, W., et al.: Identity-preserving talking face generation with landmark and appearance priors. In: Proceedings of the IEEE/CVF Conference on Computer Vision and Pattern Recognition, pp. 9729–9738 (2023)

64. Zhou, H., Liu, Y., Liu, Z., Luo, P., Wang, X.: Talking face generation by adversarially disentangled audio-visual representation. In: Proceedings of the AAAI Conference on Artificial Intelligence, vol. 33, pp. 9299–9306 (2019)
65. Zhou, H., Sun, Y., Wu, W., Loy, C.C., Wang, X., Liu, Z.: Pose-controllable talking face generation by implicitly modularized audio-visual representation. In: Proceedings of the IEEE/CVF Conference on Computer Vision and Pattern Recognition, pp. 4176–4186 (2021)
66. Zhou, Y., Han, X., Shechtman, E., Echevarria, J., Kalogerakis, E., Li, D.: Makelttalk: speaker-aware talking-head animation. ACM Trans. Graph. (TOG) **39**(6), 1–15 (2020)

Groma: Localized Visual Tokenization for Grounding Multimodal Large Language Models

Chuofan Ma[1](✉), Yi Jiang[2](✉), Jiannan Wu[1], Zehuan Yuan[2], and Xiaojuan Qi[1](✉)

[1] The University of Hong Kong, Pokfulam, Hong Kong
b20mcf@connect.hku.hk
[2] ByteDance Inc., Beijing, China

Abstract. We introduce Groma, a Multimodal Large Language Model (MLLM) with grounded and fine-grained visual perception ability. Beyond holistic image understanding, Groma is adept at region-level tasks such as region captioning and visual grounding. Such capabilities are built upon a localized visual tokenization mechanism, where an image input is decomposed into regions of interest and subsequently encoded into region tokens. By integrating region tokens into user instructions and model responses, we seamlessly enable Groma to understand user-specified region inputs and ground its textual output to images. Besides, to enhance the grounded chat ability of Groma, we curate a visually grounded instruction dataset by leveraging the powerful GPT-4V and visual prompting techniques. Compared with MLLMs that rely on the language model or external module for localization, Groma consistently demonstrates superior performances in standard referring and grounding benchmarks, highlighting the advantages of embedding localization into image tokenization. Project page: https://groma-mllm.github.io/.

1 Introduction

Multimodal Large Language Models (MLLMs) have spread the sparks of artificial general intelligence [5] from language to the visual domain [13,32,49,56,65]. Owing to the foundational capabilities of Large Language Models (LLMs) [12,35,36,44,45], MLLMs excel in vision-language tasks that require advanced understanding and reasoning, such as image captioning and visual question answering. However, despite these achievements, current MLLMs typically fall short of localization capabilities, thus cannot ground understanding to the visual context. Such limitations constrains the model from fulfilling its potential in real-world applications like robotics, autonomous driving, and augmented reality (Fig. 1).

Part of the work was done during Chuofan's internship at ByteDance.

Supplementary Information The online version contains supplementary material available at https://doi.org/10.1007/978-3-031-72658-3_24.

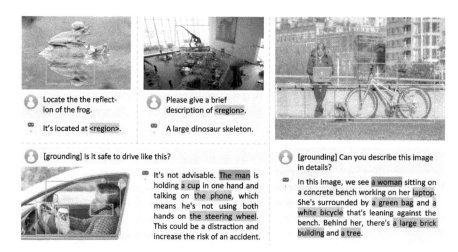

Fig. 1. Groma is a multimodal large language model with exceptional region understanding and visual grounding capabilities. It can take user-defined region inputs (boxes) as well as generate long-form responses that are grounded to visual context.

In light of the gap, one stream of research attempts to augment the LLM to directly output quantized object coordinates for localization [3,7,8,38,49,57] (Fig. 2(a)). While this method is simple in design, the substantial computational demands of LLMs make it challenging to process high-resolution image inputs, which are essential for accurate localization. Besides, the nature of sequence outputs in LLMs is not well-suited for dense prediction tasks such as segmentation. These concerns elicit another stream of research, which incorporates an external localization module (*e.g.*, SAM [22]) to decode bounding boxes or masks [25,39,42,61] (Fig. 2(b)). This approach circumvents aforementioned issues, but introduces additional model complexity.

The above motivates us to explore a new paradigm for grounded MLLMs. Drawing inspiration from open-vocabulary object detection [64], we decompose the grounding task into two sub-problems: discovering the object (localization) and relating the object to texts (recognition). We notice that localization alone requires little semantic understanding but demands perceptual skills, which is typically out of the scope of an LLM's expertise. This inspires us to decouple localization and recognition within MLLMs. But instead of using external modules, we propose exploiting the spatial understanding capability in the visual tokenizer of MLLMs for localization (Fig. 2(c)). This perceive-then-understand design also resembles human vision process.

Building upon this concept, we introduce Groma[1] (**Gro**unded **M**ultimodal **A**ssistant), an MLLM with localized and fine-grained visual perception abilities. Specifically, Groma incorporates region tokenization alongside standard image

---
[1] In Latin, Groma refers to an instrument used for accurate measurement, which implies our focus on accurate localization for MLLMs.

tokenization to identify and encode potential regions of interest (ROIs) into region tokens. During this process, location information is extracted from the image and associated with region tokens, with each region token anchored to the underlying ROI. This allows Groma to ground its textual output by simply referring to region tokens, alleviating the need for the LLM to meticulously regress object coordinates. Moreover, the tokenizer of Groma can also encode user-specified region inputs (*i.e.*, bounding boxes) into region tokens, which are directly inserted into user instructions to initiate referential dialogue.

By settling localization to the image tokenization process, Groma circumvents the heavy computation of LLMs when handling high-resolution input. Specifically, Groma uses high-resolution images for tokenizer input but downsamples image tokens for LLM input, which saves computation without sacrificing localization accuracy. Besides, unlike methods adopting separate designs for modeling grounding outputs and referring inputs [42,61], Groma seamlessly unifies the two capabilities with the use of region tokens.

From the data perspective, to improve the localized understanding of Groma, we adopt an extensive collection of datasets with region-level annotations for training, which encompasses a range of region semantics from objects and relationships to detailed region descriptions. In addition, to remedy the lack of long-form grounded data, we construct a visually grounded chat dataset called Groma Instruct for instruction finetuning. Groma Instruct is the first grounded chat dataset constructed with both visual and textual prompts, leveraging the powerful GPT-4V for data generation.

Our comprehensive experiments demonstrate the superiority of the design of Groma, with results showing that it outperforms all comparable MLLMs on established referring and grounding benchmarks. We also showcase that Groma maintains strong image-level understanding and reasoning abilities on the conversational VQA benchmark. Moreover, to assess the ability to localize multiple, diverse, and variably-sized objects, we adapt the LVIS [15] detection benchmark for object grounding evaluation. On this challenging benchmark, Groma surpasses alternative methods by a significant margin (over 10% AR), highlighting its robust and precise localization capabilities.

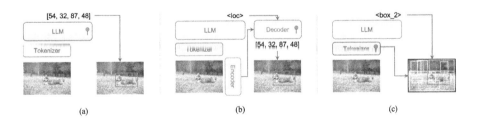

Fig. 2. Different paradigms of grounded MLLMs. We mark the modules for localization with . (a) LLM for localization (*e.g.*, Kosmos-2 [38] and Shikra [8]); (b) External modules for localization (*e.g.*, Lisa [25]); and (c) Localized visual tokenization (Ours).

2 Related Work

Image-Level MLLMs. Large language models (LLMs) such as GPT series [1, 52] and LLaMA [44,45] have recently undergone rapid development and sparked a revolution in the field of natural language processing. Such progress inspires the community to extend the foundational capabilities of LLMs to the visual domain, giving birth to multimodal large language models (MLLMs). The pioneering works [2,13,26,27,56,61,65] of MLLMs typically follow a tripartite architecture, comprising a visual encoder, a vision-language connector, and a large language model. Specifically, BLIP-2 [27] and Flamingo [2] first propose the Q-Former/Resampler to bridge vision and language. LLaVA [61] and MiniGPT4 [65] streamline this vision-language connector to a linear layer, and introduce visual instruction tuning to enhance the instruction-following ability of MLLMs. Following works [10,49] further showcase the immense potential of MLLMs by scaling up the visual components to the magnitude as LLMs. While these works have exhibited impressive visual understanding capabilities, they are predominantly constrained to image-level tasks, such as image captioning and image visual question answering. This necessitates the research into region-level MLLMs, which unlock more nuanced and granular visual-language interactions.

Region-Level MLLMs. In pursuit of fine-grained and grounded image understanding, recent studies further integrate region-level data into the training of MLLMs [7,8,38,50,51,59,63]. In particular, to model box inputs and outputs, Kosmos-2 [38] and Shikra [8] directly quantize bounding boxes into discrete location tokens or numeric representation of positions. GPT4RoI [62] and RegionGPT [14] use a simple pooling operation to extract the features within boxes or masks as the region representations. While Ferret [57] proposes a spatial-aware visual sampler to deal with free-form region inputs. Besides, to achieve more accurate localization, some works [25,42,61] resort to off-the-shelf models for pixel-level grounding. For instance, LISA [25] takes the segmentation token generated by the MLLM as the prompts for SAM [22] to produce the segmentation masks. GLaMM [42] and LLaVA-Ground [61] further advance the concept and enable grounded conversation generation. Our work shares the same focus with the aforementioned methods on region-level understanding and grounding. Yet, we distinguish ourselves from existing studies by proposing a novel perspective in enhancing the localization ability of MLLMs.

3 Method

In this section, we present Groma, a grounded multimodal large language model capable of understanding user-defined region inputs and generating visually grounded outputs. We first illustrate the model architecture of Groma in Sect. 3.1. Then we introduce how to format region input and output in Sect. 3.2. Finally, we detail the learning pipelines Sect. 3.3.

Fig. 3. Overview of Groma. Groma encodes the image input into both global image tokens and local region tokens. For region tokenization, a general-purpose region proposer is introduced to discover regions of interest, followed by a light-weight region encoder. By integrating region tokens into user instructions and model responses, Groma unlocks the referring and grounding abilities of MLLMs.

3.1 Model Architecture

As illustrated in Fig. 3, Groma primarily consists of (1) an image encoder for scene-level image tokenization, (2) a region proposer for discovering regions of interest, (3) a region encoder for region-level image tokenization, and (4) a large language model for modeling multimodal input and output. We detail each component in the following paragraphs.

Image Encoder. Groma employs a pretrained DINOv2 [37] model as the image encoder with the input image resolution set to 448×448. Compared with the commonly adopted CLIP [41] visual encoder, DINOv2 is preferred in this work for its compatibility with high-resolution inputs and fine-grained features for localization. However, the use of higher-resolution images leads to extended sequences of visual input for the language model, e.g., 1024 tokens in this case. To save computations, we further concatenate every four neighbor patch tokens into a single token following MiniGPT-v2 [7]. But slightly different from [7], we merge tokens adjacent in 2D instead of 1D, which yields better performance.

Region Proposer. To obtain localized understanding of the image, Groma innovatively incorporates a region proposer into the image tokenization process. Specifically, the region proposer is implemented as a class-agnostic detector head using the Deformable DETR (DDETR) transformer [66]. The original classification head of DDETR is replaced by a binary classifier to score region proposals based on their localization quality. Inspired by ViTDet [28], we extract feature maps from the last 4 layers of the image encoder, and rescale these feature maps to construct a hierarchical feature pyramid as the input to the region proposer. For each image, the region proposer generates 300 region proposals, which are then filtered by NMS and objectness scores before fed into the region encoder.

Region Encoder. The region encoder translates region proposals (*i.e.*, bounding boxes), coming from both user input and the region proposer, into region tokens. Akin to the previous step, we select feature maps from the last three layers of the image encoder to create a hierarchical feature pyramid. A multi-scale ROIAlign [16] module as implemented in [42,62] is utilized to crop and fuse these hierarchical features into unified region tokens. Compared with alternative ways to represent regional inputs, such as numerical representation of positions [8] and discrete location tokens [7,38], the region token representation offers distinct benefits as it is semantically aligned with the underlying region, which renders it more intuitive for the language model to comprehend.

LLM. We adopt pretrained Vicuna [12] as the language model of Groma. In particular, we instantiate Groma with the 7B version of Vicuna. Besides, we follow LLaVA v1.5 [31] to use an MLP layer to project the image tokens and region tokens into the feature space of the LLM.

3.2 Input and Output Formatting

Beyond textual only instructions and responses, Groma offers the flexibility to accept user-specified regions as input (referring) and generate visually grounded answers (grounding). Specifically, although different in task formulations, both referring and grounding are unified into one format with the use of region tokens.

Grounding Output. Remember in the tokenization process, each region token is inherently anchored to a concrete location in the image, corresponding to its region proposal. This connection allows the language model to ground its text output to particular regions in the image by simply referring to the associated region tokens. However, as region tokens are continuous embeddings, they cannot be directly integrated into the codebook of the language model and referenced in the text output. To bridge the gap, we further introduce a set of proxy tokens "$<r_1>$, $<r_2>$, ..., $<r_n>$" to register region tokens. As illustrated below, Groma can refer to any region in the output by addressing the proxy token.

> **User**: Here is an image with region crops from it. Image: `<image>`. Regions: $<r_1>$`<region>`, $<r_2>$`<region>`, ..., $<r_n>$`<region>`. [grounding] Please briefly describe the image content.
> **Groma**: `<p>`A dog`</p>` `<roi>`$<r_4>$`</roi>` is jumping to catch `<p>`a frisbee`</p>` `<roi>`$<r_7>$`</roi>` over `<p>`a fallen man`</p>` `<roi>`$<r_1>$`</roi>`.

`<image>` and `<region>` stand for placeholders of image tokens and region tokens, which are replaced by corresponding visual tokens before being fed into the LLM. `<p>` and `</p>` marks the start and end of the grounded phrase. `<roi>` and `</roi>` are used to enclose the referenced regions. In addition, we use a special token [grounding] to inform the model to generate grounded responses.

Referring Input. For a region pointed out by the user, we treat it the same as region proposals from the region proposer, i.e., encoding it into a region token and assigning a proxy token to it. This allows us to incorporate user-specified regions into our instructions by inserting corresponding region tokens. A simple example of referential dialogue in Groma is given below, where <r$_{10}$> comes from user-specified region input.

> **User**: Here is an image with region crops from it. Image: <image>. Regions: <r$_1$><region>, <r$_2$><region>, ..., <r$_n$><region>. What is <r$_{10}$><region>?
>
> **Groma**: A cute cat sleeping on a wooden bench.

3.3 Model Training

The training of Groma is partitioned into three stages: (i) detection pretraining for localization ability, (ii) alignment pretraining for image-level and region-level vision-language alignment, (iii) instruction finetuning for enhanced conversation capability. Table 1 enumerates the datasets used at different training stages. Additionally, we provide the instruction templates used to convert task-specified datasets to instruction following format in Appendix.

Table 1. Datasets used at three training stages. RefCOCO/g/+ is short for Ref-COCO, RefCOCO+, and RefCOCOg. REC means referring expression comprehension.

Training stage	Data types	Datasets
Detection pretraining	Detection	COCO, Objects365, OpenImages, V3Det, SA1B
Alignment pretraining	Image caption	ShareGPT-4V-PT
	Grounded caption	Flickr30k Entities
	Region caption	Visual Genome, RefCOCOg
	REC	COCO, RefCOCO/g/+, Grit-20m
Instruction finetuning	Grounded caption	Flickr30k Entities
	Region caption	Visual Genome, RefCOCOg
	REC	COCO, RefCOCO/g/+
	Instruction following	Groma Instruct, LLaVA Instruct, ShareGPT-4V

Detection Pretraining. This training stage only involves the image encoder and the region proposer, which collectively constitute a DDETR-like detector. The image encoder is kept frozen during training. To endow the region proposer with localization capability, an extensive collection of detection datasets, including COCO [29], Objects365 [43], OpenImages [24], and V3Det [46], is utilized for large-scale pretraining. Notably, category information is omitted from the training process, with a primary focus on box supervision.

Considering traditional detection data are typically limited to object-level annotations, we complement the training with a two million subset of SA1B [22] data filtered by GLEE [19]. Original mask annotations of SA1B are transformed into bounding boxes for consistency. The inclusion of this enriched dataset encourages the region proposer to produce region proposals encompassing not only object instances but also their constituent parts and various background stuff.

Alignment Pretraining. To align vision and language feature space of Groma, we pretrain the model on a wide range of vision-language tasks. Specifically, for image-level alignment, we leverage ShareGPT-4V-PT [9] for detailed image captioning. For region-level alignment, we engage COCO [29], RefCOCO [21], RefCOCO+ [58], RefCOCOg [34], and Grit-20m [38] for referring expression comprehension (REC), Visual Genome [23] for region captioning, and Flickr30k Entities [40] for grounded caption generation. To maintain training efficiency, we focus finetuning efforts on the MLP projection layer and the region encoder, while other modules are kept frozen throughout the training.

Instruction Finetuning. Based on alignment pretraining, we refine the training data to focus exclusively on high-quality datasets and proceed to unfreeze the language model for finetuning purposes. At this stage, LLaVA Instruct [32] and ShareGPT-4V [9] are incorporated to improve the conversational and instruction-following capabilities of Groma[2]. Besides, we curate a high-quality grounded chat dataset, named Groma Instruct (see next section for more details), to facilitate synergy of chatting and grounding abilities of Groma.

Discussions. A major difference between the training of Groma and current MLLMs is the integration of dedicated detection pretraining, which endows Groma with robust and precise localization ability. Thanks to the decoupled architecture of location and understanding within Groma, we circumvent the need to involve the LLM during detection pretraining. Such a strategic design allows Groma to benefit from pretraining on millions of bounding box annotations—a task that would be computationally prohibitive for classic MLLMs.

---

[2] Since the detailed description data in LLaVA Instruct contains severe hallucinations, we replace it with ShareGPT-4V as in [9].

4 GPT4V-Assisted Grounded Conversation Generation

Visual dialogue data have proven to be crucial in advancing the conversational capability of the MLLM as a visual chatbot. Previous methods mostly rely on coarse-grained image descriptions to derive free-form visual dialogues, which typically lack fine-grained region details and precise location information [32,65]. For grounded MLLMs, such free-form dialogue data are shown to be insufficient to enable the model to generate long-form grounded responses [61] - as the format of grounded responses significantly deviates from that of normal responses, it could be challenging for the grounded MLLM to generalize its grounding capability to long-form conversations.

To bridge the gap, we have meticulously curated a dataset containing 30k visually grounded conversations for instruction finetuning, named Groma Instruct. An illustrative example from Groma Instruct is showcased in Fig. 4. Specifically, we select images with dense region annotations from Visual Genome [23] (VG), and take the following steps to construct grounded conversations with the assistance of advanced GPT-4V model:

- First, we remove highly overlapped regions (bounding boxes) from VG annotations, normally leaving 3–10 regions of interest for each image. Then we adapt the visual prompting techniques from SoM [55] to overlay a bright numeric marker at the center of each region. Using this marked image as input unleashes the grounding capabilities of GPT-4V - it can easily make references to specific image regions by addressing the corresponding numbers.
- Besides visual input, we supply GPT-4V with rich region descriptions, image descriptions, and image-based Q&A pairs, coming from COCO [29] and VG annotations[3]. While such textual context is optional for GPT-4V input, we empirically find it useful to reduce hallucinations in generated contents and resolve potential ambiguities in visual prompts[4].
- Inspired by prior studies on visual chat data construction [9,32,47,65], we further provide GPT-4V with manually designed grounded chat as context examples. This provokes the in-context-learning ability of GPT-4V to generate grounded conversations in a uniform format. We also take a post-processing stage to filter out conversations not following the pre-defined format.

---

[3] We select VG images that also have a coco id. Thus, we can retrieve corresponding image captions from COCO Caption.
[4] There are cases where two regions highly overlap with each other and GPT-4V can hardly tell from the image which region maps to which numeric marker. For these cases, GPT-4V could rely on the numbered region descriptions to find out correspondences between regions and markers.

Region descriptions:
[1] dog walking on beach
[2] surfboard under woman's arm
[3] woman walking in wetsuit
[4] sand on ocean shore
[5] ocean with rolling waves
[6] hazy mountains on horizon

Q&A pairs:
What is the surfer holding?
Surf board.
===
What is the dog doing?
Following its master.
...

Image Descriptions:
A woman in a wetsuit carries a surfboard and walks with a dog.
A lady in a wet suit walking with a dog on a beach.
A woman carries a surf board as a dog walks beside her.
A woman and her dog are walking away from the ocean with a surfboard.
A person is walking with a dog and a surfboard.

Generated Grounded Conversations:
Question: Can you describe to me what you see in this image?
Answer: Sure, I see a beach scene with <p> a woman </p> [3] wearing a wetsuit. She's walking near the water's edge, and there's <p> a dog </p> [1] walking alongside her. <p> The woman </p> [3] is carrying <p> a surfboard </p> [2] under her arm. I can also see <p> the ocean </p> [5] with some waves and <p> faint mountains </p> [6] in the distance.
===
Question: What kind of dog is it, and what's it doing?
Answer: It looks like <p> a medium-sized, short-haired dog </p> [1], quite possibly a mixed breed. <p> The dog </p> [1] is just strolling beside <p> the woman </p> [3], seemingly following her as they both walk across <p> the sand </p> [4].
===
Question: Is the beach crowded, or does it seem deserted?
Answer: It actually looks quite deserted, with no one else in sight apart from <p> the woman </p> [3] and <p> her dog </p> [1]. It gives off a peaceful, solitary vibe - just them and the <p> vast ocean </p> [5] and <p> sandy shore </p> [4].
===
Question: What is the state of the ocean, and is it suitable for surfing?
Answer: <p> The ocean </p> [5] appears quite serene but with rolling waves, hence appropriate for surfing.

Fig. 4. A data example from Groma Instruct. The top part demonstrates the image input with visual prompts and contextual text input to GPT-4V. Notably, all numbered regions have corresponding bounding box annotations, which are not shown in the image. The bottom part is the grounded conversations generated by GPT-4V.

5 Experiments

In this section, we first quantitatively access the abilities of Groma on grounding (Sect. 5.2), referring (Sect. 5.3), and image-based conversation (Sect. 5.4) tasks. Then we provide qualitative results to exemplify the strong capabilities of Groma on a wide range of region-level tasks (Sect. 5.5). Finally, we ablate the design and training of Groma in Sect. 5.6.

5.1 Implementation Details

We adopt DINOv2-L/14 [37] as the image encoder and Vicuna-7B v1.5 [12] as the language model. The region proposer follows an encoder-decoder architecture with 6 encoder layers and 6 decoder layers. We further employ mixed query selection and look-forward-twice scheme as in [60] to accelerate convergence. We set NMS threshold to 0.6 and filter out region proposals with objectness scores lower than 0.15. Subsequently, we select the top 100 region proposals if there are more than 100 proposals left after filtering. This results in no more than 356 visual tokens in total. For training, we sequentially proceed 12 epochs of detection pretraining, 2 epochs of alignment pretraining, and 1 epoch of instruction finetuning. More training details can be found in the Appendix.

5.2 Grounding Benchmark Results

Table 2. Results on referring expression comprehension benchmarks. We report accuracy with the IoU threshold set to 0.5. We make Qwen-VL gray because it uses a much larger visual tokenizer (1.9B ViT-bigG [17]).

Method	Model type	RefCOCO			RefCOCO+			RefCOCOg		Average
		val	testA	testB	val	testA	testB	val	test	
MDETR [20]	Specialist	86.75	89.58	81.41	79.52	84.09	70.62	81.64	80.89	81.81
G-DINO [33]		90.56	93.19	88.24	82.75	88.95	75.92	86.13	87.02	86.60
UNINEXT-L [54]		91.43	93.73	88.93	83.09	87.90	76.15	86.91	87.48	86.95
VisionLLM [51]	Generalist	–	86.70	–	–	–	–	–	–	–
OFA [48]		79.96	83.67	76.39	68.29	76.00	61.75	67.57	67.58	72.65
Shikra [8]		87.01	90.61	80.24	81.60	87.36	72.12	82.27	82.19	82.93
Ferret [57]		87.49	91.35	82.45	80.78	87.38	73.14	83.93	84.76	83.91
MiniGPT-v2 [7]		88.69	91.65	85.33	79.97	85.12	74.45	84.44	84.66	84.29
Qwen-VL [4]		89.36	92.26	85.34	83.12	88.25	77.21	85.58	85.48	85.83
Groma		**89.53**	**92.09**	**86.26**	**83.90**	**88.91**	**78.05**	**86.37**	**87.01**	**86.52**

We evaluate the localization capability of Groma on visual grounding tasks. Table 2 showcases our performance on three classic referring expression comprehension benchmarks: RefCOCO [21], RefCOCO+ [58], and RefCOCOg [34]. Groma notably surpasses other generalist models of similar model size across all metrics. Moreover, as a generalist model, Groma shows competitive results with state-of-the-art specialist models [33,54]. These findings underscore the strong capability of Groma in visual grounding.

However, we notice that traditional REC benchmarks only cover a narrow range of common objects in their referring expressions, which is insufficient to thoroughly evaluate the MLLM's localization capability. Therefore, we further introduce LVIS-Ground, an object grounding benchmark converted from the LVIS [15] detection data. LVIS-Ground contains 4299 images covering 1203 categories of objects, with on average 3.7 target objects per image. Complementary to REC benchmarks, LVIS-Ground focuses on testing the model's ability to locate multiple, diverse, and variably-sized objects. For more details of LVIS-Ground, please refer to the Appendix.

Table 3 presents our results on LVIS-Ground. Notably, Groma demonstrates clear advantages over other grounded MLLMs, especially on the AR@0.75 metric. This evidences that the specialized design and training indeed bring more accurate localization for Groma. Moreover, it is noteworthy that current MLLMs all fall short of small object localization (AR@s metric). We conjecture this is mainly because the training data (e.g., RefCOCO/g/+, Flickr30k) lack annotations for small objects. We also notice that current models typically predict only one box per image, even though more than one targets are presented. This is an expected behavior as the they heavily rely on REC data for grounding training, which only has one target object per query. These findings call for the necessity of diversifying grounding data used for training in future MLLMs.

Table 3. Results on the LVIS-Ground benchmark. We report average recall (AR) to measure performances. For each model, we use the native prompt template recommended by the paper for evaluation.

Method	AR	AR@0.5	AR@0.75	AR@s	AR@m	AR@l
Shikra [8]	4.9	14.2	2.0	0.1	3.1	18.5
MiniGPT-v2 [7]	11.4	19.8	11.2	0.3	8.0	41.1
Ferret [57]	16.8	29.6	16.3	1.6	16.7	51.1
Groma	**28.8**	**37.9**	**30.3**	**8.7**	**35.6**	**64.3**

5.3 Referring Benchmark Results

We evaluate Groma on the region captioning task to assess its fine-grained region understanding capability. To prompt the model to generate region-level descriptions, we use queries like *"Please describe ¡region¿ in details."*. Table 4 presents our results on two established region captioning benchmarks, RefCOCOg and Visual Genome. Without task-specific finetuning, Groma shows comparable or improved performance over GLaMM[5] [42], which has separate designs for input referring and output grounding. This exemplifies the superiority of unified refer-and-ground formulation in Groma.

5.4 Conversational VQA Benchmark Results

In addition to region-level tasks, we further evaluate Groma on the conversational style VQA benchmark, LLaVA Bench (COCO) [32], which contains three types of questions, namely conversation, detailed description, and complex reasoning. As shown in Tab. 5, Groma surpasses the strong baseline method LLaVA [32]

---

[5] We re-evaluate the performance of GLaMM using the officially released checkpoint after fixing the bug in its original evaluation scripts.

Table 4. Results on region captioning benchmarks. We report METEOR and CIDEr scores to measure caption quality. †: with task-specific finetuning.

Method	RefCOCOg		Visual Genome	
	METEOR	CIDEr	METEOR	CIDEr
GRIT [53]	15.2	71.6	17.1	142
Kosmos-2 [38]	14.1	62.3	–	–
GPT4RoI [62]	–	–	17.4	145.2
GLaMM† [42]	16.1	101.9	**19.0**	**163.9**
ViP-LLaVA [6]	16.6	105.9	–	–
Groma	**16.8**	**107.3**	**19.0**	158.4

and achieves competitive performance among grounded MLLMs, especially in detailed image description. This demonstrates that Groma maintains decent image understanding and visual chatting abilities. For the underperformance in conversation and complex reasoning questions, we speculate this could be resulted from the DINOv2 features, as shown in recent studies [18,30]. But we believe such gap can be closed by scaling up vision-language alignment pretraining.

Table 5. Results on LLaVA-Bench (COCO).

Method	Conversation	Description	Reasoning	Average
LLaVA [32]	85.4	68.3	92.1	81.9
Kosmos-2 [38]	71.7	63.4	74.9	70.0
Shikra [8]	80.6	70.7	88.1	79.9
LLaVA-G [61]	79.3	71.2	92.8	81.2
Ferret [57]	**84.4**	79.4	**96.3**	**86.7**
Groma	82.6	**84.0**	88.8	85.2

5.5 Qualitative Results

Figure 5 presents a comparison between Groma and other grounded MLLMs on the grounded image captioning task. We choose an exemplar image that is inherently challenging with multiple and occluded instances to ground. Groma manifests exceptional grounding performance in this case with the highest recall and minimum hallucinations. In addition, we provide several visualization examples in Fig. 6 for a complementary understanding of Groma's abilities on grounded

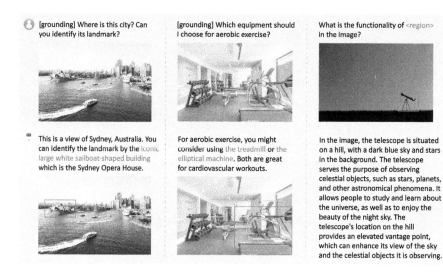

Fig. 5. A qualitative comparison on the grounded image captioning task.

Fig. 6. Examples on visually grounded chat and referential dialogue.

chat and referential dialogue. We show that Groma is capable of generating long-form, grounded and logically rich answers, which can be mainly attributed to the introduction of Groma Instruct data in finetuning.

5.6 Ablation

CLIP vs. DINOv2. To quantitatively assess the differences in localization capabilities between CLIP and DINOv2, we compare the two backbones on the COCO detection benchmark in Table 6. For this comparison, we equip each backbone with a DDETR [66] detection head and finetune only the detection head on COCO dataset. It can be seen that under the same resolution, DINOv2 backbone significantly outperforms CLIP backbone by 6.5 AP. Furthermore, by scaling the resolution to 448×448, DINOv2 backbone achieves a commendable performance of 43.6 AP. The results consolidate our choice of DINOv2 backbone in Groma.

Frozen LLM. In Table 7, we reveal that Groma retains robust localized understanding even without finetuning the LLM, *i.e.*, it demonstrates a referring ability on par with GPT4ROI [62] (148.0 *vs*. 145.2) and grounding ability comparable to Ferret [57] (84.02% *vs*. 83.91%). This finding suggests our design effectively decouples localization and understanding within Groma, such that it requires minimum 'new knowledge' from the LLM for localized understanding.

Token Merge. To save computations, Groma by default concatenates every 4 image tokens into one as LLM inputs. Through control experiments in Table 8, we find that such downsampling has negligible impacts on the grounding performances (*e.g.*, less than 0.1% average accuracy drop on the REC benchmarks). The results evidence that the decoupled design is optimal in both efficiency and localization accuracy.

Table 6. Object detection performances on COCO with different backbones and image resolutions.

Backbone	Resolution	AP
CLIP	336 × 336	32.4
DINOv2	336 × 336	38.9
DINOv2	448 × 448	43.6

Table 7. Referring and grounding abilities with the frozen LLM. We measure referring ability with CIDEr score on Visual Genome and grounding ability with average accuracy on REC benchmarks.

Stage	LLM status	Referring	Grounding
pretraining	frozen	–	82.33
finetuning	frozen	148.0	84.02
finetuning	unfrozen	158.4	86.52

Table 8. Ablation on image token downsampling on the REC benchmarks.

Downsampling	RefCOCO			RefCOCO+			RefCOCOg		Average
	val	testA	testB	val	testA	testB	val	test	
✓	89.32	92.15	85.96	84.11	88.10	78.40	86.33	87.40	86.47
✗	89.54	92.54	86.18	83.72	88.52	78.96	86.17	86.84	86.55

6 Limitations and Conclusions

In this paper, we introduce a novel paradigm, Groma, to unleash the localized perception capabilities of MLLMs. We make the pioneering attempt to embed localization into image tokenization. Our paradigm is based on a perception-then-understand mindset that separates localization from high-level understanding and reasoning. Without introducing external modules, our approach overcomes the resolution bottleneck of using LLMs as location decoders and unifies referring and visual grounding tasks. Extensive experiments showcase the superior performance of our approach in localized perception, as evidenced by its success in referring and visual grounding tasks.

However, the current implementation does not support free-form region inputs and pixel-level grounding. A promising direction to address such limitations is to re-implement the region encoder with a visual sampler as in [57,67] and replace the box region proposer by a mask region proposer like Mask2Former [11]. We leave this for future studies.

Acknowledgements. This work has been supported by Hong Kong Research Grant Council - Early Career Scheme (Grant No. 27209621), General Research Fund Scheme (Grant No. 17202422), Theme-based Research (Grant No. T45-701/22-R) and RGC Matching Fund Scheme (RMGS). Part of the described research work is conducted in the JC STEM Lab of Robotics for Soft Materials funded by The Hong Kong Jockey Club Charities Trust.

References

1. Achiam, J., et al.: GPT-4 technical report. arXiv preprint arXiv:2303.08774 (2023)
2. Alayrac, J.B., et al.: Flamingo: a visual language model for few-shot learning. In: Advances in Neural Information Processing Systems, vol. 35, pp. 23716–23736 (2022)
3. Bai, J., et al.: Qwen-VL: a frontier large vision-language model with versatile abilities. arXiv preprint arXiv:2308.12966 (2023)
4. Bai, J., et al.: Qwen-VL: a versatile vision-language model for understanding, localization, text reading, and beyond (2023)
5. Bubeck, S., et al.: Sparks of artificial general intelligence: early experiments with GPT-4. arXiv preprint arXiv:2303.12712 (2023)
6. Cai, M., et al.: Making large multimodal models understand arbitrary visual prompts. arXiv preprint arXiv:2312.00784 (2023)
7. Chen, J., et al.: Minigpt-V2: large language model as a unified interface for vision-language multi-task learning. arXiv preprint arXiv:2310.09478 (2023)
8. Chen, K., Zhang, Z., Zeng, W., Zhang, R., Zhu, F., Zhao, R.: Shikra: unleashing multimodal LLM's referential dialogue magic. arXiv preprint arXiv:2306.15195 (2023)
9. Chen, L., et al.: Sharegpt4v: improving large multi-modal models with better captions. arXiv preprint arXiv:2311.12793 (2023)
10. Chen, Z., et al.: Internvl: scaling up vision foundation models and aligning for generic visual-linguistic tasks. arXiv preprint arXiv:2312.14238 (2023)
11. Cheng, B., Misra, I., Schwing, A.G., Kirillov, A., Girdhar, R.: Masked-attention mask transformer for universal image segmentation. In: Proceedings of the IEEE/CVF Conference on Computer Vision and Pattern Recognition, pp. 1290–1299 (2022)
12. Chiang, W.L., et al.: Vicuna: an open-source chatbot impressing GPT-4 with 90%* chatgpt quality (2023). https://vicuna.lmsys.org. Accessed 14 Apr 2023
13. Dai, W., et al.: InstructBLIP: towards general-purpose vision-language models with instruction tuning. In: Thirty-Seventh Conference on Neural Information Processing Systems (2023). https://openreview.net/forum?id=vvoWPYqZJA
14. Guo, Q., et al.: Regiongpt: towards region understanding vision language model (2024)
15. Gupta, A., Dollar, P., Girshick, R.: LVIS: a dataset for large vocabulary instance segmentation. In: Proceedings of the IEEE/CVF Conference on Computer Vision and Pattern Recognition, pp. 5356–5364 (2019)

16. He, K., Gkioxari, G., Dollár, P., Girshick, R.: Mask R-CNN. In: Proceedings of the IEEE International Conference on Computer Vision, pp. 2961–2969 (2017)
17. Ilharco, G., et al.: Openclip (2021). https://doi.org/10.5281/zenodo.5143773
18. Jiang, D., et al.: From clip to dino: visual encoders shout in multi-modal large language models (2023)
19. Junfeng, W., Yi, J., Qihao, L., Zehuan, Y., Xiang, B., Song, B.: General object foundation model for images and videos at scale. arXiv preprint arXiv:2312.09158 (2023)
20. Kamath, A., Singh, M., LeCun, Y., Synnaeve, G., Misra, I., Carion, N.: MDETR-modulated detection for end-to-end multi-modal understanding. In: Proceedings of the IEEE/CVF International Conference on Computer Vision, pp. 1780–1790 (2021)
21. Kazemzadeh, S., Ordonez, V., Matten, M., Berg, T.: Referitgame: referring to objects in photographs of natural scenes. In: Proceedings of the 2014 Conference on Empirical Methods in Natural Language Processing (EMNLP), pp. 787–798 (2014)
22. Kirillov, A., et al.: Segment anything. arXiv preprint arXiv:2304.02643 (2023)
23. Krishna, R., et al.: Visual genome: connecting language and vision using crowdsourced dense image annotations. Int. J. Comput. Vision **123**, 32–73 (2017)
24. Kuznetsova, A., et al.: The open images dataset V4: unified image classification, object detection, and visual relationship detection at scale. Int. J. Comput. Vision **128**(7), 1956–1981 (2020)
25. Lai, X., et al.: LISA: reasoning segmentation via large language model. arXiv preprint arXiv:2308.00692 (2023)
26. Li, B., Zhang, Y., Chen, L., Wang, J., Yang, J., Liu, Z.: Otter: a multi-modal model with in-context instruction tuning. arXiv preprint arXiv:2305.03726 (2023)
27. Li, J., Li, D., Savarese, S., Hoi, S.: Blip-2: bootstrapping language-image pre-training with frozen image encoders and large language models. arXiv preprint arXiv:2301.12597 (2023)
28. Li, Y., Mao, H., Girshick, R., He, K.: Exploring plain vision transformer backbones for object detection. In: Avidan, S., Brostow, G., Cissé, M., Farinella, G.M., Hassner, T. (eds.) ECCV 2022. LNCS, vol. 13669, pp. 280–296. Springer, Cham (2022). https://doi.org/10.1007/978-3-031-20077-9_17
29. Lin, T.-Y., et al.: Microsoft COCO: common objects in context. In: Fleet, D., Pajdla, T., Schiele, B., Tuytelaars, T. (eds.) ECCV 2014. LNCS, vol. 8693, pp. 740–755. Springer, Cham (2014). https://doi.org/10.1007/978-3-319-10602-1_48
30. Lin, Z., et al.: Sphinx: the joint mixing of weights, tasks, and visual embeddings for multi-modal large language models. arXiv preprint arXiv:2311.07575 (2023)
31. Liu, H., Li, C., Li, Y., Lee, Y.J.: Improved baselines with visual instruction tuning. arXiv preprint arXiv:2310.03744 (2023)
32. Liu, H., Li, C., Wu, Q., Lee, Y.J.: Visual instruction tuning. In: Advances in Neural Information Processing Systems, vol. 36 (2024)
33. Liu, S., et al.: Grounding dino: marrying dino with grounded pre-training for open-set object detection. arXiv preprint arXiv:2303.05499 (2023)
34. Mao, J., Huang, J., Toshev, A., Camburu, O., Yuille, A.L., Murphy, K.: Generation and comprehension of unambiguous object descriptions. In: Proceedings of the IEEE Conference on Computer Vision and Pattern Recognition, pp. 11–20 (2016)
35. OpenAI: Chatgpt (2023). https://openai.com/blog/chatgpt/
36. OpenAI: GPT-4 technical report (2023)
37. Oquab, M., et al.: Dinov2: learning robust visual features without supervision. arXiv preprint arXiv:2304.07193 (2023)

38. Peng, Z., et al.: Kosmos-2: grounding multimodal large language models to the world. arXiv preprint arXiv:2306.14824 (2023)
39. Pi, R., et al.: Detgpt: detect what you need via reasoning. arXiv preprint arXiv:2305.14167 (2023)
40. Plummer, B.A., Wang, L., Cervantes, C.M., Caicedo, J.C., Hockenmaier, J., Lazebnik, S.: Flickr30k entities: collecting region-to-phrase correspondences for richer image-to-sentence models. In: Proceedings of the IEEE International Conference on Computer Vision, pp. 2641–2649 (2015)
41. Radford, A., et al.: Learning transferable visual models from natural language supervision. In: International Conference on Machine Learning, pp. 8748–8763. PMLR (2021)
42. Rasheed, H., et al.: GLaMM: pixel grounding large multimodal model. arXiv preprint arXiv:2311.03356 (2023)
43. Shao, S., et al.: Objects365: a large-scale, high-quality dataset for object detection. In: Proceedings of the IEEE/CVF International Conference on Computer Vision, pp. 8430–8439 (2019)
44. Touvron, H., et al.: Llama: open and efficient foundation language models. arXiv preprint arXiv:2302.13971 (2023)
45. Touvron, H., et al.: Llama 2: open foundation and fine-tuned chat models. arXiv preprint arXiv:2307.09288 (2023)
46. Wang, J., et al.: V3det: vast vocabulary visual detection dataset. arXiv preprint arXiv:2304.03752 (2023)
47. Wang, J., Meng, L., Weng, Z., He, B., Wu, Z., Jiang, Y.G.: To see is to believe: prompting GPT-4V for better visual instruction tuning. arXiv preprint arXiv:2311.07574 (2023)
48. Wang, P., et al.: OFA: unifying architectures, tasks, and modalities through a simple sequence-to-sequence learning framework. In: International Conference on Machine Learning, pp. 23318–23340. PMLR (2022)
49. Wang, W., et al.: Cogvlm: visual expert for pretrained language models. arXiv preprint arXiv:2311.03079 (2023)
50. Wang, W., et al.: The all-seeing project: towards panoptic visual recognition and understanding of the open world. arXiv preprint arXiv:2308.01907 (2023)
51. Wang, W., et al.: Visionllm: large language model is also an open-ended decoder for vision-centric tasks. In: Advances in Neural Information Processing Systems, vol. 36 (2024)
52. Wei, J., et al.: Chain-of-thought prompting elicits reasoning in large language models. In: Advances in Neural Information Processing Systems, vol. 35, pp. 24824–24837 (2022)
53. Wu, J., et al.: Grit: a generative region-to-text transformer for object understanding. arXiv preprint arXiv:2212.00280 (2022)
54. Yan, B., et al.: Universal instance perception as object discovery and retrieval. In: Proceedings of the IEEE/CVF Conference on Computer Vision and Pattern Recognition, pp. 15325–15336 (2023)
55. Yang, J., Zhang, H., Li, F., Zou, X., Li, C., Gao, J.: Set-of-mark prompting unleashes extraordinary visual grounding in GPT-4V. arXiv preprint arXiv:2310.11441 (2023)
56. Ye, Q., et al.: mPLUG-Owl: modularization empowers large language models with multimodality. arXiv preprint arXiv:2304.14178 (2023)
57. You, H., et al.: Ferret: refer and ground anything anywhere at any granularity. arXiv preprint arXiv:2310.07704 (2023)

58. Yu, L., Poirson, P., Yang, S., Berg, A.C., Berg, T.L.: Modeling context in referring expressions. In: Leibe, B., Matas, J., Sebe, N., Welling, M. (eds.) ECCV 2016. LNCS, vol. 9906, pp. 69–85. Springer, Cham (2016). https://doi.org/10.1007/978-3-319-46475-6_5
59. Yuan, Y., et al.: Osprey: pixel understanding with visual instruction tuning. arXiv preprint arXiv:2312.10032 (2023)
60. Zhang, H., et al.: DINO: DETR with improved denoising anchor boxes for end-to-end object detection. arXiv preprint arXiv:2203.03605 (2022)
61. Zhang, H., et al.: Llava-grounding: grounded visual chat with large multimodal models. arXiv preprint arXiv:2312.02949 (2023)
62. Zhang, S., et al.: GPT4RoI: instruction tuning large language model on region-of-interest. arXiv preprint arXiv:2307.03601 (2023)
63. Zhao, Y., Lin, Z., Zhou, D., Huang, Z., Feng, J., Kang, B.: BuboGPT: enabling visual grounding in multi-modal LLMs. arXiv preprint arXiv:2307.08581 (2023)
64. Zhou, X., Girdhar, R., Joulin, A., Krähenbühl, P., Misra, I.: Detecting twenty-thousand classes using image-level supervision. In: Avidan, S., Brostow, G., Cissé, M., Farinella, G.M., Hassner, T. (eds.) ECCV 2022. LNCS, vol. 13669, pp. 350–368. Springer, Cham (2022). https://doi.org/10.1007/978-3-031-20077-9_21
65. Zhu, D., Chen, J., Shen, X., Li, X., Elhoseiny, M.: MiniGPT-4: enhancing vision-language understanding with advanced large language models. arXiv preprint arXiv:2304.10592 (2023)
66. Zhu, X., Su, W., Lu, L., Li, B., Wang, X., Dai, J.: Deformable DETR: deformable transformers for end-to-end object detection. arXiv preprint arXiv:2010.04159 (2020)
67. Zou, X., et al.: Segment everything everywhere all at once. In: Advances in Neural Information Processing Systems, vol. 36 (2024)

On the Utility of 3D Hand Poses for Action Recognition

Md Salman Shamil[1](✉), Dibyadip Chatterjee[1], Fadime Sener[2], Shugao Ma[2], and Angela Yao[1]

[1] National University of Singapore, Singapore, Singapore
{salman,dibyadip,ayao}@comp.nus.edu.sg
[2] Meta Reality Labs, Menlo Park, USA
{famesener,shugao}@meta.com
https://s-shamil.github.io/HandFormer/

Abstract. 3D hand pose is an underexplored modality for action recognition. Poses are compact yet informative and can greatly benefit applications with limited compute budgets. However, poses alone offer an incomplete understanding of actions, as they cannot fully capture objects and environments with which humans interact. We propose HandFormer, a novel multimodal transformer, to efficiently model hand-object interactions. HandFormer combines 3D hand poses at a high temporal resolution for fine-grained motion modeling with sparsely sampled RGB frames for encoding scene semantics. Observing the unique characteristics of hand poses, we temporally factorize hand modeling and represent each joint by its short-term trajectories. This factorized pose representation combined with sparse RGB samples is remarkably efficient and highly accurate. Unimodal HandFormer with only hand poses outperforms existing skeleton-based methods at 5× fewer FLOPs. With RGB, we achieve new state-of-the-art performance on Assembly101 and H2O with significant improvements in egocentric action recognition.

Keywords: Skeleton-based action recognition · 3D hand poses · Multimodal transformer

1 Introduction

The popularity of AR/VR headsets has driven interest in recognizing hand-object interactions, particularly through egocentric [14,26] and multiview cameras [34,51]. Such interactions are inherently fine-grained; recognizing them requires distinguishing subtle motions and object state changes. State-of-the-art methods for hand action recognition [23,45,47,66] primarily rely on multi- or single-view RGB streams, which are computationally heavy and unsuitable for resource-constrained scenarios like AR/VR. Motivated by advancements in lightweight hand pose estimation methods leveraging monochrome

Supplementary Information The online version contains supplementary material available at https://doi.org/10.1007/978-3-031-72658-3_25.

© The Author(s), under exclusive license to Springer Nature Switzerland AG 2025
A. Leonardis et al. (Eds.): ECCV 2024, LNCS 15064, pp. 436–454, 2025.
https://doi.org/10.1007/978-3-031-72658-3_25

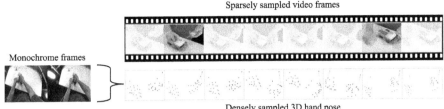

Fig. 1. We densely sample 3D hand poses to understand fine-grained hand motions and sparsely sample RGB frames to capture the scene semantics. Our 3D hand poses are acquired from low-resolution monochrome cameras but can also come from wearable sensors, facilitating an efficient understanding of hand-object interactions. Video frames and hand poses in the figure are from Assembly101 [51].

cameras [27,28,46], and the evolution of low-dimensional sensors [36,42] such as accelerometers, MMG, EMG, demonstrating real-time hand pose estimation, we advocate the use of 3D hand poses as an input modality for recognizing hand-object interactions. Hand poses are a compact yet informative representation that captures the motions and nuances of hand movements (Fig. 1).

Existing works on 3D pose-based action recognition have focused primarily on full-body skeletons [20,44,65,68]. Hand poses differ fundamentally from full-body skeletons. In full-body recognition datasets [41,52], the actions are predominantly static from a global perspective. The relative changes in joint or limb positions signify the action category, *e.g.*, 'sitting down'. Conversely, hand joints typically move together for many actions and lack a static joint as a global reference [51], *e.g.*, 'put down toy'. Full-body skeleton methods also benefit from modeling long-range spatiotemporal dependencies between joints [44], while this is less important for the hands as shown in Fig. 2 and in Sect. 3.

However, the 3D pose alone cannot encode the hands' actions. Unlike the full-body case, where actions are self-contained by the sequence of poses, the hands are often manipulating objects [34,46,51]. Hand pose is excellent for identifying motions (verbs) but struggles with associated objects [51]. Therefore, supplementing pose data with visual context from images or videos is crucial for full semantic understanding. However, as noted earlier, using dense RGB frames contradicts our motivation for using hand poses.

This paper introduces HandFormer, a novel and lightweight multimodal transformer that leverages dense 3D hand poses complemented with sparsely sampled RGB frames. To this end, we conceptualize an action as a sequence of short segments, termed *micro-actions*, which are analogous to words that form sentence-level complex actions. Each micro-action comprises a dense sequence of pose frames and a single RGB frame. As every hand joint moves in close spatial proximity, we encode the pose sequence from a Lagrangian view [49] and track each joint as an individual entity. HandFormer effectively models each micro-action whose features are then temporally aggregated for action classification. This formulation benefits from our observations on the spatiotemporal dynamics of hand movement and delivers strong performance. Additionally, limiting long-range spatiotemporal dependencies among joints and only using sparse RGB frame features improves computational efficiency, making it well-suited for always-on, low-compute AR/VR applications.

Our contributions are: *(i)* We analyze the differences between hand pose and full-body skeleton actions and design a novel pose sequence encoding that reflects hand-specific properties. *(ii)* We propose HandFormer that takes a sequence of dense 3D hand pose and sparse RGB frames as input in the form of micro-actions. *(iii)* HandFormer achieves state-of-the-art action recognition on H2O [34] and Assembly101 [51]. Unimodal HandFormer with only hand poses outperforms existing skeleton-based methods in Assembly101, incurring at least 5× fewer FLOPs. *(iv)* We experimentally demonstrate how hand poses are crucial for multiview and egocentric action recognition.

2 Related Work

Video Action Recognition. Video-based action recognition systems are well-developed with sophisticated 3D-CNN [9,23,57,67] or Video Transformer [2,3,43,47] architectures. However, they all bear significant computational expense for both feature extraction and motion modeling, either explicitly in the form of optical flow [9,33,55] or implicitly through the architecture [2,47]. Such designs are well-suited for high-facility applications but are not suitable for integration into lightweight systems. To this end, efficient video understanding is an active topic of research, with a focus on reducing expensive 3D operations [22,40,58,67], quadratic complexity of attention [32,43,47] and dropping tokens [4,21]. Additionally, *densly-sampled* RGB frames incur a high cost, yet limiting the temporal resolution will hinder fine-grained action understanding. We propose complementing 3D hand poses with *sparsely-sampled* RGB frames, developing a lightweight video understanding system.

Skeleton-Based Action Recognition. Skeleton-based action recognition has been tackled with hand-crafted features [60,62], sequence models like RNNs and LSTMs [19,69], and CNN-based methods that either employ temporal convolution on the pose sequence [56] or transform the skeleton data into pseudo-images to be processed with 2D or 3D convolutional networks [7,20,31]. Recent progress has been driven by GNN-based methods that exploit the graph structure of skeletal data to construct spatio-temporal graphs and perform graph convolutions [68,70]. These methods often model functional links between joints that go beyond skeletal connectivity [39,54] or expand the spatiotemporal receptive field [44]. Self-attention and transformer-based methods have also been proposed [48,65,71]. However, most existing methods are tailored for datasets involving full-body poses. Few works [24,30,37,50] study hand poses but are restricted to simple gestures, recognizable without complex temporal modeling.

Fusing RGB with Skeleton. Skeletal data has been used for cropping images of body parts [12], for weighting RGB patches around regions of interest [5,6],

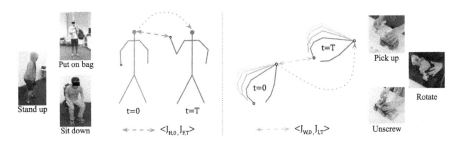

Fig. 2. Comparing skeletal changes in full-body actions from NTU RGB+D 120 [41] (left) and hand actions from Assembly101 [51] (right). Two pose frames at interval T are shown. $J_{j,t}$ indicates the 3D coordinate of joint j at timestep t, and $<J_x, J_y>$ is the correlation between two such joints. Modeling the correlation between spatiotemporally distant joints can be informative for full-body poses but does not provide a useful action cue for hand poses.

or for pooling CNN features [1,8]. Projecting into common embedding space is done in [16], enabling pose distillation [15]. Multi-stream architectures are also designed with separate paths for RGB and skeleton having lateral connections between them [38,63]. RGBPoseConv3D [20] is a two-stream model that employs 3D CNNs for both RGB and pose. Similar to skeleton-based action recognition, these multimodal approaches primarily focus on full-body poses. Some approaches simultaneously perform both hand pose estimation and action recognition, using pose data to supervise the training process [13,64]. However, it is important to note that pose estimation deals with lower-level semantics than action recognition. Actions can often be inferred even when the estimated poses are not accurate [46].

3 Modeling Full-Body vs. Hand Skeletons

Existing skeleton-based action recognition datasets primarily consist of full-body poses where actions feature significant changes over time in limb positions relative to other body parts, which highly correlate to the action category. These changes can be captured through long-range spatiotemporal modeling using graph convolutions with large receptive fields [44] or self-attention [65]. Conversely, hand actions show diverse movement patterns, as hands can move in arbitrary directions, often without significant changes in articulation. We analyze pose sequences from NTU RGB+D 120 [41] and Assembly101 [51], calculating the distance covered by each joint and identifying the least and the most active joints. We observe that, for full-body poses, these two representative joints show a significant difference in the distances covered, while for hand poses, the difference is subtle. The Pearson correlation coefficient is 0.93 for hand poses (indicating highly coupled joints) and 0.33 for full-body poses. Please refer to Suppl. Sec. A for details regarding these computations and comparisons.

An example showing the difference between body and hand poses is provided in Fig. 2, where we depict two frames separated by an interval of T frames. For the

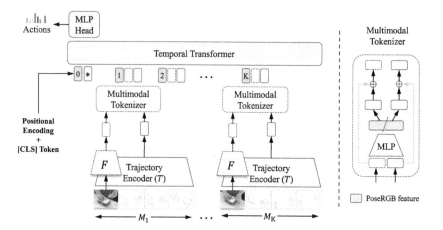

Fig. 3. Overall architecture of HandFormer. An action segment is divided into K micro-actions $\{M_1, M_2 \ldots M_K\}$. Each micro-action comprises a dense sequence of pose frames and a single RGB frame. The frame encoder F and the trajectory encoder T encode the RGB and the dense poses, respectively, after which they are passed to a Multimodal Tokenizer. The modality-mixed tokens are then fed to a Temporal Transformer. Dotted paths are optional and only required when RGB is used.

full body skeleton, the head joint J_H remains static, while the right hand joint J_F moves upward. Conversely, for the hand, all the joints move, including the wrist joint J_W and index fingertip J_I. The spatiotemporal correlation between the head joint in the first frame and the hand joint in the last frame ($<J_{H,0}, J_{F,T}>$) is a crucial action cue that captures the relative structural change. However, such a correlation between the wrist and the fingertip for hand skeleton ($<J_{W,0}, J_{I,T}>$) does not provide a stronger cue than the wrist movement itself. To avoid such redundant spatiotemporally distant correlations, we propose to divide the hand pose sequence into micro-action blocks of fixed temporal length. This formulation allows for encoding short-term movements while enabling parameter sharing.

Moreover, full-skeletal motion is a dominant feature in hand pose sequences, differentiating them from full-body poses. While the location and orientation of the human body can be trivial for most actions, this is not the case for hands. The hands do not conform to any particular 6D pose, and notably, they frequently and unpredictably alter their 6D poses over time. A comprehensive analysis of this phenomenon is provided in Suppl. Sec. A. To this end, we explicitly consider the global 6D poses of the hands during micro-action-based pose encoding.

4 HandFormer

Figure 3 illustrates our proposed HandFormer, which consists of a sequence of micro-action blocks, a novel trajectory encoder, and a temporal aggregation module. The design of our HandFormer allows us to easily incorporate semantic context by sampling a single RGB frame from certain micro-actions.

4.1 Micro-actions

Given an action segment containing $\mathcal{T}$ frames sampled at a certain fps, the input to our model comprises—*i)* a dense sequence of 3D hand poses, represented as $\mathbf{S} = \{P_1, P_2, \ldots, P_\mathcal{T}\}$, where $P_t = \{P_t^{left}, P_t^{right}\} \in \mathbb{R}^{2 \times J \times 3}$ signifying the 3D coordinates of J keypoints in the left and right hands, respectively, and *ii)* a sparse set of RGB frames sampled at intervals Δf, denoted as $\mathbf{V} = \{I_1, I_{1+\Delta f}, \ldots, I_{\mathcal{T}_r}\}$, where $I_t \in \mathbb{R}^{H \times W \times 3}$ are frame-wise RGB features and $\mathcal{T}_r = 1 + \left\lfloor \frac{\mathcal{T}-1}{\Delta f} \right\rfloor \times \Delta f$.

We factorize the raw input into a sequence of K micro-action blocks of length N frames, obtained by shifting the window across the action segment with a stride of R frames. Each block consists of two components—the initial appearance derived from the first RGB frame within the block and the hand motion characterized by the dense sequence of N pose frames. To obtain a fixed length input containing $\mathcal{T}' = (K-1) \times R + N$ pose frames, we perform linear interpolation for each joint along the temporal axis, transforming pose sequence $\mathbf{S}$ having $\mathcal{T}$ frames to $\mathbf{S}' = \{P_1', P_2', \ldots, P_{\mathcal{T}'}'\}$ having $\mathcal{T}'$ frames, where $P_t' \in \mathbb{R}^{2 \times J \times 3}$. Thus, we represent the input as a sequence of micro-actions $\mathbf{M} = \{M_1, M_2, \ldots, M_K\}$, derived from $\mathbf{V}$ and $\mathbf{S}'$ through the following equations:

$$M_k = \left[M_k^{\text{RGB}}, M_k^{\text{Pose}} \right] = \left[I_{h(k)}, \{P_{g(k)+i}'\}_{i=0}^{N-1} \right], \tag{1}$$

where $g(k) = (k-1) \times R + 1$ denotes the first pose frame in k-th Micro-action, and $h(k)$ determines the index of the nearest available RGB frame.

The dense pose sequence in a micro-action captures fine-grained hand motion crucial for recognizing verbs, whereas a single RGB frame provides semantic context for recognizing objects. To extract features from micro-actions, we use a frame encoder F and a trajectory encoder T, which operate on RGB frames and pose sequences, respectively. Consequently, the RGB and pose features for the kth micro-action is given by-

$$\left[f_k^{\text{RGB}}, f_k^{\text{Pose}} \right] = \left[F(M_k^{\text{RGB}}), T(M_k^{\text{Pose}}) \right], \tag{2}$$

where $f_k^{\text{RGB}}, f_k^{\text{pose}} \in \mathbb{R}^d$, where d denotes the common dimensionality of both RGB and pose embedding space.

4.2 Trajectory Encoder

We devise a trajectory-based pose encoder to encode dense hand pose sequences within micro-actions, illustrated in Fig. 4. Each joint is represented by its trajectory of dimension $3 \times N$, encapsulating the sequence of 3D coordinates across the N pose frames of a micro-action. This yields $2 \times J$ feature vectors for the J joints of two hands. Each joint's trajectory is passed through a TCN [35], whose parameters are shared for all joints. This produces $2 \times J$ Local Trajectory Tokens. Additionally, the full-skeletal motion of the hand during the entire action is used as a reference through an additional token named Global Wrist Token.

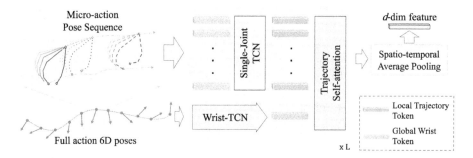

Fig. 4. Our Trajectory Encoder T, which operates on micro-actions, derives tokens with trajectory-based features and performs self-attention to encode the pose sequence into a feature vector. The Single-Joint TCN is a Temporal Convolutional Network [35] that processes the trajectories of all the joints individually with shared parameters across all joints. Wrist-TCN takes an action-wide sequence of wrist location and hand orientation (6D pose) to produce a global reference token.

This token is generated by a separate TCN operating on the sequence of 6D poses of hands, indicating wrist location and hand orientation. Subsequently, a self-attention layer is applied to these trajectory tokens, preserving the temporal dimension for subsequent stages. This iterative process culminates in spatiotemporal average pooling, summarizing the hand motion of the micro-action.

4.3 Multimodal Tokenizer

This section discusses incorporating sparsely sampled RGB frames into Hand-Former to better capture scene semantics. A single frame is sampled from each micro-action, generating an extended crop (1.25×) around the hands. While the full image provides an overall scene context, the crop focuses specifically on hand-object interaction regions [10]. Features for both are separately generated using a pre-existing image encoder and then aggregated to enrich the hand-object interaction feature with scene context. The hand-object ROI crop can be obtained using an off-the-shelf HOI detector [53] or by using the corresponding hand pose projections. We opt for the latter.

Our multimodal tokenizer in Fig. 3 generates RGB and Pose tokens enhanced via multimodal interactions. This involves concatenating each frame feature and trajectory encoding, and projecting them into a shared PoseRGB feature space using an MLP. This PoseRGB feature is then split into two parts, which are added to the original frame feature and trajectory encoding, respectively.

4.4 Temporal Transformer

The multimodal tokenizer provides RGB and pose tokens for each micro-action. Since an action segment consists of a sequence of micro-actions, these micro-action tokens are aggregated over time via a temporal transformer. A video $\mathbf{V}$, divided into K micro-actions can be represented by two sets of tokens $\{\hat{f}_k^{\text{RGB}}\}_{k=1}^K$ and $\{\hat{f}_k^{\text{Pose}}\}_{k=1}^K$, respectively, produced by the multimodal tokenizer. These $2 \times K$ tokens of dimension d form the input sequence of the temporal transformer. Positional encoding and modality embedding are added to each input token to indicate the temporal location and source modality. We use the fixed sine/cosine positional encoding [59], assigning the same position to two tokens from the same micro-action. Modality embeddings are learned and shared across the tokens from the same modality (RGB or Pose). Following standard practice [17,18], we prepend an additional learnable class token [CLS] $\in \mathbb{R}^d$, the output of which is then used for classifying actions.

4.5 Learning Objectives

HandFormer is trained end-to-end for action recognition, supervised by a cross-entropy loss: $\mathcal{L}_{cls} = -a_i \log \hat{a}_i$, where a_i represents the ground truth action label for the ith sample, and $\hat{a}_i$ denotes the predicted action category. The input pose and RGB modalities provide complementary information regarding a scene, capturing motion and interacting objects, respectively. To effectively utilize this, we employ explicit verb and object supervision ($\mathcal{L}_{verb}$, $\mathcal{L}_{obj}$) via two additional learnable class tokens. Given that pose strongly correlates with verbs, whereas objects are identifiable from RGB frames alone, we let the verb class token attend exclusively to pose encodings and the object class token to frame features.

Feature Anticipation Loss. Hand pose sequence captures the primary sources of state changes during hand-object interactions. We posit that the visual state from an initial RGB frame, in combination with the subsequent hand pose sequence, is indicative of the visual state that results from the completion of the sequence. Therefore, given an RGB feature and the corresponding pose features from a micro-action, we force our model to anticipate the RGB feature for the next micro-action by minimizing an L_1 feature loss. This loss, inspired by existing efforts [25,61], quantifies the difference between the *anticipated* image feature and the true feature extracted from a frozen image encoder. Formally,

$$\mathcal{L}_{ant} = \sum_{k=1}^{K-1} \left\| \Phi_{ant}(f_k^{\text{PoseRGB}}) - f_{k+1}^{\text{RGB}} \right\|_1 \qquad (3)$$

where Φ_{ant} denotes a linear projection layer. Hence, the total loss is-

$$\mathcal{L} = \mathcal{L}_{cls} + \lambda_1 \mathcal{L}_{verb} + \lambda_2 \mathcal{L}_{obj} + \lambda_3 \mathcal{L}_{ant} \qquad (4)$$

where $\lambda_1, \lambda_2, \lambda_3$ are hyperparameters used to balance the four losses.

5 Experiments

Datasets. We conduct experiments on two publicly available hand-object interaction datasets. **Assembly101** [51] is a large-scale multiview dataset that features videos of procedural activities for assembling and disassembling 101 take-apart toy vehicles. This dataset has 1380 fine-grained actions based on 24 verbs and 90 objects. It consists of 12 temporally synchronized views—eight static and four egocentric. The egocentric views exhibit low-resolution monochrome images, making it challenging even for human eyes to discern objects in view. Hence, we opt for fixed views as our RGB modality. 3D hand poses estimated using off-the-shelf MEgATrack [27] are provided by the dataset. **H2O** [34] features four participants performing 36 actions that involve 8 objects and 11 verbs. Ground truth 3D hand poses are provided. The dataset consists of 5 temporally synchronized RGB views—4 fixed and 1 egocentric. We use RGB frames from the egocentric view only.

Implementation Details. Similar to [65], our pose input consists of 120 frames by setting $T' = 120$. The number of frames per micro-action (N) is 15. For multimodal experiments, we set the window stride $R = N$ and take $K = 8$ non-overlapping micro-actions, which allows us to sample 8 RGB frames for each video following [51]. However, we allow a 50% overlap between consecutive micro-actions for pose-only variants. A frozen ViT [18] or ResNet [29], followed by a learnable linear layer, is used as the frame encoder F. More details are provided in Suppl. Sec. C. Each model is trained for 50 epochs using SGD with momentum 0.9, a batch size of 32, a learning rate of 0.025, and step LR decay with a factor of 0.1 at epochs $\{25, 40\}$. Loss hyperparameters $\{\lambda_1, \lambda_2, \lambda_3\}$ are chosen to be $\{1.0, 1.0, 2.0\}$.

Model Variants. We propose several variants of our model by adjusting the width d and the number of layers T_n of the transformer to balance efficiency and accuracy. Our default HandFormer, denoted HandFormer-B, has parameters $(d, T_n) = (256, 2)$. We introduce a larger variant, HandFormer-L, with $(d, T_n) = (512, 4)$. We also explore different configurations for the number of input joints J per hand. Unless otherwise mentioned, we utilize all 21 joints per hand along with the base model denoted as HandFormer-B/21 while offering a highly efficient option utilizing only six joints per hand (five fingertips and the wrist) termed HandFormer-B/6. Additionally, HandFormer-X/J×$\mathcal{N}$ denotes a multimodal variant that utilizes $\mathcal{N}$ RGB frames, where $\mathcal{N} \leq K$.

5.1 Comparison with State-of-the-Art

To evaluate the effectiveness of our proposed architecture, we compare it against several baselines. For pose-only comparisons, we choose a graph-based network MS-G3D [44] and an attention-based method ISTA-Net [65]—the two best performing skeleton-based methods on Assembly101 [51], as reported by [65]. We also employ video baselines TSM [40] and SlowFast [23], emphasizing efficiency and high temporal resolution, respectively. Additionally, we include H2OTR [13], the state-of-the-art for the H2O dataset. For Assembly101, we replicate the results of the methods above using RGB frames from *view 4* of the dataset, while results on H2O are acquired from the respective papers. Furthermore, on

Table 1. Quantitative comparison with state-of-the-art methods on Assembly101 [51] and H2O [34]. For H2O, 6D object pose is used by ISTA-Net during training and inference and by H2OTR during training. 'MS-G3D + TSM' denotes a late fusion of the corresponding unimodal architectures. Reported GFLOPs include pose estimation costs in the context of Assembly101 at 15 fps. [†]For pose-only methods, pose estimation costs are common and excluded for simpler comparison.

Method	Pose	RGB	GFLOPs	Assembly101			H2O
				Action	Verb	Object	Action
MS-G3D [44]	✓	✗	21.2[†]	28.78	63.46	37.26	50.83
ISTA-Net [65]	✓	✗	35.2[†]	28.14	62.70	36.77	89.09
SlowFast [23]	✗	✓	65.7	-	-	-	77.69
TSM [40]	✗	✓	33.0	35.27	58.27	47.45	-
H2OTR [13]	✗	✓	-	-	-	-	90.90
RGBPoseConv3D [20]	✓	✓	68.9	33.61	61.99	42.90	83.47
MS-G3D + TSM	✓	✓	66.2	39.74	65.12	51.12	-
HandFormer-B/21×8	✓	✗	4.2[†]	28.80	65.33	36.28	57.44
	✗	✓	33.0	32.07	55.61	44.89	84.71
	✓	✓	47.6	**41.06**	**69.23**	**51.17**	**93.39**

both datasets, we train and test RGBPoseConv3D [20], which is current state-of-the-art in multimodal action recognition with skeleton and RGB data.

We evaluate three variants of our method by controlling the input modalities. As shown in Table 1, our unimodal pose-only model excels in verb recognition on Assembly101. In contrast, RGB-based methods benefit from object appearance and usually perform well for actions due to the strong object recognition.

In this context, the accuracy of ISTA-Net on H2O is not directly comparable to our method, as they also use the 6D object poses, while we only use hand poses as input. RGBPoseConv3D struggles to achieve satisfactory performance, particularly on Assembly101, suggesting that generalizing to hand poses is non-trivial for skeleton-based methods. Therefore, we combine two best-performing unimodal methods with late fusion (MS-G3D + TSM) to create a stronger baseline. Our model even outperforms this, indicating the effectiveness of our proposed multimodal fusion.

5.2 Skeleton-Based Action Recognition for Hands

The compositional nature of action classes in hand-object interaction videos allows us to break down the action into a verb and an object. Recognizing such actions from 3D hand poses is an ill-posed problem, as the hand skeletons lack explicit information about the interacting objects, which are also part of the action semantics. However, as the pose data completely captures the hand motion information, it can be reliably used for verb recognition. Therefore, we evaluate a pose-only version of our method on the verb recognition task and compare

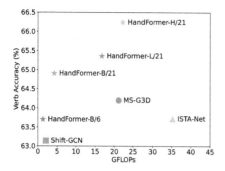
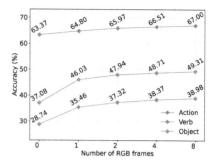

Fig. 5. Comparison of skeleton-based methods for verb recognition on Assembly101 [51]. Our method achieves state-of-the-art performance while utilizing significantly fewer FLOPs.

Fig. 6. Ablating number of RGB frames in HandFormer on Assembly101 [51]. With more RGB frames, verb recognition shows marginal gain, whereas object recognition shows improvement with diminishing returns.

the performance and efficiency metrics with other state-of-the-art skeleton-based methods in Fig. 5. Our method uses significantly fewer GFLOPs due to our spatiotemporal factorization using micro-actions. With $J = 21$, all our HandFormer variants outperform existing methods. The $J = 6$ variant is exceptionally efficient with comparable accuracy, notably surpassing the efficiency-focused Shift-GCN [11]. HandFormer-H/21 combines the HandFormer-B and HandFormer-L variants and becomes our best-performing model, improving over MS-G3D by 2.04% while maintaining comparable FLOPs.

5.3 How Many RGB Frames are Required?

In our model, the pose modality maintains a high temporal resolution to capture fine-grained hand movements, resulting in good verb recognition performance. On the contrary, RGB frames are primarily required to introduce semantic context for object recognition and does not necessitate a high temporal resolution like hand movements. The design of our model allows us to sample only a few RGB frames (as low as one) but still perform competitively at a reduced computational cost. Figure 6 shows the impact of using more RGB frames for Assembly101 [51]. For this experiment, non-overlapping micro-actions are considered. Using only one RGB frame in HandFormer (35.46) outperforms the video model TSM (35.27), as shown in Table 1. This performance gain stems primarily from a significant improvement in object accuracy, with only a slight enhancement in verb accuracy. However, including more RGB frames shows a diminishing return, which is unsurprising as additional frames are expected to provide redundant information. These results are obtained with a simplified version of our model by setting $\mathcal{L} = \mathcal{L}_{cls}$ and bypassing the multimodal tokenizer.

Table 2. Multi-view Action Recognition on Assembly101 [51]. HandFormer with 'Single View+*Pose*' has better performance than 'Single View+*Egocentric*'. Also, verb performance is comparable to using all (8) RGB views.

Views	Action	Verb	Object
Single View	35.27	58.27	47.45
Single View + *Egocentric*	37.75	61.80	49.43
Two Views	41.96	65.22	53.26
All (8) Views	47.51	70.99	57.73
Single View + *Pose*	41.06	69.23	51.17

Table 3. Cross-view performance of HandFormer in Assembly101 [51] shows its generalization capability to unseen *view 1*, outperforming the video baseline which is trained on *view 1* directly. Egocentric views are the source of hand poses here and, therefore, are included in the video models.

Method	Train on	Test on	Action	Verb	Object
TSM [40]	$v4$ + ego	$v4$ + ego	37.75	61.80	49.43
	$v4$ + ego	$v1$ + ego	35.27	59.53	47.72
	$v1$ + ego	$v1$ + ego	36.21	60.78	48.52
Our Method	$v4$ + Pose	$v4$ + Pose	41.06	69.23	51.17
	$v4$ + Pose	$v1$ + Pose	38.43	67.86	48.32

5.4 Can 3D Hand Pose Be an Efficient Alternative to Multi-View?

While multiview action recognition benefits from precise hand-movement information in 3D space, processing all views with video models is expensive and highly redundant. In Table 2, we demonstrate that combining 3D hand pose with a single RGB view (*view 4*) achieves comparable performance to multi-view action recognition on Assembly101 [51]. Specifically, our action recognition performance ('Single View + *Pose*') matches the fusion of the two most informative views—*view 1* and *view 4*. Notably, our verb recognition accuracy is on par with combining all 8 RGB views, though the latter incurs at least 5× more FLOPs despite using the efficient TSM model. Additionally, using hand pose in combination with an RGB view ('Single View + *Pose*') outperforms directly using the egocentric videos ('Single View + *Egocentric*') from which the hand poses are derived. While fusing multiple RGB views improves accuracy by ensembling multiple complementary predictions, the computational overhead increases significantly. In contrast, our model processes hand pose and single-view RGB frames, enhancing efficiency by leveraging the less redundant and low-dimensional pose data.

Cross-view Generalization. 3D hand pose offers a unique opportunity for cross-view generalization because of its universality across different viewpoints. To evaluate the effectiveness of our method for unseen views, we train our model with frame-wise RGB features from *view 4* and test it on *view 1*. We train TSM [40] as a baseline video model on both RGB views separately. As our method includes 3D hand poses obtained from the egocentric views, we include the egocentric videos in the TSM baselines for a fair comparison. As seen from Table 3, our method, trained on *view 4*, generalizes well on unseen *view 1*, outperforming the TSM model that was directly trained on *view 1*.

Table 4. Egocentric action recognition in Assembly101 [51]. TSM features from [51] are used as RGB frame features.

Method	Action	Verb	Object
TSM egocentric (fuse 4 views)	33.80	59.00	46.50
Egocentric (e3) + Pose	36.07	65.52	45.82
Egocentric (e4) + Pose	35.56	65.79	45.20
Egocentric (e3+e4) + Pose	**38.05**	**66.32**	**47.86**

Table 5. Keypoint ablation for verb recognition in Assembly101 [51]

#Joints	Global Reference	Verb Accuracy (%)
21	✗	64.17
21	✓	**64.90**
11	✓	64.77
6	✓	63.70

Table 6. Micro-action length ablation for verb recognition in Assembly101 [51]

#Frames	1	15	30	60	120
Verb Accuracy(%)	59.12	**63.70**	63.68	63.51	62.29

Table 7. Ablating unimodal temporal aggregation for verb recognition in Assembly101 [51]

Temp. Agg.	TCN	LSTM	Transformer
Verb Acc. (%)	62.95	63.34	**63.70**

5.5 Egocentric Action Recognition

Recognizing actions in egocentric videos is challenging due to camera motion and the occlusion of interacting objects by the hands. Additionally, in the case of Assembly101 [51], the egocentric cameras are similar to Oculus Quest VR headsets, which provide monochrome low-resolution frames. As a result, action recognition performance is significantly lower than fixed RGB views, as found in [51]. We address this challenging scenario with our proposed multimodal architecture and achieve state-of-the-art performance in egocentric action recognition on Assembly101. For this experiment, we used frame-wise TSM features provided by [51]. As depicted in Table 4, our model, using a single egocentric view ($e3$ or $e4$) outperforms the fusion of four egocentric views as reported in [51]. Moreover, fusing $e3$ and $e4$ significantly enhances our model, resulting in a 4.25% increase in action accuracy over the baseline.

5.6 Ablation Studies

Keypoints. Not all joints of the hands are equally informative for understanding hand actions. For instance, fingertips exhibit greater mobility compared to the inner joints. Moreover, from an egocentric viewpoint, certain joints are more prone to self-occlusion than others. In Table 5, we present the impact of incorporating varying numbers of joints on verb recognition within the Assembly101 dataset [51]. For the case of 6 joints, we consider only the wrist joint along with the five fingertips. To expand to 11 joints, we incorporate all the joints along the index and the thumb, which are least affected by self-occlusion. We also show the effect of including the Global Wrist Token, which acts as a reference to the global motion of the hands while encoding micro-actions.

Table 8. Ablating tokenization and different losses for action recognition

Multimodal Tokenizer	Feature Ant. Loss	Verb & Object Loss	Action Accuracy (%)	
			Assembly101 [51]	H2O [34]
✗	✗	✗	38.98	85.95
✓	✗	✗	40.19	88.84
✓	✓	✗	40.24	89.26
✗	✗	✓	40.56	90.50
✓	✓	✓	**41.06**	**93.39**

Micro-Action Length. As the resized input comprises a fixed number of pose frames, enlarging the window size for a micro-action decreases the number of micro-actions to aggregate, and vice-versa. In Table 6, we vary the micro-action length for verb recognition in the Assembly101 [51] using 6 joints per hand as input, *i.e.* fingertips and wrist joint. The input pose clip is temporally resized to $T' = 120$ before breaking into micro-actions. Lengths 1 and 120 represent two extreme versions with frame-based and trajectory-based encoding, respectively, while the others conform to our micro-action-based formulation.

Temporal Aggregation. After extracting micro-action features, aggregation for action recognition can be done with any sequence model. In Table 7, we evaluate the effectiveness of different temporal aggregation methods for verb recognition in the Assembly101 dataset [51]. Here, we use 6 joints per hand as input, *i.e.*, the fingertips and the wrist joint.

Loss Components. To assess the individual contributions of different components, we begin with a basic configuration. We then systematically introduce each element to understand its impact on the overall performance as observed in Table 8. Incorporating modality interaction between RGB and pose at the micro-action level through a multimodal tokenizer enhances action accuracy. The introduction of auxiliary losses also has a positive impact, resulting in an overall improvement of 2.08% for Assembly101 [51] and 7.44% for H2O [34].

6 Conclusion

With the growing interest in AR/VR and wearables, hand pose estimation has rapidly advanced, and holds promise as a compact and domain-independent modality to complement the visual input. To address the underexplored domain of using 3D hand pose as a modality for hand-object interaction recognition, we introduce HandFormer, a novel multimodal transformer that leverages dense sequences of 3D hand poses with sparsely sampled RGB frames to achieve state-of-the-art action recognition performance. Our model also reduces computational requirements, offering immediate significance across various low-resource applications in mobile devices.

Limitations. Our method relies on the availability of hand poses, which, if extracted from the visual modality with pose estimation tools [27,28], can encounter out-of-view scenarios and produce noisy poses. Our experiments reveal that these estimated poses can still achieve good accuracy, yet further research can be conducted to explicitly address this phenomenon. We also assume that a uniform sampling of RGB frames from each micro-action should provide good representations for understanding the semantic context. However, not all frames are equally important in understanding the action. In such a case, adaptive frame sampling methods can be employed, which we leave for future work.

Acknowledgements. This research is supported by A*STAR under its National Robotics Programme (NRP) (Award M23NBK0053).

References

1. Ahn, D., Kim, S., Hong, H., Ko, B.C.: Star-transformer: a spatio-temporal cross attention transformer for human action recognition. In: Proceedings of the IEEE/CVF Winter Conference on Applications of Computer Vision, pp. 3330–3339 (2023)
2. Arnab, A., Dehghani, M., Heigold, G., Sun, C., Lučić, M., Schmid, C.: Vivit: a video vision transformer. In: Proceedings of the IEEE/CVF International Conference on Computer Vision, pp. 6836–6846 (2021)
3. Bertasius, G., Wang, H., Torresani, L.: Is space-time attention all you need for video understanding? In: ICML, vol. 2, p. 4 (2021)
4. Bolya, D., Fu, C.Y., Dai, X., Zhang, P., Feichtenhofer, C., Hoffman, J.: Token merging: your VIT but faster. arXiv preprint arXiv:2210.09461 (2022)
5. Bruce, X., Liu, Y., Chan, K.C.: Multimodal fusion via teacher-student network for indoor action recognition. In: Proceedings of the AAAI Conference on Artificial Intelligence, vol. 35, pp. 3199–3207 (2021)
6. Bruce, X., Liu, Y., Zhang, X., Zhong, S.H., Chan, K.C.: MMNet: a model-based multimodal network for human action recognition in RGB-D videos. IEEE Trans. Pattern Anal. Mach. Intell. **45**(3), 3522–3538 (2022)
7. Caetano, C., Sena, J., Brémond, F., Dos Santos, J.A., Schwartz, W.R.: Skelemotion: a new representation of skeleton joint sequences based on motion information for 3D action recognition. In: 2019 16th IEEE International Conference on Advanced Video and Signal Based Surveillance (AVSS), pp. 1–8. IEEE (2019)
8. Cao, C., Zhang, Y., Zhang, C., Lu, H.: Body joint guided 3-D deep convolutional descriptors for action recognition. IEEE Trans. Cybern. **48**(3), 1095–1108 (2017)
9. Carreira, J., Zisserman, A.: Quo vadis, action recognition? A new model and the kinetics dataset. In: proceedings of the IEEE Conference on Computer Vision and Pattern Recognition, pp. 6299–6308 (2017)
10. Chatterjee, D., Sener, F., Ma, S., Yao, A.: Opening the vocabulary of egocentric actions. In: Thirty-Seventh Conference on Neural Information Processing Systems (2023)
11. Cheng, K., Zhang, Y., He, X., Chen, W., Cheng, J., Lu, H.: Skeleton-based action recognition with shift graph convolutional network. In: Proceedings of the IEEE/CVF Conference on Computer Vision and Pattern Recognition, pp. 183–192 (2020)

12. Chéron, G., Laptev, I., Schmid, C.: P-CNN: pose-based CNN features for action recognition. In: Proceedings of the IEEE International Conference on Computer Vision, pp. 3218–3226 (2015)
13. Cho, H., Kim, C., Kim, J., Lee, S., Ismayilzada, E., Baek, S.: Transformer-based unified recognition of two hands manipulating objects. In: Proceedings of the IEEE/CVF Conference on Computer Vision and Pattern Recognition, pp. 4769–4778 (2023)
14. Damen, D., et al.: Scaling egocentric vision: the epic-kitchens dataset. In: Proceedings of the European Conference on Computer Vision (ECCV), pp. 720–736 (2018)
15. Das, S., Dai, R., Yang, D., Bremond, F.: VPN++: rethinking video-pose embeddings for understanding activities of daily living. IEEE Trans. Pattern Anal. Mach. Intell. **44**(12), 9703–9717 (2021)
16. Das, S., Sharma, S., Dai, R., Brémond, F., Thonnat, M.: VPN: learning video-pose embedding for activities of daily living. In: Vedaldi, A., Bischof, H., Brox, T., Frahm, J.-M. (eds.) ECCV 2020. LNCS, vol. 12354, pp. 72–90. Springer, Cham (2020). https://doi.org/10.1007/978-3-030-58545-7_5
17. Devlin, J., Chang, M.W., Lee, K., Toutanova, K.: Bert: pre-training of deep bidirectional transformers for language understanding. arXiv preprint arXiv:1810.04805 (2018)
18. Dosovitskiy, A., et al.: An image is worth 16x16 words: transformers for image recognition at scale. arXiv preprint arXiv:2010.11929 (2020)
19. Du, Y., Wang, W., Wang, L.: Hierarchical recurrent neural network for skeleton based action recognition. In: Proceedings of the IEEE Conference on Computer Vision and Pattern Recognition, pp. 1110–1118 (2015)
20. Duan, H., Zhao, Y., Chen, K., Lin, D., Dai, B.: Revisiting skeleton-based action recognition. In: Proceedings of the IEEE/CVF Conference on Computer Vision and Pattern Recognition, pp. 2969–2978 (2022)
21. Fayyaz, M., et al.: Adaptive token sampling for efficient vision transformers. In: Avidan, S., Brostow, G., Cissé, M., Farinella, G.M., Hassner, T. (eds.) ECCV 2022. LNCS, vol. 13671, pp. 396–414. Springer, Cham (2022). https://doi.org/10.1007/978-3-031-20083-0_24
22. Feichtenhofer, C.: X3D: expanding architectures for efficient video recognition. In: Proceedings of the IEEE/CVF Conference on Computer Vision and Pattern Recognition, pp. 203–213 (2020)
23. Feichtenhofer, C., Fan, H., Malik, J., He, K.: Slowfast networks for video recognition. In: Proceedings of the IEEE/CVF International Conference on Computer Vision, pp. 6202–6211 (2019)
24. Gan, M., Liu, J., He, Y., Chen, A., Ma, Q.: Keyframe selection via deep reinforcement learning for skeleton-based gesture recognition. IEEE Robot. Autom. Lett. (2023)
25. Girdhar, R., Grauman, K.: Anticipative video transformer. In: Proceedings of the IEEE/CVF International Conference on Computer Vision, pp. 13505–13515 (2021)
26. Grauman, K., et al.: EGO4D: around the world in 3,000 hours of egocentric video. In: Proceedings of the IEEE/CVF Conference on Computer Vision and Pattern Recognition, pp. 18995–19012 (2022)
27. Han, S., et al.: Megatrack: monochrome egocentric articulated hand-tracking for virtual reality. ACM Trans. Graph. (ToG) **39**(4), 87-1 (2020)
28. Han, S., et al.: Umetrack: unified multi-view end-to-end hand tracking for VR. In: SIGGRAPH Asia 2022 Conference Papers, pp. 1–9 (2022)

29. He, K., Zhang, X., Ren, S., Sun, J.: Deep residual learning for image recognition. In: Proceedings of the IEEE Conference on Computer Vision and Pattern Recognition, pp. 770–778 (2016)
30. Hou, J., Wang, G., Chen, X., Xue, J.H., Zhu, R., Yang, H.: Spatial-temporal attention res-TCN for skeleton-based dynamic hand gesture recognition. In: Proceedings of the European Conference on Computer Vision (ECCV) Workshops (2018)
31. Hou, Y., Li, Z., Wang, P., Li, W.: Skeleton optical spectra-based action recognition using convolutional neural networks. IEEE Trans. Circuits Syst. Video Technol. **28**(3), 807–811 (2016)
32. Jaegle, A., Gimeno, F., Brock, A., Vinyals, O., Zisserman, A., Carreira, J.: Perceiver: general perception with iterative attention. In: International Conference on Machine Learning, pp. 4651–4664. PMLR (2021)
33. Kazakos, E., Nagrani, A., Zisserman, A., Damen, D.: Epic-fusion: audio-visual temporal binding for egocentric action recognition. In: Proceedings of the IEEE/CVF International Conference on Computer Vision, pp. 5492–5501 (2019)
34. Kwon, T., Tekin, B., Stühmer, J., Bogo, F., Pollefeys, M.: H2O: two hands manipulating objects for first person interaction recognition. In: Proceedings of the IEEE/CVF International Conference on Computer Vision, pp. 10138–10148 (2021)
35. Lea, C., Flynn, M.D., Vidal, R., Reiter, A., Hager, G.D.: Temporal convolutional networks for action segmentation and detection. In: proceedings of the IEEE Conference on Computer Vision and Pattern Recognition, pp. 156–165 (2017)
36. Lee, C.J., et al.: Echowrist: continuous hand pose tracking and hand-object interaction recognition using low-power active acoustic sensing on a wristband. arXiv preprint arXiv:2401.17409 (2024)
37. Li, C., Li, S., Gao, Y., Zhang, X., Li, W.: A two-stream neural network for pose-based hand gesture recognition. IEEE Trans. Cogn. Dev. Syst. **14**(4), 1594–1603 (2021)
38. Li, J., Xie, X., Pan, Q., Cao, Y., Zhao, Z., Shi, G.: SGM-Net: skeleton-guided multimodal network for action recognition. Pattern Recogn. **104**, 107356 (2020)
39. Li, M., Chen, S., Chen, X., Zhang, Y., Wang, Y., Tian, Q.: Actional-structural graph convolutional networks for skeleton-based action recognition. In: Proceedings of the IEEE/CVF Conference on Computer Vision and Pattern Recognition, pp. 3595–3603 (2019)
40. Lin, J., Gan, C., Han, S.: TSM: temporal shift module for efficient video understanding. In: Proceedings of the IEEE/CVF International Conference on Computer Vision, pp. 7083–7093 (2019)
41. Liu, J., Shahroudy, A., Perez, M., Wang, G., Duan, L.Y., Kot, A.C.: NTU RGB+D 120: a large-scale benchmark for 3D human activity understanding. IEEE Trans. Pattern Anal. Mach. Intell. **42**(10), 2684–2701 (2019)
42. Liu, Y., Zhang, S., Gowda, M.: Neuropose: 3D hand pose tracking using EMG wearables. In: Proceedings of the Web Conference 2021, pp. 1471–1482 (2021)
43. Liu, Z., et al.: Video swin transformer. In: Proceedings of the IEEE/CVF Conference on Computer Vision and Pattern Recognition, pp. 3202–3211 (2022)
44. Liu, Z., Zhang, H., Chen, Z., Wang, Z., Ouyang, W.: Disentangling and unifying graph convolutions for skeleton-based action recognition. In: Proceedings of the IEEE/CVF Conference on Computer Vision and Pattern Recognition, pp. 143–152 (2020)
45. Ma, J., Damen, D.: Hand-object interaction reasoning. In: 2022 18th IEEE International Conference on Advanced Video and Signal Based Surveillance (AVSS), pp. 1–8. IEEE (2022)

46. Ohkawa, T., He, K., Sener, F., Hodan, T., Tran, L., Keskin, C.: Assemblyhands: towards egocentric activity understanding via 3D hand pose estimation. In: Proceedings of the IEEE/CVF Conference on Computer Vision and Pattern Recognition, pp. 12999–13008 (2023)
47. Patrick, M., et al.: Keeping your eye on the ball: trajectory attention in video transformers. In: Advances in Neural Information Processing Systems, vol. 34, pp. 12493–12506 (2021)
48. Plizzari, C., Cannici, M., Matteucci, M.: Spatial temporal transformer network for skeleton-based action recognition. In: Del Bimbo, A., et al. (eds.) ICPR 2021. LNCS, vol. 12663, pp. 694–701. Springer, Cham (2021). https://doi.org/10.1007/978-3-030-68796-0_50
49. Rajasegaran, J., Pavlakos, G., Kanazawa, A., Feichtenhofer, C., Malik, J.: On the benefits of 3D pose and tracking for human action recognition. In: Proceedings of the IEEE/CVF Conference on Computer Vision and Pattern Recognition, pp. 640–649 (2023)
50. Sabater, A., Alonso, I., Montesano, L., Murillo, A.C.: Domain and view-point agnostic hand action recognition. IEEE Robot. Autom. Lett. **6**(4), 7823–7830 (2021)
51. Sener, F., et al.: Assembly101: a large-scale multi-view video dataset for understanding procedural activities. In: Proceedings of the IEEE/CVF Conference on Computer Vision and Pattern Recognition, pp. 21096–21106 (2022)
52. Shahroudy, A., Liu, J., Ng, T.T., Wang, G.: NTU RGB+D: a large scale dataset for 3D human activity analysis. In: Proceedings of the IEEE Conference on Computer Vision and Pattern Recognition, pp. 1010–1019 (2016)
53. Shan, D., Geng, J., Shu, M., Fouhey, D.F.: Understanding human hands in contact at internet scale. In: Proceedings of the IEEE/CVF Conference on Computer Vision and Pattern Recognition, pp. 9869–9878 (2020)
54. Shi, L., Zhang, Y., Cheng, J., Lu, H.: Two-stream adaptive graph convolutional networks for skeleton-based action recognition. In: Proceedings of the IEEE/CVF Conference on Computer Vision and Pattern Recognition, pp. 12026–12035 (2019)
55. Simonyan, K., Zisserman, A.: Two-stream convolutional networks for action recognition in videos. In: Advances in Neural Information Processing Systems, vol. 27 (2014)
56. Soo Kim, T., Reiter, A.: Interpretable 3D human action analysis with temporal convolutional networks. In: Proceedings of the IEEE Conference on Computer Vision and Pattern Recognition Workshops, pp. 20–28 (2017)
57. Tran, D., Bourdev, L., Fergus, R., Torresani, L., Paluri, M.: Learning spatiotemporal features with 3D convolutional networks. In: Proceedings of the IEEE International Conference on Computer Vision, pp. 4489–4497 (2015)
58. Tran, D., Wang, H., Torresani, L., Ray, J., LeCun, Y., Paluri, M.: A closer look at spatiotemporal convolutions for action recognition. In: Proceedings of the IEEE Conference on Computer Vision and Pattern Recognition, pp. 6450–6459 (2018)
59. Vaswani, A., et al.: Attention is all you need. In: Advances in Neural Information Processing Systems, vol. 30 (2017)
60. Vemulapalli, R., Arrate, F., Chellappa, R.: Human action recognition by representing 3D skeletons as points in a lie group. In: Proceedings of the IEEE Conference on Computer Vision and Pattern Recognition, pp. 588–595 (2014)
61. Vondrick, C., Pirsiavash, H., Torralba, A.: Anticipating visual representations from unlabeled video. In: Proceedings of the IEEE Conference on Computer Vision and Pattern Recognition, pp. 98–106 (2016)

62. Wang, J., Liu, Z., Wu, Y., Yuan, J.: Mining actionlet ensemble for action recognition with depth cameras. In: 2012 IEEE Conference on Computer Vision and Pattern Recognition, pp. 1290–1297. IEEE (2012)
63. Weiyao, X., Muqing, W., Min, Z., Ting, X.: Fusion of skeleton and RGB features for RGB-D human action recognition. IEEE Sens. J. **21**(17), 19157–19164 (2021)
64. Wen, Y., Pan, H., Yang, L., Pan, J., Komura, T., Wang, W.: Hierarchical temporal transformer for 3D hand pose estimation and action recognition from egocentric RGB videos. In: Proceedings of the IEEE/CVF Conference on Computer Vision and Pattern Recognition, pp. 21243–21253 (2023)
65. Wen, Y., Tang, Z., Pang, Y., Ding, B., Liu, M.: Interactive spatiotemporal token attention network for skeleton-based general interactive action recognition. arXiv preprint arXiv:2307.07469 (2023)
66. Wu, C.Y., et al.: Memvit: memory-augmented multiscale vision transformer for efficient long-term video recognition. In: Proceedings of the IEEE/CVF Conference on Computer Vision and Pattern Recognition, pp. 13587–13597 (2022)
67. Xie, S., Sun, C., Huang, J., Tu, Z., Murphy, K.: Rethinking spatiotemporal feature learning: speed-accuracy trade-offs in video classification. In: Proceedings of the European Conference on Computer Vision (ECCV), pp. 305–321 (2018)
68. Yan, S., Xiong, Y., Lin, D.: Spatial temporal graph convolutional networks for skeleton-based action recognition. In: Proceedings of the AAAI Conference on Artificial Intelligence, vol. 32 (2018)
69. Zhang, S., Liu, X., Xiao, J.: On geometric features for skeleton-based action recognition using multilayer LSTM networks. In: 2017 IEEE Winter Conference on Applications of Computer Vision (WACV), pp. 148–157. IEEE (2017)
70. Zhang, X., Xu, C., Tian, X., Tao, D.: Graph edge convolutional neural networks for skeleton-based action recognition. IEEE Trans. Neural Netw. Learn. Syst. **31**(8), 3047–3060 (2019)
71. Zhang, Y., Wu, B., Li, W., Duan, L., Gan, C.: STST: spatial-temporal specialized transformer for skeleton-based action recognition. In: Proceedings of the 29th ACM International Conference on Multimedia, pp. 3229–3237 (2021)

DG-PIC: Domain Generalized Point-In-Context Learning for Point Cloud Understanding

Jincen Jiang[1], Qianyu Zhou[2], Yuhang Li[1,3], Xuequan Lu[4(✉)],
Meili Wang[5(✉)], Lizhuang Ma[2], Jian Chang[1], and Jian Jun Zhang[1]

[1] National Centre for Computer Animation, Bournemouth University, Dorset, UK
{jiangj,jchang,jzhang}@bournemouth.ac.uk
[2] Shanghai Jiao Tong University, Shanghai, China
{zhouqianyu,lzma}@sjtu.edu.cn
[3] Shanghai University, Shanghai, China
yuhangli@shu.edu.cn
[4] La Trobe University, Victoria, Australia
b.lu@latrobe.edu.au
[5] Northwest A&F University, Yangling, China
wml@nwsuaf.edu.cn
https://github.com/Jinec98/DG-PIC

Abstract. Recent point cloud understanding research suffers from performance drops on unseen data, due to the distribution shifts across different domains. While recent studies use Domain Generalization (DG) techniques to mitigate this by learning domain-invariant features, most are designed for a single task and neglect the potential of testing data. Despite In-Context Learning (ICL) showcasing multi-task learning capability, it usually relies on high-quality context-rich data and considers a single dataset, and has rarely been studied in point cloud understanding. In this paper, we introduce a novel, practical, multi-domain multi-task setting, handling multiple domains and multiple tasks within one unified model for domain generalized point cloud understanding. To this end, we propose Domain Generalized Point-In-Context Learning (DG-PIC) that boosts the generalizability across various tasks and domains at testing time. In particular, we develop dual-level source prototype estimation that considers both global-level shape contextual and local-level geometrical structures for representing source domains and a dual-level test time feature shifting mechanism that leverages both macro-level domain semantic information and micro-level patch positional relationships to pull the target data closer to the source ones during the testing. Our DG-PIC does not require any model updates during the testing and

---

J. Jiang and Q. Zhou—Equal contributions.

Supplementary Information The online version contains supplementary material available at https://doi.org/10.1007/978-3-031-72658-3_26.

can handle unseen domains and multiple tasks, *i.e.,* point cloud reconstruction, denoising, and registration, within one unified model. We also introduce a benchmark for this new setting. Comprehensive experiments demonstrate that DG-PIC outperforms state-of-the-art techniques significantly.

Keywords: Point Cloud Understanding · Test-time Domain Generalization · In-Context Learning

1 Introduction

Recent advancements in 3D point cloud understanding, pioneered by classical works like [32,49], have revolutionized 3D vision, facilitating applications in autonomous driving, robotics, and augmented reality. Despite significant development in this line of direction, current methods mainly focus on a single domain (*i.e.,* training and testing on a single dataset) and have difficulties in addressing distribution shifts (*i.e.,* training and testing on multiple datasets), known as the domain gap in point cloud learning. For instance, models trained on well-structured synthetic data like ModelNet40 [53] may struggle to generalize to complex and noisy real-world data such as ScanObjectNN [43].

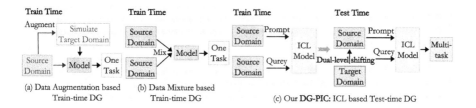

Fig. 1. (a–b) Previous train-time DG techniques for point cloud understanding are typically designed for one task and dedicated to domain-invariant features at training time while ignoring the contribution of the testing data. (c) In contrast, our **DG-PIC** aims at pulling the testing features to the source ones at dual levels to improve the pre-trained model's generalizability, which does not require any model updates at testing time, excelling at the newly proposed multi-domain and multi-task setting.

Researchers have attempted to bridge this gap using Domain Adaptation (DA) techniques, with some studies [44,62,68] utilizing generative models to synthesize diverse training data to improve the model's adaptability toward target domains. Nevertheless, these methods still face challenges in handling the unseen target domains that are infeasible to acquire in the training. Thus, Domain Generalization (DG) was introduced for 3D point cloud understanding. These DG studies [18,39,50] concentrate on learning domain-invariant representations by training models to be more adaptable to unseen domains. However, as illustrated in Fig. 1, they involve two primary limitations: *1)* They are specifically designed for one task and lack the capacity to handle multiple tasks simultaneously. *2)* They mainly emphasize learning domain-invariant features at training-time, while neglecting the potential of testing data as valuable resources in enhancing the generalizability, leading to less-desirable results.

In-Context Learning (ICL) stands out as a crucial technique for multi-task learning. It trains a single model to perform multiple tasks using contextual information from input data [22,29,37], applying the adaptive learned knowledge across various tasks and eliminating the need for distinct models for each different task. However, ICL has been rarely studied in point cloud understanding. Point In-Context (PIC) [10], a specific form of ICL on 3D point clouds, aims to handle multiple point cloud tasks using a unified model. Nonetheless, PIC suffers from its dependence on input data quality and diversity to provide meaningful contexts. Also, PIC is constrained to a single dataset, potentially hindering the model's ability to generalize across different datasets and perform effectively on new unseen data deviating significantly from the trained data. As such, while ICL including PIC present advanced strategies for multi-task learning, they are intricately tied to the nature of the training data and have limited ability in mitigating the domain gaps between diverse datasets.

Motivated by the above analysis, we propose a novel setting that is practical and important in the real world, *i.e.*, handling multiple domains and tasks within a unified model for point cloud learning. To this end, we introduce an innovative framework, namely Domain Generalized Point In-Context Learning (DG-PIC), for point cloud understanding. In particular, we employ PIC during pre-training to gather rich, generalizable information across all source domains, creating an off-the-shelf pre-trained model, and introduce test-time DG to enhance the model's generalizability by aligning the unseen target features with source domains without any model updates during the testing.

Our DG-PIC comprises two key, novel modules. Firstly, we design a dual-level module to estimate the source prototypes, considering both global-level and local-level features. This dual-context learning strategy ensures a comprehensive understanding of the source domain, through capturing rich shape patterns and fine geometric structure details. Secondly, during testing time, we design the dual-level feature shifting scheme to push the target data towards the source domains, leveraging both macro-level domain-aware semantic information and micro-level patch-aware positional structures. Subsequently, we select the most similar sample from the nearest source domain to generate the query-prompt pair. In this manner, our DG-PIC effectively bridges the gap between the source and target domains without requiring model updates at testing time, and handles multiple tasks, within a unified model, leading to strong generalization ability and superior performance in terms of the new multi-domain and multi-task setting. In addition, we introduce a new benchmark for evaluating performance in this proposed setting. We meticulously select a total of 30,954 point cloud samples from 4 distinct datasets, including 2 synthetic datasets (ModelNet40 [53] and ShapeNet [4]) and 2 real-world datasets (ScanNet [6] and ScanObjectNN

[43]), encompassing 7 same object categories, and generate corresponding ground truth for 3 different tasks (reconstruction, denoising, and registration).

Our main contributions are summarized as follows:

- We introduce a novel and practical multi-domain multi-task setting in point cloud understanding and propose the DG-PIC, the first network to handle multiple domains and tasks within a unified model for test-time domain generalization in point cloud learning.
- We devise two innovative dual-level modules for DG-PIC, *i.e.*, dual-level source prototype estimation that considers global-level shape and local-level geometry information for representing source domain, and the dual-level test-time target feature shifting that pushes the target data towards the source domain space with leveraging macro-level domain information and micro-level patch relationship information.
- We introduce a new benchmark for the proposed multi-domain and multi-task setting. Comprehensive experiments show that our DG-PIC achieves state-of-the-art performance on three different tasks.

2 Related Work

2.1 Point Cloud Understanding

Point cloud learning has become increasingly important in 3D vision [14,15,56]. Pioneering works include PointNet [32], which directly processes point clouds by learning spatial encodings with pooling operation, and PointNet++ [33], which employs hierarchical processing to capture local structures at multiple scales. In addition, DGCNN [49] updates the graph in feature space to capture dynamic local semantic features, while PCT [12] addresses global context and dependencies within point clouds using an order-invariant attention mechanism. Recently, methods like Point-BERT [63] and Point-MAE [30] have revolutionized point cloud learning by integrating strategies from Natural Language Processing (NLP). They introduced the Masked Point Modeling (MPM) framework as a pretext task for reconstructing masked point clouds, with Point-BERT employing a BERT-style pre-training approach to enhance performance in downstream tasks and Point-MAE utilizing masked autoencoders for self-supervised learning to obtain rich representations without labeled data.

2.2 In-Context Learning

In-Context Learning (ICL) [19,36,37,52,61], popular in NLP, has significantly influenced the field of deep learning, particularly with Transformer-based models like GPT [3] and BERT [7], which enables auto-regressive language models to infer unseen tasks by conditioning inputs on specific query-prompt pairs, *i.e.*, the contexts. The principles of ICL, primarily learning from the given data context without explicit task-specific instructions, have been extended beyond NLP, into computer vision and 3D data analysis. For instance, SupPR [67] delved into the importance of in-context examples in enhancing the inference abilities of large-scale vision models for unseen tasks, presenting a prompt retrieval framework that automates prompt selection. Painter [46] generalized visual ICL, facilitating the model to paint an image through various task prompts. In 3D, PIC [10] investigated the ICL paradigm for enhancing the 3D point cloud understanding, revealing the model's applicability for conducting multi-tasks concurrently. Though these methods showcase the generalizability across various tasks to some extent, they all focus on a single dataset (*i.e.*, single domain), overlooking the disparities between different domains. In fact, it is common and practical to have different domains in point cloud learning. This essentially motivates the design of our method: a unified model for multi-domain, multi-task point cloud learning.

2.3 Point Cloud Adaptation and Generalization

Domain Adaptation (DA) [11,42,55,57,58,65,66,71–73,77] and Domain Generalization (DG) [23–25,27,47,74–76] are introduced to enhance the adaptability and generalizability of models across diverse datasets. In point cloud analysis, DA [1,21,38,40,59,78] focuses on transferring knowledge from a labeled source domain to an unlabeled target domain, often employing unsupervised learning techniques to tackle challenges posed by the sparse and irregular nature of point cloud data [8,16,41,48,51,60]. PointDAN [34] addressed domain shift among the local-level geometric structures, while GLRV [9] presented a self-supervised learning strategy and a pseudo-labeling method for point cloud adaptation. Differently, DG [13,28,35,54,70] aims to train models that perform effectively across various unseen point cloud datasets, thereby enhancing generalizability [20,45]. SemanticSTF [54] used random data augmentation across geometry styles and applied contrastive learning to learn general features. DGLSS [17] simulated the unseen domain during training and built domain-invariant scene-wise class prototypes. The methods mentioned above typically introduce known target domains directly during training or mimic them, struggling to handle diverse unseen test data. In contrast, test-time DG methods [31,69] shifted testing sample features to the nearest source domain without additional model updates at test-time, improving robustness and performance on previously unseen domains. These advancements are useful in addressing the domain gap between synthetic and real-world data. However, they are usually task specific and lack generalization capabilities on multi-task learning in which a unified network model tackles different tasks. By contrast, we consider both different domains and different tasks.

3 Methodology

We propose a new multi-domain and multi-task setting for test-time domain generalized point cloud understanding aimed at attaining a universally generalized model. As shown in Fig. 2, we design an innovative method that accommodates both Domain Generalization and Point In-Context Learning, namely **DG-PIC**.

In DG-PIC, we devise dual-level source prototype estimation and dual-level test-time feature shifting modules, allowing the pre-trained model to generalize to unseen domain data by aligning the test data toward the source ones, without requiring any model updates at testing time. Lastly, we introduce a new benchmark for evaluating performance in this new setting.

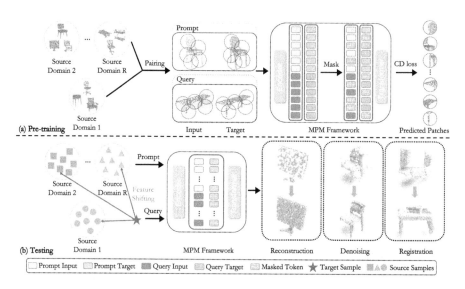

Fig. 2. The proposed DG-PIC. (a) Pre-training: we select an arbitrary sample from source domains and form a query-prompt pair with the current one. The point cloud pairs will tackle the same task. Then, we mask some patches in the target point clouds randomly through the MPM framework and reconstruct them via the Transformer model. (b) Testing: we freeze the pre-trained model, and generalize it towards unseen target sample through two key components: estimating source domain prototypes using *global-level* and *local-level* features, aligning target features with source domains by considering *macro-level* semantic information and *micro-level* positional relationships. We select the most similar sample from the nearest source as the prompt.

3.1 Revisiting Point In-Context (PIC) Learning Model

PIC is a 3D point cloud In-Context Learning (ICL) procedure inspired by 2D visual ICL principles. Through the Masked Point Modeling (MPM) framework [63], point clouds are converted into a sentence-like format, which is then encoded into tokens. Tokens are then reconstructed to generate diverse task-specific outputs.

In-Context Learning in 3D. During the training, each input sample comprises two pairs of point clouds: the input point clouds and their corresponding targets addressing the same task. Following previous works [30,63], PIC [10] employs the

MPM framework, using a Transformer model for masked point cloud reconstruction. In the inference stage, the input comprises the query point cloud and the prompt point cloud, while the output point clouds, *i.e.*, the target of the given task, include the prompt target along with the query masked tokens. Within the MPM framework, the Transformer generates masked point cloud patches based on various task prompts, and the corresponding point cloud is derived from a unified model. Thus, PIC deduces query results for various downstream point cloud tasks, achieving multi-task objectives through a unified model.

Loss Function. The Chamfer Distance (CD) serves as the training loss, comparing the predicted masked patch P with its corresponding ground truth G:

$$\mathrm{CD}(P, G) = \frac{1}{|P|} \sum_{x \in P} \min_{y \in G} \|x - y\|_2^2 + \frac{1}{|G|} \sum_{y \in G} \min_{x \in P} \|y - x\|_2^2. \tag{1}$$

3.2 Multi-domain in Point In-Context Learning

While PIC investigated the capability of a multi-task generalized model within a single domain, it remains unexplored for multiple domains. Meanwhile, significant differences exist between various data sources, like diverse data distributions and resolutions, making it challenging for the model to uniformly handle all different data, and leading to limited generalization ability.

Problem Setting: Test-time DG in PIC. Our innovation lies in test-time DG for PIC to tackle discrepancies among different domains. During the pre-training stage, we employ PIC to acquire rich information that is generalizable across all source domains, treating it as the off-the-shelf pre-trained model. The test-time DG is to facilitate the model's generalizability by aligning the unseen target domain D_t towards source domains D_s that share correlative features without any model update at testing time. We define $\{I_q, I_p\}, \{T_q^k, T_p^k\}$ as the input and target point clouds pair, where k represents the task index while q and p indicate the *query* and *prompt* sample, respectively. Suppose we have R source domains $D_s = \{D_s^1, D_s^2, \ldots, D_s^R\}$ that are available at pre-training stage. Let ν, τ denote samples from two different domains. In practice, the feature gap $\|F_\theta(\nu) - F_\theta(\tau)\|$, where $F_\theta(\cdot)$ is the shared feature encoder parameterized by θ, calculated by a unified model, might be large, especially when one new domain serves as the target domain, as the model is unfamiliar with the new data. Thereby, we aim to minimize performance degradation caused by domain gaps in the unified model and maximize its potential for testing target data.

Multi-task Setup. ICL requires maintaining the consistency between inputs and outputs within the same space, *i.e.*, the task involves XYZ coordinate regression. Thus, following PIC [10], we undertake three distinct tasks on the point cloud, including: *1) Reconstruction*, aiming at reconstructing a dense point cloud from a sparse point set. *2) Denoising*, striving to obtain a clear object shape by removing Gaussian noise from the input point cloud. *3) Registration*, focusing on restoring the original orientation from a randomly rotated point cloud.

Multi-domain Multi-task Benchmark. We propose a new setting, namely, the multi-domain and multi-task setting. To the best of our knowledge, there are no available benchmarks for performance evaluation. Therefore, we introduce a new benchmark wherein we meticulously curate and select data from 4 distinct datasets (comprising 2 synthetic and 2 real-world datasets) with 7 same object categories, and generate the corresponding ground truth based on 3 different tasks. Specifically, for the synthetic datasets, we employ ModelNet40 [53] and ShapeNet [4], with ModelNet40 choosing 3,713 samples for training and 686 for testing, and ShapeNet selecting 15,001 training samples and 2,145 testing samples. As for real-world data, we use ScanNet [6] and ScanObjectNN [43]. ScanNet provides annotations for individual objects in the real scan 3D scenes, we select 5,763 samples for training and 1,677 for testing. ScanObjectNN contains 1,577 training samples and 392 testing samples, and we merge 'desk' and 'table' in the dataset as 'table'. Following PIC, we normalize both the inputs and targets to only contain 1,024 points with XYZ coordinates, and all data are conducted with various random operations, including rotation, scaling, and perturbation.

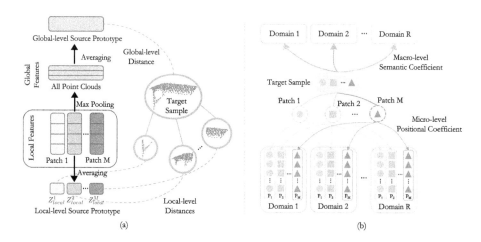

Fig. 3. Illustration of the presented method: **(a) Dual-level Source Prototype Estimation.** We estimate the prototype of the source domains at the global and local levels and consider the feature distance from the target sample to the prototypes at the dual levels. **(b) Dual-level Test-time Target Feature Shifting.** We shift the target feature by considering both macro-level semantic information across all source domains and micro-level positional relationships within corresponding patches.

3.3 Framework of DG-PIC

Test-time DG should attempt to mitigate the domain gap and shift the testing data at the testing time to boost the generalizability. Given that 3D point clouds mainly have global shape information and local geometric structures, *i.e.*, dual-level features, we propose a novel *dual-level source-domain prototype estimation* module along with a *dual-level target feature shifting* strategy, to enable the network to generalize to previously unseen target domains.

Multi-domain Prompt Pairing. During the training of PIC, the process begins with selecting another point cloud pair to serve as a prompt, guiding the model in performing a specific task. Both the prompt point cloud pair and the query point cloud pair are required to execute the same task. Unlike PIC which selects a prompt from a single training set (*i.e.*, single domain), we randomly collect samples from various source domains, reinforcing the correlation between these domains. Let $Trans(\cdot)$ denote as the Transformer blocks in the MPM framework, and query point clouds belonging to D_s^i as $\{I_i, T_i^k\}$ and prompt point clouds in other domain $D_s^j (j \neq i)$ as $\{I_j, T_j^k\}$. Then the predicted masked patch P can be depicted as:

$$P \sim (D_s^i, D_s^j) = Trans([F_\theta(I_i) \oplus F_\theta(T_i^k) \oplus F_\theta(I_j) \oplus F_\theta(T_j^k)], Mask), \quad (2)$$

where $\oplus$ represents the concatenate operation that merges all point cloud tokens, and $Mask$ denotes the masked token utilized to replace the invisible token.

Dual-Level Source Prototype Estimation. As for test-time DG, it is essential to estimate the prototype Z for each source domain. For this estimation, we propose to respectively average global and local features across all data within the domain, as shown in Fig. 3(a). Specifically, we treat the patch-wise tokens $\{F_\theta(P_1), F_\theta(P_2), \ldots, F_\theta(P_M)\}$ as the local features of each patch P_m ($m \in [1, M]$, where M is the patch number), which can be derived from the MPM encoder. Thereby, the *local-level* prototype $Z_{local} \in \mathbb{R}^{C \times M}$ of D_s^i is formulated as:

$$Z_{local}^{i,m} = \frac{1}{N_{D_s^i}} \sum_{n=1}^{N_{D_s^i}} F_\theta(P_m), \quad m \in [1, M], \quad (3)$$

where $N_{D_s^i}$ denotes the sample number of D_s^i. Following previous work [32], we use the pooling operation to extract the global feature of the point cloud, and we can obtain the *global-level* prototype $Z_{global}^i \in \mathbb{R}^C$ as:

$$Z_{global}^i = \frac{1}{N_{D_s^i}} \sum_{n=1}^{N_{D_s^i}} max(F_\theta(P_m)), \quad m \in [1, M], \quad (4)$$

where max indicates the max pooling over the point number channel. We need to pre-train the model under the multi-domain pairing settings and store all source prototypes from the final epoch. In this way, we regard the source prototypes as the shared common information available to the target domain during testing.

Given a test sample I_t in the target domain D_t, we can obtain its local features $\{F_{local}^m \mid F_\theta(P_m), m \in [1, M]\}$ along with its global feature F_{global} through the max pooling. We then calculate the feature Euclidean distances $\mathcal{E}$ between the test sample and each source domain prototype at both global and local levels:

$$\begin{aligned} \mathcal{E}_{global}^i &= ||F_{global} - Z_{global}^i||, \quad i \in [1, R], \\ \mathcal{E}_{local}^{i,m} &= ||F_{local}^m - Z_{local}^{i,m}||, \quad i \in [1, R], m \in [1, M]. \end{aligned} \quad (5)$$

Then, we retrieve the similarities between each source domain and the testing sample by comparing the global-local feature distances and consider their prototypes Z as anchors to align the testing features closer to the source domains.

Dual-Level Test-Time Target Feature Shifting. We introduce a *macro-level* semantic coefficient α, derived from the domain-aware shape information, *i.e.*, the source domain feature similarities $\mathcal{E}_{global}$, to regulate the impact from various source domains on feature shifting. Since the MPM framework models point cloud with tokens (*i.e.*, at the patch level), the shifting can be depicted as:

$$\alpha = softmax(\mathcal{E}_{global}), \quad F'_{local} = \frac{1}{R}\sum_{i=1}^{R}(\alpha_i F_{local} + (1-\alpha_i)Z_{local}^i). \quad (6)$$

To consider the *micro-level* geometric structure of the point cloud, as illustrated in Fig. 3(b), we further propose a patch-aware positional coefficient β to investigate the feature relationships within point cloud patches from different domains at each patch position in 3D space. If two point clouds have high semantic similarity, their corresponding patches occupying the same location should hold highly similar geometrical structures, even from different domains. For instance, in table point clouds derived from either real scans or synthetic data, the outer regions typically feature sharper edge structures, whereas the interior regions consist of surface planes. The designed positional coefficient is dedicated to aligning the test sample patches towards the source domains located at the same position, aiming to mitigate the discrepancies between them. Based upon this, we design the test sample feature shifting as:

$$\beta^i = softmax(\mathcal{E}_{local}^i), \quad F'_{local} = \frac{1}{R}\sum_{i=1}^{R}\alpha_i(\frac{1}{M}\sum_{m=1}^{M}\beta^{i,m}F_{local}^m) \\ + \frac{1}{R}\sum_{i=1}^{R}(1-\alpha_i)(\frac{1}{M}\sum_{m=1}^{M}(1-\beta^{i,m})Z_{local}^{i,m}). \quad (7)$$

Through shifting, the target data is pulled closer to source domains, facilitating the model to achieve comparable performance to that observed on sources.

Test-Time Prompt Selection. Upon computing the feature distances at both global and local levels between the test sample and each source domain, we can simply identify the closest source domain D_s^t by:

$$\mathcal{E}^i = \lambda \cdot \mathcal{E}_{global}^i + (1-\lambda) \cdot \frac{1}{M}\sum_{m=1}^{M}\mathcal{E}_{local}^{i,m}, \quad (8)$$

where λ, default to 0.5, serves as a balancing factor for adjusting the two parts in the equation. Therefore, the most similar sample $\{I_s, T_s^k\}$ within this source domain can be easily identified through feature distance. This similarity usually manifests similar global shapes or local geometrical structures. We select this

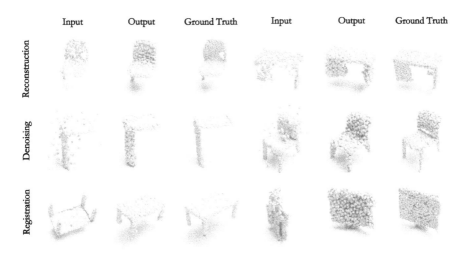

Fig. 4. Visualization results of our DG-PIC and their corresponding targets (denoted as ground truth) under 3 different tasks, including reconstruction, denoising, and registration. The real-world dataset ScanObjectNN serves as the target domain.

Comparison to Conventional Methods. Several conventional point cloud learning methods, such as PointNet [32], DGCNN [49], and PCT [12], often struggle with achieving generalization. Despite incorporating individual heads for these methods to suit diverse tasks (*i.e.*, task-specific scheme), our method consistently outperforms them across all tasks. They exhibit inferior performance with a shared network in a multi-task scheme. Our method is characterized as a fully unified model, capable of overcoming domain gaps through the multi-task in-context learning setting. Notably, these selected models suffer from significant performance drops when applying to the new dataset, demonstrating the superiority and better generalization capacity of our method. This is mainly due to our proposed test-time DG feature shifting module and ICL strategy.

Comparison to DG Methods. We compare our DGPIC with a widely-used DG strategy, specifically the mix strategy. We adopt representative works Pointmixup [5] and PointCutMix [64], by employing mixing data from various source domains during training. However, as shown in Table 1, these methods usually explore generalization between domains while overlooking task-specific differences, as indicated by the consistently high Chamfer Distance results with minimal variance across all tasks. This shows the limitation of relying solely on DG and proves insufficient for addressing multi-domain multi-task scenarios.

Comparison to ICL Method. ICL methods such as PIC [10] are effective for multi-task scenarios using a unified model but struggle with generalization across different domains. They perform poorly when faced with unseen data, failing to perform the specified task with the provided prompt in existing source domains. In contrast, our proposed test-time DG module effectively addresses this difficulty by aligning the test data features with familiar source domains. In addition, utilizing the most similar source sample as the prompt pair effectively mitigates the gap between the source and target. This enables our unified model to enjoy multi-domain generalizability.

Please refer to the supplementary for more results.

4.4 Ablation Studies

Our two novel dual-level modules, *i.e.*, global-local source prototype estimation and macro-micro target feature shifting, are pivotal for aligning the test data feature toward the source domains. To verify the effectiveness of these components, we conduct the ablation studies, and the results are shown in Table 2.

Prototype Estimation: Global-Local Levels. We consider both global and local prototypes when calculating the distance from the target sample to the source domain, allowing our model to understand sample data comprehensively by considering both global shape information and local geometric structure. We first take the prototype estimation out of our dual-level design by adopting a straightforward averaged shifting approach to align the target feature towards only one domain (Models A-D). We determine the anchor source by measuring the feature distance from the target sample to each source domain prototype. An intuitive alternative would involve randomly selecting the anchor domain. As depicted in Table 2 (Model A), the random feature shift may result in ambiguous features, leading to the loss of significant features inherent in the test sample and consequently yielding inferior results. We also compare results using only the global or local features and find that incorporating both global and local levels captures richer features, leading to superior results.

Anchor Source Domains: One or All? We observe that pushing the target sample towards only one nearest anchor source domain may result in less abundant source domain information within the model. Consequently, we delve into the superior option of shifting target samples towards all source domains to comprehensively maximize the model's generalizability. As shown in Table 2 (Models D and E), aligning target samples to all source domains leads to only a marginal improvement. We attribute this to the fact that equally pulling target sample features towards all prototypes tends to flatten features across all test data, potentially leading to less discriminative features for the test sample. By contrast, our dual-level target feature shifting mechanism enables the model to mimic prior information about new test samples, achieving much better results.

Feature Shifting: Macro-Micro Levels. To distinguish the degree of contributions from different source domains in feature shifting, as shown in Eq. (7), we implicitly utilize the feature similarity between the target samples and the global-local prototype of each source domain, allowing the model to derive informative coefficients that facilitate shifting. The comparison in Table 2 (Models E-G and ours) manifests that incorporating semantic information among all domains at the macro-level and considering the positional relationships of the point cloud patches at the micro-level both contribute to the final model, assisting in alleviating the domain discrepancy between sources and targets.

Table 2. Ablations on dual-level prototype estimation and test-time feature shifting.

Models	Prototype	→ Shifting	Anchors	Recon.	Denoi.	Regis.
Model A	random	→ average	one domain	8.4	40.5	6.7
Model B	only global	→ average	one domain	7.2	38.3	6.4
Model C	only local	→ average	one domain	7.3	36.7	6.7
Model D	global-local	→ average	one domain	6.8	35.1	6.2
Model E	global-local	→ average	all domains	6.3	32.4	6.5
Model F	global-local	→ only macro	all domains	5.2	22.7	6.0
Model G	global-local	→ only micro	all domains	4.9	25.6	6.2
Our choice	global-local	→ macro-micro	all domains	**4.1**	**15.2**	**5.8**

5 Conclusion

In this paper, we introduce the multi-domain multi-task setting that handles multiple domains and tasks within a unified model. We propose Domain Generalized Point-In-Context (DG-PIC), boosting the generalizability across different domains and tasks. Specifically, we devise dual-level source prototype estimation, capturing global and local-level features to comprehensively represent source domains, and a dual-level test-time feature shifting strategy that aligns target data with source domains at both macro and micro-level during testing. Extensive experiments demonstrate that DG-PIC outperforms state-of-the-art techniques.

Acknowledgments. Jincen Jiang is supported by the China Scholarship Council (Grant Number 202306300023), and the Research and Development Fund of Bournemouth University. Meili Wang is supported by the National Key Research and Development Program of China (Grant Number 2022YFD1300201).

References

1. Achituve, I., Maron, H., Chechik, G.: Self-supervised learning for domain adaptation on point clouds. In: Proceedings of the IEEE/CVF Winter Conference on Applications of Computer Vision, pp. 123–133 (2021)
2. Bar, A., Gandelsman, Y., Darrell, T., Globerson, A., Efros, A.: Visual prompting via image inpainting. In: Advances in Neural Information Processing Systems, vol. 35, pp. 25005–25017 (2022)
3. Brown, T., et al.: Language models are few-shot learners. In: Advances in Neural Information Processing Systems, vol. 33, pp. 1877–1901 (2020)
4. Chang, A.X., et al.: Shapenet: an information-rich 3D model repository. arXiv preprint arXiv:1512.03012 (2015)
5. Chen, Y., et al.: PointMixup: augmentation for point clouds. In: Vedaldi, A., Bischof, H., Brox, T., Frahm, J.-M. (eds.) ECCV 2020. LNCS, vol. 12348, pp. 330–345. Springer, Cham (2020). https://doi.org/10.1007/978-3-030-58580-8_20

6. Dai, A., Chang, A.X., Savva, M., Halber, M., Funkhouser, T., Nießner, M.: Scannet: richly-annotated 3D reconstructions of indoor scenes. In: Proceedings of the IEEE/CVF Conference on Computer Vision and Pattern Recognition, pp. 5828–5839 (2017)
7. Devlin, J., Chang, M.W., Lee, K., Toutanova, K.: Bert: pre-training of deep bidirectional transformers for language understanding. arXiv preprint arXiv:1810.04805 (2018)
8. Ding, R., Yang, J., Jiang, L., Qi, X.: DODA: data-oriented sim-to-real domain adaptation for 3D semantic segmentation. In: Avidan, S., Brostow, G., Cissé, M., Farinella, G.M., Hassner, T. (eds.) ECCV 2022. LNCS, vol. 13687, pp. 284–303. Springer, Cham (2022). https://doi.org/10.1007/978-3-031-19812-0_17
9. Fan, H., Chang, X., Zhang, W., Cheng, Y., Sun, Y., Kankanhalli, M.: Self-supervised global-local structure modeling for point cloud domain adaptation with reliable voted pseudo labels. In: Proceedings of the IEEE/CVF Conference on Computer Vision and Pattern Recognition, pp. 6377–6386 (2022)
10. Fang, Z., Li, X., Li, X., Buhmann, J.M., Loy, C.C., Liu, M.: Explore in-context learning for 3D point cloud understanding. arXiv preprint arXiv:2306.08659 (2023)
11. Gu, Q., et al.: PIT: position-invariant transform for cross-FoV domain adaptation. In: Proceedings of the IEEE/CVF International Conference on Computer Vision, pp. 8761–8770 (2021)
12. Guo, M.H., Cai, J.X., Liu, Z.N., Mu, T.J., Martin, R.R., Hu, S.M.: PCT: point cloud transformer. Comput. Visual Media **7**, 187–199 (2021)
13. Huang, S., Zhang, B., Shi, B., Li, H., Li, Y., Gao, P.: SUG: single-dataset unified generalization for 3D point cloud classification. In: Proceedings of the 31st ACM International Conference on Multimedia, pp. 8644–8652 (2023)
14. Jiang, J., Lu, X., Zhao, L., Dazaley, R., Wang, M.: Masked autoencoders in 3D point cloud representation learning. IEEE Trans. Multimedia (2023)
15. Jiang, J., Zhao, L., Lu, X., Hu, W., Razzak, I., Wang, M.: DHGCN: dynamic hop graph convolution network for self-supervised point cloud learning. In: Proceedings of the AAAI Conference on Artificial Intelligence, vol. 38, pp. 12883–12891 (2024)
16. Katageri, S., De, A., Devaguptapu, C., Prasad, V., Sharma, C., Kaul, M.: Synergizing contrastive learning and optimal transport for 3D point cloud domain adaptation. In: Proceedings of the IEEE/CVF Winter Conference on Applications of Computer Vision, pp. 2942–2951 (2024)
17. Kim, H., Kang, Y., Oh, C., Yoon, K.J.: Single domain generalization for lidar semantic segmentation. In: Proceedings of the IEEE/CVF Conference on Computer Vision and Pattern Recognition, pp. 17587–17598 (2023)
18. Lehner, A., et al.: 3D-VField: adversarial augmentation of point clouds for domain generalization in 3D object detection. In: Proceedings of the IEEE/CVF Conference on Computer Vision and Pattern Recognition, pp. 17295–17304 (2022)
19. Li, L., Peng, J., Chen, H., Gao, C., Yang, X.: How to configure good in-context sequence for visual question answering. In: Proceedings of the IEEE/CVF Conference on Computer Vision and Pattern Recognition, pp. 26710–26720 (2024)
20. Li, M., Zhang, Y., Ma, X., Qu, Y., Fu, Y.: BEV-DG: cross-modal learning under bird's-eye view for domain generalization of 3D semantic segmentation. In: Proceedings of the IEEE/CVF International Conference on Computer Vision, pp. 11632–11642 (2023)
21. Liu, F., et al.: Cloudmix: dual mixup consistency for unpaired point cloud completion. IEEE Trans. Visual. Comput. Graph. (2024)
22. Liu, J., Shen, D., Zhang, Y., Dolan, B., Carin, L., Chen, W.: What makes good in-context examples for GPT-3? arXiv preprint arXiv:2101.06804 (2021)

23. Long, S., et al.: Dgmamba: domain generalization via generalized state space model. arXiv preprint arXiv:2404.07794 (2024)
24. Long, S., Zhou, Q., Ying, C., Ma, L., Luo, Y.: Diverse target and contribution scheduling for domain generalization. arXiv preprint arXiv:2309.16460 (2023)
25. Long, S., Zhou, Q., Ying, C., Ma, L., Luo, Y.: Rethinking domain generalization: discriminability and generalizability. IEEE Trans. Circ. Syst. Video Technol. 1 (2024)
26. Loshchilov, I., Hutter, F.: Decoupled weight decay regularization. In: International Conference on Learning Representations (2019)
27. Lu, H., Yu, Z., Niu, X., Chen, Y.C.: Neuron structure modeling for generalizable remote physiological measurement. In: Proceedings of the IEEE/CVF Conference on Computer Vision and Pattern Recognition, pp. 18589–18599 (2023)
28. Lu, H., Zhang, Y., Lian, Q., Du, D., Chen, Y.: Towards generalizable multi-camera 3D object detection via perspective debiasing. arXiv preprint arXiv:2310.11346 (2023)
29. Min, S., et al.: Rethinking the role of demonstrations: what makes in-context learning work? arXiv preprint arXiv:2202.12837 (2022)
30. Pang, Y., Wang, W., Tay, F.E., Liu, W., Tian, Y., Yuan, L.: Masked autoencoders for point cloud self-supervised learning. In: Avidan, S., Brostow, G., Cissé, M., Farinella, G.M., Hassner, T. (eds.) ECCV 2022. LNCS, vol. 13662, pp. 604–621. Springer, Cham (2022). https://doi.org/10.1007/978-3-031-20086-1_35
31. Park, J., Han, D.J., Kim, S., Moon, J.: Test-time style shifting: handling arbitrary styles in domain generalization. In: International Conference on Machine Learning (2023)
32. Qi, C.R., Su, H., Mo, K., Guibas, L.J.: Pointnet: deep learning on point sets for 3D classification and segmentation. In: Proceedings of the IEEE/CVF Conference on Computer Vision and Pattern Recognition, pp. 652–660 (2017)
33. Qi, C.R., Yi, L., Su, H., Guibas, L.J.: Pointnet++: deep hierarchical feature learning on point sets in a metric space. In: Advances in Neural Information Processing Systems, vol. 30 (2017)
34. Qin, C., You, H., Wang, L., Kuo, C.C.J., Fu, Y.: Pointdan: a multi-scale 3D domain adaption network for point cloud representation. In: Advances in Neural Information Processing Systems, vol. 32 (2019)
35. Qu, S., Pan, Y., Chen, G., Yao, T., Jiang, C., Mei, T.: Modality-agnostic debiasing for single domain generalization. In: Proceedings of the IEEE/CVF Conference on Computer Vision and Pattern Recognition, pp. 24142–24151 (2023)
36. Radford, A., et al.: Learning transferable visual models from natural language supervision. In: International Conference on Machine Learning, pp. 8748–8763. PMLR (2021)
37. Rubin, O., Herzig, J., Berant, J.: Learning to retrieve prompts for in-context learning. arXiv preprint arXiv:2112.08633 (2021)
38. Saleh, K., et al.: Domain adaptation for vehicle detection from bird's eye view lidar point cloud data. In: Proceedings of the IEEE/CVF International Conference on Computer Vision Workshops (2019)
39. Sanchez, J., Deschaud, J.E., Goulette, F.: Domain generalization of 3D semantic segmentation in autonomous driving. In: Proceedings of the IEEE/CVF International Conference on Computer Vision, pp. 18077–18087 (2023)
40. Shen, Y., Yang, Y., Yan, M., Wang, H., Zheng, Y., Guibas, L.J.: Domain adaptation on point clouds via geometry-aware implicits. In: Proceedings of the IEEE/CVF Conference on Computer Vision and Pattern Recognition, pp. 7223–7232 (2022)

41. Sinha, A., Choi, J.: Mensa: mix-up ensemble average for unsupervised multi target domain adaptation on 3D point clouds. In: Proceedings of the IEEE/CVF Conference on Computer Vision and Pattern Recognition, pp. 4766–4776 (2023)
42. Song, Y., Zhou, Q., Li, X., Fan, D.P., Lu, X., Ma, L.: BA-SAM: scalable bias-mode attention mask for segment anything model. In: Proceedings of the IEEE/CVF Conference on Computer Vision and Pattern Recognition (2024)
43. Uy, M.A., Pham, Q.H., Hua, B.S., Nguyen, T., Yeung, S.K.: Revisiting point cloud classification: a new benchmark dataset and classification model on real-world data. In: Proceedings of the IEEE/CVF International Conference on Computer Vision, pp. 1588–1597 (2019)
44. Wang, F., Li, W., Xu, D.: Cross-dataset point cloud recognition using deep-shallow domain adaptation network. IEEE Trans. Image Process. **30**, 7364–7377 (2021)
45. Wang, S., et al.: Towards domain generalization for multi-view 3D object detection in bird-eye-view. In: Proceedings of the IEEE/CVF Conference on Computer Vision and Pattern Recognition, pp. 13333–13342 (2023)
46. Wang, X., Wang, W., Cao, Y., Shen, C., Huang, T.: Images speak in images: a generalist painter for in-context visual learning. In: Proceedings of the IEEE/CVF Conference on Computer Vision and Pattern Recognition, pp. 6830–6839 (2023)
47. Wang, X., et al.: TF-FAS: twofold-element fine-grained semantic guidance for generalizable face anti-spoofing. In: European Conference on Computer Vision. Springer, Cham (2024)
48. Wang, Y., Yin, J., Li, W., Frossard, P., Yang, R., Shen, J.: SSDA3D: semi-supervised domain adaptation for 3D object detection from point cloud. In: Proceedings of the AAAI Conference on Artificial Intelligence, vol. 37, pp. 2707–2715 (2023)
49. Wang, Y., Sun, Y., Liu, Z., Sarma, S.E., Bronstein, M.M., Solomon, J.M.: Dynamic graph CNN for learning on point clouds. ACM Trans. Graph. **38**(5), 1–12 (2019)
50. Wei, X., Gu, X., Sun, J.: Learning generalizable part-based feature representation for 3D point clouds. In: Advances in Neural Information Processing Systems, vol. 35, pp. 29305–29318 (2022)
51. Wu, B., Zhou, X., Zhao, S., Yue, X., Keutzer, K.: Squeezesegv2: improved model structure and unsupervised domain adaptation for road-object segmentation from a lidar point cloud. In: 2019 International Conference on Robotics and Automation, pp. 4376–4382. IEEE (2019)
52. Wu, Y., Yang, X.: A glance at in-context learning. Front. Comp. Sci. **18**(5), 185347 (2024)
53. Wu, Z., et al.: 3D shapenets: a deep representation for volumetric shapes. In: Proceedings of the IEEE/CVF Conference on Computer Vision and Pattern Recognition, pp. 1912–1920 (2015)
54. Xiao, A., et al.: 3D semantic segmentation in the wild: Learning generalized models for adverse-condition point clouds. In: Proceedings of the IEEE/CVF Conference on Computer Vision and Pattern Recognition, pp. 9382–9392 (2023)
55. Xiao, A., et al.: CAT-SAM: conditional tuning network for few-shot adaptation of segmentation anything model. arXiv preprint arXiv:2402.03631 (2024)
56. Xie, S., Gu, J., Guo, D., Qi, C.R., Guibas, L., Litany, O.: PointContrast: unsupervised pre-training for 3D point cloud understanding. In: Vedaldi, A., Bischof, H., Brox, T., Frahm, J.-M. (eds.) ECCV 2020. LNCS, vol. 12348, pp. 574–591. Springer, Cham (2020). https://doi.org/10.1007/978-3-030-58580-8_34
57. Xiong, Y., et al.: PYRA: parallel yielding re-activation for training-inference efficient task adaptation. arXiv preprint arXiv:2403.09192 (2024)

58. Xiong, Y., Chen, H., Lin, Z., Zhao, S., Ding, G.: Confidence-based visual dispersal for few-shot unsupervised domain adaptation. In: Proceedings of the IEEE/CVF International Conference on Computer Vision, pp. 11621–11631 (2023)
59. Xu, Q., Zhou, Y., Wang, W., Qi, C.R., Anguelov, D.: SPG: unsupervised domain adaptation for 3D object detection via semantic point generation. In: Proceedings of the IEEE/CVF International Conference on Computer Vision, pp. 15446–15456 (2021)
60. Yang, Q., Liu, Y., Chen, S., Xu, Y., Sun, J.: No-reference point cloud quality assessment via domain adaptation. In: Proceedings of the IEEE/CVF Conference on Computer Vision and Pattern Recognition, pp. 21179–21188 (2022)
61. Yang, X., Wu, Y., Yang, M., Chen, H., Geng, X.: Exploring diverse in-context configurations for image captioning. In: Advances in Neural Information Processing Systems, vol. 36 (2024)
62. Yi, L., Gong, B., Funkhouser, T.: Complete & label: a domain adaptation approach to semantic segmentation of lidar point clouds. In: Proceedings of the IEEE/CVF Conference on Computer Vision and Pattern Recognition, pp. 15363–15373 (2021)
63. Yu, X., Tang, L., Rao, Y., Huang, T., Zhou, J., Lu, J.: Point-BERT: pre-training 3D point cloud transformers with masked point modeling. In: Proceedings of the IEEE/CVF Conference on Computer Vision and Pattern Recognition, pp. 19313–19322 (2022)
64. Zhang, J., et al.: Pointcutmix: regularization strategy for point cloud classification. Neurocomputing **505**, 58–67 (2022)
65. Zhang, Y., Deng, B., Tang, H., Zhang, L., Jia, K.: Unsupervised multi-class domain adaptation: theory, algorithms, and practice. IEEE Trans. Pattern Anal. Mach. Intell. **44**(5), 2775–2792 (2020)
66. Zhang, Y., Zhu, W., Tang, H., Ma, Z., Zhou, K., Zhang, L.: Dual memory networks: a versatile adaptation approach for vision-language models. In: Proceedings of the IEEE/CVF Conference on Computer Vision and Pattern Recognition, pp. 28718–28728 (2024)
67. Zhang, Y., Zhou, K., Liu, Z.: What makes good examples for visual in-context learning? arXiv preprint arXiv:2301.13670 (2023)
68. Zhao, S., et al.: ePointDA: an end-to-end simulation-to-real domain adaptation framework for lidar point cloud segmentation. In: Proceedings of the AAAI Conference on Artificial Intelligence, vol. 35, pp. 3500–3509 (2021)
69. Zhao, X., Liu, C., Sicilia, A., Hwang, S.J., Fu, Y.: Test-time fourier style calibration for domain generalization. In: The International Joint Conference on Artificial Intelligence (2022)
70. Zhao, Y., Zhao, N., Lee, G.H.: Synthetic-to-real domain generalized semantic segmentation for 3D indoor point clouds. arXiv preprint arXiv:2212.04668 (2022)
71. Zhou, Q., et al.: Uncertainty-aware consistency regularization for cross-domain semantic segmentation. Comput. Vis. Image Underst. **221**, 103448 (2022)
72. Zhou, Q., et al.: Context-aware mixup for domain adaptive semantic segmentation. IEEE Trans. Circuits Syst. Video Technol. **33**(2), 804–817 (2023)
73. Zhou, Q., Gu, Q., Pang, J., Lu, X., Ma, L.: Self-adversarial disentangling for specific domain adaptation. IEEE Trans. Pattern Anal. Mach. Intell. **45**(7), 8954–8968 (2023)
74. Zhou, Q., Zhang, K.Y., Yao, T., Lu, X., Ding, S., Ma, L.: Test-time domain generalization for face anti-spoofing. In: Proceedings of the IEEE/CVF Conference on Computer Vision and Pattern Recognition, pp. 175–187 (2024)

75. Zhou, Q., Zhang, K.Y., Yao, T., Lu, X., Yi, R., Ding, S., Ma, L.: Instance-aware domain generalization for face anti-spoofing. In: Proceedings of the IEEE/CVF Conference on Computer Vision and Pattern Recognition, pp. 20453–20463 (2023)
76. Zhou, Q., Zhang, K.Y., Yao, T., Yi, R., Ding, S., Ma, L.: Adaptive mixture of experts learning for generalizable face anti-spoofing. In: Proceedings of the 30th ACM International Conference on Multimedia, pp. 6009–6018 (2022)
77. Zhou, Q., et al.: Generative domain adaptation for face anti-spoofing. In: Avidan, S., Brostow, G., Cissé, M., Farinella, G.M., Hassner, T. (eds.) ECCV 2022. LNCS, vol. 13665, pp. 335–356. Springer, Cham (2022). https://doi.org/10.1007/978-3-031-20065-6_20
78. Zou, L., Tang, H., Chen, K., Jia, K.: Geometry-aware self-training for unsupervised domain adaptation on object point clouds. In: Proceedings of the IEEE/CVF International Conference on Computer Vision, pp. 6403–6412 (2021)

Operational Open-Set Recognition and PostMax Refinement

Steve Cruz[1](✉), Ryan Rabinowitz[2], Manuel Günther[3], and Terrance E. Boult[2]

[1] University of Notre Dame, Notre Dame, USA
stevecruz@nd.edu
[2] University of Colorado Colorado Springs, Colorado Springs, USA
{rrabinow,tboult}@uccs.edu
[3] University of Zurich, Zürich, Switzerland
guenther@ifi.uzh.ch

Abstract. Open-Set Recognition (OSR) is a problem with mainly practical applications. However, recent evaluations have largely focused on small-scale data and tuning thresholds over the test set, which disregard the real-world operational needs of parameter selection. Thus, we revisit the original goals of OSR and propose a new evaluation metric, Operational Open-Set Accuracy (OOSA), which requires predicting an operationally relevant threshold from a validation set with known and a surrogate set with unknown samples, and then applying this threshold during testing. With this new measure in mind, we develop a large-scale evaluation protocol suited for operational scenarios. Additionally, we introduce the novel PostMax algorithm that performs post-processing refinement of the logit of the maximal class. This refinement involves normalizing logits by deep feature magnitudes and utilizing an extreme-value-based generalized Pareto distribution to map them into proper probabilities. We evaluate multiple pre-trained deep networks, including leading transformer and convolution-based architectures, on different selections of large-scale surrogate and test sets. Our experiments demonstrate that PostMax advances the state of the art in open-set recognition, showing statistically significant improvements in our novel OOSA metric as well as in previously used metrics such as AUROC, FPR95, and others.

1 Introduction

The original Open-Set Recognition (OSR) formulation [1,38,39] gained attention due to the flaws in existing recognition systems when handling inputs from outside the training classes. These systems lack control over inputs, and any unknown sample that is not rejected leads to a misclassification, thereby reducing accuracy. Therefore, early open-set work [1,3,35] introduced protocols tai-

---

Supplementary Information The online version contains supplementary material available at https://doi.org/10.1007/978-3-031-72658-3_27.

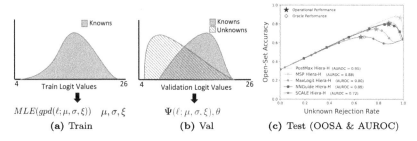

Fig. 1. POSTMAX PIPELINE. During train (a), post-processed maximum logits of correctly classified train samples are normalized by feature magnitude ℓ; the set is then used in a Maximum Likelihood Estimation (MLE) to obtain Generalized Pareto Distribution (GPD) [33] with parameters μ, σ, and ξ. During validation (b), we use a validation and a surrogate set where the cumulative distribution of the GPD (Ψ) gives the probability of the normalized max logit being from the train distribution and predict an operational threshold. During operation (c), we deploy each method's threshold and apply it to determine OSR and compute Operational Open Set Accuracy (OOSA). For evaluation we compare against Maximum Softmax Probability (MSP) [16,44], Maximum Logit (MaxLogit) [15,44], Nearest Neighbor Guidance (NNGuide) [32], and SCALE [50]. Operational Open-Set Accuracy (OOSA) performance is ($\star$) and ($\diamond$) is the optimal Open-Set Accuracy (OSA) for an oracle that knows the final test set (shown to highlight differences from OOSA). AUROC scores are shown for comparison.

lored to emulate real-world systems, allowing for a proper evaluation of open-set performance.

For a decade, researchers explored techniques that enhanced OSR performance [1,5,7,10,11,20,25,27,28,35,39,55,57]. While contributing novel ideas, recent works have diverged from the original goals of OSR. Instead of leveraging large-scale data as in early research on open-set deep networks [1], recent OSR evaluations have focused on small-scale datasets with few classes, with only a few notable exceptions [30,35,44]. Moreover, the prevalent reliance on metrics such as the Area Under the Receiver Operating Characteristics (AUROC) curve for performance evaluation has led to inopportune assessments since such metrics do not offer a realistic perspective on OSR. Figure 1 illustrates an example where AUROC scores exhibit one ranking, yet in an operational setting, there is a slight difference. Inherently, common measures overlook the real-world operational needs of threshold selection. Some may argue for their use with a dedicated validation set; however, it is not clear how one would extract a threshold from these metrics and deploy it at test time. This is problematic as methods have largely ignored a fundamental practice of machine learning systems – having separate training, validation, and test sets to ensure that models generalize, avoid overfitting, and provide unbiased evaluation and reliable performance prediction on new data.

Initial aspirations behind OSR aimed to enhance *real systems* [4]. Imagine an engineer tasked with implementing an OSR algorithm into their system. As they analyze results from various algorithms through tables and plots to inform decisions, a formidable challenge emerges. While AUROC and similar metrics assess a model's discriminative capability across all possible classification thresholds, they do not guide the engineer toward an optimal threshold for decision-making. There is a pressing need for a metric that enables confident and statistically rigorous evaluations with anticipated data, while also demonstrating robustness against varying numbers of unknowns. When deploying an open-set system, engineers must, at a minimum, consider two factors: **(1)** how an approach scales with operational data, and **(2)** how to choose an operating point that balances their desired known classification accuracy and unknown rejection. These considerations underscore the importance of achieving overall accuracy with a predicted operational threshold. Therefore, we introduce a new metric called Operational Open-Set Accuracy (OOSA), illustrated in Fig. 1, which facilitates algorithm and operational threshold selection. To ensure real-world applicability, we implement a large-scale evaluation protocol with proper training/validation/test splits. We determine a threshold on the validation set and deploy it during testing on unseen unknowns. Additionally, for completeness, fairness, and robustness, we evaluate state-of-the-art algorithms on ImageNet-scale open-set recognition using common metrics (AUROC, FPR95, AUOSCR [6], and OpenAUC [46]), many of which are detailed in the supplemental.

Before incorporating an operational threshold, it is natural to ask about the distribution of scores above this threshold. This naturally leads to a Peak-over-Threshold (POT) formulation of the distribution, best modeled via Extreme Value Theory (EVT). The Pickands-Balkema-De Haan theorem of EVT [33] yields a Generalized Pareto Distribution (GPD) over normalized scores, as depicted in Fig. 1. This GPD-based score transformation provides theoretically grounded probabilities rather than raw, network-dependent ad-hoc confidence scores (see supplemental for details). While GPD can be applied directly to Softmax confidences or logit values, our approach goes further. We introduce a novel algorithm, **PostMax**, which applies GPD to post-processed normalized maximum logits. Here, normalization by the deep feature magnitude enhances the separation of known and unknown classes. Prior observations and theoretical explanations [10,13,31,44] suggest that deep feature magnitudes extracted from inputs that the network was not trained on are generally smaller than those for known samples. However, our findings show the exact opposite, *i.e.*, that modern networks trained on the large-scale ILSVRC2012 dataset generally exhibit larger deep feature magnitudes for unknown samples (details in supplemental). Therefore, applying GPD to logit values *divided* by the deep feature magnitude can enhance OSR beyond the strong baselines of using raw Softmax confidence [16], maximum logit values [15], overconfidence reduction with features and logits [32], or scaled logit values [50]. The contributions of this paper are as follows:

- We design a novel evaluation metric, Operational Open-Set Accuracy (OOSA), that emphasizes real-world usage by predicting an operational threshold on a validation set (knowns and unknowns surrogate), which is then deployed during testing on unseen data.

- We introduce a novel algorithm, PostMax, that post-processes maximum logits by normalizing them by deep feature magnitude and then applies GPD to provide probabilities. Our code is publicly available.[1]
- We develop new large-scale evaluation protocols suited for assessing algorithms in operational scenarios.
- We showcase that PostMax advances the state of the art in large-scale open-set recognition with statistical significance on OOSA and prior metrics.

2 Related Work

Improving OSR, distinct from OOD detection, anomaly detection, or novel category discovery, can be approached in two ways: providing better features through improved network training, or through post-processing where a pre-trained network is trained for closed-set tasks and then adapted for OSR. Our work is rooted in the post-processing approaches.

2.1 Related Problems

A common question regarding open-set is, *What about a two-stage system with OOD followed by classification, does that not solve OSR?* This hypothesis is rejected with an example from one of our operational systems where we have to recognize objects in novel contexts, *e.g.*, in snowy or foggy conditions. Such conditions are OOD with respect to training samples, but they should **not** be rejected from an open-set point of view. Out-Of-Distribution (OOD) detection, though distinct from OSR, has been more consistent in its use of large-scale experimentation as were used in early OSR work [15,16,32,45,47,50,52,54]. Even when OOD is restricted to out-of-class, the two-stage system needs to be evaluated in that context, including a process for selecting an OOD threshold and then evaluating the resulting classification performance. The OpenOOD Benchmark [52,54] has implemented various OSR and OOD techniques such as OpenGan [20], MOS [19], ReAct [41], ViM [45], GEN [23], NNGuide [32], and SCALE [50]. While we caution the use of OOD algorithms in OSR evaluations/settings, based on the OpenOOD Benchmark leaderboard, we compare with the state-of-the-art, NNGuide & SCALE.

2.2 Closed-Set Classifiers

Recently, Vaze *et al.* [44] argued that closed-set classifiers are sufficient. This aligns with Hendrycks *et al.* [15,16], who demonstrated thresholding on Softmax confidences or, especially, on logits provide unreasonably good baselines. Our work shows that such approaches/networks can be improved with normalization.

We evaluate several leading pre-trained architectures trainined on ILSVRC2012-1K with no additional data. Particularly, Meta's Vision Transformer Hiera-H [37], which strips non-essential components making it faster and

---

[1] https://github.com/Vastlab/PostMax-OOSA

more accurate during training/inference. Also, we use their Masked Autoencoders (MAE) model, ViT-H [14], which masks random patches of an image and reconstructs the missing pixels. To show generalization, we also utilize CNNs, including Meta's ConvNeXtV2-H [48] and ConvNeXt-L [24] which modernize a standard ResNet. Other networks are found in the supplemental material.

2.3 Post-processing Approaches

Besides thresholding softmax scores or logits [15,16,44], OSR methods take features and use Weibull-calibrated Support Vector Machines (W-SVM) [38] or Extreme Value Machines (EVM) [35] to estimate a probability of unknown. OpenMax [1] was the first method to add an artificial probability of unknown to closed-set networks by computing a logit score for the unknown class based on deep feature similarities to features of known classes. Other approaches include a small adaptor network [13,42] that adds open-set capabilities [10] to features extracted from closed-set networks.

2.4 Learning-Based Approaches

Most methods use *negative* aka. *known unknown* samples in training, which represent samples of no interest to the classifier and should be classified as unknown, with a few exemptions such as replacing the final Softmax classifier with a set of binary classifiers [40,43]. While some use real negative samples [10,13,30], the majority of research tries to artificially create them. For example, combinations of knowns [57], added noise to knowns [47], or Generative Adversarial Networks (GANs) [11,20]. In our approach, we avoid the use of any negative samples.

2.5 Evaluation Techniques

Evaluating OSR correctly in a real-world setting is difficult and current measures do not satisfy operational requirements. For example, the widely used AUROC metric [5,7,27,28,44,51] only looks at the binary decision of known *vs.* unknown, but ignores the task of assigning the correct class for known samples. Such metrics are often combined with closed-set classification accuracy in order to evaluate OSR. These evaluations are reasonable, but provide no information under realistic scenarios, *i.e.*, an operational threshold classifying samples as known or unknown.

Another metric often employed is the macro-averaged F1 score [1,27,35,53], which treats unknown as a separate class. This measure has many counterintuitive properties that make it difficult to interpret the results. For example, a different threshold is required for each class so a sample could be classified as several known classes and unknown (at the same time). Thus, incorrect classification of unknown samples can easily be overlooked.

Recently, Dhamija *et al.* [10] introduced the Open-Set Classification Rate (OSCR) curve. This curve handles knowns and unknowns separately, evaluating

Fig. 2. MOTIVATION FOR NORMALIZATION. We illustrate why PostMax divides by the feature magnitude norm before applying GPD. The comparison above shows squared deep feature values extracted from a pre-trained ILSVRC2012-1K Transformer (ViT-H [14]). We sorted the deep features of 20 randomly sampled images from class 990 based on increasing mean square magnitude. Next, we applied this sorted order to the features of 20 images misclassified as class 990 from each unknown dataset. Unknowns exhibit high feature values (bright colors) in regions where known sample activations are low. These large, uncorrelated responses accumulate to give it a high score for the class but also result in unknown dataset features having, on average, larger L_2 norms than those of the correct class, depicted in the magnitude histogram on the right. Therefore, dividing by magnitude enhances separation.

the Correct Classification Rate (CCR) at various thresholds corresponding to specific False Positive Rates (FPR). While valuable for high-risk applications like open-set face recognition, its practicality in general OSR tasks is questionable. In scenarios where a threshold is chosen based on a specific FPR, the curve utilizes unknowns from the test set, raising doubts about its applicability to unseen classes. Moreover, the curve faces the challenge of not providing a single value for comparing different methods. Additionally, the proposed metrics, Area Under the OSCR (AUOSCR) [6] and Open Area Under the ROC Curve (OpenAUC) [46], do not provide an intuition on how to select an operating threshold.

3 Approach

Several approaches advocate selecting the maximum class and then applying thresholding to certain scores, e.g., Softmax [16] or logits [15], to address OSR. While such thresholding proves effective in some recognition tasks [44], its relevance often extends only to establishing a ranked order of classes for individual samples. We focus on leveraging information from the features themselves and the formal distribution of these maxima. The result is PostMax, an effective approach with theoretically grounded probabilities to address OSR through (1) normalization and (2) score distribution.

3.1 Normalization and GPD

From Fig. 2, unknown samples activate high dimensions in the feature space, which goes against our intuition that deep networks would only learn features necessary to classify known classes and disregard those belonging to other uninteresting ones. We attribute these findings to high logit values and Softmax scores, where positions with high feature magnitudes align with larger weights of certain classes, even when an input exhibits a large feature magnitude and the high-scoring class has a small or negative weight. The observed effect may depend on the network's loss function and the training dataset; in small-scale datasets such as MNIST or CIFAR, which only differentiate between 10 classes, the opposite has been observed [10,18,31,44]. Further details regarding the normalizations in these works and how they differ from our findings are provided in the supplemental.

As we classify samples based on their "maximum score" threshold, a key question arises: *Does this score belong to the training distribution of maximum scores?* Statistical Extreme Value Theory (EVT) provides a grounded approach to answering this question. Unlike ad-hoc Softmax confidences, transforming logits into real probabilities allows probabilities to be more interpretable compared to Softmax, probabilities do not need to sum up to one (images may contain multiple objects), and EVT models probabilities on are based on thousands/millions of samples rather than the current sample. Given that thresholdling maximum logits or Softmax scores serves as a reasonable baseline [15,32,44], applying EVT to this problem is both theoretically justified and intuitive. Note, prior OSR research [1,26,35,38] utilizes the Fisher-Tippett-Gnedenko EVT [12], resulting in a Weibull distribution. However, practical effectiveness was limited due to challenges in parameter selection affecting EVT modeling. In contrast, we employ the Generalized Pareto Distribution (GPD) derived from the Peak-over-threshold (POT) approach [33], which models extreme values above a threshold.

3.2 PostMax

Using these intuitions and observations, we define our Postnormalization of Maxima (PostMax) algorithm. We aim to find a general distribution Ψ of logit values that is valid for all of our known classes $c \in \{1, \ldots, C\}$. As observed in Fig. 2, modeling raw logit values might not be fruitful since unknown samples have generally higher activation of deep features, which often leads to high logit values. For a sample, we utilize *normalized* logit values by dividing the original logit values by their deep feature magnitude via (1).

Based on POT EVT [33], we understand that if we consider all maximum values above a threshold (such as the smallest correct logit observed during training), the resulting distribution follows a Generalized Pareto Distribution (GPD). Let $D_{\text{train}} = \{(x_i, t_i)\}$ be the collection of all samples x_i with their respective ground-truth target label t_i in the training set. Let $\text{FV}(x)$ be the function returning the feature vector of a sample x. Let $\text{L}(x)$ be the function returning the logit vector of a sample. We collect the normalized logits from the target class for all correctly classified training samples, which are then used to model a Generalized Pareto Distribution $\Psi_{\mu,\sigma,\xi}$. Details can be found in Algorithm 1.

At inference for test sample x we use $\text{FV}(x)$ and $\text{L}(x)$ to extract the deep features and logits of x and then for class c assign probability $p_c(x)$ using the

Algorithm 1. POSTMAX FITTING. This implements the probability distribution estimation of the proposed PostMax algorithm.

Require: D_{train}, FV(x), L(x)
  $M \leftarrow \{\}$
  for each $(x,t) \in D_{\text{train}}$ **do**
    if $\arg\max \text{L}(x) = t$ **then**
      $\ell \leftarrow \frac{\max \text{L}(x)}{\|\text{FV}(x)\|}$
      $M \leftarrow M \cup \{\ell\}$
    end if
  end for
  $\mu, \sigma, \xi \leftarrow$ `scipy.stats.genpareto.fit`(M)

---

normalized logit value:

$$\ell_c = \frac{\text{L}(x)[c]}{\|\text{FV}(x)\|} \qquad p_c(x) = \Psi_{\mu,\sigma,\xi}(\ell_c) \qquad (1)$$

where $\Psi_{\mu,\sigma,\xi}(\ell)$ is the cumulative distribution of the GPD with location μ, scale σ and shape ξ computed via Algorithm 1. If we just want the top class (as all experiments), we compute the class with the max logit and apply the above to that logit.

Note that we do not utilize any unknowns or negatives in this process, nor do we explicitly model a probability of unknowns. Instead, we threshold based on the cumulative GPD probability of known classes, computing the maximum value of normalized logits.

3.3 Operational Open-Set Accuracy

By designing the OpenAUC metric, Wang *et al.* [46] proposed four different conditions that a good open-set evaluation metric should fulfill, which we rephrase:

P1 The metric needs to check that known samples are classified correctly with high probability.
P2 The metric needs to evaluate if unknown samples are assigned to known classes with low probability.
P3 The metric should be insensitive to a score threshold.
P4 The metric should be a single number.

We totally agree with **P1**, **P2** and **P4**. However, we reject **P3** because it means that the metric cannot be used to select a classifier that works well in a specific operational setting, which requires obtaining an operational threshold. In our view, a useful open-set algorithm must include that step. Therefore, we reformulate condition **P3'** to be operationally relevant:

P3' The metric should indicate an operational threshold that is optimal under specified circumstances and can be applied to unseen data and unseen unknown classes.

These four conditions (**P1**, **P2**, **P3'**, **P4**) exclude all existing open-set evaluation metrics; we need a new metric to satisfy them. As discussed in Sect. 2.5, metrics such as AUROC, AUOSCR, and OpenAUC simply average over all possible operational thresholds, while F1 and normalized accuracy fail to use a single threshold [46]. On the other hand, the Open-Set Classification Rate (OSCR) curve [10] does not provide a single value for comparison, which makes it hard to compare different algorithms. Therefore, we introduce Operational Open-Set Accuracy (OOSA), which is inspired by OSCR but fulfills all four conditions.

From an engineer's view, the real question is how well a specific method can perform under the presence of both known and unknown samples in their data. Most applications of open-set classification are not security-sensitive, so there is no need to have very low False Positive Rates (FPR) as typically evaluated in the OSCR [10,30].

Usually, an engineer has a rough estimate of how many known and unknown samples the system will observe – so they are rather interested in the total accuracy the system will achieve and which kind of operational threshold will lead to this performance. To provide both, we exploit the FPR and CCR calculations used in OSCR [10] to define the *Unknown Rejection Rate* (URR) and the *Open-Set Accuracy* (OSA), both of which rely on the operational threshold θ and only make use of the probabilities of known classes $P_c(x)$:

$$\mathrm{CCR}(\theta) = \frac{|\{x \mid x \in D_c \wedge \arg\max_c P_c(x) = \hat{c} \wedge P_c(x) \geq \theta\}|}{|D_c|}$$

$$\mathrm{URR}(\theta) = \frac{|\{x \mid x \in D_a \wedge \max_c P_c(x) < \theta\}|}{|D_a|} \quad (2)$$

$$\mathrm{OSA}(\theta) = \alpha \cdot \mathrm{CCR}(\theta) + (1 - \alpha) \cdot \mathrm{URR}(\theta)$$

where D_c is the collection of all known test samples with class labels $\hat{c}$, while D_a collects all unknown test samples [10]. Please note that $\mathrm{URR}(\theta) = 1 - \mathrm{FPR}(\theta)$ provides a positive view on the unknown samples, *i.e.*, it calculates how many of them are correctly rejected under threshold θ. OSA includes a weighted average over CCR and URR to provide a holistic view of the expected total accuracy, where the application-dependent α can be used to provide a stronger focus on correctly classifying known or unknown samples. When both types of samples should be treated similarly, $\alpha = \frac{|D_c|}{|D_c|+|D_a|}$ can be selected, which we do in most of our experiments. In that case, OSA can be thought of as evaluating a C class classifier using $C+1$ class accuracy where the unknown class $C+1$ is classified when the maximal probability does not exceed our operational threshold θ.

When plotting OSA over URR, by varying the operational threshold, we arrive at the plots seen in Fig. 1. This plot is *not monotonic*, but rather provides a peak where the OSA is maximized. The maximum value can be used to select a threshold θ^* on a given set of scores S effectively:

$$\theta^* = \arg\max_{\theta \in S} \mathrm{OSA}_S(\theta)$$
$$S = \{P_{\hat{c}}(x) \mid x \in D_c\} \cup \{\max_c P_c(x) \mid x \in D_a\} \quad (3)$$

Table 1. OPERATIONAL OPEN-SET ACCURACY (OOSA). The mean OOSA ($\uparrow$) of all methods. To compute, we validate methods on ImageNetV2 [34] ($10K$ images) as knowns and 20% of our surrogate 21K-P *Hard* [44] ($9.8K$ images) as unknowns and predict an operational threshold. Then, we deploy each method's threshold and test on five different ILSVRC2012 *val* [36] splits (each $10K$ images) and specified unknowns. Standard deviation is omitted as it is < 0.005 for all methods/unknowns. OSR is performed on extractions from two state-of-the-art pre-trained architectures on ILSVRC2012 - (1) Transformer Hiera-H [37] (2) CNN ConvNeXtV2-H [48]. The best scores are in **bold**.

Unknowns (# images)	OOSA $\uparrow$									
	Hiera-H [37]					ConvNeXtV2-H [48]				
	PostMax (Ours)	MSP [16,44]	MaxLogit [15,44]	NNGuide [32]	SCALE [50]	PostMax (Ours)	MSP [16,44]	MaxLogit [15,44]	NNGuide [32]	SCALE [50]
iNaturalist [17] ($10K$)	**0.825**	0.815	0.780	0.782	0.727	**0.812**	0.796	0.786	0.713	0.772
NINCO [2] ($\sim 5.8K$)	**0.740**	0.739	0.707	0.625	0.667	**0.732**	0.723	0.718	0.597	0.705
OpenImage-O [45] ($\sim 17.6K$)	**0.864**	0.814	0.745	0.796	0.663	**0.849**	0.820	0.792	0.745	0.767
Places [56] ($10K$)	**0.785**	0.754	0.691	0.745	0.625	**0.779**	0.749	0.710	0.693	0.676
SUN [49] ($10K$)	**0.795**	0.758	0.709	0.749	0.651	**0.783**	0.751	0.721	0.693	0.690
Textures [8] ($\sim 5.1K$)	**0.764**	0.756	0.736	0.660	0.715	**0.744**	0.736	0.740	0.601	0.740
21K-P *Easy* [44] ($50K$)	**0.818**	0.740	0.662	0.744	0.568	**0.813**	0.763	0.691	0.783	0.633

Here, S is the collection of all maximum scores for known and unknown samples. When applying this calculation on the test set S_{test}, we will find the optimal threshold θ^*_{test}, indicated by $\diamond$ in Fig. 1. However, since test set samples are unknown in operational settings, we make use of the validation set D_{val} of known samples as D_c and an additional surrogate set D_{sur} of unknown samples to replace D_a in (3) for obtaining the operational threshold θ^*_{sur}, which we will then apply to the test set: $\text{OSA}_{S_{test}}(\theta^*_{sur})$, which is indicated by $\star$ in Fig. 1. How to optimally select this surrogate set is discussed below.

4 Experiments

Although a few OSR methods have explored large-scale datasets [1,30,35,44], many recent techniques [5–7,10,11,20,25,27,28,55,57] still lack comprehensive large-scale evaluations. The performance on small-scale data such as MNIST [22], CIFAR [21], or SVHN [29] does not reflect operational scenarios and lacks generalizability due to small image crops and a limited number of classes [30].

An open-set protocol is essential and necessitates a large-scale setting for the proper evaluation of real-world scenarios [30,44]. Additionally, it is paramount to employ proper machine-learning methodology, including a validation set (and a surrogate set) for operational threshold selection. Thus, we evaluate the robustness and generalization of PostMax on multiple large-scale datasets and splits with realistic and high-quality images. We also leverage the recently introduced large-scale ImageNet open-set splits [44]. Lastly, we conduct ablations to understand the performance impact of our approach.

Table 2. COMMON METRICS. The mean AUROC (↑) and FPR95 (↓) of all methods. To compute, we test each method on five different ILSVRC2012 *val* [36] splits (each $10K$ images) as knowns and specified unknowns. Standard deviation is omitted as it is < 0.01 for all methods/unknowns. OSR is performed on extractions from two state-of-the-art pre-trained architectures on ILSVRC2012-1K - (1) Transformer Hiera-H [37] (2) CNN ConvNeXtV2-H [48]. The best scores are in **bold**.

Unknowns (# images)	AUROC ↑ / FPR95 ↓									
	Hiera-H [37]					ConvNeXtV2-H [48]				
	PostMax (Ours)	MSP [16,44]	MaxLogit [15,44]	NNGuide [32]	SCALE [50]	PostMax (Ours)	MSP [16,44]	MaxLogit [15,44]	NNGuide [32]	SCALE [50]
iNaturalist [17] ($10K$)	**.96 / .21**	.92 / .37	.88 / .36	.90 / .50	.81 / .51	**.95 / .26**	.91 / .42	.90 / .37	.86 / .58	.88 / .44
NINCO [2] (~$5.8K$)	**.87 / .58**	.83 / .62	.79 / .61	.69 / .78	.73 / .67	**.86 / .56**	.83 / .64	.82 / .60	.71 / .81	.80 / .62
OpenImage-O [45] (~$17.6K$)	**.95 / .25**	.88 / .49	.80 / .49	.88 / .45	.72 / .61	**.94 / .29**	.88 / .49	.87 / .44	.81 / .70	.85 / .47
Places [56] ($10K$)	**.89 / .49**	.83 / .60	.75 / .61	.85 / .55	.67 / .71	**.89 / .50**	.84 / .62	.79 / .60	.82 / .64	.76 / .64
SUN [49] ($10K$)	**.90 / .46**	.84 / .57	.78 / .57	.85 / .56	.71 / .66	**.89 / .47**	.84 / .59	.81 / .57	.83 / .60	.77 / .61
Textures [8] (~$5.1K$)	**.93 / .32**	.88 / .47	.86 / .42	.74 / .86	.83 / .46	**.91 / .34**	.87 / .49	.88 / .40	.73 / .85	.88 / .38
21K-P *Easy* [44] ($50K$)	**.86 / .50**	.79 / .64	.72 / .64	.80 / .68	.65 / .73	**.85 / .51**	.79 / .65	.75 / .64	.78 / .72	.72 / .68

4.1 Details

Architectures. We use state-of-the-art pre-trained models on ILSVRC2012 [36] with standard 224×224 image crops - Meta's Hiera-H [37] and ConvNeXtV2-H [48]. Meta fine-tuned these models with no additional/external data. Note, we do not perform additional training or fine-tuning.

Datasets. For knowns, we utilize the large-scale ImageNet [9] subset ILSVRC2012 [36]. This dataset provides an advantage due to the extensive variety of available pre-trained architectures, enabling us to compare with more diverse deep networks. Our PostMax model is trained on the ILSVRC2012 *train* split, which contains $\sim 1.28M$ images and $1K$ classes (PostMax training requires no additional negative/unknown data). For validation, we employ ImageNetV2 [34], which contains $10K$ images. For testing, we split ILSVRC2012 *val* into five equal splits of $10K$ images each.

For unknowns, we use iNaturalist [17], NINCO [2], OpenImage-O [45], Places [56], SUN [49], Textures [8], and ImageNet-21K-P Open-Set splits [44]. The vision community [15,16,32,44,45,50,52,54] has utilized these datasets as they present unique challenges not seen in small-scale settings. For validation, to predict an operational threshold, we utilize 20% of ImageNet-21K-P *Hard* ($9.8K$ images) as our surrogate unknowns. For testing, NINCO, OpenImage-O, Textures, and ImageNet-21K-P *Easy* are entirely used, while others sample $10K$ images from a manual selection of classes. Further descriptions of all datasets and splits are provided in the supplemental.

Evaluation Metrics. To align with existing literature, we utilized common metrics in Table 2, namely, Area Under the Receiver Operating Curve (AUROC) and False Positive Rate at a fixed True Positive Rate of 95% (FPR95). Additional metrics like Area Under Open-Set Classification Rate (AUOSCR) [6] and Open Area Under ROC Curve (OpenAUC) [46] are provided in the supplemental. Alongside these, we introduce Operational Open-Set Accuracy (OOSA) in

Table 1, where we predict an operational threshold on a validation and a surrogate set and deploy it at test time. This measure, described in detail in Sect. 3.3, serves as a more realistic alternative to existing measures. When evaluating OSR methods in real-world scenarios, other measures are insufficient as they do not provide an operating point for deployment and are usually only evaluated with a test set.

4.2 Results

Our main results are presented in Tables 1 and 2, where we report performance on each unknown dataset mentioned in Sect. 4.1. We compare our approach with two commonly seen methods, Maximum Softmax Probability (MSP) [16,44] and Maximum Logit (MaxLogit) [15,44], along with the recently introduced Nearest Neighbor Guidance (NNGuide) [32] and SCALE [50]. We omit comparisons with older techniques such as OpenMax [1] and EVM [35] as they are much worse. According to the OpenOOD [52,54] Benchmark leaderboard[2] for ImageNet-1K, NNGuide stands out as the leader in the OOD space, closely followed by SCALE. We caution against relying solely on OOD methods and metrics, as they may not accurately reflect OSR performance. Note, some OSA and ROC curves are found in the supplemental.

When utilizing both architectures (Hiera-H & ConvNeXtV2-H) as feature extractors, PostMax outperforms all other methods in every measure (OOSA, AUROC, FPR95) on each unknown dataset. PostMax demonstrates superior performance, emphasizing robustness and generalizability. Effectively, PostMax achieves an excellent trade-off between classification accuracy and unknown rejection. Overall, the results suggest that a system operator can seamlessly integrate PostMax, enhance the performance beyond good closed-set classifiers, and confidently select the operational threshold that best suits its operation.

Statistical Testing. To underscore the significance of performance differences between algorithms in our experiments, we prioritized the assessment of statistical significance. As a result, we performed paired, two-tailed t-tests, with Bonferroni corrections, to compare PostMax with the four methods (MSP, MaxLogit, NNGuide, and SCALE) using both architectures (Hiera-H and ConvNeXtV2-H) across 5 fold runs for each metric (OOSA, AUROC, and FPR95). On Hiera-H, PostMax **(1)** OOSA scores improvements range from **2.5%** to **12.4%** with corresponding P-values all less than $1.0E^{-7}$, **(2)** AUROC scores improvements range from **5.7%** to **17.9%** with corresponding P-values all less than $1.0E^{-8}$, and **(3)** FPR95 scores improvements range from **12.7%** to **22.5%** with corresponding P-values all less than $1.0E^{-8}$. Similarly, on ConvNeXtV2-H, PostMax **(1)** OOSA scores improvements range from **2.3%** to **10.4%**, **(2)** AUROC scores improvements range from **4.7%** to **10.6%**, and **(3)** FPR95 scores improvements range from **9.6%** to **28.1%**, where all corresponding P-values are less than $1.0E^{-8}$ for all tests on all three metrics. In summary, the t-tests conducted for PostMax,

---
[2] https://zjysteven.github.io/OpenOOD.

Table 3. DIFFERENT VALIDATION - KNOWNS & UNKNOWNS SURROGATE. The Operational Open-Set Accuracy (OOSA) (↑) of all methods on a different validation set. To compute, we validate methods on ILSVRC2012 *val* [36] (50K images) as knowns and 21K-P *Easy* [44] (50K images) as unknowns surrogate and predict an operational threshold. Then, we deploy each method's threshold and test on ImageNetV2 [34] (10K images) and specified unknowns. OSR is performed again on extractions from Hiera-H [37] and ConvNeXtV2-H [48]. The best scores are in **bold**.

Unknowns (# images)	OOSA ↑									
	Hiera-H [37]					ConvNeXtV2-H [48]				
	PostMax (Ours)	MSP [16,44]	MaxLogit [15,44]	NNGuide [32]	SCALE [50]	PostMax (Ours)	MSP [16,44]	MaxLogit [15,44]	NNGuide [32]	SCALE [50]
iNaturalist [17] (10K)	**0.863**	0.808	0.780	0.808	0.716	**0.842**	0.806	0.794	0.759	0.763
NINCO [2] (~5.8K)	**0.733**	0.730	0.704	0.655	0.654	**0.759**	0.733	0.726	0.643	0.695
OpenImage-O [45] (~17.6K)	**0.885**	0.810	0.736	0.809	0.649	**0.865**	0.821	0.786	0.746	0.757
Places [56] (10K)	**0.801**	0.748	0.684	0.766	0.611	**0.795**	0.753	0.708	0.731	0.664
SUN [49] (10K)	**0.812**	0.752	0.702	0.771	0.635	**0.802**	0.757	0.722	0.738	0.679
Textures [8] (~5.1K)	**0.813**	0.746	0.739	0.691	0.706	**0.784**	0.750	0.757	0.657	0.732
21K-P *Hard* [44] (49K)	0.634	**0.673**	0.583	0.535	0.515	0.647	**0.672**	0.593	0.559	0.566

Table 4. DIFFERENT TESTING - PRIOR METRICS. The AUROC (↑) and FPR95 (↓) of all methods on different test sets. To compute, we test each method on ImageNetV2 [34] (10K images) as knowns and specified unknowns. OSR is performed again on extractions from Hiera-H [37] and ConvNeXtV2-H [48]. The best scores are in **bold**.

Unknowns (# images)	AUROC ↑ / FPR95 ↓									
	Hiera-H [37]					ConvNeXtV2-H [48]				
	PostMax (Ours)	MSP [16,44]	MaxLogit [15,44]	NNGuide [32]	SCALE [50]	PostMax (Ours)	MSP [16,44]	MaxLogit [15,44]	NNGuide [32]	SCALE [50]
iNaturalist [17] (10K)	**.96 / .16**	.92 / .37	.88 / .37	.91 / .40	.80 / .55	**.95 / .23**	.91 / .41	.90 / .37	.87 / .53	.87 / .44
NINCO [2] (~5.8K)	**.88 / .52**	.83 / .61	.77 / .62	.71 / .72	.70 / .71	**.87 / .54**	.83 / .63	.81 / .60	.73 / .78	.79 / .63
OpenImage-O [45] (~17.6K)	**.96 / .20**	.87 / .48	.79 / .49	.90 / .38	.70 / .64	**.94 / .27**	.88 / .49	.86 / .43	.82 / .65	.84 / .48
Places [56] (10K)	**.90 / .44**	.83 / .60	.74 / .62	.87 / .48	.65 / .74	**.89 / .48**	.83 / .61	.78 / .60	.84 / .59	.74 / .64
SUN [49] (10K)	**.91 / .41**	.84 / .57	.77 / .58	.87 / .49	.69 / .70	**.90 / .44**	.84 / .59	.80 / .57	.84 / .55	.76 / .61
Textures [8] (~5.1K)	**.94 / .27**	.88 / .47	.85 / .42	.77 / .79	.81 / .49	**.92 / .32**	.87 / .49	.88 / .39	.75 / .82	.87 / .39
21K-P *Hard* [44] (49K)	**.77 / .68**	.73 / .73	.68 / .73	.68 / .76	.62 / .82	**.76 / .72**	.74 / .75	.70 / .74	.67 / .82	.67 / .79

considering the combined scores of all metrics across all datasets and architectures, revealed very statistically significant differences compared to every other method.

4.3 Ablation Study

In Tables 3 and 4, we investigate the effects of varying the validation datasets (knowns and unknowns surrogate). We interchanged ImageNetV2 with ILSVRC2012 *val* for knowns and 21K-P *Hard* with 21K-P *Easy* for the unknowns surrogate. Unlike in Table 1, this involves validating on a larger number of knowns and unknowns, with an easier unknowns surrogate. Performance remained consistent regardless of the datasets.

Also, we investigated the impact of other architectures as extractors and interchanged pre-trained models (ILSVRC2012-1K only w/ 224 crops) with Meta's

ViT-H [14] and ConvNeXt-L [24]. Results showcasing PostMax's robustness are in the supplemental.

Lastly, we explored the impact of varying the application-dependent α from Eq. (2) and the normalization (L_2) within PostMax. Details and results are in the supplemental.

5 Discussion

This paper introduces the new evaluation metric Open-Set Accuracy (OSA) (2); we framed the evaluation of an open-set problem as predicting a threshold and then evaluating open-set accuracy at that threshold. This supports an engineer selecting an algorithm and a threshold to be applied in some Operational setting.

In real-world scenarios, unknown samples can be drawn from an unlimited variety, unlike known samples, which are assumed to be related to the training data. Thus, the evaluation must consider the computation of the predictive threshold on validation data and a separate evaluation on test data. The quality of the predicted threshold will depend on many factors with the two most influential being (1) how representative are the surrogate samples of the real unknowns, and (2) the relative proportion of knowns and unknowns. In Sect. 4.3 and our supplemental, we explore some variations of those factors.

Maybe not surprisingly, if we train with "harder" unknowns, we generally see better performance. As shown in Tables 1 and 2, the novel PostMax algorithm is better on unknowns across two networks (other networks in the supplemental). However, as shown in Sect. 4.3, when 21K-P *Easy* is used as our unknowns surrogate, PostMax does not exhibit an advantage on 21K-P *Hard*. In general, 21K-P *Hard* poses the most significant challenge for all algorithms. Notably, the sets close visual and semantic similarity to known samples suggests it represents a fine-grained categorization challenge rather than a typical open-set scenario.

One can adjust the importance of knowns *vs.* unknowns either by explicitly weighting them or implicitly through their relative percentages in the data. Although performance degrades when there is a significant disparity in their importance during threshold prediction stages, PostMax consistently outperforms and demonstrates greater stability across all thresholds. Refer to the supplemental for further details.

Finally, to account for inherent random variations in predictions and tests, we developed training/validation/test splits, which are available on the GitHub repo. This protocol enables rigorous experimentation to assess the statistical significance of our evaluation. While the paper demonstrates statistically significant improvements across all results, some ablations in the supplemental do not.

6 Conclusion

This paper contributes to OSR research by shifting focus from common metrics on small-scale experiments to more realistic scenarios and datasets, thereby fostering more practically relevant research. We revisited and revised the criteria for effective open-set metrics as suggested by Wang et al. [46], making them more applicable to real-world scenarios. Our proposed Operational Open-Set Accuracy (OOSA), which incorporates known and unknown samples, offers a comprehensive metric and facilitates the selection of the most effective method and operating threshold for system operators.

Our investigation of deep feature magnitudes of unknown samples from various ILSVRC2012 pre-trained networks yielded surprising findings. Contrary to smaller-scale studies [10,31], we observed that these magnitudes are often larger for unknown than for known samples. This observation was consistent across multiple networks and types of unknown samples. Thus, we introduced our novel PostMax algorithm, refining normalized logit values from closed-set architectures into real probabilities of class inclusion using a Generalized Pareto Distribution (GPD). This distribution method offers advantages over previous extreme-value-based techniques, including negating the need to select a tail size and minimizing the influence of outliers. Also, it eliminates the need to define negative training samples as representatives of unknown classes, a common approach in much of the OSR literature that has not yielded a sufficient solution.

While a good closed-set classifier is an important place to start [44], our novel **PostMax** algorithm shows that one can do better than "all you need" and that for operational usage, one needs to predict a threshold to use. With this paper, we empower engineers to operationalize their usage of OSR.

References

1. Bendale, A., Boult, T.E.: Towards open set deep networks. In: Conference on Computer Vision and Pattern Recognition (CVPR). IEEE (2016)
2. Bitterwolf, J., Mueller, M., Hein, M.: In or out? Fixing ImageNet out-of-distribution detection evaluation. In: ICLR Workshop on Trustworthy and Reliable Large-Scale Machine Learning Models (2023)
3. Bodesheim, P., Freytag, A., Rodner, E., Denzler, J.: Local novelty detection in multi-class recognition problems. In: Winter Conference on Applications of Computer Vision (WACV) (2015)
4. Boult, T.E., Cruz, S., Dhamija, A.R., Günther, M., Henrydoss, J., Scheirer, W.J.: Learning and the unknown: surveying steps toward open world recognition. In: AAAI Conference on Artificial Intelligence, vol. 33, pp. 9801–9807 (2019)
5. Cevikalp, H., Uzun, B., Salk, Y., Saribas, H., Köpüklü, O.: From anomaly detection to open set recognition: bridging the gap. Pattern Recogn. **138**, 109385 (2023)
6. Chen, G., Peng, P., Wang, X., Tian, Y.: Adversarial reciprocal points learning for open set recognition. Trans. Pattern Anal. Mach. Intell. (TPAMI) **44**(11), 8065–8081 (2021)
7. Chen, G., et al.: Learning open set network with discriminative reciprocal points. In: Vedaldi, A., Bischof, H., Brox, T., Frahm, J.-M. (eds.) ECCV 2020. LNCS, vol. 12348, pp. 507–522. Springer, Cham (2020). https://doi.org/10.1007/978-3-030-58580-8_30
8. Cimpoi, M., Maji, S., Kokkinos, I., Mohamed, S., Vedaldi, A.: Describing textures in the wild. In: Conference on Computer Vision and Pattern Recognition (CVPR) (2014)

9. Deng, J., Dong, W., Socher, R., Li, L.J., Li, K., Fei-Fei, L.: Imagenet: a large-scale hierarchical image database. In: Conference on Computer Vision and Pattern Recognition (CVPR), pp. 248–255. IEEE (2009)
10. Dhamija, A.R., Günther, M., Boult, T.: Reducing network agnostophobia. In: Advances in Neural Information Processing Systems (NeurIPS), vol. 31 (2018)
11. Ge, Z., Demyanov, S., Garnavi, R.: Generative OpenMax for multi-class open set classification. In: British Machine Vision Conference (BMVC) (2017)
12. Gumbel, E.J.: Statistical Theory of Extreme Values and Some Practical Applications: A Series of Lectures, vol. 33. US Government Printing Office (1954)
13. Günther, M., Dhamija, A.R., Boult, T.E.: Watchlist adaptation: protecting the innocent. In: International Conference of the Biometrics Special Interest Group (BIOSIG) (2020)
14. He, K., Chen, X., Xie, S., Li, Y., Dollár, P., Girshick, R.: Masked autoencoders are scalable vision learners. In: Conference on Computer Vision and Pattern Recognition (CVPR), pp. 16000–16009 (2022)
15. Hendrycks, D., et al.: Scaling out-of-distribution detection for real-world settings. In: International Conference on Machine Learning (ICML). PMLR (2022)
16. Hendrycks, D., Gimpel, K.: A baseline for detecting misclassified and out-of-distribution examples in neural networks. In: International Conference on Learning Representations (ICLR) (2017)
17. van Horn, G., et al.: The INaturalist species classification and detection dataset. In: Conference on Computer Vision and Pattern Recognition (CVPR) (2018)
18. Huang, R., Geng, A., Li, Y.: On the importance of gradients for detecting distributional shifts in the wild. In: Advances in Neural Information Processing Systems (NeurIPS), vol. 34, pp. 677–689 (2021)
19. Huang, R., Li, Y.: MOS: towards scaling out-of-distribution detection for large semantic space. In: Conference on Computer Vision and Pattern Recognition (CVPR), pp. 8710–8719 (2021)
20. Kong, S., Ramanan, D.: Opengan: open-set recognition via open data generation. In: International Conference on Computer Vision (ICCV), pp. 813–822 (2021)
21. Krizhevsky, A., Hinton, G.: Learning multiple layers of features from tiny images. Technical report, University of Toronto (2009)
22. LeCun, Y., Bottou, L., Bengio, Y., Haffner, P.: Gradient-based learning applied to document recognition. Proc. IEEE **86**(11), 2278–2324 (1998)
23. Liu, X., Lochman, Y., Zach, C.: GEN: pushing the limits of softmax-based out-of-distribution detection. In: Conference on Computer Vision and Pattern Recognition (CVPR), pp. 23946–23955 (2023)
24. Liu, Z., Mao, H., Wu, C.Y., Feichtenhofer, C., Darrell, T., Xie, S.: A convnet for the 2020s. In: Conference on Computer Vision and Pattern Recognition (CVPR), pp. 11976–11986 (2022)
25. Lu, J., Xu, Y., Li, H., Cheng, Z., Niu, Y.: PMAL: open set recognition via robust prototype mining. In: AAAI Conference on Artificial Intelligence, vol. 36:2, pp. 1872–1880 (2022)
26. Lyu, Z., Gutierrez, N.B., Beksi, W.J.: Metamax: improved open-set deep neural networks via weibull calibration. In: Winter Conference on Applications of Computer Vision Workshops (WACVW) (2023)
27. Moon, W., Park, J., Seong, H.S., Cho, C.H., Heo, J.P.: Difficulty-aware simulator for open set recognition. In: Avidan, S., Brostow, G., Cissé, M., Farinella, G.M., Hassner, T. (eds.) ECCV 2022. LNCS, vol. 13685, pp. 365–381. Springer, Cham (2022). https://doi.org/10.1007/978-3-031-19806-9_21

28. Neal, L., Olson, M., Fern, X., Wong, W.K., Li, F.: Open set learning with counterfactual images. In: European Conference on Computer Vision (ECCV) (2018)
29. Netzer, Y., Wang, T., Coates, A., Bissacco, A., Wu, B., Ng, A.Y.: Reading digits in natural images with unsupervised feature learning. In: NIPS Workshop on Deep Learning and Unsupervised Feature Learning (2011)
30. Palechor, A., Bhoumik, A., Günther, M.: Large-scale open-set classification protocols for imagenet. In: Winter Conference on Applications of Computer Vision (WACV), pp. 42–51. CVF/IEEE (2023)
31. Park, J., Chai, J.C.L., Yoon, J., Teoh, A.B.J.: Understanding the feature norm for out-of-distribution detection. In: International Conference on Computer Vision (ICCV), pp. 1557–1567 (2023)
32. Park, J., Jung, Y.G., Teoh, A.B.J.: Nearest neighbor guidance for out-of-distribution detection. In: International Conference on Computer Vision (ICCV), pp. 1686–1695 (2023)
33. Pickands, J., III.: Statistical inference using extreme order statistics. Ann. Stat. **3**(1), 119–131 (1975)
34. Recht, B., Roelofs, R., Schmidt, L., Shankar, V.: Do Imagenet classifiers generalize to Imagenet? In: International Conference on Machine Learning (ICML), pp. 5389–5400. PMLR (2019)
35. Rudd, E.M., Jain, L.P., Scheirer, W.J., Boult, T.E.: The extreme value machine. Trans. Pattern Anal. Mach. Intell. (TPAMI) **40**(3), 762–768 (2017)
36. Russakovsky, O., et al.: ImageNet large scale visual recognition challenge. Int. J. Comput. Vision (IJCV) **115**(3), 211–252 (2015)
37. Ryali, C., et al.: Hiera: a hierarchical vision transformer without the bells-and-whistles. In: International Conference on Machine Learning (ICML) (2023)
38. Scheirer, W.J., Jain, L.P., Boult, T.E.: Probability models for open set recognition. Trans. Pattern Anal. Mach. Intell. (TPAMI) **36**(11), 2317–2324 (2014)
39. Scheirer, W.J., de Rezende Rocha, A., Sapkota, A., Boult, T.E.: Towards open set recognition. Trans. Pattern Anal. Mach. Intell. (TPAMI) **35**(7) (2013)
40. Shu, L., Xu, H., Liu, B.: DOC: deep open classification of text documents. In: Conference on Empirical Methods in Natural Language Processing (EMNLP). Association for Computational Linguistics (2017)
41. Sun, Y., Guo, C., Li, Y.: React: out-of-distribution detection with rectified activations. In: Advances in Neural Information Processing Systems (NeurIPS), vol. 34, pp. 144–157 (2021)
42. Vareto, R.H., Linghu, Y., Boult, T.E., Schwartz, W.R., Günther, M.: Open-set face recognition with maximal entropy and Objectosphere loss. Image Vis. Comput. (IMAVIS) **141** (2024)
43. Vareto, R.H., Schwartz, W.R., Günther, M.: Toward open-set face recognition with neural ensemble, maximal entropy loss and feature augmentation. In: Conference on Graphics, Patterns and Images (SIBGRAPI) (2023)
44. Vaze, S., Han, K., Vedaldi, A., Zissermann, A.: Open-set recognition: a good closed-set classifier is all you need? In: International Conference on Learning Representations (ICLR) (2022)
45. Wang, H., Li, Z., Feng, L., Zhang, W.: Vim: out-of-distribution with virtual-logit matching. In: Conference on Computer Vision and Pattern Recognition (CVPR), pp. 4921–4930 (2022)
46. Wang, Z., Xu, Q., Yang, Z., He, Y., Cao, X., Huang, Q.: OpenAUC: towards AUC-oriented open-set recognition. In: Advances in Neural Information Processing Systems (NeurIPS), vol. 35, pp. 25033–25045 (2022)

47. Wilson, S., Fischer, T., Dayoub, F., Miller, D., Sünderhauf, N.: Safe: sensitivity-aware features for out-of-distribution object detection. In: International Conference on Computer Vision (ICCV), pp. 23565–23576 (2023)
48. Woo, S., et al.: Convnext v2: co-designing and scaling convnets with masked autoencoders. In: Conference on Computer Vision and Pattern Recognition (CVPR), pp. 16133–16142 (2023)
49. Xiao, J., Hays, J., Ehinger, K.A., Oliva, A., Torralba, A.: Sun database: large-scale scene recognition from abbey to zoo. In: Conference on Computer Vision and Pattern Recognition (CVPR), pp. 3485–3492. IEEE (2010)
50. Xu, K., Chen, R., Franchi, G., Yao, A.: Scaling for training time and post-hoc out-of-distribution detection enhancement. In: International Conference on Learning Representations (ICLR) (2024)
51. Yang, H.M., Zhang, X.Y., Yin, F., Yang, Q., Liu, C.L.: Convolutional prototype network for open set recognition. Trans. Pattern Anal. Mach. Intell. (TPAMI) **44**(5), 2358–2370 (2020)
52. Yang, J., et al.: OpenOOD: benchmarking generalized out-of-distribution detection. In: Advances in Neural Information Processing Systems (NeurIPS), vol. 35, pp. 32598–32611 (2022)
53. Yoshihashi, R., Shao, W., Kawakami, R., You, S., Iida, M., Naemura, T.: Classification-reconstruction learning for open-set recognition. In: Conference on Computer Vision and Pattern Recognition (CVPR) (2019)
54. Zhang, J., et al.: OpenOOD v1.5: enhanced benchmark for out-of-distribution detection. In: NeurIPS Workshop on Distribution Shifts: New Frontiers with Foundation Models (2023)
55. Zhang, X., Cheng, X., Zhang, D., Bonnington, P., Ge, Z.: Learning network architecture for open-set recognition. In: Proceedings of the AAAI Conference on Artificial Intelligence, vol. 36, no. 3, pp. 3362–3370 (2022)
56. Zhou, B., Lapedriza, A., Khosla, A., Oliva, A., Torralba, A.: Places: a 10 million image database for scene recognition. Trans. Pattern Anal. Mach. Intell. (TPAMI) **40**(6), 1452–1464 (2017)
57. Zhou, D.W., Ye, H.J., Zhan, D.C.: Learning placeholders for open-set recognition. In: Conference on Computer Vision and Pattern Recognition (CVPR) (2021)

Author Index

B
Bartolomei, Luca 125
Bharadwaj, Tejas 324
Bi, Mengxiao 398
Bian, Liheng 54
Boult, Terrance E. 475

C
Cao, Chunshui 380
Chang, Haoran 324
Chang, Jian 455
Chari, Pradyumna 324
Chatterjee, Dibyadip 436
Chen, Jiacheng 90
Chen, Kai 216
Chen, Tianyi 71
Chen, Xiangyu 108
Chen, Yun 305
Chi, Hyung-gun 361
Chilimbi, Trishul 146
Chung, Woojeh 18
Conti, Andrea 125
Cruz, Steve 475
Cui, Ruodai 252
Cygert, Sebastian 181

D
Ding, Tianyu 71
Dong, Weisheng 286
Duan, Haodong 216

F
Fan, Zhiwen 324
Feng, Mingtao 286
Furukawa, Yasutaka 90

G
Gao, Ge 234
Gong, Yiming 269

Gu, Ming 234
Günther, Manuel 475

H
He, Conghui 216
He, Zhiqiang 380
Hoiem, Derek 269
Hou, Saihui 380
Hu, Guosheng 252
Hua, Gang 163
Huang, Lujie 198
Huang, Panjian 380
Huang, Yongzhen 380

J
Jayasuriya, Suren 18
Ji, Bin 398
Ji, Rongrong 38
Jiang, Jincen 455
Jiang, Yi 417
Jin, Yeying 1
Jung, Haeji 361

K
Kadambi, Achuta 324
Kim, Jinkyu 361
Kim, Sangpil 361

L
Li, Bo 216
Li, Nianyi 18
Li, Shengtao 234
Li, Xin 1
Li, Yuhang 455
Liang, Luming 71
Lim, Hyerin 361
Lin, Dahua 216
Lin, Mingbao 38
Lin, Zhihang 38
Liu, Daizong 198

Liu, Ligang 198
Liu, Xina 108
Liu, Xu 380
Liu, Yuan 216
Liu, Yudong 234
Liu, Yu-Shen 234
Liu, Ziwei 216
Lu, Xuequan 455
Luo, Lei 343

M
Ma, Chuofan 417
Ma, Hang 90
Ma, Lizhuang 455
Ma, Shugao 436
Ma, Yang 198
Mahjourian, Reza 361
Manivasagam, Sivabalan 305
Marczak, Daniel 181
Mattoccia, Stefano 125
Mian, Ajmal 286
Miao, Bo 286
Moon, Seokha 361

N
Neiman, Tal 146
Nevatia, Ram 71
Niu, Li 252

P
Pan, Ye 398
Park, Hongbeen 361
Peng, Lintao 54
Peng, Weilong 198
Peng, Yunjie 380
Poggi, Matteo 125

Q
Qi, Xiaojuan 417
Qin, Dehao 18

R
Rabinowitz, Ryan 475
Rizve, Mamshad Nayeem 146

S
Saha, Ripon Kumar 18
Sener, Fadime 436
Shah, Mubarak 146

Shamil, Md Salman 436
Shi, Haoyue 163
Sirnam, Swetha 146
Song, Binbin 108
Sun, Yanguang 343

T
Tan, Jiaqi 90
Tan, Shuai 398
Tang, Keke 198
Tang, Wei 163
Tian, Zhihong 198
Tran, Son 146
Trzciński, Tomasz 181
Twardowski, Bartłomiej 181

U
Urtasun, Raquel 305

W
Wang, Jiadong 1
Wang, Jiaqi 216
Wang, Jingkang 305
Wang, Le 163
Wang, Meili 455
Wang, Xiaofei 198
Wang, Yaonan 286
Wang, Zhangyang 324
Woo, Hyun 361
Wu, Jiannan 417
Wu, Yuefan 90
Wu, Zijie 286

X
Xie, He 286
Xie, Siyu 54
Xu, Chunyan 343
Xu, Dejia 324
Xu, Shuning 108
Xuan, Hanyu 343

Y
Yang, Jian 343
Yang, Jinyu 146
Yang, Ze 305
Yao, Angela 436
Yao, Benjamin 146
Ye, Jinwei 18
You, Suya 324

Author Index

Yuan, Yike 216
Yuan, Zehuan 417

Z

Zhang, Jian Jun 455
Zhang, Malu 1
Zhang, Songyang 216
Zhang, Yan 1
Zhang, Yuanhan 216

Zhao, Meng 38
Zhao, Wangbo 216
Zharkov, Ilya 71
Zhou, Jiantao 108
Zhou, Qianyu 455
Zhou, Sanping 163
Zhou, Shijie 324
Zhu, Haidong 71
Zhu, Zhen 269

Printed in the USA
CPSIA information can be obtained
at www.ICGtesting.com
CBHW062023131024
15807CB00011B/205